THE HETEROCYCLIC DERIVATIVES OF PHOSPHORUS, ARSENIC, ANTIMONY AND BISMUTH

This is the second edition of the first volume in the series
THE CHEMISTRY OF HETEROCYCLIC COMPOUNDS

THE CHEMISTRY OF HETEROCYCLIC COMPOUNDS

A SERIES OF MONOGRAPHS

A. WEISSBERGER and E. C. TAYLOR, *Editors*

The Heterocyclic Derivatives of
PHOSPHORUS, ARSENIC, ANTIMONY
and BISMUTH

SECOND EDITION

Frederick George Mann

(Fellow of the Royal Society)

Cambridge University, England

1970

WILEY–INTERSCIENCE

a division of John Wiley & Sons

New York - London - Sydney - Toronto

Library of Congress Catalog Card Number: 75-110402

ISBN 0 471 37489 X

PRINTED IN GREAT BRITAIN BY SPOTTISWOODE, BALLANTYNE AND CO. LTD.
LONDON AND COLCHESTER

The Chemistry of Heterocyclic Compounds

The chemistry of heterocyclic compounds is one of the most complex branches of organic chemistry. It is equally interesting for its theoretical implications, for the diversity of its synthetic procedures, and for the physiological and industrial significance of heterocyclic compounds.

A field of such importance and intrinsic difficulty should be made as readily accessible as possible, and the lack of a modern, detailed and comprehensive presentation of heterocyclic chemistry was therefore keenly felt. It is the intention of the present series to fill this gap by expert presentation of the various branches of heterocyclic chemistry. The subdivisions have been designed to cover the field in its entirety by monographs which reflect the importance and the interrelations of the various ring systems, and accommodate the specific interests of the authors.

In order to keep *The Chemistry of Heterocyclic Compounds* up to date, two methods will be applied in accordance with the nature of the respective compound group and the timeliness of the respective Volume of the First Edition. For those areas where the Volumes have become obsolete by overwhelming progress, new editions are planned. For those areas where the necessary changes are not too great, supplements to the respective volumes will be published. In this way, we will achieve our purpose with a minimum of new investments for the subscriber to the whole set.

The series started in 1950 with a slim volume by Frederick George Mann on the Heterocyclic Derivatives of Phosphorus, Arsenic, Antimony, Bismuth, and Silicon. Great progress in these fields has been made and we are, therefore, pleased to publish a new edition of Dr. Mann's book.

December 1968

A. WEISSBERGER
E. C. TAYLOR

Yet all experience is an arch wherethro'
Gleams that untravell'd world, whose margin fades
For ever and for ever when I move.

TENNYSON

Preface

In the First Edition of this work, published in 1950, an attempt was made to give 'a systematic account of the heterocyclic organic derivatives of phosphorus, arsenic, antimony, bismuth (and silicon), that is to say, of those derivatives which contain these elements and carbon (with or without other elements) in the ring system', and this object was achieved in 162 pages of text. The subsequent vast expansion of our knowledge of heterocyclic organic chemistry has presented the author with a massive task: for example, by 1963 there were about 150 different ring systems containing phosphorus and about 160 such systems containing arsenic, and these numbers are rapidly increasing.

In these circumstances some selection has been inevitable: in this Second Edition I have had to omit a number of systems of which very little is known and which appear to be (at present) of small interest. My original intention, moreover, was to deal only briefly with rings in which the Group VB element, X, was present as —O—X(R)—C— or, more particularly, as —O—X(R)—O—, with the atom X often oxidized, because such systems are really cyclic esters, frequently vulnerable to hydrolysis, and thus markedly different from the highly stable ring systems containing solely carbon and the heteroatoms. This intention has often failed, however, because our knowledge of several such systems is based on a considerable volume of work, much of which is of great interest and importance.

It should be emphasized that this volume is one of a series on organic heterocyclic compounds, and consequently purely inorganic rings, i.e., those devoid of carbon as a ring member, are omitted; this would in any case have been essential, for certain inorganic ring systems such as the phosphonitriles have already received separate volumes in their own right.

In the First Edition of this book, the heterocyclic derivatives of silicon occupied two pages: our current knowledge of this subject would also require a separate volume, and the cyclic derivatives of silicon have, therefore, also had to be omitted from this Edition.

The order in which the heterocyclic systems are presented is shown in the Table of Contents (p. xv): taking phosphorus as an example, rings containing carbon and one phosphorus atom are discussed in order of increasing size of ring; rings containing carbon and two or more phosphorus atoms follow in the same order; finally rings containing phosphorus and another heteroatom, usually in order of increasing atomic weight of this atom, are considered, each group also in order of increasing size of ring.

An important factor in a book of this type is the nomenclature. It is, I think, abundantly clear that the only acceptable system is that of the Ring Index (2nd Edition) and its Supplements (p. xiii). The authors of the Ring Index, working in consultation with the Nomenclature Committees of the International Union of Pure and Applied Chemistry (IUPAC), have produced a logical and flexible system for the naming, numbering, and presentation of more than 14,000 homocyclic and heterocyclic rings (p. xiii).

In the present Edition, each ring system is discussed in its own section, which is headed by the Ring Index name and gives the number, mode of presentation of the ring, and its notation: the ring (with certain exceptions) is shown in its lowest state of hydrogenation, and any modification of the name for hydrogenated derivatives is stated.

The adoption of Ring Index names is rapidly becoming general. The older names were, however, widely used until recently and some still persist: these names are usually given in each section, so that a chemist having occasion to consult one of the older chemical indexes can find the entry he seeks.

The Ring Index in certain cases retains a familiar and/or long established trivial name for a ring system: all other names are based on the rules fully explained in the Second Edition of that work. The Ring Index has one minor disadvantage, namely, that a hydrogenated and heavily substituted ring system may develop a rather long name. This is offset by the fact that every ring system now has its own name and number. In particular, when derivatives of a ring system have been prepared or investigated by organic chemists, biochemists, pharmacologists, physiologists, and workers in other fields, a remarkable variety of names may have been employed (cf. for example 1,3,2-dioxaphosphorinanes, Part I, pp. 295–301): this confusion of names should be rapidly reduced by adoption of the Ring Index system.

In the Foreword to the First Edition, I briefly indicated certain differences in the English and the American nomenclature of some simple compounds of the Group VB elements. These differences were removed shortly afterwards, but some years later the chemists of both countries united in the adoption of a comprehensive system of nomenclature for phosphorus compounds which was so extensive and often so involved that only the names of the more common compounds have come into wide use. It covered, moreover, only compounds containing one phosphorus atom, and the promised extension of this remarkable system to compounds containing two or more phosphorus atoms has never appeared. It does not, of course, affect the names of cyclic compounds and appears in this book only in loyal application to the names of various reagents used in the synthetic work.

Many of the earlier volumes in the Heterocyclic Series have had complete tables of all known members of each type of compound, together with melting points, yields, etc. The adoption of this practice in the present work would have taken up much space: such tables have, therefore, been strictly limited both in number and in size. On the other hand, the melting points of most compounds are given immediately after their first mention in the text: apart from the reference value of these constants, they indicate immediately to the reader whether the compound is solid or liquid at room temperature.

During recent years there has been an increasing tendency in the larger chemical volumes, reviews, reports, etc., to amplify the text solely by reference numbers, usually as small superscripts, leaving the reader, who may want to know when and by whom (and even where) the work was done, to grope through the distant pages to the complete references. When this practice is allied to a text that consists almost solely of a series of rather brief factual statements, the printed page can have a rather dreary, sterile, and utterly impersonal atmosphere, well calculated to subdue, if not to kill, the reader's interest. The result is that most students, junior and senior, now use a chemical book almost solely to check individual facts, and only rarely to read. I have tried to write the present book with sufficient detail to ensure that, without in any way diminishing its value as a book of reference, whole passages may actually be *read* by those to whom the substance of the passage has some interest. I have also tried to give some human interest to the material by citing the names of authors, though frequently in parentheses for brevity. Each of the four Parts is followed by its own list of references, which are therefore readily available as a comprehensive whole.

All references to papers in languages unfamiliar to most American and Western European chemists, and all patent specifications, are accompanied by their *Chemical Abstracts* references.

Whilst writing this book, I have been repeatedly and deeply indebted to Dr. Robert S. Cahn for so patiently and so clearly answering my many collections of queries regarding nomenclature. The last Supplement to the Ring Index covers the field up to 1963: since then a steady stream of new heterocyclic compounds, tending to be of increasing complexity, has been recorded. I am singularly fortunate in having the help and

advice of so great an authority on nomenclature as Dr. Cahn, to whom
I tender my warmest thanks. If this book enables many chemists to get
a clear and accurate knowledge of the names of these Group VB heterocyclic rings, they will be indebted primarily to Dr. Cahn. It must be
emphasized, however, that I am entirely responsible for any errors in
nomenclature that may still remain.

I wish also to thank Dr. Arnold Weissberger, the Editor of this
Series, both for the patience with which over a very long period he bore
my many delays in starting this book, and for the help and advice he
gave me when at last I had sufficient time to undertake this task.

I must also thank Dr. Reinhard Schmutzler who, whilst I was fully
engaged in writing, kept a sharp look-out through a wide range of
chemical journals for papers dealing with my subjects, and then ensured
that I received a reprint or a copy of them. Without this aid I should have
missed a number of recent and important papers.

I am also much indebted to Professor Lord Todd for the excellent
facilities for compilation and writing which he kindly provided for me
in the University of Cambridge Chemical Laboratory and which I have
greatly appreciated.

It would be difficult adequately to express my thanks to my wife
for all the help which she has given me during the progress of this book,
a progress which at times would have seemed slow indeed without her
cheerful encouragement.

A book of this size, although carefully checked and scrutinized, must
suffer from some errors and omissions. I should be grateful if readers
would notify me of any such faults that they may detect.

F. G. MANN

Note on Nomenclature

Prefixes and Suffixes of Fundamental Heterocyclic Systems

Table 1. Prefixes

Element	Valency	Prefix	Element	Valency	Prefix
Phosphorus	III	Phospha	Oxygen	II	Oxa
Arsenic	III	Arsa	Sulphur	II	Thia
Antimony	III	Stiba	Nitrogen	III	Aza
Bismuth	III	Bisma			

When immediately followed by '-in' or '-ine', the prefixes 'phospha-', 'arsa-' and 'stiba-' should be replaced by 'phosphor-', 'arsen-' and 'antimon-', respectively.

Table 2. Suffixes

No. of atoms in the ring	Rings containing no nitrogen		Rings containing nitrogen	
	Unsaturated	Saturated	Unsaturated	Saturated
3	-irene	-irane	-irine	-iridine
4	-ete	-etane	-ete	-etidine
5	-ole	-olane	-ole	-olidine
6	-in	-ane	-ine	*
7	-epin	-epane	-epine	*
8	-ocin	-ocane	-ocine	*
9	-onin	-onane	-onine	*
10	-ecin	-ecane	-ecine	*

* State position and number of hydrogen atoms, e.g., '3,4-dihydro' before name of unsaturated member (cf. pp. 220–221).

The Ring Index

By (the late) Austin M. Patterson, Leonard T. Capell and Donald F. Walker. Published by the American Chemical Society, Washington, D.C., U.S.A., 1960.

	Prefix and Numbers of Ring Index Systems	Covers abstracted literature
2nd Edition, 1960	RRI 1–7727	Up to Jan. 1st 1957
Supplement I, 1963	RIS 7728–9734	Jan. 1st 1957–Jan. 1st 1960
Supplement II, 1964	RIS 9735–11524	Jan. 1st 1960–Jan. 1st 1962
Supplement III, 1965	RIS 11525–14265	Jan. 1st 1962–Dec. 31st 1963

In the following pages, the Ring Index Prefix and Number are usually given after the name of each ring system encountered. The entries [RRI —] and [RIS —] indicate that the ring system has not been traced or has not yet been published.

Contents

(Centre headings denote the heteroatoms in individual rings)

Part I. Heterocyclic Derivatives of Phosphorus*

Phosphorus only

1P

* Certain polycyclic compounds are classified for convenience of discussion on the basis of one of their smaller rings, in both this and other Parts.

Phosphorus and Nitrogen

Phosphorus and Oxygen

Part III. Heterocyclic Derivatives of Antimony

Antimony only

1Sb

Antimony and Oxygen

Antimony, Oxygen and Sulfur

Antimony and Sulfur

Part IV. Heterocyclic Derivatives of Bismuth

Bismuth only

1Bi

1Bi + 2O

Bismuth and Sulfur

1Bi + 2S

PART I

Heterocyclic Derivatives of Phosphorus

Three-membered Ring Systems containing only Carbon and One Phosphorus Atom

1H-Phosphirene and Phosphirane

The ring system (1) is termed 1H-phosphirene, [RIS 11527], and its dihydro derivative (2) is phosphirane.

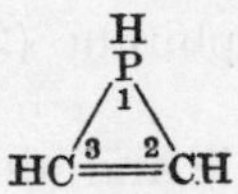

$$
\begin{array}{cc}
\overset{\displaystyle H}{\overset{\displaystyle P}{\underset{\displaystyle HC \!=\!=\! CH}{\triangle}}} & \overset{\displaystyle H}{\overset{\displaystyle P}{\underset{\displaystyle H_2C \!-\! CH_2}{\triangle}}} \\
(1) & (2)
\end{array}
$$

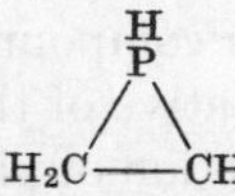

Sodium dissolves in liquid ammonia to give a deep blue solution and, when one equivalent of phosphine, PH_3, or an alkyl- or aryl-phosphine, RPH_2, is added, the blue color is converted into the yellow color of the sodio derivative, $NaPH_2$ or $RPHNa$; indeed the change in color can be used as an indicator for the addition of a primary phosphine.

Wagner[1] has claimed that, when half a molecular equivalent of an alkylene dihalide, $X(CH_2)_nX$, where $n = 1$–7, is now added, two products are usually formed, namely, the open-chain diphosphine (3A) or (3B),

$$
\begin{array}{cccc}
H_2P(CH_2)_nPH_2 & HRP(CH_2)_nPRH & (CH_2)_n\,PH & (CH_2)_n\,PR \\
(3A) & (3B) & & \\
& & (4A) & (4B)
\end{array}
$$

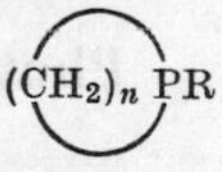

and the corresponding cyclic secondary phosphine (4A) or tertiary phosphine (4B), according to whether phosphine itself or a primary phosphine has initially been used.

Wagner[1] states that, when phosphine (1 molar equivalent) is thus added to a solution of sodium (1 equivalent) in ammonia in a tube at $-78°$, 0.5 molar equivalent of 1,2-dichloroethane then added at $-196°$, and the tube sealed and allowed to warm to room temperature for 20 minutes, subsequent distillation of the product affords the liquid phosphirane (2), m.p. $-121.4°$ to $-120.9°$, molecular weight 60.0 (theoretical, 60.0). It slowly gives a non-volatile viscous liquid at room temperature.

3

The identity of this compound must at present be accepted with considerable reserve. The patent specification gives only the above values, without analysis or chemical properties. The compound could very well be vinylphosphine, $CH_2:CHPH_2$, formed by loss of hydrogen chloride from the intermediate (2-chloroethyl)phosphine, $ClCH_2CH_2PH_2$. It may be significant that the use of methylphosphine in the above experiment is stated to give 1,2-bis(methylphosphino)ethane [or ethylenebis(methylphosphine)] (**3B**; $n = 2$, $R = CH_3$), m.p. $-20°$ to $-13°$, molecular weight 102.2 (theoretical 108.1), and no mention is made of 1-methylphosphirane (**4B**; $n = 2$, $R = CH_3$) although the latter might be more stable than the phosphirane (**2**) itself. Furthermore, the next higher homologue (**4B**; $n = 3$, $R = CH_3$), which should also have greater stability, is not mentioned.

If Wagner's compound is the true phosphirane (**2**), it is the only known representative of this class.

When the action of phosphorus trichloride and aluminum chloride on unhindered simple olefins (see p. 5) has been more fully investigated compounds having the ring system (**4B**; $n = 2$) may be discovered.

The phosphirane system, when incorporated in an appropriate bicyclic system, may acquire a sufficiently enhanced stability to allow the bicyclic compound to be isolated in the crystalline state, although the general stability may not be high; for an example, see p. 158.

Four-Membered Ring Systems containing only Carbon and One Phosphorus Atom

Phosphetes, Phosphetenes, and Phosphetanes

This four-membered system could theoretically exist as phosphete (**1**), as the isomeric 1-phosphetene (**2**), and 2-phosphetene (**3**), and as

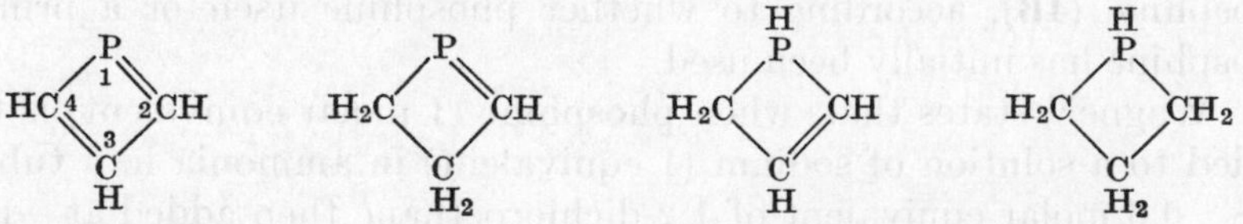

the fully hydrogenated phosphetane (**4**), [RIS 7730]. Derivatives of phosphete and the phosphetenes are at present unknown, and phosphetane is known only as substituted derivatives.

Jungermann, McBride, *et al.*,[2, 3, 4] in an investigation of the action of phosphorus trichloride in the presence of aluminum chloride on various unsaturated fatty compounds such as methyl oleate, obtained evidence of the addition of the phosphorus trichloride at the double bond, for the product after careful treatment with water appeared to contain the P(O)Cl group. Purification of products was difficult, however, and the exact nature of the reaction with these compounds was not elucidated.

When, however, a sterically hindered olefin, 2,4,4-trimethyl-2-pentene, $(CH_3)_2C{=}CHC(CH_3)_3$, was used in this reaction, a chloride, $C_8H_{16}POCl$, m.p. 74–75°, was isolated: this compound was hydrolyzed to an acid $C_8H_{16}PO_2H$, m.p. 75–76°, [dihydrate, m.p. 54–58° (from water)], and was also converted by sodium methoxide in methanol into the methyl ester, $C_8H_{16}PO_2CH_3$, b.p. 78°/0.7 mm, m.p. 35–36°. The composition of these compounds was confirmed by analysis; the molecular weight of the methyl ester in boiling acetone was normal, and that of the acid indicated a dimer.

To prepare the chloride, equimolecular quantities of phosphorus trichloride, aluminum chloride, and the olefin were added in turn to stirred dichloromethane at 0–10°; after 1 hour water was added dropwise, the temperature being kept below 25°. Distillation of the dried organic layer left a residue of the white solid chloride.

(5)   (6)   (7)   (8)

The physical and chemical properties gave strong evidence that the chloro compound was 1-chloro-2,2,3,4,4-pentamethylphosphetane 1-oxide (5), *i.e.*, a cyclic phosphinic chloride, and that the acid and the methyl ester therefore had the corresponding structures (6) and (7): (*a*) The compounds (5), (6), and (7) do not decolorize bromine or aqueous permanganate, therefore a C=C group or a PH group is absent. (*b*) The acid (6) has very great stability. It was recovered unchanged after treatment with boiling concentrated nitric acid or hot concentrated sulfuric acid, and after prolonged boiling with concentrated sodium hydroxide solution; therefore a P—O—C linkage is absent. Boiling thionyl chloride converted the acid back into the chloride (5). (*c*) These findings are confirmed by the infrared spectra of three compounds, which all show a band near $8\,\mu$ (1250 cm^{-1}), indicating a P=O bond. (d) Potentiometric titration shows that (6) is a strong monobasic acid, p$K' = 2.85$. (e) The proton-nmr spectra of the acid showed a single peak at $\tau -2.48$ for the OH proton, and when this is used as a standard area equivalent to one proton, two multiplets at τ 8.70 and 9.00 were integrated to 7.1 (two CH$_3$ groups and one CH) and 8.9 (three CH$_3$ groups) protons. The spectrum of the methyl ester showed the OCH$_3$ doublet at τ 3.72 (J_{PH} 10 c/sec) and two groups equivalent to 7.2 and 9.3 protons.

The ring structure of the methyl ester (7) should enable this compound to exist as *cis–trans*-isomers. High-resolution gas chromatography showed only one peak when a highly purified sample was chromatographed on a 6-ft. column, but two distinct peaks when put through a 100-ft. capillary column.

This evidence does not, however, distinguish decisively between the ring system in (5) and the isomeric ring system (8; R = Cl, OH, or OCH$_3$).

The mechanism suggested for the formation of the ring system involves the stages: (i) Formation of the PCl$_2^+$ ion by acid–base reaction

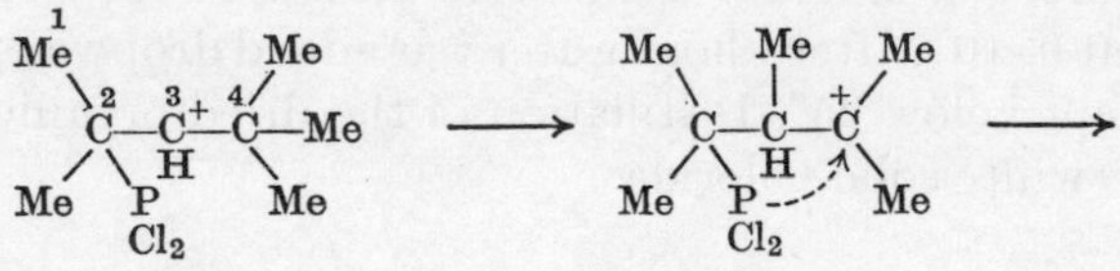

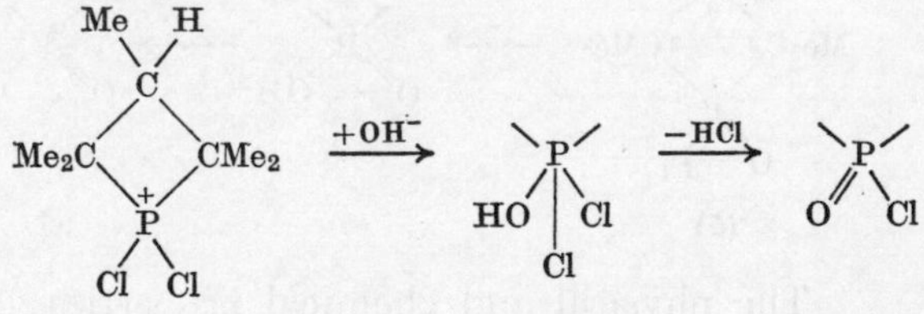

between the PCl$_3$ and the AlCl$_3$. (ii) Attack of the PCl$_2^+$ ion at the C-2 atom rather than at the C-3 atom, giving the carbonium ion (9); for steric reasons it is considered that the bulky PCl$_2^+$ ion is more likely to attack the C-2 atom than the neopentyl type C-3 atom. (iii) Shift of a methyl group from C-4 to C-3 (cf. 10) to convert the secondary into the more stable tertiary carbonium ion. (iv) Cyclization to (11) in preference to loss of a proton, possibly because of the very favorable geometry of the intermediate (10). (v) Reaction of the cyclic cation (11) with a solvent anion to give the neutral compound (12), followed by loss of hydrogen chloride to give the chloride (5).

This mechanism is reasonable, whereas any mechanism to give the alternative ring system (8) would involve the migration of at least two methyl groups; on this basis, the structure (5) is the more probable.

To obtain decisive evidence for the structure (5) or (8), Bergesen [5] has converted the chloride (5) into the ethyl ester, which could be (13; $R = C_2H_5$) or (14; $R = C_2H_5$); each of these esters could exist as

(13)   (14)

cis–trans-isomers. It is recorded that the product did consist of a mixture of such isomers, of which the cis-member underwent alkaline hydrolysis more rapidly than the trans-member, the latter being thus obtained isomerically pure. The cis-acid, isolated from the alkaline solution, was then converted via the chloride into the cis-ethyl ester; both esters had b.p. 112°/10 mm and on hydrolysis gave the trans-acid, m.p. 74.6°, and the cis-acid, m.p. 74.1°, severally.

The nmr spectra of the 'pure acid isomers' showed decisively that they had the structure (13; R = H).

The allocation of cis and trans to these compounds was based on the rate of hydrolysis of the esters; the one which underwent hydrolysis ca. 7 times more rapidly than the other was deemed to be the cis-form.

This hydrolysis portion of Bergesen's work has been criticized by Hawes and Trippett,[6] who repeated the preparation of the ethyl ester (13; $R = C_2H_5$), which from its nmr spectrum appeared to be homogeneous; its hydrolysis rate constants agree with those of Bergesen's trans-ester. They also note that the isomeric acids (13; R = H) would have a common anion and might thus be identical.

X-Ray crystal analysis of the monohydrated acid has now shown decisively that it has the structure (13; R = H), with a planar ring system: the oxygen atoms of the PO group and of the OH group are both hydrogen-bonded to water molecules, and the OH group is depicted as trans to the 3-CH_3 group (Swank and Caughlan,[6a] who term the acid 2,3,4-trimethylpentane-2,4-phosphinic acid).

The alkaline hydrolysis of certain phosphetanium salts has been investigated by Fishwick, Flint, Hawes, and Trippett,[7] who emphasize three characteristics of such hydrolysis of quaternary phosphonium salts, (a) the rate varies with $[^-OH]^2$, (b) inversion occurs at the phosphorus atom, and (c) the group which is lost is that which is the most stable as an anion. The mechanism of the process involves an intermediate trigonal bipyramidal phosphorane (15), which loses from an apical position the group forming the stable anion. If the phosphorus

(15)   (16)   (17)   (18)

atom is part of a small strained ring, the considerable difference in bond angles between a group linked to an apical and an equatorial position (90°) and to two equatorial positions (120°) could produce constraint on the conformation of bipyramid (15) and thus lead to unusual behavior. These considerations are markedly supported by their results.

They repeated the preparation of Jungerman et al.,[3] except that a mixture of phenylphosphonous dichloride and aluminum chloride was treated with the 2,4,4-trimethylpent-2-ene, and the product on hydrolysis furnished 2,2,3,4,4-pentamethyl-1-phenylphosphetane 1-oxide (16). This was reduced with lithium aluminum hydride to the tertiary phosphine, which with methyl iodide gave 1,2,2,3,4,4-hexamethyl-1-phenylphosphetanium iodide (17; R = H, X = I). Alkaline hydrolysis gave a phosphine oxide, $C_{15}H_{25}OP$ (88%), m.p. 155°, the infrared, ultraviolet, and nmr spectra of which showed that it had the spirocyclic structure (18) with a five-membered ring (cf. p. 9). The mass spectrum showed the molecular ion at m/e 252, with a major cracking pattern consistent with (18). The oxide absorbed 2 molar equivalents of hydrogen over platinum, giving the saturated oxide, m.p. 133–135°, having a molecular ion m/e 256.

The above tertiary phosphine was also quaternized with diiodomethane to give 1-(iodomethyl)-2,2,3,4,4-pentamethyl-1-phenylphosphetanium iodide (17; R = I, X = I), which on alkaline hydrolysis gave a phosphine oxide, $C_{15}H_{23}OP$ (30%), m.p. 141°, identified by its infrared and nmr spectra as 2,2,3,4,4-pentamethyl-1-phenylphospholane 1-oxide (19). The mass spectrum showed the molecular ion at m/e 250, also with a confirmatory major cracking pattern.

The hydrolysis of both the phosphonium iodides (17; R = H, X = I) and (17; R = I, X = I) thus followed unusual directions. It is suggested that in the respective intermediates (**20**; R = CH_3) and (**21**; R = CH_3), the four-membered rings have been constrained to occupy apical-equatorial positions: this blocks the expected loss of the phenyl group

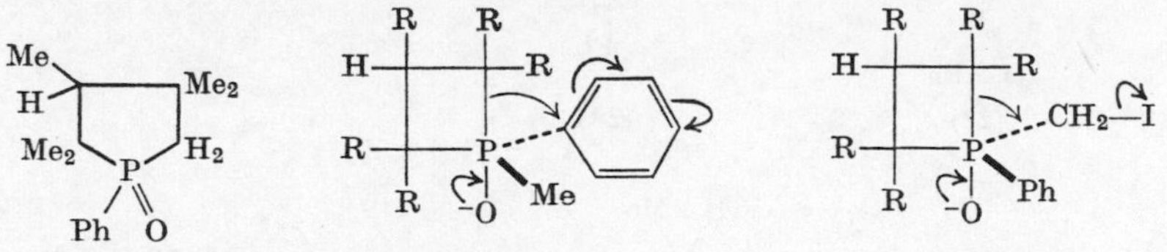

(**19**)   (**20**)   (**21**)

from (**20**), which must occur from an apical position, and also blocks the migration of the phenyl group in (**21**). The apical $(CH_3)_2C$ group therefore migrates either to the phenyl group (**20**) with protonation in the *para*-position, or to the CH_2I group (**21**) with expulsion of the iodine atom and ring expansion.

[*Note on Nomenclature*. The compound (**18**) is a phospholane, but on Ring Index method it would probably be termed 1,2,2,3,4,4-hexamethyl-1-phosphaspiro[4.5]deca-6,9-diene 1-oxide (**18A**), where [4.5] indicates

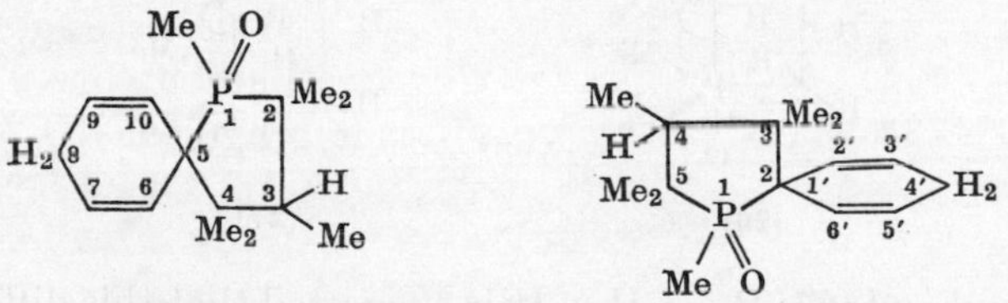

(**18A**)   (**18B**)

the number of atoms in the two cyclic links which are joined to the spiro atom, in the order of the numbering. On the phospholane basis (**18B**), the compound could be termed 2-(2',5'-cyclohexadienylidene)-1,3,3,4,5,5-hexamethylphospholane 1-oxide, but this is not entirely satisfactory.]

Fishwick and Flint[8] have similarly treated phenylphosphonous chloride and aluminum chloride with 3,3-dimethyl-1-butene, and hydrolyzed the product to give 2,2,3-trimethyl-1-phenylphosphetane 1-oxide (**22**). This was reduced by trichlorosilane (cf. p. 50) to the tertiary phosphine (**23**), which with methyl iodide gave 1,2,2,3-tetra-methyl-1-phenylphosphetanium iodide (**24**). This salt on alkaline hydrolysis gave a phosphine oxide, $C_{13}H_{21}OP$ (63%), m.p. 74–76°, which was identified by infrared, nmr and mass spectroscopy as *P*-methyl-*P*-phenyl-*P*-(1,1,2-trimethylpropyl)phosphine oxide (**25**).

2

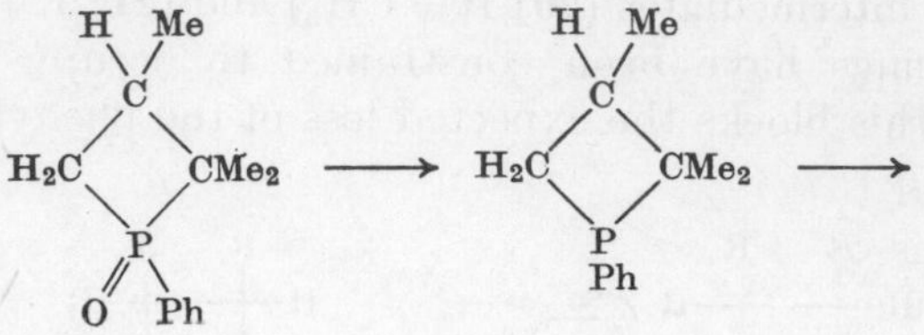

(22)   (23)

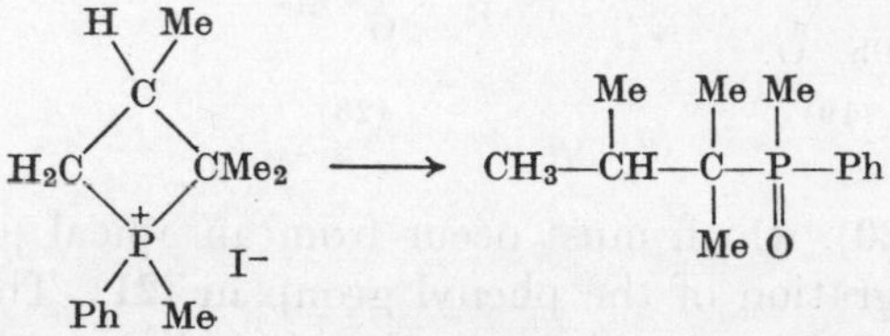

(24)   (25)

The formation of this oxide indicates that the trigonal bipyramidal phosphorane intermediate has undergone cleavage at the P—CH_2 bond (26; R = CH_3), followed by protonation, and not at the

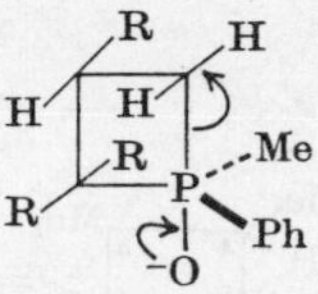

(26) 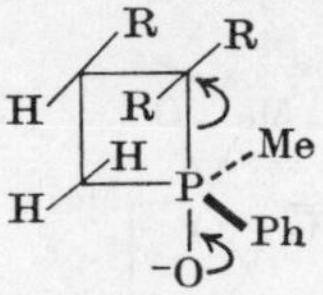   (27)

P—$C(CH_3)_2$ bond (27; R = CH_3). It is suggested that this differentiation between the two possible reactions is based on the greater stability of the CH_2 carbanion from (26) compared with that of the $C(CH_3)_2$ carbanion from (27). This less stable $C(CH_3)_2$ carbanion is also formed in the rearrangement of (20) where, however, without separation it attacks the phenyl group with protonation to form the compound (18).

Hawes and Trippett[8a] have also recorded that 2,2,3,4,4-penta-methyl-1-phenylphosphetane 1-oxide (16) exists as two geometric isomers, m.p. 118° and 127°, which can be separated chromatographi-cally. Reduction of a mixture of these oxides by lithium aluminum hydride gives the phosphetane, which when quaternized with benzyl bromide gives the homogeneous crystalline 1-benzyl-2,2,3,4,4-penta-methyl-1-phenylphosphetanium bromide (17; R = C_6H_5, X = Br), m.p. 202–203°.

This salt on alkaline hydrolysis gives solely the oxide (16), m.p. 127°. Alternatively, this salt is converted by ethanolic sodium ethoxide into the Wittig olefinic $\geq$P=CHC$_6$H$_5$ phosphorane, which reacts with benzaldehyde also to give the oxide (16(, m.p. 127°, and stilbene. The Wittig olefin synthesis occurs with retention of configuration at the phosphorus atom, and this must have occurred therefore also during the alkaline hydrolysis of the phosphetanium bromide (17; R = C$_6$H$_5$, X = Br). The probable mechanism of this hydrolysis supports the earlier thesis that the presence of a small ring, which prefers the equatorial-apical 90° bond angle to the diequatorial 120° angle in a trigonal-bipyramidal intermediate, can direct the course of the subsequent reaction. Substituents in the phosphetane ring may thus determine whether retention or inversion of the phosphorus configuration occurs.[501]

In contrast with the work of Fishwick et al.,[7] Cremer and Chorvat[9] record that 1,2,2,3,4,4-hexamethyl-1-phenylphosphetanium bromide (17; R = H, X = Br) on treatment with 1N-NaOH solution at 10° gives

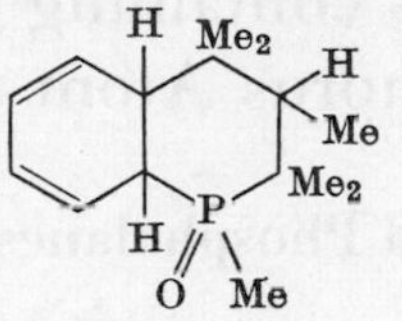

(28) 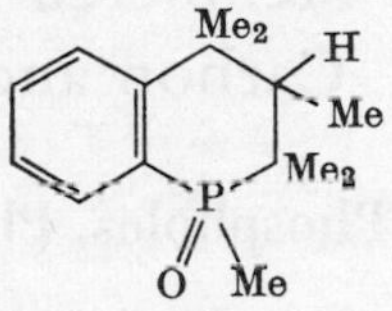   (29)

a phosphine oxide, C$_{15}$H$_{25}$OP (70–80%), m.p. 133–143°, which they identify, largely on the nmr spectra, as the compound (28); it undergoes aromatization (palladium–charcoal in boiling decalin) to give (29), i.e., 1,2,3,4-tetrahydro-1,2,2,3,4,4-hexamethylphosphinoline 1-oxide (cf. p. 124).

It is also recorded that the corresponding iodide (17; R = H, X = I), when treated with phenyllithium in ether at 25° and then with methyl iodide, gave dimethyl-(1,1,2,3-tetramethyl-3-phenylbutyl)phenylphos-phonium iodide (30), m.p. 223–227° (38%), and that the oxide (22)

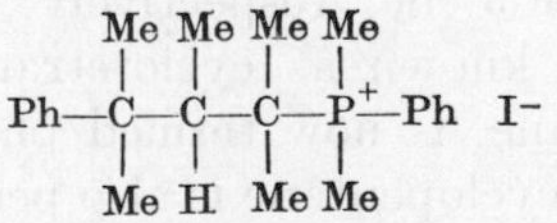

(30) 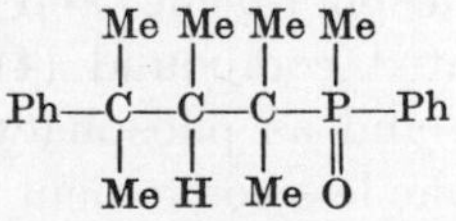   (31)

when similarly treated gave methyl-(1,1,2,3-tetramethyl-1-phenyl-butyl)phenylphosphine oxide (31), m.p. 108–120° (81%). For the mechanism whereby these compounds are formed see Cremer and Chorvat.[9]

Cremer [9a] has later reinvestigated the alkaline hydrolysis of 1,2,2,3,4,4-hexamethyl-1-phenylphosphetanium bromide, and by deuterium labelling has reaffirmed the formation of the phosphine oxide (**18**): rearrangement of this oxide to the substituted phosphinoline oxide (**29**) presumably occurs during the aromatization process in boiling decalin.

Kosolapoff and Struck[10] have described the synthesis of 1-hydroxyphosphetane 1-oxide (trimethylenephosphinic acid), melting 'unsharply' at 65°, from 3-bromopropylphosphonous dichloride; but, since the yield was minute (12 mg of acid from 93.1 g of dichloride) and the acid, isolated by evaporation of the final solution, was identified solely by a phosphorus analysis, its identity must remain uncertain.

The phosphetane ring occurs in other compounds in which it is fused to other ring systems (cf. p. 154).

Five-Membered Ring Systems containing only Carbon and One Phosphorus Atom

Phospholes, Phospholenes, and Phospholanes

Phosphole has the formula (**1**), [RRI 148, in which the molecule is depicted inverted, with the phosphorus atom at the top]; it can be

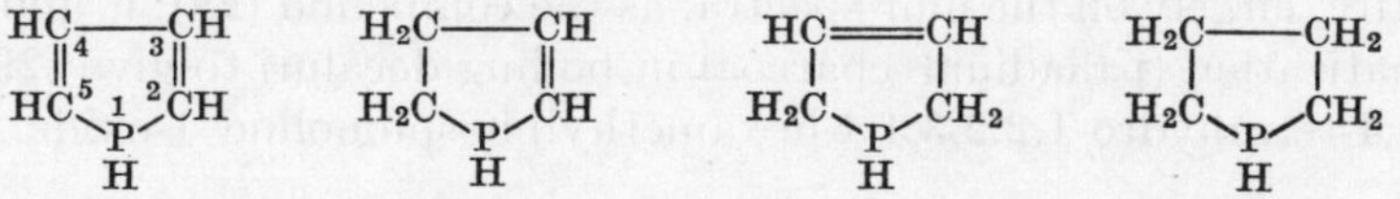

regarded as the phosphorus analogue of pyrrole, but it has many markedly different properties. Two isomeric dihydro derivatives should exist, the 2-phospholene (**2**) and the 3-phospholene (**3**), alternatively named phosphol-2-ene and phosphol-3-ene, respectively. The fully hydrogenated compound (**4**), earlier known as cyclotetramethylenephosphine and as phosphacyclopentane, is now termed phospholane, although the less convenient phosphacyclopentane is also permitted by IUPAC rules. The name phospholidine (analogous to pyrrolidine) is still occasionally encountered.

Each of the three classes, phospholes, phospholenes, and phospholanes, has its own extensive chemistry, and will therefore be discussed in turn.

Phospholes

Phosphole (**1**), [RRI 148], has not been isolated, and it is probably very reactive and therefore in most circumstances an unstable compound. The preparation, properties, and reactions of substituted phospholes will be considered first, with the evidence that these factors provide for the nature of the phosphole ring system.

(**1**)

The preparation of aryl-substituted phospholes was reported almost simultaneously from two sources (1959). Smith and Hoehn[11] had earlier (1941) shown that diphenylacetylene (**2**), when treated in ether with lithium, underwent dimerization with the formation of the 1,4-dilithio derivative (**3**) of 1,2,3,4-tetraphenylbutadiene; an alternative product, however, under these conditions was 1,2,3-triphenylnaphthalene. Leavitt, Manuel, and Johnson[12] in a preliminary Note recorded the

$$2C_6H_5C{\equiv}CC_6H_5 \rightarrow C_6H_5C(Li){=}C(C_6H_5){-}C(C_6H_5){=}C(Li)C_6H_5$$
$$\quad\quad (2) \quad\quad\quad\quad\quad\quad\quad\quad\quad (3)$$

reaction of the dilithio compound (**3**) with phenylphosphonous dichloride to give 1,2,3,4,5-pentaphenylphosphole (**4**; R = C_6H_5), m.p. 261–262°. In a later paper, with Matternas and Lehmann,[13] they point out that, although the reaction of lithium with diphenylacetylene (**2**) (followed by hydrolysis) gives either 1,2,3,4-tetraphenylbutadiene or 1,2,3-triphenylnaphthalene, a mixture is never obtained; in the course of the dimerization of (**2**), the concentration of some intermediate apparently catalyzes the ring closure to the naphthalene. By carrying out the dimerization to *ca.* 60% conversion into (**3**) before adding the phenylphosphonous dichloride, a maximum yield of 68% of the yellow crystalline phosphole (**4**; R = C_6H_5), m.p. 256–257°, was obtained. In this reaction, the compound (**3**) is presumably the *cis–cis*-isomer. The

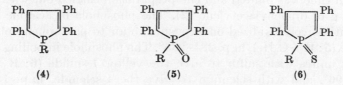

(**4**)   (**5**)   (**6**)

similar use of the dilithio compound (**3**) to form five-membered ring systems containing arsenic, antimony, tin, and germanium, and the corresponding spirocyclic derivatives of tin and germanium are recorded. Leavitt and Johnson, in a patent specification,[14] recommend that the ether solution of (3) should be chilled to $-40°$ before addition of the phosphonous dichloride.

Hübel and Braye meanwhile reported that various iron carbonyl complexes, such as $Fe_2(CO)_6C_{28}H_{20}$, where $C_{28}H_{20}$ represents 1,2,3,4-tetraphenylbutadiene linked at both double bonds to one of the iron atoms, react with sulfur to give tetraphenylthiophene,[15] and with phenylphosphonous dichloride at $140°$ to give pentaphenylphosphole (4; $R = C_6H_5$), m.p. 255–256° (66%).[16] To confirm the identity of the latter, they also independently used the interaction of the dilithio compound (3) with phenylphosphonous dichloride to prepare the same phosphole (4; $R = C_6H_5$). This paper,[17] although short, gives much information about derivatives of pentaphenylphosphole, its reactions with iron carbonyls, and with the normal Diels–Alder reagents; more information was given later (Braye, Hübel, and Caplier[18, 19]), and the contents of the two studies may be summarized together. These workers showed that the dilithio derivative (3) when treated with iodine, both in ethereal solution, deposited 1,4-diiodo-1,2,3,4-tetraphenylbutadiene, $C_6H_5C(I){=}C(C_6H_5){-}C(C_6H_5){=}C(I)C_6H_5$, m.p. 202°, and that this compound would condense with many compounds of type R_nMX_2, where R is an alkyl or aryl group, M is the heteroatom, and X a univalent element, to give the corresponding five-membered ring: for example, the diiodo compound condensed with phenyldisodiophosphine, to give pentaphenylphosphole (4; $R = C_6H_5$), thus providing a third synthetic route.

The method using the dilithio compound (3) has the most extensive use: for example, the five-membered arsenic and antimony compounds could not be prepared from the 1,4-diiodo intermediate, but the use of one or both methods afforded the corresponding five-membered derivatives of mercury, gold, boron, silicon (single ring and spirocyclic), zirconium, sulfur, arsenic, antimony, and tellurium.

Pentaphenylphosphole was obtained in 88% yield from the 1,4-diiodo compound, and in 84% from the 1,4-dilithio compound (3) if the reaction time was limited and the preparation carried out with not less than 5 g of diphenylacetylene (2). The phosphole in acetone solution is quantitatively oxidized on exposure to air to pentaphenylphosphole 1-oxide (5; $R = C_6H_5$), m.p. 284–285°. The phosphole in boiling benzene readily unites with sulfur to give the yellow 1-sulfide (6; $R = C_6H_5$), m.p. 196°, and with selenium to give the 1-selenide, m.p. 186–188°. It is noteworthy that the dilithio derivative (3) reacts with phenylthiophosphonic dichloride, $C_6H_5P(S)Cl_2$, to give pentaphenylphosphole

(45%) and its 1-sulfide (1%); this P–S fission occurs also in the penta-carbonyliron reaction (see below).

p-Bromophenylphosphonous dichloride reacts with (3) to form the 1-p-bromophenyl compound (4; R = p-BrC_6H_4), m.p. 236–239°. Benzyl-phosphonous dichloride reacts similarly with (3) to give the greenish-yellow 1-benzyl-2,3,4,5-tetraphenylphosphole (4; R = $CH_2C_6H_5$), m.p. 203–213° (38%); the 1-oxide (5; R = $CH_2C_6H_5$), m.p. 228–230°, is formed as a 'by-product' in this reaction, and also (in 10% yield) when certain iron carbonyls (p. 14) are heated with benzylphosphonous dibromide in benzene at 170° for 16 hours. Methylphosphonous di-chloride reacts with (3) to give 1-methyl-2,3,4,5-tetraphenylphosphole (4; R = CH_3) which, however, is so readily oxidized that only the 1-oxide (5; R = CH_3), m.p. 240–241°, has been isolated.

Pentaphenylphosphole has a strong green fluorescence, its oxide has a bright yellow fluorescence, and its sulfide and selenide have none: the fluorescence of this phosphole (4; R = C_6H_5) is stronger than that of analogous phospholes in which the 1-phenyl group is replaced by an alkyl group.

The ready conversions of pentaphenylphosphole into the oxide, sulfide, and selenide show clearly that its phosphorus atom has a reactive lone pair of electrons. This is confirmed by its reaction with pentacarbonyliron to give the orange-red tetracarbonylphospholeiron (7), m.p. 180–185°, the infrared spectrum of which is closely similar to

(7)   (8)

that of other carbonyls of general structure R_3P—$Fe(CO)_4$ (cf. Reppe et al.[20]). The phosphole reacts, however, with $Fe_3(CO)_{12}$ in boiling isooctane to give the tetracarbonyl (7) and also a yellow tricarbonyl (8), m.p. 205–215° (dec.), having a structure similar to that of the cyclo-pentadienone iron tricarbonyls (cf. Weiss and Hübel[21]). In the com-pound (8), the four π-electrons of the conjugated double bonds of the phosphole system are donated to the iron atom; this illustrates the resemblance between this system and the wide variety of non-aromatic conjugated dienes which also react very readily with iron carbonyls.

It is noteworthy that pentaphenylphosphole 1-oxide reacts even more readily than the parent phosphole with pentacarbonyliron, giving a compound of composition $C_{34}H_{25}PO,Fe(CO)_3$, yellow crystals, m.p. 226–235°, which has the same structure as (8), with of course an oxygen atom on the phosphorus. Pentaphenylphosphole 1-sulfide reacts about as readily as the parent phosphole with pentacarbonyliron, but with the formation solely of the tetracarbonyl (7), the sulfur having been split off.

Pentaphenylphosphole acts as a diene in the Diels–Alder reaction. When it was heated with maleic anhydride in benzene in a sealed tube at 150° for 20 hours, 83.5% of unchanged phosphole was recovered, but some of the adduct, presumably (9), m.p. 260–264° (dec.), was isolated. The phosphole 1-oxide, similarly treated, gave the P-oxide adduct (67%), m.p. 245–255° (impure). The two adducts are apparently

Ph————Ph   Ph————Ph   Ph
Ph\ P /Ph   Ph\ P /Ph   Ph——COOMe
    Ph   Ph   Ph——COOMe
HC————CH   Ph
OC   CO   MeOOC   COOMe
   O
  (9)   (10)   (11)

different,[18] contrary to an earlier statement,[17] but analytical results are not recorded. The phosphole, heated with dimethyl acetylene-dicarboxylate in dioxan at 150° for 15 hours, apparently gave the adduct (10), which then decomposed to form dimethyl tetraphenylphthalate (11) (88%), the evicted $P—C_6H_5$ units probably combining to form one of the polyphenyl-substituted cyclic polyphosphines.

From a consideration of all these reactions, these workers conclude that pentasubstituted phospholes act as tertiary phosphines having a conjugated diene system, and that they possess little (if any) aromatic character.

Two other syntheses of substituted phospholes should be mentioned before further discussion of properties.

Phenylphosphonous dichloride when heated with 1,4-diphenyl-butadiene at 226–230° for 10 hours gives the yellow 1,2,5-triphenyl-phosphole (13), m.p. 186.5–187.5° (50–66%) with vigorous evolution of hydrogen chloride (Campbell, R. C. Cookson, and Hocking[22]); presumably this process initially involves a normal McCormack reaction (p. 31) with the formation of the phosphorane (12), which at the high temperature loses hydrogen chloride with formation of the phosphole (13). The structure of the phosphole is confirmed by (a) its conversion by hydrogen peroxide into the 1-oxide, m.p. 237.5–238°, and (b) its reaction

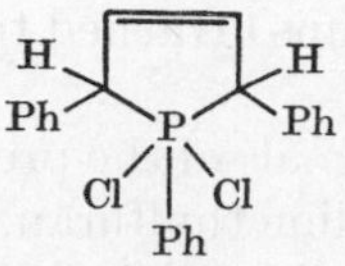

(12) 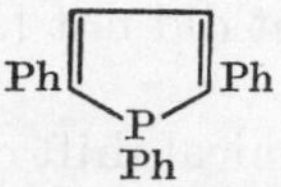   (13) 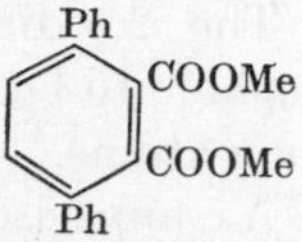   (14)

with dimethyl acetylenedicarboxylate to form a normal adduct of type
(10), which also spontaneously loses the P—C_6H_5 unit to form dimethyl
3,6-diphenylphthalate (14), m.p. 189–191°.

A mixture of phenylphosphonous dichloride and 1,2,3,4-tetra-
phenylbutadiene, when similarly heated at 210–230° for 10 hours, gave
pentaphenylphosphole (4; R = C_6H_5) in 40% yield—probably the
simplest method for preparing this phosphole (Campbell, R. C. Cookson,
Hocking, and Hughes [23]).

A recent simple synthesis of the phosphole system (Märkl and
Potthast [24]) has two variations. (a) When various 1,4-butadiynes (15;
where R is an aryl or an alkyl group) are added to a solution of bis-
(hydroxymethyl)phenylphosphine, $C_6H_5P(CH_2OH)_2$, in anhydrous pyri-

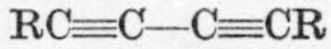

RC≡C—C≡CR

(15) 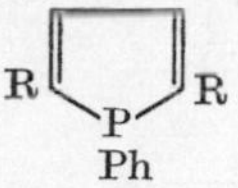   (16)

dine, which is then boiled for 5 hours, the corresponding 2,5-disubstituted
1-phenylphosphole (16) usually crystallizes out. (b) When the butadiyne
(15) is added to a cold solution of phenylphosphine in benzene or
benzene–tetrahydrofuran containing a catalytic quantity of phenyllithium, a vigorous initial reaction may occur; the reaction mixture,
after 18 hours at room temperature, gives the phosphole (16) on distillation.

The substituents in the 2,5-position in (16) and the yield by method
(a) or (b) are listed in Table 1.

Table 1. 2,5-Disubstituted-phenylphospholes

R in (16)	M.p. or b.p. (°)	Yield (%)	
		(a)	(b)
C_6H_5	184–186	20.5	60
2-$C_{10}H_7$	230–231	17.5	89.5
p-$CH_3C_6H_4$	194–196	29	59
p-BrC_6H_4	200–202	16.5	—
CH_3	66–69°/0.2	0	51

The 2,5-dimethyl-1-phenylphosphole is claimed to be the first phosphole to be prepared that did not have aryl groups attached to the carbon atoms.

A comparison of the chemical shift of the nmr signals of the protons on C-3 and C-4 in this compound with that in 2,5-dimethylfuran, 2,5-dimethylthiophen, and 1,2,5-trimethylpyrrole is considered to be 'not inconsistent with aromatic bonding in phospholes'.

The conversion of 1-phenyl-3-phospholene 1-oxide and its 3-methyl and 3,4-dimethyl analogues into the corresponding 1-phenylphosphole 1-oxide derivatives has been briefly recorded (Howard and Donadio [25]).

The 1-phenyl group in pentaphenylphosphole can be evicted by certain metals to give the reddish-violet anion (17), although the degree of ionization of the metallic derivative may vary in different solvents; reaction with potassium proceeds to 50–85% when the phosphole and the metal are (a) in boiling dioxan or toluene for 3 hours or (b) in tetrahydrofuran at 25° for 1 hour; reaction with lithium proceeds to 60–70%

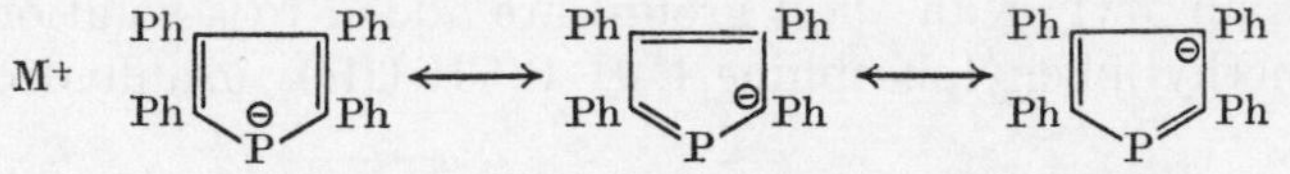

in tetrahydrofuran at room temperature. The deep color of the anion is undoubtedly due to contributions from (17A) (two forms) and (17B) (two forms). 1,2,5-Triphenylphosphole, treated with potassium in boiling dioxan for 3 hours, gives the reddish-violet anion (18) (40%), which will have similar canonical forms (Braye [26]).

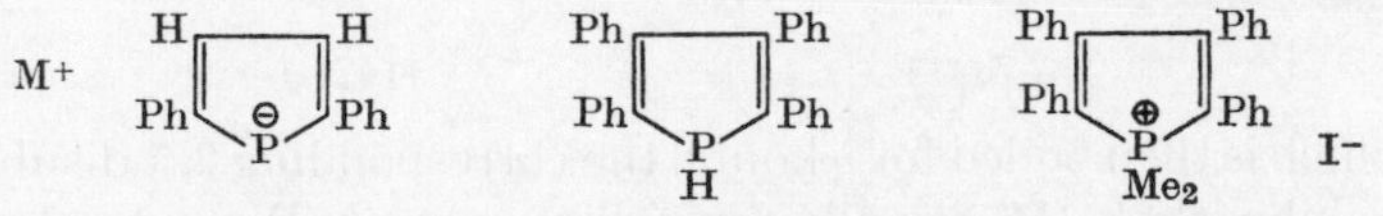

The lithium or potassium salt (17) is a ready source of many derivatives, examples of which are listed:

(i) Water gives 2,3,4,5-tetraphenylphosphole (19), m.p. 147–150° (24%): its infrared spectrum shows a marked P—H band at 2309 cm^{-1}.

(ii) Methyl iodide gives, first, 1-methyl-2,3,4,5-tetraphenylphosphole (4; R = CH$_3$), m.p. 195–196°, and then 1,1-dimethyl-2,3,4,5-tetraphenylphospholium iodide (20), m.p. 297°. [Similarly, the 2,5-diphenyl anion (18) with methyl iodide gives 1-methyl-2,5-diphenylphosphole, m.p. 110–111° (1-oxide, m.p. 163–165°).]

(iii) Ethyl bromoacetate gives 1-(ethoxycarbonylmethyl)-2,3,4,5-tetraphenylphosphole (**20A**; $R = C_2H_5$), m.p. 157–158°, which on hydrolysis gives the 1-carboxymethyl derivative (**20A**; $R = H$), m.p. 210–212°.

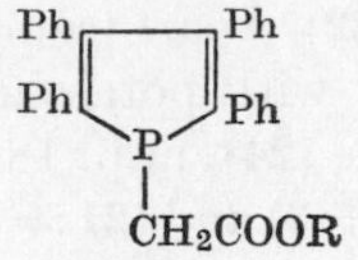

(iv) When the alkali salt (**17**) was added to a large (often 10 molar) excess of $\alpha\omega$-dibromoalkanes, the corresponding 1-(ω-bromoalkyl)-phosphines (**21**) were obtained. These compounds are listed in Table 2.

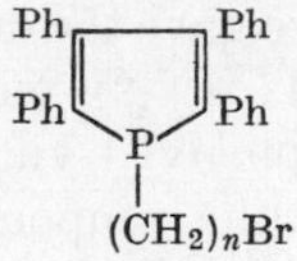

Table 2. 1-(ω-Bromoalkyl)-2,3,4,5-tetraphenylphospholes

n in (**21**)	2	3	4	5	6
M.p.	219°	205–206°	147–150°	132–135°	160–163°
Yield (%)	73	56	19	51	44

All these compounds should be capable of self-quaternization. The actual results are of considerable interest. The compound (**21**; $n = 2$) in

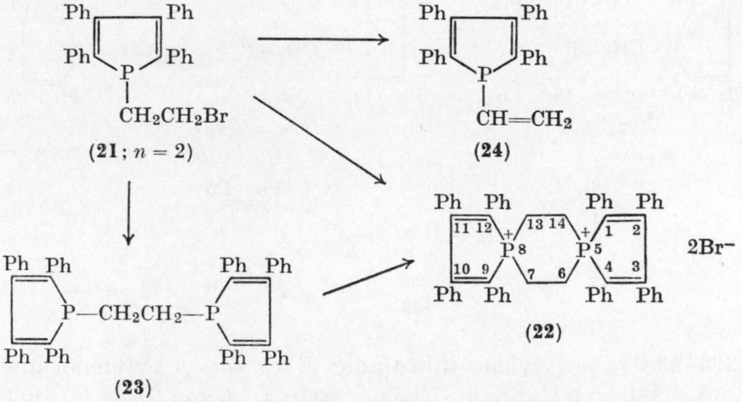

boiling benzonitrile gives the tricyclic dibromide (**22**), i.e., 1,2,3,4,9,10,-11,12-octaphenyl-5,8-diphosphoniadispiro[4.2.4.2]tetradeca-1,3,9,10-tetraene dibromide, [RIS —], m.p. 320–325°, with a polyphosphonium polymer formed almost certainly by linear condensation: the yield of (**22**) varied from 0 to 61%. Reversal of the conditions for the preparation of (**21**; $n = 2$), *i.e.*, the addition of 1,2-dibromoethane to an excess of the salt (**17**), gave 2,2′,3,3′,4,4′,5,5′-octaphenyl-1,1′-ethylenedi(phosphole) (**23**), m.p. 325–330°, which when heated with an equivalent of 1,2-dibromoethane also gave the dibromide (**22**). Apart from these reactions, the compound (**21**; $n = 2$), on treatment with potassium *tert*-butoxide gave 2,3,4,5-tetraphenyl-1-vinylphosphole (**24**), m.p. 181–183°.

When heated, the compounds (**21**; $n = 3$) and (**21**; $n = 6$) gave solely polymers, but the intermediates (**21**; $n = 4$) and (**21**; $n = 5$) underwent intramolecular quaternization with the formation of 1,2,3,4-tetraphenyl-5-phosphoniaspiro[4.4]nona-1,3-diene bromide (**25**), m.p. 333–335°

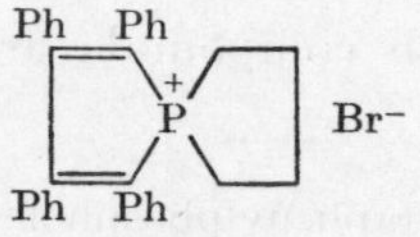

(**25**) 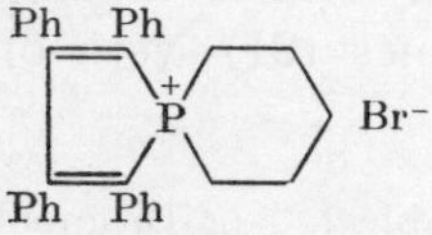   (**26**)

(p. 63), and 1,2,3,4-tetraphenyl-5-phosphoniaspiro[4.5]deca-1,3-diene bromide (**26**), m.p. 357° (p. 120); the yield of (**25**) was almost quantitative, but that of (**26**) was only 31% since much polymeric material was also formed.

The conditions employed for the preparation of (**23**), when used for the interaction of the salt (**17**) and dibromomethane give methylenebis-1,1′-(2,3,4,5-tetraphenylphosphole) (**27**), m.p. 328–329° (1,1′-dioxide,

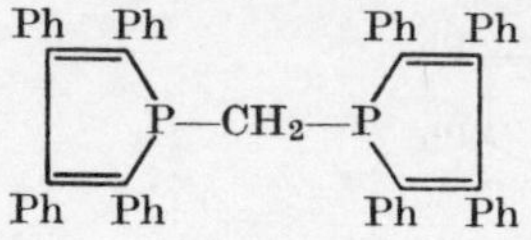

(**27**) 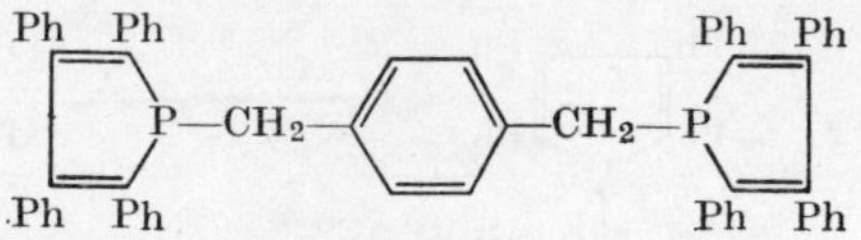   (**28**)

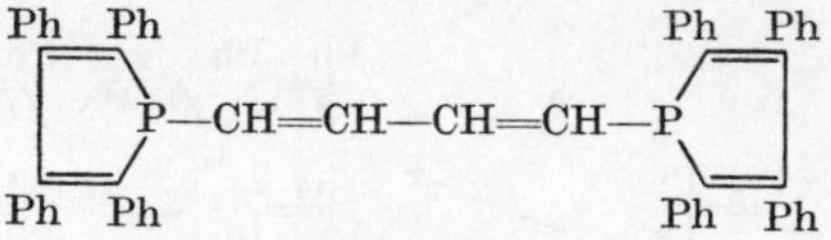

(**29**)

m.p. 334–338°); *p*-xylylene dibromide gives the *p*-xylylene analogue (**28**), m.p. 280° (1,1′-dioxide, m.p. >370°); *trans-trans*-1,4-dichlorobutadiene gives 2′,2″,3′,3″,4′,4″,5′,5″-octaphenyl-1′,1″-(*trans*-1,*trans*-3-butadienylene)di(phosphole) (**29**), m.p. 280° (1,1′-dioxide, m.p. 350–357°) (Braye [27]).

An entirely different type of compound arises when the salt (**17**) is treated with sulfur in the presence of water, namely 2,2′,3,3′,4,4′,5,5′-octaphenyl-1,1′-thiodiphosphole 1,1′-disulfide (**30**; $n = 1$), m.p. 252–258°, and the dithio analogue (**30**; $n = 2$), m.p. 245–247°.

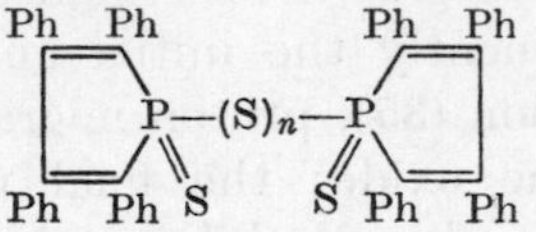

(**30**)

Braye and Joshi [28] have shown that the lithium salt (**17**) reacts with dicarbonylcyclopentadienyliron bromide (**31**), to form dicarbonylcyclo-pentadienyl-1-(2,3,4,5-tetraphenylphosphole)iron (**32**), m.p. 175–180°

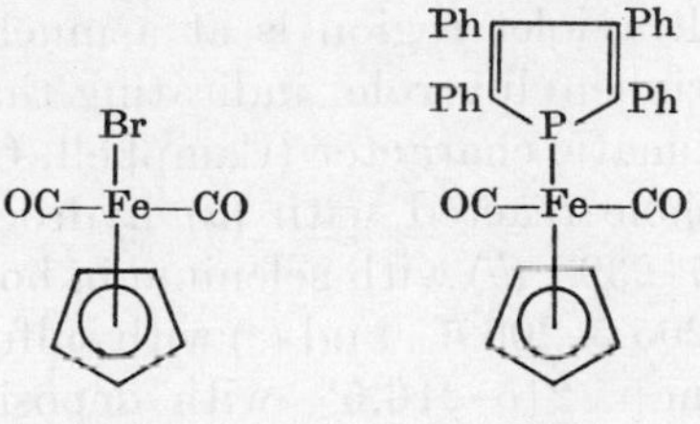

(**31**)   (**32**)

(dec.), which readily gives the corresponding P-oxide, m.p. 250° (dec.); the infrared spectrum of (**32**) shows C=O bands at 2012 and 1969 cm^{-1}, whereas that of the oxide shows them at 2024 and 1972 cm^{-1}. Attempts to remove the CO groups from (**32**) and thus obtain a 'phosphaferrocene' failed, however.

Hydrolysis of Quaternary Salts. The alkaline hydrolysis of 1-methyl-1,2,5-triphenylphospholium iodide (**33**; R = H) and of 1-methyl-1,2,3,4,5-pentaphenylphospholium iodide (**33**; R = C$_6$H$_5$) shows second-order

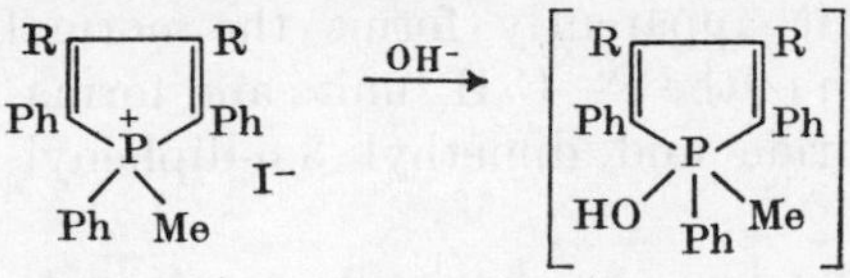

(**33**)   (**34**)

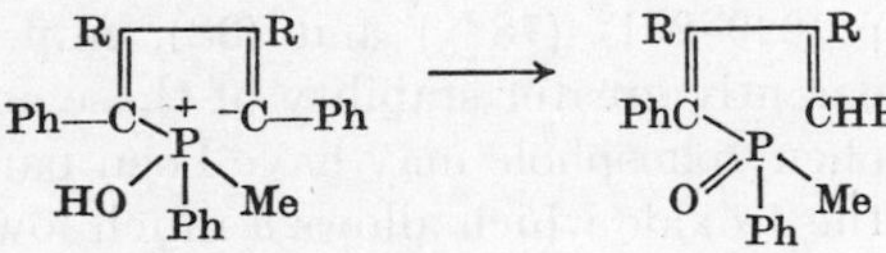

(**35**)   (**36**)

kinetics—first-order dependence on the concentration of the phospholium ion and on that of the hydroxyl ion (Bergesen [29]). No cyclic compound was found among the products. This was attributed to the conjugation between the ring system and the phenyl groups attached to it giving rise to an electronically more stabilized anion than that of the $P—C_6H_5$ group. Consequently the initial unstable phosphorane (34) gives rise to the zwitterion (35), proton migration then producing the usual tertiary phosphine oxide; the final products were therefore (1,4-diphenyl-1,3-butadienyl)methylphenylphosphine oxide (36; $R = H$), m.p. 169–170°, and methylphenyl-(1,2,3,4-tetraphenyl-1,3-butadienyl)-phosphine oxide (36; $R = C_6H_5$), m.p. 204°.

The properties and reactions of 1,2,5-triphenylphosphole (16; $R = C_6H_5$) have been examined in some detail. The compound is bright yellow and shows a strong blue fluorescence in solution; its absorption in the ultraviolet region is at a much longer wave-length than that of 1,2,5-triphenylpyrrole, indicating that the phosphole has (as expected) less aromatic character (Campbell, Cookson, et al.[23])

Triphenylphosphole reacted with (a) hydrogen peroxide to give the 1-oxide, m.p. 237–239°, (b) with selenium in boiling benzene to form the 1-selenide, m.p. 205.5–206.5°, and (c) with sulfur in boiling xylene to give the 1-sulfide, m.p. 215–216.5°, with deposition of an insoluble polymer, m.p. 350–351° (after crystallization from dimethylformamide). All these compounds were yellow except the colorless polymer. The methiodide formed a yellow monohydrate, m.p. 218.5–222°.

The 1-oxide in ethyl acetate–acetic acid with platinum oxide was hydrogenated to 1,2,5-triphenylphospholane 1-oxide, m.p. 205–207°; the phosphole itself could not be hydrogenated, for it poisoned the catalyst.

1,2,5-Triphenylphosphole is not regarded as a characteristic diene, for it does not react with maleic anhydride or with acrylonitrile in benzene at 80°, but with maleic anhydride and with dimethyl acetylene-dicarboxylate in benzene at 150° it apparently forms the normal adducts which decompose with fission of the $P—C_6H_5$ units and formation of 3,6-diphenylphthalic anhydride and dimethyl 3,6-diphenyl-phthalate, respectively.

The phosphole 1-oxide in boiling benzene, however, reacts with maleic anhydride and acrylonitrile to give the stable adducts (37), m.p. 229–231° (78%) and (38), m.p. 180° (40%), respectively; the apparently greater stability of these adducts than of those from 1,2,5-triphenylphosphole may have been partly due to the greater reactivity of the 1-oxide which allows a much lower temperature of reaction. The 1-oxide, (a) when heated with dimethyl acetylenedicarboxylate at 156°

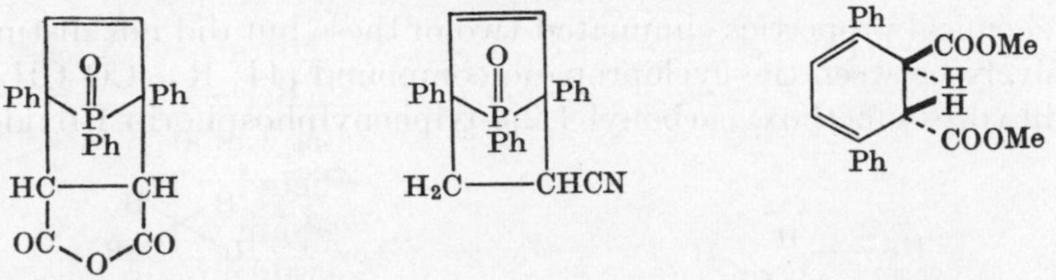

(37)   (38)   (39)

for 2 minutes, gave a dark residue from which only dimethyl 3,6-diphenylphthalate was isolated, and (b) reacted vigorously with dimethyl fumarate at 160–230° (15 minutes) to give dimethyl 3,6-diphenyl-3,5-cyclohexadiene-*trans*-1,2-dicarboxylate, m.p. 169–171° (39); in each case the bridging —P(O)(C_6H_5)— group of the initial adduct has been lost, as in the unstable adducts formed from the tertiary phosphine.

The triphenylphosphole, when heated with ethyl bromoacetate in boiling benzene for 47 hours, gave 1-ethoxycarbonylmethyl-1,2,5-triphenylphospholium bromide (40), m.p. 184.5–185°, which when

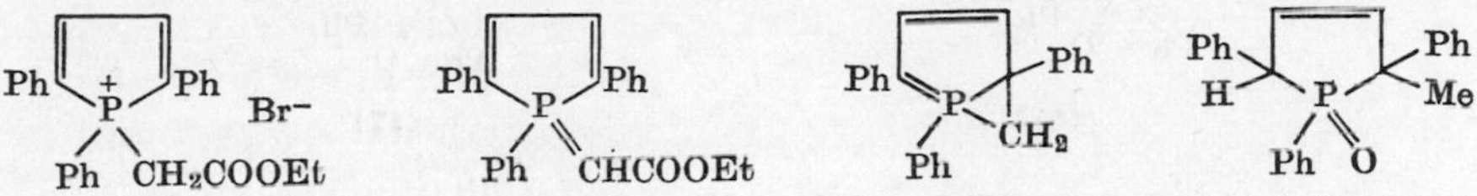

(40)   (41)   (42)   (43)

treated with 4N-sodium hydroxide until just alkaline gave the phosphorane (41), m.p. 161–163°; this did not undergo a Wittig reaction with cyclohexanone. On one occasion, this treatment of (40) with alkali gave the phosphine oxide and a colorless compound, m.p. 175.5°–176.5°, identified by its nmr spectrum as 2-methyl-1,2,5-triphenyl-3-phospholene 1-oxide (43), which, it is suggested, arose from the unstable three-membered ring system (42); catalytic hydrogenation of (43) gave the corresponding phospholane.

The products formed by the action of the triphenylphosphole and its 1-oxide with methyl diazoacetate varied considerably with the conditions. (a) The phosphole in boiling dioxan containing copper powder gave the dimethyl ester (39), presumably by intermediate decomposition of the diazoacetate to dimethyl fumarate. (b) Repetition of (a) but without the copper gave a colorless methyl ester, $C_{25}H_{21}O_3P$, m.p. 217–218°, which on hydrolysis gave the acid $C_{24}H_{19}O_3P$, m.p. 285–287°. The yield of the ester from the phosphole was very small, but rather larger from the phosphole oxide. The ester could have one of four possible structures: the combination of ultraviolet and nmr spectra and chemical properties eliminated two of these but did not distinguish decisively between the cyclopropane compound (44; R = CO$_2$CH$_3$) and 1,4-dihydro-3-methoxycarbonyl-1,2,5-triphenylphosphorin 1-oxide (45;

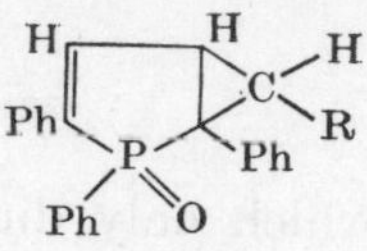

(44)

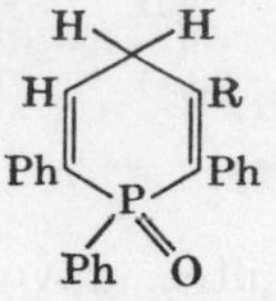

(45)

R = CO$_2$CH$_3$). (c) The phosphole and the diazoacetate in dioxan at room temperature for 42 days afforded the pale-yellow pyrazoline (46), m.p. 233–235° (34%); the phosphole oxide, similarly treated for 3 months,

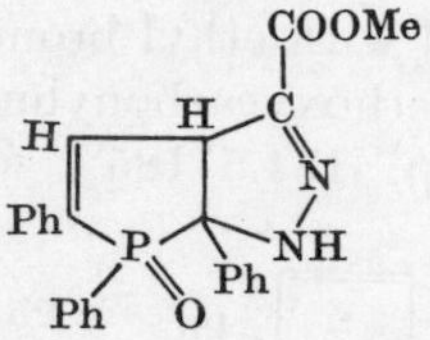

(46)

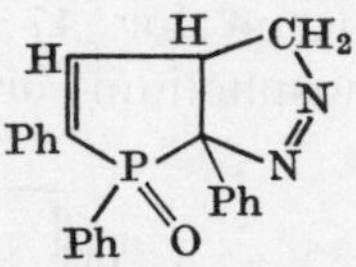

(47)

also gave (46) (89%). The —NH—N= structure was assigned to (46) rather than the azo structure, for the infrared spectrum showed an —NH— band at 3130 cm^{-1}. This compound, in boiling bis-(2-methoxyethyl) ether containing copper powder, decomposed giving the methyl ester, m.p. 217.5–219°, identical with that above, i.e., either (44; R = CO$_2$CH$_3$) or (45; R = CO$_2$CH$_3$).

A solution of the phosphole 1-oxide and diazomethane in dioxan–ether, when kept at −15° to −18° for 6 days, gave the pale yellow pyrazoline (47), which decomposed at 190°. This with sodium borohydride gave a dihydro derivative, m.p. 256–259°; no NH band was present in the infrared spectrum of (47), but it appeared in that of the dihydro derivative.

The pyrazoline (47), heated in boiling bis-(2-methoxyethyl) ether with copper powder, readily lost nitrogen, giving the pale yellow cyclopropane compound (44; R = H), m.p. 234–235°; the same compound was formed when the phosphole oxide and diazomethane in dioxan–ether was subjected to ultraviolet light. The structure of (44; R = H) is shown by its nmr spectrum and its catalytic dihydrogenation; it is a remarkably stable compound, being unaffected by bromine, boiling hydrobromic acid, or warm concentrated sulfuric acid.

These authors (Campbell, Cookson, Hocking, and Hughes [23]) also conclude that the physical and chemical evidence indicates that in 1,2,5-triphenylphosphole the heterocyclic ring has little or no aromatic character. This is in general agreement with theoretical predictions (Brown [30]) and with estimates of the resonance energy of similar systems (Bedford, Heinekey, Millar, and Mortimer [31]).

It is noteworthy that 1,2,5-triphenylphosphole in dichloromethane coordinates readily with certain metallic halides: with potassium tetrachloro- and tetrabromo-palladate(II) it forms orange and red crystals, respectively, of composition $(C_{22}H_{17}P)_2PdX_2$, where X = Cl or Br; the corresponding dichloro- and dibromo-platinum(II) derivatives are both yellow. The phosphole reacts with mercuric chloride to give only the yellow bridged derivative $(C_{22}H_{17}P)ClHgCl_2HgCl(C_{22}H_{17}P)$ (cf. p. 51), and with rhodium(III) chloride it gives orange crystals having the exceptional composition $(C_{22}H_{17}P)RhCl_3$. The infrared spectra of these compounds has been recorded in detail (Walton [32]). For the metal-carbonyl derivatives of this phosphine see R. C. Cookson, Fowles, and Jenkins.[33]

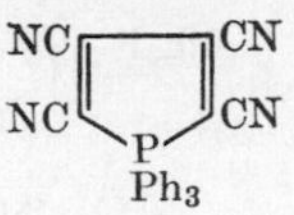

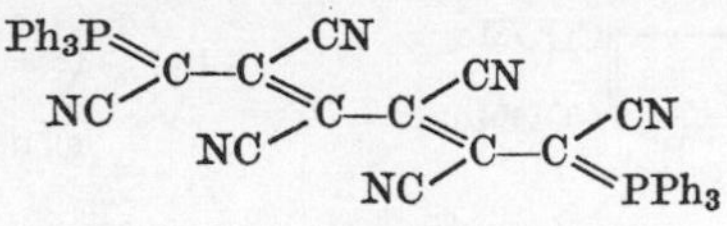

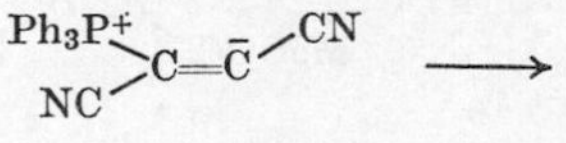

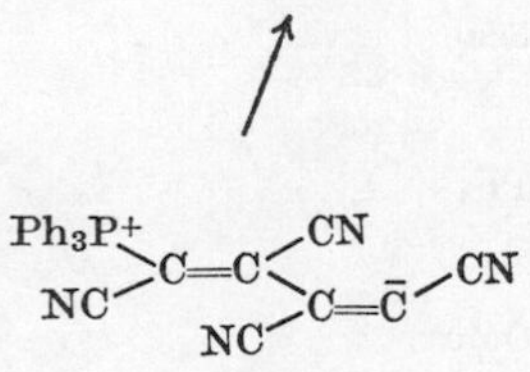

It has been claimed that the addition of an acetonitrile solution of dicyanoacetylene (2 molar equivalents) to a similar solution of triphenyl-phosphine (1 equivalent) at 25–30° gives a deep purple solution which deposits 2,3,4,5-tetracyano-1,1,1-triphenylphosphole (48), orange crystals, m.p. 237–239° (from pyridine) (Reddy and Weis [34]) The structure of this compound was subsequently investigated in considerable detail (Shaw, Tebby, Ward, and Williams [35]). The mass spectrum shows a molecular ion at m/e 752, of composition $C_{48}H_{30}N_6P_2$, determined by high resolution; the compound must be an adduct of three molecules of dicyanoacetylene and two of triphenylphosphine. Consideration of the full mass-spectrometer results, of the ultraviolet spectrum, and of the chemical analysis indicates that the compound is 1,6-bis(triphenyl-phosphoranylidene) - 2,4 - hexadiene - 1,2,3,4,5,6 - hexacarbonitrile **(49)**. This is probably formed by the combination of the initial reactants to give the 1:1 dipolar intermediate **(50)**, which then combines with a second molecule of the acetylene to give an extended dipolar intermediate **(51)**; combination of **(50)** and **(51)** then gives the final product **(49)**. This compound could have four contributing structures: its structure and its ultraviolet spectrum are in harmony with its yellow color, whereas the phosphole **(48)** would be expected to be colorless.

Tetramethyl 1,1,1-triphenylphosphole-2,3,4,5-tetracarboxylate **(52)** is stated to be formed when equimolecular quantities of triphenyl-phosphine and dimethyl acetylenedicarboxylate in ether are mixed

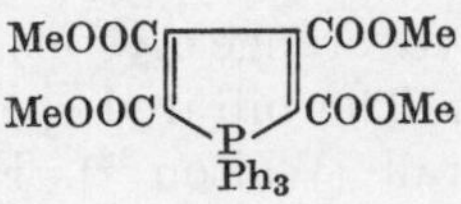

(52) 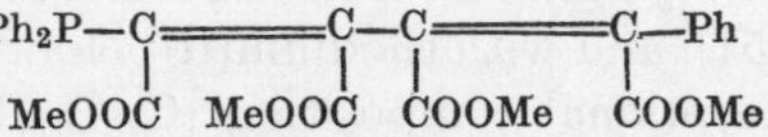   (53)

under nitrogen at −50°. The precipitated yellow ester **(52)** is collected at −50°: it is very unstable and in less than 1 hour forms a brown gum. A solution of the fresh compound **(52)** in methanol, when treated with a drop of methanolic hydrogen chloride, gives an immediate deposit apparently of the rearranged yellow crystalline tetramethyl-1-phenyl-4-diphenylphosphinobutadiene-1,2,3,4-tetracarboxylate **(53)** (Hendrickson, Spenger, and Sims [36]).

This preparation had been carried out earlier by Johnson and Tebby,[37] who analyzed the yellow product, which they considered was probably the isomeric zwitterion **(54)**. By carrying out the reaction under carbon dioxide, they were able to trap an unstable intermediate

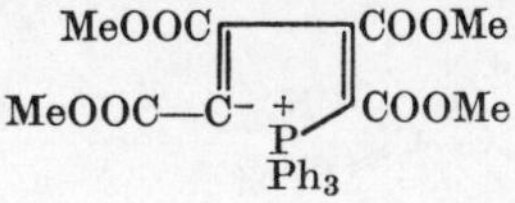

(54) 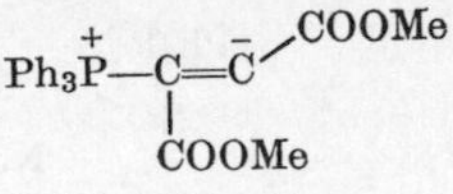   (55) 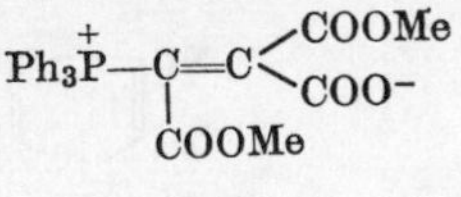   (56)

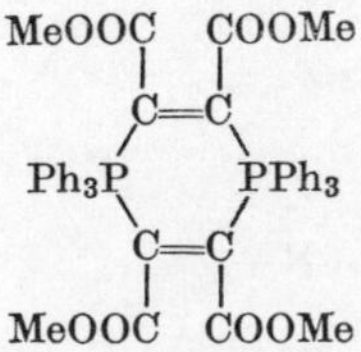

(57) 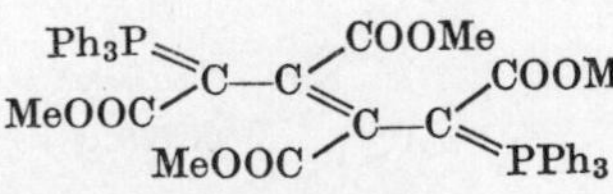   (58)

(55), formed from 1 mole of the phosphine and 1 mole of the ester, as the colorless adduct (56), m.p. 78–80°; when this compound was boiled in methyl iodide (or other solvents) it lost carbon dioxide with dimerization to a more stable orange compound, m.p. 266°, considered to be the diphosphorane (57), *i.e.*, tetramethyl 1,4-dihydro-1,1,1,4,4,4-hexaphenyl-1,4-diphosphorin-2,3,5,6-tetracarboxylate. A later, detailed investigation has shown, however, that this compound almost certainly has the structure (58), *i.e.*, that it is 1,4-bis(triphenylphosphoranylidene)-2-butene-1,2,3,4-tetracarboxylate, although the geometric configuration has not been established (Shaw, Tebby, Ronayne, and Williams [38]).

The phospholes so far considered have all carried one or more aryl substituents. In contrast, two of the simplest substituted phospholes, namely, 1-methylphosphole and 1-ethoxyphosphole (as the 1-oxide), have been synthesized: both proved to be highly reactive, and their isolation in the pure state has offered considerable difficulty.

The McCormack reaction (p. 31), when applied to butadiene and methylphosphonous dichloride, gave 1,1-dichloro-1-methyl-3-phospholene (59), which on hydrolysis gave the 1-methyl-3-phospholene 1-oxide

(59)   (60)   (61)   (62)   (63)

(60), a deliquescent solid, b.p. 59°/0.16 mm, which with bromine gave 3,4-dibromo-1-methylphospholane 1-oxide (61), m.p. 145–146°. The pmr spectra of (60) and of various derivatives showed that the double bond in (59) had not migrated during the hydrolysis to (60) (Quin, Peters, Griffin, and Gordon [39]) (cf. p. 41)

The oxide (61), when treated in boiling benzene with trichlorosilane, HSiCl$_3$ (cf. Fritzsche, Hasserodt, and Korte [40]), was reduced to 3,4-dibromo-1-methylphospholane (62), m.p. 47.5–49°, characterized as the 1-benzyl-1-methylphospholanium bromide, m.p. 171.5–172.5°. The compound (62) in pentane solution on treatment with potassium *tert*-butoxide (3 equivalents) at room temperature underwent dehydrobromination to 1-methylphosphole (63); this was contaminated with some 1-methyl-3-phospholene which was extracted from the pentane solution with 0.5 N-hydrochloric acid, which did not affect the phosphole (63) (24%). The phosphole, alone or in concentrated solution, is affected by prolonged heating, and considerable decomposition occurred on attempted fractionation. After removal of almost all the solvent under reduced pressure, a rapid distillation gave a sample containing 87.8% of the phosphole (**63**), the remainder being mainly pentane. The sample, which is best preserved at low temperatures, reacted readily with methyl iodide to give 1,1-dimethylphospholium iodide, m.p. 190–194° (dec.) (Quin and Bryson [41]).

The pmr spectrum (60 Mc; external standard) of the phosphole (**63**) consists of the methyl signal as an apparent singlet at δ 1.36 ppm (any PCH coupling being less than *ca.* 1 cps) and a 4H multiplet for the vinyl protons at 6.51–7.52 ppm. This α- and β-proton multiplet, centered at 7.09 ppm, is at a lower field than in vinylphosphines. Results for trivinyl-phosphine (calculated to δ) allocate the α-proton to 6.66, and the β-protons to 5.69 (*trans* to phosphorus) and 5.56 ppm (*cis* to phosphorus) (Anderson, Freeman, and Reilly [42]). In 3-methyl-1-phenyl-2-phos-pholene, the vinyl proton appears at δ 5.68 ppm [internal tetramethyl-silane (TMS)] (Quin and Barket [43]).

The ring protons of (**63**) resonate in the range characteristic of analogous 'aromatic' systems: *e.g.*, the α- and β-protons in thiophene are at 7.04 and 6.92 ppm, respectively (internal TMS) (Page, Alger, and Grant [44]).

The ultraviolet spectrum of (**63**) in isooctane has a band at λ_{max} 286 mμ (log ε 3.88) with intense end absorption at 200 mμ; that of 1-methylpyrrole has bands at λ_{max} 280 mμ (log ε 2.06) and 214 mμ (log ε 3.77). The mass spectrum of (**63**) shows a strong molecular ion at m/e 98, and a disintegration pattern similar to that of 1-methylpyrrole.

The chemical properties of (**63**) in general harmonize with these physical properties. The phosphole has a very low basicity (cf. 1-methyl-pyrrole) and it is not extracted from pentane solution by 0.5N-hydro-chloric acid, as noted above. It is extracted by 4N-hydrochloric acid but then undergoes decomposition; even in a dilute solution (10^{-4}M) in 0.01N-hydrochloric acid it slowly decomposes. It does not form an addition product with carbon disulfide. These properties, therefore, may indicate an approach to the extensive electron delocalization of an aromatic system[41] (see also Quin *et al.*[502]).

The strong phosphine odor of the phosphine (**63**) may be partly due to its ready volatility.

1-Ethoxyphosphole 1-oxide synthesized by Usher and West-heimer,[45] who dehydrogenated 1-ethoxy-2-phospholene 1-oxide (**64**) by the four classical steps: (*a*) N-bromosuccinimide gave the 4-bromo derivative (**65**), which (*b*) with dimethylamine gave 1-ethoxy-4-di-methylamino-2-phospholene 1-oxide (**66**); (*c*) this with methyl iodide gave the quaternary iodide (**67**), m.p. 141–141.5°; (*d*) the iodide was converted into the chloride, treatment of which with sodium ethoxide at 25° completed the exhaustive methylation, giving 1-ethoxyphosphole 1-oxide (68).

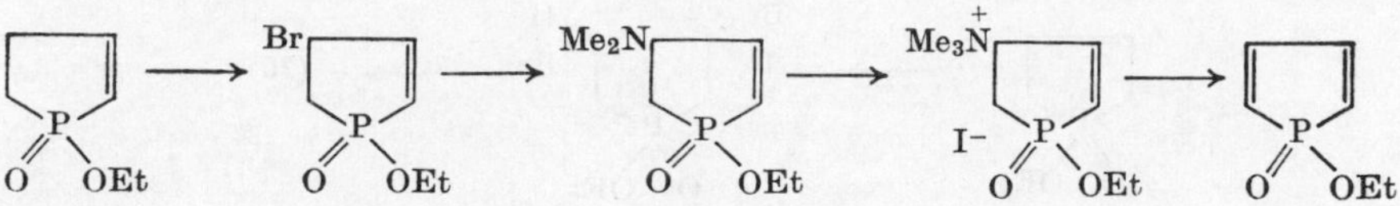

(64)   (65)   (66)   (67)   (68)

The ester (68) dimerized so readily that initially it defied isolation: it was identified by its ultraviolet absorption spectrum, λ_{max} 293 mμ (ϵ 1050), which is closely similar to that of thiophen 1,1-dioxide, λ_{max}

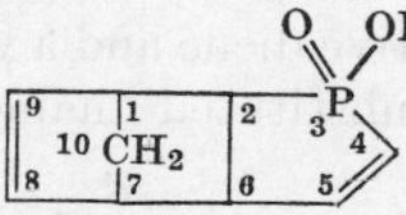

(69) 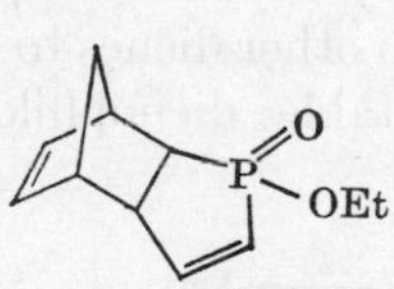   (69A)

289 mμ (ϵ 1230), and by its adduct with cyclopentadiene. For nomenclature purposes, this adduct can be shown as (69) and termed 3-ethoxy-3-phosphatricyclo[5.2.1.0^{2,6}]deca-4,8-diene 3-oxide [RIS —]; its structure can be better depicted as (69A), which is consistent with its complex nmr spectrum although its stereochemistry is uncertain.

The crystalline dimer, m.p. 125–126°, of (68) was subsequently isolated; its physical properties indicated that it had the structure (70), i.e., 3,10-diethoxy-3,10-diphosphatricyclo[5.2.1.0^{2,6}]deca-4,8-diene 3,10-dioxide. Catalytic hydrogenation in absolute ethanol gave a tetrahydro derivative (71), m.p. 80–81°, but the stereochemistry of the ring junction and of the substituents on the phosphorus atoms remains uncertain in

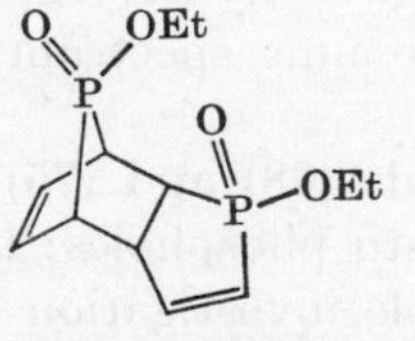

(70) 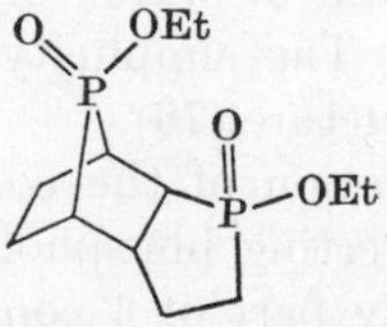   (71)

both compounds. The nmr spectra of the two compounds are compatible with the assigned structures.

The dimer (70) is, however, best prepared by brominating 1-ethoxy-3-phospholene 1-oxide (72) in chloroform at 0° to the 3,4-dibromo

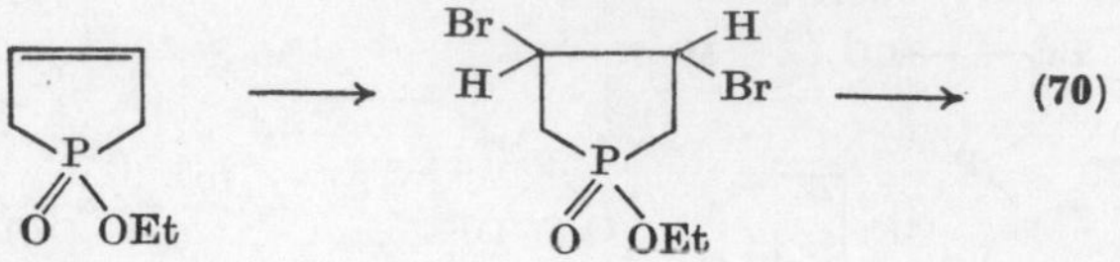

(72)   (73)   (70)

derivative (73), m.p. 47–48°, which when treated in carbon tetrachloride with anhydrous triethylamine at 0° gives the soluble dimer (70) and insoluble triethylammonium bromide (Kluger, Kerst, Lee, and Westheimer [46]).

1-Ethoxy-3,4-dimethylphosphole 1-oxide (75) was expected by analogy with other dienes to be a more reactive diene and a weaker (and thus more stable) dienophile than the unsubstituted analogue (68). It

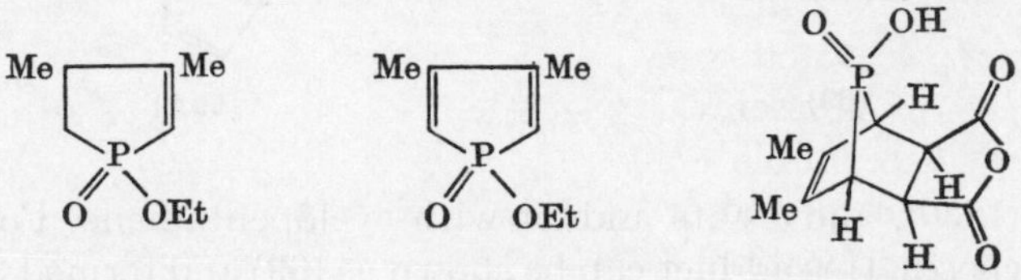

(74)   (75)   (76)

was obtained by N-bromosuccinimide bromination of 1-ethoxy-3,4-dimethyl-2-phospholene 1-oxide (74) (cf. p. 29), followed by treatment of the crude product with sodium ethoxide in ethanol. The solution, made just acid with acetic acid, undoubtedly contained the phosphole (75) for it showed absorption at λ_{max} 298 mμ (unchanged at room temperature for several days), and when treated with maleic anhydride gave the adduct (76), m.p. 261–263°. The ethoxy group was unchanged if this preparation was carried out in 'superdry' solvents, but the presence of traces of water or acid gave hydrolysis to the P—OH compound (76). The simplicity of the nmr spectrum of the adduct confirms the structure (76).

The preparation of the compounds (68) and (75) illustrates one method of converting phospholenes into phospholes; the above work is, however, only part of a considerable investigation by Westheimer and his co-workers on the mechanism of the hydrolysis of the esters of cyclic phosphates and cyclic phosphinic acids, the latter including those having the phosphole, phospholene, and phospholane systems (cf. pp. 277, 278).

Phospholenes

It has already been stated that the phospholenes can exist in two isomeric forms, (1) and (2): these are usually termed 2-phospholene (1)

(1)   (2)

and 3-phospholene (2), respectively; in certain types of derivative in which these names would be immediately preceded by other numbers, ambiguity is avoided by the alternative names phosphol-2-ene and phosphol-3-ene for (1) and (2), respectively.

Our knowledge of the phospholenes is comparatively recent and has been developed by three main streams of work, (a) the pioneering work of McCormack in the U.S.A. (1953–1954), (b) the Russian work, mainly by N. A. Razumova and by B. A. Arbuzov (1961–1963), and (c) the work of Hasserodt, Hunger, and Korte (1963–1964) on the migration of the double bond in the phospholene series. (These dates indicate the main period of work in each case, and do not necessarily include every publication.) From the chemical aspect, the work in (c) follows more directly after that in (a), with much of the work in (b) arriving by a new approach and joining the main stream. Current work—for example, that of Quin and his school—is largely concerned with the decisive identification of the nature of the double-bond migration, the specific stage at which it occurs, and the effect of substituents on the direction and completeness of the migration.

McCormack's work, which has appeared almost solely in detailed patent specifications, can be regarded as the application of a special type of Diels–Alder reaction, in that he showed that various conjugated dienes would condense with arylphosphonous dichlorides to give 1-aryl-1,1-dichloro-3-phospholenes.[47, 48] Thus butadiene and $C_6H_5PCl_2$ gave 1,1-dichloro-1-phenyl-3-phospholene (3; R = R′ = H); isoprene

$$CH_2=CH—CH=CH_2 + PhPCl_2 \longrightarrow$$

(3)   (3A)   (4)

gave the 3-methyl analogue (**3**; $R = CH_3$, $R' = H$); and 1,4-dimethyl-butadiene gave the 3,4-dimethyl analogue (**3**; $R = R' = CH_3$). These products are conveniently represented as cyclic phosphor(v)anes (**3**), but the ionic structure (**3A**) may make a significant contribution to the structure. These dichloro products are readily affected by damp air, and were therefore usually hydrolyzed with cold water (sometimes with the addition of hydrogen peroxide) to give the stable 1-phenyl-3-phospholene 1-oxides (**4**).

The above condensation can be carried out by using a dibromide such as $C_6H_5PBr_2$, which gives a more rapid reaction than $C_6H_5PCl_2$, or by using an alkylphosphonous dichloride such as $C_2H_5PCl_2$; phenyl-phosphonous dichloride is quoted, however, in most of the specific examples given in the specifications.

The reaction is usually carried out at room temperature without a solvent; the onset of the reaction, indicated by a cloudiness or even deposition of a solid, may often be seen within 24 hours but the mixture is usually left undisturbed for several days to ensure complete reaction. The formation of the phospholene adduct is usually accompanied by that of glutinous polymeric material, and the proportion of the latter is increased if the reaction mixture is warmed. McCormack usually added a small proportion of copper stearate to reduce the degree of polymerization, but this was omitted in some later experiments [cf. the preparation of 3-methyl-1-phenyl-3-phospholene 1-oxide (later shown to be the 2-phospholene, p. 42) (McCormack[49])]. Trinitrobenzene, Methylene Blue, and phenothiazine were also tested as inhibitors.

Certain limitations on the structure of the conjugated diene were noted by McCormack. (*a*) The conjugated diene portion (**5**) must not

$$-CH=\overset{|}{C}-\overset{|}{C}=CH-$$

(**5**)

have more than one substituent on the terminal atoms of this portion. (*b*) The diene should also be free from substituent CN and CO groups, for these groups cause considerable polymerization. (*c*) A 1,4-disubsti-tuted butadiene, $RCH=CH-CH=CHR'$, can exist in four geometrically

trans,trans   cis,cis   trans,cis   cis,trans

isomeric forms. The 1,4-addition of the phosphorus to form a planar five-membered ring should proceed with negligible hindrance at the *trans,trans*-form, but should meet considerable obstruction at the *cis,cis*-form; the *trans,cis*- and the *cis,trans*-forms should offer intermediate obstruction. This explains factor (*a*), for two substituents on each of the 1- and 4-carbon atoms would always have two in the above *cis,cis*-position.

(*d*) No carbon atom of the butadiene skeleton (5) may be part of an aromatic ring, and no three carbon atoms may be part of the same cyclo-aliphatic ring. Compounds in which two adjacent carbon atoms of the skeleton (5) are part of the same cyclo-aliphatic ring, or in which each of the double bonds forms part of a polymethylene ring can be employed in the reaction. The example of the last case, 1,1'-bicyclo-hexenyl, cited by McCormack, has later been again used in the synthesis of hydrogenated 1*H*-dibenzophosphole derivatives (p. 80).

Although the stereochemical considerations discussed in (*c*) above probably give a reasonably accurate picture, there is no quantitative evidence for their accuracy. It is noteworthy, however, that commercial piperylene (1,3-pentadiene), which is a mixture of *cis*- and *trans*-forms, gives incomplete reaction with the highly reactive methylphosphonous dichloride, whereas the pure *trans*-form should give complete reaction; further, in support of (*a*), 2,5-dimethyl-1,4-hexadiene (1,1,4,4-tetra-methylbutadiene) does not react with methylphosphonous dichloride (Quin and Mathewes [50]).

For illustration, the following (Table 3) are some of the examples cited by McCormack [47, 48] for interaction of the diene and the phosphonous dihalide, to give after hydrolysis the 3-phospholene 1-oxides (6).

(6)

Some of the boiling points recorded in Table 3 refer to the crude product and others to the refractionated product, which is probably the main cause of occasional discrepancies.

It should be noted that the older nomenclature is used in the above specifications, [47, 48] *e.g.*, the compound (6; R = CH_3, R^1 = H, R^2 = C_6H_5) is termed 3-methyl-1-phenyl-1-phospha-3-cyclopentene *P*-oxide.

Table 3. Preparation of 3-phospholene 1-oxides* (6) [47, 48]

Cpd. no.	Diene	Phosphonous dihalide	1-Oxide (6)			B.p. (°/mm)†	Yield (%)
			R	R^1	R^2		
1	$CH_2=CH-CH=CH_2$	$C_6H_5PCl_2$	H	H	C_6H_5	158–160/0.4	37
2	$CH_2=CH-CH=CH_2$	$C_6H_5PBr_2$	H	H	C_6H_5	139–142/0.5	61
3	$CH_2=CH-C(CH_3)=CH_2$	$C_6H_5PCl_2$	CH_3	H	C_6H_5	172–174/0.2	67.5
4	$CH_2=CH-C(CH_3)=CH_2$	$C_6H_5PBr_2$	CH_3	H	C_6H_5	143–144/0.3	78
5	$CH_2=CHCl-CH=CH_2$	$C_6H_5PCl_2$	Cl	H	C_6H_5	158–164/0.1	37
6	$CH_2=CHCl-CH=CH_2$	$C_6H_5PBr_2$	Cl	H	C_6H_5	161–170/0.5	53
7	$CH_2=CHBr-CH=CH_2$	$C_6H_5PCl_2$	Br	H	C_6H_5	160–165/0.5	53
8	$CH_3-CH=CH-C(CH_3)=CH_2$	$C_6H_5PCl_2$	CH_3	CH_3	C_6H_5	202–208/8	55
9	$CH_2=CH-C(CH_3)=CH_2$	$p\text{-}CH_3OC_6H_4PCl_2$	CH_3	H	$p\text{-}CH_3OC_6H_4$	210–212/0.7	22
10	$CH_2=CH-C(CH_3)=CH_2$	$C_2H_5PCl_2$	CH_3	H	C_2H_5	116–117/0.55	59

*Some of these products may be 2-phospholene 1-oxides (see pp. 40, 42).
†Product from example (1) had m.p. 67–75°; those from (3) and (5) solidified but no m.p. was recorded.

For extensive tables of McCormack's products, and an account of the action of trivalent phosphorus compounds on dienophiles (up to 1964), see Quin.[51]

In patent specifications immediately following those already noted, McCormack [52, 53] showed that the first obvious products of his reaction namely, the 1,1-dichloro derivatives (**3**), if suspended in benzene and treated with a stream of hydrogen sulfide, gave the corresponding 1-sulfides (**7**): thus 1,1-dichloro-3-methyl-1-phenyl-3-phospholene (**3**;

(7)   (8)

R = CH_3, R' = H) gives 3-methyl-1-phenyl-3-phospholene 1-sulfide, m.p. 69–70°.[52]

Further, the 1-oxides of type (**4**), when treated with hydrogen in the presence of Raney nickel, underwent hydrogenation of the ring system without reduction of the oxide, thus giving the 1-oxides (**8**) of the corresponding phospholanes.[53] A platinum catalyst can be used for this purpose, provided the oxide is entirely free from tertiary phosphines; palladium is almost useless. The following are some of the substituted phospholane 1-oxides thus prepared: 1-phenyl, b.p. 136–137°/0.3 mm; 3,4-dimethyl-1-phenyl, b.p. 160–168°/0.7 mm; 1-ethyl-3-methyl, b.p. 95–100°/0.6 mm; 1-p-methoxyphenyl-3-methyl, b.p. 178–185°/0.3 mm.

McCormack's reaction was of great value in providing ready access to phospholenes, but this value was initially limited by the fact that the final hydrolyzed products were the highly stable phospholene 1-oxides. This disadvantage was first overcome by Balon,[54] who showed that the intermediate 1,1-dichloro product (**3**), when reduced with lithium aluminum hydride, gave the tertiary phospholene, i.e., the oxygen was removed without hydrogenation of the phospholene ring. Balon claims that these substituted 3-phospholenes and also substituted phospholanes (p. 61) are of value for catalyzing the conversion of isocyanates into carbodiimides:

$$2C_6H_5NCO \rightarrow C_6H_5N{=}C{=}NC_6H_5 + CO_2$$

After testing various members in each series, he decided that 3-methyl-1-phenyl-3-phospholene and 1-ethyl-3-methylphospholane were the most effective members in their respective series for this purpose.

Various phospholene 1-oxides were also investigated for this purpose by Campbell, Monagle, and Foldi,[55] who found that 1-ethyl-3-methyl-3-phospholene 1-oxide was the most effective, but that 3-methyl-1-phenyl-3-phospholene 1-oxide, although somewhat less effective, was more readily available.

A later reductive method, simpler in manipulation, consists of the portion-wise addition of dry magnesium to a solution of the 1,1-dichloro product (3) in cold anhydrous tetrahydrofuran; when the initial reaction subsided, the mixture was boiled for *ca.* 1 hour, cooled, and hydrolyzed. After working up the product, the 3-phospholene was distilled in nitrogen, for it is very readily oxidized in air. A number of 1-substituted 3-phospholenes were thus isolated, and they were characterized by quaternization with benzyl bromide to give the crystalline 1-substituted 1-benzyl-3-phospholenium bromides (Quin and Mathewes[50]).

Since the McCormack reaction was regarded as a particular type of Diels–Alder reaction, it was natural to assume that the product was a 3-phospholene, and they have been so depicted in the foregoing pages to avoid confusion. Fresh light was thrown on this subject some 10 years later by the work of Hasserodt, Hunger, and Korte,[56, 57] who showed that 1,3-dienes would also combine with phosphorus trichloride (with the usual addition of copper stearate), although at 20° the reaction was sometimes not complete after 60 days. Thus butadiene gave 1,1,1-trichloro-3-phospholene (9; X = Cl), 2-methylbutadiene (isoprene)

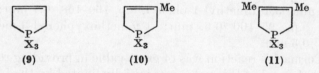

(9)   (10)   (11)

gave the 3-methyl analogue (10), and 2,3-dimethylbutadiene gave the 3,4-dimethyl analogue (11). 2-Methylbutadiene reacted with phosphorus tribromide to give 1,1,1-tribromo-3-methyl-3-phospholene (10; X = Br) in 66% yield after 5 days at 20°, while 2,3-dimethylbutadiene reacted almost explosively with phosphorus tribromide, and the reaction had to be carried out in petroleum solution at −10°, an 85.3% yield of the 1,1,1-tribromo-3,4-dimethyl-3-phospholene (11; X = Br) being obtained after 1 hour.

The trichloro compounds (9), (10), and (11) and their tribromo analogues can, by virtue of their reactive halogen atoms, readily give rise to other 1-substituted derivatives, but considerable evidence was adduced to show that the formation of these derivatives from the trichloro compounds was very frequently accompanied by migration of the double bond to give a 2-phospholene product, whereas the formation of such derivatives from the tribromo compounds was unaccompanied by double-bond migration and thus gave the isomeric 3-phospholene products.

The general conditions for the preparation of these derivatives are briefly noted, but minor variations are made for the preparation of members of any one class according to the nature and position of substituents in the ring. The trihalogeno compound (9; X = Cl) is used as an example in the following summary of the methods for preparing the chief types of derivative.

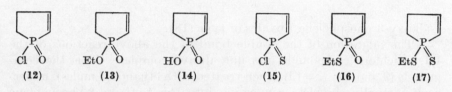

(a) The trichloro compound (9) (1 mole), when treated with acetic anhydride (1 mole) at −20°, gives low-boiling material (chiefly acetyl chloride) and the higher-boiling 1-chloro-2-phospholene 1-oxide (12). Alternatively the compound (9) in dichloromethane solution, when treated at −10° with sulfur dioxide (1.5 moles), is reduced to the chloro oxide (12).

(b) The compound (9) (1 mole) in dry dichloromethane solution at −10° to −15°, when treated with triethylamine (1.6 moles) and absolute ethanol (1.6 moles), gives triethylamine hydrochloride and 1-ethoxy-2-phospholene 1-oxide (13), which is distilled.

It is noteworthy that (9), when treated with triethyl orthoformate, $HC(OC_2H_5)_3$ (1 mole), gives (12) (46.5%) but with the orthoformate (2.2 moles) it gives (13) (61%); many other orthoesters show this distinctive behavior (Hunger and Korte [58]).

The reverse change (13) → (12) can be readily achieved by slowly adding a solution of (13) in tetrachloromethane to a saturated solution of phosgene in tetrachloromethane at 0°. After several hours, distillation gives (12) in high yield and purity.

(c) When added to an excess of ice, the compound (9) gives a vigorous reaction and ultimately a clear solution, which on removal of the water leaves the residual crude 1-hydroxy-2-phospholene 1-oxide (14) in almost theoretical yield. The 3-methyl and 3,4-dimethyl members have respectively b.p.s 240–260°/0.002 mm and 202–222°/0.001 mm and m.p.s 95–97° and 69–70°.

(d) Compounds of type (15), (16), and (17) are all obtained by treating (9) with ethanethiol under various conditions. A solution of (9) in dry dichloromethane, treated with the thiol (1 mole) at room temperature and distilled after several hours, gives the 1-chloro 1-sulfide (15). A solution of (9) in benzene, treated at room temperature simultaneously with benzene solutions of the thiol (1 mole) and triethylamine (1 mole), followed by removal of the amine hydrochloride and distillation, gives the 1-ethylthio 1-oxide (16). Finally, a solution of (9) in dry dichloromethane at −10° to −20°, treated dropwise with the thiol (3 moles), followed by distillation after 2 hours at room temperature, gives the 1-ethylthio 1-sulfide (**17**).

In contrast to these results, the 1,1,1-tribromo compounds (**9**, **10**, and **11**; X = Br) react with methanol or ethanol to give the corresponding

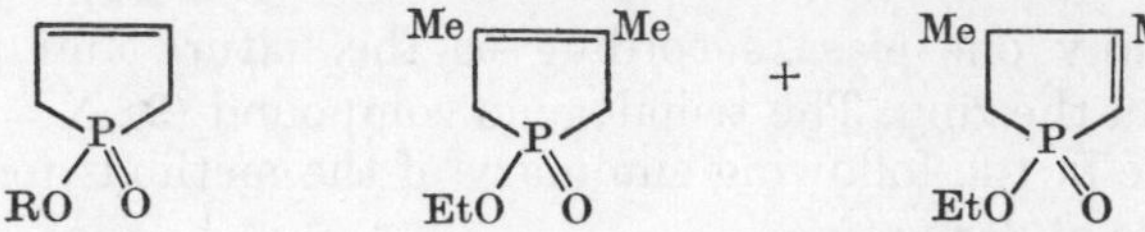

1-alkoxy-3-phospholene 1-oxides of type (**18**).

The migration of the double bond in the above reactions of the 1,1,1-trichloro compounds was not always complete. Thus the compounds (**9** and **10**; X = Cl), when treated with ethanol (2 moles), underwent virtually complete conversion into the 1-ethoxy-2-phospholene 1-oxide of type (**13**), yet the compound (**11**; X = Cl) when similarly treated gave a mixture of 65% of 1-ethoxy-3,4-dimethyl-3-phospholene 1-oxide (**19**) and 35% of the 2-phospholene isomer (**20**).

The 1-alkoxy-3-phospholene 1-oxides will themselves undergo conversion into the 2-phospholene isomers under suitable conditions. Thus 1-ethoxy-3-phospholene 1-oxide was unaffected by exposure to ultraviolet light for 200 hours, but when heated at 200° for 2 hours gave 17% of the 2-phospholene isomer (determined by gas chromatography). When heated at 80° (*a*) in 0.5N-ethanolic potassium hydroxide (1 equivalent) for 2.5 hours, or (*b*) in a sodamide–ethanol mixture for 20 hours, the recovered ester contained 25% and 75%, respectively, of the 2-phospholene isomer. 1-Ethoxy-2-phospholene 1-oxide, similarly heated with ethanolic potassium hydroxide, retained 35% of the 2-phospholene ester.

The early evidence for the position of the double bond in the above compounds was mainly spectroscopic, and the detailed results were published later (Weitkamp and Korte[59]).

The pmr spectra provide the most decisive evidence. In the absence of nuclear substituents, the 3-phospholene ring will show two equivalent CH_2 groups and two equivalent vinylene protons: the 2-phospholene ring will clearly not possess equivalent CH_2 groups or vinylene protons. The presence of substituents in the ring changes this simple picture but a more detailed analysis of the spectra (taken if necessary under various conditions) will usually provide the required differentiation. Some specific detailed results are given (p. 41) for illustration.

The infrared spectra may give some indication, as the $\nu C{=}C$ band of the 2-phospholene ring is in general about 25 cm^{-1} below that of the 3-phospholene ring. This difference must be interpreted with reserve, however; the values quoted by Hunger et al.,[57] range from 1580 to 1608 cm^{-1} for the 2-phospholene ring, and from 1602 to 1635 for the 3-phospholene ring, whereas the values for the much greater number of compounds quoted by Weitkamp et al.,[59] range from 1589 to 1619 cm^{-1} (with one at 1635) and from 1614 to 1651 cm^{-1}, respectively. The infrared spectra are therefore most reliable for distinguishing between the members of a pair of 2- and 3-phospholene isomers.

The ultraviolet spectra of both series show marked absorption at ca. 200 mμ, but the extinction value in the 2-phospholene series is much greater: for example, 1-ethoxy-3-phospholene 1-oxide has λ_{max} 200 (ϵ 305) and the 2-phospholene isomer has λ_{max} 199 (ϵ 4565); the 1-ethoxy-3-methyl-3-phospholene 1-oxide has λ_{max} 200 (ϵ 2524) and the 2-phospholene isomer has λ_{max}197 (ϵ 8970).[57]

The position of the double bond in 2- and 3-phospholenes can also be confirmed by chemical means. When the isomerically pure 1-methoxy-3,4-dimethyl-2-phospholene 1-oxide (21) is ozonized in ethyl acetate at $-80°$ and the product subsequently hydrolyzed, carbon dioxide is liberated with the formation of the phosphine (22), which however

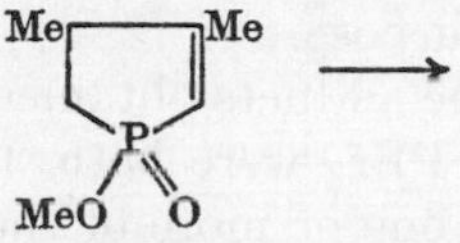

(21)

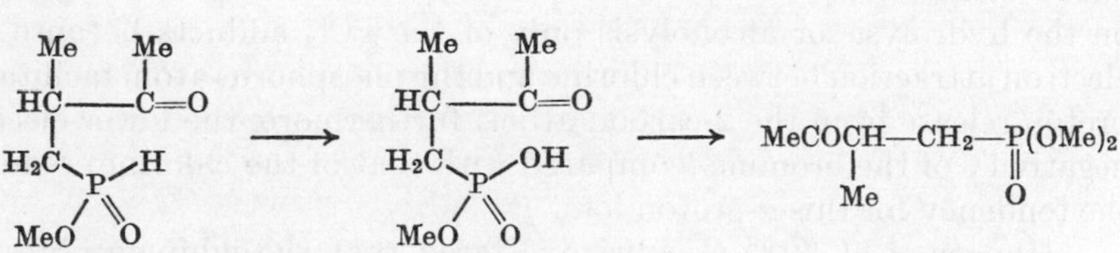

(22)   (23)   (24)

undergoes very ready oxidation to the P–OH compound (23), which when esterified with diazomethane gives dimethyl 2-methyl-3-oxo-butylphosphonate (24).

The isomeric 3-phospholene (25), prepared from the 1,1,1-tribromo adduct, when similarly ozonized, gives mainly methyl bis(acetonyl)-phosphinate (26), with a small proportion of dimethyl acetonylphosphonate (27). The compounds (24) and (26) were fully analyzed, and (27) was identified by its mass spectrum and molecular fragmentation.[57]

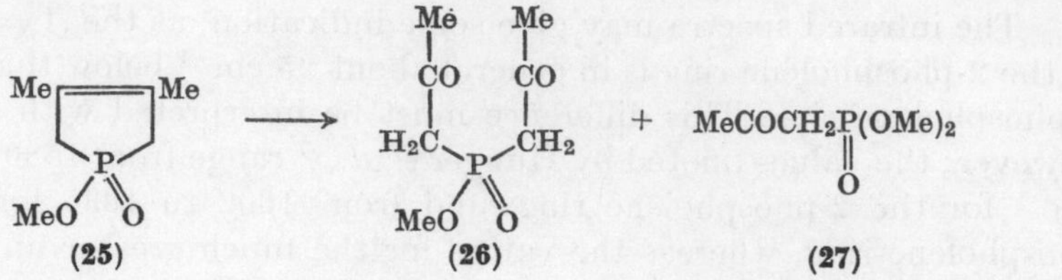

(25)   (26)   (27)

The mechanism of the double-bond migration remains uncertain for lack of adequate evidence. Hunger et al.,[57] suggest the annexed scheme for the intermediate condition of the 1,1,1-trichloro adduct as a

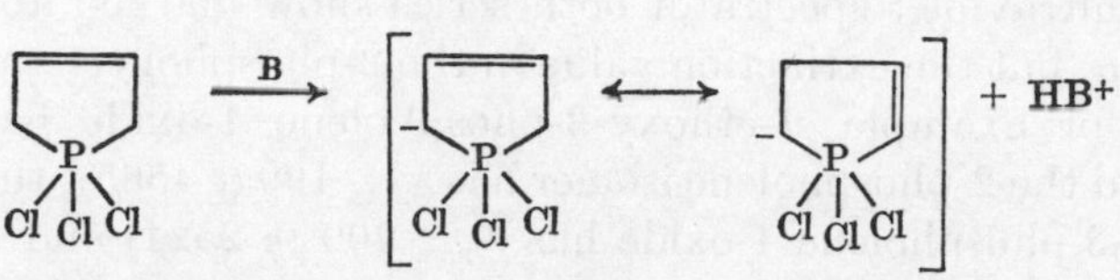

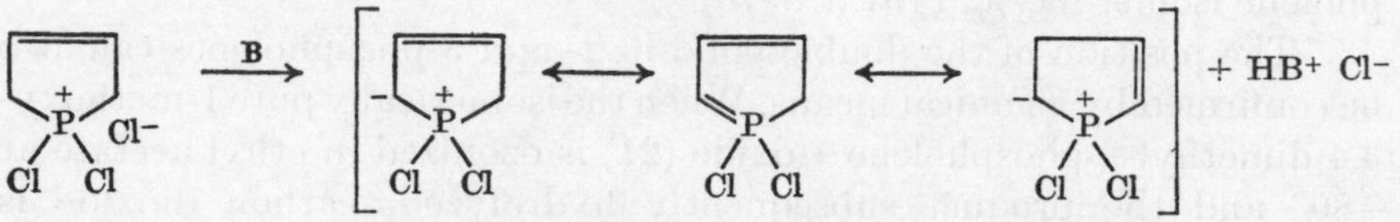

phosphorane and as an ionic salt. Here B represents a base, ethanol being considered a sufficiently strong base for this purpose.

At this stage it was apparently assumed that the adducts obtained from the union of a 1,3-diene with PCl_3 and with PBr_3 were both 3-phospholene derivatives (possibly reacting through one or more of the above contributory forms), and that double-bond migration occurred on the hydrolysis or alcoholysis only of the PCl_3 adducts because the electron attraction between chlorine and the phosphorus atom facilitated proton release from the 2-carbon atom: furthermore, the lower electronegativity of the bromine (compared with that of the chlorine) reduced the tendency for this α-proton loss.

Hunger et al.,[57, 58] also briefly stated that the adducts obtained from the union of phenylphosphonous dichloride with butadiene and with isoprene gave on hydrolysis the corresponding 1-phenyl-2-phospholene 1-oxide, but that the adduct from 2,3-dimethylbutadiene gave the corresponding 1-phenyl-3-phospholene 1-oxide. This result, supported by later confirmation (p. 42), will entail the re-examination of the hydrolysis products of McCormack's numerous dichloro adducts.

A series of investigations by Quin and his co-workers have given very interesting results regarding the structure and reaction of the adducts and other factors in phospholene chemistry.

In a comparatively early investigation, it was shown that butadiene and methylphosphonous dichloride in cyclohexane solution containing copper stearate, after 3 months at room temperature, gave the crystalline

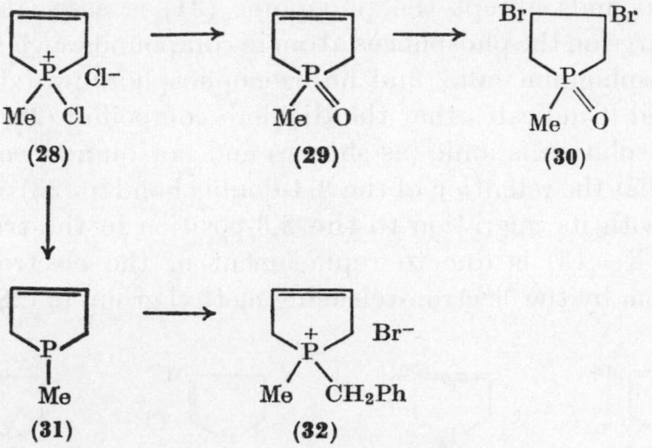

Me Cl
(28)

Me O
(29)

Br   Br
Me   O
(30)

Me
(31)

Me   CH₂Ph
(32)

dichloro adduct (28)—shown here in the ionic form—in 72% yield. This on hydrolysis gave the 1-methyl-3-phospholene 1-oxide (29), b.p. 59°/0.16 mm (80%), characterized as the 3,4-dibromo-1-methyl-phospholane 1-oxide (30), m.p. 145–146° (cf. p. 27).

The dichloro compound (28) was also reduced with magnesium to the 1-methyl-3-phospholene (31), b.p. 114–115° (73%), characterized as the quaternary benzyl bromide (32), m.p. 185–186.5° (Quin, Peters, Griffin, and Gordon [39]).

Details of the pmr spectra of these compounds in deuteriochloroform solution are given to illustrate the distinction between 2- and 3-phospholenes.

The pmr spectrum of the oxide (29) showed three 1:1 doublets: τ 8.39 ppm, J 12.8 cps (3 methyl protons); τ 6.69 ppm, J 10.9 cps (4 methylene protons); τ 4.17, J 27.2 cps (2 vinylene protons). These show that (29) must have the 3-phospholene structure, for the 2-phospholene structure would have two-non-equivalent CH₂ groups and two non-equivalent vinylene protons.

The spectrum of the dichloro compound (28) also showed three doublets: τ 6.13 ppm, J 14 cps (3 methyl protons); τ 5.53 ppm, J 8 cps (4 methylene protons); τ 3.18 ppm, J 36 cps (2 vinylene protons).

The spectrum of the methylphospholene (31) showed a 1:1 doublet, τ 9.19 ppm, J 3.4 cps (3 methyl protons); a complex multiplet, τ 7.25–8.25 ppm (4 methylene protons); and a 1:1 doublet at τ 4.28 ppm, J 6.5 cps (2 vinylene protons).

The spectrum of the quaternary bromide (32) showed four 1:1 doublets and a phenyl multiplet: τ 7.60 ppm, J 14.3 cps (3 methyl protons); τ 6.68 ppm, J 9.9 cps (4 methylene protons); τ 5.40 ppm, J 16.5 cps (2 protons of benzyl CH₂); τ 4.17, J 28.0 cps (2 vinylene protons); τ 2.35–2.75 (5 aromatic protons).

3

Double resonance experiments (^{31}P decoupling) confirmed the observed couplings in (**29**), (**31**), and (**32**) to be ^{31}P–H couplings. These results are consistent only with the 3-phospholene nucleus throughout this series. The large vinyl–^{31}P coupling which is common to all the above compounds except the phosphine (**31**) is associated with the positive charge on the phosphorus atom in compounds such as phosphine oxides, phosphonium salts, and halogenophosphoranes (cf. Wiley and Stine [60]), and it indicates that the dichloro compound (**28**) in deuterio-chloroform solution is ionic (as shown) and not quinquecovalent. It is suggested that the retention of the 3,4-double bond in (**28**) on hydrolysis compared with its migration to the 2,3-position in the trichloro compound (**9**; X = Cl) is due to replacement of the electron-attracting chlorine atom by the electron-releasing methyl group in (**28**).[39]

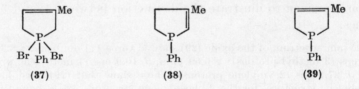

Another investigation (Quin and Barket [43]) provided confirmation that the isoprene–phenylphosphonous dichloride adduct and its hydrolysis product, originally thought to have the 3-phospholene structures (**33**) and (**34**), respectively, have in fact the 2-phospholene structures (**35**, covalent or ionic) and (**36**), respectively. The isoprene–phenylphosphonous dibromide adduct and its hydrolysis product,

however, retain the 3-phospholene structures (**37**) and (**34**), respectively. The dichloro adduct (**35**) and the dibromo-adduct (**37**), when reduced with magnesium to the tertiary phosphine, retain their ring systems, giving respectively 3-methyl-1-phenyl-2-phospholene (**38**) (benzyl bromide salt, m.p. 178–180°) and 3-methyl-1-phenyl-3-phospholene (**39**) (benzyl bromide salt, m.p. 184–185°); a mixture of the two quaternary benzyl bromide salts had m.p. 166–176°.

Three points in particular arise from these investigations: (*a*) the migration of the double bond may occur during the formation of the adduct itself, (b) the double bond retains a remarkably stable position in reactions that occur after hydrolysis or alcoholysis (and in some cases from the adduct onwards), and (c) the presence of chloro groups in the adduct is not essential for 2-phospholene formation, for their influence may be nullified by other groups attached to the phosphorus atom: a methyl group can nullify, and a phenyl group promote, this influence of the chloro groups.

A later investigation (Quin, Gratz, and Barket[61]), extending the preliminary account, records the adducts formed by the union of methylphosphonous dichloride with butadiene (40; $R = R^1 = R^2 = H$), isoprene (40; $R = R^2 = H$, $R^1 = CH_3$), 2,3-dimethylbutadiene (40; $R = H$, $R^1 = R^2 = CH_3$), and piperylene (40; $R = CH_3$, $R^1 = R^2 = H$),

and the corresponding 1-methyl 1-oxides (41) formed by their hydrolysis. The highly reactive adducts (40) were generally unsuitable for spectroscopic investigation, but the butadiene adduct (40; $R = R^1 = R^2 = H$) was shown to have the 3-phospholene structure; all the above four 1-oxides (41), and the corresponding phospholenes (42) obtained by magnesium reduction of the adducts, and the quaternary benzyl bromide salts of these phospholenes, also had this structure.

The isomeric stability of these oxides (41) (a) when boiled under reflux, (b) in boiling 3N-sodium hydroxide solution, and (c) in boiling 3N-hydrochloric acid was investigated (see Table 4), and the ratio of 3-phospholene oxide/2-phospholene oxide was determined.

Table 4. Stability of 3-phospholene oxides (41)

Compound (41)	Ratio of isomers*		
	(a)	(b)	(c)
($R = R^1 = R^2 = H$)	1:1	2:3	No change
($R = R^2 = H$, $R^1 = CH_3$)	1:1	1:2	1:9
($R = CH_3$, $R^1 = R^2 = H$)	1:9	d	No change
($R = H$, $R^1 = R^2 = CH_3$)	9:1	9:1	9:1

* For a, b, c see text; the values in any one vertical column are not strictly comparable, as the period of heating varied for the four compounds. (d) Almost complete conversion.

The significance of these results, and also the spectral properties of the 3-phospholene 1-oxides (41), their 2-phospholene isomers (43), the 3-phospholenes (42) and their benzyl bromide derivatives are discussed in detail.[61]

It is noteworthy that the above conversions afford the only known route to 1-methyl-2-phospholene 1-oxide (43; $R = R^1 = R^2 = H$), 1,3-dimethyl-2-phospholene 1-oxide (43; $R = R^2 = H$, $R^1 = CH_3$), and 1,2-dimethyl-2-phospholene 1-oxide (43; $R = CH_3$, $R^1 = R^2 = H$), each of which was obtained pure by distillation; the last member (43; $R = H$, $R^1 = R^2 = CH_3$) was obtained in insufficient quantity for purification.

The preliminary work[43] on the identity of the products obtained by the hydrolysis of the phenylphosphonous dichloride and dibromide adducts with certain 1,3-dienes was also extended.[61] On hydrolysis, the $C_6H_5PCl_2$–isoprene adduct gave 3-methyl-1-phenyl-2-phospholene 1-oxide (44), identified unmistakably by its pmr spectrum, with a trace

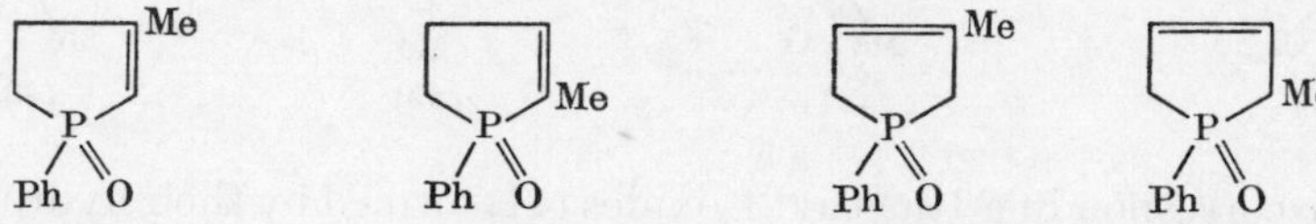

of the 3-phospholene isomer. The $C_6H_5PCl_2$–piperylene adduct similarly gave 2-methyl-1-phenyl-2-phospholene 1-oxide (45); gas chromatography indicated the presence of ca. 4.5% of the 3-phospholene isomer.

The $C_6H_5PBr_2$–isoprene adduct gave the expected 3-methyl-1-phenyl-3-phospholene 1-oxide (46) with only ca. 0.5% of the isomer, and the $C_6H_5PBr_2$–piperylene adduct gave solely the 3-phospholene product (47).

The identities of the 2-phospholene product (44) and the 3-phospholene product (46) were confirmed by synthesis. 1-Methoxy-3-methyl-2-phospholene 1-oxide (48), the identity of which is well established,[56, 57]

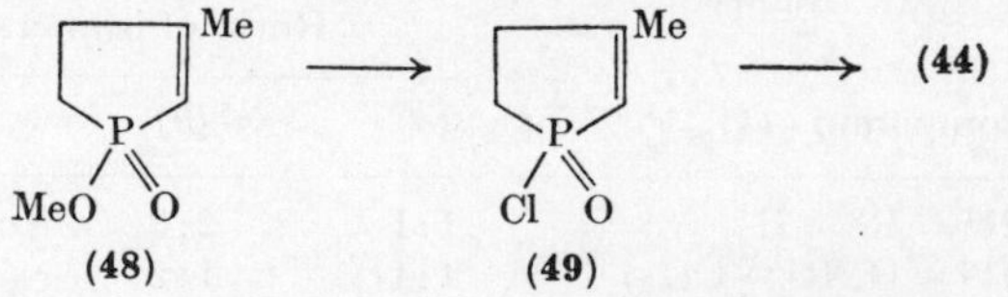

when treated with a saturated solution of phosgene in benzene, gave 1-chloro-3-methyl-2-phospholene 1-oxide (49), which when treated in turn with phenylmagnesium bromide gave the corresponding 1-phenyl derivative, identical with (44). 1-Methoxy-3-methyl-3-phospholene

1-oxide, also of established identity, when similarly treated gave the corresponding 1-phenyl derivative, identical with (46).

It follows that the differentiation made by Hasserodt and his colleagues[56, 57] between PCl_3–diene and PBr_3–diene adducts, namely, that on hydrolysis they give 2-phospholene and 3-phospholene derivatives, respectively, also applies similarly to the $C_6H_5PCl_2$–diene and the $C_6H_5PBr_2$–diene adducts. It should be emphasized, however, that this is a generalization, to which there are exceptions; for example, the substituents in the ring and those directly joined to the phosphorus may affect the nature of the hydrolysis product.

A stereochemical feature of some interest arose in the above investigations. In view of the configurational stability of the phosphorus in phospholenes and in their 1-oxides, it is clear that a substituent in the ring, other than on a double-bonded carbon atom, would allow the formation of *cis*- and *trans*-geometric isomers relative to the group on the phosphorus atom.

Quin, Gratz, and Montgomery[62] condensed methylphosphonous dichloride with commercial *cis,trans*-piperylene in heptane at room temperature in the usual way, giving the adduct (50). Hydrolysis gave the 1,2-dimethyl-3-phospholene 1-oxide (62%), which could exist in

(50)   (51A)   (51B)   (52A)   (52B)

the *cis*-dimethyl form (51A) and the *trans*-dimethyl form (51B). The oxide, b.p. 58–64° (0.17–0.25 mm), gave on preparative gas chromatography one pure isomer, and on distillation with a spinning-band column gave a second isomer which was not completely purified. The infrared spectra of the two isomers showed $\nu C{=}C$ 1610 and 1613 cm^{-1}, respectively; their pmr spectra were almost identical and confirmed the 3-phospholene structure for each compound.

Reduction of the adduct (50) in the usual way gave 1,2-dimethyl-3-phospholene (54%), which was easily separated by preparative gas chromatography into two isomers. The pmr spectra again showed both isomers to be 3-phospholenes (52A) and (52B), and the infrared spectra showed $\nu C{=}C$ 1615 cm^{-1} for both compounds. The two isomers were quaternized with benzyl bromide, giving the corresponding 1-benzyl-1,2-dimethyl-3-phospholenium bromides, m.p. 200–201° and 202–203°, respectively, mixed m.p. 183–185°.

The ratio of the 1-oxide isomers was *ca.* 2:1, and that of the 3-phospholenes *ca.* 3:1; it is suggested that the adduct (**50**) may consist of isomers in a ratio approximating to these values.

There is at present no decisive evidence for the configuration of each member of the pair of isomers, *i.e.*, as to which has the *cis-* and which the *trans*-dimethyl groups.

A novel approach to phospholene chemistry came originally from the Russian chemists. Razumova and Petrov[63] showed that when 2-chloro-4-methyl-1,3,2-dioxaphospholane (**53**) (p. 272) mixed with

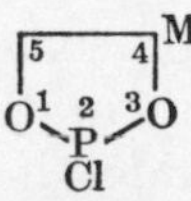

(53)

some zinc chloride was heated at 100–120° with butadiene, isoprene, piperylene, or chloroprene, adducts were formed which could be distilled. The molecular formulae and b.p. (at 10 mm) of these adducts from the above four 1,3-dienes were:

(*a*) $C_7H_{12}O_2PCl$, b.p. 147–148°   (*b*) $C_8H_{14}O_2PCl$, b.p. 155–156°
(*c*) $C_8H_{14}O_2PCl$, b.p. 144–146°   (*d*) $C_7H_{11}O_2PCl_2$, b.p. 169–170°

These compounds were unaffected by oxygen, sulfur, or cuprous chloride but readily added bromine (1 molar equivalent). These workers concluded that the adducts were esters of dialkylphosphinic acids.

B. A. Arbuzov, Shapshinskaya, and Erokhina[64] decided that the above adducts were cyclic compounds. This was illustrated experimentally by heating 2,3-dimethylbutadiene (**54**) with 2-chloro-1,3,2-dioxaphospholane (**55**) and a trace of zinc chloride and hydroquinone at

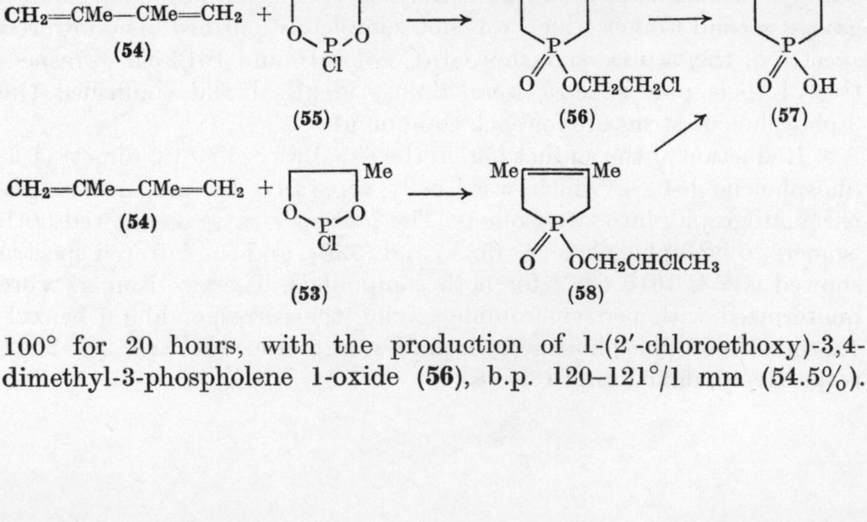

$CH_2{=}CMe{-}CMe{=}CH_2$ +
(54)

(55)   **(56)**   **(57)**

$CH_2{=}CMe{-}CMe{=}CH_2$ +
(54)

(53)   **(58)**

100° for 20 hours, with the production of 1-(2′-chloroethoxy)-3,4-dimethyl-3-phospholene 1-oxide (**56**), b.p. 120–121°/1 mm (54.5%).

This in turn when heated with 20% hydrochloric acid at 120–130° underwent hydrolysis to 1-hydroxy-3,4-dimethyl-3-phospholene 1-oxide (57), m.p. 119–121°. The diene (54) when similarly heated with 2-chloro-4-methyl-1,3,2-dioxaphospholane (53) gave 1-(2′-chloropropoxy)-3,4-dimethyl-3-phospholene 1-oxide (58), b.p. 135–136°/3.5 mm (51%), which on hydrolysis also gave the 1-hydroxy 1-oxide (57).

It was suggested that in both cases the initial reaction was a 1,4-addition of the diene (54) to the phosphorus atom to give a compound of type (59), in which the chlorine can readily ionize and attack the P—O—alkyl group (the last stage of a normal Arbuzov rearrangement) to give the ester (56) or (58).

(59)   (60)

Note re Nomenclature: The 'parent' compound (60) [RIS —], may be termed spiro[1,3,2-dioxaphosphole-2,1′-phosphol-3-ene], and (59) is therefore 2-chloro-3′,4′-dimethylspiro[1,3,2-dioxaphospholane-2,1′-phosphol-3-ene].

The formation of the intermediate (59) is supported by the union of 1,3-dienes with 2-chloro-1,3,2-benzodioxaphosphole to give similar spirocyclic compounds, which in this case are stable compounds and do not give the Arbuzov rearrangement.

The above acid hydrolysis of compounds of type (56) and (58) to the 1-hydroxy 1-oxide derivative (57) can be replaced by the action of phosphorus pentachloride, which gives the corresponding 1-chloro 1-oxide, the chloroalkyl groups in (56) and (58) coming off as 1,2-dichloroethane and 1,2-dichloropropane, respectively (Razumova and Petrov[65]).

The lithium aluminum hydride reduction of these 1-chloroalkoxy 1-oxides gives the tertiary phosphine; thus the compound (61) is said

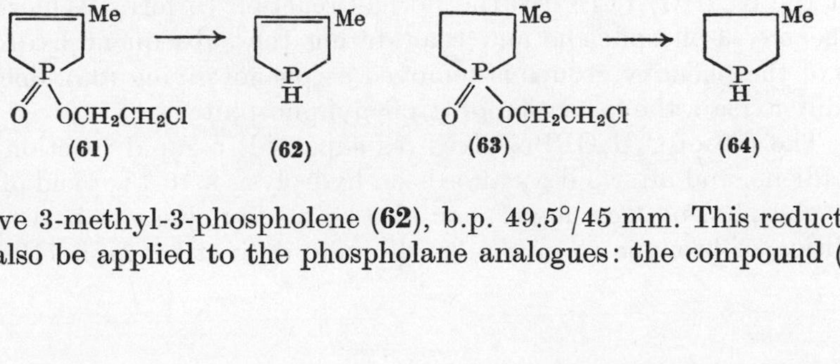

(61)   (62)   (63)   (64)

to give 3-methyl-3-phospholene (62), b.p. 49.5°/45 mm. This reduction can also be applied to the phospholane analogues: the compound (63) gives 3-methylphospholane (64), b.p. 119°. The PH group in both (62) and (64) appears as a band at 2270 cm^{-1} (Bogolyubov *et al.*[66]). It is stated that 'halogenation and dehydrohalogenation of (62) can give 3-methylphosphole among other products'.

Hunger *et al.*[57] have investigated some of the above 1-(chloroalkoxy) 1-oxides and find that the 3,4-double bond is retained.

Arbuzov and Shapshinskaya[67] had earlier repeated much of McCormack's work, using butadiene and piperylene, and the phosphonous dichlorides p-RC$_6$H$_4$PCl$_2$ (where R = CH$_3$, Cl, or Br) and C$_2$H$_5$PCl$_2$, the reaction being carried out at room temperature in the presence of a trace of copper stearate and the main product being hydrolyzed to the corresponding 1-oxide.

They also showed that phenyl phosphorodichloridite, C$_6$H$_5$OPCl$_2$, combined with butadiene, the main reaction product on hydrolysis giving 1-phenoxy-3-phospholene 1-oxide (65), b.p. 163–163.5°/3 mm m.p. 60–62° (14%), and the liquid portion of the reaction product containing unchanged C$_6$H$_5$OPCl$_2$ and also (C$_6$H$_5$O)$_2$PCl, *i.e.*, diphenyl phosphorochloridite. When, however, (C$_6$H$_5$O)$_2$PCl and butadiene were allowed to react for 90 days, the solid deposit on hydrolysis again gave (65), whereas the liquid portion contained (C$_6$H$_5$O)$_2$PCl and triphenyl

(65)   (66)   (67)

phosphite, (C$_6$H$_5$O)$_3$P. This strongly suggests that during the long period of reaction an equilibrium of the three simple phenoxy compounds

$$2(C_6H_5O)_2PCl \rightleftarrows C_6H_5OPCl_2 + (C_6H_5O)_3P$$

begins to be attained and that the C$_6$H$_5$OPCl$_2$ is the most active component in the above McCormack reaction. The alternative explanation, that the (C$_6$H$_5$O)$_2$PCl gives the normal reaction to form 1-chloro-1,1-diphenoxy-3-phospholene and that during the subsequent hydrolysis one of the phenoxy groups is removed as phenol giving (65), does not readily explain the formation of triphenyl phosphite.

The use of C$_6$H$_5$OPBr$_2$ gives (as expected) a rapid reaction with butadiene, and after 5 days affords on hydrolysis a 70.7% yield of (65). The phospholene 1-oxide (65) can also be readily characterized by the addition of bromine to give a 3,4-dibromo derivative, m.p. 119–126°.

The similar addition of $C_6H_5OPCl_2$ to isoprene, 2,3-dimethyl-butadiene, and piperylene gives products that on hydrolysis give the corresponding substituted 1-phenoxy 1-oxides (66), namely, (66; $R = R^2 = H$, $R^1 = CH_3$), b.p. 156–157°/2–2.5 mm, m.p. 56–58° (51.5%); (66; $R = H$, $R^1 = R^2 = CH_3$), b.p. 146–147°/2.5 mm (46%); and (66; $R = CH_3$, $R^1 = R^2 = H$), b.p. 162–164°/3 mm (9%).

Ethyl phosphorodichloridite, $C_2H_5OPCl_2$, reacts with 2,3-dimethyl-butadiene to give a liquid product which on direct distillation affords 1-chloro-3,4-dimethyl-3-phospholene 1-oxide (67), b.p. 121–122°/8 mm, m.p. 78–82° (55%); this probably arises by thermal decomposition of the initial 1,1-dichloro-1-ethoxy-3,4-dimethyl-3-phospholene with loss of ethyl chloride. The compound (67) could be hydrolyzed to the 1-hydroxy 1-oxide, m.p. 121–123° (67%).

The 3-phospholene structure was given to the compounds prepared in this investigation, which preceded that of Hasserodt et al.[56] revealing the possibilities of double-bond migration.

The use of magnesium for removing the 1-halogen atoms from the initial adducts, of hydrogen–nickel for reducing the phospholene 1-oxides to phospholane 1-oxides, and of lithium aluminum hydride for reducing 1-alkoxyphospholene 1-oxides to the parent phospholene has been noted. Various silanes are known to reduce tertiary phosphine oxides to the phosphines and their use for the reduction of phospholene 1-oxides to the corresponding phospholenes without affecting the double bond has been introduced by Fritzsche, Hasserodt, and Korte.[68, 40, 60]

Noteworthy examples of these are methylpolysiloxane (68) (preparation: Sauer, Scheiber, and Brewer[70]) and mono-, di-, and triphenyl-silane. A mixture of 3-methyl-1-phenyl-2-phospholene 1-oxide and

$$\left[\begin{array}{c} H \\ | \\ -O-Si- \\ | \\ Me \end{array}\right]_n \qquad R_3PO + 2Ph_3SiH \longrightarrow R_3P + Ph_3Si-O-SiPh_3$$

(68)   (69)   (70)

methylpolysiloxane ('MPS'), when shaken for 2 hours at 250° (without a solvent), gives 3-methyl-1-phenyl-2-phospholene, b.p. 79–80°/0.05 mm (88%).

Triphenylsilane (69) reduces tertiary phosphine oxides to tertiary phosphines with the formation of hexaphenylsiloxane (70).[68, 69]

Trichlorosilane[40] mixed with triethylamine gives the reduction:

$$R_3PO + Cl_3SiH + (C_2H_5)_3N \rightarrow R_3P + (Cl_2SiO)_n + (C_2H_5)_3N,HCl$$

3-Methyl-1-phenyl-2-phospholene 1-oxide reacts with the tri-chlorosilane–triethylamine mixture in boiling benzene to give 3-methyl-1-phenyl-2-phospholene, b.p. 64–67°/0.02 mm (24%).

Diphenylsilane, $(C_6H_5)_2SiH_2$, is active in the following types of reduction:

$$RPCl_2 \rightarrow RPH_2; \quad R_2PCl \rightarrow R_2PH; \quad R_2P(O)OH \rightarrow R_2PH$$

A mixture of 1-hydroxy-3-methyl-2-phospholene 1-oxide and diphenylsilane (1.5 molar equivalents) when heated at 150–190° for 4 hours gives 3-methyl-2-phospholene, b.p. 80°/760 mm (78%). Similarly, 1-ethoxy-2-phospholene 1-oxide is reduced by phenylsilane, $C_6H_5SiH_3$, to 2-phospholene, b.p. 75°/760 mm (73%).[68,69]

The conversion of a phospholene into the corresponding phosphole may be achieved by two methods, depending on the number and nature of nuclear substituents: (a) addition of bromine to the 3,4-double bond and subsequent removal of hydrogen bromide (2 equivalents); (b) mono-bromination of a nuclear methylene group by N-bromosuccinimide, followed by removal of hydrogen bromide either directly or by replacement of the bromine by a $N(CH_3)_2$ group, with subsequent quaternization with methyl iodide followed by heating with alkali or silver oxide. For example, see Quin and Bryson (pp. 27, 28),[41] and Usher and Westheimer (p. 28).[45]

Phospholanes

Preparation

(I) In the earliest method, Grüttner and Krause[71] brought the di-Grignard reagent from 1,4-dibromobutane, $(—CH_2CH_2MgBr)_2$, into reaction with phenylphosphonous dichloride, $C_6H_5PCl_2$, to form 1-phenylphospholane (**1**; $R = C_6H_5$), a colorless oil, b.p. 132–133°/17 mm, having a strong 'phosphine odor'; it is only slowly oxidized by air. It shows the normal chemical properties of a tertiary phosphine. Thus with ethyl iodide it gives a quaternary ethiodide (**2**; $R = C_6H_5$, $R' = C_2H_5$), m.p. 122°, formally termed 1-ethyl-1-phenylphospholanium

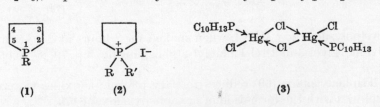

(1)   (2)   (3)

iodide, and with n-propyl iodide a similar salt (**2**; $R = C_6H_5$, $R' = C_3H_7$), m.p. 153–154°. It also gives a colorless mercuric chloride 'addition product', of composition $C_{10}H_{13}P,HgCl_2$, m.p. 143–144°. The earlier chemists frequently characterized their tertiary phosphines by these 1:1 adducts, but later work with simpler tertiary phosphines (Evans, Mann, Peiser, and Purdie[72]) shows that they are bridged compounds of structure (3).

(II) In the reverse preparation (Issleib and Häusler[73]) the dilithio derivative of a primary phosphine in tetrahydrofuran or benzene is treated with 1,4-dibromo (or dichloro)butane (one equivalent): an excess of the dihalide must be avoided, otherwise quaternary halides

$$RPLi_2 + Br(CH_2)_4Br \rightarrow (1) + 2LiBr$$

are formed. In this way they synthesized the above 1-phenylphospholane (1; $R = C_6H_5$), b.p. 97°/3 mm (30%), which gave a methiodide (2; $R = C_6H_5$, $R' = CH_3$), m.p. 130°, and with sulfur in boiling benzene gave a crystalline sulfide (4; $R = C_6H_5$), m.p. 77°. The 1-ethylphospholane, b.p. 145° (22%), gave a methiodide, m.p. 289–291°; 1-cyclohexylphospholane, b.p. 94°/3.5 mm (23%), gave a methiodide, m.p. 249°, and a sulfide (4; $R = C_6H_{11}$), m.p. 81°.

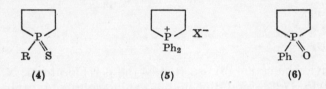

(4)   (5)   (6)

It is noteworthy that this synthetic route using 1,5-dibromo(or dichloro)pentane similarly gives the corresponding phosphorinanes (p. 100), but the attempted use of 1,3- or 1,6-dihalogenoalkanes gives solely polymeric products. Similar limitations apply to other synthetic routes.

(III) A third method (Issleib, K. Krech, and Gruber[74]) is intermediate in type between the two former methods. When equal weights of cyclohexylphosphine, $C_6H_{11}PH_2$, and 1,4-dibromobutane in ethanol are boiled for 15 hours, treatment with sodium ethoxide gives 1-cyclohexylphospholane (1; $R = C_6H_{11}$), b.p. 90°/3 mm (32%). There is little doubt that these reactants first form 4-bromobutylcyclohexylphosphine, $Br(CH_2)_4PH(C_6H_{11})$, which then undergoes intramolecular condensation to form the phospholane. This preparation, like the previous one, is limited in its scope to the synthesis of phospholanes and phosphorinanes (p. 101).

Diphenylphosphine, $(C_6H_5)_2PH$, when similarly boiled with ethanolic 1,4-dibromobutane, gives 1,1-diphenylphospholanium bromide

(5; X = Br), m.p. 162° (24%), again almost certainly with the similar intermediate formation of 4-bromobutyldiphenylphosphine.[74] Diphenyl-potassiophosphine in dioxane–tetrahydrofuran, when slowly added to boiling 1,4-dibromobutane, also gives the same cation, isolated as the tetraphenylborate [5; X = $B(C_6H_5)_4$], m.p. 185–187° (Märkl[75]).

The bromide (5; X = Br), when treated in solution with sodium iodide, gives the iodide, m.p. 163°, and when treated in boiling aqueous solution with silver oxide gives 1-phenylphospholane 1-oxide (6), b.p. 176–180°/3 mm (60%).[74] This eviction of an aryl or alkylaryl group from a quaternary phosphonium salt by the action of alkali or other metallic hydroxide, with the formation of a tertiary phosphine oxide and liberation of the corresponding hydrocarbon, is of course a recognized reaction; its mechanism is indicated on p. 176.

(IV) Märkl[76] has shown that an o-dichlorobenzene solution of 1,4-dibromobutane (2 equivalents) and tetraphenyldiphosphine, $(C_6H_5)_2P—P(C_6H_5)_2$ (1 equivalent), when added dropwise to the boiling

solvent, gives the cation (5), isolated as the perchlorate (5; X = ClO_4), m.p. 114–115° (72%), and diphenylphosphinous bromide; this reaction probably proceeds through the intermediate (7), which decomposes as shown, the 4-bromobutyldiphenylphosphine then undergoing the normal intramolecular quaternization to give the cation (5). This reaction can also be applied to the synthesis of phosphorinanes (p. 101), phosphepanes (p. 153) and certain other cyclic phosphines (p. 69).

(V) A novel synthesis of cyclic phosphines, applicable also to phospholanes, phosphorinanes, and phosphepanes, has recently been described by Davies, Downer, and Kirby.[77] A solution of phenyl-phosphine in 2-methylheptane at 80° under nitrogen is treated over 4 hours with sodium (one equivalent) dispersed in the same solvent, to give phenylsodiophosphine (8). 1-Bromo-3-butene (9) (0.8 equivalent) is then added at 75–81° over 8 hours, to give 3-butenylphenylphosphine (10), b.p. 109–110°/12 mm (58%). A 5% solution of this phosphine in

$$C_6H_5PNaH + CH_2{=}CH(CH_2)_2Br \rightarrow CH_2{=}CH(CH_2)_2PHC_6H_5 \rightarrow (1; R = C_6H_5)$$
$$(8) \qquad\qquad (9) \qquad\qquad\qquad (10)$$

petroleum (b.p. 40–60°) under nitrogen is boiled under reflux for 80 hours whilst being irradiated with ultraviolet light: during this process the phosphine (**10**) undergoes intramolecular cyclization by addition of the PH group to the double bond, and distillation of the reaction mixture gives 1-phenylphospholane (**1**; $R = C_6H_5$), b.p. 125°/14 mm (38%), with a residue of polymeric material. The phospholane was further characterized by conversion into the 1-sulfide (**4**; $R = C_6H_5$), and into the 1:1 $HgCl_2$ adduct, m.p. 199–200°, a value which is strikingly different from that found by Grüttner and Krause,[71] although analytical values are recorded for both products. 1-Phenylphospholane 1-oxide (**6**) was obtained by manganese dioxide oxidation of the phospholane (**1**; $R = C_6H_5$) in 87% yield, or similar oxidation of the phosphine (**10**) in 27% yield, the latter process involving cyclization and oxidation: the oxide (**6**) was obtained as a very hygroscopic solid, m.p. 56–57°, b.p. 99–100°/0.15 mm.

(VI) Wagner[1] has claimed (for details, see p. 3) that sodium phosphide, $NaPH_2$, reacts with 1,4-dichlorobutane in liquid ammonia to give 1,4-diphosphinobutane, $H_2P(CH_2)_4PH_2$, and phospholane (**1**; $R = H$), b.p. 0°/9 mm, and that the use of alkyl- or aryl-sodiophosphines, NaPHR, gives phospholanes having as 1-substituents CH_3, C_2H_5, iso-C_3H_7, n-C_8H_{17}, p-$CH_3C_6H_4$, p-ClC_6H_4, (? 2,4-)$(CH_3)_2C_6H_3$. No physical constants or analyses are given for identification.

(VII) The double bond in the ring system of 3-phospholenes is chemically stable to many reagents, but it is readily hydrogenated in the presence of Raney nickel to form the corresponding phospholanes (McCormack[78]) (p. 35). This would, however, rarely be used as a preparative method for phospholanes, in view of the other methods available.

Derivatives

Sulfides. The monosulfides of 1-substituted phospholanes are usually prepared by direct addition of sulfur as noted above. They can, however, be obtained without formation of the tertiary phosphine; equimolecular quantities of the di-Grignard reagent from 1,4-dibromobutane and of methylthiophosphonic dibromide, $CH_3P(S)Br_2$, interact in ethereal solution giving the liquid 1-methylphospholane 1-sulfide (**4**; $R = CH_3$), b.p. 73–75°/0.01 mm, infrared spectral details of which are recorded (Maier[79]).

Schmutzler[80] has shown that the above di-Grignard reagent reacts with thiophosphoryl chloride, $SPCl_3$, also in ether under nitrogen at −78°, to form much viscous polymeric material and in addition the highly crystalline 1,1'-biphospholane 1,1'-disulfide (**11**), m.p. 185°, which can be readily isolated. This disulfide is a valuable intermediate for the synthesis of several types of phospholane derivatives. It is readily reduced by iron powder to give the liquid 1,1'-biphospholane (**12**), b.p. 50°/0.05 mm, which is spontaneously inflammable in air. The disulfide, when heated with ethylene and a trace of iodine in an autoclave at 275–300° for 48 hours, gives *sym*-ethylenebis-(1,1'-biphospholane 1,1'-disulfide) (**13**), m.p. 174.5°. The use of (**11**) for the preparation of 1-fluoro-derivatives is noted later (p. 56). Finally, the disulfide (**11**), when oxidized by 30% nitric acid at 70–80°, gives the crystalline phosphinic acid (1-hydroxyphospholane 1-oxide) (**14**), m.p. 53–54°.

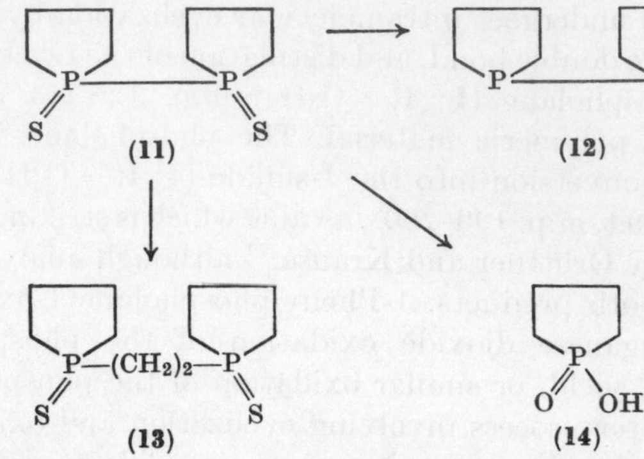

This acid had been previously prepared (Helferich and Aufderhaar[81]) by treating ethereal 1-bromo-4-chlorobutane in turn with magnesium to form 4-chlorobutylmagnesium bromide, then with cadmium chloride, and finally under nitrogen at −30°, with phosphorus trichloride, thus obtaining a mixture of 4-chlorobutylphosphonous dichloride (**15**) and 1,8-dichlorooctane (**16**). This mixture, treated in

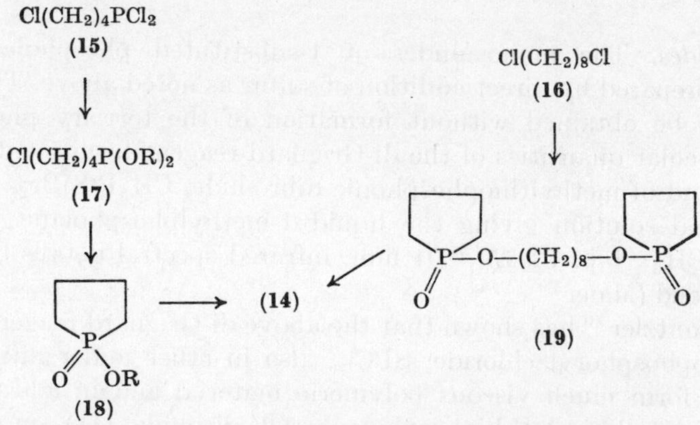

pyridine with 1-butanol and then fractionally distilled, gave di-*n*-butyl 4-chlorobutylphosphinite (**17**; R = C$_4$H$_9$), b.p. 99°/0.05 mm, which when heated at 165–180° underwent cyclization to the n-butyl ester (**18**; R = C_4H_9), b.p. 76°/0.05 mm, of the phosphinic acid (**14**), to which it was readily hydrolyzed by hot hydrochloric acid. When the original mixture of (**15**) and (**16**) was treated with ethanol in dimethylaniline, it gave diethyl 4-chlorobutylphosphinite (**17**; R = C_2H_5), which could not be separated from the dichloro derivative (**16**). The mixture was therefore heated at 145°, and fractional distillation of the product gave the ethyl ester (**18**; R = C_2H_5), b.p. 107°12 mm, of the phosphinic acid and the 1,8-octylene diester (**19**), b.p. 218–220°/0.01 mm; both esters on hydrolysis gave the acid (**14**), m.p. 53–54.5°. The m.p.s of the acid prepared by this route and by Schmutzler's are identical.

Kosolapoff and Struck[10] attempted a third synthesis in which the di-Grignard reagent, $BrMg(CH_2)_4MgBr$, and N,N-diethylamido-phosphoryl dichloride (or N,N-diethylphosphoramidic dichloride), $(C_2H_5)_2N$—$P(O)Cl_2$, were brought into reaction in boiling ether. The crude product was hydrolyzed and then subjected to several processes to remove polymeric material and to purify and isolate the final product; an acid, m.p. 99–100°, was thus obtained in 0.1% yield and identified as the phosphinic acid (**14**) on the basis of a phosphorus analysis.

It is noteworthy, however, that Hilgetag, Henning, and Gloyna[82] state that, although the above dichloride reacts with two equivalents of a monofunctional Grignard reagent to give after hydrolysis a phosphinic acid, yet its reaction with the difunctional reagent $BrMg(CH_2)_4MgBr$

$$(C_2H_5)_2NP(O)Cl_2 + 2RMgX \rightarrow (C_2H_5)_2NP(O)R_2 \rightarrow R_2P(O)OH$$

stops at the intermediate stage, even when the reaction is carried out in boiling tetrahydrofuran to obtain a higher reaction temperature,

$$(C_2H_5)_2NP(O)(Cl)—(CH_2)_4MgBr \rightarrow (HO)_2(O)P(CH_2)_3CH_3$$

and cyclization therefore does not occur: subsequent hydrolysis gives n-butylphosphonic acid. They add, as a similar case, that although the di-Grignard reagent reacts with $C_6H_5PCl_2$ to give 1-phenylphospholane (p. 50), it reacts with $C_6H_5P(O)Cl_2$ to give only the intermediate stage,

$$C_6H_5P(O)Cl(CH_2)_4MgBr \rightarrow CH_3(CH_2)_3(C_6H_5)PO(OH)$$

which therefore on hydrolysis furnishes n-butylphenylphosphinic acid.

Halogenophospholanes. 1-Chlorophospholane (**21**) has been synthesized by Burg and Slota,[83] the first stage being the simultaneous addition of ethereal solutions of the di-Grignard derivative of 1,4-dibromobutane and of dichloro(dimethylamino)phosphine (also named N,N-dimethylphosphoramidous dichloride), $(CH_3)_2NPCl_2$, to ether at −78°, which gave the 1-(dimethylamino)phospholane (**20**) in 8% yield. Treatment of this compound with just less than one equivalent of

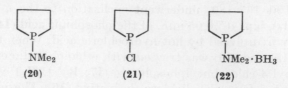

hydrogen chloride gave 1-chlorophospholane (**21**). The compound (**20**) was also converted into the boron hydride derivative (**22**), which when heated gave a mixture of products from which the parent phospholane (**1**; R = H), m.p. −88°, was ultimately isolated. Its infrared spectrum gave a strong P—H band at 2280 cm⁻¹. All these products were identified by their physical properties, no analyses of the compounds being recorded.

Schmutzler[84] has shown that antimony trifluoride reacts with 1-chlorophospholane (**21**), giving 1,1,1-trifluorophospholane (**23**), b.p. 62°/90 mm (45%); it was to avoid the laborious preparation of (**21**) that

he developed the synthesis of the disulfide (**11**),[80] which also when heated with antimony trifluoride gives the trifluorophospholane (**23**). The latter compound is of some theoretical interest, for its ¹⁹F-nmr spectra indicate that it has a trigonal bipyramidal structure, in which there is a rapid intramolecular exchange of the equatorial and apical fluorine atoms (Muetterties *et al.*,[85] Nixon and Schmutzler[86]).

A third method of making polyfluoro derivatives is by treating, for example, 3-methyl-1-phenylphospholane 1-oxide (**24**) with sulfur trifluoride, giving the 1,1-difluoro-3-methyl-1-phenylphospholane (**25**).[80]

2,2,3,3,4,4,5,5-Octafluoro-1-iodophospholane (**26**) has been obtained by a novel synthesis (Krespan and Langkammerer[87, 88]). A mixture of red phosphorus, iodine, and tetrafluoroethylene was heated in a sealed vessel at 220° for 8 hours. Distillation of the reaction mixture gave the phospholane (**26**), b.p. 116–119° (4% based on the iodine used). The undistilled residue afforded 2,2,3,3,5,5,6,6-octafluoro-1,4-di-iodo-1,4-diphosphorinane (27%) (p. 174).

The structure of (**26**) was confirmed by nmr spectra and also by oxidative hydrolysis with cold water. This process gave a colorless solution which was partly evaporated in a stream of air, cooled to 0°, treated with 12% hydrogen peroxide, and similarly evaporated, leaving the oily 1,1,2,2,3,3,4,4-octafluorobutylphosphonic acid, $H(CF_2)_4PO(OH)_2$. The identity of this acid was also confirmed by its nmr spectra and by conversion into the crystalline diammonium salt. A simpler example of this oxidative hydrolysis, which entails ready cleavage of a P—C link

$$(F_3C)_2PI \rightarrow F_3C \cdot PO(OH)_2$$

followed by oxidation of the phosphonous acid so formed, is provided by bis(trifluoromethyl)phosphinous iodide which when similarly treated yields trifluoromethylphosphonic acid (Bennett, Eméleus, and Haszeldine [89]).

Hydroxyphospholanes. The 2,5-dihydroxy derivatives of the quaternary phospholanium salts can be readily prepared, although the reaction conditions are critical. If concentrated hydrochloric acid is added to a solution of a secondary phosphine and 25% aqueous succinaldehyde in methanol–tetrahydrofuran at room temperature, and the reaction mixture is then boiled for 2 hours and concentrated, the crystalline chloride (27) is obtained (Buckler and Epstein [90, 91]). Thus di-*n*-butylphosphine gives 1,1-di-*n*-butyl-2,5-dihydroxyphospholanium chlor-

(27)

ide [27; $R = CH_3(CH_2)_3^-$], m.p. 107–108°; diisobutylphosphine gives the 1,1-diisobutyl analogue, m.p. 149–150°. The nmr shifts in the phosphorus region, in ppm relative to 85% H_3PO_4, are −46 and −39, respectively; these values are of interest compared with those of the analogous phosphorinanium salts (p. 111).

Reddy and Weis [34] have recorded that triphenylphosphine in acetonitrile reacts exothermally with two equivalents of tetracyanoethylene, $(NC)_2C{=}C(CN)_2$, to give the colorless crystalline 2,2,3,3,4,4,5,5-octacyano-1,1,1-triphenylpholane (28), m.p. 168.5–170°. The use of tri-*o*-tolylphosphine and tri-*p*-tolylphosphine gave the 1,1,1-tri-*o*-tolyl derivative, m.p. 235–237°, and the 1,1,1-tri-*p*-tolyl derivative, m.p. 205–206°, respectively.

The chemical evidence for the structure (28) is that the compound (a) in boiling methanol gave 2,3,3,4,4,5-hexacyano-2,5-bis[imido-(methoxy)methyl]-1,1,1-triphenylpholane (29), m.p. 215–220°, although this was not decisively identified, and (b) when hydrolyzed with hydrochloric acid at 140–150° yielded triphenylphosphine oxide and a

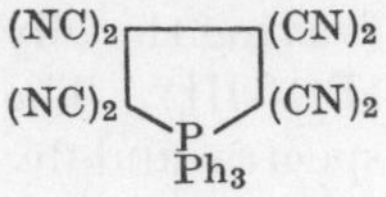

(28) 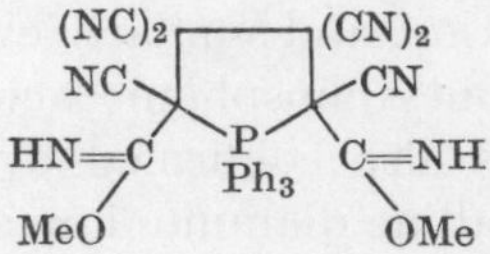   (29) 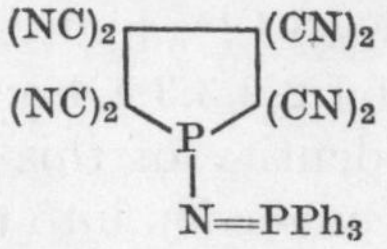   (30)

mixture of the low- and high-melting forms of 1,2,3,4-butanetetra-
carboxylic acid. This mixture treated with diazomethane gave the mixed
tetramethyl esters; the mixed acids, moreover, in boiling acetic anhydride
gave the dianhydride, m.p. 248°. The identity of the mixed acids, their
methyl ester, and the dianhydride was confirmed by comparison of their
infrared spectra with those of the authentic compounds.

The physical evidence for the structure (28) was mainly based on
the nmr spectra. In particular, the $(C_6H_5)_3P$ unit might (improbably)
have migrated to an adjacent CN group to give (30), which in turn might
on hydrolysis also give the butanetetracarboxylic acid. In dioxan
solution, with 85% orthophosphoric acid as a reference, the ^{31}P chemical
shift of (28) was -22.0 ± 1.0 ppm, whereas that for the compound
$(C_6H_5)_3P{=}NC_6H_5$, akin in structure to (30),[34] was 0.0 ± 1.0 ppm.

This identification of (28) has been criticized (Shaw, Tebby, Ward,
and Williams[35]) on the ground that the ^{31}P shift of (28) was compared
with the ^{31}P shift (-17.0 ppm) of the compound considered at that time
to be dimethyl 4,4,4,6-tetraphenyl-4H-1,4-oxaphosphorin-2,3-dicarb-
oxylate (31) (Hendrickson[92]), but later shown to have the structure
(32) (Gough and Trippett[93]; Hendrickson et al.[94]). Considerable evidence

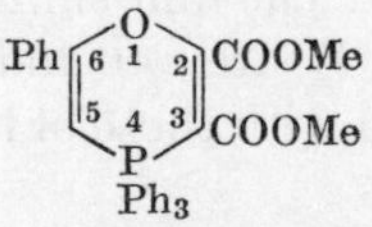

(31) 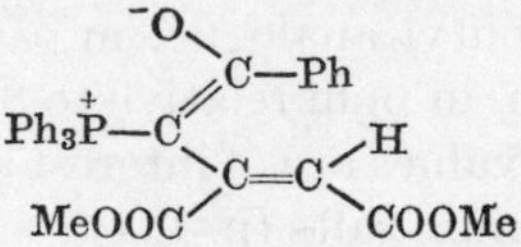   (32)

has been adduced that phosphoranes that have five single bonds from
the phosphorus atom to carbon atoms, have ^{31}P shifts in the region
$+80$ to $+100$ ppm (Hellwinkel).[95]

The identification of the compound (28) must await additional
evidence.

Keto-phospholane Derivatives. The first phospholane derivative
having a keto group in the ring system has recently been synthesized
and has very interesting properties (Quin and Caputo[95a]). The interaction
of chloroprene and methylphosphonous dichloride gave 1,1,3-trichloro-
1-methyl-3-phospholene (33), a compound originally prepared by

McCormack.[47, 48] The crude product was hydrolyzed by the addition of ice and when, after 12 hours, the solution was neutralized with sodium carbonate, a 9:1 mixture of 3-chloro-1-methyl-2-phospholene 1-oxide (**34**) and the 3-phospholene isomer (**35**) was obtained. The ratio of the isomers is determined by the length of exposure of the hydrolyzed product in the acid solution: a short exposure gives a higher proportion of the 3-phospholene (**35**). The former, when treated with methanolic

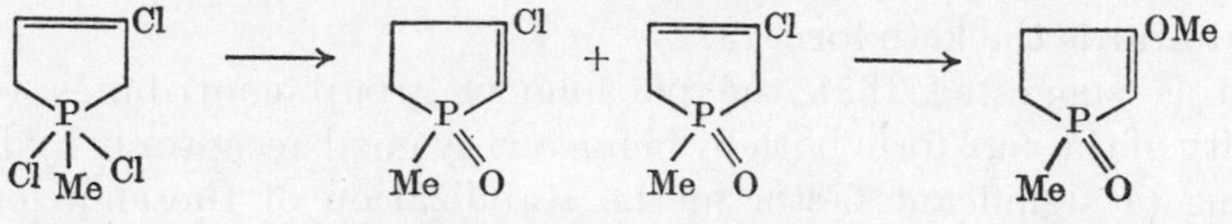

sodium methoxide, gave 3-methoxy-1-methyl-2-phospholene 1-oxide (**36**), b.p. 96°/0.15 mm (58%), which in turn readily underwent hydrolysis in warm aqueous solution containing a trace of hydrochloric acid to give 1-methylphospholan-3-one 1-oxide (**37**), m.p. 89–91° (53%).

The structure of this compound in dilute chloroform solution was confirmed by its infrared spectrum which showed a strong CO band at

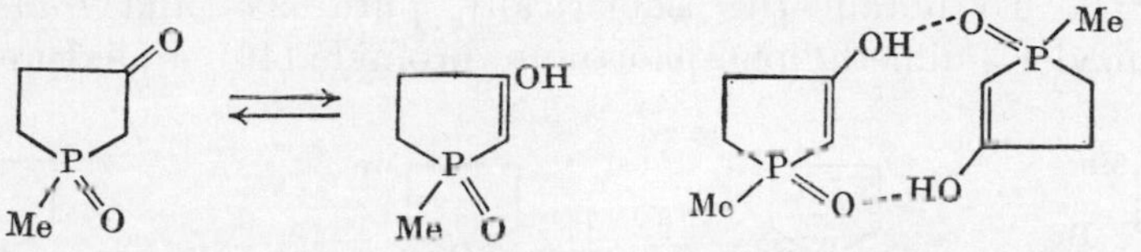

1730 cm^{-1} and no appreciable OH or C=C absorption. In a Nujol mull, however, the spectrum showed only a very weak CO band, but had a broad band at 2440–2275 cm^{-1} attributed to hydrogen-bonded OH, and a very strong peak in the C=C region (1558 cm^{-1}). The spectrum of a KBr pellet of (**37**) of low concentration resembled that in the dilute chloroform solution, but the spectrum of a KBr pellet of high concentration resembled that in the Nujol mull. These results strongly suggest that the keto form (**37**) and the enol form (**38**) give a tautomeric mixture, the keto form predominating in low concentration and the enol form in high concentration.

The broad band at 2440–2275 cm^{-1} is attributed to a dimeric hydrogen-bonded form (**39**) similar in type to those formed by carboxylic acids.

The nmr spectra provide further evidence of this tautomerism. The ^{31}P spectrum of a 3.4-molal solution of (**37**) shows two signals,

−51.0 and −60.5 ppm (H_3PO_4 reference), the latter signal (*ca.* 25% of
the total area) being attributed to the enol form (**38**). The [1]H-nmr
spectrum of the same solution shows a vinyl proton doublet (J 22 c/sec)
at δ 5.07 and two P-CH_3 doublets (δ 1.82, J 13.5 c/sec; δ 1.67, J 13.5
c/sec). The area ratio of vinyl to total P-CH_3 signals is *ca.* 1:12, indicating the presence of 20% of the enol (**38**), in reasonably close accord
with the ^{31}P assignment. Dilution of the chloroform solution causes the
vinyl signal and one P—CH_3 doublet to disappear as the equilibrium
shifts towards the keto form (**37**).

It is suggested that the phosphoryl group contributes to the
stability of the enol form both by being a very good acceptor in hydrogen
bonding (a significant factor in the stabilization of the enol form of
dicarbonyl compounds) and by the resultant *d–p* resonance with the
enolic double bond. The operation of the latter effect among phospholene
oxides has been earlier suggested by the greater thermodynamic
stability of 2-phospholene oxides relative to that of isomeric 3-phos-
pholene oxides.[56, 61]

Alkaline Hydrolysis of Quaternary Phospholanium Salts. This
process, unlike hydrolysis of phosphetanium salts (p. 11), normally
occurs with retention of the configuration of the phosphorus atom
(Marsi[95b]). To obtain the isomerically pure *cis-* and *trans*-forms
of 1-benzyl-1,3-dimethylphospholanium bromide (**40**), a mixture of the

(**40**)   (**41**)   (**42**)   (**43**)

isomeric 1,3-dimethylphospholane 1-oxides (**41**) was reduced with
trichlorosilane (p. 50) to the corresponding mixed 1,3-dimethylphos-
pholanes (**42**). Careful fractional distillation gave a pure isomer (**42a**),
b.p. 141°, and a fraction of lower b.p. rich in the second isomer (**42b**).
The configurational stability of (**42a**) was shown by heating a mixture
of the isomers at 150° for three days without detectable change in
the nmr spectra of the cold liquid. The isomer (**42a**) when treated with
benzyl bromide gave the phospholanium bromide (**40a**), m.p. 168.5–
169.5°. The fraction of lower b.p., when similarly treated, gave a mixture
of the isomeric bromides, from which fractional crystallization furnished
the pure second isomeric bromide (**40b**), m.p. 180–181°. The nmr
spectra of the two bromides were almost identical.

The two isomeric bromides, when separately treated with boiling aqueous 1N-NaOH solution, gave the respective isomeric 1,3-dimethyl-phospholane 1-oxide (**41a**), m.p. 22°, and (**41b**), m.p. 70–71°, toluene being the only other product detected.

The melting point and nmr spectrum of (**41a**) were identical with those of the oxide obtained by the oxidation of (**42a**) with *tert*-butyl hydroperoxide, which oxidizes phosphines stereospecifically with retention of configuration at the phosphorus atom. Furthermore, the 1-oxide (**41b**) was identical with the phospholane oxide isolated by fractional crystallization of the mixture of (**41a**) and (**41b**) prepared by catalytic hydrogenation of 1,3-dimethyl-3-phospholene 1-oxide (**43**)[61] (p. 35).

These results show that there is no common intermediate between the *cis*- and the *trans*-isomer of the quaternary bromide (**40**) during their alkaline conversions into the 1-oxides (**41**).

The probable mechanism of this conversion is discussed, with the nmr spectral data confirming the structure of the various products.[95b]

Several phospholanes, such as the 1-ethyl and the 1-ethyl-3-methyl members, can be used to catalyze the conversion of isocyanates into carbodiimides, a property also shown by similar phospholenes (p. 35) (Balon[54]).

5-Phosphoniaspiro[4.4]nonane Cation

This name is given to the quaternary phosphonium cation (**1**), [RIS 7918], in which each ring system is a phospholane, with one phosphorus atom common to both.

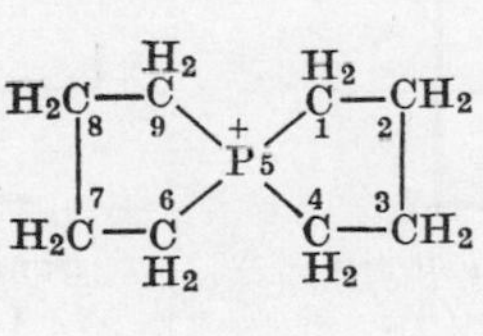

(**1**) 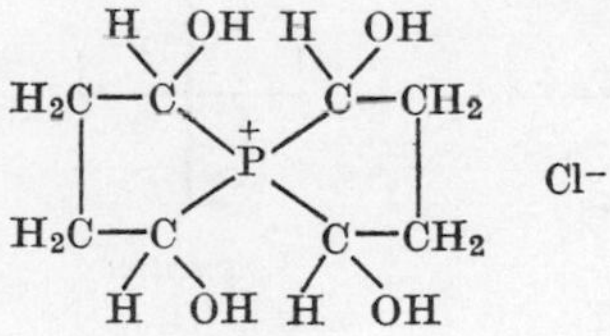   (**2**)

It has been known for many years (Messinger and Engels[96]) that acetaldehyde and propionaldehyde react readily with phosphine in an ethereal solution of hydrogen chloride to give tetrakis(1-hydroxyalkyl)-phosphonium chlorides:

$$4RCHO + PH_3 + HCl \rightarrow [RCH(OH)]_4P^+ \, Cl^-$$

Formaldehyde reacts similarly with phosphine in concentrated hydro-chloric acid to give tetrakis(hydroxymethyl)phosphonium chloride

(Hoffmann [97]). α-Branched aliphatic aldehydes may, however, give substituted 1,3,5-dioxaphosphorinanes (p. 309).

Buckler and Wystrach [98, 99] have studied the action of succindialdehyde, using for convenience, however, 2,5-diethoxytetrahydrofuran which as a cyclic acetal readily hydrolyzes to the dialdehyde in aqueous acid. The gaseous phosphine was added at atmospheric pressure to a vigorously stirred mixture of 2,5-diethoxytetrahydrofuran and concentrated hydrochloric acid at 25–30°; evaporation under reduced pressure gave a gummy residue, which when stirred with 1-pentanol at 0° deposited 1,4,6,9-tetrahydroxy-5-phosphoniaspiro[4.4]nonane chloride (**2**) (34%), m.p. 94–95° (dec.) after recrystallization from 6:1 1-pentanol–concentrated hydrochloric acid.

The stereochemistry of this compound, which possesses four asymmetric carbon atoms, might appear to be complex. Buckler and Wystrach [99] draw attention, however, to the apparent stereochemical similarity of 2,3,7,8-tetramethyl-5-azoniaspiro[4.4]nonane toluene-*p*-sulfonate (**3**; $X = CH_3C_6H_4SO_3$), which McCasland and Proskow[100] have

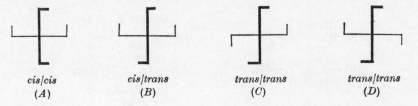

(3)

obtained in four isomeric forms, and which they represent formally by (*A*), (*B*), (*C*), and (*D*). These diagrams represent the cation viewed down

| *cis/cis* | *cis/trans* | *trans/trans* | *trans/trans* |
| (*A*) | (*B*) | (*C*) | (*D*) |

the axis running across both pyrrolidine rings and the tetrahedral nitrogen atom: in each case the two long lines represent the planes of the upper and the lower pyrrolidine rings; the short terminal lines represent the relative position of the methyl groups and indicate whether the two methyl groups in any one ring are *cis* or *trans* to one another. The forms (*A*), (*B*), and (*C*) are racemic and susceptible to optical resolution, but (*D*) is a *meso* form, being unique in that it owes its optical inactivity to the possession of a four-fold alternating axis of symmetry and no other element of symmetry.

The cation (**2**) can be represented by four precisely similar forms (*A*), (*B*), (*C*), and (*D*). This cation, however, differs fundamentally from (**3**) in that it has an OH group on each of the four carbon atoms directly linked to the phosphorus atom, and models indicate that (*D*) is the only form in which the OH groups from different rings do not clash with one another. No isomeric forms of the salt (**2**) were detected.

Braye[101] has found that potassium or lithium 2,3,4,5-tetraphenyl-phospholide (**4**) when added slowly to a considerable excess of 1,4-dibromobutane gives 1-(4′-bromobutyl)-2,3,4,5-tetraphenylphosphole

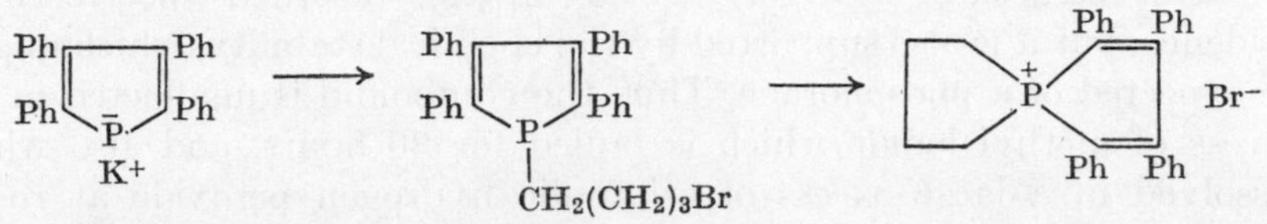

(4) (5) (6)

(**5**), m.p. 147–150° (19%). This compound when heated at 175° for 2 hours without a solvent undergoes cyclization to give 1,2,3,4-tetraphenyl-5-phosphoniaspiro[4.4]nona-1,3-diene bromide (**6**), m.p. 333–335°, in almost quantitative yield.

5-Phospha(v)spiro[4.4]nonane

The above compound (**1**), [RIS —], could have various dehydrogenated derivatives, the extreme case being 5-phospha(v)spiro[4.4]nona-1,3,6,8-tetraene (**2**). In all the compounds (**1**)—(**2**), the 5H atom would be replaced by an alkyl or an aryl group, thus forming a phosphorane.

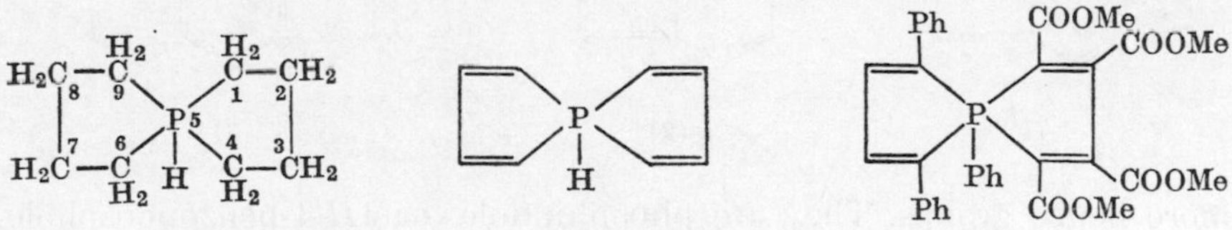

(1) (2) (3)

It has been noted (p. 22) that 1,2,5-triphenylphosphole reacts with dimethyl acetylenedicarboxylate in benzene at 150° apparently to form a normal but unstable adduct which decomposes to form 3,6-diphenylphthalic anhydride and dimethyl 3,6-diphenylphthalate; 1,2,5-triphenylphosphole 1-oxide reacts very readily with the acetylenedicarboxylate at 156° to form a residue from which only dimethyl 3,6-diphenylphthalate has been isolated.

Hughes and Uaboonkul[102] have recorded that, when a mixture of

1,2,5-triphenylphosphole and an excess of dimethyl acetylenedicarboxylate was shaken under nitrogen for 2 days at room temperature, the products isolated were (a) dimethyl 3,4-diphenylphthalate (9%), (b) a yellow crystalline 1:2 adduct, m.p. 182° (37%), and (c) a red crystalline 1:4 adduct, m.p. 218–220° (2.5%), which was not investigated further.

The yellow compound might be one of a number of isomers, but evidence is adduced that it is most probably tetramethyl 5,6,9-triphenyl-5-phospha(v)spiro[4.4]nona-1,3,6,8-tetraene-1,2,3,4-tetracarboxylate (**3**).

This structure is largely based on the recorded spectroscopic evidence, but it is also supported by the chemical stability, which might be expected of a phosphorane. Thus the compound is unaffected in an excess of methyl iodide which is boiled for 30 hours, and also when dissolved in a large excess of ethanolic hydrogen peroxide at room temperature for 24 hours (and much is left unchanged after boiling for 2.5 hours).

On the other hand, in boiling chloroform or ethanol it undergoes a slow rearrangement over several hours to give a colorless isomer, m.p. 220°. This isomer is a tertiary phosphine which readily gives a P-oxide, and the ultraviolet absorption shows that the degree of conjugation has been considerably reduced during the rearrangement. This isomer was not further investigated.

Benzo Derivatives of Phosphole. There are three simple compounds which can be regarded systematically as phosphole fused to one or

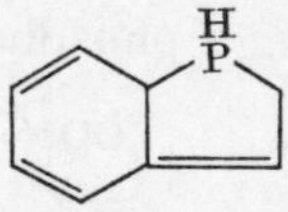

(1) 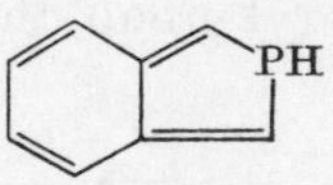   (2) 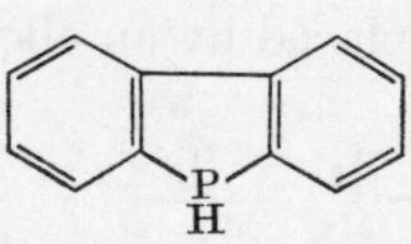   (3)

more benzo groups. They are phosphindole, or 1*H*-1-benzophosphole, (**1**) [RRI 1334]; isophosphindole, or 2*H*-2-benzophosphole, (**2**) [RRI 1335] (p. 62); and 5*H*-dibenzophosphole or 9-phosphafluorene, (**3**) [RRI 3032] (p. 72).

These three fundamental compounds and their derivatives will be considered in turn.

Phosphindole and Phosphindolines

The parent compound, [RRI 1334], is depicted and numbered as in (**1**); both phosphindole (which is understood to be the 1*H*-compound, like indole itself) and the isomeric $3H$-phosphindole (**1A**) are unknown, but the 2,3-dihydro derivative (**2**), named phosphindoline, is known as its 1-substituted products.

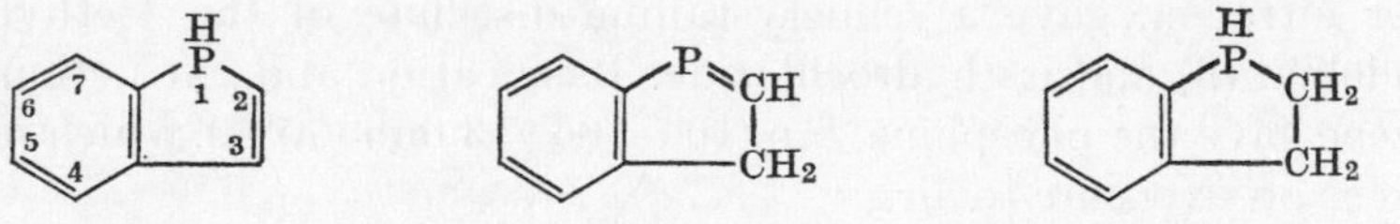

(**1**)   (**1A**)   (**2**)

1-Ethylphosphindoline has been synthesized (Mann and Millar[103]) by a method closely similar to that used earlier in the synthesis of 1-ethyl-1,2,3,4-tetrahydrophosphinoline (p. 124). For this purpose, o-bromobenzyl bromide (**3**) was converted into o-bromobenzylmagnesium bromide, which reacted with chlorodimethyl ether, $ClCH_2OCH_3$,

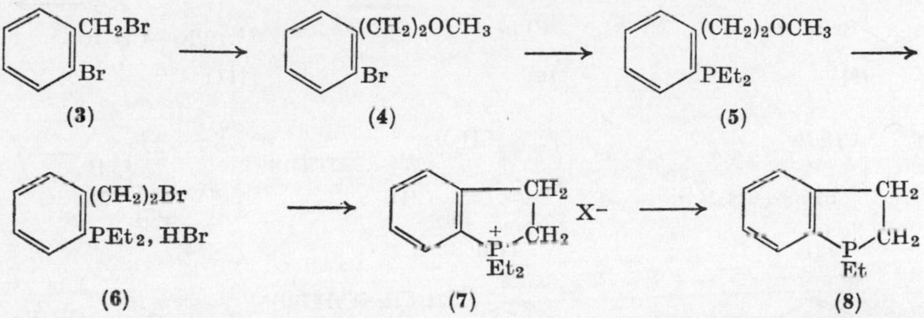

(**3**)   (**4**)   (**5**)

(**6**)   (**7**)   (**8**)

to give o-bromophenethyl methyl ether (**4**), b.p. 64–67°/0.7 mm (46%). By the use of activated magnesium, this compound was converted into the Grignard reagent, which on treatment with diethylphosphinous chloride, $(C_2H_5)_2PCl$, furnished diethyl-(o-2-methoxyethylphenyl)phosphine (**5**), b.p. 119–133°/26 mm (67%); persistent frothing on distillation precluded ready purification, but the phosphine gave a methiodide, m.p. 95.5–96°, and with ethanolic potassium tetrabromopalladium(II) gave the orange dibromobisphosphinepalladium, $(C_{13}H_{21}OP)_2PdBr_2$, m.p. 178°.

The phosphine (**5**) in acetic acid–hydrobromic acid was saturated with hydrogen bromide and boiled for 3 hours, giving the water-soluble hydrobromide (**6**) of diethyl-(o-2-bromoethylphenyl)phosphine. The addition of aqueous sodium hydrogen carbonate to the concentrated solution deposited the oily phosphine; subsequent addition of chloroform with shaking now extracted the free phosphine, which underwent an exothermic cyclization to 1,1-diethylphosphindolinium bromide (**7**;

X = Br), which promptly returned to the aqueous solution. The latter, when treated with aqueous sodium picrate, deposited the picrate (**7**; X = $C_6H_2N_3O_7$), m.p. 145–145.5° (29%). The pure picrate was converted into the chloride which, when dried and heated at 350–370°/12 mm under nitrogen, gave a semicrystalline distillate of the 1-ethylphosphindoline (**8**) and its hydrochloride. Basification and extraction with benzene gave the phosphine, b.p. 104–106°/13 mm (75%), which readily darkens on exposure to air.

With ethyl iodide it gave 1,1-diethylphosphindolinium iodide, m.p. 123°, and it also gave the dibromodiphosphinepalladium, $(C_{10}H_{13}P)_2PdBr_2$, m.p. 147°.

Märkl[104] has extended this synthesis by introducing a substituent in position 2 of the phosphindoline ring. The compound (**3**) was converted

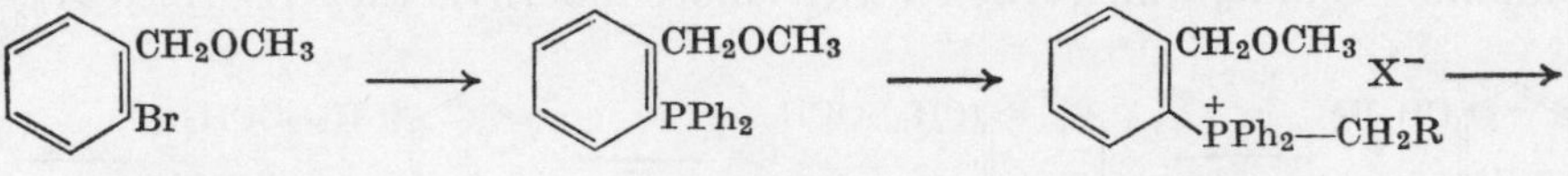

(**9**)   (**10**)   (**11**)

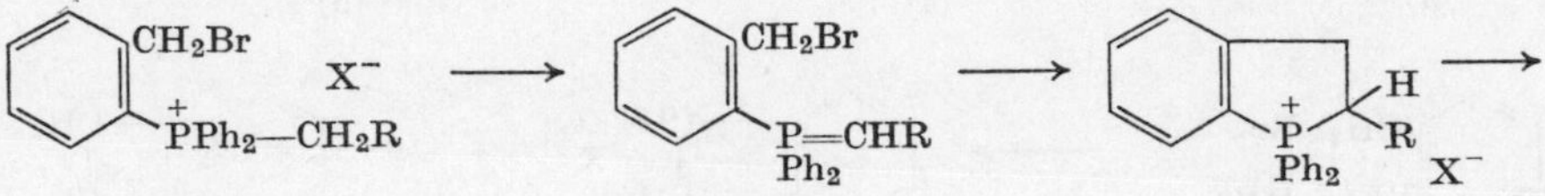

(**12**)   (**13**)   (**14**)

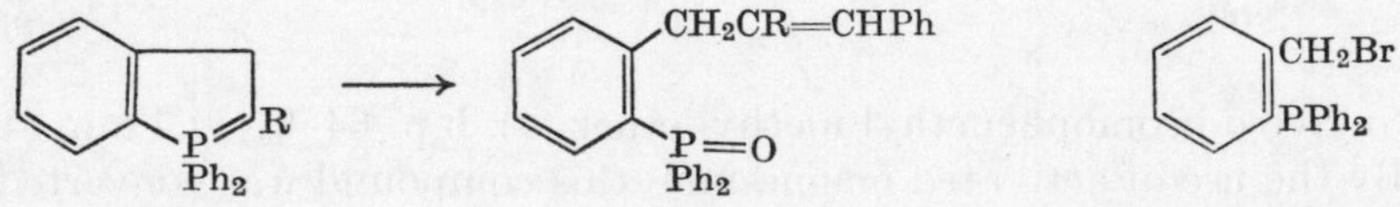

(**15**)   (**16**)   (**17**)

into o-bromobenzyl methyl ether (**9**), and then as above into [o-(methoxymethyl)phenyl]diphenylphosphine (**10**), m.p. 94–96°. This was then quaternized, for example, with benzyl chloride to form benzyl-[o-(methoxymethyl)phenyl]diphenylphosphonium bromide (**11**; R = C_6H_5, X = Cl), m.p. 254–256°. This was then converted as above into benzyl-[o-(bromomethyl)phenyl]diphenylphosphonium bromide (**12**; R = C_6H_5, X = Br), m.p. 247–249°. Phenyllithium in toluene, or potassium *tert*-butoxide in dimethylformamide converted this bromide into the red benzylidene-[o-(bromomethyl)phenyl]diphenylphosphorane (**13**; R = C_6H_5), which when treated with one equivalent of a base underwent cyclization by *C*-alkylation to form the 1,1,2-triphenylphosphindolinium cation, isolated as the tetrafluoroborate (**14**; R = C_6H_5,

$X = BF_4$), m.p. 203–205°. The addition of one equivalent of base to this salt, or of two equivalents to (**13**; $R = C_6H_5$), produced the phosphorus ylide, 1,1,2-triphenyl-3H-phosph[v]indole (**15**; $R = C_6H_5$).

This compound shows the Wittig reaction, giving a product having the C=C and the P=O groups in one compound: for example, with benzaldehyde it gives (*o*-diphenylphosphinobenzylstilbene) *P*-oxide (**16**; $R = C_6H_5$), consisting mainly of the substituted *cis*-stilbene, an amorphous powder (λ_{max} 280, 273 mμ in methanol), with some *trans*-stilbene, colorless crystals, m.p. 167–169° (λ_{max} 310 mμ).

Quaternization of (**10**) with methyl bromide gives the corresponding series of compounds: (**11**; $R = H$, $X = Br$), m.p. 158–160°; (**12**; $R = H$, $X = Br$), isolated as the iodide, m.p. 192–194°; (**14**; $R = H$, $X = BF_4$), m.p. 91–93°.

It is clear that, if an alkyl halide containing an ester group were used to form the quaternary salt (**11**), hydrolysis would occur during the next stage. Therefore the salt (**12**; $R = CO_2C_2H_5$, $X = Br$), m.p. 152–153°, must be prepared by the action of ethyl bromoacetate at room temperature on [*o*-(bromomethyl)phenyl]diphenylphosphine (**17**); if the latter is heated, it readily undergoes bimolecular quaternization to 5,6,11,12-tetrahydro-5,5,11,11-tetraphenyldibenzo[*b,f*][1,5]diphospho-cinium dibromide (p. 199).

Isophosphindole and Isophosphindolines

Isophosphindole is the name given to the (probably non-existent)

(1)   (2)

compound (**1**) [RRI 1335]: the 1,3-dihydro compound (**2**) is isophosphindoline, which has been encountered only as its 2-substituted derivatives.

The isophosphindoline system was first synthesized by McCormack[48] who, in the course of his wide exploration of the McCormack reaction (p. 31), showed that a mixture of 1,2-dimethylenecyclohexane (**3**) and phenylphosphonous dibromide, $C_6H_5PBr_2$, gave a 'complete reaction' in 30 minutes at 60°, to form 2,2-dibromo-4,5,6,7-tetrahydro-2-phenyl-isophosphindoline (**4**), which on hydrolysis gave the stable 4,5,6,7-tetrahydro-2-phenylisophosphindoline 2-oxide (**5**).

(3)   (4)   (5)

In an attempt (Mann, Millar, and Stewart[105]) to synthesize the
simpler 2-phenylisophosphindoline (**6**), o-xylylene dibromide (**7**) was

(6)   (7)   (8)

treated with dimethylphenylphosphine; rapid quaternization occurred
giving o-xylylenebis(dimethylphenylphosphonium)dibromide (**8**; R =
CH_3); the use of diethylphenylphosphine gave the analogous diethyl
salt (**8**; R = C_2H_5). These bisphosphonium dibromides were formed
irrespective of the relative quantities of the tertiary phosphines and
the dibromide (**7**) employed. Precisely similar arsonium salts are formed
from analogous tertiary arsines and the dibromide (**7**), but whereas
these bisarsonium dibromides when heated give 2-phenylisoarsindoline
in yields up to 60% (p. 369), the bisphosphonium dibromides (**8**) when
heated undergo general decomposition without evidence of cyclization.

The dibromide (**7**) reacted readily with phenylphosphinebis-
(magnesium bromide), $C_6H_5P(MgBr)_2$, in ether–benzene, but the
organic product was almost entirely an ivory-colored glass, indicating a
very extensive linear condensation of the reagents. This material, when
heated under nitrogen at 0.3 mm, underwent decomposition but gave a
distillate of 2-phenylisophosphindoline (**6**) in only 4% yield. The pure
phosphine, b.p. 110–112°/0.2 mm, reacted readily with methyl iodide,
giving 2-methyl-2-phenylisophosphindolinium iodide, m.p. 207–209°
(picrate, m.p. 88–90°).[105] This iodide decomposes when heated above
its m.p., giving a mixture of the phosphine (**6**) and its hydriodide, thus
illustrating the thermal stability of the heterocyclic ring, for simple
quaternary phosphonium halides containing both a benzyl and a methyl
group tend, when heated, to lose the benzyl rather than the methyl
group (Meisenheimer et al.[106]).

An effective synthesis of (**6**) was achieved using o-(methoxymethyl)-
benzyl chloride (**9**), which can be readily prepared by the action of acetyl
chloride with a trace of zinc chloride on o-xylylene dimethyl ether,

$C_6H_4(CH_2OCH_3)_2$, but requires careful manipulation because of its toxic properties (Mann and Stewart[107]).

$$\text{(benzene ring)}\begin{matrix}CH_2OCH_3\\CH_2Cl\end{matrix} \longrightarrow \text{(benzene ring)}\begin{matrix}CH_2OCH_3\\CH_2PPhR\end{matrix} \longrightarrow \text{(bicyclic)}\ \overset{H_2}{\underset{H_2}{C}}\ \overset{+}{\underset{R}{P}}\ \overset{Ph}{}\quad Br^-$$

(9)   (10)   (11)

Initially, diphenylphosphine was added to a solution of sodium (one equivalent) in liquid ammonia to form $(C_6H_5)_2PNa$, which on the addition of the chloride (9) gave o-(methoxymethylbenzyl)diphenylphosphine (10; $R = C_6H_5$), a viscous oil (methiodide, m.p. 130–131°). Hydrogen bromide was passed through a solution of this phosphine in boiling acetic acid–hydrobromic acid for 3 hours; this converted the methoxymethyl group into a bromomethyl group, which quaternized the phosphine group, giving 2,2-diphenylisophosphindolinium bromide (11; $R = C_6H_5$), characterized as the picrate, m.p. 180–181°.

This synthetic route was readily modified. Phenylphosphine in ammonia was treated in turn with Na, C_2H_5Br, and Na (one equivalent throughout) to give the sequence

$$C_6H_5PH_2 \rightarrow C_6H_5PHNa \rightarrow C_6H_5P(C_2H_5)H \rightarrow C_6H_5P(C_2H_5)Na$$

Addition of the chloride (9) now gave ethyl-o-(methoxymethylbenzyl)phenylphosphine (10; $R = C_2H_5$), b.p. 202–206°/13 mm [72% calculated on (9)]. This phosphine, treated as above, gave the 2-ethyl-2-phenylisophosphindolinium bromide (11; $R = C_2H_5$), m.p. 273° (picrate, m.p. 102–102.5°). The bromide, when heated under nitrogen at 15 mm, decomposed giving a mixed distillate of the phosphine (6) and its hydrobromide, which on basification and working up furnished the phosphine (6), b.p. 182–186°/15 mm, 110–113°/0.2 mm (67%) (Mann, Millar, and Watson[108]).

Two recent syntheses of isophosphindolinium salts should be mentioned. Märkl[76] has recorded the reaction of the dibromide (7) with tetraphenyldiphosphine, $(C_6H_5)_2P–P(C_6H_5)_2$, to give the cation (11; $R = C_6H_5$), isolated as the perchlorate, m.p. 180–181° (42%), this being a special example of his general reaction (p. 52). If symmetrical dialkyldiaryl- or tetraalkyl-diphosphines give the same reaction, it might afford a preparative method for 2-aryl- and 2-alkyl-isophosphindolines, respectively.

It is claimed that the acid-catalyzed condensation of secondary phosphines with succindialdehyde and glutardialdehyde to give quaternary 2,5-dihydroxyphospholanium and 2,6-dihydroxyphosphorinanium salts, respectively (pp. 57, 110), when similarly applied to phthal-aldehyde, gives the analogous 1,3-dihydroxyisophosphindolinium salts

$$R_2PH \ + \quad R'\!\!\begin{array}{c}CHO\\CHO\end{array} \longrightarrow \quad R'\!\!\begin{array}{c}CHOH\\^+PR_2 \quad X^-\\CHOH\end{array}$$

$$(12) \qquad\qquad\qquad (13)$$

(Buckler and Epstein[91]). For example, di-(n-dodecyl)phosphine reacts with phthalaldehyde (**12**; R' = H) in tetrahydrofuran–sulfuric acid at 5° to give 2,2-bisdodecyl-1,3-dihydroxyisophosphindolinium hydrogen sulfate (**13**; R = $C_{12}H_{25}$, R' = H, X = HSO_4). Similarly diphenethyl-phosphine reacts with 4-carboxyphthalaldehyde (**12**; R' = CO_2H) in dioxan–sulfuric acid at 25° to give 5-carboxy-1,3-dihydroxy-2,2-diphenethylisophosphindolinium sulfate (**13**; R = $CH_2CH_2C_6H_5$, R' = CO_2H, X = $\frac{1}{2}H_2SO_4$). Di(hydroxymethyl)phosphine in butanol–hydriodic acid with the aldehyde (**12**; R' = CO_2H) at 15° gives the 5-carboxy-1,3-dihydroxy-2,2-bis(hydroxymethyl)isophosphindolinium iodide (**13**; R = CH_2OH, R' = CO_2H, X = I). Unfortunately these and other 1,3-dihydroxy compounds are quoted solely as examples in the patent specification and no physical constant or chemical property is recorded.

Pure 2-phenylisophosphindoline is not readily oxidized in air, and for experimental purposes can conveniently be stored in a calibrated glass syringe, preferably in a nitrogen-filled desiccator.

It has many of the normal properties of a tertiary phosphine, and with metallic halides will form normal stable complexes; for example, with dihalogenopalladium(II) it forms stable covalent compounds of composition $(C_{14}H_{13}P)_2PdX_2$, where X = Cl, Br, or I. It has in addition a striking ability to stabilize complex metallic compounds having compositions which are either unusual or often associated with considerable instability: this property is exemplified in the following brief outline.

With the dihalogenopalladiums, the phosphine forms complexes of type (**14**), of which the dichloride exists in two colorless and two orange-red modifications, the dibromide in a yellow and a red modification, and the diiodide only in a crimson form (Mann and H. R. Watson[109]). These

$$[(C_{14}H_{13}P)_3PdX_2] \ \rightleftarrows \ [(C_{14}H_{13}P)_3PdX^+]X^- \qquad [(C_{14}H_{13}P)_3IrHCl_2] \qquad [(C_{14}H_{13}P)_3IrCl]_3$$
$$(14) \qquad\qquad\qquad (15) \qquad\qquad\qquad (16) \qquad\qquad\qquad (17)$$

$$[C_{14}H_{13}P,CuI]_4 \qquad [C_{14}H_{13}P,AuX] \qquad [(C_{14}H_{13}P)_2MX]$$
$$(18) \qquad\qquad (19) \qquad\qquad\quad (20)$$

$$[(C_{14}H_{13}P)_3MX] \qquad [(C_{14}H_{13}P)_4Cu]X$$
$$(21) \qquad\qquad\quad (22)$$

compounds are covalent in the crystalline state; the red form of the dibromide, for example, has the structure shown in Figure 1, where P represents one molecule of the phosphine. They are covalent also in non-polar solvents, but in polar solvents exist in equilibrium with the ionic form (15). The phosphine forms compounds similar in composition to (14) with PtX_2, NiX_2, and with CoX_2, but the cobalt compounds do not ionize in solution (Collier, Mann, D. G. Watson, and H. R. Watson[110]). With sodium hexachloroiridate(IV), $Na_2IrCl_6, 6HO_2$, the phosphine gives the dichlorohydridotris(phosphine)iridium (16), which when

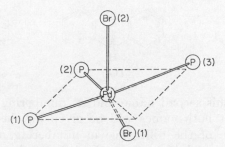

Figure 1. The structure of the red form of dibromotri-(2-phenylisophosphindo-line)palladium (14; X = Br); the symbol P represents one molecule of the tertiary phosphine. (Reproduced, by permission, from J. W. Collier, F. G. Mann, D. G. Watson, and H. R. Watson (*J. Chem. Soc.*, 1964, 1803).

treated with chlorine or hydrogen chloride gives the trichlorotris(phosphine)iridium (17) (Collier and Mann[111]). The phosphine and cuprous iodide form the tetrameric 1:1 complex (18), with silver iodide forms a similar complex, but with aurous iodide gives the expected monomeric 1:1 complex (19). The phosphine also combines with these three metallic iodides to form the covalent complexes (20), in which the metal is three-co-ordinate: the AuI member is remarkable in that in organic solvents it has a covalent trigonal plane structure but in the crystalline state it has an ionized structure. The phosphine combines further with the three iodides to form covalent complexes (21), in which the metal is four-co-ordinate, but only cuprous iodide gives ionic four-co-ordinate salts (22; X = NO_3 or ClO_4) (Collier, Fox, I. G. Hinton, and Mann[112]).

The ability to form complexes of type (14) with various metallic dihalides is also shown by 5-alkyl-5*H*-dibenzophospholes (p. 80), but such complexes do not apparently undergo ionization in solution.

5H-Dibenzophosphole (9-Phosphafluorene)

This fundamental member (1), [RRI 3032], of an important class of heterocyclic phosphorus compounds is given the name 5H-dibenzophosphole and the numbering shown in (1) by the IUPAC Nomenclature Committee and the Ring Index. The alternative name 9-phosphafluorene with the numbering shown in (1A) is, however, very widely used; it is almost universal in European literature and is still often used in the U.S.A. (1967). The authority of the Ring Index will probably ultimately

(1)   (1A)   (1B)

oust the use of this second name and its numbering (1A), but in the following account both names will be given for the more important members because of the difference in numbering of the ring. (The compounds are indexed, however, under Dibenzophosphole.)

This unfortunate plurality of names also arises in the case of the arsenic (p. 373) and the antimony analogue (p. 596) of the compound (1).

Certain dodecahydro derivatives of dibenzophosphole are known in which the two aryl groups are linked by a double bond (p. 81); for purposes of nomenclature these are regarded as hydrogenated derivatives of 1H-dibenzophosphole (1B).

An earlier name for (1), diphenylenephosphine, is rapidly falling into disuse.

Compounds having the ring system (1) are almost invariably 5-substituted derivatives. The various syntheses, some of practical value and others of theoretical interest, will be classified mainly by the type of substituent present.

5-Alkyl-(or Aryl)-5H-dibenzophospholes. The earliest syntheses are due to Wittig and Geissler[113] who showed (a) that pentaphenylphosphorane, $(C_6H_5)_5P$, decomposed when heated under nitrogen at 130° forming triphenylphosphine, biphenyl, and 5-phenyl-5H-dibenzophosphole, i.e., 9-phenyl-9-phosphafluorene, (2; R = C_6H_5), m.p. 93–94°, in very small yield; (b) that a mixture of triphenylphosphine and phenylsodium in benzene under nitrogen, when shaken at 20° for 10 hours and then at 70° for 30 hours, also gave (2; R = C_6H_5) in low yield; and (c) that an ethereal solution of 2,2′-diiodobiphenyl, when gently boiled with lithium under nitrogen, gave 2,2′-dilithiobiphenyl (3), which reacted with phenylphosphonous dichloride, $C_6H_5PCl_2$, to form (**2**; $R = C_6H_5$) which they isolated solely as the oxide (**4**; $R = C_6H_5$), m.p. 162–164° (6.6%).

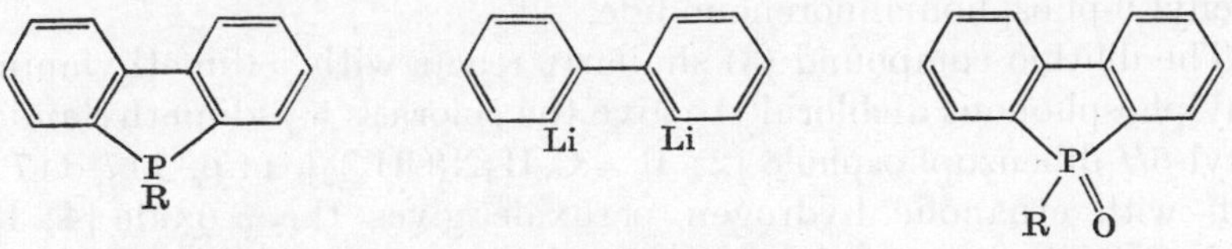

(**2**)   (**3**)   (**4**)

These three pioneering methods can be amplified.

(a) Razuvaev and Osanova[114, 115, 116] claim that pentaphenyl-phosphorane, when decomposed in benzene, gives triphenylphosphine (25%), biphenyl (20%), and (**2**; $R = C_6H_5$) in very small yield, but that the phosphorane in pyridine, when shaken for 150 hours at room temperature, gives (**2**; $R = C_6H_5$) (60%), benzene (50%), and traces of biphenyl and 4-phenylpyridine.

They adduce evidence that $(C_6H_5)_5P$ in solution gives rise to free phenyl radicals; these may abstract an *ortho*-hydrogen atom from one of the remaining phenyl groups, giving in effect the intermediate (**10**), the cyclization of which to (**2**; $R = C_6H_5$) with release of benzene is discussed on p. 77.

(b) The interaction of triphenylphosphine and phenylsodium is very probably an example of *ortho*-metalation which will be discussed

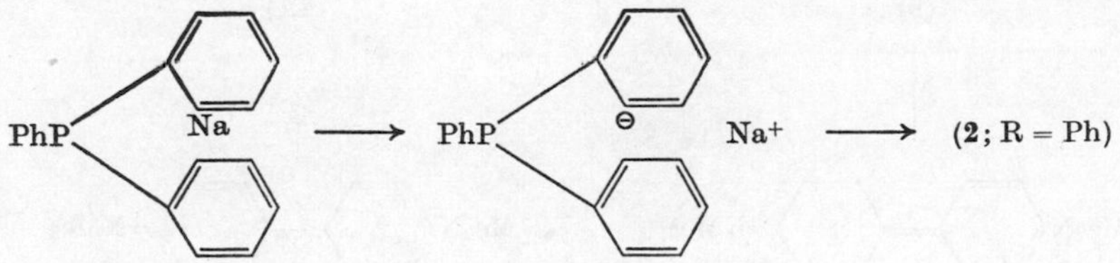

$PhP \quad Na \longrightarrow PhP^{\ominus} \quad Na^+ \longrightarrow (\mathbf{2}; R = Ph)$

(**5**)   (**6**)

more fully later. The o-sodio derivative (**5**), which can be written as (**6**), undergoes direct cyclization to the five-membered ring system (**2**; $R = C_6H_5$), apparently with the formation of sodium hydride.

(c) The reaction of the dilithiobiphenyl (**3**) with $C_6H_5PCl_2$ has been shown to yield (**2**; $R = C_6H_5$) in 45% yield (Bedford, Heinekey, Millar, and Mortimer[31]). Wittig and Maercker[117] have later shown that the dilithio compound (**3**), now prepared by the action of butyllithium on

4

2,2′-dibromobiphenyl, reacts with $C_6H_5PCl_2$ to give (**2**; $R = C_6H_5$) in yield of 78–88% (crude) and 54–57% (pure). It gives a methiodide (**7**; $R = C_6H_5$), m.p. 204–206° (dec.); this salt would systematically be named 5-methyl-5-phenyl-5H-dibenzophospholium iodide or 9-methyl-9-phenyl-9-phosphoniafluorene iodide.

The dilithio compound (**3**) similarly reacts with p-dimethylamino-phenylphosphonous dichloride to give the colorless 5-p-dimethylamino-phenyl-5H-dibenzophosphole [**2**; $R = C_6H_4N(CH_3)_2$], m.p. 117–117.5°, which with ethanolic hydrogen peroxide gives the 5-oxide [**4**; $R = C_6H_4N(CH_3)_2$], m.p. 177–177.5°.[117]

4,4′-Bis(dimethylamino)-2,2′-dilithiobiphenyl reacts with $C_6H_5PCl_2$ to give the yellow 3,7-bis(dimethylamino)-5-phenyl-5H-dibenzophos-phole (**8**), m.p. 226–227°, which gives a yellow 5-oxide, m.p. 271–272°, and an orange-red methiodide, m.p. 299.5–300°. The color of the

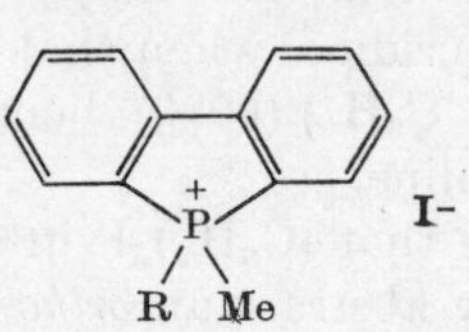

(7)

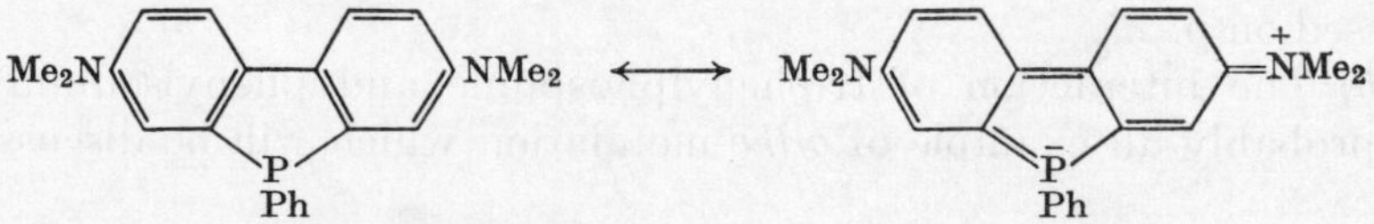

(8) ⟷ (8A)

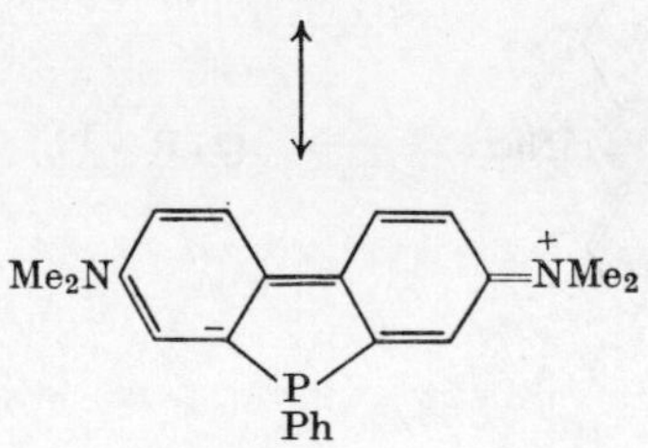

(8B) 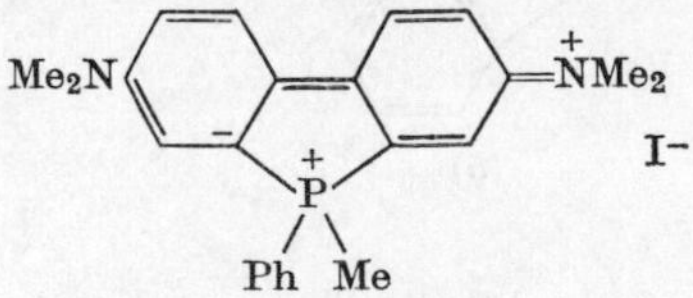 (9)

compound (**8**) has been attributed to contribution by forms such as (**8A**) in which the phosphorus is participating in the conjugation of the biphenyl system. This type of charge separation is, however, not altogether satisfactory and the resonance form (**8B**), an extended ylide type, may better represent the polar form; the darker-colored methiodide could be similarly represented in one contributory form as (9).

The reaction of 2,2'-dilithiobiphenyl (3) with alkyl- and aryl-phosphonous dichlorides is one of the most general methods of synthesizing the corresponding 5-substituted 5H-dibenzophospholes.

Wittig and Geissler[113] state that when an ethereal solution of tetraphenylphosphonium bromide and methyllithium (1:1 equivalents) is shaken under nitrogen for 4 months, and then treated with water, it yields (2; R = C_6H_5) (37%), triphenylphosphine (2%), and dimethyl-diphenylphosphonium bromide, isolated as the iodide (20%).

The mechanism of this and allied reactions has been investigated by Seyferth *et al.* In a preliminary investigation (Seyferth, Eisert, and Heeren[118]), it was pointed out that *tert*-butyltriphenylphosphonium bromide could react with an alkyllithium, RCH_2Li, in two ways:

(*a*) Attack of the Li reagent at the C—H linkage of the $(CH_3)_3C$— group, giving β-elimination:

$$(CH_3)_3C \cdot P^+(C_6H_5)_3 \; Br^- + RCH_2Li \; \rightarrow \; (C_6H_5)_3P + (CH_3)_2C{=}CH_2 + RCH_3 + LiBr$$

(*b*) Attack of the Li reagent at the P atom, giving an alkylidene-*tert*-butyldiphenylphosphorane and benzene:

$$(CH_3)_3C(C_6H_5)_2P{=}CHR + C_6H_6 + LiBr$$

It was found that methyllithium reacted with the phosphonium bromide:

$$(CH_3)_3C \cdot P^+(C_6H_5)_3 \; Br^- + CH_3Li \; \rightarrow \; (CH_3)_3C(C_6H_5)_2P{=}CH_2 + C_6H_6 \; (68\%)$$

The methylenephosphorane was identified by reaction with cyclohexanone, which gave methylenecyclohexane (46%) and *tert*-butyldiphenylphosphine oxide. The methyllithium is in this case giving reaction (*b*). Ethyllithium reacted similarly but more slowly.

When, however, the phosphonium bromide was treated with lithium piperidide, it gave 5-phenyl-5H-dibenzophosphole (2; R = C_6H_5) (42%), isobutene (16%, identified as the dibromide), and triphenylphosphine (8.4%); this is reaction (*a*) (with *ca.* 16% of β-elimination), a result to be expected since lithium piperidide cannot in these

$$[(CH_3)_3C{-}P(C_6H_5)_3]Br + LiN(CH_2)_5 \; \rightarrow \; (2; R = C_6H_5) + (CH_3)_2C{=}CH_2 + (C_6H_5)_3P$$

circumstances form an alkylidenephosphorane. The authors[118] suggest, therefore, the annexed mechanism for this reaction (*a*), based on *ortho*-metalation by the lithium piperidide.

The tetraphenylphosphonium cation resembles the *tert*-butyltri-phenylphosphonium cation in having no CH groups joined to the

$$\text{Ph}\!-\!\overset{+}{\underset{(CH_3)_3C}{P}}\quad + \quad LiN\!\!\bigcirc \quad \longrightarrow \quad \text{Ph}\!-\!\overset{+}{\underset{(CH_3)_3C}{P}}\quad \ominus \quad + \ HN(CH_2)_5 + Li^+$$

$$\text{Ph}\!-\!\underset{(CH_3)_3C}{P}\!\!\diagdown H \quad \longrightarrow \quad PhP \qquad + (CH_3)_3CH$$

phosphorus atom. Seyferth, Hughes, and Heeren,[119] working therefore on similar lines, added methyllithium in ether to tetraphenylphosphonium bromide in tetrahydrofuran, a deep red-brown solution being formed. After being stirred first in the cold and then whilst boiling, cyclohexanone was added and, after 6 hours' boiling, volatile components were distilled from the reaction product. The distillate yielded benzene (95%) and methylenecyclohexane (70%, identified as before). The residue was extracted with benzene, which when treated with methyl bromide gave 5-methyl-5-phenyldibenzophospholium bromide, identified as the tetraphenylborate $[B(C_6H_5)_4]$ salt, m.p. 172–173.5° (8% crude). Triphenylphosphine oxide was also recovered.

When in a similar experiment hydrogen bromide was passed over the reddish-brown solution until it was colorless, evaporation gave a residue which was treated with aqueous sodium hydroxide until the pH was *ca.* 6 and then extracted with ether. This extract when (i) evaporated gave (**2**; R = C_6H_5), and (ii) treated with methyl iodide gave 5-methyl-5-phenyldibenzophospholium iodide.

These results are interpreted by the annexed mechanism. The *ortho*-metalation gives the zwitterion (**10**), which undergoes cyclization to (**11**), followed by expulsion of benzene and formation of the dibenzophosphole.

Confirmation of this mechanism was obtained by using tetrakis-(2,4,6-trideuteriophenyl)phosphonium bromide. Expulsion of an *ortho*-deuterium atom would now be required to form the corresponding zwitterion (**10A**), which would cyclize to the dodecadeuterium intermediate (**11A**); this by the above mechanism should give tetradeuteriobenzene and the heptadeuteriophenyldibenzophosphole. The former was

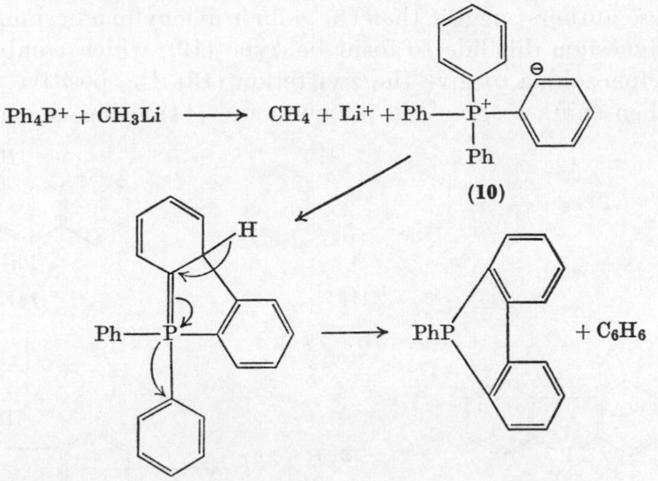
$Ph_4P^+ + CH_3Li \longrightarrow CH_4 + Li^+ + Ph—P^+$
Ph
(10)
H
Ph—P
PhP
$+ C_6H_6$
(11)

obtained in 12% yield, and the latter was identified by quaternization with methyl iodide to form deuteriated 5-methyl-5-phenyldibenzophospholium iodide, isolated as the tetraphenylborate salt, m.p. 173–176°.

Wittig and Benz[120] have prepared 5-phenyl-5*H*-dibenzophosphole by the intermediate use of benzyne. A solution of *o*-bromofluorobenzene

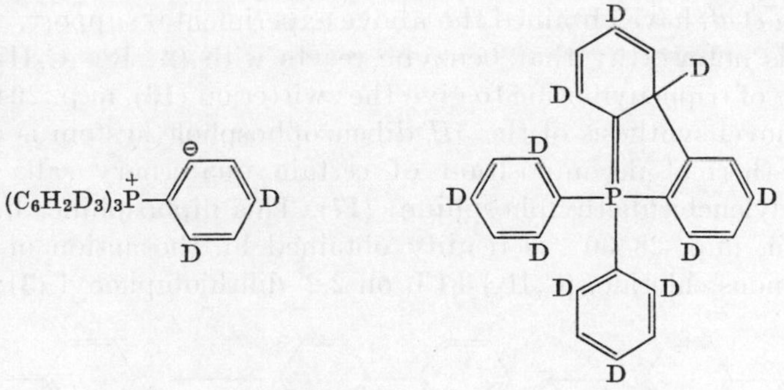
$(C_6H_2D_3)_3\overset{+}{P}$
D
(10A)
(11A)

and triphenylphosphine in tetrahydrofuran was added slowly at room temperature to magnesium under ether. The reaction mixture was then heated for 30 minutes, cooled, and decomposed with aqueous ammonium chloride. Triphenylphosphine (41%) was recovered from the ethereal layer, and the residue after chromatographic purification gave (**2**; $R = C_6H_5$), b.p. 160–165°/0.01 mm, m.p. 93.5–94.5°.

These authors suggest that the *o*-fluorophenylmagnesium bromide loses magnesium dihalide to form benzyne (**12**), which combines with triphenylphosphine to give the zwitterion (**13**), the positive charge of which then shifts to give the phosphorane (**14**). Ring closure follows

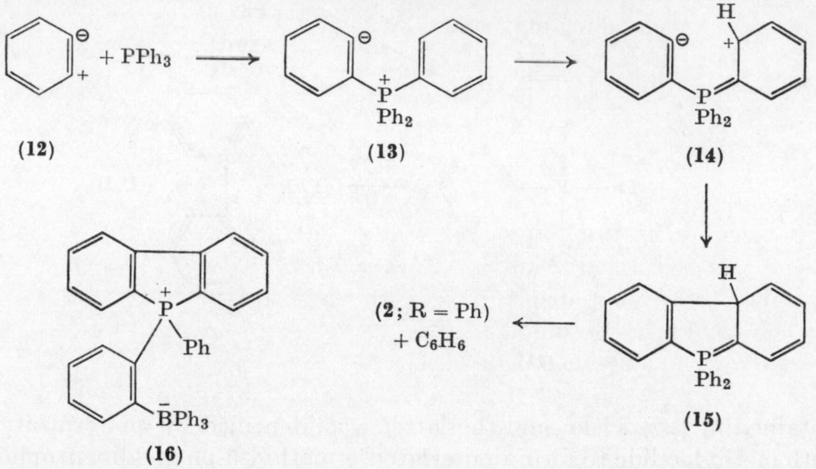

to give the isomeric phosphorane (**15**), which undergoes aromatization by splitting off benzene and forming (**2**; $R = C_6H_5$). This mechanism was in its essentials confirmed by the later work of Zbiral[121]; its progress from stage (**13**) to (**2**; $R = C_6H_5$) is closely similar to that for which Seyferth *et al.* have obtained the above experimental support.

It is noteworthy that benzyne reacts with (**2**; $R = C_6H_5$) in the presence of triphenylborine to give the zwitterion (**16**), m.p. 264–264.5°.

A novel synthesis of the 5*H*-dibenzophosphole system is achieved by the thermal decomposition of certain quaternary salts of 2,2′-biphenylylenebis(diethylphosphine) (**17**). This diphosphine, b.p. 152°/0.25 mm, m.p. 28–30°, is readily obtained by the action of diethylphosphinous chloride, $(C_2H_5)_2PCl$, on 2,2′-dilithiobiphenyl (**3**); it gives

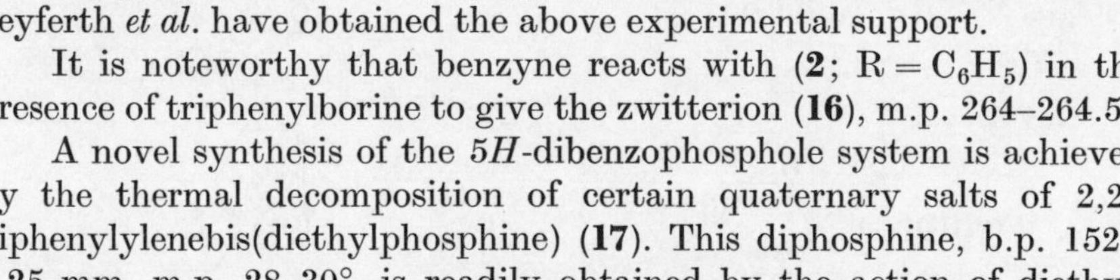

a dioxide, m.p. 193–195°, and a dimethiodide (**18**), *i.e.*, 2,2′-biphenylylenebis(diethylmethylphosphonium) diiodide, m.p. 255–256° (dipicrate, m.p. 137–138°).

The thermal decomposition of the diiodide (18) at $250°/0.15$ mm gives rise to 5-methyl-5H-dibenzophosphole (2; R = CH$_3$), and 5-ethyl-5H-dibenzophosphole (2; R = C$_2$H$_5$) in the proportions of $1:4.8$ (Allen, Mann, and Millar[122, 123]). It is suggested that this cyclization is an additional example of a general internal nucleophilic displacement ($S_N i$) mechanism which may be generalized as:

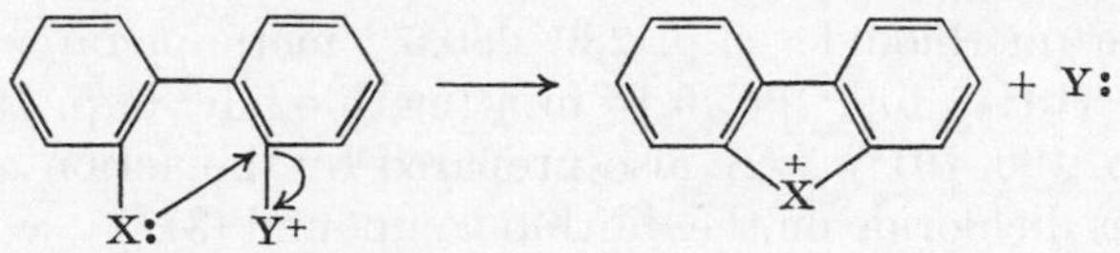

Other cyclizations apparently of this type are known. The conversion of diazotized 2-amino-2′-halogenobiphenyls into cyclic halogenonium salts with displacement of nitrogen (Mascarelli[124]; Sandin and Hay[125]), the decomposition of diazotized 2-amino-2′-aryloxybiphenyl to form a cyclic oxonium salt, and the analogous decomposition of diazotized 2-amino-2′-(diphenylamino)biphenyl to form a cyclic quaternary ammonium salt (Nesmeyanov[126]) may also be $S_N i$ displacements of this type.

For the thermal decomposition of the dimethiodide (18), it is suggested that initial loss of ethyl (or methyl) iodide occurs yielding the monophosphonium salts (19) and (20), which then undergo an $S_N i$ displacement of $(C_2H_5)_2(CH_3)P$ to form the quaternary derivatives (21) and (22). These would on further heating undergo the usual loss of an

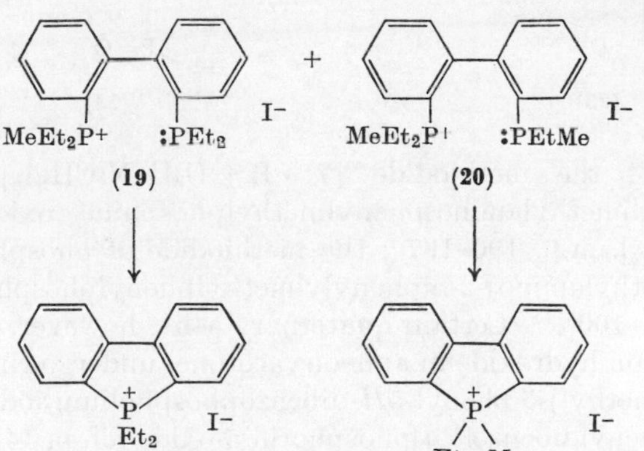

alkyl group to give the 5H-dibenzophospholes (2; R = CH$_3$) and (2; R = C$_2$H$_5$). The authentic compound (22), i.e., 5-ethyl-5-methyl-5H-dibenzophospholium iodide, on thermal decomposition does in fact give the 5-methyl- and the 5-ethyl-5H-dibenzophospholes in the proportion of 1:1.3.

The formation of the 5H-dibenzophosphole system by the thermal decomposition of heterocyclic quaternary salts of the diphosphine (17), apparently by essentially the same mechanism, is described on pp. 193, 198, 200, 201.

For identification purposes, the compounds (2; R = CH$_3$), b.p. 103°/0.2 mm (methiodide, m.p. 280–280.5°; methopicrate, m.p. 217°), and (2; R = C$_2$H$_5$), b.p. 106°/0.15 mm (methiodide, m.p. 195°; methopicrate, m.p. 190–191°), were also prepared by the action of the alkylphosphonous dichloride on the dilithio compound (3).[123]

It is noteworthy that these 5-alkyl derivatives readily give stable five-coordinate complexes with many metallic salts, e.g., the phosphine (2; R = CH$_3$) gives the complex [(C$_{13}$H$_{11}$P)$_3$NiBr$_2$], whereas the phenyl derivative gives only the normal four-coordinate complex, e.g., [(C$_{18}$H$_{13}$P)$_2$NiBr$_2$] (Allen, Mann, and Millar[127, 128]). The 5-alkyl- and 5-aryl-phenoxaphosphines (p. 295) show the same difference.

The quaternary salts of type (7) when warmed with aqueous sodium hydroxide usually undergo the expected rupture of the heterocyclic ring with formation of the P-oxide. For example, the methiodide (7; R = CH$_3$) gives 2-biphenylyldimethylphosphine oxide (23; R = CH$_3$),[123]

Me
R ᐳP=O

(23)

Me₂N NMe₂
Me
Ph ᐳP=O

(24)

m.p. 92–93°; the methiodide [7; R = C$_6$H$_4$N(CH$_3$)$_2$] gives 2-biphenylyl-p-(dimethylamino)phenylmethylphosphine oxide [23; R = C$_6$H$_4$N(CH$_3$)$_2$], m.p. 196–197°; the methiodide of phosphine (8) gives 4,4′-bis(dimethylamino)-2-biphenylylmethylphenylphosphine oxide (24), m.p. 97–100°.[117] Certain quaternary salts, however, when treated with potassium hydroxide in aqueous acetone, undergo ring expansion; e.g., 5-(iodomethyl)-5-phenyl-5H-dibenzophospholium iodide gives 5,6-dihydro-5-phenyldibenzo[b,d]phosphorin 5-oxide (cf. p. 149).

In another novel route to the dibenzophosphole system (Campbell, Cookson, Hocking, and Hughes[23]), 1,1′-bicyclohexenyl (25) was converted by the McCormack reaction (p. 33), using phenylphosphonous dichloride, into the 5,5-dichloro-decahydro-5-phenyl-1H-dibenzophosphole (26), which on hydrolysis with aqueous sodium hydrogen carbonate

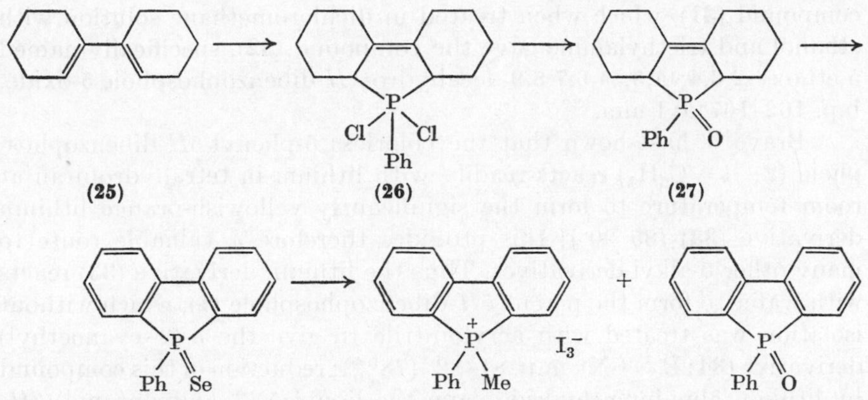

(25)   (26)   (27)

(28)   (29)   (30)

gave the corresponding 5-oxide (27), a viscous oil, b.p. 210–220°/0.5 mm. This oxide resisted all the normal methods for dehydrogenating the six-membered rings. Ultimately the oxide (27) was dispersed in a mixture of selenium and potassium dihydrogen phosphate, which was heated at 270–370° for $5\frac{1}{2}$ hours, i.e., until evolution of hydrogen selenide ceased. Working up the residue yielded 5-phenyl-5H-dibenzophosphole 5-selenide (28), m.p. 162–164° (25%).

To show that dehydrogenation was complete, a solution of (28) in methyl iodide was boiled for 3 hours, trimethylselenonium iodide, $(CH_3)_3SeI$, m.p. 150–152°, being deposited; the mother-liquor, after concentration to half-bulk, deposited 5-methyl-5-phenyl-5H-dibenzophospholium triiodide (29), red crystals, m.p. 105–108°; complete evaporation gave a residue which after treatment with aqueous sodium hydroxide gave 5-phenyl-5H-dibenzophosphole 5-oxide (30), m.p. 166–168°.

The same series of reactions, starting with 4-methyl-1,1'-bicyclohexenyl, gave 3-methyl-5-phenyl-5H-dibenzophosphole 5-selenide, m.p. 162.5–164° (31%), which was not further investigated.[23]

It is noteworthy that when the diene (25) was similarly treated with PCl_3, addition of chlorine to the product gave the 5,5,5-trichloro

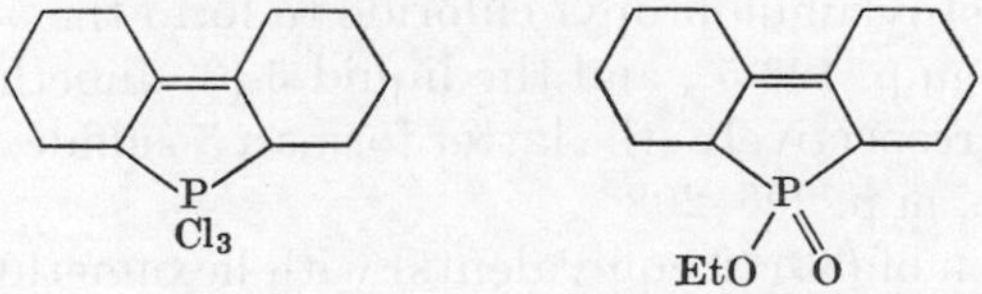

(31)   (32)

compound (31), which when treated in dichloromethane solution with
ethanol and triethylamine gave the compound (32), specifically named
5-ethoxy-2,3,4,4a,5,5a,6,7,8,9-decahydro-1H-dibenzophosphole 5-oxide,
b.p. 162–167°/0.1 mm.[129]

Braye[130] has shown that the colorless 5-phenyl-5H-dibenzophos-
phole (2; R = C$_6$H$_5$) reacts readily with lithium in tetrahydrofuran at
room temperature to form the significantly yellowish-orange lithium
derivative (33) (85–90°); this provides therefore a valuable route to
many other 5-alkyl derivatives. Thus the lithium derivative (33) reacts
with water to form the parent 5H-dibenzophosphole (1), which without
isolation was treated with acrylonitrile to give the 5-(2′-cyanoethyl)
derivative (34; R = CN), m.p. 81–82° (78%); reduction of this compound
by lithium aluminum hydride gave the liquid 5-(3′-aminopropyl)-5H-
dibenzophosphole (34; R = CH$_2$NH$_2$).

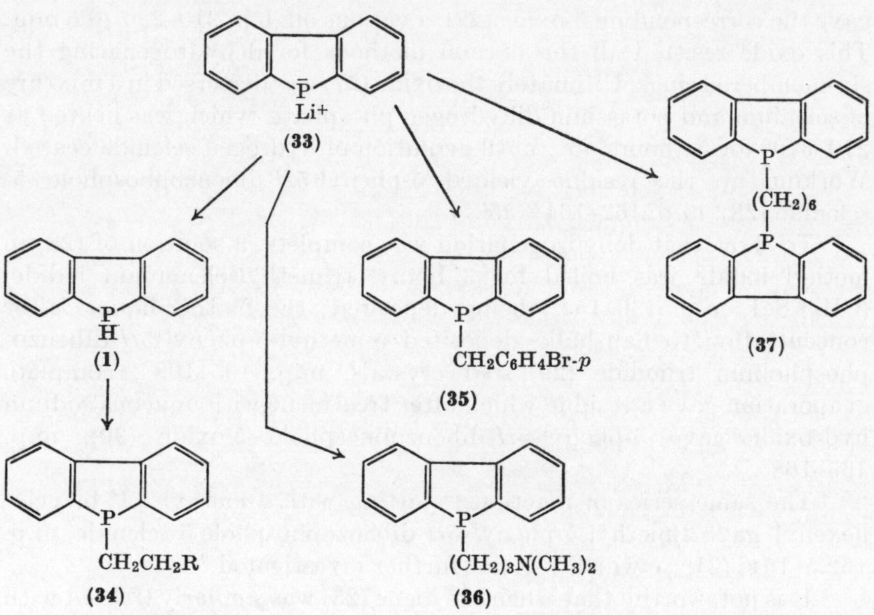

The lithio compound reacts similarly with p-bromobenzyl bromide
and with 3-dimethylaminopropyl chloride to form the 5-p-bromobenzyl
derivative (35), m.p. 123.5°, and the liquid 5-(3′-dimethylaminopropyl)
derivative (36), respectively; the latter forms a 5-sulfide, which furnishes
a hydrochloride, m.p. 206–208°.

The reaction of (33) (2 equivalents) with hexamethylene dibromide
gives hexamethylenebis(5H,5′H-dibenzophosphole) (37), m.p. 142–143°.

5-*Hydroxy*-5H-*dibenzophosphole* 5-*oxide* (**38**). This phosphinic acid was initially termed phosphafluorinic acid, with its structural formula numbered as in (**1A**); this name corresponds to the name arsafluorinic acid (p. 374), but both these very convenient names must now be replaced by systematic names based on 5*H*-dibenzophosphole(arsole) and notation (**1**).

The arsenic analogue of (**38**) can readily be prepared by the cyclization of 2-biphenylylarsonic acid using sulfuric acid (p. 374), but this method cannot be applied to the corresponding phosphonic acid. This

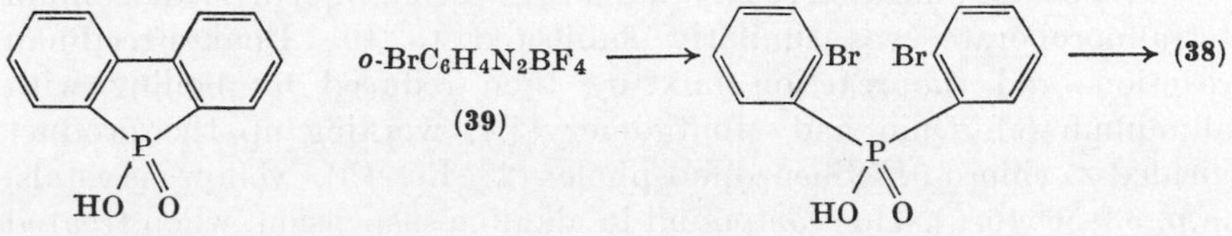

acid (**38**) was first prepared by Freedman and Doak,[131] who heated a mixture of dry *o*-bromophenyldiazonium tetrafluoroborate (**39**), phosphorus trichloride, ethyl acetate, and a small quantity of cuprous bromide; after the vigorous reaction (the Doak–Freedman reaction[132, 133]), the solution was concentrated and treated with water, yielding bis(*o*-bromophenyl)phosphinic acid (**40**). A solution of this acid in aqueous-methanolic potassium hydroxide was boiled with palladized calcium carbonate for 48 hours to remove bromine and thus couple up the *o*-phenylene groups; working up gave the acid (**38**), m.p. 239–248° (41%). This method of ring closure was based on Busch and Weber's method[134] of preparing biphenyl and its derivatives. The acid (**38**) is readily precipitated when an aqueous solution of the sodium salt is made just acid to Congo Red.

The similar use of bis(2-chloro-5-tolyl)phosphinic acid and of bis(2-bromo-4-tolyl)phosphinic acid gave the 3,7-dimethyl acid (**41**; R = CH₃, R′ = H), m.p. 325–328° (5.5%), and the 2,8-dimethyl acid (**41**; R = H, R′ = CH₃), m.p. 303–305° (57%), respectively. Oxidation of the latter acid by pyridine–permanganate (cf. Morgan and Herr[135])

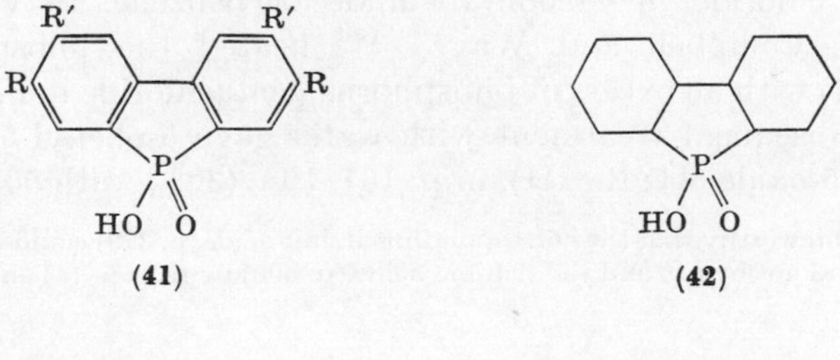

gave the 2,8-dicarboxylic acid (**41**; R = H, R′ = CO_2H), m.p. above 300° (44%). Nitration of the parent acid (**38**) gave the 3,7-dinitro acid (**41**; R = NO_2, R′ = H) (93%) which on reduction with Raney nickel gave the 3,7-diamino acid (**41**; R = NH_2, R′ = H); both these acids decompose above 260°. Reduction of the parent acid in ethanolic solution containing rhodium-on-alumina as a catalyst gave dodecahydro-5-hydroxy-5*H*-dibenzophosphole 5-oxide (**42**), m.p. 153–154.5°. The ultraviolet spectra of these acids, their diarylphosphinic acid precursors, and other kindred acids have been recorded (Freedman and Doak[136]).

A second cyclization route was shown when 2-biphenylyldiazonium tetrafluoroborate was similarly subjected to the Doak–Freedman reaction, and the reaction mixture then reduced by boiling with aluminum (cf. Quin and Montgomery[137]); working up the product yielded 5-chloro-5*H*-dibenzophosphole (**2**; R = Cl), yellow crystals, m.p. 53–56° (6%). This compound in alkaline suspension, when treated with hydrogen peroxide, gave the acid (**38**) (94%) (Doak, Freedman, and Levy[138]). Were it not for the low yield, the chloro compound (**2**; R = Cl) would be useful for the preparation of 5-alkyl and 5-aryl derivatives by the action of the appropriate Grignard or lithium reagents.

The difficulty in cyclizing 2-biphenylylphosphonic acid was also initially encountered in cyclizing 2-biphenylylphenylphosphinic acid (**43**; R = H) to the 5-phenyl-5*H*-dibenzophosphole 5-oxide (**44**; R = H).

(**43**) (**44**)

The phosphinic acid is unaffected by (a) acetic anhydride containing 1% of sulfuric acid at 90° or (b) by polyphosphoric acid at 160°.* Trifluoroacetic anhydride does not cyclize the phosphinic acid, and concentrated sulfuric acid at 100° apparently sulfonates it. The acid chloride does not undergo cyclization when heated with aluminum chloride in carbon disulfide or benzene. In view of this experience, Campbell and Way[139, 140] heated the phosphinic acid (**43**; R = H) with an excess of phosphorus pentachloride in nitrobenzene at 180°; subsequent treatment with water gave 5-phenyl-5*H*-dibenzophosphole 5-oxide (**44**; R = H), m.p. 167–168° (30%, with 50% recovery

* It is noteworthy that the corresponding arsinic acids (p. 379) could be cyclized by reagent (b) and not by (a), and the stibinic acids (p. 600) by reagent (a) and not by (b).

of unchanged acid). The 4'-nitro acid (43; R = NO$_2$) gave the 3-nitro 5-oxide (44; R = NO$_2$), m.p. 203°, but more slowly (20%, with 60% recovery of acid).

Maximum yields in this cyclization were obtained when three molecular proportions of phosphorus pentachloride were used. Campbell and Way[140] suggest that the phosphinyl chloride Ar$_2$POCl, initially formed from the phosphinic acid, is converted in the above conditions into the trichloride, Ar$_2$PCl$_3$, which in the presence of the PCl$_5$ ionizes as [Ar$_2$PCl$_2$]$^+$[PCl$_6$]$^-$, and that an electrophilic cyclization by attack of the positive phosphorus on the adjacent benzene ring of the biphenylyl system then follows. This suggested ionization is closely comparable to the known ionization of PCl$_5$ to [PCl$_4$]$^+$[PCl$_6$]$^-$. It is noteworthy that the cyclization was facilitated by a methyl group and impeded by a nitro group in the biphenylyl system, a characteristic feature of an electrophilic substitution.

It is of interest that 5-bromo-2-biphenylylphenylphosphinic acid when cyclized as above gave the corresponding 2-chloro-5-phenyldibenzophosphole 5-oxide (65%) with 10% recovery of the 5-bromo acid. This halogen exchange may occur when the phosphorus atom becomes positively charged and attack on the halogen in the para-position is facilitated.[140]

Direct cyclization to the tertiary dibenzophosphole has been achieved (Levy, Doak, and Freedman[141]) by treating 2-biphenylyl-diazonium tetrafluoroborate and cuprous bromide in ethyl acetate with phenylphosphonous dichloride; after the vigorous reaction, the solution was boiled with aluminum, concentrated, and the residue distilled. In these circumstances the 2-biphenylylphenylphosphinous chloride, (C$_6$H$_4$—C$_6$H$_5$)(C$_6$H$_5$)PCl, which is almost certainly formed undergoes cyclization and the distillation gives 5-phenyldibenzophosphole (2; R = C$_6$H$_5$), b.p. 180°/0.005 mm, m.p. 92–94° (21% purified product).

The acid (43; R = H) when heated at 350°/0.2 mm for 1 hour decomposes with the formation of biphenyl, phenylphosphonic acid, and the 5-oxide (44; R = H) in yield too low to be of practical value (Lynch[142]).

Stereochemistry. Campbell and Way,[140] in their studies of optical stability of three-covalent phosphorus compounds, attempted to synthesize and resolve a 5-phenyl-5H-dibenzophosphole having a suitable substituent in the biphenylylene portion. Since the tricyclic system is planar, any optical activity must be due to the pyramidal configuration of the phosphorus atom, which would lift the 5-phenyl group above or below this plane.

For this purpose they prepared the 3-methyl 5-oxide (44; R = CH$_3$), and oxidized it to the 3-carboxy oxide (**44**; R $= CO_2H$) by permanganate-pyridine (cf. Morgan and Herr[135]). By heating this ($\pm$)-acid with (+)-1-phenylethylamine, they obtained the ($\pm$)-3-(+)-(1′-phenylethyl)-amide 5-oxide [**44**; R $= CO_2NHCH(C_6H_5)CH_3$], which by fractional crystallization was separated into the less soluble (+)-acid–(+)-amide, $[\alpha]_D + 38.6°$, and the more soluble (−)-acid–(+)-amide, $[\alpha]_D - 212.0°$. These amides, when hydrolyzed in boiling ethanolic hydrochloric acid, gave the (+)-acid, m.p. 250–251°, $[\alpha]_D + 126.0°$ in 0.1n-NaOH, and the (−)-acid, m.p. 250–251°, $[\alpha]_D -126.1°$, respectively. Treatment of the (+)-acid with lithium aluminum hydride in benzene–dibutyl ether reduced both the CO_2H and the P=O group, giving the optically inactive 3-hydroxymethyl-5-phenyl-5*H*-dibenzophosphole mono-ethanolate, m.p. 52–54°, which was readily oxidized in boiling benzene to the 5-oxide (**44**; R $= CH_2OH$), m.p. 195–196°, and also gave a methiodide, m.p. 226–230°.

These results may be contrasted with the same workers' optical resolution of 5,6-dihydro-6-*p*-dimethylaminophenyldibenz[*c,e*][1,2]aza-phosphorine 6-oxide (p. 228), which when similarly reduced gave the optically active phosphine; the latter compound had considerable optical stability, but exposure to the air rapidly oxidized it back to the active 6-oxide. Possible reasons for this difference in behavior on reduction are discussed.[140]

5,5′-*Bi(dibenzophosphole)* or 9,9′-*Bi(9-phosphafluorene)* (**45**). This stable crystalline diphosphine was first obtained as a by-product in the synthesis of the phosphanthrene system (Davis and Mann[143]) (p. 184).

Di-(*o*-chlorophenyl)ethylphosphine (**46**) in tetrahydrofuran was treated with lithium metal at −35° to give the dilithio derivative (**47**);

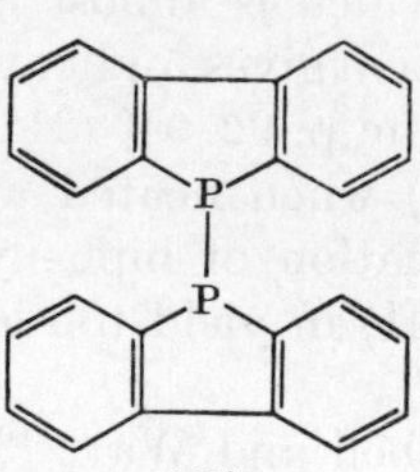

(**45**)

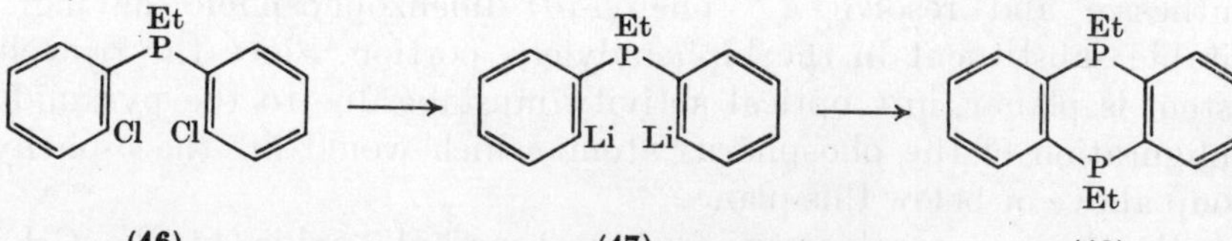

(**46**) → (**47**) → (**48**)

the mixture at $-65°$ was then treated with ethylphosphonous dichloride to form 5,10-diethyl-5,10-dihydrophosphanthren (**48**). The final reaction mixture yielded, on hydrolysis, ethyldiphenylphosphine, the phosphanthrene (**48**), and the 5,5-bi(dibenzophosphole) (**45**), pale yellow crystals, m.p. 242–243°, almost insoluble in cold ethanol.

The identification of this diphosphine was based on the following evidence. (*a*) Analysis and molecular weight determination. (*b*) Aerial oxidation of a suspension in boiling chloroform gave the acid (**38**). (*c*) An excess of hot benzyl bromide gave 5,5-dibenzyl-5*H*-dibenzophospholium bromide, m.p. 263–264° (picrate, m.p. 188–189°). (*d*) An ethereal suspension of (**45**) when shaken with iodine (one molar equivalent) gave a yellow solution of 5-iodo-5*H*-dibenzophosphole, which with phenylmagnesium bromide yielded the 5-phenyl derivative (**2**; $R = C_6H_5$); oxidation of the latter gave the 5-phenyl 5-oxide (**4**; $R = C_6H_5$).

The precise origin of the diphosphine is uncertain. The ethyldiphenylphosphine in the reaction mixture must have arisen from hydrolysis of some unused dilithio compound (**47**). A precursor of (**47**)

(**49**)   (**50**)   (**51**)

would almost certainly be the chloro-lithio compound (**49**), some of which might undergo direct cyclization to 5-ethyl-5*H*-dibenzophosphole (**50**). Although, in general, lithium cleaves a P—Aryl bond more readily than a P—Alkyl bond, Issleib and Völker[144] have shown that potassium in dioxane will readily cleave the P—Alkyl bond in a tertiary phosphine provided that the latter contains at least one aryl group. It is possible, therefore, that the two aryl groups in (**50**) facilitate a similar action by the lithium to give the 5-lithio compound (**51**). This, on hydrolysis, would yield the parent phosphine (**1**); it is highly unlikely that the phosphine (**1**) could undergo oxidation to the diphosphine (**45**), because all the reactions were carried out under nitrogen, and secondary phosphines on atmospheric oxidation give secondary phosphine oxides, $R_2P(H){=}O$. No indication of the formation of the phosphine (**1**) or of the secondary phosphine oxide was obtained.

The mechanism of the conversion of the lithio compound (**51**) into the diphosphine (**45**) is almost certainly revealed by the work of Britt and Kaiser,[145] who showed that 5-phenyl-5*H*-dibenzophosphole in tetrahydrofuran reacts at room temperature with lithium, sodium, potassium, and cesium to give a yellow solution. This contained unchanged material and the yellow diphosphine (45); addition of hydrogen peroxide converted the former into the 5-oxide and meanwhile caused fading of the yellow colour with formation of the acid (38). A detailed investigation of this reaction system, in which free radical formation was followed step-wise by electron spin resonance determinations, showed that the first detectable reaction was (A), in which M represents the

(A) + 2M ⟶   + MC_6H_5

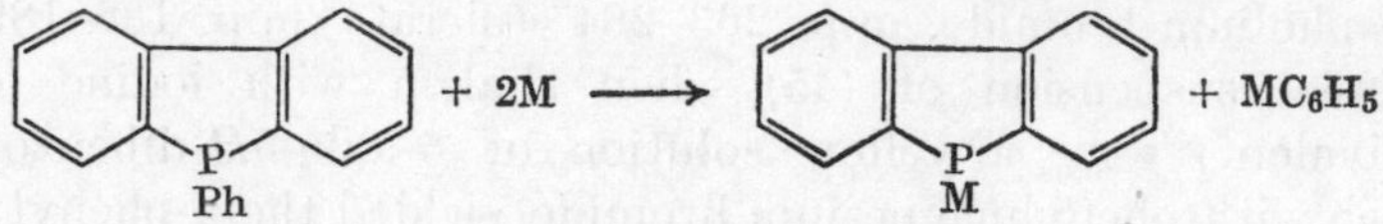

(52)

metal atom, the solution now being yellow. The probable sequence then was a second reaction which was sensitive to temperature and was studied at $-50°$; it consisted of the further action of the metal on (52) to give the

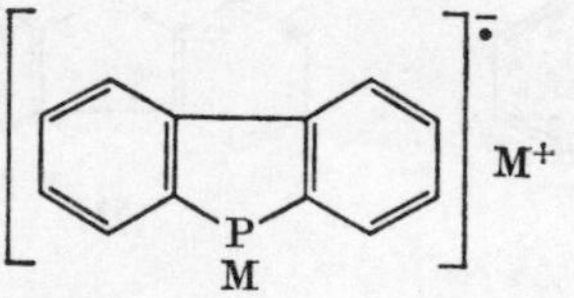

(53)

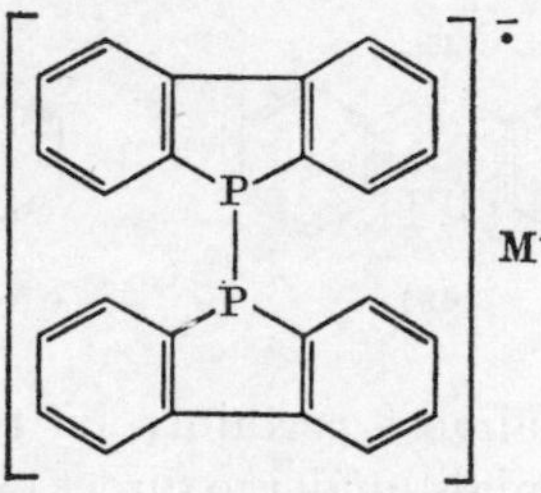

(54)

radical-anion (53). The final reaction between (52) and (53) gave the temperature-stable radical-anion (54). These results show that the formation of the lithium compound (51) in the presence of some unchanged metal could readily form the diphosphine (45) without the intervention of other reagents.

The comparatively high stability of the diphosphine may be due to its considerable steric protection.

Phosphoranes containing Dibenzophosphole Units

The syntheses of these phosphoranes and the investigation of their reactions have been very largely the work of G. Wittig and his school. A general outline of this very interesting work is given here.

A ring system which occurs in many of these compounds is 5,5'-spirobi[dibenzophosphole] (**55**), [RIS —], which can give rise to the corresponding 5,5'-spirobi(dibenzophospholium) cation (**56**), [RIS —], which is similarly numbered.

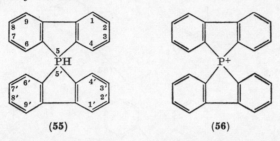

(55)   **(56)**

The names initially given to the following phosphoranes are not sufficiently specific, and names consonant with Ring Index practice have now been used. These names are reasonably clear; it should be borne in mind, however, that compounds such as (**73**) and (**75**), which have a p-dimethylaminophenyl group joined to the phosphorus atom have by nomenclature rules to be regarded as fundamentally substituted anilines, and named as such.

In 1958 Wittig and Kochendörfer[146] showed that an extension of Standinger's very early work[147, 148] readily led to pentaphenylphosphorane. Triphenylphosphine reacted with phenyl azide, $C_6H_5N_3$, to give triphenylphosphine phenylimine (**57**), m.p. 131–132°, which with methyl bromide gave quaternary bromide (**58**), of which (**58A**) is

$(C_6H_5)_3P{=}NC_6H_5 \rightarrow [(C_6H_5)_3P{=}N^+CH_3(C_6H_5)] \ Br^- \leftrightarrow [(C_6H_5)_3P^+{-}NCH_3(C_6H_5)] \ Br^-$
    (57)   **(58)**   **(58A)**

$$\Big\downarrow 2LiC_6H_5$$

$(C_6H_5)_5P + LiNCH_3(C_6H_5)$
(59)

probably a contributory form (m.p. 235.7–237°). This salt, when treated with phenyllithium (2 equivalents), gave pentaphenylphosphorane (**59**) and, after hydrolysis, monomethylaniline.

This series of reactions could be applied to 5-phenyl-5H-dibenzophosphole (**60**), which similarly gave the phenylimine (**61**), m.p. 194–195°, and its methobromide (**62**), m.p. 219–221°. The latter reacted with 2,2'-dilithiobiphenyl (**63**) to give the phosphorane (**64**), m.p. 201.5–202.5°.

Modern names of these compounds could be: (**61**) 5-phenyl-5-phenylimino-5H-dibenzophosph(v)ole; (**62**) 5-(N-methyl-N-phenyliminia)-5-phenyl-5H-dibenzophosph(v)ole bromide; and (**64**) 5-phenyl-5,5'-spirobi(dibenzophosph(v)ole).

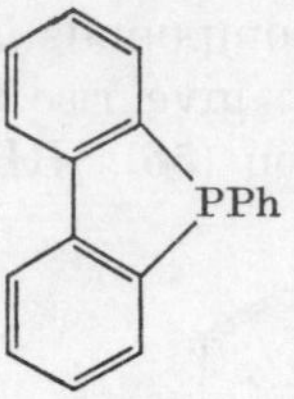

(60) 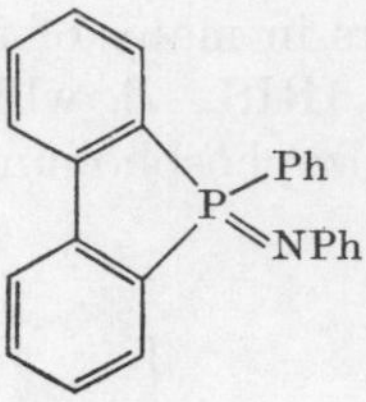   (61) 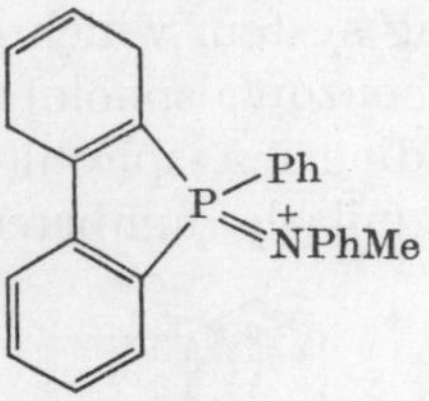   (62)

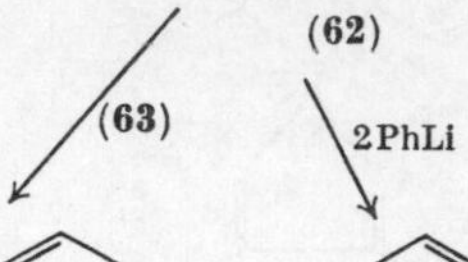

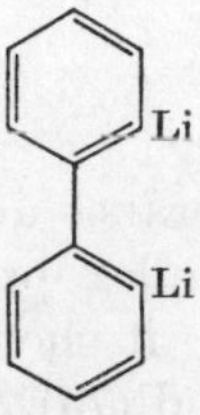

(63)   2 PhLi

(63) 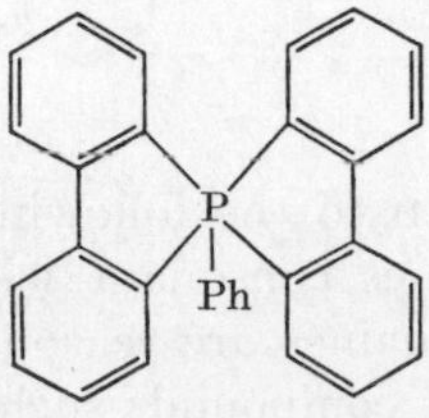   (64) 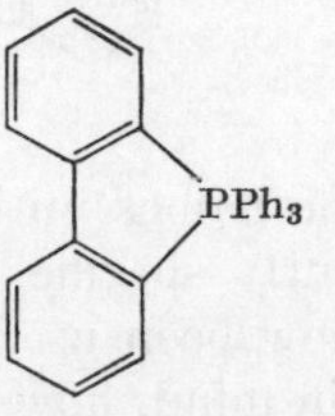   (65)

Later work (Wittig and Kochendörfer[149]) showed that the metho-bromide (62) reacted readily with phenyllithium to give the phosphorane (65), *i.e.*, 5,5,5-triphenyl-5*H*-dibenzophosph(v)ole, m.p. 155.5–156.5°, which however could not be obtained by the interaction of the metho-bromide (58) and the dilithio derivative (63). This phosphorane (65) when shaken in ether with the dilithio derivative (63) for 3 days also gave the phosphorane (64). It is noteworthy that when the phosphorane (64) was boiled with hydrochloric acid, P—C fission occurred in the heterocyclic ring and not at the P—C_6H_5 position, the product being

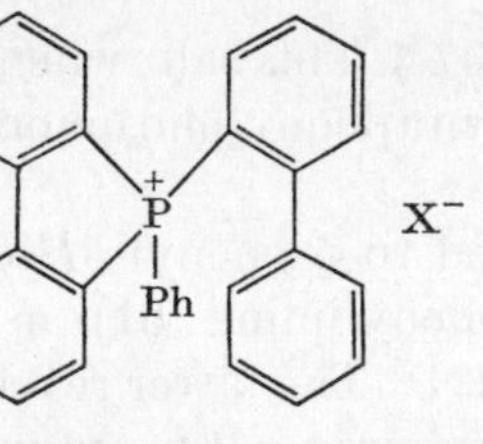

(66) 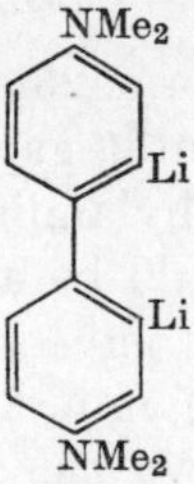   (67) 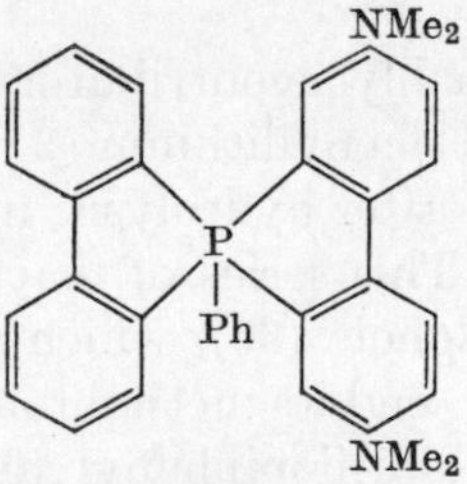   (68)

5-(2-biphenylyl)-5-phenyl-5*H*-dibenzophospholium chloride (66; X = Cl), isolated as the iodide, m.p. 253.5–254.5°, and as the tetraphenyl-borate [66; X = $B(C_6H_5)_4$], m.p. 200–210°.

Similarly 2,2'-dilithio-4,4'-bisdimethylaminobiphenyl, *i.e.*, 2,2'-dilithio-N,N,N',N'-tetramethylbenzidine (**67**), reacts with the metho-bromide (**62**) to give the phosphorane (**68**), now named N,N,N',N'-tetramethyl-5-phenyl-5,5'-spirobi(dibenzophosphole)-3,7-diamine, m.p. 212.5–213° (55%), and 5-phenyl-*5H*-dibenzophosphole 5-oxide, m.p. 164–164.5° (30%) as a by-product.

The above method of preparing pentaarylphosphoranes cannot be used for the preparation of the corresponding arsoranes, because triarylarsines do not react with phenyl azide. To overcome this difficulty, Wittig and Hellwinkel[150] developed a different synthesis, which was found to be applicable to the corresponding derivatives of phosphorus, arsenic, and antimony. Mann and Chaplin[151] had much earlier shown that triaryl-phosphines and -arsines reacted with anhydrous 'Chlor-amine-T', p-$CH_3C_6H_4SO_2NNaCl$, in anhydrous ethanol to give com-pounds of type $R_3X{=}NSO_2C_6H_4CH_3$, where $X = P$ or As. The compound (**69**) would, for example, be termed p-(tolylsulphonylimino)-triphenylphosphorane: this compound reacted readily with phenyl-lithium in ether to give the pentaphenylphosphorane (**59**).

$$(C_6H_5)_3P{=}NSO_2C_7H_7 + 2C_6H_5Li \rightarrow (C_6H_5)_5P + Li_2NSO_2C_7H_7$$
(69)   (59)

This synthetic method has been considerably developed by Wittig and Maercker,[117] who showed that 5-phenyl-5-(p-tolylsulphonylimino)-*5H*-dibenzophosphole (**70**), m.p. 176–177°, reacted with the dilithio derivative (**63**) to give the 5-phenyl-5,5'-spirobi(dibenzophosphole) (**64**)

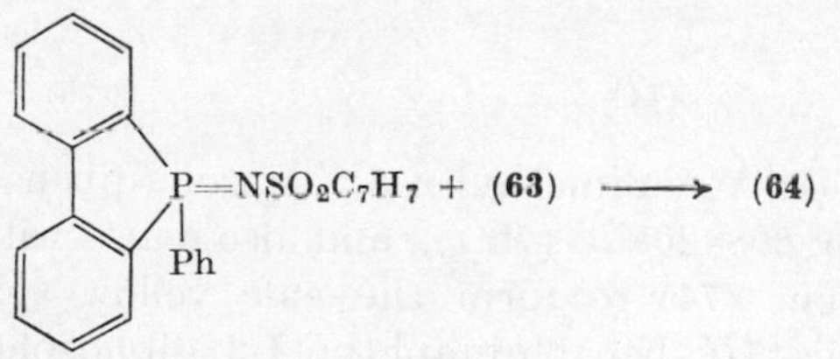

(**70**)

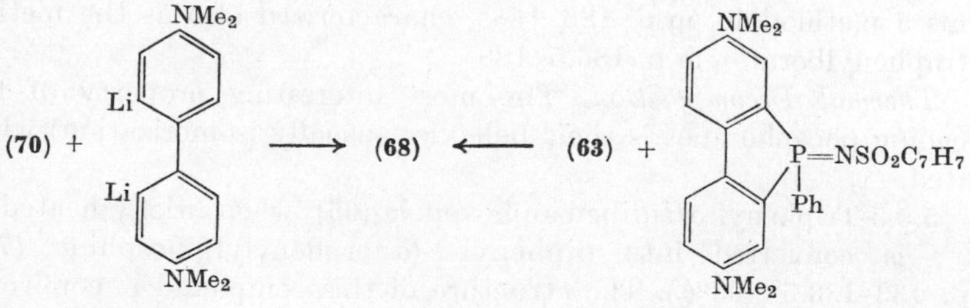

(**70**) +   ⟶ (**68**) ⟵ (**63**) +

(**67**)   (**71**)

(77%). The sulphonylimino compound (70) also reacted with the dilithio derivative (67) to give the above bisdimethylamino phosphorane (68) (48%), which was additionally obtained by the interaction of the pale yellow N,N,N',N'-tetramethyl-5-phenyl-5-(p-tolylsulphonylimino)-5H-dibenzophosphole-3,7-diamine (71), m.p. 255–259°, and the unsubstituted dilithio derivative (63) (61%).

Other examples of this synthetic route are provided by 5-(p-dimethylaminophenyl)-5-(p-tolylsulphonylimino)-5H-dibenzophosphole (72), m.p. 198–200°, which reacts with the dilithio derivative (63) to

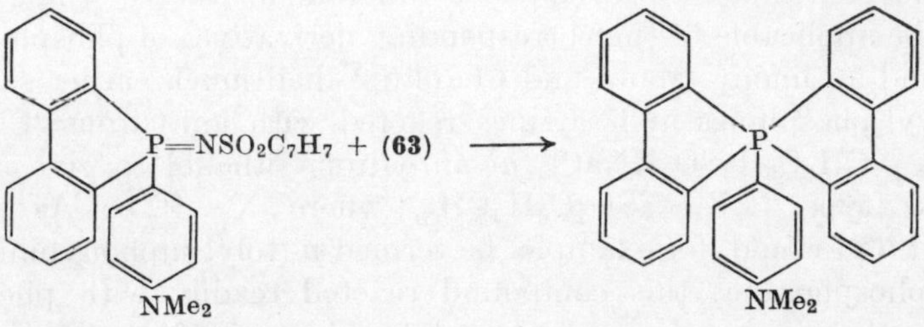

P=NSO$_2$C$_7$H$_7$ + (63) $\longrightarrow$

NMe$_2$
(72)

P

NMe$_2$
(73)

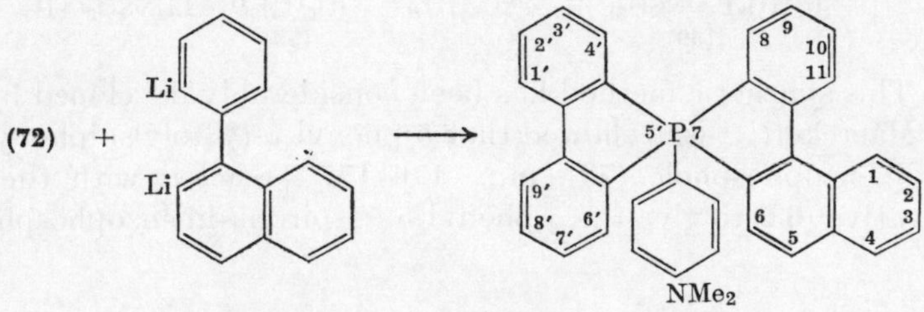

(72) +

Li

Li

(74)

$\longrightarrow$

2 3
1' 4'
5' P 7
9'
8' 7' 6'

8 9
10
11

1 2
3
6 5 4

NMe$_2$
(75)

form the colorless N,N-dimethyl-p-5,5'-spirobi(dibenzophosphol)-5-yl-aniline (73), m.p. 208–208.5° (79%), and also reacts with 2,2'-dilithio-1-phenylnaphthalene (74) to form the pale yellow spiran (75) (NN'-dimethyl-p-spiro[7H-benzo[b]naphtho[1,2-d]phosphole 7,5'-5[H]dibenzophosphol]-7-yl-aniline), m.p. 167–168° (60%). The latter compound forms a methiodide, m.p. 186–188°, characterized also as the metho-(tetraphenylborate), m.p. 186.5–188°.

Thermal Decomposition. The most interesting property of the foregoing phosphoranes is their behavior (usually isomerization) when heated.

5,5,5-Triphenyl-5H-dibenzophosphole (65), when briefly heated at 200°, is converted into diphenyl-2-(o-terphenylyl)phosphine (76), m.p. 133–133.5° (93%). The structure of this compound is confirmed by the fission of its pale yellow methiodide, m.p. 313–315°, in meth-

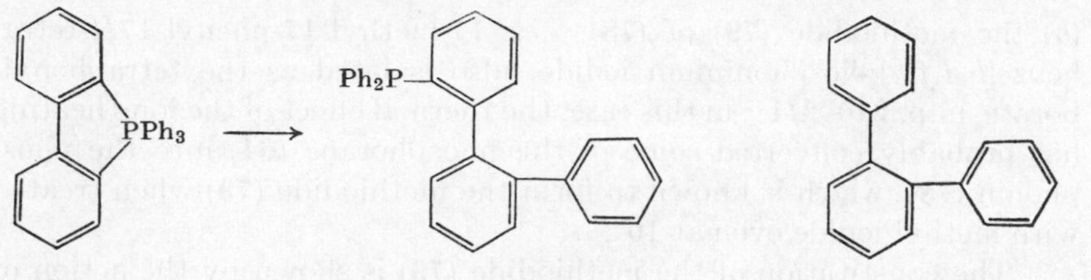

(65)   (76)   (77)

anolic potassium hydroxide to give methyldiphenylphosphine oxide, $CH_3(C_6H_5)_2PO$, o-terphenyl (77), and benzene.

5-Phenyl-5,5'-spirobi[dibenzophosphole] (64), when heated at 210° for 5 minutes, is converted into 17-phenyl-17H-tetrabenzo[b,d,f,h]phosphonin (78), m.p. 131–132° (86%). [For the names and numbering of

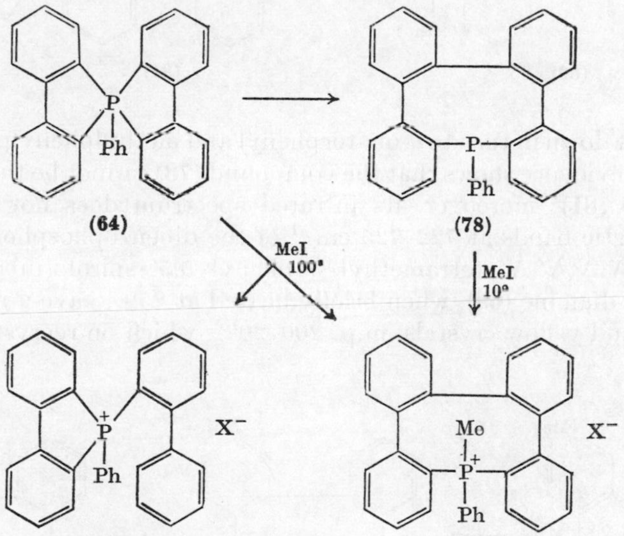

(64)   (78)

MeI
100°

MeI
10°

(66)   (79)

the phosphonin ring system and its benzo derivatives, see pp. 156–158.] The phosphorane (64) is unaffected by methyl iodide at room temperature for several weeks, but in 14 days at 100° gives rise to two compounds: (a) 5-(2-biphenylyl)-5-phenyl-5H-dibenzophospholium iodide (66; X = I), isolated as the tetraphenylborate (66; X = B(C_6H_5)_4), m.p. 205–206°; this cation may well arise by the action of hydrogen iodide (generated during the long heating) on the phosphorane (64), which, as noted earlier, readily gives this cation with methanolic hydrochloride;

(b) the methiodide (79) of (78), i.e., 17-methyl-17-phenyl-17H-tetra-
benzo[b,d,f,h]phosphoninium iodide, also isolated as the tetraphenylborate, m.p. 240–241°; in this case, the thermal effect of the long heating
has probably converted some of the phosphorane (64) into the phosphonin (78), which is known to form the methiodide (79) when treated
with methyl iodide even at 10°.

The constitution of the methiodide (79) is shown by the action of
boiling aqueous sodium hydroxide, which converts it into methylphenyl-
2-(o-quaterphenylyl)phosphine oxide (80); molten sodium hydroxide

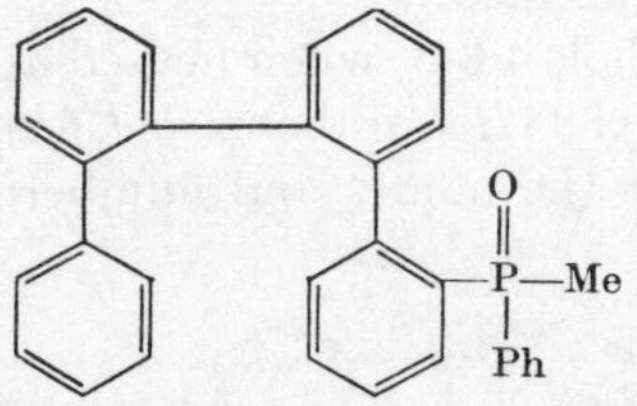

                    (80) 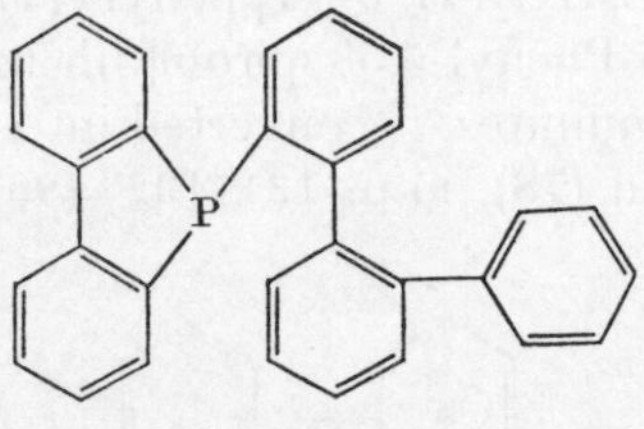   (81)

breaks this down in turn to o-quaterphenyl and methylphenylphosphinic
acid. This evidence shows that the compound (78) cannot be the isomeric
phosphine (81); moreover, its infrared spectrum does not show the
characteristic bands at 722–725 cm^{-1} of the dibenzophosphole system.

The N,N,N′,N′-tetramethyl-5-phenyl-5,5′-spirobi(dibenzophosphole)-3,7-diamine (68), when briefly heated at 250°, gave a mixture of
colorless and yellow crystals, m.p. 200–207°, which on recrystallization

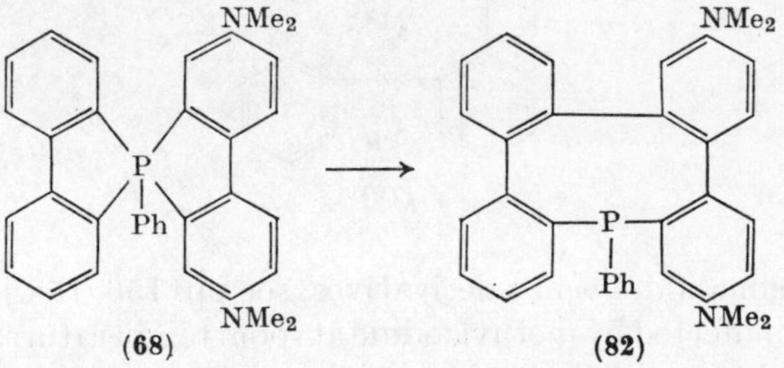

        (68)   (82) 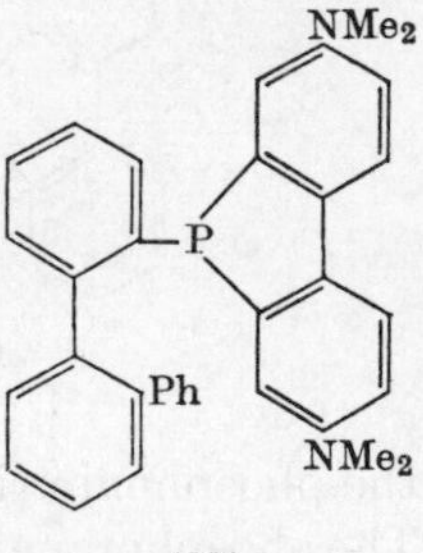   (83)

from ethanol gave the colorless NNN′N′-tetramethyl-17-phenyl-
17H-tetrabenzo[b,d,f,h]phosphonin-2,7-diamine (82), m.p. 207–208°; its
infrared spectrum gives no indication of the dibenzophosphole system.
It is suggested that the yellow crystals may have the structure (83), but
they were not investigated.

The phosphorane (**73**) when similarly heated at 240° gave the isomeric   N,N-dimethyl-p-(17H-tetrabenzo[b,d,f,h]phosphonin-17-yl)aniline (**84**), m.p. 235–236° (92%), which gives a hydrochloride, m.p.

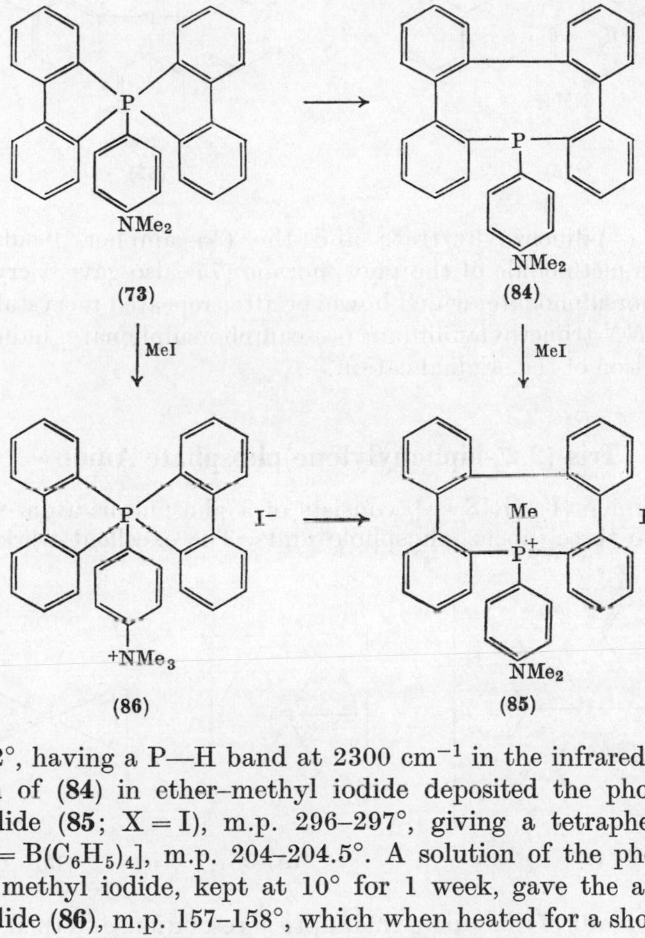

223–232°, having a P—H band at 2300 cm^{-1} in the infrared region. A solution of (**84**) in ether–methyl iodide deposited the phosphonium methiodide (**85**; X = I), m.p. 296–297°, giving a tetraphenylborate [**85**; X = B(C$_6$H$_5$)$_4$], m.p. 204–204.5°. A solution of the phosphorane (**73**) in methyl iodide, kept at 10° for 1 week, gave the ammonium methiodide (**86**), m.p. 157–158°, which when heated for a short time at 240° was converted into the isomeric iodide (**85**; X = I).

The phosphorane (**75**), when heated at 175° for 5 minutes, was converted into   NN'-dimethyl-p-(17H-tribenzo[b,d,f]naphtho[1,2-h]phosphonin-17-yl)aniline (**87**), m.p. 168–170°; this phosphine is sufficiently basic to be soluble in half-concentrated hydrochloric acid. Its infrared spectrum shows no indication of the dibenzophosphole system.

The above phosphoranes must almost certainly have a dissymmetric configuration. Attempts to resolve the methiodide (**86**) of the phosphorane (**73**) by fractional recrystallization of the corresponding

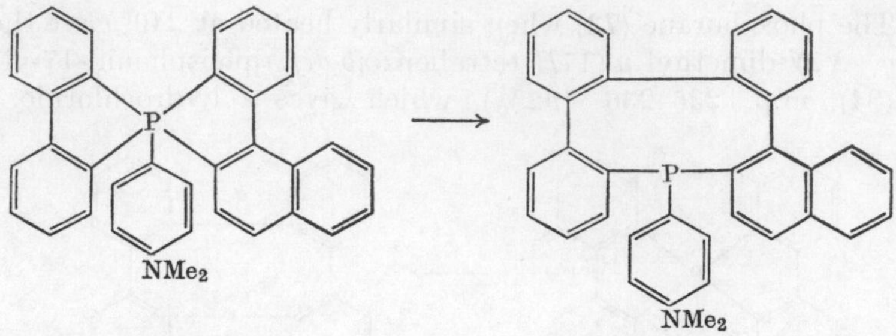

(75)   (87)

hydrogen (−)-dibenzoyltartrate and the (+)-camphor-10-sulphonate failed. The methiodide of the phosphorane (75) also gave a crystalline (+)-camphorsulphonate, which however after repeated recrystallization yielded NNN-trimethylanilinium (+)-camphorsulphonate, indicating a decomposition of the original cation.

Tris-(2,2′-biphenylylene)phosphate Anion

This anion (1), [RIS —], consists of a phosphorus atom which is common to three dibenzophosphole units. The excellent work on the

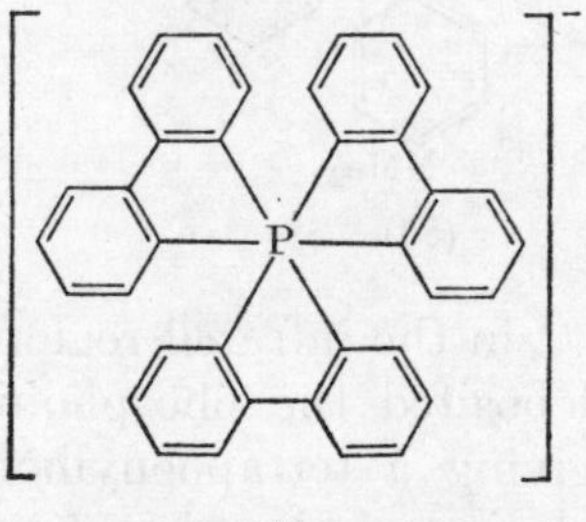

(1) 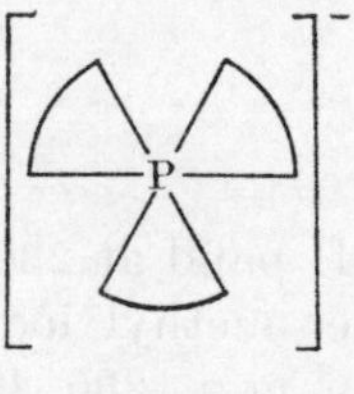   (1A) 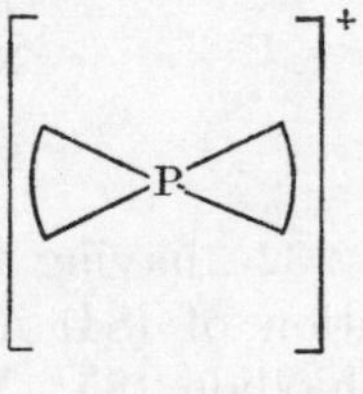   (2)

synthesis of salts of this anion, their optical resolution, and their thermal decomposition by Hellwinkel [95, 152] follows logically the previous section on derivatives of $5H$-dibenzophosphole.

For brevity, the three 2,2′-biphenylylene units in the anion (1) will be depicted as in (1A): the same units in the 5,5′-spirobi(dibenzophospholium) cation [cf. (8), (9), p. 98] will be similarly depicted as (2).

Hellwinkel [95] has shown that an ice-cold ethereal solution of 2,2′-diiodobiphenyl, when treated with butyllithium (two equivalents), and then after 4 hours at room temperature chilled to −70° and treated with powdered phosphorus pentachloride, deposits yellow crystals, m.p.

254–256° (dec.), of 5,5'-spirobi(dibenzophospholium) tris-(2,2'-biphenyl-ylene)phosphate (3) (48%): these can be recrystallized from ethanolic dimethylformamide, but the salt, like others containing the anion (1), then requires heating at 120° in a high vacuum for several days to eliminate traces of solvent. The formation of this salt is, of course, equivalent to the direct replacement of pairs of chlorine atoms in the salt $[PCl_4][PCl_6]$ by 2,2'-biphenylylene groups.

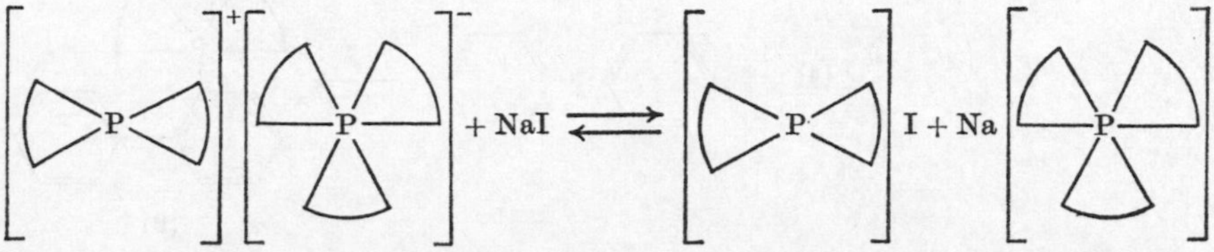

(3)   (4)   (5)

A suspension of the salt (3) in acetone, when warmed with a large excess of sodium iodide, gives 5,5'-spirobi(dibenzophospholium) iodide (4), m.p. 295–297° (82%), and sodium tris-(2,2'-biphenylylene)phosphate (5). Conversely, a solution of (4) and (5) in warm tetrahydrofuran deposits the salt (3) (92%) when diluted with ethanol.

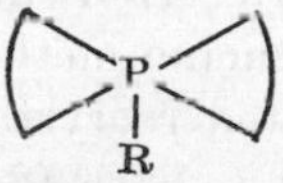

(6) 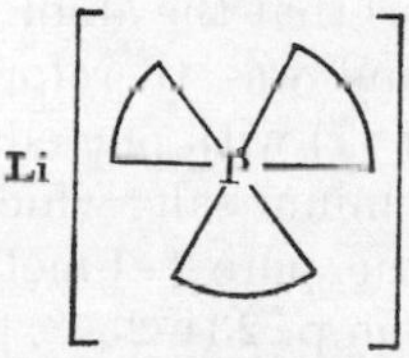   (7)

The iodide (4) in ether when treated with an excess of butyllithium gives 5-butyl-5,5'-spirobi(dibenzophosphole) (6; $R = C_4H_9$), and with methyllithium gives the 5-methyl analogue (6; $R = CH_3$). It is not surprising therefore that the salt (3) when treated in ether with an excess of methyllithium gives the phosphorane (6; $R = CH_3$) and an almost quantitative yield of lithium tris-(2,2'-biphenylylene)phosphate (7), and that phenyllithium behaves similarly. When the iodide (4) is stirred in ether with an excess of 2,2'-dilithiobiphenyl for 20 hours, the reaction product on hydrolysis gives the salt (3) (65%): the use, however, of an equimolecular quantity of the dilithio derivative in ether–tetra-hydrofuran gives the lithium salt (7) (71%), which crystallizes from tetrahydrofuran or from acetone with various proportions of the solvent which (as previously noted) are difficult to remove.

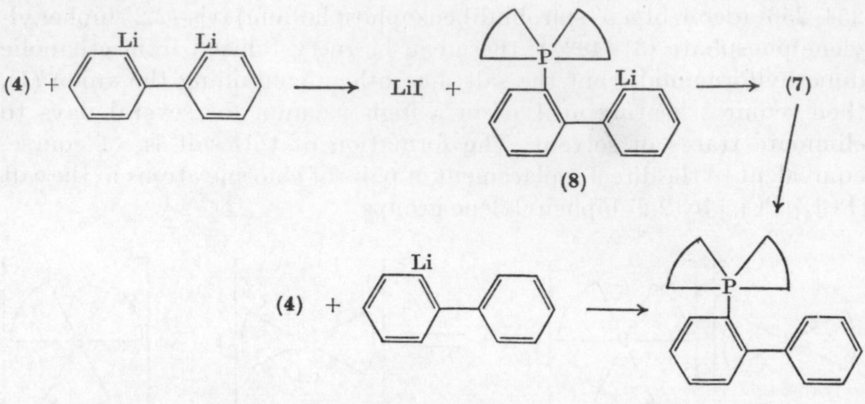

This reaction is interpreted as involving the intermediate formation of the 2'-lithio-2-biphenylyl-5,5'-spirobi(dibenzophosphole) (8), which then undergoes cyclization to the lithium salt (7). This is supported by the fact that when the salt (7) in methanol is treated with hydrochloric acid it rapidly gives the phosphorane (9) corresponding to the derivative (8), and that this phosphorane, m.p. 244–249°, can in turn be prepared by the action of 2-lithiobiphenyl on the iodide (4).

On the basis that the anion (1) had almost certainly the octahedral configuration and was therefore dissymmetric, Hellwinkel[152] treated the lithium salt (7) with one equivalent of brucine methiodide to form the methylbrucinium salt, which after repeated recrystallization from acetone gave the pure (−)-methylbrucinium (−)-tris-(2,2'-biphenylylene)phosphate, m.p. 234–236°, $[\alpha]_{578}$ −1250 ± 15° in dichloromethane. The mother-liquor furnished the (−)-methylbrucinium (+)-tris-(2,2'-biphenylylene)phosphate, m.p. 230–234°, $[\alpha]_{578}$ +950 ± 50°.

The two diastereoisomers, when treated with potassium iodide in acetone, gave the respective potassium tris-(2,2'-biphenylylene)phosphates, m.p. 295–298°, $[\alpha]_{578}$ ±1930 ± 20°, $[M]_{578}$ ±10150 ± 100°, in acetone solutions. The potassium or sodium salts, treated in methanol with the iodide (4), gave the two optically active forms of 5,5'-spirobi-(dibenzophospholium) tris-(2,2'-biphenylylene)phosphate (3), m.p. 247–250° (dec.), $[\alpha]_{578}$ ±1265 ± 15°, $[M]_{578}$ ±10410 ± 120° in dimethyl-formamide.

The thermal decomposition of certain of the above compounds, like that of the phosphoranes noted earlier (p. 93), gives rise to derivatives of the nine-membered phosphonin ring system. The lithium salt (7), when heated at 270° for 10 minutes, gives rise to 17-(2-biphenylyl)-17H-tetrabenzo[b,d,f,h]phosphonin (10), m.p. 193–195°; the same compound

is obtained (not unexpectedly) when the phosphorane (9) is heated at
270° for 5 minutes. The latter conversion is identical in type with that
of the methylphosphorane (6; R = CH₃), m.p. 217–218°, which when

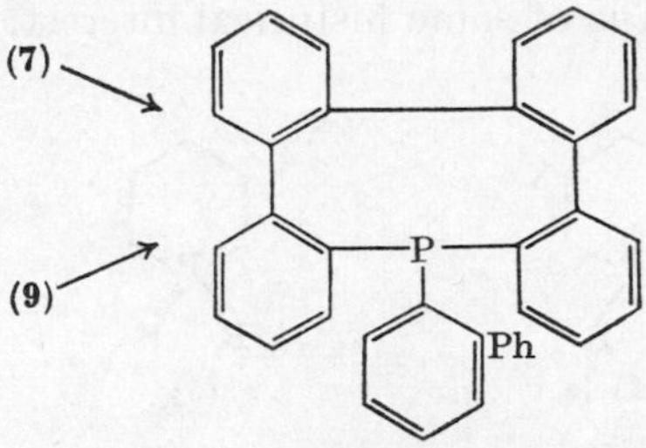

(7) ⟶

(9) ⟶

(10) 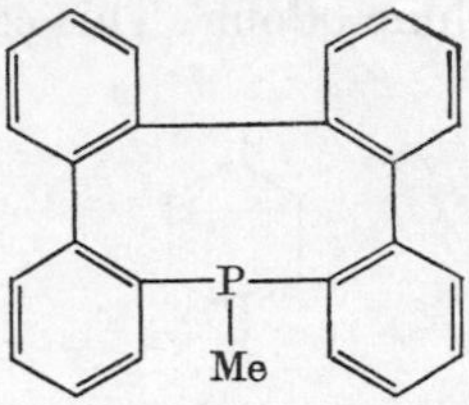   (11)

similarly heated gives the isomeric 17-methyl-17H-tetrabenzo[b,d,f,h]-
phosphonin (11), m.p. 130–131° [methiodide, m.p. 370° (dec.) [95]].

Six-membered Ring Systems Containing Only Carbon and One Phosphorus Atom

Phosphorin and Phosphorinanes

Phosphorin (1), [RRI 281], is—as one would expect—unknown, but
certain substituted derivatives which retain the phosphorin ring system
are known; its hexahydro derivative is phosphorinane (2), and our

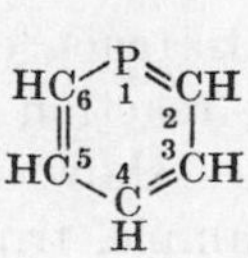

(1) 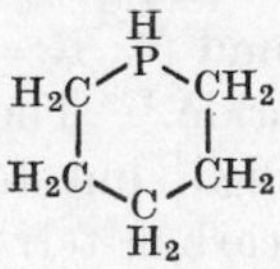   (2)

knowledge of the synthesis and reactions of phosphorinane and its
substituted derivatives is comparatively extensive. The chemistry of
these phosphorinanes will therefore be discussed first.

The system (2) was originally termed (cyclo)pentamethylene-
phosphine or phosphacyclohexane.

The main syntheses of phosphorinanes devoid of substituents on
the carbon atoms follow closely those given for similar phospholanes
(p. 50) and can therefore be briefly described in the same order.

(I) Grüttner and Wiernik[153] showed that the di-Grignard reagent prepared from 1,5-dibromopentane, $CH_2(CH_2CH_2MgBr)_2$, reacted with phenylphosphonous dichloride to give 1-phenylphosphorinane (**3**; $R = C_6H_5$), a liquid, b.p. 143–144°/16–18 mm, having a 'characteristic phosphine odour'. This compound is of some historical interest, as it is

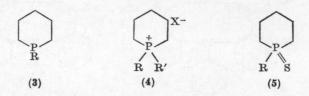

(**3**)   (**4**)   (**5**)

apparently the first heterocyclic phosphorus compound having a ring consisting solely of carbon and phosphorus to be synthesized. It gives a quaternary methiodide, named 1-methyl-1-phenylphosphorinanium iodide, m.p. 188°, and a mercuric chloride addition product of composition $C_{11}H_{15}P,HgCl_2$, m.p. 172° (cf. p. 51).

1-p-Tolylphosphorinane (**3**; $R = p\text{-}C_6H_4CH_3$), b.p. 167–168°/24 mm, similarly prepared, forms an ethiodide, m.p. 163–164°, and a 1:1 mercuric chloride addition product,[153] m.p. 157°.

More recently, benzylphosphonous dichloride has been used in this reaction to give 1-benzylphosphorinane (**3**; $R = CH_2C_6H_5$), b.p. 153–156°/30 mm [methiodide, m.p. 123°; 1-oxide, m.p. 149–150° (Slota[154])]; the dichloride, b.p. 103°/8 mm, was prepared by Fox's method,[155] ethereal benzylmagnesium chloride being treated first at −20° with cadmium chloride to form dibenzylcadmium and then with the appropriate quantity of phosphorus trichloride.

The compound (**3**; $R = C_6H_5$) is stated to react 'explosively' with carbon tetrachloride.[153] The products of this reaction have apparently not been identified, but tertiary phosphines show a wide range of reactivity with carbon tetrachloride; for example, triphenylphosphine (Rabinowitz et al.[156]) gives the products:

$$2(C_6H_5)_3P + CCl_4 \rightarrow (C_6H_5)_3PCl_2 + (C_6H_5)_3P{=}CCl_2$$

(II) Issleib and Häusler have shown that a suspension of an aryl-, alkyl- or cyclohexyl-dilithiophosphine, $RPLi_2$, in tetrahydrofuran–benzene reacts with 1,5-dibromopentane to give the corresponding 1-substituted phosphorinanes; these were characterized by formation of their methiodides (**4**) and, on treatment with sulfur in boiling benzene, their crystalline sulfides (**5**). The products thus obtained are listed in **Table 5**.

Table 5. Phosphorinanes (**3**) and their derivatives

Phosphorinanes	B.p. (°/mm)	Yield (%)	Methiodide, m.p. (°)	Sulfide, m.p. (°)
(**3**; R = C_6H_5)	119/3	31	176	86
(**3**; R = C_2H_5)	170	20	293–296	67
(**3**; R = C_6H_{11})	115/3	20	230–232	153

(III) Cyclohexylphosphine, $C_6H_{11}PH_2$, reacts with 1,5-dibromo-pentane vigorously at 130° to give the crystalline 1-cyclohexylphos-phorinane hydrobromide, which with alkalis liberates the phosphorinane (**3**; R = C_6H_{11}), b.p. 112°/2 mm (57%); this in turn gives a sulfide (**5**; R = C_6H_{11}), m.p. 152–153° (Issleib, K. Krech and Gruber[74]).

Diphenylphosphine, being a secondary phosphine, reacts with 1,5-dibromopentane in boiling ethanol to form 1,1-diphenylphosphor-inanium bromide (**4**; R = R' = C_6H_5, X = Br), m.p. 246° (28%), which with silver oxide gives 1-phenylphosphorinane 1-oxide (**6**; R = C_6H_5), m.p. 128° (65%).[74]

Diphenylpotassiophosphine also reacts with the dibromopentane, giving the bromide (**4**; R = R' = C_6H_5; X = Br), m.p. 262–263° (49%) (Märkl[75]). The rather special case of the interaction of 1,5-dibromo-3-methoxypentane with the potassiophosphine is discussed later (p. 111).

A solution of diphenylphosphine and 1,5-diiodopentane in aceto-nitrile, when boiled for 12–16 hours, gives diphenylphosphonium iodide, $[(C_6H_5)_2PH_2]I$ and 1,1-diphenylphosphorinanium iodide (**4**; R = R' = C_6H_5, X = I), m.p. 278–279° (45%); the latter, on reduction with lithium aluminum hydride, gives 1-phenylphosphorinane and benzene (Grim and Schaaff[157]). It is noteworthy that diphenylphosphine when similarly treated with 1,3-diiodopropane gives the eight-membered octahydro-1,1,5,5-tetraphenyl-1,5-diphosphocanium diiodide (p. 195).

The intermediate compounds formed in the above reactions are undoubtedly the 5-halogenopentyl analogues of those noted in the corresponding phospholane syntheses (p. 51 *et seq.*).

(IV) 1,5-Dibromopentane reacts with tetraphenyldiphosphine, $(C_6H_5)_2P$—$P(C_6H_5)_2$, also to give the bromide (**4**; R = R' = C_6H_5; X = Br), m.p. 261–262° (66%), and the bromide, $(C_6H_5)_2PBr$ (Märkl[76]; cf. p. 52).

(V) 1 Bromo-4-pentene (**7**) reacts with phenylsodiophosphine, both in 2-methylheptane, to give (4-pentenyl)phenylphosphine (**8**), b.p. 90–92°/1.5 mm [61% from (**7**)], which when irradiated with ultraviolet light in the absence of a solvent gives 1-phenylphosphorinane (**3**;

R = C_6H_5), b.p. 94°/1 mm (64%); it gives a 1:1 $HgCl_2$ adduct, m.p. 180°, a sulfide (**5**; R = C_6H_5), m.p. 83°, and an oxide (**6**; R = C_6H_5),

$$CH_2{=}CH(CH_2)_3Br \;\rightarrow\; CH_2{=}CH(CH_2)_3PH(C_6H_5) \;\rightarrow\; (\mathbf{3}; R = C_6H_5)$$
$$\qquad\quad (\mathbf{7}) \qquad\qquad\qquad\qquad (\mathbf{8})$$

b.p. 138°/0.05 mm, m.p. 130° after sublimation. This oxide was also obtained by the oxidation of (**8**) (cf. p. 52) (Davies, Downer, and Kirby[77]).

The high yield of (**3**; R = C_6H_5) compared with that of the phenyl-phospholane (38%) and the phenylphosphepane (41%) (p. 152) is attributed to the freedom from strain of the six-membered ring and the favorable conformation of the alkenylphosphine (**8**).

(VI) Wagner[1] has claimed that the addition of 1,5-dichloropentane to sodium phosphide, $NaPH_2$, in liquid ammonia gives the diphosphine, $H_2P(CH_2)_5PH_2$, and phosphorinane (**3**; R = H) 'in good yields' and that the use of sodium isobutyl- and n-pentyl-phosphide similarly gives 1-isobutyl- and 1-n-pentyl-phosphorinane, respectively. Physical constants or analyses are not given in the specification.

Derivatives. Simple sulfides of type (**5**) can be prepared by Maier's method (p. 53); for example, the di-Grignard reagent, $BrMg(CH_2)_5MgBr$, and methylthiophosphonic dibromide, $CH_3P(S)Br_2$, react in ether to give 1-methylphosphorinane 1-sulfide (**5**; R = CH_3), b.p. 113–130°/0.2 mm, m.p. 51–52° (19%).[79]

1-*Hydroxyphosphorinane* 1-*oxide* ('pentamethylenephosphinic acid') (**6**; R = OH). Three syntheses of this acid have been recorded.

(1) In the first (Kosolapoff[158]), *N,N*-diethylamidophosphoryl di-chloride, $(C_2H_5)_2N{-}P(O)Cl_2$, was treated with the usual di-Grignard reagent, $BrMg(CH_2)_5MgBr$, in boiling ether to obtain the 1-(diethyl-amino)phosphorinane 1-oxide [**6**; R = $(C_2H_5)_2N$] which, after hydrolysis

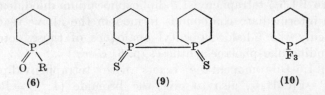

and a purification involving several stages, gave the acid (**6**; R = OH), m.p. 128–129°. This acid was converted by phosphorus pentachloride into the 1-chloro derivative (**6**; R = Cl), b.p. 151–152°/30 mm, which when treated with sodium butoxide in butanol gave the n-butyl ester (**6**; R = OC_4H_9), b.p. 80°/1 mm.

(2) Howard and Braid[159, 160] prepared four esters of type $X(CH_2)_5P(O)(OR)_2$, where $X = Cl$ or Br, and $R = C_2H_5$ or C_4H_9, by adding triethyl or tri-n-butyl phosphite to a large excess of 1,5-dichloro- or 1,5-dibromo-pentane at 149–163°, the Arbusov rearrangement thus giving the diethyl or dibutyl ester of the 5-chloro- or 5-bromo-pentyl-phosphonic acid. Slow addition of these esters to magnesium in a large volume of anisole at 135° caused formation of the Grignard reagent followed by cyclization to give the ethyl ester (**6**; $R = OC_2H_5$) and the butyl ester (**6**; $R = OC_4H_9$) of the phosphinic acid. These esters when hydrolysed with boiling hydrochloric or hydrobromic acid gave the phosphinic acid (**6**; $R = OH$), m.p. 127.5–128.5° after removal of traces of n-pentylphosphonic acid by crystallization from benzene and sublimation. The phosphinic acid was converted into the chloride (**6**; $R = Cl$), b.p. 115–116°/2 mm, m.p. 80°, which in turn with butanol gave the n-butyl ester.

(3) Schmutzler,[80] following his previous work (p. 53), treated the reagent $BrMg(CH_2)_5MgBr$ with thiophosphoryl chloride, $SPCl_3$, to obtain 1,1'-diphosphorinane 1,1'-disulfide (**9**) which, when 'analytically' pure, still melted at 185–225°. This disulfide was oxidized by nitric acid to the phosphinic acid (**6**; $R = OH$), m.p. 128–129°. The disulfide when treated with antimony trifluoride gave 1,1,1-trifluoro-phosphorinane (**10**), b.p. 64–65°/40 mm.

The virtually identical melting points of the acid obtained by these independent methods indicate strongly that the product in each case was the phosphinic acid (**6**; $R = OH$), and that the reaction of $(C_2H_5)_2N\!-\!P(O)Cl_2$ with the di-Grignard reagent does not stop at the intermediate stage, as that with $BrMg(CH_2)_4MgBr$ is claimed to do (Hilgetag et al.,[82] p. 55).

Braid[160] has reduced the pure chloride (**6**; $R = Cl$) with lithium aluminum hydride to phosphorinane (**11**), b.p. 121°, m.p. 19° (35%), which was also obtained in lower yield by the similar reduction of the n-butyl ester (**6**; $R = OC_4H_9$). Phosphorinane is described as a colorless, malodorous, toxic liquid, which inflames on exposure to air. Careful oxidation in the absence of air, obtained by passing oxygen into a dilute petroleum solution of a small quantity of (**11**), gives 1,1'-diphosphor-inane 1,1'-dioxide (**12**), described (surprisingly) as a liquid, whereas a similar oxidation in the presence of water gives a hydrated phosphorinane 1-oxide (**13**). This compound on attempted distillation undergoes disproportionation to the acid (**6**; $R = OH$) and phosphorinane (**11**), whereas cautious hydrogen peroxide oxidation of (**13**) gives solely the phosphinic acid.

The acid chloride (**6**; $R = Cl$) reacts with a secondary amine, e.g.,

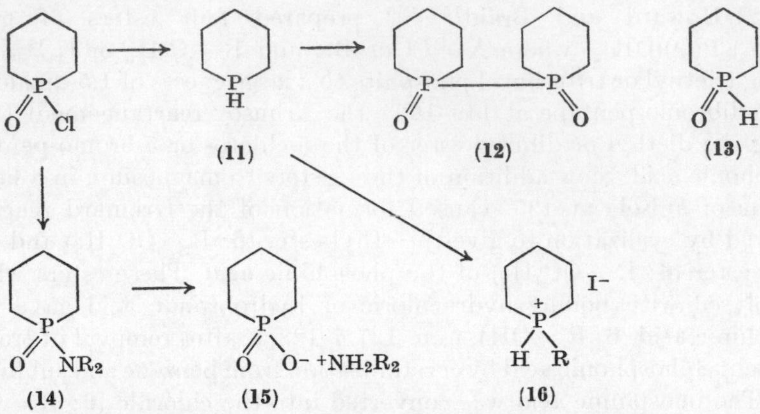

(11)   (12)   (13)

(14)   (15)   (16)

dicyclohexylamine, to give 1-(dicyclohexylamino)phosphorinane 1-oxide
(**14**; $R = C_6H_{11}$), which undergoes ready hydration to the dicyclohexylamine salt (**15**; $R = C_6H_{11}$) of the phosphinic acid.

Phosphorinane (**11**) with concentrated hydriodic acid gives a
crystalline hydriodide (**16**; $R = H$), m.p. 239–240° (dec.), and with
methyl iodide forms 1-methylphosphorinane hydriodide (**16**; $R = CH_3$),
m.p. 232° (dec.); 1-methylphosphorinane is apparently too weakly basic
to form a methiodide.

Phosphorinane (**11**) reacts with sulfur in boiling benzene, apparently
to form 1-mercaptophosphorinane 1-sulfide (**17**), which has not been
isolated as the pure compound, because it undergoes such ready oxidation

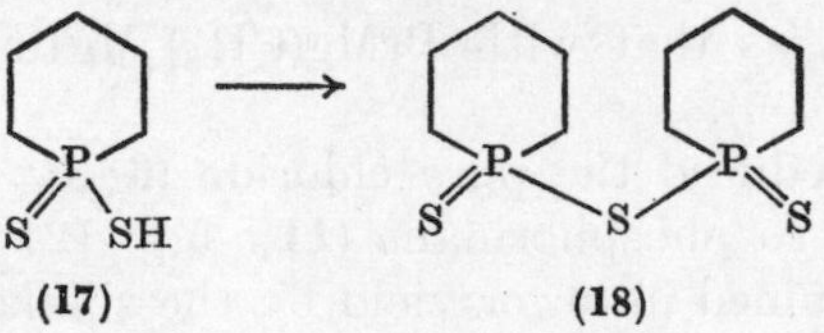

(17)   (18)

to the crystalline 1,1′-thiodiphosphorinane 1,1′-disulphide (**18**), m.p.
126.5–127°. Infrared data are given for most of Braid's compounds.[160]

   4-*Phosphorinanones and* 4-*Phosphorinanols.* The first synthesis of a
phosphorinane having a nuclear carbonyl group was achieved by
Welcher, Johnson, and Wystrach,[161, 162] starting with bis(2-cyano-
ethyl)phenylphosphine (**19**; $R = C_6H_5$), which had been prepared by
the interaction of phenylphosphine and acrylonitrile when heated
without a solvent (Mann and Millar[163]) or in acetonitrile containing
potassium hydroxide (Rauhut *et al.*[164]). The phosphine (**19**; $R = C_6H_5$),
when treated with sodium *tert*-butoxide in boiling toluene, undergoes a

$$RP(CH_2CH_2CN)_2 \longrightarrow$$

(19)   **(20)**   **(21)**   **(22)**

Thorpe cyclization to form 4-amino-3-cyano-1,2,5,6-tetrahydro-1-phenylphosphorin (**20**; $R = C_6H_5$), m.p. 139.5–140° (80%). The infrared and the nmr spectra confirm that this compound has the structure assigned, and is not the isomer (**22**; $R = C_6H_5$) or in solution a tautomeric mixture of the two compounds. The infrared spectrum[161, 165] shows significant bands at 2170 (attributed to a β-amino unsaturated nitrile), at 3400 and 3330 (sharp, indicating an NH_2 group), at 3240 (NH_2 overtone band), and 1645 cm^{-1} (NH_2 deformation band); the NH group of (**22**) would have given only one band at 3400–3300 cm^{-1}. The compound (**20**) in boiling 6N-hydrochloric acid undergoes hydrolysis and decarboxylation to give 1-phenyl-4-phosphorinanone (**21**; $R = C_6H_5$), b.p. 185–190°/1 mm, m.p. 43–44° (21%): it gives a methiodide, m.p. 155–156°, and a semicarbazone, m.p. 155.5–156.5°.

Bis(2-cyanoethyl)ethylphosphine (**19**; $R = C_2H_5$) similarly gives 4-amino-3-cyano-1-ethyl-1,2,5,6-tetrahydrophosphorin (**20**; $R = C_2H_5$), m.p. 74.5–75° (83%), which on hydrolysis gives 1-ethyl-4-phosphorin-anone (**21**; $R = C_2H_5$), b.p. 92°/7 mm (methiodide, m.p. 213–214°; semicarbazone, m.p. 167–169°).

The compounds (**19** and **20**; $R = C_6H_5$) were named 3,3'-(phenylphosphinidene)dipropionitrile and 4-amino-1,2,5,6-tetrahydro-1-phenylphosphorin-3-carbonitrile, respectively; it is claimed that the addition of members of each class to gasoline prevents misfire and surface ignition.[160]

The eutropic series of 1-phenyl-4-piperidone (**23**), 1-phenyl-4-phosphorinanone (**21**; $R = C_6H_5$), and 1-phenyl-4-arsenanone (**24**)

(23)   **(24)**

(p. 394) has been prepared for comparison of their properties (Gallagher and Mann[165]). The phosphorinanone is sufficiently basic to give a stable

5

hydrochloride dihydrate, m.p. 231–233°, giving on prolonged drying an intensely hygroscopic anhydrous salt of unchanged m.p., and a stable hydrobromide dihydrate, m.p. 240.5–241.5°. The methiodide is also a stable compound and gives no indication of the formation of a 4,4-dihydroxy or a 4,4-dimethoxy derivative, which is such a striking feature of the methiodides of the nitrogen and arsenic analogues[165] (p. 395). The phosphorinanone gives a phenylhydrazone which in boiling ethanol containing hydrogen chloride undergoes the Fischer indolization reaction to form 2,3,4,5-tetrahydro-2-phenyl-1H-phosphorino[4,3-b]indole (p. 121), whereas the phenylhydrazones of (23) and (24) are apparently too unstable to give a similar reaction.

The compounds (23) and (24) do, however, show two reactions in common with the phosphorinanone (21; R = C$_6$H$_5$). The latter condenses with o-aminobenzaldehyde (the Friedländer reaction) to give 1,2,3,4-tetrahydro-2-phenylphosphorino[4,3-b]quinoline, whose salts show some interesting color relationships (p. 122), and it also condenses with alkaline isatin (the Fitzinger reaction) to form 1,2,3,4-tetrahydro-2-phenylphosphorino[4,3-b]quinoline-10-carboxylic acid (p. 123), which unlike the analogous derivative of (23) does not form a zwitterion.[165]

The range of available 4-phosphorinanones has been considerably increased by a simple synthesis which readily allows the introduction of substituents into the heterocyclic ring (Welcher and Day[166, 167]). It consists of the direct addition of a primary phosphine to a conjugated dienone, $i.e.$, to a substituted divinyl ketone. Thus a mixture of phenylphosphine and dibenzylideneacetone (C$_6$H$_5$CH=CH)$_2$CO, when heated at 120–125° for 13 minutes under nitrogen, gives 1,2,6-triphenyl-4-phosphorinanone (25; R = C$_6$H$_5$), m.p. 176.5–177.5°: a mixture of the phosphine and phorone, [(CH$_3$)$_2$C=CH]$_2$CO, similarly heated at 115–130° for 6 hours, crystallizes on cooling, giving 2,2,6,6-tetramethyl-1-phenyl-4-phosphorinanone (26; R = C$_6$H$_5$), m.p. 91–92°. The reaction with dibenzylideneacetone is clearly more rapid than that with phorone,

(25) (26) (27) (28)

probably because the four methyl groups of phorone exert greater obstruction than the two phenyl groups of dibenzylideneacetone. This reaction between the phosphine and the ketone is markedly sensitive to the reaction conditions, and the use of a solvent or a free-radical catalyst is not desirable.

The 4-phosphorinanones of type (25) and (26) sublime very readily, and an accurate determination of their boiling points is therefore difficult. Compounds of type (25) are usually less basic than the corresponding compounds of type (26); this is shown by their pK_a values, determined by titration with perchloric acid in nitromethane. Both show the normal properties of tertiary phosphines and of ketones and therefore form quaternary methiodides and semicarbazones. These points are summarized in Table 6.

Table 6. Properties and derivatives of the 4-phosphorinanones of type (25) and (26)

R	B.p. (°/mm)	M.p. (°)	pK_a	CH_3I salt, m.p. (°)	Semi- carbazone, m.p. (°)
In (25)					
C_6H_5	Subl. 200/0.5	176.5–177.5	4.61	137	>270
C_6H_{11}	Subl. 190/1	120–121	7.00		
$(CH_2)_2CN$	Subl. 180/0.6	126–128			
iso-C_4H_9	Subl. 150/1	118–119			
n-C_8H_{17}	Subl. 190/1	80–81			
In (26)					
C_6H_5	130–140/0.5 Subl. 100/0.3	91–92	1.85	229–230	198–199.5
C_6H_{11}	135–140/1 Subl. 110/1	63–65	3.86	275–278	
$(CH_2)_2CN$	140–150/1 Subl. 100/0.8	48–50	0.79	290–295	
iso-C_4H_9	90–97/0.3	—	3.58	263–265	
n-C_8H_{17}	146–148/0.8	—		150–152	

The 4-phosphorinanones are readily reduced by lithium aluminum hydride in tetrahydrofuran to the corresponding 4-phosphorinanols. Thus the compound (26; R = C_6H_5) gives 2,2,6,6-tetramethyl-1-phenyl-4-phosphorinanol (27; R = C_6H_5). This product when sublimed and recrystallized from hexane had m.p. 98–100°; however, repeated recrystallization from hexane gave a product of identical composition but having m.p. 123–124.5°; these values indicate that the low-melting product was probably a mixture of geometric isomers, differing in the relative positions of the 4-hydroxyl group and the phenyl group attached to the pyramidal phosphorus atom.

The more vigorous Wolff–Kishner reduction, when applied to the 4-phosphorinanones, regenerates the 4-phosphorinanes. A mixture of the compound (**25**; $R = C_6H_5$), 85% aqueous hydrazine, sodium hydroxide, and diethylene glycol, when heated under nitrogen for 2 hours at 168° and then for 3 hours at 197°, gives 1,2,6-triphenylphosphorinane (**28**), m.p. 167–168° (82%).

A considerable number of compounds of the types (**20**), (**21**), and (**27**) are cited in the patent specifications.[162, 167, 168]

Stereoisomerism of 4-*Phosphorinanols.* Quin and Shook[169] have converted bis(2-cyanoethyl)methylphosphine (**19**; $R = CH_3$) into 1-methyl-4-phosphorinanone (**21**; $R = CH_3$), b.p. 57–58°/0.7 mm, which with ethylmagnesium bromide gave 4-ethyl-1-methyl-4-phosphorinanol

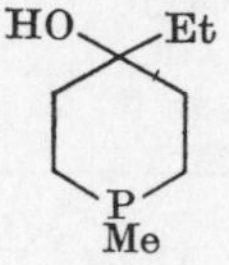

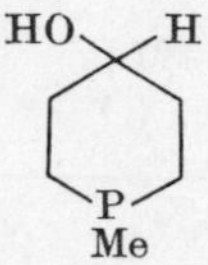

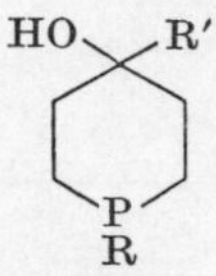

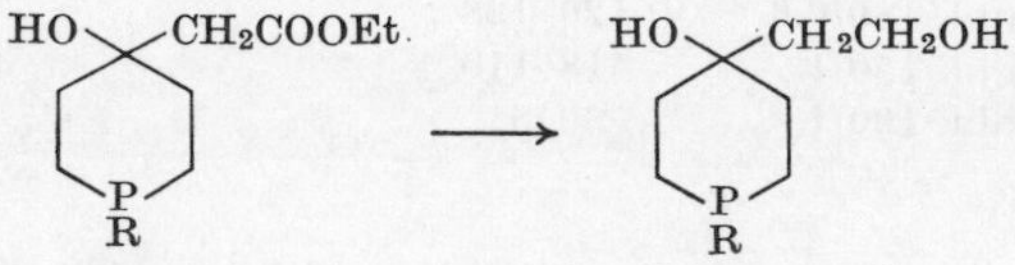

(**29**), b.p. 45–62°/0.2 mm. Gas chromatography revealed the presence in this distillate of two isomers, which were separated by fractionation on a spinning-band column, and the two isomers, each chromatographically homogeneous, were isolated, (**29a**), b.p. 62°/0.55 mm, and (**29b**), b.p. 68–69°/0.6 mm. The pmr spectra were in general very similar and in accordance with the structure (**29**). The infrared spectra were almost identical, each having an OH stretching band at 3350 cm^{-1}. The C—O stretching band for (**29a**) appeared at 1105 cm^{-1}, and that for (**29b**) at 1100 cm^{-1}; this indicates that the OH group is in a similar conformational situation in both isomers, as significant differences in the position of this band usually appear in axial and equatorial alcohols (Pickering and Price[170]). Quaternization of each isomer with benzyl bromide, followed by crystallization of the cation as the perchlorate gave 1-benzyl-4-ethyl-4-hydroxy-1-methylphosphorinanium perchlorate, m.p. 199–200° (from **29a**), 166–168° (from **29b**). No change in the proportion of (**29a**) and (**29b**) occurred when a mixed sample in benzene was boiled for 64 hours, or in toluene for 28 hours, indicating that the phosphorus atom in such isomers has considerable configurational stability.

A later more detailed investigation (Shook and Quin[171]) has been made of the compound (29) and of various derivatives prepared from 1-methyl-4-phosphorinanone (21; $R = CH_3$): reduction of this compound with lithium aluminum hydride, or by aluminum isopropoxide in 2-propanol, gave 1-methyl-4-phosphorinanol (30); application of phenylmagnesium bromide gave (31; $R = CH_3$, $R' = C_6H_5$); a Reformatsky reaction with zinc and ethyl bromoacetate gave 4-(ethoxycarbonylmethyl)-4-hydroxy-1-methylphosphorinane (32; $R = CH_3$), which on reduction with lithium aluminum hydride gave 4-hydroxy-4-(2'-hydroxyethyl)-1-methylphosphorinane (33; $R = CH_3$). The compound (31; $R = R' = C_2H_5$) was also prepared.

Gas-chromatographic evidence for the existence of isomers in each of these compounds was obtained, although separation of the isomers was carried out only with (31; $R = CH_3$, $R' = C_6H_5$), which when fractionated through a spinning-band column yielded two fractions, (a) b.p. 112–114°/0.2 mm and (b) b.p. 114–116°/0.2 mm, (a) being predominantly the cis- and (b) the trans-isomer. In one preparation the mixed isomers, when set aside for 2 weeks, deposited the crystalline (predominantly) cis-isomer, m.p. 92–100°. A mixture of these isomers in a xylene solution containing a small quantity of concentrated hydrochloric acid, when boiled for 19 hours, underwent dehydration to give 1,2,3,6-tetrahydro-1-methyl-4-phenylphosphorin (34; $R = C_6H_5$), b.p. 84–85°/0.1 mm. This structure was confirmed by infrared and proton magnetic resonance spectra. The compound reacted with benzyl bromide to give 1-benzyl-1,2,3,6-tetrahydro-1-methyl-4-phenylphosphorinium bromide, m.p. 225–228°.[171]

The proton magnetic resonance spectra of the above compounds (30)–(33), in addition to confirming the presence of isomers, gave some evidence for the configuration of the molecule, and also for example

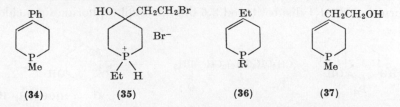

(34)   (35)   (36)   (37)

indicated in which isomer the P-substituent was axial and in which equatorial. The full discussion with its diagrams should be consulted.

It should be noted that Quin and Mathewes[172] had earlier converted

1-ethyl-4-phosphorinanone (**21**; R = C_2H_5) into the 4-(ethoxycarbonyl-methyl)-4-hydroxyl derivative (**32**; R = C_2H_5), b.p. 97–101°/0.05 mm, and thence by the usual reduction to the 4-hydroxy-4-(2-hydroxyethyl) derivative (**33**; R = C_2H_5), b.p. 128–130°/0.15–0.20 mm. This compound when heated with 62% hydrobromic acid for 12 hours appeared to give the hydrobromide (**35**) of 4-(2-bromoethyl)-1-ethyl-4-hydroxyphosphor-inane, for this phosphine when liberated by potassium carbonate and heated apparently underwent an internal quaternization by the bromo-ethyl group reacting with the tertiary phosphine group. The unlikelihood of the tertiary OH group surviving this treatment was reinvestigated (Quin and Shook[173]). The analogous compound (**29**) was similarly treated with hydrobromic acid, and the only product was 4-ethyl-1,2,3,6-tetrahydro-1-methylphosphorin (**36**; R = CH_3), which gave a quaternary benzyl bromide characterized as the corresponding perchlorate, m.p. 108–110°. The 1,4-diethyl-4-hydroxyphosphorinane (**31**; R = R′ = C_2H_5) with hydrobromic acid similarly gave 1,4-diethyl-1,2,3,6-tetra-hydrophosphorin (**36**; R = C_2H_5), characterized as the benzyl per-chlorate, m.p. 134°.

It thus became evident that the earlier experiments had not given the 4-hydroxy derivative (**35**). Furthermore when 4-hydroxy-4-(2-hydroxyethyl)-1-methylphosphorinane (**33**; R = CH_3) and 1,2,3,6-tetrahydro-4-(2-hydroxyethyl)-1-methylphosphorin (**37**) were each in turn treated with hot hydrobromic acid precisely as before, they gave identical crude products which were not the 1-methyl analogue of (**35**) or of its thermal quaternization product. The phosphorus analogue of quinuclidine apparently cannot therefore be prepared in this way.

2,6-*Phosphorinandiols.* A number of these 2,6-diols as quaternary salts have been prepared (Buckler and Epstein[90]) by the method described for the corresponding 2,5-phospholanediols (p. 57). When concentrated hydrochloric acid is added at room temperature to a solution of a secondary phosphine and glutaraldehyde in methanol–tetrahydrofuran, which is then boiled at room temperature, the solution furnishes the 1,1-disubstituted 2,6-dihydroxyphosphorinanium chloride

(**38**)   (**39**)   (**40**)

(**38**). Some of the compounds thus obtained are listed (Table 7), with their m.p.s, and also the nmr shift in the P region in ppm relative to 85%

H_3PO_4. These shifts are markedly smaller than those of the corresponding 2,5-dihydroxyphospholanium chlorides (p. 57).

Table 7.   2,6-Dihydroxyphosphorinanium
chlorides (38)

R in (38)	M.p. (°)	Nmr shift
n-C_4H_9	115–117	−20
iso-C_4H_9	150–152	−23
n-C_8H_{17}	100–105	−25
cyclo-C_6H_{11}	160–164	−20

Buckler and Epstein[174] have also shown that phosphine will react with ketones in the presence of concentrated hydrochloric acid to give primary phosphine oxides, R_2CH—$P(O)H_2$. One example of 2,6-phosphorinanediol formation which they regard as exceptional is the interaction of 4-heptanone, $(CH_3CH_2CH_2)_2CO$, with phosphine and hydrochloric acid to give the (1-propylbutyl)phosphine oxide (39), which undergoes cyclization with glutaraldehyde to give 2,6-dihydroxy-1-(1-propylbutyl)phosphorinane 1-oxide (40), m.p. 226–227°.

A number of very interesting reactions showing the interrelationship of various substituted phosphorins, dihydro- and tetrahydro-phosphorins, and phosphorinanes, and often involving compounds which contain five-covalent phosphorus and are therefore systematically phosphoranes, have been elucidated mainly by Märkl. This work is next summarized.

Märkl, using the reaction[75] noted earlier (p. 101), has treated 1,5-dibromo-3-methoxypentane, $CH_3OCH(CH_2CH_2Br)_2$, with diphenylpotassiophosphine to obtain 4-methoxy-1,1-diphenylphosphorinanium bromide (41; R = CH_3), m.p. 220–221° (46%), which when heated with hydrobromic acid–acetic acid undergoes demethylation to the 4-hydroxy analogue (41; R = H), m.p. 148–149° (58%). This secondary alcohol, when heated with potassium hydrogen sulfate at 210–220°, is dehydrated to the 1,2,3,6-tetrahydro-1,1-diphenylphosphorinium cation, isolated as the perchlorate (42), m.p. 177–179° (60%). Addition of bromine in acetic acid gives 3,4-dibromo-1,1-diphenylphosphorinanium perchlorate (43), m.p. 206–208° (68%), which in turn, in quinoline–dimethylformamide at room temperature, loses hydrogen bromide to form 4-bromo-1,2,3,4-tetrahydro-1,1-diphenylphosphorinium perchlorate (44), m.p. 175–176° (88%). Hot anhydrous phosphoric acid repeats this process, with the formation of 1,2-dihydro-1,1-diphenylphosphorinium perchlorate (45;

(41) (42) (43) (44) (45) (46)

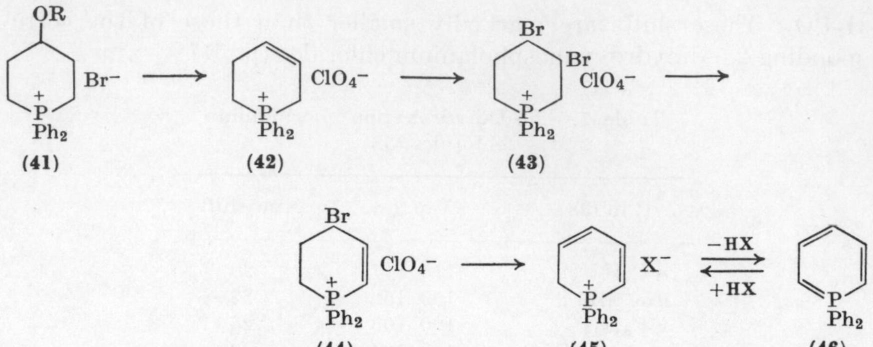

$X = ClO_4$), m.p. 117–119°. The addition of dilute sodium hydroxide to an aqueous solution of this salt causes the deposition of 1,1-diphenylphosphorin (**46**) as an amorphous yellow powder, stable in water in the absence of oxygen which converts it into red and violet phosphoranes of unknown structure. The compound (**46**), which Märkl terms 1,1-diphenyl-1-phosphabenzene, reacts readily with acids to re-form the cation (**45**).

Infrared spectra[75] show the C=C group of (**42**) by a band at 1637 cm^{-1}, and those of (**45**) by bands at 1626, 1572, and 1550 cm^{-1}.

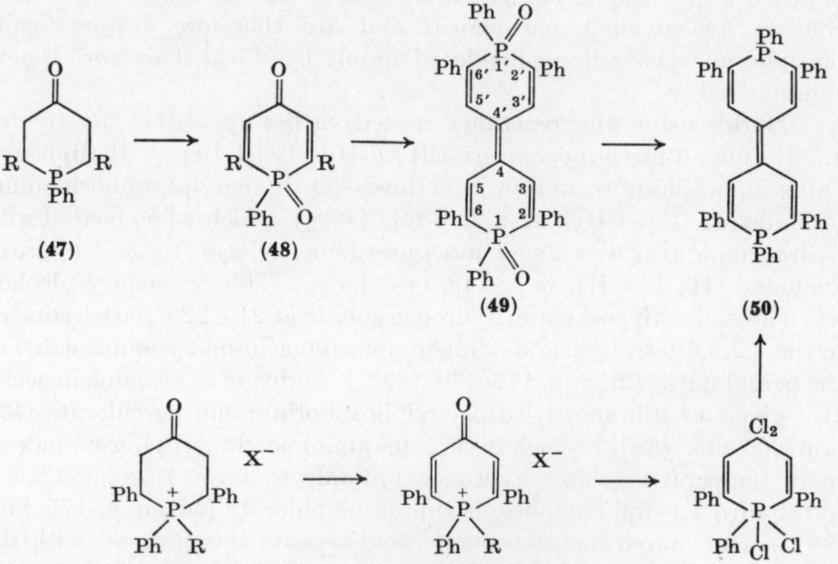

In another investigation (Märkl and Olbrich[175]), 1-phenyl-4-phosphorinanone (**47**; R = H) was treated in boiling ethanol with selenium

(47) (48) (49) (50) (52) (53) (51)

dioxide for 36 hours: oxidation of the tertiary phosphine and dehydrogenation of the phosphorinane ring occurred, with the formation of 1,4-dihydro-4-oxo-1-phenylphosphorin 1-oxide (48; R = H), m.p. 130–131° (10%), $\nu(CO)$ 1650 cm^{-1}; 1,2,6-triphenyl-4-phosphorinanone (47; R = C$_6$H$_5$), similarly treated, gave 1,4-dihydro-4-oxo-1,2,6-triphenylphosphorin 1-oxide (48; R = C$_6$H$_5$), m.p. 162° (67–80%), $\nu(CO)$ 1627 cm^{-1}. These phosphine oxides could not be reduced to the tertiary phosphines by other active phosphines or phosphites, or by the use of silanes (see p. 49). When, however, the compound (48; R = C$_6$H$_5$) was heated with triethyl phosphite at 80–100°, removal of the ketonic oxygen occurred, giving 1,1′,4,4′-tetrahydro-1,1′,2,2′,6,6′-hexaphenyl-4,4′-biphosphorinylidene 1,1′-dioxide (49), m.p. 425–428°. Treatment of this compound with phosphorus pentachloride gave the corresponding deep red 1,1,1′,1′-tetrachloride, which on reduction with lithium aluminum hydride gave bis-4,4′-(1,4-dihydro-1,2,6-triphenyl-4-phosphorinylidene) (50), brownish-red crystals, m.p. 335–337° (90%); the molecular weight was confirmed by mass spectrometry.

The diphosphine (50) resists quaternization, even by vigorous reagents such as trialkyloxonium tetrafluoroborates, [R$_3$O][BF$_4$]; it does, however, readily combine with sulfur in boiling benzene to give the 1,1′-disulfide, m.p. 403–405°, analogous to the 1,1′-dioxide (49). 1,1,4,4-Tetrachloro-1,4-dihydro-1,2,6-triphenylphosphorin (51), when reduced with lithium aluminum hydride also gives the diphosphinylidene (50), but when reduced with hydrogen sulfide or thioacetic acid in benzene at 25° gives the above 1,1′-disulfide.

Dehydrogenation of the phosphorinane system is also shown by quaternary salts (52) of 1,2,6-triphenyl-4-phosphorinanone, which in boiling ethanol with selenium dioxide for 24 hours give the corresponding quaternary salts (53) of the 4(1H)-phosphorinone, e.g., (53; R = CH$_2$C$_6$H$_5$, X = ClO$_4$, m.p. 218–220°, and R = C$_2$H$_5$, X = ClO$_4$, m.p. 173–175°).

The reactivity of 1,2,6-triphenyl-4(1H)-phosphorinone 1-oxide (48; R = C$_6$H$_5$) is further shown by its condensation with diphenylketene (as the quinoline adduct) at 170° to give the lemon-yellow 4-(diphenylmethylene)-1,4-dihydro-1,2,6-triphenylphosphorin 1-oxide (54), m.p. 319–321° (80%). This compound, when treated with phosphorus pentachloride and then reduced with lithium aluminum hydride, gives the corresponding yellow phosphine (55A). Märkl and Olbrich[176] consider that (55A) may have a dipolar limiting structure (55B) containing a 'phosphabenzene' system comparable with aromatic pyrylium salts. If (55B) made a significant contribution to the resonance structure, alkylating agents should attack the negative carbon atom of the

Ph Ph
\   Ph Ph
 C   \
 ‖   C
Ph   Ph   ‖   I⁻
 \   /   Ph   Ph
  P   \   /
 / \   P⁺
Ph   O   / \
                          Ph   Me
(54)   (56)

Ph Ph   Ph Ph
 \   \ ..
  C   C⁻
  ‖ |
Ph   Ph   Ph   Ph
 \   /   ⟷   \   /
  P   P⁺
 /   /
Ph   Ph
(55A)   (55B)

diphenylmethylene group; in practice, however, methyl iodide gave the
normal 1-methylphosphorinium iodide (56), orange-yellow needles,
m.p. 257–258°.

The phosphorinone 1-oxide (48; R = C₆H₅) will also condense with
ethyl cyanoacetate to give 4-[1-cyano-2-(ethoxycarbonyl)ethylidene]-
1,4-dihydro-1,2,3-triphenylphosphorin 1-oxide (57), m.p. 205–206°,

NC   CH₂COOEt   NC   CN   NC   CN
 \   /   \   /   \   /
  C   C   C
  ‖   ‖   ‖
Ph   Ph   Ph   Ph   Ph   Ph
 \   /   \   /   \   /
  P   P   P
 / \   / \   / \
Ph   O   Ph   O   Ph   C(CN)₂
(57)   (58)   (59)

λ_{max} 277, 374 mμ (54%), and with malononitrile to give the 4-(dicyanomethylene) analogue (58), m.p. 259–260°, λ_{max} 285, 388 mμ (31%). The
use of an excess of malononitrile gives additional condensation with the
P=O group, thus forming 1,4-bis(dicyanomethylene)-1,4-dihydro-
1,2,6-triphenylphosphorin (59), deep red leaflets with a green metallic
sheen, m.p. 185–187°, λ_{max} 361, 533.5 mμ (40–60%); this compound is
in general type the phosphorus analogue of a quinodimethane.[176]

Price et al.,[177, 178] when investigating methods for synthesizing
compounds having the 'phosphabenzene' system, found that a mixture

of phenylphosphine and 2,4,6-triphenylpyrylium fluoroborate (60) in boiling pyridine rapidly reacted to form two isomers of composition $C_{29}H_{25}O_2P$, one forming colorless crystals, m.p. 256–257°, and the other

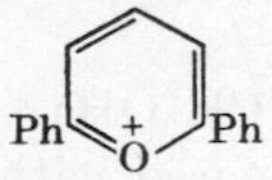

(60) 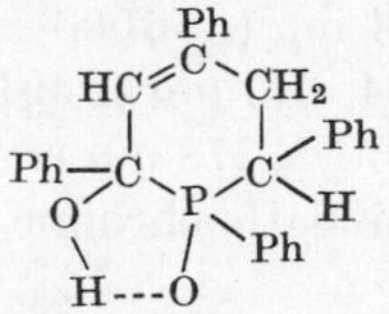   (61) 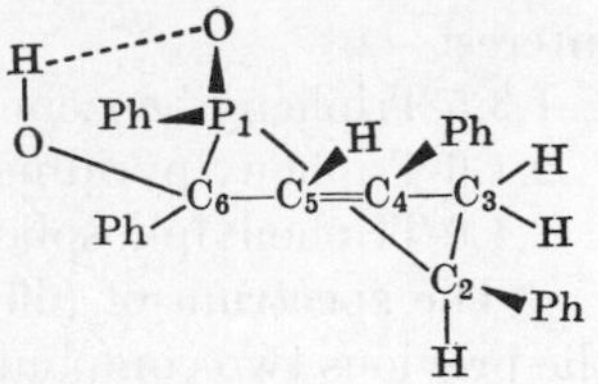   (61A)

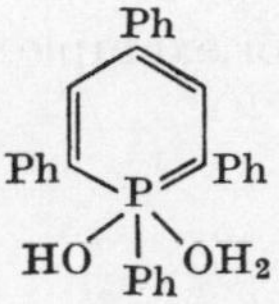

(62) 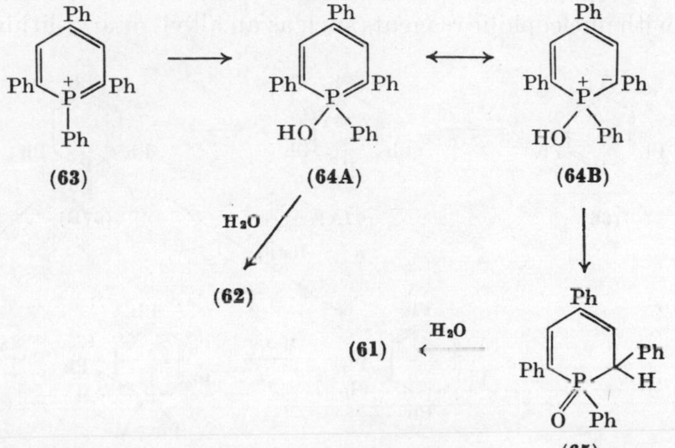   (63)   (64A)   (64B)

H_2O

(62)

(61) H_2O (65)

being an amorphous solid, m.p. 70–80°. A considerable (but not decisive) volume of evidence indicated that the first of these isomers had the hydrogen-bonded structure (61), shown in conformation as (61A), and that the second had the structure (62). It was suggested that the initial action of the phenylphosphine on the compound (60) was to give the cation (63), which united with an OH anion to give the compound of which (64A) and (64B) are contributory forms. The latter form (64B) could rearrange to (65), in which a proton and an OH ion could add on to the C-3 and C-6 atoms of the diene group, respectively, to give the first isomer (61) stabilized by the hydrogen bonding; the form (64A) would apparently unite directly with water to give the second isomer (62).

Märkl[179] has shown that a mixture of the compound (60) and tris(hydroxymethyl)phosphine, $(HOCH_2)_3P$, in 1:1.25 molar proportions, reacts in boiling pyridine to form the pale yellow crystalline 2,4,6-triphenylphosphorin (66), m.p. 172–173°; the molecular weight of this compound (which is stable to air) has been determined by mass spectrometry: the ^{31}P signal in the nmr spectrum at δ -178.2 ppm (δ H_3PO_4 as 0) occurs at an unexpectedly low field.

The ultraviolet spectra of three analogous compounds are of interest:

1,3,5-Triphenylbenzene, λ_{max} 254 mμ (ϵ 56 000)

2,4,6-Triphenylpyridine, λ_{max} 254, 317 mμ (ϵ 49 500, 9390)

2,4,6-Triphenylphosphorin (66), λ_{max} 278 mμ (ϵ 41 000) (in CH_3OH)

The spectrum of (66) shows a bathochromic shift compared with the previous two compounds.

The compound (66) does not undergo alkylation with methyl iodide or even with triethyloxonium tetrafluoroborate, but it reacts readily with nucleophilic reagents such as an alkyl- or aryl-lithium; for example,

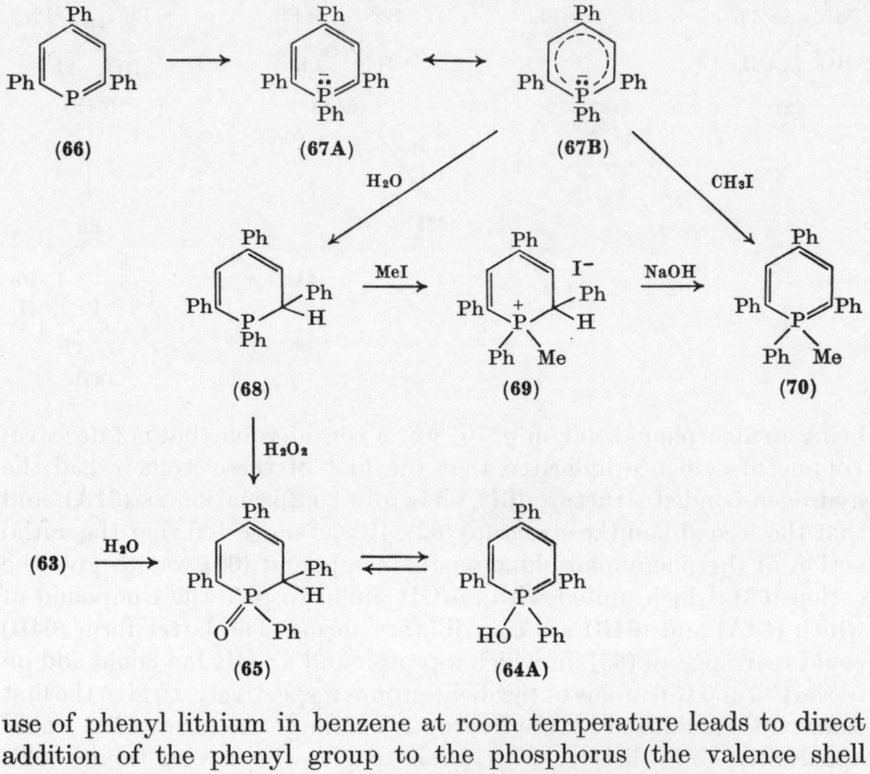

use of phenyl lithium in benzene at room temperature leads to direct addition of the phenyl group to the phosphorus (the valence shell apparently increasing to a decet) and the product, represented as the canonical forms (67A) and (67B), forms a deep blue solution. The addition of water gives 1,2-dihydro-1,2,4,6-tetraphenylphosphorin (68), m.p. 144–145°, λ_{max} 327 mμ, ϵ 7300 in benzene. This product readily undergoes quaternization, methyl iodide thus giving 1,2-dihydro-1-methyl-1,2,4,6-tetraphenylphosphorinium iodide (**69**), m.p. 160–162°. Treatment of this iodide in aqueous-ethanol with 2N-aqueous sodium hydroxide gives the red non-crystalline ylide, 1-methyl-1,2,4,6-tetraphenylphosphorin (**70**), which can also be obtained directly by the action of methyl iodide on the benzene solution of (**67A–67B**) (Märkl, Lieb, and Merz[180]).

It is noteworthy that oxidation of the phosphine (**68**) with hydrogen peroxide gives 1,2-dihydro-1,2,4,6-tetraphenylphosphorin 1-oxide (**65**), colorless crystals, m.p. 156–158°, λ_{max} 332 mμ (ϵ 6600) in benzene; this is the compound suggested as an intermediate in Price's work, and it can also be isolated by the action of water on the products of the action of phenylphosphine on the pyrylium salt (**60**), probably arising from the action of the water on the cation (**63**).

The ultraviolet spectrum in methanol shows that (**65**) is in tautomeric equilibrium with the form (**64A**). The addition of weak bases to this tautomeric mixture produces a brilliant red resonance-stabilized anion.[180]

Märkl, Lieb, and Merz[181] have developed an improved synthesis of aryl-substituted phosphorins, in which, for example, 2,4,6-triphenylpyrylium iodide (**71**) and tris(trimethylsilyl)phosphine (**72**) in acetonitrile are boiled under nitrogen for 20 hours.

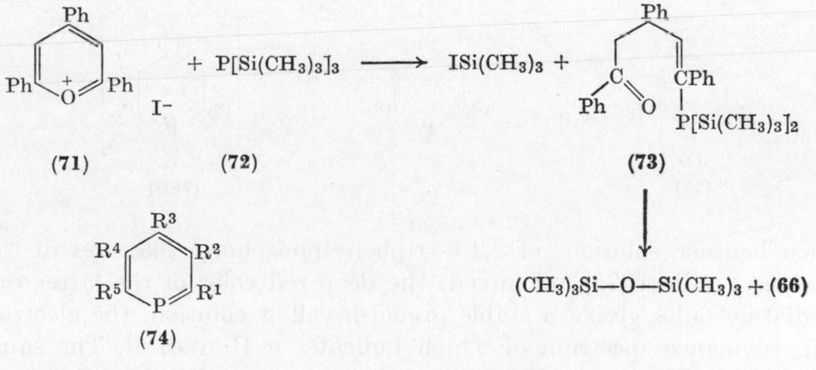

Ph
Ph
Ph
I⁻
+ P[Si(CH3)3]3
ISi(CH3)3 +
Ph
CPh
Ph
P[Si(CH3)3]2
(71)
(72)
(73)
(CH3)3Si—O—Si(CH3)3 + (66)
R3
R4
R2
R5
R1
(74)

It is suggested that two intermediates are thus formed, namely, iodotrimethylsilane and the phosphine (**73**), which in turn gives rise to 2,4,6-triphenylphosphorin (**66**) and hexamethyldisiloxane. The solvent is removed from the reaction mixture, and chromatographic separation of the residue gives the phosphorin (**66**). This synthetic method has the advantages that no basic solvent is required and no water is formed, and it is claimed that the yields in all cases are higher than those by the $P(CH_2OH)_3$ method.

The compounds thus isolated are given in Table 8, based on the general formula (**74**).

Table 8. Aryl-substituted phosphorins (Märkl[181])

$R^1 = R^5$	R^2	R^3	R^4	M.p. (°)	Yield (%)	λ_{max} (ϵ) in CH_3CH_2OH
C_6H_5	H	C_6H_5	H	172–173	45.0	278 (41,000)
C_6H_5	C_6H_5	C_6H_5	H	209–210	41.0	270 (28,000)
C_6H_5	C_6H_5	C_6H_5	C_6H_5	253–254	50.6	258 (32,700)
						283 (25,400)[a]
p-$CH_3C_6H_4$	H	C_6H_5	H	133–134	62.5	283 (40,700)
C_6H_5	H	p-$CH_3OC_6H_4$	H	106	35.2	283 (36,200)
p-$CH_3OC_6H_4$	H	C_6H_5	H	136–137	44.7	292 (40,200)
p-$CH_3OC_6H_4$	H	p-$CH_3OC_6H_4$	H	105–106	36.5	299 (46,600)

[a] Shoulder.

Dimroth and his coworkers have studied in some detail the properties[182] of the free-radical 2,4,6-triphenylphenoxyl (**75**) and certain deuterated derivatives,[183] its spin-density distribution,[184] and its reaction with certain aryl-substituted phosphorins.[185] They find that,

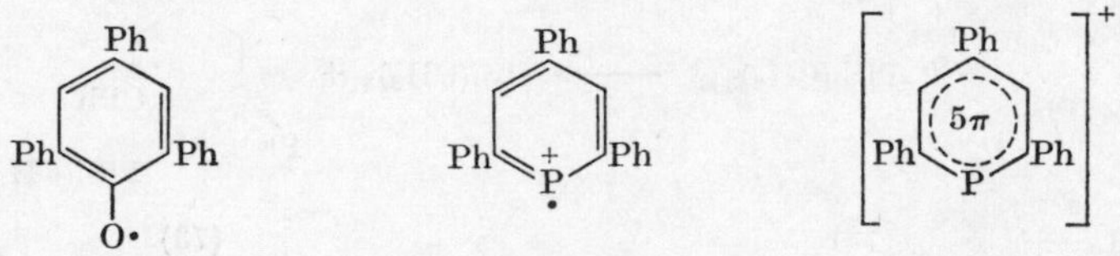

(**75**)   (**76B**)

when benzene solutions of 2,4,6-triphenylphosphorin (**66**) and of the phenoxyl radical (**75**) are mixed, the deep red color of the latter immediately fades giving a stable greenish-yellow solution, the electron spin resonance spectrum of which indicates a P· radical. The same spectrum is shown when (**66**) is treated with various deuteriated-phenyl analogues of (**75**) or with the ^{17}O derivative of (**75**); it is clearly independent of the particular oxidizing phenoxyl radical and must be due to the P radical cation shown as (**76A**) or (**76B**).

In extending the scope of this work,[186] they have used the $P[Si(CH_3)_3]_3$ method to prepare the substituted phosphorins shown in Table 9.

Table 9. Derivatives of the phosphorin (**74**) (Dimroth[186])

R^1	R^2	R^3	R^4	R^5	M.p. (°)	λ_{max} (nm) in n-hexane
C_6D_5	H	C_6D_5	H	C_6D_5	168–171	278
C_6D_5	H	C_6H_5	H	C_6D_5	167	278
C_6H_5	C_6H_5	C_6H_5	H	C_6H_5	188.5–189.5	271
C_6H_5	C_6H_5	C_6H_5	C_6H_5	C_6H_5	216–217	264
p-ClC_6H_4	H	C_6H_5	H	C_6H_5	166–167	281
p-ClC_6H_4	H	p-ClC_6H_4	H	p-ClC_6H_4	181–182	285
C_6H_5	H	p-$CH_3OC_6H_4$	H	C_6H_5	110.5–112	283
p-$CH_3OC_6H_4$	H	C_6H_5	H	C_6H_5	161.5–163	287
p-$CH_3OC_6H_4$	H	p-$CH_3OC_6H_4$	H	C_6H_5	134–136	291
p-$CH_3OC_6H_4$	H	C_6H_5	H	p-$CH_3OC_6H_4$	132–133	291
p-$CH_3C_6H_4$	H	C_6H_5	H	C_6H_5	155–156.5	281
p-$CH_3OC_6H_4$	H	p-$C_6H_5C_6H_4$	H	C_6H_5	148–150.5	293
α-$C_{10}H_7$	H	C_6H_5	H	C_6H_5	163–164	272
C_6H_5	H	$tert$-C_4H_9	H	C_6H_5	87.5–88.5	267
$tert$-C_4H_9	H	C_6H_5	H	$tert$-C_4H_9	104–105	288
$tert$-C_4H_9	H	p-$CH_3OC_6H_4$	H	$tert$-C_4H_9	116–116.5	300
$tert$-C_4H_9	H	$tert$-C_4H_9	H	$tert$-C_4H_9	88	263

In the course of this work, they synthesized 2,4,6-tri-*tert*-butyl-phosphorin (**74**; $R^2 = R^4 = H$, $R^1 = R^3 = R^5 = tert$-C_4H_9), m.p. 88° (Dimroth and Mach[187]), the first known phosphorin containing only alkyl substituents.

The phosphorins in Table 9 were oxidized by the phenoxyl radical (**75**) or by Pb(IV) acetate or benzoate, or by Hg(II) acetate.[186] The esr spectra of all the radicals so formed were very similar in type, showing a doublet due to the interaction of the free electron with the phosphorus; the P coupling constants a_P ranged between 21.5 and 27 gauss.

2,4,6-Triphenylphosphorin (**66**) also gives stable radicals by reduction.[188] Treatment of the phosphorin in tetrahydrofuran with potassium or potassium–sodium alloys gives three reduction stages:

(*a*) A solution with a green fluorescence: the esr spectrum shows a doublet, with a P coupling constant a_P 32.4. If this solution is treated with a benzene solution of an equimolecular amount of the radical (**76A**), the phosphorin (**66**) is re-formed.

(*b*) Further action of potassium on the above solution causes the production of a deep red color and the esr signal disappears.

(*c*) Further reaction with potassium gives a new esr signal.

If this final solution is treated (i) with 0.5 equivalent of the phos-phorin (**66**), the red solution without the signal is re-formed, (ii) with

2 equivalents of the phosphorin, the solution having a_P 32.4 gauss is reformed.

These three reduction stages are considered to indicate the stepwise addition of three electrons with the formation of a monoanion radical, a non-radical dianion, and a trianion radical, respectively.[188]

5-Phosphoniaspiro[4.5]decane Cation

This name applies to the spirocyclic cation (1), [RRI —], which is still unknown in the form of its salts; one derivative of the 1,2,3,4-tetradehydro cation (4) is however known. The cation (1) lies systematically between the similar cation consisting of two phospholane units

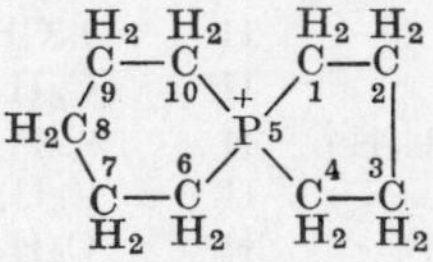

$$
\begin{array}{c}
\text{H}_2 \ \text{H}_2 \qquad \text{H}_2 \ \text{H}_2 \\
\text{C--C} \qquad \text{C--C} \\
{}_{/9}\quad {}_{10}\!\!\diagdown \ +\diagup{}_1\quad {}_2| \\
\text{H}_2\text{C}\,8 \qquad \text{P}\,5 \\
{}_{\diagdown 7}\quad {}_6\diagup \quad {}_{4}\quad {}_3| \\
\text{C--C} \qquad \text{C--C} \\
\text{H}_2 \ \text{H}_2 \qquad \text{H}_2 \ \text{H}_2
\end{array}
$$

(1)

with one phosphorus atom in common (p. 61), and that consisting of two phosphorinane units similarly oriented (see below).

Braye,[101] continuing his study of the cyclization of 1-(ω-bromoalkyl)phospholes, has found that the 1-potassio (or lithio) derivative of 2,3,4,5-tetraphenylphosphole (2), when added to a considerable excess

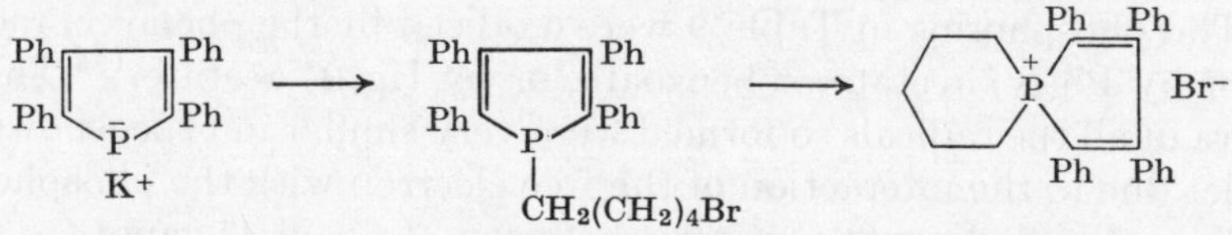

(2)   (3)   (4)

of 1,5-dibromopentane, gives 1-(5'-bromopentyl)-2,3,4,5-tetraphenylphosphole (3), m.p. 132–135° (51%), which when heated at 175–200° for 10 minutes undergoes cyclization to form 1,2,3,4-tetraphenyl-5-phosphoniaspiro[4.5]deca-1,3-diene bromide (4), m.p. 357° (31%), and much polymeric material (p. 20).

6-Phosphoniaspiro[5.5]undecane Cation

This name is given to the spirocyclic phosphonium cation (1), [RIS 8114], which consists of two phosphorinane rings with the phosphorus atom in common.

```
    H2  H2        H2  H2                    H  OH  H  OH
    C — C         C — C                 H2 \  /   \  / H2
   /10  11\+    /1   2\                    C — C  +  C — C
 H2C9      P6      3CH2              H2C        P        CH2    Cl⁻
   \8   7/    \5   4/                    C — C     C — C
    C — C         C — C                H2 /  \   /  \ H2
    H2  H2        H2  H2                    H  OH  H  OH
         (1)                                    (2)
```

The study of this system follows on from the work of Buckler and Wystrach[98, 99] (p. 61), who also showed that phosphine, when passed into a stirred mixture of tetrahydrofuran, concentrated hydrochloric acid and 25% aqueous glutaraldehyde at 20–30°, gave 6-phosphonia-spiro[5.5]undecane-1,5,7,11-tetraol chloride (2) (65%), m.p. 167–168° (dec.).

The stereochemistry of the cation (2) should also follow the general lines of that described for the lower homologue (p. 61), and again the *meso* form (*d*) is probably the only form free from steric hindrance by the OH groups. Only one form of (2) was in fact detected.

1*H*-Phosphorino[4,3-*b*]indole

The parent compound having the above name is represented formally as (1), [RIS 12507], and is unknown.

```
         H2
         C  P
   9    1  2                                                     
 8          0                  PPh                       PPh
 7       4                 N                          O
 6   5  N                  H
 
      (1)                      (2)                     (3)
```

1-Phenyl-4-phosphorinanone (3) (p. 106) gives an oily phenyl-hydrazone, which when boiled in ethanol previously half-saturated with hydrogen chloride gives the pale yellow 2,3,4,5-tetrahydro-2-phenyl-1*H*-phosphorino[4,3-*b*]indole (2), m.p. 113–114°; the structure of this compound is confirmed by a strong NH band at 3470 cm⁻¹ in its infrared spectrum (Gallagher and Mann[165]). The indole (2) with hot methyl iodide forms a gum which when heated to 120°/15 mm gives the pale yellow methiodide, m.p. 205–206°; quaternization has almost certainly occurred here at the phosphorus atom, and the salt should be named 2,3,4,5-tetrahydro-2-methyl-2-phenyl-1*H*-phosphorino[4,3-*b*]indolium iodide.

The structure of the compound (2) required verification, because the phenylhydrazones of some cyclic amino ketones, such as 1,2,3,4-tetrahydro-1-phenyl(or methyl)-4-quinolones,[189, 190, 191] when similarly

subjected to the Fischer indolization reaction, lose hydrogen to give
ψ-indoles (3H-indolenines) having the =N— structure; this process is
at present unknown in cyclic keto phosphines or keto arsines (cf. pp.
121, 133, 395, 411, 433).

It is noteworthy that the nitrogen and the arsenic analogues of
the phosphorinanone (3) give unstable phenylhydrazones which on
attempted indolization give only tars; in this reaction therefore the
phosphorus member differs from the other two members of the series.

The compound (2) was earlier named 1,2,3,4-tetrahydro-3-phenyl-
3-phospha-9-azafluorene.[165]

Phosphorino[4,3-b]quinoline

The above compound (1), [RIS 12701], is unknown and because of
the 'aromatic' nature of the phosphorus ring, is probably non-existent.
2-Substituted 1,2,3,4-tetrahydro derivatives have, however, been
prepared (Gallagher and Mann[165]).

A solution of 1-phenyl-4-phosphorinanone (3) (p. 106) and o-aminobenzaldehyde in ethanol containing a trace of sodium hydroxide, when
left at room temperature for 5 days and then saturated with hydrogen

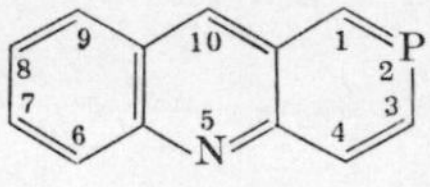

(1) 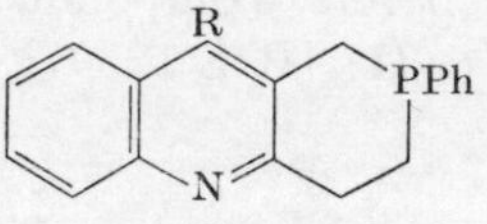   (2)

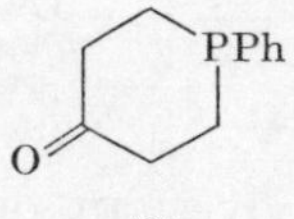

(3) 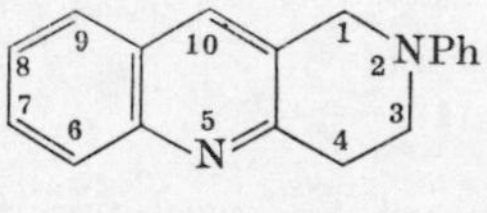   (4)

chloride, deposited the colorless dihydrated dihydrochloride of 1,2,3,4-
tetrahydro-2-phenylphosphorino[4,3-b]quinoline (2; R = H), m.p. 171–
172°, which on basification gave the parent compound (2; R = H),
m.p. 66.5–67°. The latter gave a monomethiodide, m.p. 212–213°,
quaternization occurring almost certainly on the phosphorus and not
the nitrogen atom.

Grinding the dihydrated dihydrochloride with water gave the
yellow dihydrated monohydrochloride, m.p. 150–151°, which when
heated at 60°/0.4 mm for 24 hours gave the anhydrous cream-colored
monohydrochloride, m.p. 178–179.5°; this reverted to the dihydrate
slowly on exposure to air and immediately on contact with water. This
difference in color between the dihydrochloride and the monohydro-

chloride is more pronounced in the corresponding salts of the analogous 1,2,3,4-tetrahydro-2-phenylbenzo[b][1,6]naphthyridine (4), [RRI 3473], where a stable yellow dihydrated dihydrochloride forms a stable red dihydrated monohydrochloride when recrystallized from water.

Essentially the same relationship holds, therefore, for the salts of the phosphorus and the nitrogen compounds, (2; R = H) and (4), respectively:

$$\text{Base} \xrightarrow{\text{Aq.HCl}} \text{Base,2HCl,2H}_2\text{O} \underset{\text{HCl}}{\overset{\text{H}_2\text{O}}{\rightleftharpoons}} \text{Base,HCl,2H}_2\text{O} \rightleftharpoons \text{Base,HCl}$$

The benzonaphthyridine (4) is similarly prepared by the condensation of o-aminobenzaldehyde and 1-phenyl-4-piperidone, and its structure has been confirmed by its nmr spectrum;[165] the three bases, (4), (2; R = H), and the analogous arsenic compound (similarly prepared, p. 398), have almost identical ultraviolet spectra and must have identical structures; only the members (4) and (2; R = H), however, have two basic groups. Monoprotonation of (4) or of (2; R = H) could not give a colored cation of the cyanine salt type, and diprotonation would suppress resonance even if it could exist. The ultraviolet spectra of the monohydrochlorides of (4) and (2; R = H) show that they cannot have structures markedly different from those of the parent bases, whereas the spectra of the dihydrochlorides have only a general resemblance to those of the parent bases.

The cause of the color of the hydrochlorides remains obscure; the interested reader should consult the fuller discussion.[165]

A solution of 1-phenyl-4-phosphorinanone (3) and isatin in aqueous ethanol containing sodium hydroxide, when boiled in nitrogen under reflux, gave a product which, on working up with final sublimation at 210–220°/0.05 mm, gave 1,2,3,4-tetrahydro-2-phenylphosphorino[4,3-b]-quinoline-10-carboxylic acid (2; R = CO₂H), m.p. 248–249°; it gave a benzylthiuronium salt, $C_{19}H_{16}NO_2P,C_8H_{10}N_2S$, which was dimorphic, having m.p. 138.5–139.5° and 212.5–213.5°. The infrared spectrum of the acid was closely similar to that of its arsenic analogue (p. 398) and of the 10-carboxylic acid of (4). The spectrum showed a broad band of low intensity centered at ca. 2000 cm⁻¹, attributable to strongly hydrogen-bonded OH groups, presumably arising from intermolecular interaction of the CO_2H groups and the quinoline-nitrogen atoms, and C=O absorption at 1625 cm⁻¹, this low value being influenced by the above bonding. No evidence of the CO_2^- ion, caused by zwitter-ion formation, could be detected.

The compounds (2; R = H) and (2; R = CO₂H) were earlier termed 1,2,3,4-tetrahydro-2-phenyl-10-aza-2-phospha-anthracene and its 9-

carboxylic acid, the exceptional numbering of anthracene being retained.[165]

Benzo Derivatives of Phosphorin. There are four compounds that can be regarded systematically as consisting of phosphorin fused to one or two benzo groups. They are phosphinoline or 1-benzophosphorin **(1)**

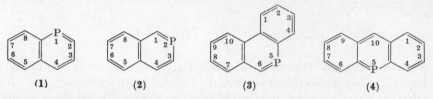

(1)   **(2)**   **(3)**   **(4)**

[RRI 1741]; isophosphinoline or 2-benzophosphorin **(2)** [RRI 1742]; dibenzo[*b,d*]phosphorin **(3)** [RIS 12714]; dibenzo[*b,e*] phosphorin **(4)** [RIS —]. Since most reactions of phosphinoline and isophosphinoline occur at the phosphorus atom, it is convenient in practice to invert the formulae **(1)** and **(2)**.

Phosphinoline and Tetrahydrophosphinolines

Phosphinoline **(1)**, [RRI 1741], is so named to indicate its relationship to quinoline. It is unknown, and almost all known compounds with this ring system are derivatives of 1,2,3,4-tetrahydrophosphinoline **(2)**.

(1)   **(2)**

The synthesis of 1-ethyl-1,2,3,4-tetrahydrophosphinoline (Beeby and Mann[192]), like those of 1-ethylphosphindoline (p. 65) and the spirocyclic salt (p. 131), is based essentially on the earlier synthesis of a 2-substituted 1,2,3,4-tetrahydroisophosphinoline (pp. 138, 139).

o-Bromobenzyl bromide **(3)** was converted under nitrogen into the *o*-bromobenzylmagnesium bromide, which was treated at room temperature with an excess of ethylene oxide in ether. Hydrolysis with dilute sulfuric acid, followed by fractional distillation of the ether residue, gave 3-(*o*-bromophenyl)-1-propanol (**4**; R = OH), b.p. 106–108°/0.5 mm (47%); 2,2'-dibromobibenzyl, [*o*-BrC$_6$H$_4$CH$_2$—]$_2$, b.p. 139–145°/0.5 mm, m.p. 84°, and 2,2'-di-(2'-hydroxyethyl)bibenzyl, [*o*-(HOCH$_2$CH$_2$)C$_6$H$_4$CH$_2$—]$_2$, m.p. 98–99°, were isolated from small higher fractions.

The propanol (**4**; R = OH) with phosphorus tribromide gave the 3-(*o*-bromophenyl)propyl bromide (**4**; R = Br), b.p. 84–85°/0.3 mm (83%), which with methanolic sodium methoxide formed 3-(*o*-bromophenyl)propyl methyl ether (**4**; R = OCH$_3$), b.p. 127–129°/15 mm (84%). The bromine atom in this compound has very low reactivity, but its solution in ether containing ethyl bromide (the entrainment method) reacted with activated magnesium powder giving a Grignard reagent which in turn reacted with diethylphosphinous chloride to give diethyl-[*o*-(3′-methoxypropyl)phenyl]phosphine (**5**), b.p. 155–158°/15 mm (73%).

A solution of (**5**) in hydrobromic acid–acetic acid, maintained at 120–130° for 3 hours and then evaporated, gave the hydrobromide (**6**). Basification liberated the phosphine, which in chloroform readily cyclized to 1,1-diethyl-1,2,3,4-tetrahydrophosphinolinium bromide (**7**; X = Br) (cf. p. 139); this was converted *via* the picrate into the pure chloride, which at 350–370°/20 mm decomposed smoothly to give 1-ethyl-1,2,3,4-tetrahydrophosphinoline (**8**), b.p. 141–143°/18 mm (78%). The initial distillate from the thermal decomposition contained the phosphine and its hydrochloride, indicating that the chloride (**7**; X = Cl) had given the phosphine, ethylene, and hydrogen chloride.

The phosphine (**8**) was characterized as its methiodide, m.p. 184–185°, and methopicrate, m.p. 121°, and as orange bis(phosphine)-dibromopalladium, m.p. 153.5–154°.

Märkl,[193] working later on similar lines, showed that [*o*-(3-methoxypropyl)phenyl]diphenylphosphine (**9**), treated as (**5**) above, gives the 1,2,3,4-tetrahydro-1,1-diphenylphosphinolinium cation (**11**; R = H), isolated as the tetrafluoroborate (X = BF$_4$), m.p. 193–195° (74%). Similarly benzyl-[*o*-(2′-bromoethyl)phenyl]diphenylphosphonium bromide (**10**), m.p. 225–227°, undergoes intramolecular *C*-alkylation to give the 1,2,3,4-tetrahydro-1,1,2-triphenylphosphinolinium cation (**11**; R = C$_6$H$_5$), also isolated as the tetrafluoroborate, m.p. 248–250° (68%).

These salts, when treated in chloroform with N-bromosuccinimide, give, respectively, the 4-bromo derivatives (**12**; R = H, X = BF$_4$), m.p. 192° (91%), and (**12**; R = C$_6$H$_5$, X = BF$_4$), m.p. 195–197° (86%).

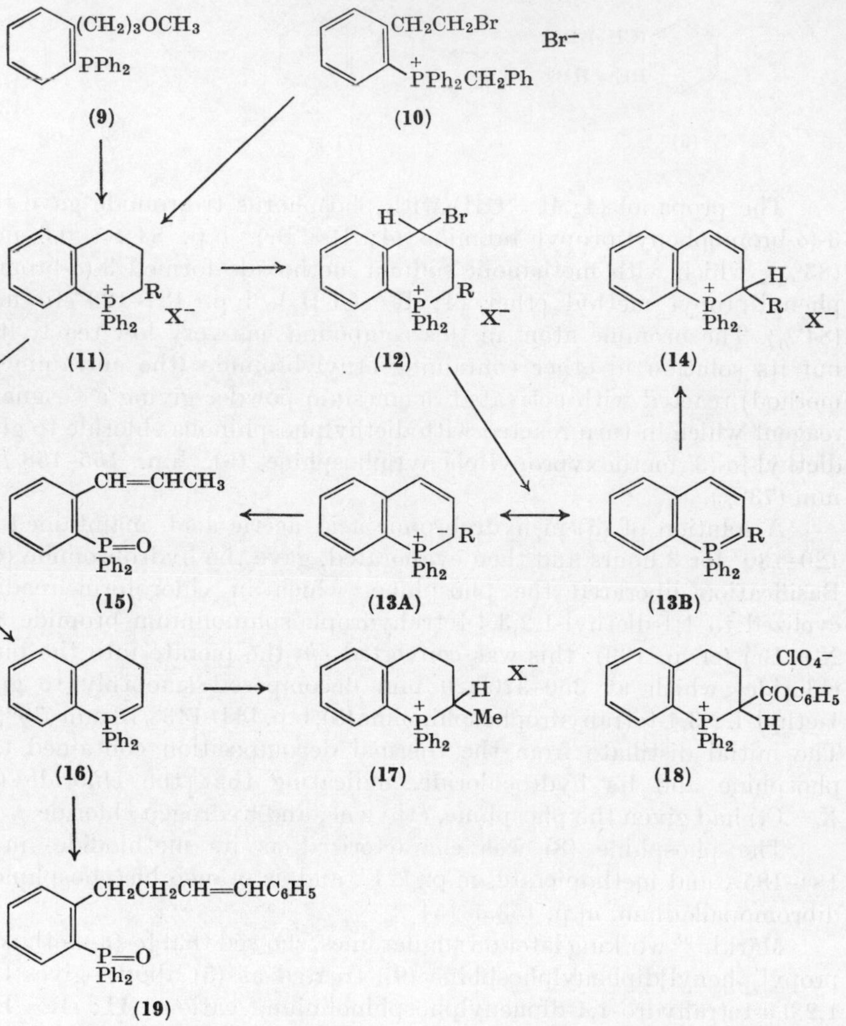

(CH$_2$)$_3$OCH$_3$
PPh$_2$

(**9**)

CH$_2$CH$_2$Br
$\overset{+}{\text{P}}$Ph$_2$CH$_2$Ph   Br$^-$

(**10**)

$\overset{+}{\text{P}}$Ph$_2$ R   X$^-$

(**11**)

H   Br
$\overset{+}{\text{P}}$Ph$_2$ R   X$^-$

(**12**)

H
$\overset{+}{\text{P}}$Ph$_2$ R   X$^-$

(**14**)

CH=CHCH$_3$
P=O
Ph$_2$

(**15**)

$\overset{+}{\text{P}}$Ph$_2$ $^-$ R

(**13A**)

$\overset{+}{\text{P}}$Ph$_2$ R

(**13B**)

$\overset{+}{\text{P}}$Ph$_2$

(**16**)

H
$\overset{+}{\text{P}}$Ph$_2$ Me   X$^-$

(**17**)

COC$_6$H$_5$
H   ClO$_4^-$
$\overset{+}{\text{P}}$Ph$_2$

(**18**)

CH$_2$CH$_2$CH=CHC$_6$H$_5$
P=O
Ph$_2$

(**19**)

These bromo salts in turn when treated, in the absence of water or oxygen, either with sodamide in liquid ammonia or with potassium *tert*-butoxide in dimethylformamide undergo dehydrobromination, to give, respectively, the yellow 1,1-diphenylphosphinoline, which has the two contributory structures (**13A** ↔ **13B**; R = H), and the orange-red

1,1,2-triphenylphosphinoline, having the two structures (**13A** ↔ **13B**; R = C_6H_5). The phosphorane (**13A** ↔ **13B**; R = H) forms a yellow amorphous powder, λ_{max} 362, 420 mμ in benzene, and (**13A** ↔ **13B**; R = C_6H_5) forms a similar orange-red powder, λ_{max} 387, 426, 488 mμ. Treatment of these compounds in solution with hydrogen chloride gives, respectively, 1,2-dihydro-1,1-diphenylphosphinolinium chloride, isolated as the tetrafluoroborate (**14**; R = H, X = BF_4), m.p. 220–221°, and 1,2-dihydro-1,1,2-triphenylphosphinolinium chloride, similarly isolated as (**14**; R = C_6H_5, X = BF_4), m.p. 178–180°.

The phosphorane (**13A** ↔ **13B**; R = H) is unexpectedly stable towards hydrolysis and requires hot water to convert it into (o-1-propenylphenyl)diphenylphosphine oxide (**15**), m.p. 140–142°; when exposed in methanolic solution to the air, however, it is readily oxidized to various red-violet methylenephosphoranes.

The bromide (**11**; R = H) reacts with *tert*-butoxide in dimethylformamide to give 3,4-dihydro-1,1-diphenylphosphinoline (**16**). This reacts as a normal ylide: with methyl iodide it forms the 1,2,3,4-tetrahydro-2-methyl-1,1-diphenylphosphinolinium cation (**17**), isolated as the perchlorate, m.p. 225–227°; it undergoes acylation, *e.g.*, to the corresponding 2-benzoyl cation (**18**) (perchlorate, m.p. 243–245°); and it reacts smoothly with benzaldehyde with ring opening to form the olefin-phosphine oxide (**19**).[194]

Märkl[193] suggests that, considering the properties of other comparable methylene-phosphoranes, the ready formation and stability towards water of the compounds (**13**) indicate an increase in resonance energy resulting from the attainment in the heterocyclic ring of cyclic conjugation involving the p_π–d_π double bond of the yline; however, the participation of the ylide form (**13A**) is small, and the formation of betaines with benzaldehyde or p-nitrobenzaldehyde with subsequent production of olefins therefore does not occur.

1,2,3,4-*Tetrahydro-1-phenyl-4-phosphinolone* (**20**). Few derivatives of 1,2,3,4-tetrahydrophosphinoline containing substituents in other than the 1- and 2-positions have been prepared; these may require special synthetic methods. Several routes to the 4-ketone (**20**), alternatively named 1,2,3,4-tetrahydro-4-oxo-1-phenylphosphinolane, were investigated by Gallagher, Kirby, and Mann,[195] who achieved the following synthesis.

o-Bromoaniline was converted into o-bromobenzenediazonium tetrafluoroborate (**21**), which when mixed with ethyl acetate and phenylphonous dichloride and a trace of cuprous bromide gave on gentle heating the usual vigorous Doak–Freedman reaction[132, 133]; the resulting solution was treated with magnesium below 50° and the ethyl acetate removed under reduced pressure (cf. Quin and Humphrey[196, 197]). The residue yielded o-bromophenylphenylphosphinous chloride (22), b.p. 145–146°/0.5 mm, which when reduced with lithium aluminum hydride afforded (o-bromophenyl)phenylphosphine (23), b.p. 128–130°/0.1 mm

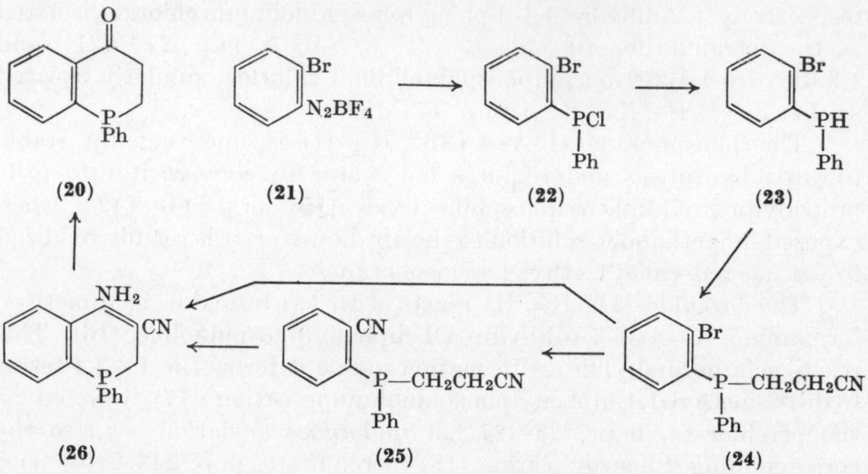

(82%). A mixture of this phosphine, acrylonitrile, and acetic acid, when boiled for $2\frac{1}{2}$ hours, gave o-bromophenyl-(2-cyanoethyl)phenylphosphine (24), b.p. 188–191°/0.03 mm, m.p. 61–62° (88%). It is noteworthy that the phosphine (23) and its o-chloro analogue give unusually stable secondary phosphine oxides, $R_2P(O)H$, on exposure to air.

Attempts to cyclize either the nitrile (24) or its apparent o-lithio derivative failed. A mixture of the nitrile with cuprous cyanide and dimethyl sulfoxide was therefore heated at 170°, then poured into water, and the precipitated material was digested with boiling aqueous potassium cyanide; extraction of the cyanide solution with benzene gave the crude 2-cyanoethyl-(o-cyanophenyl)phenylphosphine (25). This phosphine when treated in warm xylene with sodium tert-butoxide deposited a yellow sodium derivative, which with water yielded 4-amino-3-cyano-1,2-dihydro-1-phenylphosphinoline (26), m.p. 180–181° after sublimation.

When a more concentrated solution of the above mixture of the nitrile (24) and cuprous cyanide in dimethyl sulfoxide was heated, the benzene extract on distillation gave the 4-amino derivative (26) directly (28%), and the isolation of the intermediate (25) is therefore unnecessary.

The constitution of the amine (26) was confirmed by its infrared spectrum.

The hydrolysis of the amine with hydrochloric acid, when carried out under specific conditions, gave the colorless 1,2,3,4-tetrahydro-1-phenyl-4-phosphinolone (**20**), b.p. 143–145°/0.05 mm, m.p. 46–47° (55%), characterized by its 4-phenylsemicarbazone, m.p. 225–226°, and by its methiodide which gives the crystalline methopicrate, m.p. 153–154°. The molten phosphine when exposed as a thin layer to air rapidly forms the 1-oxide (**27**), m.p. 124–126°; a warm ethanolic solution of the phosphine when treated with perchloric acid deposited the colorless perchlorate (**28**), m.p. 179.5–181°.

(**27**)   (**28**)   (**29**)   (**30**)

The evidence that the phosphine has the true ketonic structure (**20**) and that the polar form (**29**) makes no detectable contribution to its structure, and furthermore that the perchlorate has the cation (**28**) and not the 4-hydroxy cation (**30**) can be summarized. (*a*) If the phosphine and perchlorate had the structures (**29**) and (**30**), respectively, they would almost undoubtedly be colored. (*b*) The infrared spectrum of the phosphine (**20**) shows a strong CO band at 1680 cm^{-1} (normal for aryl-conjugated ketones), whereas that of the 1-oxide shows this band at 1700 cm^{-1}, the change being apparently caused by the P=O group, which shows strong absorption at 1190 cm^{-1}; the spectrum of the perchlorate shows a CO band at 1687 cm^{-1} but no OH absorption. (*c*) The ultraviolet spectra of the phosphine (and its 1-oxide) in ethanol were unaffected by the addition of hydrochloric acid, whereas that of the phosphine would have been markedly affected by the formation of the cation (**30**). This spectrum of the phosphine is markedly different from that of 1,2,3,4-tetrahydro-1-methyl-4-quinolone, but has a general resemblance to that of 1,2,3,4-tetrahydro-7-methoxy-1-methyl-4-arsinolone (p. 404).

The 4-phosphinolone (**20**) does not condense with malonodinitrile, and in this respect again resembles its arsenic analogue and differs from the nitrogen analogue.

The 4-phosphinolone (**20**) gives an unstable phenylhydrazone which, however, readily undergoes the Fischer indolization reaction (p. 132); the phosphine also condenses with *o*-aminobenzaldehyde to give the quinolino derivative (p. 134). These two types of compound containing indole and quinoline rings, respectively, 'fused' to the phosphinoline system, are discussed separately on the pages indicated.

Earlier attempts to synthesize the ketophosphine (**20**) by the apparently simple method of cyclizing 3-(diphenylphosphino)propionic

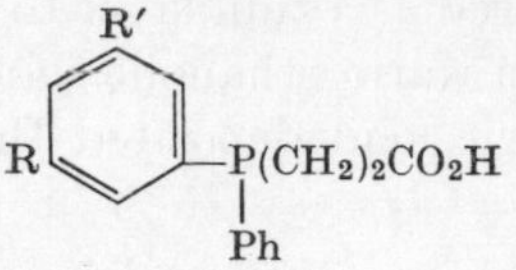

acid (**31**; R = R′ = H) had failed, although the corresponding diphenyl-aminopropionic acid readily gave 1,2,3,4-tetrahydro-1-phenyl-4-quinolone.[189] To investigate the effect of activating groups in appropriate positions in one of the benzene rings, the following acids were prepared: 3-[(m-methoxyphenyl)phenylphosphino]propionic acid (**31**; R = OCH₃, R′ = H); the 3-[(m-tolyl)phenylphosphino]propionic analogue (**31**; R = CH₃, R′ = H); and the 3-[(3′,5′-dimethylphenyl)phenylphosphino]-propionic analogue (**31**; R = R′ = CH₃). These attempts, in which a variety of reagents was used, failed, except that the last acid (**31**; R = R′ = CH₃) when heated with polyphosphoric acid gave a viscous syrup that formed a deliquescent methiodide, identified by conversion into the crystalline methopicrate, i.e., 1,2,3,4-tetrahydro-1,5,7-trimethyl-4-oxo-1-phenylphosphinolinium picrate (**32**; X = C₆H₂N₃O₇). The very low yield made this route of negligible practical value.[195]

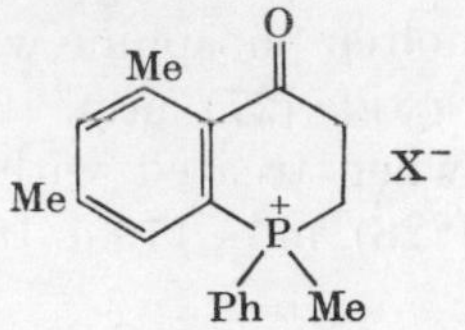

1,1′(2H,2′H)-Spirobiphosphinolinium Cation

One of the simplest derivatives of phosphinoline—simplest in the absence of substituents—would be a spirocyclic cation formed from two phosphinoline systems having one phosphorus atom in common. Such a cation, in its lowest degree of hydrogenation, is termed 1,1′(2H, 2′H)-spirobiphosphinolinium (**1**) [RRI 5242], numbered as shown. Salts

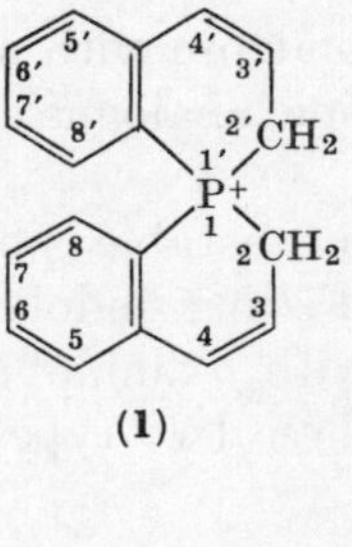

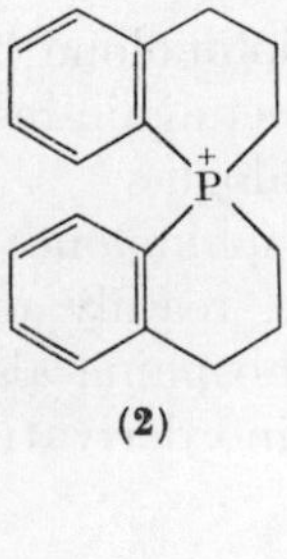

of this cation are unknown, but those of the 3,3′,4,4′-tetrahydro-1,1′-
(2H,2′H)-spirobiphosphinolinium cation (**2**) have been synthesized by
Hart and Mann,[198, 199] who gave the simpler name P-spiro-bis-1,2,3,4-
tetrahydrophosphinolinium to this cation.

The comparatively ready synthetic routes to comparable spirocyclic
arsonium salts (pp. 372, 407, 419) could not be applied to phosphonium
analogues, and salts of the cation (**2**) were ultimately prepared by the
following method.

3-(*o*-Bromophenyl)propyl methyl ether (**3**) (p. 125) was converted
by *n*-butyllithium into the *o*-lithio derivative, which reacted with
diethylphosphinous chloride, $(C_2H_5)_2PCl$, to give ethyldi[*o*-(3′-methoxy-
propyl)phenyl]phosphine (**4**; R = OCH_3), b.p. 215–218°/2 mm (65%):

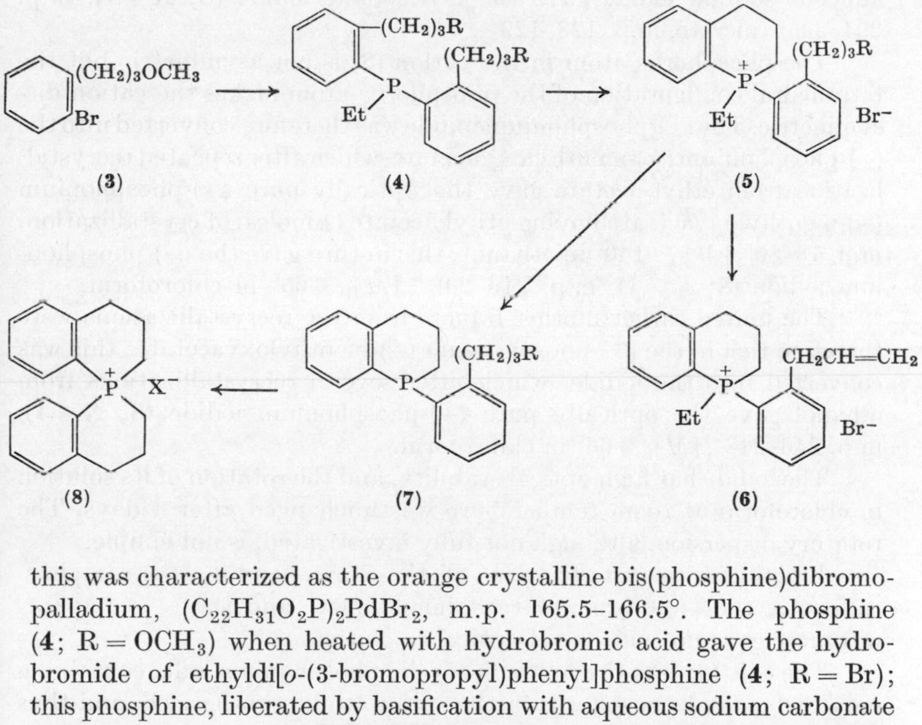

(**3**)   (**4**)   (**5**)

(**8**)   (**7**)   (**6**)

this was characterized as the orange crystalline bis(phosphine)dibromo-
palladium, $(C_{22}H_{31}O_2P)_2PdBr_2$, m.p. 165.5–166.5°. The phosphine
(**4**; R = OCH_3) when heated with hydrobromic acid gave the hydrobromide of ethyldi[*o*-(3-bromopropyl)phenyl]phosphine (**4**; R = Br);
this phosphine, liberated by basification with aqueous sodium carbonate
and cyclized in hot chloroform, furnished 1-[*o*-(3′-bromopropyl)phenyl]-
1-ethyl-1,2,3,4-tetrahydrophosphinolinium bromide (**5**; R = Br), m.p.
110–112° (81%) (picrate, m.p. 83–85°).

An attempt to break off the ethyl group by heating this bromide
and thus obtain a second cyclization direct to the bromide (**8**; X = Br)
failed, for the 3-bromopropyl group lost hydrogen bromide giving

1-(o-allylphenyl)-1-ethyl-1,2,3,4-tetrahydrophosphinolinium bromide (**6**), characterized as the picrate, m.p. 130–131°.

The bromide (**5**; R = Br) was then converted by methanolic sodium methoxide into the corresponding 1-o-(3′-methoxypropyl) analogue (**5**; R = OCH$_3$), m.p. 144–146° (picrate, m.p. 84–86°), which when heated under nitrogen at 250–320°/1 mm lost ethylene and hydrogen bromide, giving a viscous distillate of 1,2,3,4-tetrahydro-1-[o-(3′-methoxypropyl)phenyl]phosphinoline (**7**; R = OCH$_3$) and its hydrobromide. This crude product, heated with 48% hydrobromic acid, gave the water-soluble hydrobromide of the 1-[o-(3′-bromopropyl)phenyl] analogue (**7**; R = Br). Basification as before, with cyclization of the liberated phosphine in chloroform gave the 3,3′,4,4′-tetrahydro-1,1′-(2H,2′H)-spirobiphosphinolinium bromide (**8**; X = Br), which with aqueous sodium iodide gave the less soluble iodide (**8**; X = I), m.p. 294–295° (picrate, m.p. 128–129°).

The phosphorus atom in the cation (**8**) is not asymmetric but the tetrahedral configuration of the phosphorus atom makes the cation dissymmetric. The (±)-phosphonium iodide was therefore converted into the (±)-phosphonium (−)-menthyloxyacetate, which after repeated recrystallization from ethyl acetate gave the optically pure (−)-phosphonium (−)-menthyloxyacetate having ethyl acetate (4 moles) of crystallization, m.p. 78–80°, $[M]_D$ −140° in ethanol; this in turn gave the (−)-phosphonium iodide (**8**; X = I), m.p. 246–248°, $[M]_D$ −65° in chloroform.

The united earlier mother-liquors from the recrystallizations were therefore rich in the (+)-phosphonium (−)-menthyloxyacetate; this was converted into the iodide, which after several recrystallizations from ethanol gave the optically pure (+)-phosphonium iodide (**8**; X = I), m.p. 246–248°, $[M]_D$ +66° in chloroform.

The iodide has high optical stability, and the rotation of its solution in chloroform at room temperature was unchanged after 4 days. The rotatory dispersion, although not fully investigated, is not simple.

Fractional recrystallization of the corresponding (+)-camphorsulfonate, the (+)-3-bromo-8-camphorsulfonate and the (+)-camphornitronate gave no evidence of optical resolution.

The salt (**8**) was the first spirocyclic phosphonium salt, containing only carbon and phosphorus in the cation, to be synthesized, and thus also the first to be isolated in optically active forms.

5H-Phosphinolino[4,3-b]indole

A solution of the phenylhydrazone of 1,2,3,4-tetrahydro-1-phenyl-4-phosphinolone (**1**) (p. 129) in boiling ethanolic hydrogen chloride undergoes the Fischer indolization reaction, giving the compound (2), originally termed 1,2-dihydro-1-phenylindolo(3′,2′:3,4)phosphinoline (Gallagher and Mann [200]). The parent ring system in (2) is now termed

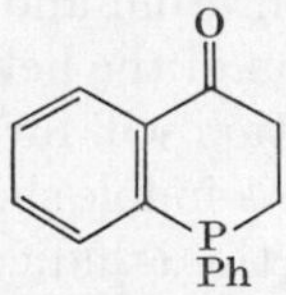

(1) 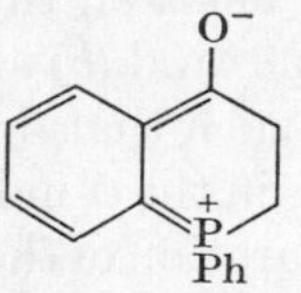   (1A) 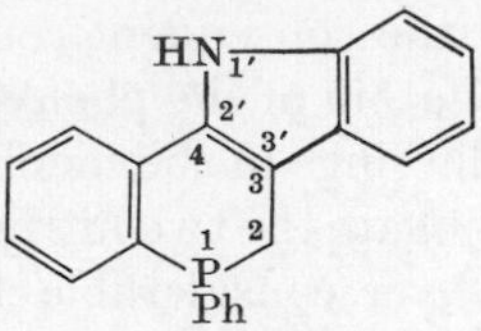   (2)

5H-phosphinolino[4,3-b]indole and numbered as in (3), [RIS 13145], and the compound (2) is 5,6-dihydro-5-phenyl-5H-phosphinolino[4,3-b]-indole.

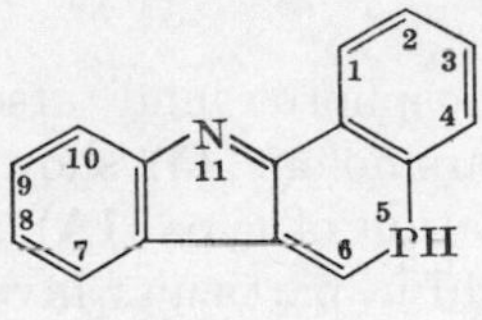

(3) 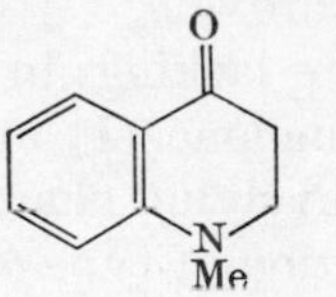   (4) 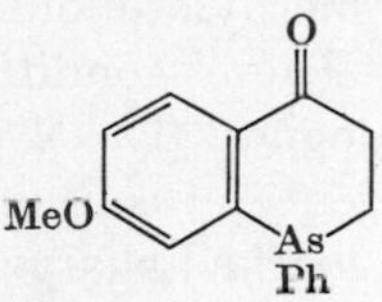   (5)

The structure of the compound (2) is beyond doubt: its infrared spectrum shows a sharp :NH band at 3420 cm^{-1}, indicating a true indole, and the compound forms a methiodide in which quaternization must therefore have occurred normally on the tertiary phosphorus atom. This structure is of some theoretical importance, because the phosphinolone (1) occupies an intermediate position between 1,2,3,4-tetrahydro-1-methyl-4-quinolone (4) and 1,2,3,4-tetrahydro-7-methoxy-1-methyl-4-arsinolone (5) (p. 403). The phenylhydrazone of the quinolone (4)

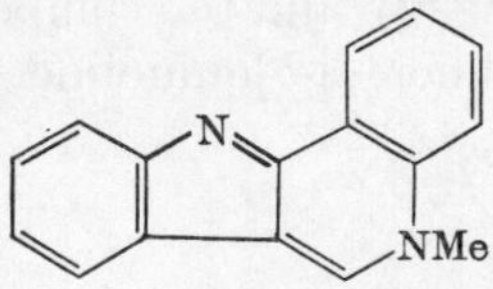

(6) 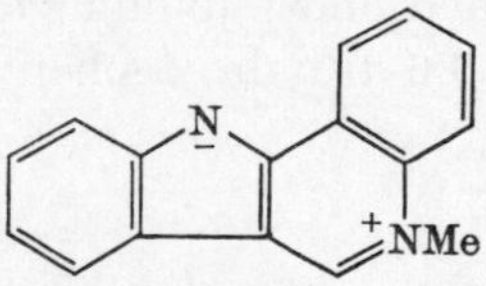   (6A)

when similarly subjected to the Fischer indolization loses two atoms of hydrogen with the formation of the ψ-indole (or 3H-indolenine) (6),[191] whereas the phenylhydrazone of the 4-arsinolone (5) gives the normal arsinolino[4,3-b]indole analogous to (2) (p. 410).

A number of cyclic keto amines give polycyclic ψ-indoles similar to (6), and it has been suggested (Braunholtz and Mann[191]) that the formation of a ψ-indole is dependent on (a) an initial conjugation between, for example, the CO group and the 1-nitrogen of the 4-quinolone (4) and the consequent conjugation between this 1-nitrogen atom and the =N-atom of the phenylhydrazone, and (b) an orientation of the heterocyclic rings in the indole system such that dehydrogenation will further conjugate the two nitrogen atoms in the ψ-indole (6). The ψ-indole shows, however, considerable charge separation of the type (6A); this is indicated by the yellow color of the ψ-indole, and by the avidity with which the ψ-indole unites with hydrogen chloride or methyl iodide to give salts in which the proton and the methyl group respectively have united with the indole-N atom, the positive charge remaining on the quinoline-N atom; in some cases the ψ-indole will combine firmly with water, the proton again uniting with the indole-N atom with the production of the methylquinolinium hydroxide.

These conditions are lacking in the phosphorus and arsenic analogues. The 4-phosphinolone (1) and the arsinolone (5) show no evidence of conjugation involving charge separation of type (1A), and the marked charge separation of type (6A) would in particular involve an 'internal' positive charge on the phosphorus or arsenic atom which they very rarely show.

Phosphinolino[4,3-b]quinolines

1,2,3,4-Tetrahydro-1-phenyl-4-phosphinolone (1) when condensed with alkaline o-aminobenzaldehyde by the Friedländer reaction gives the compound (2), m.p. 125–127°, which was originally termed 1,2-dihydro-1-phenylquinolino(3′,2′:3,4)phosphinoline (Gallagher and Mann[200]).

The parent system, [RIS 13327], presented and numbered as in (3), is now called phosphinolino[4,3-b]quinoline (3), and the compound (2) is therefore 5,6-dihydro-5-phenylphosphinolino[4,3-b]quinoline.

(1)   (2)   (3)

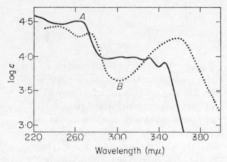

(4)   (4A)

The particular interest of the colorless compound (2) is that it forms a bright yellow, very stable monohydrochloride (4); the color of this salt indicates that the form (4A) must be making a small contribution to the structure, although the positively charged phosphorus atom does not thereby become a member of an 'aromatic-type' ring. The fact that protonation on the nitrogen atom of (2) has involved a change in structure is clearly shown by the different ultraviolet spectra (Figure 2)

Figure 2. Ultraviolet spectra of (*A*) the phosphinolinoquinoline (2) (2.242 mg in 100 ml of ethanol), and (*B*) the same solution diluted with 2% of concentrated hydrochloric acid (Reproduced, by permission, from M. J. Gallagher and F. G. Mann, *J. Chem. Soc.*, **1963**, 4855)

of the compound (2) in ethanol and in ethanol containing hydrochloric acid. These conclusions are confirmed by the facts that (*a*) if protonation of (2) occurred on the phosphorus atom no change in structure could have occurred, and (*b*) the phosphine oxide, *i.e.*, 5,6-dihydro-5-phenyl-phosphinolino[4,3-*b*]quinoline 5-oxide (5), also forms an almost colorless but stable hydrochloride; the ultraviolet spectra (Figure 3) of (5) also in ethanol and in ethanol–hydrochloric acid are closely similar to one

(5)

another and to that of the base (**2**), and show that no significant change
in structure has occurred on protonation; clearly a canonical form of
type (**4A**) could not be formed from the phosphine oxide. Furthermore,

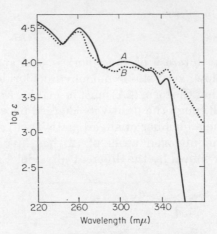

Figure 3. Ultraviolet spectra of (*A*)
the phosphinolinoquinoline oxide (**5**)
(2.394 mg in 100 ml of ethanol), and
(*B*) the same solution diluted with 2%
of concentrated hydrochloric acid.
(Reproduced, by permission, from
M. J. Gallagher and F. G. Mann,
J. Chem. Soc., **1963**, 4855)

the spectral changes which occur on salt formation of the base (**2**) are
not simply those observed when simple quinolines are protonated.

The small contribution made by the form (**4A**) to the structure of
the hydrochloride (**4**) is also indicated by the fact that the quinolinoquinoline analogue (**6**; R = C$_6$H$_5$) of the compound (**2**) is colorless,
whereas its hydrochloride, analogous to (**4** ↔ **4A**), is deep red.

The quinolinoquinoline (**6**; R = CH$_3$) also gives a deep red hydrochloride, which when boiled with hydrochloric acid undergoes an allylic
rearrangement to the yellow hydrochloride of the 5,7-dihydro isomer
(**7**; R = CH$_3$); the two bases, m.p. 98–100° and 155°, respectively, have
been isolated, and both in cold benzene solution undergo a rapid
atmospheric oxidation to the 6-oxo compound (**8**) (Braunholtz and
Mann[191]). The phosphinolinoquinoline hydrochloride (**4** ↔ **4A**) was
unaffected by boiling hydrochloric acid, and the crystalline base (**2**) was
stable to the air; in these respects the compound (**2**) behaves like its
arsinolino analogue (p. 412). In view, however, of the probable mechanism
of the above allylic rearrangement,[191] it is possible that the *P*-methyl

analogue of (2), in which the tertiary phosphine would have maximum basic properties, might also show this rearrangement.

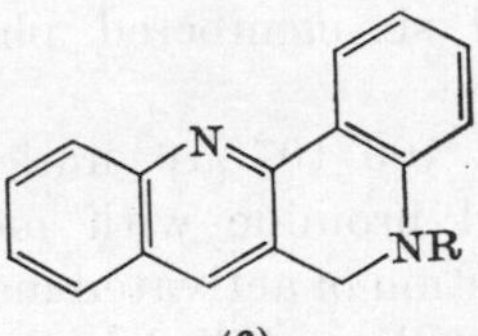

(6)

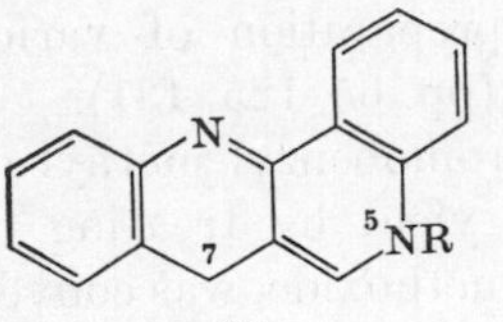

(7)

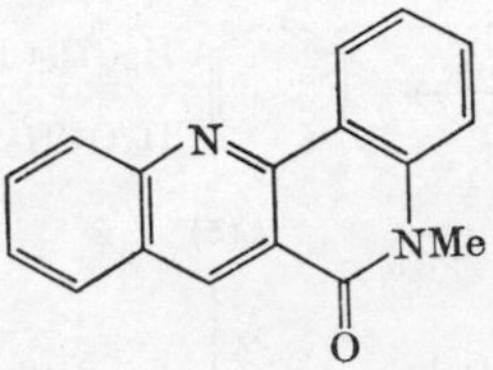

(8)

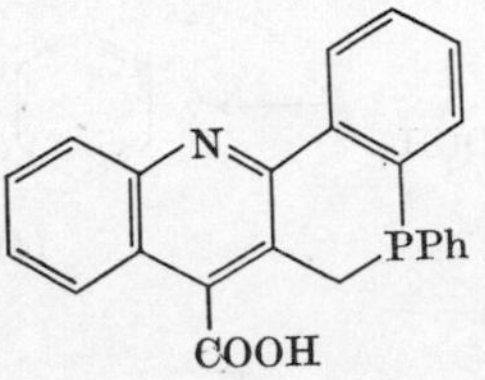

(9)

A suspension of the 4-phosphinolone (1) and of isatin in aqueous potassium hydroxide, when boiled under nitrogen for 48 hours, gave in small yield the crude 5,6-dihydro-5-phenylphosphinolino[4,3-b]quinoline-7-carboxylic acid (9), which on sublimation at 220–240°/0.0005 mm gave the compound (2) and some crystals which, on the basis of analysis and infrared spectra, were the still impure acid (9). It is not known if this acid exists as the covalent form (9), or as a zwitterion similar to that of its arsinolino[4,3-b]quinoline analogue (p. 413).

Isophosphinoline and Tetrahydroisophosphinolines

Isophosphinoline (1), [RRI 1742], is unknown and the recorded compounds having this ring system are mainly derivatives of 1,2,3,4-tetrahydroisophosphinoline (2).

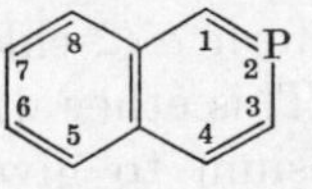

(1)

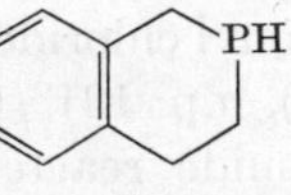

(2)

The tetrahydroisophosphinoline system was first synthesized (Holliman and Mann[201, 202]) in a stereochemical investigation of quaternary phosphonium salts, in which it was hoped to obtain maximum stability of the phosphonium cation by having the phosphorus

6

atom in a suitable stable ring system, with its remaining valencies
linked to two unlike aryl groups.

The following general method of synthesis has since been applied
to the preparation of various five- and six-membered phosphorus
systems (pp. 65, 125, 131).

o-Bromobenzyl methyl ether (3), b.p. 106–107°/16 mm, prepared
in 90% yield by treating o-bromobenzyl bromide with methanolic
sodium methoxide, was converted by the action of activated magnesium
in ether into the Grignard reagent, which when chilled in an ice–salt

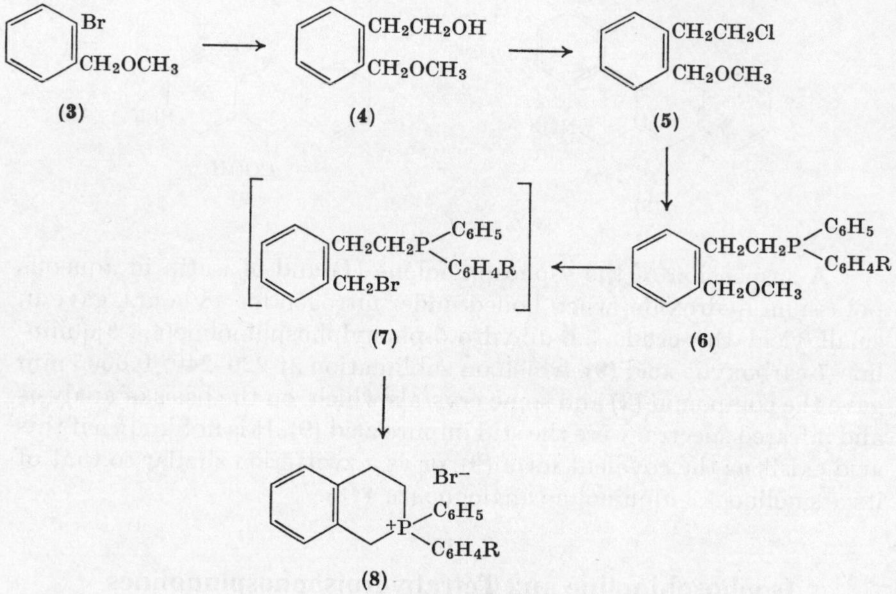

mixture and treated with ethereal ethylene oxide gave o-(2′-hydroxy-
ethyl)benzyl methyl ether (4), b.p. 152–154°/14 mm (53%); this ether
was characterized by the action of phenyl isocyanate, giving (o-methoxy-
methyl)phenethyl N-phenylcarbamate, m.p. 64–65°. The ether (4)
reacted with thionyl chloride in pyridine to form o-(2-chloroethyl)benzyl
methyl ether (5), b.p. 131°/14 mm (72%). This ether when mixed with
some ethyl bromide reacted with magnesium to give the Grignard
reagent, which in turn reacted with (p-bromophenyl)phenylphos-
phinous chloride, $(BrC_6H_4)C_6H_5PCl$, to give p-bromophenyl-[o-(methoxymethyl)phenethyl]phenylphosphine (6; R = Br), a viscous oil,
b.p. 214–216°/0.1 mm (59%) (methiodide, m.p. 167–168°).

A solution of the phosphine (6; R = Br) in acetic acid–hydrobromic
acid was heated to 120° for 2 hours whilst a stream of hydrogen bromide

was passed through it to maintain the concentration of this acid and to remove methyl bromide. The methoxymethyl phosphine (6) was thus converted into the bromomethyl phosphine (7) which then underwent cyclization to form 2-*p*-bromophenyl-1,2,3,4-tetrahydro-2-phenyliso-phosphinolinium bromide (8; R = Br), m.p. 137–149° with resolidification and melting at 218–221° (picrate, m.p. 186–187°). Repeated recrystallization of the pure bromide did not change its behavior on heating.

The Grignard reagent from the chloride (5), when similarly treated with (*p*-methoxyphenyl)phenylphosphinous chloride, gave [*o*-(methoxy-methyl)phenethyl](*p*-methoxyphenyl)phenylphosphine (6; R = OCH₃), b.p. 208°/0.05 mm. When this phosphine was treated with acetic acid–hydrobromic acid (as above), but with 5 hours' heating, the same changes occurred, but the *p*-methoxyl group was also demethylated; the product was therefore 1,2,3,4-tetrahydro-2-(*p*-hydroxyphenyl)-2-phenylisophos-phinolinium bromide (8; R = OH), m.p. 287–287.5°. The presence of the hydroxyl group was confirmed by treating the bromide with acetic anhydride in pyridine, giving the deliquescent 2-*p*-acetoxyphenyl analogue (8; R = OCOCH₃), m.p. 100–103°.

The above synthesis could be readily modified to obtain a 2-substituted tetrahydroisophosphinoline (Beeby and Mann[192]). The Grignard reagent from the chloride (5) reacted with diethylphosphinous chloride to give diethyl-*o*-(methoxymethyl)phenethylphosphine (9), b.p. 117–118°/0.2 mm (74%). This phosphine, when treated with acetic

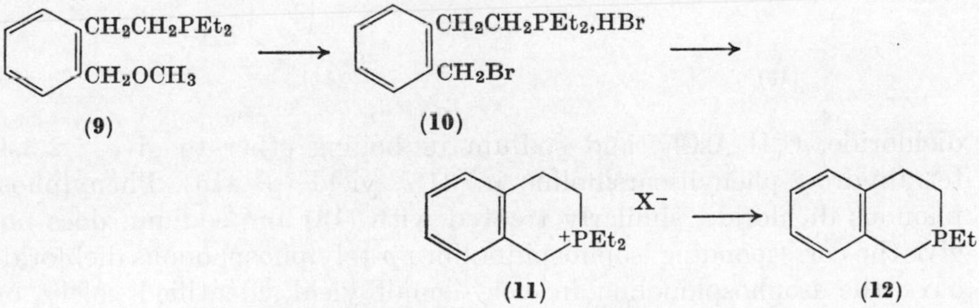

(9)   (10)

(11)   (12)

acid–hydrobromic acid (as before), gave the water-soluble [*o*-(bromo-methyl)phenethyl]diethylphosphine hydrobromide (10); the cold concentrated solution when basified with sodium carbonate liberated the free phosphine which immediately cyclized to form 2,2-diethyl-1,2,3,4-tetrahydroisophosphinolinium bromide (11; X = Br). This salt was converted *via* the recrystallized picrate, m.p. 98–98.5°, into the pure chloride, which when heated under nitrogen to 350–370°/20 mm decomposed smoothly to give a distillate that rapidly solidified, being apparently the pure 2-ethyl-tetrahydroisophosphinoline hydrochloride. Basification, extraction, and distillation gave the pure 2-ethyl-1,2,3,4-tetrahydroisophosphinoline (12), b.p. 129–132°/15 mm. It was characterized as the ethiodide, m.p. 93–94°, and as the deep orange crystalline bis(phosphine)dibromopalladium, m.p. 165.5–166°.

The isophosphinoline (12), is, as expected, more strongly basic and more reactive than the isomeric 1-ethyl-1,2,3,4-tetrahydrophosphinoline (p. 125). It is noteworthy that the quaternary chloride (11; X = Cl) on thermal decomposition gave solely the isophosphinoline hydrochloride (by recombination of the phosphine and the acid), whereas the corresponding phosphinolinium chloride gave a mixture of the free phosphine and its hydrochloride; furthermore, the isophosphinoline (12) is markedly more rapidly oxidized in air than the phosphinoline. The fact that, when liberated from its hydrobromide (10), the phosphine immediately cyclized to the bromide (11; X = Br), whereas the isomeric phosphine in the phosphinoline synthesis, when similarly liberated, formed an oil that underwent rapid cyclization only when dissolved in chloroform, is another aspect of this difference.

Two other attempted synthetic routes may be recorded. o-(2-Bromoethyl)benzyl bromide (13) condenses with phenylarsonous

(13)   (14)

dichloride, $C_6H_5AsCl_2$, and sodium in boiling ether to give 1,2,3,4-tetrahydro-2-phenylisoarsinoline in 31% yield (p. 415). Phenylphosphonous dichloride, similarly treated with (13) and sodium, does not give the corresponding isophosphinoline; p-tolylphosphonous dichloride gave the isophosphinoline in very small yield, identified solely by quaternization with p-chlorophenacyl bromide to form 2-p-chlorophenacyl-1,2,3,4-tetrahydro-2-p-tolylisophosphinolinium bromide (14), m.p. 227–230°. This difference between the phosphorus and the arsenic compounds may result from two factors: (a) the above Michaelis-type reaction with sodium proceeds far more slowly with phosphorus halides than with arsenic halides; for example, a mixture of bromobenzene, phosphorus trichloride, and sodium in ether, even after 48 hours boiling, gives only a small yield of triphenylphosphine, whereas the analogous reaction with arsenic trichloride proceeds rapidly to give triphenylarsine in high yield (Michaelis and Gleichman [203]; Michaelis and Reese [204]); (b) quaternary salt formation from tertiary phosphines is considerably faster than that from corresponding tertiary arsines; for example, under identical conditions the reaction of diethylphenylphosphine with ethyl iodide in acetone is over 90% complete in 60 hours, whereas that with diethylphenylarsine is only 20% complete in this time (Davies and Lewis [205]). It is possible, therefore, that the reaction of (13) with sodium and an aryldichlorophosphine is comparatively slow, and that the product undergoes rapid quaternization with unchanged (13), possibly giving various complex products.

In the second route, ethylphenyl-p-tolylphosphine was quaternized with one equivalent of the dibromide (13) to give a crystalline salt, m.p. 146–148°, which was almost certainly [o-(2-bromoethyl)benzyl]-ethylphenyl-p-tolylphosphonium bromide (15; R = C_2H_5) and not the

$$C_6H_5 \quad + \quad CH_2C_6H_4(CH_2)_2Br$$
$$CH_3C_6H_4 \overset{P}{\diagup} \diagdown R \qquad Br^-$$
(15)

$$C_6H_5 \quad + \quad (CH_2)_2C_6H_4CH_2Br$$
$$CH_3C_6H_4 \overset{P}{\diagup} \diagdown R \qquad Br^-$$
(16)

$$\overset{+}{P} \diagup^{C_6H_5} \diagdown_{C_6H_4CH_3} \qquad Br^-$$
(17)

isomeric bromide (16). The possibility that this bromide on thermal decomposition would lose ethyl bromide or ethylene and hydrogen bromide with subsequent cyclization to the bromide (17) could not be experimentally realized; similar failure attended the use of the methyl analogue (15; R = CH_3), m.p. 185–186°.

Stereochemistry. The cations in the salts (8; R = Br or OH) contain an asymmetric phosphorus atom and therefore should be resolvable into optically active forms. The bromide (8; R = Br) was therefore converted into the corresponding (+)-camphorsulfonate, the (+)-3-bromo-8-camphorsulfonate, the (+)-hydrogen tartrate, and the (−)-N-(α-methylbenzyl)phthalamate. The last three of these salts were syrups or glasses, and only the camphorsulfonate was obtained crystalline, but fractional recrystallization gave no evidence of resolution.

The bromide (8; R = OH) was also converted into a crystalline (+)-camphorsulfonate, initially having m.p. 153–158° and $[M]_D$ +102° in ethanolic solution; it was repeatedly recrystallized from ethanol–ethyl acetate until it had m.p. 174–175° and $[M]_D$ +113.5° in ethanol, these

values being unchanged by further crystallization. Ethanolic solutions
of this salt and of calcium bromide, when mixed at 0°, rapidly deposited
the bromide (8; R = OH), m.p. 268–270°, $[M]_D$ +32.9° in aqueous
ethanol (1:2 v/v). A solution of this bromide in methanol had $[M]_D$
+4.8°; this low value was not caused by racemization, for the bromide
recovered by evaporation of the methanol still had $[M]_D$ +32.9° in
ethanol.

The rotation of the (+)-bromide was determined over the range of
wave-lengths λ 6104–4358; these values indicated that the cation had
simple dispersion, although this range is too short to provide decisive
evidence.

The salt (8; R = OH) was the first quaternary phosphonium salt
to be resolved into optically active forms (Holliman and Mann[202]).
Hitherto, the failure to resolve such salts had been attributed to the
possibility of the formation in solution of a 'dissociation equilibrium',
$[PR^1R^2R^3R^4]X \rightleftharpoons PR^1R^2R^3 + R^4X$, where X was a halogen ion and
R^4 an alkyl group. There was in fact little evidence for the occurrence
of such dissociation in phosphonium salts, although there was significant
evidence for it in analogous quaternary ammonium and arsonium salts
(for details, see p. 417).

Much more recent work by Horner et al.[206] has shown, however,
that quaternary phosphonium salts having alkyl groups joined to the
phosphorus atom can be resolved into optically stable forms, and that
these can furthermore be reduced to tertiary phosphines possessing
quite reasonable optical stability.

Dibenzo[*b,e*]phosphorin

The parent compound in this series, dibenzo[*b,e*]phosphorin (1),
[RIS —], has recently been prepared but not isolated (de Koe and

(1)   (2)

Bickelhaupt[207]). Our previous knowledge of this ring system was limited
to 5-substituted derivatives of 5,10-dihydrodibenzo[*b,e*]phosphorin (2).

Doak, Freedman, and Levy[138] first converted *o*-aminodiphenyl-
methane (3; R = NH₂) into the diazonium fluoroborate (3; R = N₂BF₄),

which when dry was heated with a mixture of phosphorus trichloride, ethyl acetate, and cuprous bromide under the usual conditions for the Doak–Freedman reaction.[132, 133] Powdered aluminum (cf. Quin and Montgomery[137]) was then added as a reducing agent to the reaction product, which when worked up furnished (o-benzylphenyl)phosphonous

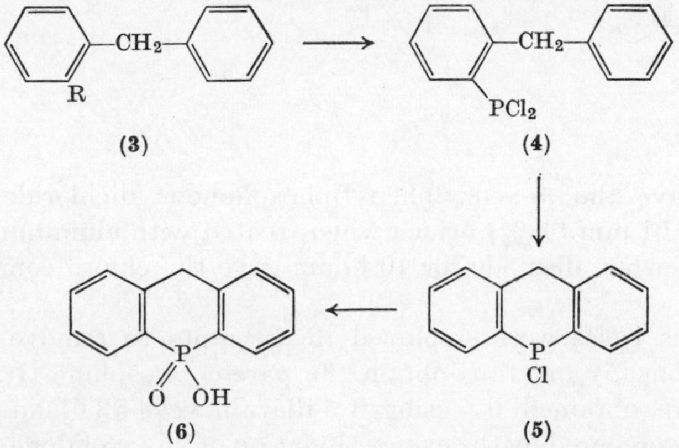

(3) → (4)

(6) ← (5)

dichloride (4), b.p. 132–137°/0.2 mm (6.4%). This dichloride, when heated at 100° with zinc chloride under nitrogen for 24 hours with final heating under reduced pressure, underwent a cyclodehydrohalogenation to give 5-chloro-5,10-dihydrodibenzo[b,e]phosphorin (5), yellow crystals, m.p. 78–86° (25%). This compound in turn when oxidized with alkaline hydrogen peroxide furnished the corresponding phosphinic acid, i.e., 5,10-dihydro-5-hydroxydibenzo[b,e]phosphorin 5-oxide (6) (81%), which decomposes above 225°.

The ultraviolet spectrum of the acid (6) shows the maxima:

λ_{max} (mμ)	205	227*	264*	268.5	275.5
ϵ_{max}	36,600	9,180	1,260	1,630	1,610

* Shoulders.

de Koe and Bickelhaupt[207] used two similar methods to prepare the 5-chloro compound (5). 2,2′-Dibromodiphenylmethane (7) was converted into a di-Grignard reagent in tetrahydrofuran and added dropwise to a solution of N,N-diethylphosphoramidous dichloride, $(C_2H_5)_2NPCl_2$ (one molar equivalent) in tetrahydrofuran at −80°. The solvent was evaporated and the residue extracted with cyclohexane; treatment of the filtered extract with hydrogen chloride deposited the chloro compound (5), (47%), m.p. 102–109° after sublimation. The identity of the product was confirmed by oxidation to the phosphinic acid (6).

Alternatively, 2-bromodiphenylmethane (8) was similarly converted
into a Grignard reagent and added to tetra-N-ethylphosphordiamidous
chloride, $[(C_2H_5)_2N]_2PCl$, at $-80°$. Treatment as before with hydrogen

(7)   (8)   (9)   (10)

chloride gave the (o-benzylphenyl)phosphonous dichloride (4), b.p.
142–143°/0.01 mm (68%), which when treated with aluminum chloride
in boiling carbon disulfide for 10 hours gave the chloro compound (5)
(40%).

Various bases were employed in attempts to remove hydrogen
chloride from (5) and thus obtain the parent phosphine (1), but best
results were obtained by using 1,5-diazabicyclo[4.3.0]non-5-ene (9).
This base is prepared by the cyanoethylation of 2-pyrrolidone, reduction
to the 3-aminopropyl derivative, and condensation of the amino with
the keto group (Oediger, Kabbe, Möller, and Eiter[208, 209]).

A slight excess of this base (9) in dry toluene is added to a solution
of the chloro compound (5) in dry 'degassed' toluene at $-196°$, and the
reaction vessel is sealed in a high vacuum and allowed to warm to room
temperature. During this period the solution slowly becomes pale
yellow and deposits 60–75% of the hydrochloride of (9); the solution
is stable for several days. If the reaction is carried out in dimethylformamide, the formation of (1) is complete in about an hour and the
base hydrochloride remains in solution, but largely decomposes in the
course of one day.

The ultraviolet spectrum of the filtered toluene solution resembles
that of anthracene far more than that of acridine in the number of the
bands and their relative positions and intensities, apart from a marked
bathochromic shift, the long-wave peak being at $\lambda = 429$ mμ (ϵ unknown).
This spectrum disappears immediately on admission of atmospheric
oxygen, very rapidly on addition of sodium hydroxide or hydrogen
chloride, and slowly on addition of dilute acid or when maintained at
room temperature. Evaporation of the filtrate, or cooling it to $-196°$,
causes decomposition.

Under mass spectroscopy conditions, the $(5)^+$ ion (at m/e 232,
intensity 43%), very rapidly splits off hydrogen chloride to give the
more stable $(1)^+$ ion (m/e 196, intensity 100%), which loses phosphorus

to give the (9-fluorenyl)$^+$ ion (m/e 165, intensity 57.7%); the appearance of the (1)$^{2+}$ ion ($m/2e$ 98, intensity 34%) is also noteworthy.

Oediger and Möller[210] have more recently reported that 1,5-diazabicyclo[5.4.0]undec-5-ene (10) (similarly prepared from caprolactam) is considerably superior to (9) for the removal of hydrogen halides from bromoalkanes to form alkenes, and its use in the above reactions might well give a more ready and effective production of the parent compound (1). [It has been used subsequently in the attempted preparation of dibenzo[b,d]phosphorin (p. 148).]

Dibenzo[b,d]phosphorin

Dibenzo[b,d]phosphorin (1), [RIS 12714], is presented and numbered as shown; the only stable members of this series are derivatives of 5,6-dihydrobenzo[b,d]phosphorin (2). The parent compound (1) is however

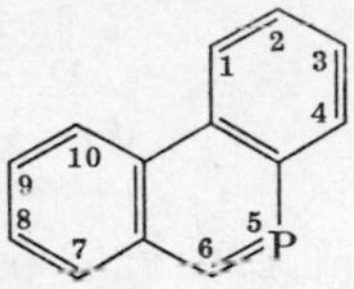

(1) 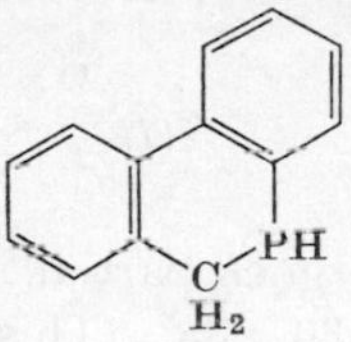   (2) 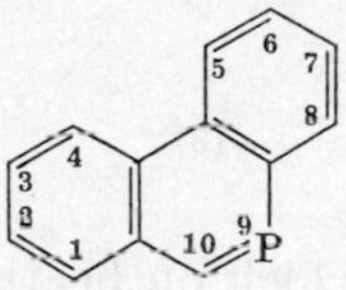   (1A)

still frequently numbered as in (1A), based on the 'exceptional' numbering of phenanthrene, [RRI 3619], and the dihydro derivative (2) is consequently then termed 9,10-dihydro-9-phosphaphenanthrene; this numbering and nomenclature should now become obsolete.

Although (1) is the correct presentation of this ring system, it is often more convenient to depict the system as shown below.

Our earliest knowledge of the derivatives of this system is due to Lynch[211, 212] who carried out an Arbusov rearrangement between o-phenylbenzyl chloride (2-chloromethylbiphenyl) (3) and triethyl phosphite to obtain the diethyl ester, b.p. 148–152°/0.3 mm, of (o-phenylbenzyl)phosphonic acid (4); hydrolysis of the ester with hydrochloric acid furnished the phosphonic acid (4), m.p. 167–169°.

Initial attempts to cyclize this acid or its derivatives to the required acid (5) failed. For example, phosphorus pentachloride converted the acid (4) into its dichloride, which however was unaffected when distilled, when heated with aluminum chloride, or when heated in nitrobenzene at 180° (cf. Campbell and Way,[139, 140] p. 84). When, however, the acid (4) was heated under nitrogen at 350°/0.2 mm for 6 hours, it gave a mixture of the phosphinic acid (5) and the anhydride of this acid: dissolution of this mixture in hot aqueous sodium hydroxide followed by acidification gave the acid, 5,6-dihydro-5-hydroxydibenzo[b,d]phosphorin 5-oxide

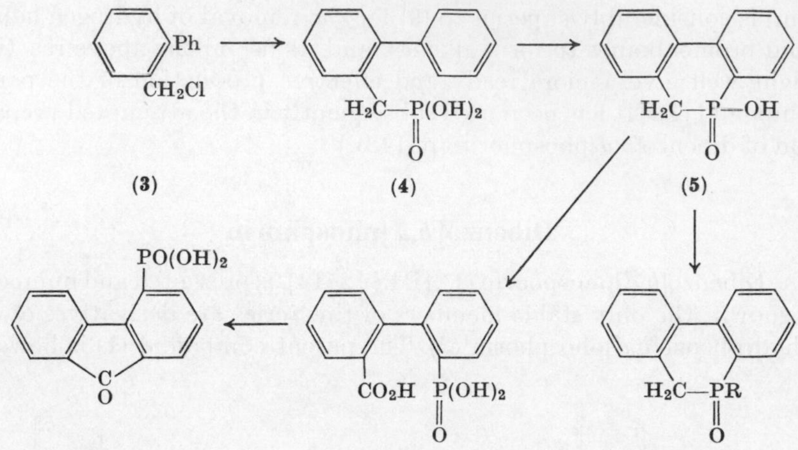

(5), which on heating undergoes partial melting at 220–225°, resolidification, and remelting at 236–238°. (This acid (5) was initially termed 9,10-dihydro-9-hydroxy-9-phosphaphenanthrene 9-oxide.[211])

The acid (4), when heated at 410°/0.3 mm for 1 hour, gave a brown melt, from which soluble components were removed with ethanol, leaving a residue of the anhydride, m.p. 272–274°, of the acid (5).

A mixture of the acid (4) and thionyl chloride, when boiled under reflux for 10–12 hours, gave mainly the anhydride of (4); removal of the thionyl chloride gave a residue which when strongly heated gave a mixture of the acid (5) and its anhydride.

The acid (5) when heated with thionyl chloride or phosphorus pentachloride gave the corresponding chloride, i.e., 5-chloro-5,6-dihydrodibenzo[b,d]phosphorin (6; R = Cl), pale yellow crystals, m.p. 125–126°. This compound in turn gave the following derivatives: (a) Heated with a mixture of methanol, N,N-dimethylaniline, and benzene, it gave the methyl ester (6; R = OCH$_3$), m.p. 171.5–172.5°. (b) Heated with benzene, phenol, and pyridine, it afforded the phenyl ester (6; R = OC$_6$H$_5$), b.p. 197–198°/0.1 mm, m.p. 75–77°. (c) Heated with benzene and aniline, it gave the anilide (6; R = NHC$_6$H$_5$), m.p. 202.5–204°. (d) Reduction in ether with lithium aluminum hydride gave 5,6-dihydrodibenzo[b,d]phosphorin 5-oxide (6; R = H), m.p. 99–100°, which showed a P—H band at 2330 cm^{-1} in its infrared spectrum.

Oxidation of the acid (5) with alkaline permanganate cleaved the heterocyclic ring with the formation of 2′-phosphobiphenyl-2-carboxylic acid (7). Attempted decarboxylation of (7) to form biphenylyl-2-phosphonic acid by brief heating with copper powder at 260–300° failed, giving a mixture of fluorenone and 9-oxofluorene-4-phosphonic acid (8), m.p. 306–307.5°, a compound which could be readily obtained by the action of sulfuric acid on the acid (7).

The infrared spectrum of the acid (5) has a trio of bands, at ~2703, 2326 and 1667 cm^{-1}, characteristic of phosphinic acids; the ultraviolet

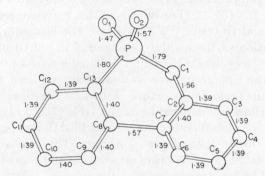

Figure 4. The structure and bond lengths (in Å) of 5,6-dihydro-5-hydroxydibenzo[b,d]phosphorin 5-oxide (5). (Reproduced, by permission, from P. J. Wheatley, J. Chem. Soc., 1962, 3733)

spectrum shows a band at 268 mμ (ϵ_{max} 15,000) which is probably the characteristic biphenyl band (normally at 252 mμ) shifted because the two o-phenylene rings in (5) are coaxial but not coplanar. It is noteworthy that the methiodides of the structurally analogous 5,6-dihydro-5-methyl- and -5-phenylarsanthridine also have one band, at 268 and 269 mμ, respectively (p. 427).

An X-ray crystal analysis of the acid (5) by Wheatley[213] has fully confirmed the above structure. This is shown in Figure 4, in which all the atoms are numbered for crystallographic reference, as distinct from the 'chemical' numbering in formula (1).

The molecule has a 'skew' configuration, the planes of the two benzene rings being at 22°. One plane contains the atoms C-1 to C-8, and the other the atoms C-7 to C-13 and the phosphorus atom. These two planes therefore contain all the atoms except the two oxygen atoms. The mean C–C distances in the benzene rings is 1.394 Å. The length of the C-7/C-8 bond indicates that there is no conjugation between the benzene rings; moreover, the equality in length of the P—C-1 bond

(1.79 Å) and the P—C-13 (1.80 Å) also indicates that there is little conjugation between the phosphorus atom and the adjacent benzene ring.

de Koe, van Veen, and Bickelhaupt[214] have recently (1968) recorded an attempt to convert this 5,6-dihydro-5-hydroxybenzo[b,d]-phosphorin 5-oxide (5) into the parent compound (1). For this purpose, the acid (5) was reduced by diphenylsilane (p. 50) to the 5,6-dihydro compound (2), b.p. 132°/0.01 mm (92%), which when treated in dichloromethane solution under nitrogen with phosgene at −30° and later at room temperature gave 5-chloro-5,6-dihydrobenzo[b,d]phosphorin, m.p. 58–62° (35%). (For details of this reaction, see Henderson, Buckler, Day, and Grayson.[215]) To remove the elements of hydrogen chloride, a solution of the 5-chloro compound in anhydrous 'degassed' ether at −196° was treated with 1,5-diazabicyclo[5.4.0]undec-5-ene (p. 145), and the apparatus was sealed under reduced pressure and allowed to attain room temperature. Filtration in absence of oxygen removed the yellow hydrochloride of the base (73%). The ultraviolet spectrum of the colorless filtrate closely resembled that of phenanthridine but much less closely that of phenanthrene. The pure compound (1) could not be isolated from the ethereal solution and was obviously very rapidly affected by air.

The mass spectrum of the 5-chloro compound strongly confirmed its structure.

Two other syntheses of the dibenzo[b,d]phosphorin system have one feature in common, namely, the migration of an aryl group from phosphorus to carbon. A simple example of this process is afforded by the alkaline hydrolysis of (bromomethyl)triphenylphosphonium bromide (9), which gives benzyldiphenylphosphine oxide (10) (Schlosser[216]).

$$[(C_6H_5)_3PCH_2Br]^+ \ Br^- \ \rightarrow \ (C_6H_5)_2P(=O)CH_2C_6H_5$$
$$(9) \hspace{4cm} (10)$$

Allen, Tebby, and Williams[217] found that the phenyl group in phenylacetylene activated the triple bond towards nucleophilic attack by triphenylphosphine in boiling wet diethylene glycol, $O(CH_2CH_2OH)_2$, with the resultant formation of diphenyl-(1,2-diphenylethyl)phosphine oxide (11), m.p. 233–234°.

$$(C_6H_5)_3P + C_6H_5C\equiv CH \ \rightarrow \ (C_6H_5)_2P(=O)—CH(C_6H_5)—CH_2C_6H_5$$
$$(11)$$

Further investigation confirmed that the most probable mechanism of this process was the initial union of the two reactants to give the dipolar product (12), which, in the presence of water, added a proton to give triphenylstyrylphosphonium hydroxide (13); addition of water then gave (β-hydroxyphenethyl)triphenylphosphonium hydroxide (**14**), which could readily form the betaine or zwitterion (**15**), and then by a

$$Ph_3\overset{+}{P}-CH=\overset{-}{C}-Ph \quad \xrightarrow{H_2O} \quad Ph_3\overset{+}{P}-CH=CH-Ph \quad ^-OH$$

(**12**)   (**13**)

$$\Big\downarrow H_2O$$

(**11**) $Ph_2\overset{+}{P}\underset{\underset{-O-CHPh}{\big\downarrow}}{\overset{\overset{Ph\ H}{|\ \ |}}{\curvearrowright C}}-H \quad \xleftarrow{-H_2O} \quad Ph_3\overset{+}{P}-CH_2-CH(OH)-Ph \quad ^-OH$

(**14**)

(**15**)

type of internal Wittig rearrangement give the phosphine oxide (**11**). Tri-*p*-tolylphosphine reacted similarly with phenylacetylene to form 2 - phenyl - 1 - (*p* - tolylethyl)di - *p* - tolylphosphine oxide, m.p. 218–219° (Allen and Tebby [218]).

The use of this type of reaction was greatly facilitated by the replacement of the phenylacetylene by methyl propiolate, $HC\equiv CCO_2CH_3$, which reacted with triphenylphosphine in ether at room temperature to form [α-(methoxycarbonylmethyl)benzyl]diphenylphosphine oxide (**16**)

Ph_2P—CH—CH_2COOMe
‖ |
O Ph

                                        H—C—P—Me
                                        | ‖
                              CH_3COO—H_2C   O

(**16**)   (**17**)   (**18**)

(Richards and Tebby [219]). This 1,2-aryl shift was then applied to 5-methyl-5*H*-dibenzophosphole (**17**), which with methyl propiolate underwent ring expansion to 5,6-dihydro-6-(methoxycarbonylmethyl)-5-methyldibenzo[*b,d*]phosphorin 5-oxide (**18**), m.p. 145–148°, termed by these authors 9-methyl-10-methoxycarbonylmethyl-9,10-dihydro-9-phosphaphenanthrene 9-oxide. It was identified by analysis, by its nmr spectra, and by the close similarity of its ultraviolet spectrum to that of the bromide (**26**) (below).

Schlosser's rearrangement was applied (Allen and Millar [220, 221]) to the action of sodium hydroxide at room temperature on an aqueous-acetone solution of 5-(iodomethyl)-5-phenyl-5*H*-dibenzophospholium iodide (**19**; R = C_6H_5), m.p. 219–220°, which is readily prepared by the quaternization of 5-phenyl-5*H*-dibenzophosphole with diiodomethane.

(19)   (20)   (21)

A comparable aryl migration occurred, presumably through the phosphorane (**20**) to give finally 5,6-dihydro-5-phenyldibenzo[*b,d*]phosphorin 5-oxide (**21**; $R = C_6H_5$), m.p. 127–130° (58%), with traces of 5-phenyldibenzophosphole and its 5-oxide.

The analogous 5-methyl salt (**19**; $R = CH_3$), m.p. 235–236°, similarly gave the 5,6-dihydro-5-methyl 5-oxide (**21**; $R = CH_3$) (71%) as a hygroscopic gum which was not obtained crystalline: its characteristic ultraviolet spectrum was closely similar to those of other compounds having the ring system (**2**). The infrared spectra of the compounds (**21**; $R = C_6H_5$ or CH_3) showed P=O absorption by a strong band at 1197 and 1190 cm^{-1}, respectively; the structure of each compound was confirmed by its pmr spectrum.

Derivatives of the ring system (**1**) have been prepared in a study of the factors determining the stability of methylenephosphoranes (E. A. Cookson and Crofts[222, 222a]). Many methylenephosphoranes are too readily hydrolyzed to permit preparation in an aqueous medium;

$$R_3P{=}CR_2 \leftrightarrow R_3\overset{+}{P}{-}\overset{-}{C}R_2$$
(**22A**)   (**22B**)

others are stabilized to an extent greater than that expected by the normal resonance indicated by (**22A ↔ 22B**). This stabilization is usually attributed to delocalization of the negative charge in the form (**22B**). Stabilization might also result, however, from the inclusion of the P=C bond of (**22A**) in a fully conjugated six-membered ring, if this led to aromaticity in this ring.

To obtain evidence on this point, 2'-bromobiphenyl-2-carboxylic acid (**23**) was reduced with lithium aluminum hydride to 2-bromo-2'-(hydroxymethyl)biphenyl (**24**; $R = OH$), m.p. 83–83.5°, which was then subjected to the familiar stages (p. 125) to obtain the desired cyclization. The compound (**24**; $R = OH$) when treated in benzene solution with hydrogen bromide gave 2-bromo-2'-(bromomethyl)biphenyl (**24**; $R = Br$), m.p. 55–57°, which in boiling methanol afforded 2-bromo-2'-(methoxymethyl)biphenyl (**24**; $R = OCH_3$), b.p. 140–150°/2 mm, m.p. 40–42°. This was converted into a Grignard reagent, which with diphenylphosphinous chloride gave 2-(diphenylphosphino)-2'-(methoxy-methyl)biphenyl (**25**), m.p. 104.5–105.5°. This compound, when treated

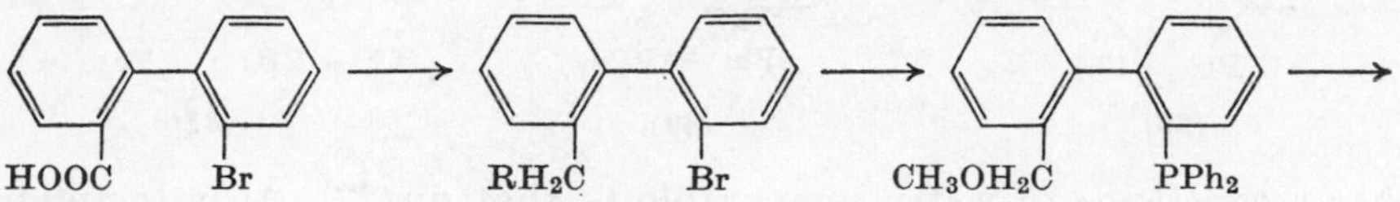

in boiling acetic acid with hydrogen bromide, gave the intermediate 2'-bromomethyl derivative which underwent spontaneous cyclization to 5,6-dihydro-5,5-diphenyldibenzo[*b,d*]phosphorinium bromide (**26**), m.p. 335–337°.

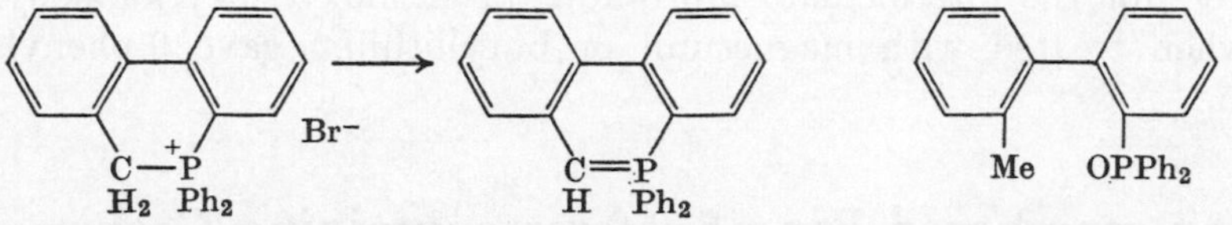

The addition of a solution of (**26**) in ice-cold water to a similar solution of sodium hydroxide precipitated a bright orange solid, undoubtedly 5,5-diphenyldibenzo[*b,d*]phosph(v)orin (**27**), which when collected, washed with water, and dried remained unchanged for several hours before the color faded. Repetition of this experiment at 100° caused rapid fading of the color and the formation of 2-(diphenylphosphinyl)-2'-methylbiphenyl or (2'-methyl-2-biphenylyl)diphenylphosphine oxide (**28**), m.p. 154–156°: this compound was also formed when the moist phosphorane (**27**) was dissolved in acetone.

The addition of benzaldehyde or *p*-nitrobenzaldehyde to an ethereal solution of (**27**) gave no evidence of a normal Wittig reaction: the product from the benzaldehyde reaction was the hemibenzoate, m.p. 138–139°, of the phosphine oxide (**28**), identical with that prepared by the direct action of benzoic acid on the authentic (**28**).

[*Note.* In the original communication [222] the compounds (**26**) and (**27**) were termed 9,10-dihydro-9,9-diphenyl-9-phosphoniaphenanthrene bromide and 9,9-diphenyl-9-phospha(v)phenanthrene, respectively.]

It is noteworthy that Märkl's two yellow amorphous phosphoranes, 1,1-diphenylphosphorin (p. 112) and 1,1-diphenylphosphinoline (p. 126), also showed a moderate resistance to hydrolysis and did not undergo the Wittig reaction.

Although benzylidenetriphenylphosphorane (**29**; R = H) is rapidly hydrolyzed, (diphenylmethylene)triphenylphosphorane (**29**; R = C_6H_5)

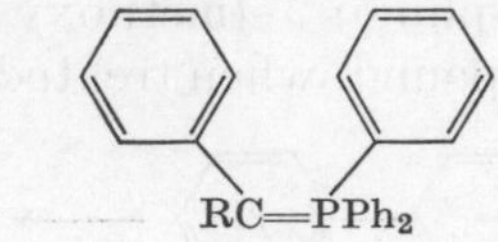

RC=PPh$_2$	PhC=PPh$_2$	CH$_3$OCHPh Br
(29)	(30)	(31)

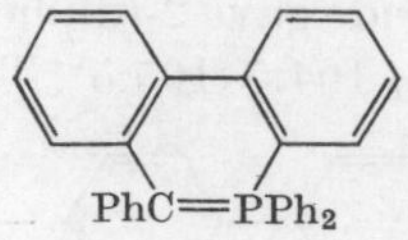

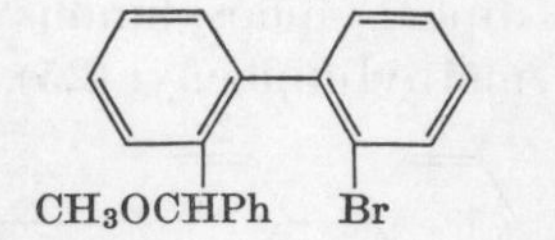

has a resistance to water comparable to that of (27). An attempt to prepare the analogous 5,5,6-triphenyldibenzo[b,d]phosphorin (30) failed however, for the intermediate 2-bromo-2'-(α-methoxybenzyl)biphenyl (31) when treated with magnesium or butyllithium gave 9-phenyl-fluorene.

Seven-membered Ring Systems containing Carbon and One Phosphorus Atom

1H-Phosphepin and Phosphepanes

The unsaturated system (1) is known as 1H-phosphepin, [RIS 11634], and its fully saturated derivative (2) as phosphepane.

Compound (1) is unknown. Wagner[1] has claimed that sodium phosphide, NaPH$_2$, prepared by passing phosphine into a solution of

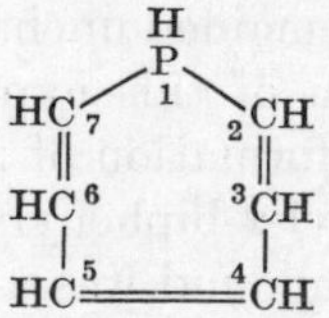

(1)

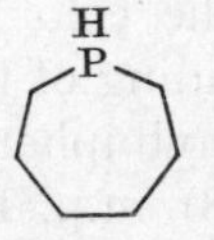

(2)

sodium in ammonia, reacts with 1,6-dibromohexane to form a mixture of the disecondary phosphine, H$_2$P(CH$_2$)$_6$PH$_2$, and phosphepane (2), from which each component can be isolated by fractional distillation. The use of phenylphosphine similarly gives 1-phenylphosphepane. No physical constants are quoted in the patent specification.

Davies, Downer, and Kirby[77] have employed their synthetic method for the preparation of 1-phenylphosphepane. 1-Bromo-5-hexene (3) was added during several hours to phenylsodiophosphine, C$_6$H$_5$PNaH,

CH$_2$:CH(CH$_2$)$_4$Br   CH$_2$:CH$_2$(CH$_2$)$_4$PHC$_6$H$_5$
(3)   (4)

in 2-methylheptane at 80° to obtain 5-hexenylphenylphosphine (4) in 37% yield. Cyclization of the latter by irradiation with ultraviolet light gave 1-phenylphosphepane (41%), b.p. 106–120°/1.5 mm, which however could not be purified and was therefore converted into the pure crystalline 1-phenylphosphepane 1-sulphide (5), m.p. 90.5–91.5° (37%). This method of cyclization, which worked so well in the synthesis of the corresponding phospholane (p. 52) and phosphorinane (p. 101) appears to be reaching its practical limit at phosphepane: many other cyclization reactions have the same practical limits.

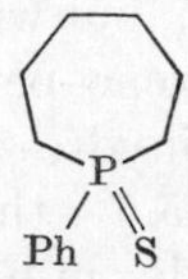

(5) 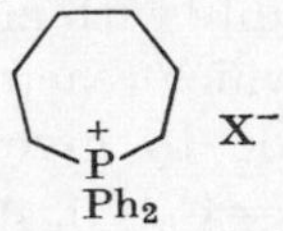   (6)

Märkl[75] has briefly noted that the slow addition of diphenylpotassio-phosphine, $(C_6H_5)_2PK$, in dioxane–tetrahydrofuran to a boiling solution (in tetrahydrofuran) of 1,6-dibromohexane gives (6-bromohexyl)di-phenylphosphine, which immediately undergoes intramolecular quaternization to form 1,1-diphenylphosphepanium bromide (6; X = Br), isolated as the perchlorate, m.p. 208° (36%) (p. 52).

Märkl[76] has also applied his other method (pp. 52, 101) for the synthesis of the cation (6), a mixture of 1,6-dibromohexane (2 equivalents) and tetraphenyldiphosphine, $(C_6H_5)_2P$—$P(C_6H_5)_2$ (1 equivalent) in o-dichlorobenzene being added dropwise to boiling o-dichlorobenzene. The solution, on cooling, deposits the crystalline bromide (6; X = Br), m.p. 245–247° (54%), the other product, diphenylphosphinous bromide, $(C_6H_5)_2PBr$, remaining in solution.

5H-Dibenzo[b,f]phosphepin

The compound (1) is clearly a dibenzophosphepin and is termed 5H-dibenzo[b,f]phosphepin, [RRI 3705], and the compound (2; R = H) is 10,11-dihydro-5H-dibenzo[b,f]phosphepin.

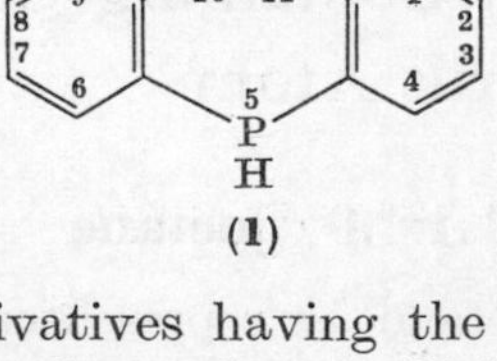

(1) 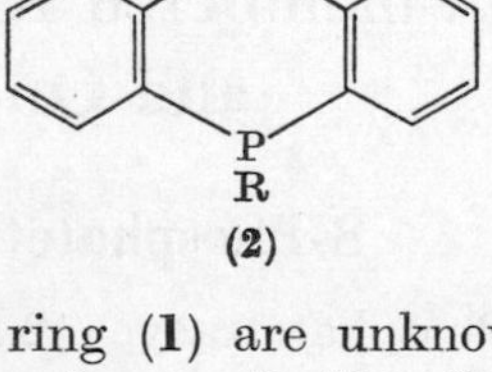   (2)

Derivatives having the unsaturated ring (1) are unknown, but 10,11-dihydro-5-phenyl-5H-dibenzo[b,f]phosphepin (2; R = C_6H_5) has been prepared (Mann, Millar, and Smith[223]). For this purpose, a modified version of Kenner and Wilson's preparation of 2,2'-dibromobibenzyl[224] was employed, o-bromobenzyl bromide (3) being converted by hydrazine

$$\text{BrC}_6\text{H}_4\text{CH}_2\text{Br} \qquad (\text{BrC}_6\text{H}_4\text{CH}_2)_2\text{N}\text{---NH}_2 \qquad \text{BrC}_6\text{H}_4\text{CH}_2\text{CH}_2\text{C}_6\text{H}_4\text{Br}$$
$$(\textbf{3}) \qquad\qquad\qquad (\textbf{4}) \qquad\qquad\qquad\qquad (\textbf{5})$$

into 1,1-di-o-bromobenzylhydrazine (4), which on oxidation with mercuric oxide furnished the 2,2'-dibromobibenzyl (5). The bromine atoms in (5) did not react readily with magnesium, but with butyllithium gave 2,2'-dilithiobibenzyl, which reacted in benzene–petroleum with phenylphosphonous dichloride to give 10,11-dihydro-5-phenyl-5H-dibenzo-[b,f]phosphepin (2; R = C_6H_5), m.p. 75–75.5°; this compound formed a 5-oxide, m.p. 173.5–174.5°, a methiodide, m.p. 251–252°, a yellow bisphosphinedichloropalladium having a molecule of dioxane of crystallization, $(C_{20}H_{17}P)_2PdCl_2,C_4H_8O_2$, m.p. 278–280°, and a brown bridged diiodobisphosphine-$\mu\mu'$-diiodopalladium, $(C_{20}H_{17}P)_2(PdI_2)_2$, m.p. 283–285°.

In many cases an aryl group attached to an arsenic atom forming part of a ring may be replaced by iodine, without fission of the ring, when the compound is boiled with hydriodic acid (cf. pp. 371, 416). A similar experiment with the phosphepin (2; R = C_6H_5) left it chemically unaffected, but when isolated it had m.p. 94.5–95° (instead of 75–75.5°). The phosphine is thus dimorphic: the low-melting form is stable in the absence of the high-melting form, but when ground or treated just above its melting point with a trace of the latter it very readily passes over to the high-melting form.

The arsenic analogue of (2; R = C_6H_5) also exists in dimorphic forms, each isomorphous with the corresponding form of the phosphepin. The phosphorus ring system is considerably more stable than the arsenic system; a comparison of the physical and chemical properties of the two compounds is given on pp. 436, 437.

The dihydrophosphepin (2; R = C_6H_5) was originally named 1-phenyl-1-phospha-2,3:6,7-dibenzocyclohepta-2,6-diene.[223]

Eight-membered Ring System Containing Carbon and One Phosphorus Atom

8-Phosphatetracyclo[2.2.1.1^{2,6}.0^{3,5}]octane

8-Phosphatetracyclo[2.2.1.1^{2,6}.0^{3,5}]octane is represented as (1), [RIS 12839], although the ring system is more clearly depicted as (2): the numbers carrying superscripts in the name indicate the position and

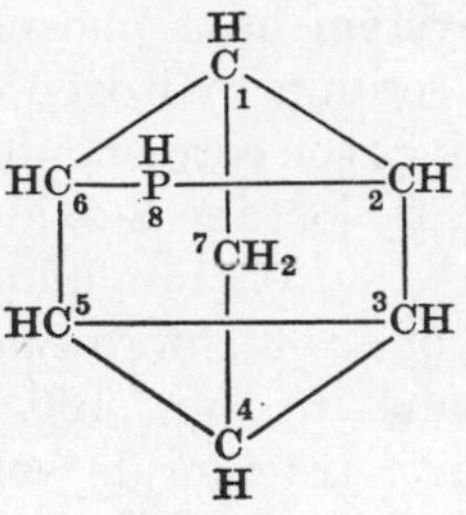

H
C
1
HC 6 — P — 2 CH
8
^{7}CH$_2$
HC^5 — ^{3}CH
4
C
H

(1)

7
4
5   3
1
6   2
8
P
H

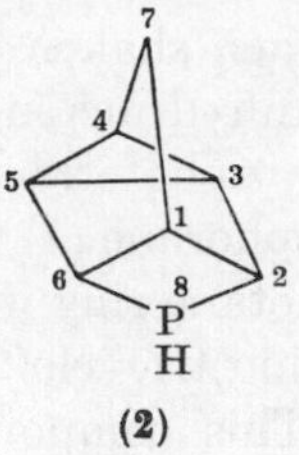

(2)

the number of atoms in those bridges which do not link the two tertiary C-1 and C-4 atoms.

The synthesis and reactions of derivatives of (1) illustrate the formation of the phosphetane ring (p. 4) as part of the tetracyclo system (1) and its subsequent degradation.

Bicyclo[2.2.1]heptadiene (3), or 2,5-norbornadiene [cf. RRI 1031], is known to react with various Lewis acids, such as boron trichloride, phenylboron dichloride, and stannic chloride to form substituted nortricyclenes.

Green [225, 226] has shown that when the diene (3), is treated with methylphosphonous dichloride, CH_3PCl_2, under dry oxygen-free nitrogen, it combines to form the crystalline 1:1 adduct (4), namely 8,8-dichloro 8 methyl 8 phosphatetracyclo[2.2.1.1^{2,6}.0^{3,5}]octane, which

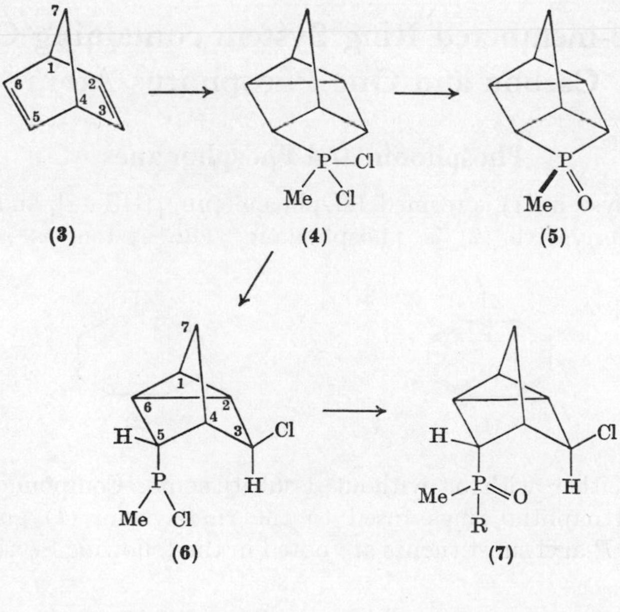

7
1
6   2
5   4   3

$\longrightarrow$

P—Cl
Me Cl

$\longrightarrow$

P
Me O

(3)   (4)   (5)

7
1
6   2
H — 5   4   3 — Cl
P   H
Me   Cl

(6)

$\longrightarrow$

H   Cl
Me — P=O   H
R

(7)

contains the fused phosphetane ring system as a phosphorane. This compound, when shaken with aqueous sodium hydrogen carbonate or treated with anhydrous sulfur dioxide, gives the corresponding oxide (**5**), 8-methyl-8-phosphatetracyclo[2.2.1.1^{2,6}.0^{3,5}]octane 8-oxide, m.p. 157°.

The phosphorane (**4**) when heated at 180°/0.1 mm undergoes fission of the phosphetane ring with formation of *exo*-3-chloro-*endo*-5-(methyl-chlorophosphino)tricyclo[2.2.1.0^{2,6}]heptane (**6**), b.p. 100°/0.2 mm [cf. RRI 2058]. This compound, in carbon tetrachloride solution, when treated in succession with chlorine and sulfur dioxide, undergoes oxidation at the phosphorus atom to give *exo*-3-chloro-*endo*-5-(methyl-chlorophosphinyl)tricyclo[2.2.1.0^{2,6}]heptane (**7**; R = Cl), b.p. 130°/0.2 mm, which in dry ether reacts with methanol at 0° to give the corresponding methyl ether, *exo*-3-chloro-*endo*-5-(methoxymethylphos-phinyl)tricyclo[2.2.1.0^{2,6}]heptane (**7**; R = OCH$_3$), b.p. 121°/0.1 mm.

Considerable evidence for the above structures has been adduced from their infrared and proton nuclear magnetic resonance spectra and, in the case of the phosphine oxide (**5**) and the phosphinous chloride (**6**), from their mass spectra: the original paper should be consulted for the detailed evidence. The infrared spectra of (**5**), (**7**; R = Cl), and (**7**; R = OCH$_3$) show a P=O band at 1190, 1235, and 1212 cm^{-1}, respectively.

The compounds (**4**) and (**5**) were initially termed 2,2-dichloro-2-methyl-2-phospha(v)tetracyclo[3.2.1.0^{3,6}.0^{4,7}]octane and 2-methyl-2-phospha(v)tetracyclo[3.2.1.0^{3,6}.0^{4,7}]octane 2-oxide, respectively.[226]

Nine-membered Ring System containing Only Carbon and One Phosphorus Atom

Phosphonin and Phosphonanes

This system (**1**) is termed 1*H*-phosphonin, [RIS —], and the fully reduced derivative (**2**) is phosphonane. The system is apparently

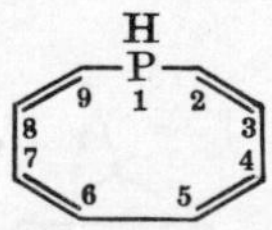

(1) 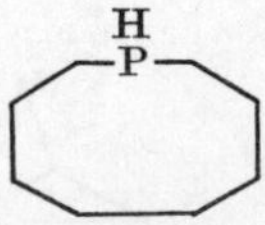   (2)

unknown either with or without 1-substituents. Compounds having benzo and naphtho rings fused to the ring-system (**1**) and having *P*-alkyl or *P*-aryl substituents are noted in the following Section.

17H-Tetrabenzo[b,d,f,h]phosphonin and
17H-Tribenzo[b,d,f]naphtho[1,2-h]phosphonin

17H-Tetrabenzo[b,d,f,h]phosphonin (**1**), [RIS —], has the structure and numbering shown. 17H-Tribenzo[b,d,f]naphtho[1,2-h]phosphonin

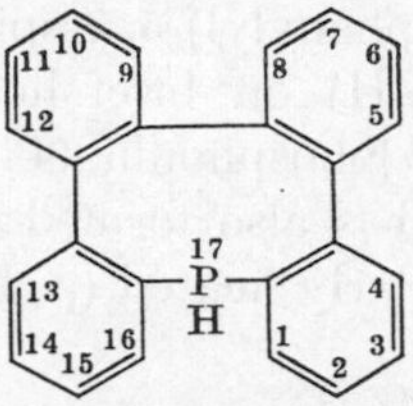

(1) 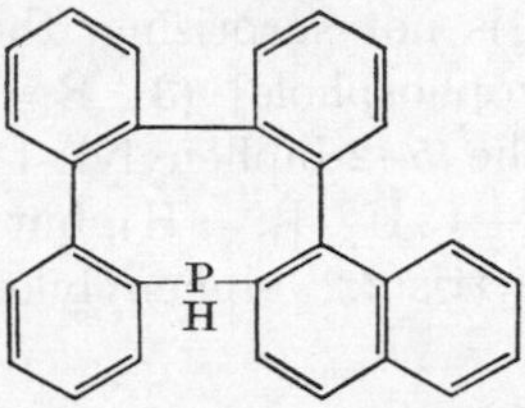   (2)

(**2**), [RIS —], is regarded as a tribenzonaphtho derivative of the same nine-membered ring. [In the RIS, both will probably be depicted with the phosphorus atom uppermost (hence the numbering shown in (**1**)), but it is more convenient to depict them as above.]

They are listed together, because both systems are formed by the same thermal isomerization process, which has been already described (pp. 93–96 *et seq.*); the specific members are briefly mentioned here for quick systematic reference.

Wittig and Maercker[117] first showed that 5-phenyl-5,5'-spirobi[dibenzophosphole] (**3**; R = C_6H_5, R' = H) when briefly heated at 210° gave the isomeric 17-phenyl-17H-tetrabenzo[b,d,f,h]phosphonin (**4**; R = C_6H_5, R' = H), m.p. 131–132°, which readily formed a methiodide.

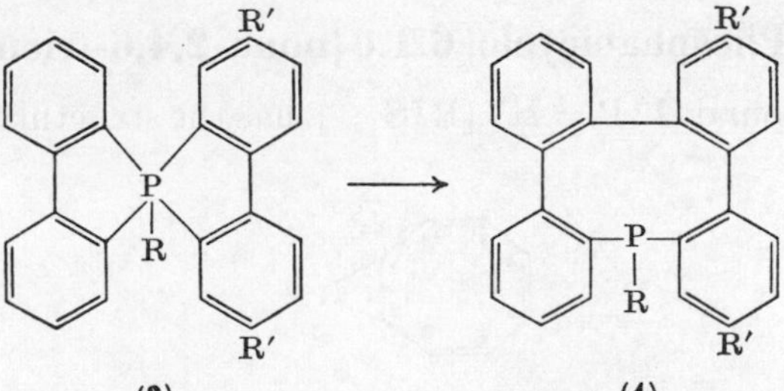

(3)   (4)

Hellwinkel[95] later described the simplest case of the 5-methyl-phosphorane (**3**; R = CH_3, R' = H) being similarly converted into the 17-methyl isomer (**4**; R = CH_3, R' = H), m.p. 130–131°.

The *p*-dimethylaminophenyl phosphorane (**3**; R = $C_6H_4N(CH_3)_2$-*p*,

R′ = H) gives the compound (**4**; R = $C_6H_4N(CH_3)_2$-*p*), named *N,N*-dimethyl-*p*-(17*H*-tetrabenzo[*b,d,f,h*]phosphorin-17-yl)aniline, m.p. 235–236°; and the phosphorane [**3**; R = C_6H_5, R′ = $N(CH_3)_2$] similarly gives the compound [**4**; R = C_6H_5, R′ = $N(CH_3)_2$], m.p. 207–208°, named *N,N,N′,N′*-tetramethyl-17-phenyl-17*H*-tetrabenzo[*b,d,f,h*]phosphorin-2,7-diamine.[117]

It is not surprising, therefore, that 5-(2-biphenylyl)-5,5′-spirobi-[dibenzophosphole] (**3**; R = 2-C_6H_4·C_6H_5, R′ = H) on brief heating gives the 5-(2-biphenylyl)-17*H*-tetrabenzo[*b,d,f,h*]phosphonin (**4**; R = 2-C_6H_4—C_6H_5, R′ = H), but the latter compound is also formed when lithium tris(2,2′-biphenylylene)phosphate is similarly heated (p. 98).[95]

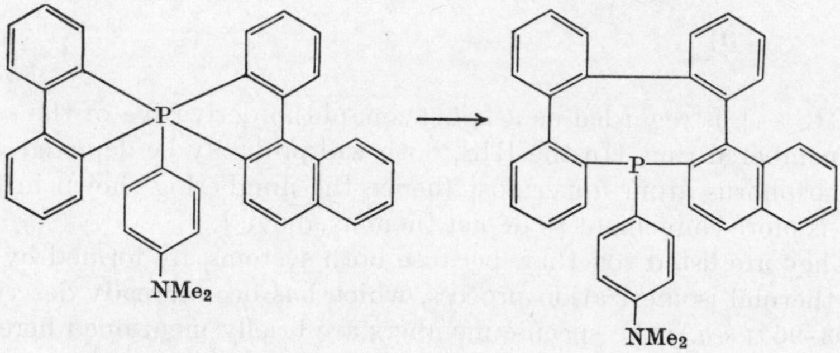

The yellow phosphorane (**5**) undergoes a similar isomerization, to give the compound (**6**), named *N,N*-dimethyl-*p*-(17*H*-tribenzo[*b,d,f*]-naphtho[1,2-*h*]phosphonin-17-yl)aniline, m.p. 168–170° (p. 95).[117]

9-Phosphabicyclo[6.1.0]nona-2,4,6-triene

This compound (**1**; R = H), [RIS —], has the structure and notation shown.

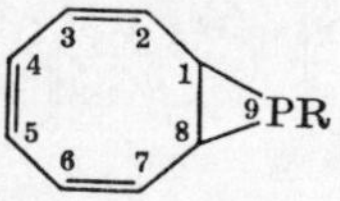

A three-membered ring system consisting of one phosphorus and two carbon atoms must be unstable and highly reactive; by incorporating this phosphirane system (p. 3) in an appropriate bicyclic system, however, it may acquire a sufficiently enhanced stability to allow the bicyclic compound to be isolated in the crystalline state, although the 'overall' stability may still not be high.

A comparable case in the carbocyclic series arises when cyclo-octatetraene (2) is treated in hot tetrahydrofuran with potassium (two equivalents), whereby the pale yellow crystalline dipotassium cyclo-octatetraenide, conveniently depicted as (3), is deposited (Katz[227]).

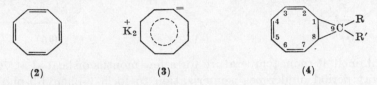

(2)   (3)   (4)

This compound, which has a flat anion, is stable in solution but readily explodes when dry. It combines with various alkyl chlorides in tetra-hydrofuran, even at $-30°$, to give bicyclo[6.1.0]nona-2,4,6-triene (4; $R = R' = H$) or 9-substituted derivatives (Katz and Garratt[228]): for example, with dichloromethane, it gives the parent compound (4; $R = R' = H$); with chloroform the 9-chloro derivative (4; $R = Cl$, $R' = H$); with carbon tetrachloride the 9,9-dichloro derivative (4; $R = R' = Cl$); with dichloromethyl methyl ether the 9-methoxy deriva-tive (4; $R = OCH_3$, $R' = H$). These compounds, when heated, may undergo a rearrangement to the corresponding 1-substituted 3a,7a-dihydroindenes.

Katz, Nicholson, and Reilly[229] have recently extended this reaction to the phosphorus field by showing that the compound (3) similarly combines with phenylphosphonous dichloride, $C_6H_5PCl_2$, in tetra-hydrofuran at $0°$ to give 9-phenyl-9-phosphabicyclo[6.1.0]nona-2,4,6-triene (1; $R = C_6H_5$). The crude product can be sublimed at $55°/1 \times 10^{-6}$ mm, giving the pure compound as white crystals, which are stable at $-78°$ but become oily at room temperature and darken rapidly on exposure to air. Attempts to isolate a P-oxide or a methiodide failed. Proton nmr spectra were consistent with the structure (1; $R = C_6H_5$) and with the existence of only one isomeric form in the sample.

The compound in chloroform solution undergoes a rearrangement, slowly at room temperature and rapidly at $70°$, to form the isomeric 9-phenyl-9-phosphabicyclo[4.2.1]nona-2,4,7-triene, which is discussed below.

9-Phosphabicyclo[4.2.1]nona-2,4,7-triene

The above compound, [RIS —], has the structure (1) and is at present known only as its 9-phenyl derivative.

Katz, Nicholson and Reilly [229] have shown (p. 159) that dipotassium cyclooctatetraenide in tetrahydrofuran combines with phenylphosphonous dichloride at 0° to give 9-phenyl-9-phosphabicyclo[6.1.0]nona-2,4,6-triene (2). This compound, when in chloroform solution which is

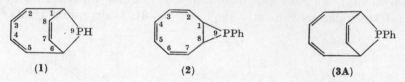

(1)   (2)   (3A)

maintained at room temperature for a few months or heated at 70° for a short period, undergoes isomerization to form 9-phenyl-9-phosphabicyclo[4.2.1]nona-2,4,7-triene (3A), which after sublimation at 80°/0.1 mm forms white crystals, m.p. 85.5–86.5°, which alone or in solution darken rapidly on exposure to air.

The isolation of the phosphine (2) in the preparation of (3A) is not necessary. Dipotassium cyclooctatraenide in tetrahydrofuran is added under nitrogen to $C_6H_5PCl_2$ at 0°, and the dark mixture is stirred at room temperature for 2 hours and then treated with water, pentane, and sodium carbonate until neutral. After two more extractions with pentane, the united dried extracts are distilled, giving a final fraction, b.p. 160–170°/0.4 mm, which forms a pale yellow solid and gives the phosphine (3A), m.p. 85.5–86.5° after sublimation at 80°/0.1 mm. Under these conditions the intermediate phosphine (2) would almost certainly be entirely converted into the phosphine (3A).

The structure of (3A) is established primarily by the proton nmr spectra. It is noteworthy that the intensities of the resonances of the phenyl, olefinic, and allylic protons are in the ratio 5.16:5.93:1.92, indicating a bicyclic skeleton; this was confirmed by the preparation of the phosphine methiodide, m.p. 239–240°, which on catalytic hydrogenation gave the hexahydrophosphine methiodide, m.p. 244.5–245°, characterized as the methopicrate, m.p. 187.1–187.6°. There are six possible such bicyclic structures in which each carbon is linked to one hydrogen atom; of these, the nmr evidence eliminated five and confirmed the structure (3A).

When the phosphine (3A) was subjected to rapid gas-phase pyrolysis at 480°, or was heated in chloroform solution containing free hydrogen chloride at 100°, it was converted into another isomer (3B), which when purified also formed white crystals, m.p. 84.5–85.5°. The nmr evidence showed that (3A) and (3B) were epimers, *i.e.*, they differed solely in the relative positions of the phenyl group about the pyramidal phosphorus atom.

Table 10. Derivatives of the epimeric 9-phenyl-9-phosphabicyclo[4.2.1]nona-2,4,7-trienes (3)

Reagent	Product	Derivative of	
		(3A), m.p. (°)	(3B), m.p. (°)
CH₃I	Methiodide	239–240	219–220
	Methopicrate	186.0–186.5	
H₂O₂–CHCl₃	P-Oxide	182.8–183.4	197.5–197.7
Air–CHCl₃	P-Oxide	←——— 197.5–197.7 ———→	
(C₆H₅CN)₂PdCl₂	2(Phosphine)·PdCl₂	Dec. 264–269	Dec. 260–268

The main reactions of these two isomers are summarized (Table 10). The following points may be noted. (a) No evidence could be obtained for the conversion of the methiodide of (3A) into that of (3B), or vice-versa. (b) The methopicrate of (3A) and that of the hexahydro-phosphine, i.e., 9-phenyl-9-phosphabicyclo[4.2.1]nonane, have melting points that are almost identical, but a mixture had m.p. 182.5–184.5°. (c) The P-oxides prepared by the H_2O_2–CHCl₃ oxidation of (3A) and (3B) are quite distinct, but (3A) when oxidized by air always gave the P-oxide of (3B). (d) The yellow complex 2(phosphine)PdCl₂ obtained from (3B), when treated with aqueous KOH solution, liberated the pure (3B) phosphine. This was a valuable method of purification, for the above direct conversions (3A) → (3B) were never complete. (e) The phosphine (3A) when treated with an excess of (C₆H₅CN)₂PdCl₂ gave the orange crystalline bridged complex, (phosphine)₂(PdCl₂)₂, i.e., dichlorobisphosphine-$\mu\mu'$-dichlorodipalladium, dec. 228–235°. (f) No Diels–Alder adducts could be obtained by the action of maleic anhydride or dimethyl acetylene-dicarboxylate on the phosphine (3A), which was recovered unchanged.

The configuration of the two phosphines (3A) and (3B) can be deduced from their proton nmr spectra. The general principle is that the difference in the chemical shifts in the olefin proton resonances of the two phosphines (and also that in the corresponding resonances in the spectra of the two phosphine oxides) show that the protons on atoms C-1 and C-7 are more shielded, and those on atoms C-2 and C-3 are less shielded, in the (3B) series than in the (3A) series—a similarity which incidentally confirms the retention of configuration during H_2O_2 oxidation of the tertiary phosphines to their oxides. If the major contribution to these differences is the long-range shielding based on the magnetic anisotropy of the benzene ring, it follows that the phosphines (3A) and (3B) have the configurations (3A') and (3B'), respectively.

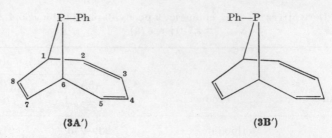

(3A')   (3B')

It is noteworthy that the conversion of the phosphine (**2**) into (**3A**) is stereospecific, since only (**3A**) is formed, although this phosphine is less stable than its epimer (**3B**) under other conditions. It is suggested that the conversion (**2**) → (**3A**) is an intramolecular rearrangement,

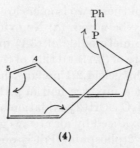

(**4**)

which probably proceeds as indicated in (**4**); as a bond forms between P and C-4, the phenyl group moves from a position equidistant to C-4 and C-5 towards C-5, with the formation of (**3A'**). [The phosphine (**2**) is shown as the *anti*-form in (**4**), because a molecular model of the *syn*-form, epimeric at the P atom and permitting an intramolecular rearrangement to the phosphine (**3A**), could not be constructed.]

The mechanism of the acid-catalyzed conversion of (**3A'**) into (**3B'**) remains uncertain; the authors discuss several possible mechanisms, but at present there is no decisive evidence for any one of them.

The original paper[229] should be consulted for the complete nmr and ultraviolet data, and their discussion.

Four-membered Ring Systems containing only Carbon and Two Phosphorus Atoms

1,2-Diphosphete, 1,2-Diphosphetins, and 1,2-Diphosphetane

This ring system (**1**), [RRI —], could theoretically give rise to three isomeric dihydro derivatives, namely, the 1,2-, 1,4- and 3,4-dihydro derivatives [(2), (3), and (4) respectively] collectively known as 1,2-diphosphetins, and to the tetrahydro derivative, 1,2-diphosphetane (5). The three isomeric 1,2-diphosphetins can be distinguished by naming

```
HC₄——P₁      HC——PH      H₂C——PH      H₂C——P       H₂C——PH
  ‖ ‖ ‖        ‖  |         |   |         |   ‖        |    |
HC³——P²      HC——PH      HC===P       H₂C——P       H₂C——PH
   (1)          (2)          (3)          (4)          (5)
```

the position of the double bond, e.g., in (2), by the prefix Δ^3- or (less clearly) 3-, the full name thus being Δ^3-1,2-diphosphetin or 3-1,2-diphosphetin.

Only one member of the above systems is known, and in preparation and properties it is closely allied to the analogous derivative in the 1H-1,2,3-triphospholene series (p. 202). Mahler[230] has shown that when tetrakis(trifluoromethyl)cyclotetraphosphine $(F_3C—P)_4$, or pentakis-(trifluoromethyl)cyclopentaphosphine $(F_3C—P)_5$, is heated with an excess of bis(trifluoromethyl)acetylene, $F_3C—C\equiv C—CF_3$, a 55% yield of tetrakis(trifluoromethyl) 1,2-diphosphetin (6) is obtained, together with a 31% yield of 1,2,3,4,5-pentakis(trifluoromethyl)-1,2,3-triphospholene (p. 202).

```
F₃C—C——P—CF₃
     ‖   |
F₃C—C——P—CF₃
      (6)
```

The structure of the diphosphetin (6) has been established by analysis, molecular-weight determinations (mass spectrometer and gas density), and by its infrared spectrum, which shows the C=C group by a band at 1625 cm^{-1}; its nmr spectrum confirms this structure. It is a liquid, which has b.p. 110° (estimated from vapor pressure) and which does not freeze at −120°; it is spontaneously inflammable in air, but is otherwise stable at 200°. Its stability to iodine is shown by the fact that when 1,2,3,4,5-pentakis(trifluoromethyl)-1,2,3-triphospholene (p. 202) is heated with iodine at 170° it gives equivalent amounts of diiodo(trifluoromethyl)phosphine, $F_3C—PI_2$, and the diphosphetin (6), the latter being resistant to further action of iodine. The stability of the P–P bond in the diphosphetin (6) is thus considerably greater under these conditions than that in the homocyclic polyphosphines $(F_3C—P)_4$ and $(F_3C—P)_5$, both of which, when similarly treated with iodine, give solely the diiodophosphine, $F_3C—PI_2$ (Mahler and Burg[231]).

Five-membered Ring Systems containing Carbon and Two Phosphorus Atoms

Diphospholes and Diphospholanes

There are theoretically two isomeric diphospholes, namely, $1H$-1,2-diphosphole (1), [RIS 11565], and $1H$-1,3-diphosphole (2), [RRI —]. These systems are at present known only as derivatives of the tetrahydro compounds, 1,2-diphospholane (3) and 1,3-diphospholane.

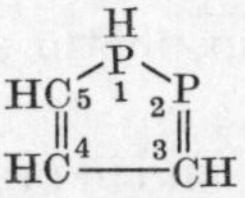

(1) 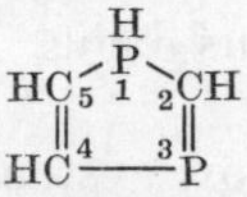   (2) 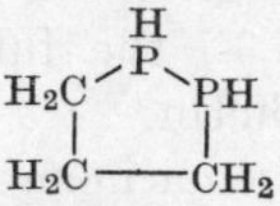   (3)

1,2-Diphospholanes

Issleib and F. Krech [232, 233] have shown that phenylsodiophosphine, C_6H_5NaPH, will react in ether with tri-, tetra-, penta-, or hexa-methylene dibromide to give the corresponding disecondary phosphines (4), where

$$C_6H_5PH(CH_2)_nPHC_6H_5 \qquad\qquad C_6H_5PLi(CH_2)_nPLiC_6H_5$$
$$(4) \qquad\qquad\qquad\qquad (5)$$

$n = 3, 4, 5,$ or 6. These diphosphines in turn will react in warm ether with phenyllithium to give the dilithio compounds (5) which, on addition of dioxane to the final solution, may be readily isolated as the crystalline dioxanates: thus the compounds (5; $n = 3, 4,$ and 6) form di(dioxanates) and (5; $n = 5$) forms a tetra(dioxanate) which, when heated at 100° under reduced pressure, gives a more stable mono(dioxanate).

The compound (5; $n = 3$) in dilute ethereal solution reacts normally with ethyl chloride to give 1,3-bis(ethylphenylphosphino)propane (6), b.p. 196–200°/3 mm. When, however, the compound (5; $n = 3$) is treated

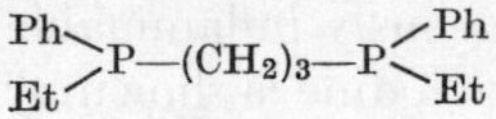

(6) 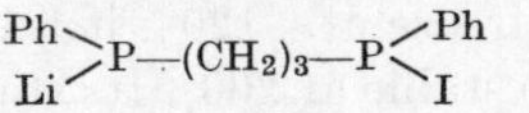   (7)

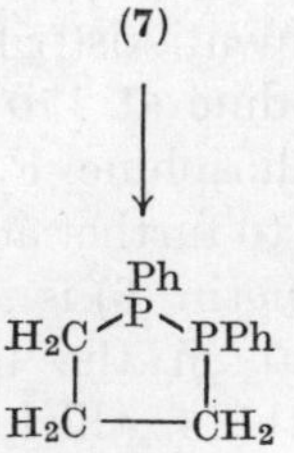

(9)   ←   (8)

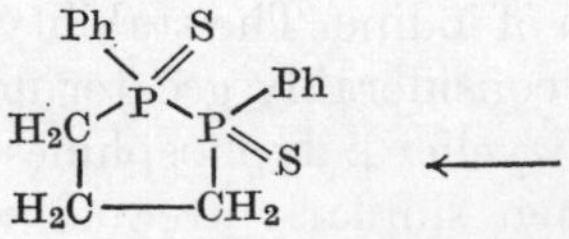

with ethyl iodide, an exceptional metal–halogen exchange occurs to give the lithio iodo derivative (7) and ethyllithium, which presumably reacts with more ethyl iodide to give n-butane. The compound (7) undergoes cyclization to form 1,2-diphenyl-1,2-diphospholane (8), b.p. 184–190°/4 mm.

This process can be followed more definitely by adding a benzene solution of 1,2-dibromoethane dropwise over 3 hours to a suspension of (5; $n = 3$) in hot benzene–tetrahydrofuran. The lithio bromo analogue of (7) is now formed, together with 2-bromoethyllithium, $Br(CH_2)_2Li$, which breaks down liberating ethylene in 90% yield. Removal of the solvent and distillation then gives the diphospholane (8).

The diphospholane (8) forms a 1,2-disulfide (9), m.p. 178–180°, which could theoretically exist in *cis*- and *trans*-isomers: it is possible that steric obstruction prevents the formation of one of the isomers (*cf.* the analogous case of the following disubstituted 1,2-diphosphorinane disulfides; these also do not show this isomerism, which is shown, however, by the 1,4-diphosphorinane disulfides).

It is noteworthy that the lithium–halogen exchange shown by the dilithio compound (5; $n = 3$) is also shown by the corresponding diarsenic compound (p. 444).

In a later synthesis (Issleib and K. Krech[234]), tetraphenylcyclotetraphosphine, $[C_6H_5\text{-}P]_4$, m.p. 154–156° (3 molar equivalents) was treated with potassium (8 atomic equivalents) in tetrahydrofuran to form 1,2,3-triphenyl-1,3-dipotassiotriphosphine (10), which crystallizes

(10)   (11)

with two molecules of tetrahydrofuran. A suspension of these crystals in the same solvent, treated slowly with 1,3-dichloropropane, presumably forms the unstable cyclic triphosphine (11): evaporation of the solvent from the filtered reaction mixture, followed by distillation, gives however 1,2-diphenyl-1,2-diphospholane (8), b.p. 180°/1.5 mm (77%), and the undistilled residue, when crystallized from benzene, yields the initial tetraphosphine. It is suggested therefore that the triphosphine (11) under these conditions breaks down to form the diphosphine (8), the expelled C_6H_5P units combining to form the cyclotetraphosphine.

Note. Two compounds of composition $[C_6H_5P]_n$, melting at 150° and 190°, respectively, have been identified by workers as isomeric forms of $[C_6H_5\text{-}P]_4$. Daly has shown by X-ray crystal analysis that the former is a cyclic pentaphosphine[235] and the latter a cyclic hexaphosphine[236] (p. 205); since their condition in solution is uncertain, the tetraphosphine name for the low-melting compound has for convenience been retained above.

1,3-Diphospholanes

This system, [RIS —], is at present represented by only one member which has been briefly recorded.

Horner, Bercz, and Bercz[237] have shown that benzylmethylphenylphosphine combined with 1,2-dibromoethane to form ethylenebis-(benzylmethylphenylphosphonium) dibromide, which can be separated

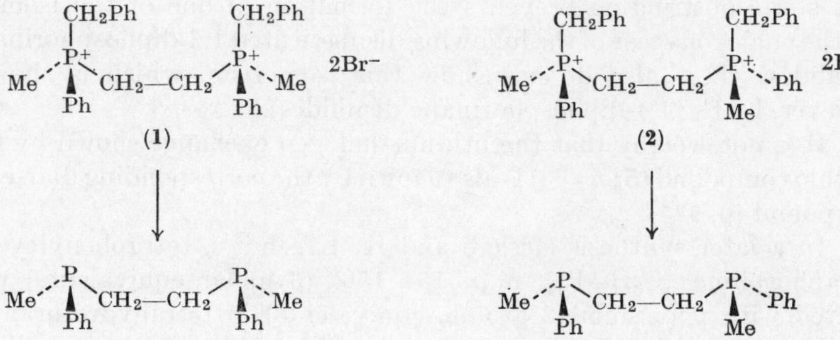

into the *meso-* (1) and the racemic form (2), having m.p.s 278° and 293°, respectively. (The two forms can be readily separated by recrystallization of the corresponding diperchlorates from methanol.) The configurational identity of (2) has been confirmed by optical resolution of the corresponding di[hydrogen (−)-benzoyltartrates]. Cathodic reduction of the two dibromides, by removing the benzyl groups, gives the *meso-*diphosphine (3), m.p. 90°, and the oily racemic diphosphine (4).

The *meso-*diphosphine, when heated with dibromomethane at 100°, gives the *meso-* or *cis-*1,3-dimethyl-1,3-diphenyl-1,3-diphospholanium dibromide (5), m.p. 328°.

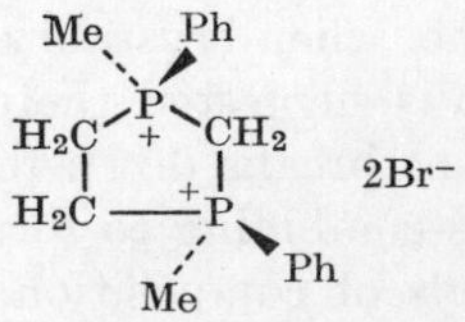

This synthetic route could be widely extended. For example, to obtain a 1,3-disubstituted 1,3-diphospholane, condensation of ethylene-bis(benzylphenylphosphine) (p. 171), with dibromomethane should give the 1,3-dibenzyl-1,3-diphenyl analogue of the dibromide (5); anodic reduction would then presumably give 1,3-diphenyl-1,3-diphospholane.

Six-membered Ring Systems containing Carbon and Two Phosphorus Atoms

Diphosphorins and Diphosphorinanes

The three isomeric compounds (1), (2), and (3), named 1,2-diphosphorin, [RRI —], 1,3-diphosphorin, [RRI —], and 1,4-diphosphorin, [RIS 7755], respectively, are unknown; their hexahydro derivatives are

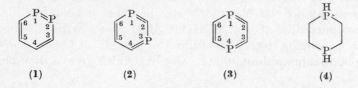

(1)   (2)   (3)   (4)

the corresponding diphosphorinanes, e.g., 1,4-diphosphorinane (4). Derivatives of each of the partly or fully hydrogenated diphosphorins are known.

1,2-Diphosphorin and 1,2-Diphosphorinanes

Issleib and K. Krech[238] have shown that tetraethylcyclotetraphosphine, $(C_2H_5\text{-}P)_4$, reacts exothermally with potassium in cold tetrahydrofuran to give the crystalline 1,2,3,4-tetraethyl-1,4-dipotassiotetraphosphine (1). In the presence of an excess of potassium and with heating for 36 hours, this is converted into 1,2-diethyl-1,2-dipotassiodiphosphine (2). In ethereal suspension at −10° this diphosphine reacts with 1,4-dichlorobutane forming 1,2-diethyl-1,2-diphosphorinane (3; $R = C_2H_5$), b.p. 70–71°/2 mm (70%), which unites readily with sulfur in hot benzene to give the disulfide, m.p. 115° (79%).

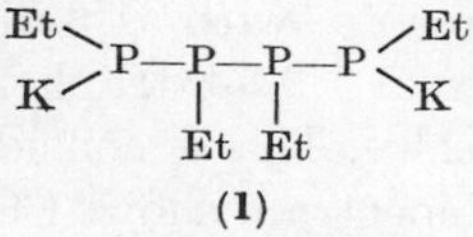

Et⟍   ⟋Et
   P—P—P—P
K⟋ | | ⟍K
    Et Et

(1)

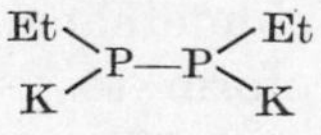

Et⟍   ⟋Et
   ⟍P—P⟋
K⟋   ⟍K

(2)

      R
      P⟍
     ⟋   PR

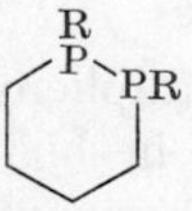

(3)

The same authors[234] have shown that tetraphenylcyclotetraphosphine, $(C_6H_5-P)_4$, reacts similarly with potassium to give 1,2-diphenyl-1,2-dipotassiodiphosphine, which in turn reacts with 1,4-dichlorobutane in tetrahydrofuran at room temperature to give 1,2-diphenyl-1,2-diphosphorinane (**3**; $R = C_6H_5$), m.p. 90° (91%). This also gives a disulfide, m.p. 171°. The disulfides of each of these phosphorinanes (**3**) could theoretically exist in *cis*- and *trans*-forms, although such forms have not been recorded. The sharp melting point of each disulfide indicates that it is very probably one form only; steric hindrance may prevent the existence of a second form.

The two compounds (**3**; $R = C_2H_5$ or C_6H_5) were originally given the alternative names based on 1,2-diphosphacyclohexane.[238, 234]

1,3-Diphosphorin and 1,3-Diphosphorinanes

Some very interesting derivatives of the above compounds (p. 167) have been studied by Märkl,[239] who has shown that bis(diphenylphosphino)methane, $(C_6H_5)_2PCH_2P(C_6H_5)_2$ reacts with 1,3-dibromopropane in boiling tetrahydrofuran to give 1,1,3,3-tetraphenyl-1,3-diphosphorinanium dibromide (**1**; X = Br), which gives a diperchlorate

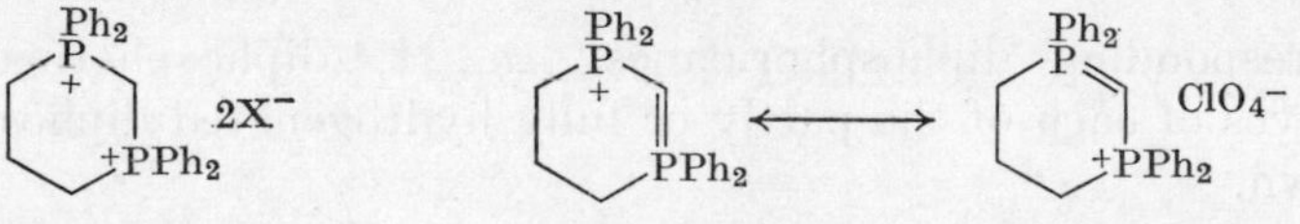

(**1**)   (**2A**)   (**2B**)

(**1**; X = ClO_4), m.p. 345–347°. This dibromide reacts quantitatively with aqueous sodium hydroxide to give the 1,4,5,6-tetrahydro-1,1,3,3-tetraphenyl-1,3-diphosphorinium cation, stabilized as a resonance hybrid (**2A** ↔ **2B**), and isolated as the perchlorate, m.p. 255–257°.

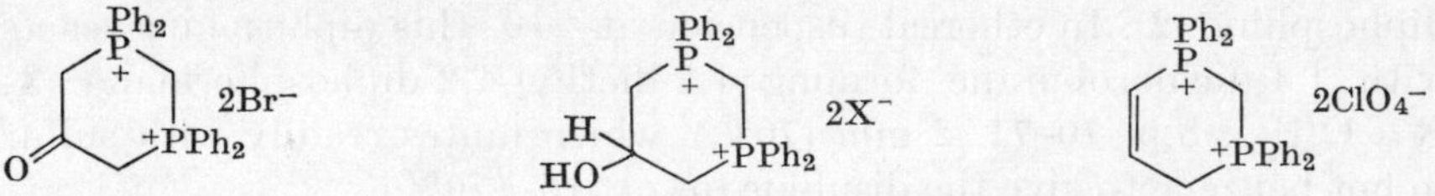

(**3**)   (**4**)   (**5**)

Bis(diphenylphosphino)methane also reacts with 1,3-dibromoacetone in boiling chloroform to give 5-oxo-1,1,3,3-tetraphenyl-1,3-diphosphorinanium dibromide (**3**), m.p. 230–250°. This is considered to be predominantly the enol form; its infrared spectrum shows CO bands at 1706 and 1725 cm^{-1} and an OH band at 2800 cm^{-1}, and the salt gives an orange-red ferric chloride color.

The dibromide (3) when reduced with sodium boron hydride in methanol gives 5-hydroxy-1,1,3,3-tetraphenyl-1,3-diphosphorinanium dibromide (4; X = Br), also isolated as the diperchlorate (4; X = ClO$_4$), m.p. 287–288° (68%). The dibromide in boiling o-dichlorobenzene breaks down to the original diphosphinomethane and 1,3-dibromo-2-propanol, but when heated with anhydrous phosphoric acid at 180–200° for 15–20 minutes undergoes dehydration to the 1,2,3,4-tetrahydro-1,1,3,3-tetraphenyl-1,3-diphosphorinium cation, isolated as the diperchlorate (5), m.p. 320–321° (86%); its infrared spectrum shows a C=C band at 1600 cm^{-1}.

The diperchlorate (5) when treated with aqueous sodium carbonate loses a proton from its cation to give the 1,2-dihydro-1,1,3,3-tetraphenyl-1,3-diphosphorinium cation (6A), of which (6B) is another canonical

(6A)   (6B)   (7A)   (7B)

form, isolated as the perchlorate, m.p. 236–237° (λ_{max} 340 mμ in methanol; C=C, 1620 cm^{-1}). Aqueous sodium hydroxide takes this process one stage further, with the formation of 1,1,3,3-tetraphenyl-1,3-diphosphorin (7A ↔ 7B), or '1,1,3,3-tetraphenyl-1,3-diphosphabenzene',[239] a yellow voluminous precipitate which is stable in cold water, and when dry is easily soluble in ether, benzene, and other organic solvents (λ_{max} 281 and 384 mμ in ether).

Each of the phosphorus atoms in (2A ↔ 2B) and in (6A ↔ 6B) is thus partly of phosphonium and partly of phosphorane nature, whereas those in (7A ↔ 7B) are both phosphorane in type.

1,3-Benzodiphosphorin

This ring system (1), [RRI —], is at present known only as quaternary salts of 1,2,3,4-tetrahydro-1,3-benzodiphosphorin (2). Davis and Mann [240] have condensed [2-(diethylphosphino)-5-methylbenzyl]diethylphosphine (3) with dibromomethane to form 1,1,3,3-tetraethyl-1,2,3,4-tetrahydro-6-methyl-1,3-benzodiphosphorinium dibromide (4), m.p. 308–309° (dipicrate, m.p. 189°). The compound (4) was initially termed 1,1,3,3 - tetraethyl - 1,2,3,4 - tetrahydro - 6 - methyl - 1,3 - diphosphonianaphthalene dibromide; the ring system was not further investigated.[240]

7

$$(1) \qquad (2)$$

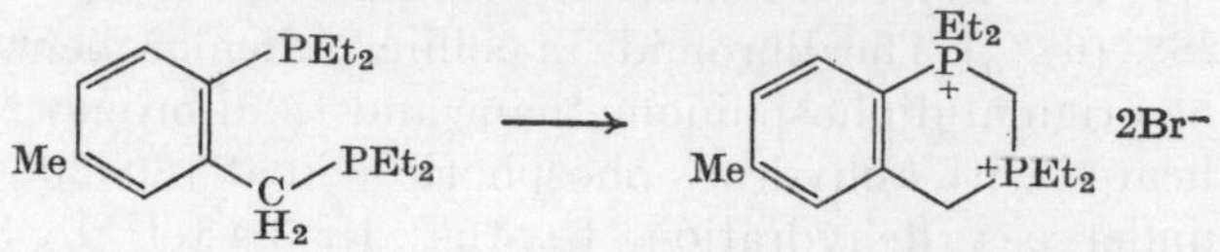

$$(3) \longrightarrow (4) \quad 2Br^-$$

1,4-Diphosphorin and 1,4-Diphosphorinanes

The various syntheses of the 1,4-diphosphorinane system (p. 167) are listed as follows.

(I) The first synthesis of this system was achieved by Hitchcock and Mann,[241] who treated phenylphosphine in liquid ammonia with

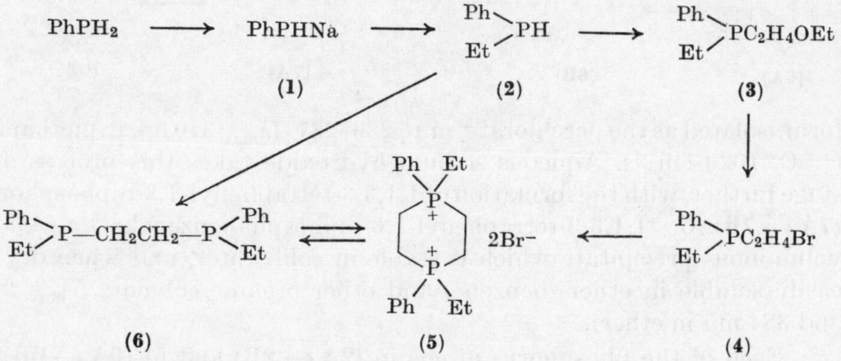

$$PhPH_2 \longrightarrow PhPHNa \longrightarrow \underset{Et}{\overset{Ph}{>}}PH \longrightarrow \underset{Et}{\overset{Ph}{>}}PC_2H_4OEt$$

$$(1) \qquad\qquad (2) \qquad\qquad (3)$$

$$\underset{Et}{\overset{Ph}{>}}P—CH_2CH_2—P\underset{Et}{\overset{Ph}{<}} \rightleftharpoons (5) \quad 2Br^- \longleftarrow \underset{Et}{\overset{Ph}{>}}PC_2H_4Br$$

$$(6) \qquad\qquad (5) \qquad\qquad (4)$$

sodium to give phenylsodiophosphine (**1**); treatment *in situ* with ethyl bromide (1 equivalent) formed ethylphenylphosphine (**2**), then with sodium gave the sodio derivative, which with ethyl 2-iodoethyl ether gave (2-ethoxyethyl)ethylphenylphosphine (**3**). The latter, after evaporation of the ammonia, was treated in boiling acetic acid with hydrogen bromide to give (2-bromoethyl)ethylphenylphosphine (**4**). This oily phosphine underwent self-quaternization erratically and vigorously, but in boiling toluene solution smoothly gave the 1,4-diethyl-1,4-diphenyl-1,4-diphosphorinanium dibromide (**5**), dec. at *ca.* 370° (dipicrate, m.p. 240–245°; dichloride, dec. at *ca.* 350°). [The salt (**5**) was initially termed 1,4-diethyl-1,4-diphenyl-1,4-diphosphoniacyclohexane dibromide[241]]. Alternatively the phosphine (**2**) was treated again in liquid ammonia with sodium and then with 1,2-dibromoethane to give ethylenebis(ethylphenylphosphine) (6), which when isolated and heated with 1,2-dibromoethane gave the dibromide (5), in yield inferior to that of the first method.

Therapeutic tests showed that the dichloride was ineffective against a *Trypanosoma congolese* in mice and against *Streptococcus haemolyticus in vitro*; it showed a feeble activity against a mouse nematode infection.

The 1,4-diphenyl-1,4-diphosphorinane could not be obtained from the dibromide (5) by thermal decomposition, which regenerated the linear diphosphine (6).

To overcome this difficulty, R. C. Hinton and Mann [242] condensed dibenzylphenylphosphine with 1,2-dibromoethane to give ethylenebis(dibenzylphenylphosphonium) dibromide (7; $R = C_6H_5$), which

$$[(C_6H_5CH_2)_2RP^+\!\!-\!C_2H_4\!-\!P^+R(CH_2C_6H_5)_2]Br_2 \rightarrow$$
$$(7)$$

$$(C_6H_5CH_2)RP\!-\!C_2H_4\!-\!P\!-\!R(CH_2C_6H_5)$$
$$(8)$$

when reduced in tetrahydrofuran with lithium aluminum hydride (the method of Bailey and Buckler [243]) gave ethylenebis(benzylphenylphosphine) (8; $R = C_6H_5$); condensation with 1,2-dibromoethane now gave 1,4-dibenzyl-1,4-diphenyldiphosphorinanium dibromide (9; $R = C_6H_5$), m.p. 375° (dec.) (dipicrate, m.p. indefinite), which when similarly reduced gave 1,4-diphenyl-1,4-diphosphorinane (10; $R = C_6H_5$), m.p. 92–95°. This compound gave a dioxide (11; $R = C_6H_5$), m.p. 250°, which showed the normal molecular weight in chloroform.

Repetition of this synthesis starting with tribenzylphosphine gave in turn the dibromide (7; $R = CH_2C_6H_5$), m.p. 275–279° (dipicrate, m.p. 220–222°), the diphosphine (8; $R = CH_2C_6H_5$), (identified as the diphosphine-dibromopalladium, $C_{30}H_{32}P_2,PdBr_2$, m.p. 215–217°), the dibromide (9; $R = CH_2C_6H_5$), m.p. 287–292° (dipicrate, m.p. 214°), and finally 1,4-dibenzyl-1,4-diphosphorinane (10; $R = CH_2C_6H_5$), m.p. 128–130°. This compound, treated in ether with a stream of air, gave the monooxide, m.p. 302–305°, and with hydrogen peroxide in acetone gave the insoluble dioxide (11; $R = CH_2C_6H_5$). It is noteworthy that cold

benzyl bromide converted the 1,4-dibenzyl-1,4-diphosphorinane into
the diquaternary dibromide (**9**; R = CH$_2$C$_6$H$_5$), but hot benzyl bromide
rapidly caused rupture of the heterocyclic ring with formation of the
linear dibromide (**7**; R = CH$_2$C$_6$H$_5$).[242]

(II) A second synthesis, developed by Issleib and Standtke,[244] and
Issleib and Döll,[245] has the advantage of avoiding intermediate diquaternary salts.

$$\text{RPH—C}_2\text{H}_4\text{—PHR} \rightarrow \text{RPLi—C}_2\text{H}_4\text{—PLiR} \rightarrow (\textbf{10}; \text{R} = \text{C}_6\text{H}_5)$$
$$\quad (\textbf{12}) \qquad\qquad\qquad (\textbf{13})$$

For the synthesis of 1,4-diphenyl-1,4-diphosphorinane (**10**; R =
C$_6$H$_5$), phenylsodiophosphine (**1**) was treated in ether solution at −20°
with 1,2-dichlorocthane to give ethylenebis(phenylphosphine) (**12**;
R = C$_6$H$_5$); this, when again treated in ether with phenyllithium at room
temperature with subsequent boiling, gave the dilithio derivative (**13**;
R = C$_6$H$_5$), which on the addition of dioxane crystallized as the bisdioxanate. The latter, on treatment with 1,2-dichloroethane in ether,
gave 1,4-diphenyl-1,4-diphosphorinane.[244]

Similarly, ethylpotassiophosphine gave ethylenebis(ethylphos-
phine) (**12**; R = C$_2$H$_5$), which formed the dilithio derivative (**13**;
R = C$_2$H$_5$), which also crystallized as a bisdioxanate. Addition of 1,2-
dichlorethane gave[245] 1,4-diethyl-1,4-diphosphorinane (**10**; R = C$_2$H$_5$),
b.p. 135–145°/0.4 mm.

Cyclohexylpotassiophosphine, subjected to the same process,[245]
gave 1,4-dicyclohexyl-1,4-diphosphorinane (**10**; R = C$_6$H$_{11}$), b.p. 225–
230°/0.2 mm.

In the above synthesis, to avoid the formation of polymeric
material the dilithio compound (**13**) must be added to the 1,2-dichloro-
ethane, and not *vice-versa*.

(III) A novel reaction was employed by Rauhut, Borowitz, and
Gillham,[246] who showed that (2-acetoxyethyl)dialkylphosphines (**14**)

$$2\text{R}_2\text{P—CH}_2\text{CH}_2\text{—O—COCH}_3 \longrightarrow \begin{array}{c} \text{R}_2 \\ \text{P} \\ + \\ \\ + \\ \text{P} \\ \text{R}_2 \end{array} \quad 2\text{X}^-$$

$$\qquad\qquad (\textbf{14}) \qquad\qquad\qquad\qquad\qquad (\textbf{15})$$

$$[\text{CH}_3(\text{CH}_2)_3]_2\text{P—(CH}_2)_3\text{—O—COCH}_3$$
$$(\textbf{16})$$

when heated to 80° underwent, often after an induction period, a vigorous
self-quaternization to give the corresponding 1,1,4,4-tetrasubstituted

diphosphorinanium diacetates (15; $X = OCOCH_3$). These were the main products if the rise in temperature was uncontrolled: when, however, this rise was controlled by the use of a solvent or by cooling, a polymeric material was the chief product. The addition of a trace of this material to the phosphine (14) caused the self-quaternization to the salt (15) to proceed smoothly.

The reactivity of the phosphines (14) in this reaction was dependent on the group R and fell off steadily in the order $R = (CH_2)_2CN$, $(CH_2)_3CH_3$, $(CH_2)_7CH_3$, and $(CH_2)_2CH(CH_3)_2$. A mechanism for the

Table 11. Quaternary 1,4-diphosphori-
nanium salts of type (15)

R	X	M.p. (°)*
n-C_4H_9	OAc	195
n-C_4H_9	Br	316–319
n-C_4H_9	I	>335
iso-C_4H_9	OAc	157–158
iso-C_4H_9	I	337
n-C_8H_{17}	I	298–299
C_6H_{11}†	OAc	221–223
$(CH_2)_2CN$	OAc	149–150
$(CH_2)_2CN$	Cl	286–288
$(CH_2)_2CN$	Br	303–304
$(CH_2)_2CN$	I	320
$(CH_2)CO_2H$	Cl	280–282

* All melting with decomposition except
the last.
† Cyclohexyl.

formation of various polymeric products, and their breakdown at higher temperatures to give the diacetates (15) was suggested, based partly on the observation that a tertiary phosphine readily displaces an acetoxy group located in the β-position to a quaternary phosphonium group. It is noteworthy that (3-acetoxypropyl)-di-n-butylphosphine (16), in which such a group in this location does not occur, is unchanged when heated to 300°.

The quaternary salts of type (15) that were thus prepared are recorded in Table 11.

The compound [15; $R = (CH_2)_2CN$; $X = Br$] was independently prepared by the action of 1,2-dibromoethane on ethylenebis[di-(2-cyanoethyl)phosphine] and on bis-(2-cyanoethyl)phosphine, and by the

action of vinyl bromide on bis-(2-cyanoethyl)phosphine in the presence
of αα'-azobisisobutyronitrile, added as an initiating agent; in the last
two reactions the intermediate (2-bromoethyl)bis-(2-cyanoethyl)phos-
phine was not isolated.

(IV) A synthetic method akin to the above reaction with vinyl
bromide, and of considerable potential value, has been briefly announced
by Märkl.[247] Diphenylvinylphosphine, $(C_6H_5)_2PCH{=}CH_2$, in the
presence of a proton donor such as acetic acid undergoes a type of
dimerization with addition of two protons to form 1,1,4,4-tetraphenyl-
1,4-diphosphorinanium diacetate; other proton donors are triethylammonium hydrochloride and α-halohydrins.

This ready reaction may indicate that the exothermic dimerization
of (2-bromoethyl)ethylphenylphosphine (4) to give the 1,4-diphosphor-
inanium dibromide (5) does not proceed by a simple quaternization, but
by a preliminary decomposition of the phosphine (4) to hydrogen
bromide and ethylphenylvinylphosphine, which undergo Märkl's
reaction so rapidly that they are in effect transient intermediates.

(V) A very interesting synthesis (although of limited application
in phosphorus work) has been developed by Krespan and Lang-
kammer,[87, 88] who heated a mixture of tetrafluoroethylene, red
phosphorus, and iodine in a closed vessel at 220° for 8 hours. Distillation
of the product gave 2,2,3,3,4,4,5,5-octafluoro-1-iodophospholane (p. 156)
(4% based on the iodine). The undistilled residue, when sublimed at

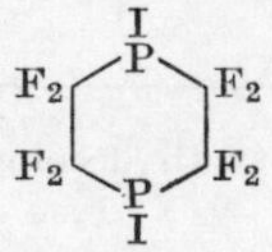

(17)

50–60°/1 mm, gave pale yellow crystals of 2,2,3,3,5,5,6,6-octafluoro-1,4-
diiodo-1,4-phosphorinane (17) (27% based on the iodine). The compound
can be crystallized from petroleum and is stable in cold water.

A brief nmr spectral investigation indicated that the ring has a
puckered form, and that the two fluorine atoms attached to each carbon
are non-equivalent 'because of the lack of symmetry in the ring'. The
eight fluorine atoms are distributed in two non-equivalent sets of four
atoms each; the inequality of the coupling of the two sets is the result
of two different F–C–P angles. The spectral pattern did not change
when taken at 100°, and it is concluded that there is a high barrier to
axial–equatorial inversion.

Properties of the 1,4-*diphosphorinanes*

Stereochemistry. The diquaternary salts of the types (**5**) and (**9**; R = $CH_2C_6H_5$) showed no indication of *cis–trans*-isomerism of the cation. Horner, Bercz, and Bercz[237] have recently shown, however, that the *meso*-ethylenebis(methylphenylphosphine),

$$CH_3(C_6H_5)PCH_2CH_2P(C_6H_5)CH_3,$$

(p. 166) undergoes cyclic diquaternization with 1,2-dibromoethane to give the *cis*-1,4-dimethyl-1,4-diphenyl-1,4-diphosphorinanium dibromide, m.p. 332°, while the racemic diphosphine similarly gives the corresponding *trans*-dibromide, m.p. 350°.

1,4-Diphenyl-1,4-diphosphorinane 1,4-disulfide (**18**) has been obtained in two forms, m.p. 154° and 253°, undoubtedly *cis–trans*-isomers; the low-melting form is not converted into the high-melting form even at 200° (Isslieb and Standtke[244]).

Similarly 1,4-dicyclohexyl-1,4-diphosphorinane forms two isomeric disulfides, m.p. 250–255° and 325–326°, respectively; 1,4-diethyl-1,4-diphosphorinane gave apparently only one disulfide, m.p. 225–235°,

(18)   (19)   (20)   (21)

but this indefinite m.p. may suggest a mixture.[245] No isomerism of the dioxides (**11**), corresponding to that of the disulfides, has been detected: their low solubility makes investigation difficult.

The 1,4-diphosphorane ring system can readily assume the boat conformation. Thus the 1,4-dibenzyl derivative forms the orange crystalline monomeric palladium dibromide complex (**19**; R = $CH_2C_6H_5$), showing a normal molecular weight in chloroform, and an insoluble buff-colored product of identical composition, almost certainly the salt (**20**); the first compound is readily soluble, and the second insoluble, in ethanol (R. C. Hinton and Mann[242]).

Further evidence of this conformation is the ready condensation of the 1,4-diphenyl- and the 1,4-dibenzyl-1,4-diphosphorinane with 1,2-dibromoethane to give the corresponding diquaternary salt (**21**; R = C_6H_5 or $CH_2C_6H_5$) of 'triethylenediphosphine', (1,4-diphospha-bicyclo[2.2.2]octane) (p. 180).[242]

Hydrolysis of quaternary salts. The action of alkali hydroxides on unhindered quaternary phosphonium salts is normally to give a tertiary phosphine oxide and a hydrocarbon:

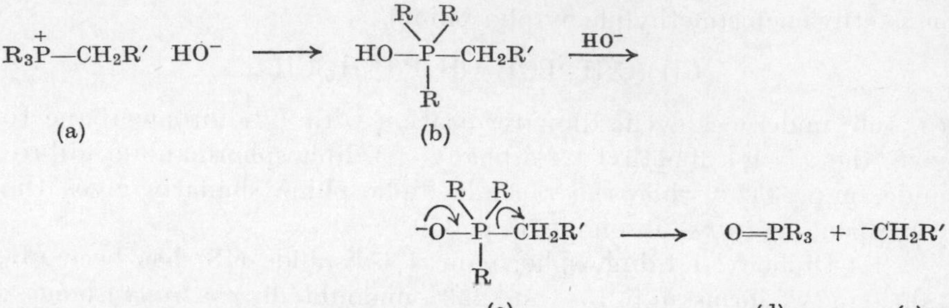

These products almost certainly arise from the initial attack of hydroxyl ion on the phosphonium cation (a) to give the phosphorane-type intermediate (b), which is deprotonated by further hydroxyl-ion attack to the anion (c); this breaks down as shown, to give the phosphine oxide (d) and the carbanion (e), which by proton addition gives the hydrocarbon (Zanger, VanderWerf, and McEwen [248]).

Aguiar, Aguiar, and Daigle [249] have prepared 1,1,4,4-tetraphenyl-1,4-diphosphorinanium dibromide (**22**), m.p. 324–325°, by the cyclic diquaternization of ethylenebis(diphenylphosphine) with 1,2-dibromo-ethane. They point out that the above 'normal' action of alkali should

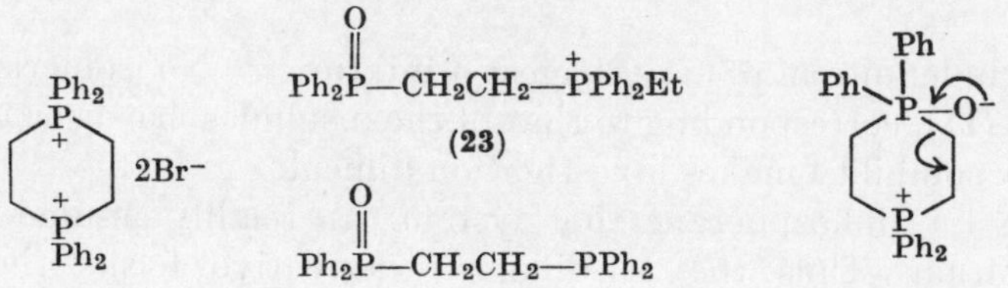

give the cation (**23**), which in turn should decompose normally into $(C_6H_5)_2P(O)CH_2CH_2P(O)(C_6H_5)(C_2H_5)$ or even into $(C_6H_5)_2(C_2H_5)PO$: the actual product is ethylenebis(diphenylphosphine) monooxide (**24**), presumably formed by β-elimination from the normal intermediate (**25**) with liberation of ethylene, although the final stages of this process remain uncertain.

This reaction recalls the action of aqueous sodium hydroxide on the di(benzylobromide) of 5,10-dihydro-5,10-diphenylphosphanthrene, whereby both the monooxide and the expected dioxide are formed (p. 189), but in this case without fission of the heterocyclic ring.

1,2,3,4-*Tetrahydro*-1,4-*diphosphorins*

Compounds of this type, intermediate in degree of hydrogenation between the diphosphorins and the diphosphorinanes, have been prepared by Aguiar and Aguiar.[250] *cis*-Vinylenebis(diphenylphosphine) (**26**) (cf. Aguiar and Daigle[251]) reacts with boiling 1,2-dibromoethane

(**26**)　　　　　(**27**)　　　　　(**28**)

to give 1,2,3,4-tetrahydro-1,1,4,4-tetraphenyl-1,4-diphosphorinium dibromide (**27**), m.p. 298–300° (dipicrate, m.p. 273–275°). The nmr spectrum of (**27**) in trifluoroacetic acid at 60 Mc shows two peaks centered at τ 1.12 ppm, 51 cps apart with a broad complex at the center, a complex centered at τ 2.15 ppm, and a doublet centered at τ 6.11 (J_{PCH} 8 cps), in the ratio 1:10:2, assigned to *cis*-vinyl, the phenyl and the CH_2 protons, respectively. The significance of this spectrum and those of various other quaternary phosphonium salts is discussed by the authors.[250]

Hot dilute aqueous sodium hydroxide reacts with the cation of the salt (**27**) to give the monooxide (**24**), which indicates that the intermediate form (**28**) may break down as shown, although the evolution of acetylene, which this mechanism would presumably entail, has not been detected.

The compound (**27**) was initially termed 1,1,4,4-tetraphenyl-1,4-diphosphoniacyclohexene-2 dibromide.[250]

1,4-*Dihydro*-1,4-*diphosphorins*

Derivatives of this type were obtained (Aguiar, Hansen, and Reddy[252]) by dissolving diphenyl(phenylethynyl)phosphine (**29**) in glacial acetic acid containing hydrogen bromide: after 100 hours at room temperature, the solution yielded 1,4-dihydro-1,1,2,4,4,5-hexaphenyl-1,4-diphosphorinium dibromide (**30**), m.p. 286–290° (dipicrate, m.p. 246–248°). The nmr spectrum of the dibromide in trifluoroacetic acid at 60 Mc shows two sharp peaks at τ 0.6 and 1.5, with a shallow broad complex centered between them; this is very similar to the spectrum of the dibromide (**27**).

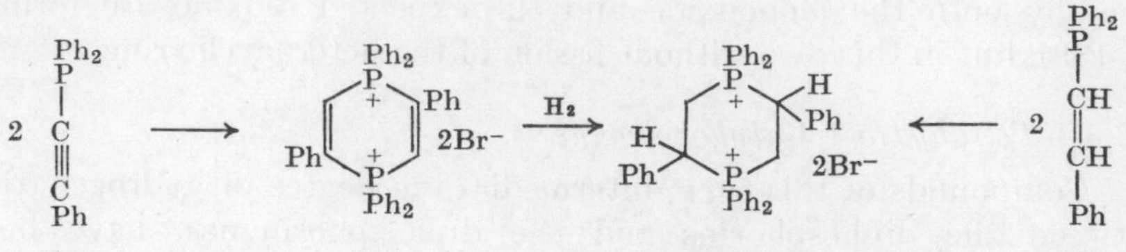

(29)   (30)   (31)   (32)

The dibromide (**30**) in methanolic solution can be hydrogenated over a rhodium–alumina catalyst to give 1,1,2,4,4,5-hexaphenyl-1,4-diphosphorinanium dibromide (**31**), m.p. 269–272° (dipicrate, m.p. 230–233°). This salt can also be obtained by similarly dissolving either the *cis-* or the *trans-*form of diphenylstyrylphosphine (**32**) at room temperature in acetic acid containing hydrogen bromide.

At 40.5 Mc/sec, the ^{31}P nmr shift (ppm of methanolic solution relative to external standard of 85% H_3PO_4) is +3.5 for (**30**) and −20.4 for (**31**). It is claimed that the dibromide (**30**) is the first recorded phosphonium salt having a positive ^{31}P nmr shift; most such salts have

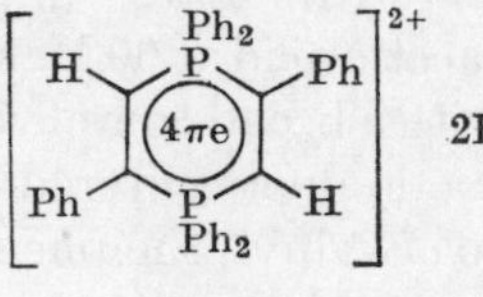

(33)

shifts in the region −20 to −30 cps. These spectra and the ultraviolet spectra of (**30**) are considered to be in accord with delocalization of the π-electronic charge over the phosphorus atoms, so that the dibromide (**30**) can be more adequately represented as (**33**).

The dibromides (**30**) and (**31**) were initially termed 1,1,2,4,4,5-hexaphenyl-1,4-diphosphonia-cyclohexadiene-2,5 dibromide, and -cyclohexane dibromide, respectively; the ring system in (**30**), was termed (incorrectly) 1,4-dihydrophospha(v)pyrazine.[252]

The above synthetic route was later used (Aguiar and Hansen[253]) for the similar condensation of ethynyldiphenylphosphine (**34**) to form 1,4-dihydro-1,1,4,4-tetraphenyl-1,4-diphosphorinium dibromide (**35**;

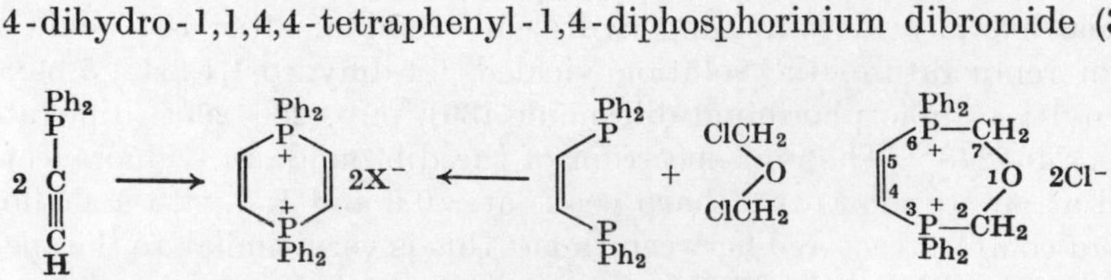

(34)   (35)   (26)   (36)   (37)

X = Br), m.p. 306–308° (dipicrate, m.p. 235–237°). This dibromide was of value in the study of the unexpected product which resulted from the interaction of *cis*-vinylenebis(diphenylphosphine) (26) and bis-(chloromethyl) ether (36). This reaction was expected to proceed by cyclic diquaternization to give 2,3,6,7-tetrahydro-3,3,6,6-tetraphenyl-1,3,6-oxadiphosphepinium dichloride (37) (cf. p. 178), precisely as ethylenebis(diphenylphosphine) combines with the ether (36) to form the 4,5-dihydro derivative of (37), *i.e.*, the dichloride (40). The product from the interaction of (26) and (36), although having the composition of (37), was shown to be 1,4-dihydro-1,1,4,4-tetraphenyl-1,4-diphosphorinium dichloride monohydrate (35; X = Cl), m.p. 251–253°. This identification is based on the evidence: (*a*) the dichloride gives an anhydrous dipicrate, m.p. 235–237°, unchanged when mixed with the dipicrate from (35; X = Br); (*b*) the infrared spectrum of the dichloride is devoid of the C–O single-bond stretching band; (*c*) the dibromide and the dichloride have identical infrared and proton nmr spectra, and the latter give no indication of CH_2 protons.

Methanolic solutions of the dichloride (35; X = Cl) showed a ^{31}P nmr shift of +16.0 ppm, relative to 85% H_3PO_4 (external standard), which the authors claim is consistent, as in hexaphenyl analogue (30), with delocalization of the four π-electrons over the phosphorus $3d$ orbitals.

The mechanism whereby the dichloride (35; X = Cl) arises from the interaction of (26) and (36) is unknown; this salt may be the end-product of one definite sequence of reactions, or the end-product of one of several competing reactions. A complex series of reactions following the initial formation of the dichloride (37) has been suggested. Alternatively, initial monoquaternization of (26) may give the monochloride

(38), which undergoes cleavage as shown, regenerating the ethynyl-phosphine (34) with formation of hydrogen chloride and (chloromethoxy-methyl)diphenylphosphine (39): combination of (34) and the HCl would then give the dichloride (35; X = Cl). It is noteworthy that when, after the interaction of (26) and (36) and the removal of the crystalline dichloride (35; X = Cl), the mother-liquor on hydrogenation furnished

3,3,6,6-tetraphenyl-1,3,6-oxadiphosphepanium dichloride (**40**), which gave the same dibromide as that prepared directly (p. 318) (Aguiar[254]); a certain amount of the dichloride (**37**) must thus have been formed in the early stages of the reaction.

1,4-Diphosphabicyclo[2.2.2]octane

This system, [RIS 10049], is represented in the Ring Index as (**1**): the general conformation, with the pyramidal phosphorus atoms at the two apices of the molecule, is represented more clearly by (**1A**). The

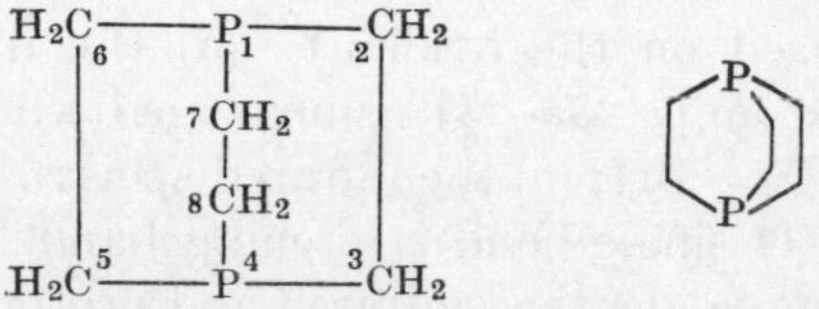

(**1**)   (**1A**)

diphosphine (**1**), which was initially and conveniently also termed triethylenediphosphine, can be regarded as 1,4-diphosphorinane with the terminal hydrogen atoms replaced by a bridging —CH_2CH_2— group; historically it is the first known organic phosphine with phosphorus in a bridge-head position.

The preparation of 1,4-dibenzyl-1,4-diphosphorinane (**2**; R = $CH_2C_6H_5$) has been described (p. 171). This compound when heated with

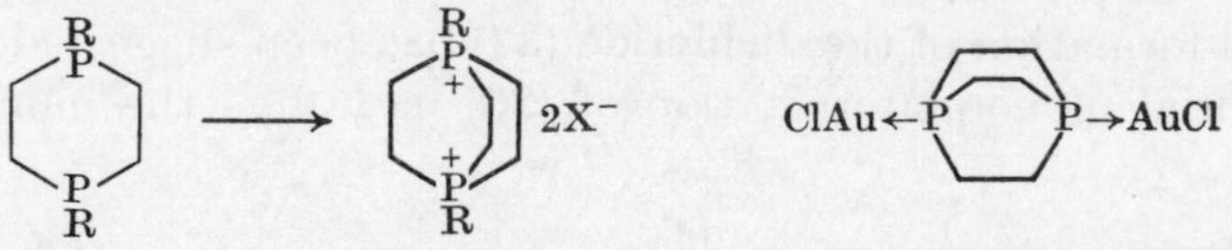

(**2**)   (**3**)   (**4**)

1,2-dibromoethane gave the highly deliquescent 1,4-dibenzyl-1,4-diphosphoniabicyclo[2.2.2]octane dibromide (**3**; R = $CH_2C_6H_5$, X = Br), m.p. 340–344°, which was also characterized as the dipicrate (**3**; R = $CH_2C_6H_5$, X = $C_6H_2N_3O_7$); these salts formed a stable hemihydrate and a diethanolate, respectively. The dibromide suspended in tetrahydrofuran was reduced with lithium aluminum hydride to remove the benzyl groups, and after evaporation of the solvent the residue in ether was hydrolyzed. Evaporation of the dried ethereal layer followed by sublimation of the residue gave the diphosphine (**1**), colorless crystals, m.p. 252° (under nitrogen in a sealed tube) after repeated sublimation;

it showed a normal molecular weight in boiling chloroform (R. C. Hinton and Mann [242]).

The diphosphine (1) sublimes very readily under reduced pressure. This property of ready volatility, often linked with a reasonably high melting point, is shared by 1,4-diazabicyclo[2.2.2]octane,[255] by quinuclidine,[256] and by adamantane, and is often associated with a highly symmetric structure.

The diphosphine and certain derivatives show a remarkable capacity for tenaciously retaining traces of solvent: evaporation of a benzene solution of the diphosphine gave a residue which required repeated sublimation at $100°/0.2$ mm to yield the pure diphosphine.

This compound (1) shows many of the normal properties of a tertiary phosphine, forming a dimethiodide (3; $R = CH_3$, $X = I$), m.p. 375–380°, and a dimethopicrate, m.p. 270°, with benzyl bromide reforming the dibromide (3; $R = CH_2C_6H_5$, $X = Br$), and when treated cautiously in ether with tetrachloroauric acid forming the yellow crystalline unstable dichloro(diphosphine)digold (4), the gold being reduced to the monovalent condition by an excess of the phosphine.

A cold ethanolic solution of the diphosphine on exposure to air followed by evaporation gave the 1,4-dioxide, which required heating at $220°/1$ mm for 4 hours to remove traces of ethanol. In contrast, the diphosphine in benzene solution is remarkably protected from oxidation, which does not occur when air is passed through even the hot solution, or when the warm solution is shaken with acetone–hydrogen peroxide. This protection does not apply to the action of sulfur, which readily gives the 1,4-disulfide in hot benzene solution.

The infrared spectrum of the diphosphine (1) is very simple, and shows two sharp bands at 643 and 700 cm^{-1}, indicating the C–P stretching frequency of a tertiary phosphine; that of the 1,4-dioxide shows the expected strong band of the $\equiv$PO group at 1165 cm^{-1}.

The yield of the diphosphine (1) in the above synthesis was necessarily low owing to the formation of by-products. The ethereal solution after the hydrolysis also contained an unidentified, yet apparently simple, phosphine, whereas the residue from the first sublimation of the diphosphine yielded a colorless, rather greasy, crystalline hydrocarbon of composition $CH_3(CH_2)_nCH_3$, where n is ca. 34. The lithium aluminum hydride clearly caused considerable breakdown of the cyclic system, and a more detailed investigation of the reduction under milder conditions might achieve a higher yield of the diphosphine.

The unknown compound (5) is termed 1,4-diphosphabicyclo[2.2.2]-octa-2,5,7-triene. A remarkable derivative of (5), having all six hydrogen atoms replaced by trifluoromethyl groups, has been recorded briefly by

Krespan, McKusick, and Cairns [257] and in detail by Krespan.[258] Hexa-fluoro-2-butyne, F_3C—$C{\equiv}C$—CF_3, is heated with red phosphorus and a small quantity of iodine at 200° under pressure for 8 hours, and the product then briefly heated at 90°/10 mm to remove the more volatile

(5)   (6)

material. The solid residue is shaken with mercury, sublimed at 100°/10 mm, recrystallized from acetic acid, and again sublimed, giving the pure 2,3,5,6,7,8 - hexakis(trifluoromethyl) - 1,4 - diphosphabicyclo[2.2.2] - octa-2,5,7-triene, m.p. 118–119° (43%); this compound can be most clearly depicted as (6).

The same compound can be similarly obtained 'in good yield' when hexafluoro-2,3-diiodo-2-butene, F_3C—$CI{=}CI$—CF_3, is heated with red phosphorus at 210°.

The structure of (6) has been determined by the cumulative evidence of analysis, molecular weight, and physical properties. The ultraviolet spectrum in acetonitrile shows absorption at 262 mμ (ϵ 450) and 316 mμ (ϵ 640), and the infrared spectrum shows C=C absorption at 6.22 μ; in both spectra these bands are at longer wavelengths than those in similar bicyclooctatrienes having no hetero-atoms (Krespan, McKusick, and Cairns [259]), and indicate a P—C=C linkage. The nmr spectra in tetra-hydrofuran shows only a doublet at −1210 and −1170 cps, arising from one type of F_3C group split by adjacent phosphorus; the same splitting occurs at 40 megacycles.

The compound (6) is (not unexpectedly) unreactive to atmospheric oxygen, to methyl iodide at room temperature, to benzyl chloride at 100°, and to bromine in chloroform. This very low reactivity is probably the joint result of steric hindrance and of the inductive withdrawal of electrons from the phosphorus atoms by the vinyl groups, a process strongly reinforced by the highly electronegative F_3C groups.

The compound (6) and its arsenic analogue (p. 450) show a ready volatility, which again as in (1) and the other examples earlier cited, is associated with a highly symmetric structure.

1,4-Benzodiphosphorin

1,4-Benzodiphosphorin (**1**), [RIS 8106], is known at present only as quaternary salts derived from 1,2,3,4-tetrahydro-1,4-benzodiphosphorin (**2**).

o-Phenylenebis(diethylphosphine) (**3**; R = H) combines vigorously with 1,2-dibromoethane to give 1,1,4,4-tetraethyl-1,2,3,4-tetrahydro-1,4-benzodiphosphorinium dibromide (**4**; R = H), initially named

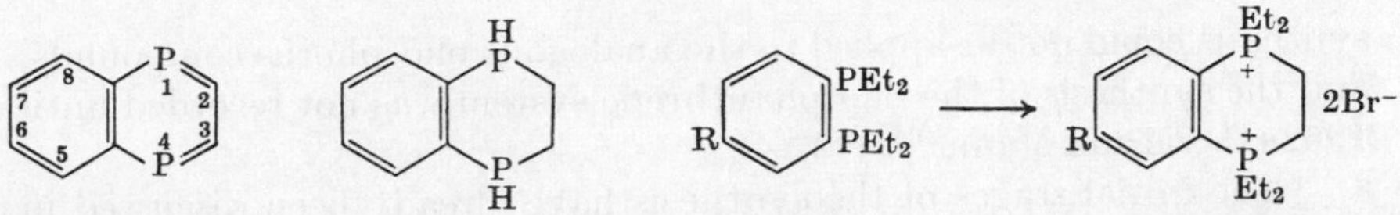

(**1**)   (**2**)   (**3**)   (**4**)

1,1,4,4-tetraethyl-1,4-diphosphonianaphthalene dibromide (dipicrate, m.p. 230–232°) (Hart [260]). 4-Methyl-o-phenylenebis(diethyl)phosphine (**3**; R = CH₃) had earlier been similarly converted into the cyclic 6-methyl homologue (**4**; R = CH₃), m.p. >340° (dipicrate, m.p. 178–179°), apparently the first recorded compound having a ring consisting solely of carbon atoms and two phosphorus atoms (Hart and Mann [261, 262]).

It is possible that these diquaternary dibromides, when heated, might lose ethylene and hydrogen bromide, giving 1,4-diethyl-1,2,3,4-tetrahydro-1,4-benzodiphosphorin and its 6-methyl derivative, respectively: this mode of decomposition cannot be predicted with certainty however (cf. p. 170) and does not occur with the analogous phosphorus–arsenic dibromides (p. 333).

Phosphanthrene

Phosphanthrene has the structure and notation shown in (**1**), [RIS 12689], but all known compounds having this ring system are 5,10-disubstituted derivatives of 5,10-dihydrophosphanthrene (**2**).

(**1**)   (**2**)

The diarsenic analogue of (**1**) is similarly termed arsanthrene (**3**) (p. 460); the intermediate phosphine-arsine (**4**), originally termed

5-phospha-10-arsaanthracene, is now termed dibenzo[1,4]phospharsenin (p. 335). It is noteworthy that the synthesis of the arsanthrene ring system was first recorded in 1921 (Kalb[263]); the various stages of this

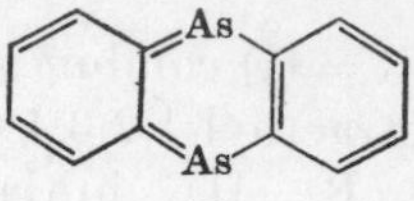

synthesis could not be applied to the analogous phosphorus compounds, and the synthesis of the phosphanthrene system was not recorded until 1962 (Davis and Mann[264, 143]).

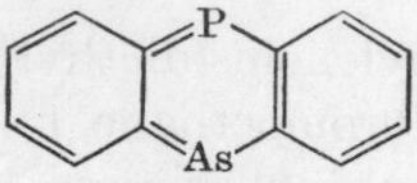

The initial stages of the synthesis have already been discussed in connection with that of the by-product 5,5'-bi(dibenzophosphole) (p. 86). *o*-Chlorophenylmagnesium bromide (5) when treated in ether with ethylphosphonous dichloride (0.5 equivalent) gave bis-(*o*-chlorophenyl)ethylphosphine (6), m.p. 85° (39%).

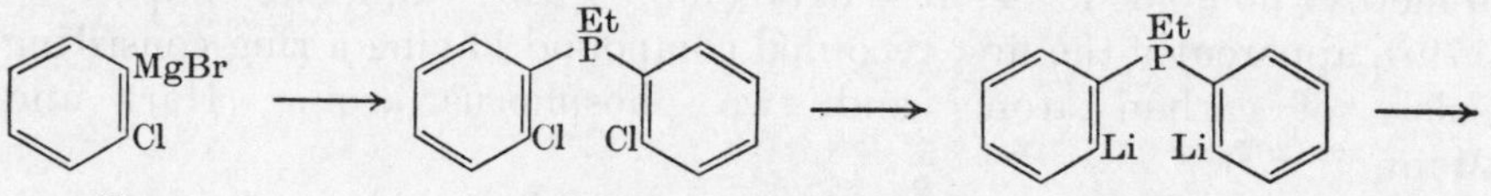

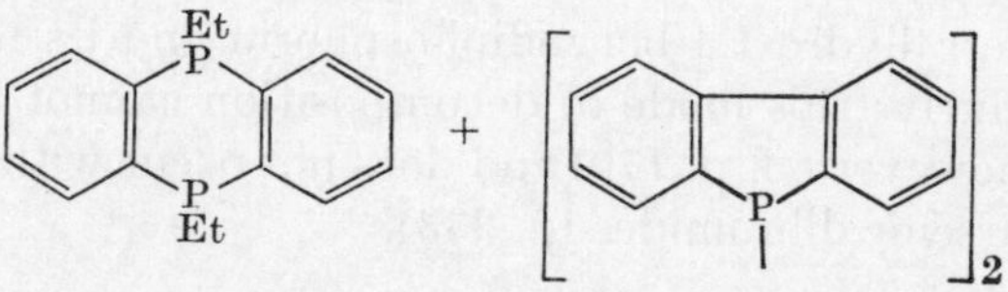

This phosphine in tetrahydrofuran at −35° was stirred with lithium until the formation of the 2,2'-dilithio compound (7) was apparently complete, and the black solution was treated at *ca.* −65° with the phosphonous dichloride (1 equivalent) also in tetrahydrofuran, the color of the mixture fading considerably. The reaction mixture was stirred at room temperature with water and benzene, and the dried organic layer on distillation gave an unbroken fraction, b.p. 100–200°/0.4 mm. Redistillation gave some ethyldiphenylphosphine, probably arising from the hydrolysis of unchanged (7): the two other main constituents were 5,10-diethyl-5,10-dihydrophosphanthrene (8) and 5,5'-bi(dibenzophosphole) (9), which apparently co-distilled. These compounds could be readily separated, for (9) is almost completely insoluble in cold ethanol, in which (8) is reasonably soluble: the latter, however, gives diquaternary salts which are also insoluble in ethanol. The addition of ethanol therefore precipitated the insoluble yellow (9); the filtrate, when warmed with benzyl iodide, deposited the di(benzyl iodide) of the phosphanthrene (8), i.e., 5,10-dibenzyl-5,10-diethyl-5,10-dihydrophosphanthronium diiodide (10; X = I).

For a discussion of the formation and structure of (9), see p. 87 et seq.

The tricyclic system in the cation (10) must be almost flat, in view of the tetrahedral angle at the phosphorus atoms; this is confirmed by

(10)   (11)

models, which, although not completely flat, 'flex' readily about the planar position. A salt of this cation, such as the diiodide (10; X = I) should exist in a cis- and a trans-form. This diiodide was separated into two forms, one, m.p. 326°, being insoluble in boiling water, and the other, m.p. 320–321°, moderately soluble; the low-melting isomer was formed in very small amount. These two forms gave isomeric dipicrates (10; X = $C_6H_2N_3O_7$), m.p. 300° and 241–244°, respectively. The quaternary dibromides (10; X = Br) were separated by extraction with boiling ethanol into two forms, m.p. 346° and 319°, respectively, which gave the above-mentioned different dipicrates.

The isomeric diiodides were further characterized by the action of warm aqueous sodium hydroxide with which they gave the normal reaction of a quaternary phosphonium salt. The high-melting form gave toluene and 5,10-diethyl-5,10-dihydrophosphanthren 5,10-dioxide (11), m.p. 235°; the low-melting form similarly gave toluene and the isomeric 5,10-dioxide (11), m.p. 257°. The dipole moment of the symmetric trans-dioxide must be zero, and that of the cis-dioxide must have a finite value. The dipole moment of the dioxide, m.p. 235°, determined in benzene solution, was 4.0 D and it must therefore be the cis-isomer. A similar measurement on the dioxide, m.p. 257°, was precluded by insufficiency of material.

The 5,10-diethyl-5,10-dihydrophosphanthrene (8) was regenerated from the diiodide, m.p. 326°, by reduction in tetrahydrofuran suspension with lithium aluminum hydride: after distillation and crystallization from ethanol and then methanol, it had m.p. 52–53°; the final mother-liquor slowly deposited a second form, m.p. 96–97°, a mixture having m.p. 43–50°. The second form was obtained in very small amount, and the subsequent investigation was limited to the form of m.p. 52–53°.

This diethyl-phosphanthrene has moderate basic properties, being soluble in cold 5N- but not in 2N-hydrochloric acid; in ethanolic solution it deposits a deep reddish-purple picrate, which dissociates on exposure to air or attempted recrystallization.

In ethanolic solution it readily gives a dimethiodide, m.p. 400° (dimethopicrate, m.p. 250°); in warm ethanol it gives a monoethiodide, m.p. 159–160°, and after several hours' boiling a diethiodide, m.p. above 400° (diethopicrate, m.p. 267°). The monoethiodide is exceptional in that with aqueous sodium picrate it gives an ethopicrate-sodium picrate salt, $C_{24}H_{25}N_3O_7P_2,C_6H_2N_3NaO_7$, m.p. 209–210°, and with aqueous picric acid the corresponding ethopicrate-hydrogen picrate, m.p. 120–121°; these salts crystallize unchanged from aqueous ethanol containing a trace of sodium picrate and picric acid, respectively. These are the first recorded examples of such double picrates from quaternary phosphonium salts, whereas the methiodides of pyridine, 2-, 3-, and 4-picoline, quinoline, quinaldine, lepidine, and isoquinoline all form methopicrate-sodium picrates and some also form the double hydrogen picrate (Mann and Baker [265]).

The diethyl-phosphanthrene (8), m.p. 52–53°, when exposed to the air in ethanolic solution deposited the monooxide, m.p. 140°, and when oxidized with hydrogen peroxide gave the dioxide, m.p. 234–235°, identical with that obtained by the action of alkalis on the di(benzyl iodide) (10; X = I), m.p. 320°.

The stereochemistry of the 5,10-dihydrophosphanthrene system must be closely similar to that of the 5,10-dihydroarsanthrene system, deduced initially by Chatt and Mann,[266] who separated 5,10-dihydro-5,10-di-p-tolylarsanthrene into two isomeric forms (cf. p. 463). In simple symmetric phosphines, PR_3, where R is alkyl or halogen, the average intervalency angle at the phosphorus atom is ca. 98°. This angle would not be possible in the dihydrophosphanthrene system (8) if the tricyclic system were flat. The phosphorus atoms could maintain this 'natural' angle, however, if the molecule were folded about the P—P axis, giving the ring system a 'butterfly' conformation. The maintenance of this angle would, however, cause the 5,10-substituents to adopt one of two possible positions: they could both be within the angle (calculated to be 121°) subtended by the two o-phenylene groups (the cis-position) or one substituent could be within this angle and one outside (the trans-position). This is shown in Figure 5 in which the diagrams are based on scale models of 5,10-dihydro-5,10-dimethylphosphanthrene, this being chosen to ensure minimum steric hindrance by the substituents. The *cis*-form is shown in (A1), and the *trans*-form in (B1).

Theoretical evidence has been adduced (Mislow, Zimmerman, and Melillo[267]) to show that both the *cis*- (A1) and the *trans*-form (B1)

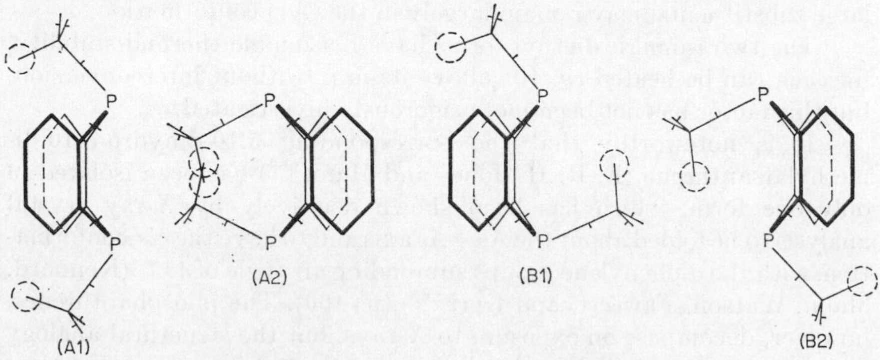

(A2)   (B1)

(A1)   (B2)

Figure 5. 5,10-Dihydro-5,10-dimethylphosphanthrene

The phosphorus atoms, the carbon atoms of the methyl groups, and one C—H bond of each methyl group are depicted in the plane of the paper, the size of the hydrogen atom in these C—H groups being indicated by broken circles. The molecule is folded about the P—P axis, one *o*-phenylene group (shown in thick lines) projecting towards, and the other away from, the observer. The boundary of the phosphorus atoms along the P—C bonds, and that of the carbon atoms along the C—H bonds of the methyl groups, are indicated by a notch in each bond. The C—H bonds of the central methyl groups in (A2) are (exceptionally) not in the plane of the paper because of mutual obstruction; in the model of (A2) these crossed bonds are in contact, but the hydrogen atoms would force them further apart. (Reproduced, by permission, from M. Davis and F. G. Mann, *J. Chem. Soc.*, **1964**, 3770)

can 'flex' readily about the planar conformation of the tricyclic system, and consequently that this flexing may occur in solution, where the degree of flexing would be determined mainly by the size of the substituent groups. Figure 5 is based on accurate Dreiding models having an intervalency angle of 98° around each phosphorus atom and constructed to allow rotation of the P—aryl and the P—CH$_3$ bonds without change of intervalency angles. These models show that, when pressure is suitably applied, the tricyclic system can be brought into the planar position, but release of the pressure causes the system to spring forwards or backwards into the almost strainless folded conformation. Furthermore, when the *trans*-(B1) is flexed completely over to (B2), it still has the *trans*-form, and therefore no configurational change *cis* $\rightleftarrows$ *trans*

occurs in this process. Similar flexing of the *cis*-(A1) form over the mean
planar position produces the form (A2); in this case, however, the process
is always incomplete because of the mutual obstruction of the two
methyl groups.

If this flexing does indeed occur in solution, the *trans*-form is
essentially unaffected, and the *cis*-form, particularly if it is carrying
large substituents, may remain largely in the (A1) conformation.

The two isomeric diethyl forms have reasonable thermal stability,
for each can be heated *ca.* 10° above its m.p. without interconversion,
but this factor has not been more rigorously investigated.

It is noteworthy that the corresponding 5,10-dihydro-5,10-di-
methylarsanthrene (E. R. H. Jones and Mann[268]) has been isolated in
only one form, which has been shown decisively by X-ray crystal
analysis to be folded about the As—As axis and to have the *cis*-conforma-
tion, with the *o*-phenylene groups subtending an angle of 117° (Kennard,
Mann, Watson, Fawcett, and Kerr[269]) (p. 466). The phosphanthrenes,
however, decompose on exposure to X-rays, but the structural analogy
between the arsanthrenes and the phosphanthrenes is so close that they
almost certainly have the same conformation.

There is at present no decisive evidence to show which of the
isomeric diethyl-phosphanthrenes is *cis* and which is *trans*, but the
following chemical evidence indicates that the form, m.p. 52–53°, has
almost certainly the *cis*-configuration.

(i) Both 1,2-dibromoethane and *o*-xylylene dibromide combine
readily in 1:1 molecular proportion with the phosphanthrene (cf. pp.
191 and 192). The resulting salts are very probably formed from one
molecule of each reagent giving (12) and (13), respectively, although

(12)   (13)

they could be formed from two molecules of each reagent to give a
'dimer'. In either case, a *cis*-conformation of the phosphanthrene would
be required.

(ii) Similarly the diethyl-phosphanthrene coordinates with palladium dibromide to form a deep yellow, crystalline complex of composition
$C_{16}H_{18}P_2,PdBr_2$. In this compound it is uncertain whether the palladium

atom can link the two phosphorus atoms of one molecule or if the
compound has a 2:2 molecular composition. Its insolubility and its
crystalline character have precluded molecular-weight determinations
either in solution or by X-ray crystal analysis. Again, however, both a
1:1 and a 2:2 composition would require a *cis*-configuration of the
phosphanthrene.

5,10-Dihydro-5,10-diphenylphosphanthrene (**14**) has been similarly
prepared by the reaction of the Grignard reagent (**5**) (2 equivalents) with
phenylphosphonous dichloride (1 equivalent), giving di-(*o*-chlorophenyl)-
phenylphosphine, $(C_6H_4Cl)_2PC_6H_5$, m.p. 133–134° (30%), and some
o-chlorophenylphenylphosphine, $(C_6H_4Cl)_2PH$. The tertiary phosphine
was converted into the 2,2'-dilithio derivative, which reacted with the
phosphonous dichloride (1 equivalent) to form the diphenyl-phosphanthrene (**14**), accompanied by some diphenylphosphine, $(C_6H_5)_2PH$, and
a very small proportion of the diphosphine (**9**). [This formation of a
secondary phosphine in the above Grignard and dilithio reactions is
interesting, particularly as in the latter case it must involve a cleavage
of the P—C_6H_5 bond.]

As before, the crude phosphanthrene was isolated as the insoluble
diquaternary benzyl bromide (**15**), m.p. 387° after thorough extraction
with organic solvents (dipicrate, m.p. 281°). When reduced with lithium

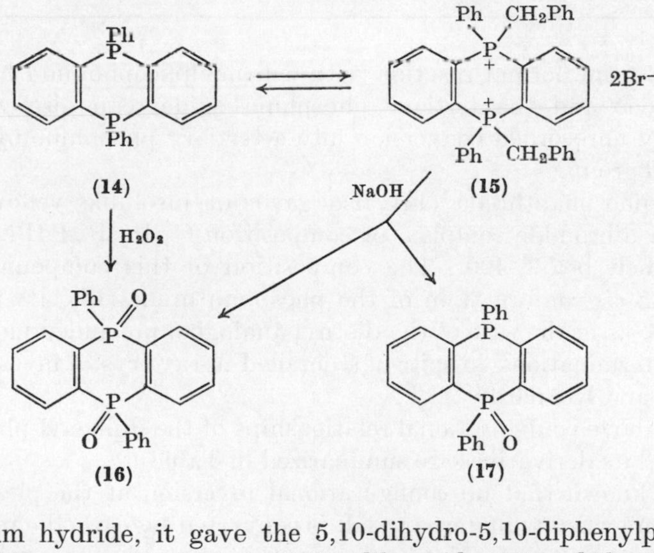

aluminum hydride, it gave the 5,10-dihydro-5,10-diphenylphosphan-
threne (**14**), m.p. 184–187°; both the dibromide (**15**) and the phosphanthrene (**14**) had sharp melting points, and no evidence for the existence
of isomeric forms in either compound could be obtained.

The phosphanthrene (**14**) with hydrogen peroxide gave a 5,10-dioxide (**16**), m.p. 276–278°. The dibromide (**15**) on treatment with aqueous sodium hydroxide gave an isomeric 5,10-dioxide (**16**), m.p. above 400°, and a monooxide (**17**), m.p. 231–232.5°. The isomeric dioxides had identical infrared spectra. The isomer, m.p. 276–278°, has a dipole measurement of 3.4 D in benzene solution and is almost certainly the *cis*-isomer; the high-melting isomer was too insoluble for dipole-moment determination.

The formation of both the dioxide and the monooxide in the above reaction is remarkable, for it shows that the sodium hydroxide, in

---

Table 12. Interrelations of the diphenylphosphanthrene
derivatives (**14**)–(**16**)

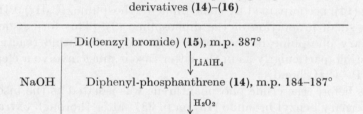

```
        ┌──Di(benzyl bromide) (15), m.p. 387°
        │              │ LiAlH₄
        │              ▼
  NaOH  │     Diphenyl-phosphanthrene (14), m.p. 184–187°
        │              │ H₂O₂
        │              ▼
        │     5,10-Dioxide (16), m.p. 276–278° (cis)
        └───▶ 5,10-Dioxide (16), m.p. >400°
```

---

addition to the normal reaction with a benzylphosphonium halide to give toluene and the tertiary phosphine oxide, can also give the apparently unrecorded conversion into a tertiary phosphine by loss of the benzyl group.

The phosphanthrene (**14**) also gave an insoluble yellowish-red palladium dibromide complex, of composition $C_{24}H_{18}P_2,PdBr_2$, which did not melt below 400°. The composition of this compound again indicates a *cis*-conformation of the phosphanthrene (**14**). Its physical properties, as in the case of the diethyl analogue, precluded molecular-weight determinations, in spite of a detailed X-ray crystal investigation (de Wolff and Kennard[270]).

The above configurational relationships of the diphenyl-phosphanthrene and its derivatives are summarized in Table 12.

It is known that no configurational inversion at the phosphorus atom occurs when a quaternary salt is converted by cathodic reduction into the tertiary phosphine, or when the phosphine is oxidized by hydrogen peroxide to the oxide or converted by benzyl halide into the quaternary salt; conversion of this salt by sodium hydroxide into the phosphine oxide is, however, accompanied by inversion (Kumli, McEwen, and VanderWerf[271, 272]; Horner et al.[206, 273]).

The 5,10-dioxide (16), m.p. 276–278°, has the *cis*-configuration (dipole moment) and the phosphanthrene (14) must also be *cis*; and, if the reduction by lithium aluminum hydride follows the course of cathodic reduction, the di(benzyl bromide) (15) must also be *cis*. But this bromide (15) with aqueous sodium hydroxide gives the *trans*-dioxide (16), m.p. above 400°; this would entail an inversion at only one phosphorus atom, for inversion at both would leave the configuration unchanged. There is at present no decisive evidence for the retention or inversion of configuration when a quaternary salt is reduced by lithium aluminum hydride; if this process causes inversion at one phosphorus atom, the bromide (15) may have the *trans*-configuration and the apparent anomaly would be explained. It should be borne in mind, however, that the above 'rules' regarding retention or inversion were based on compounds in which the phosphorus atom was not part of a ring and was thus largely unrestricted; acceptance of the above configurational relationships in the phosphanthrene series should, therefore, await more decisive evidence.

The diphenylphosphanthrene gives an insoluble dimethiodide, m.p. *ca.* 385° (dipicrate, m.p. 238–242°), and with bromine gives an unstable phosphine bromide which readily undergoes hydrolysis. This behavior is to be expected: it contrasts strongly, however, with that of 5,10-dihydro-5,10-dimethylarsanthrene which with bromine forms a colorless stable dibromide, which is covalent in the crystalline state and ionized in certain organic solvents (pp. 468, 469).

A second synthesis of 5,10-dihydro-5,10-diphenylphosphanthrene (14) starts with the interaction of dilithiophenylphosphine, $C_6H_5PLi_2$, and certain o-dihalogenobenzenes (Mann and Pragnell[274]). This process furnishes a gum and a mixture of (14) and 1,2,3-triphenyl-1*H*-1,2,3-benzotriphospholene (cf. p. 204); the latter can be extracted with acetone, and the phosphanthrene again isolated as the di(benzyl bromide) (15). The yield of the latter is small, but this preparative route may prove to be the quicker. The method and the probable mechanisms of the formation of the above products are discussed on p. 205 *et seq*.

5,10-Ethanophosphanthrene

This compound, [RIS —], has the structure and numbering (1); it is known only as the diethyl quaternary salts.

A mixture of 5,10-diethyl-5,10-dihydrophosphanthrene and an excess of 1,2-dibromoethane, when gently warmed, combined vigorously.

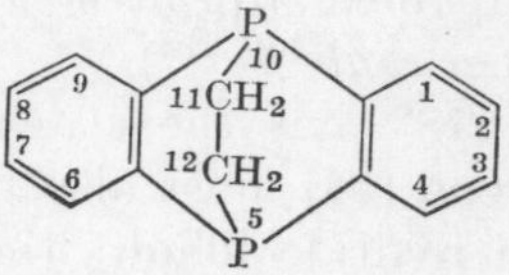

(1) 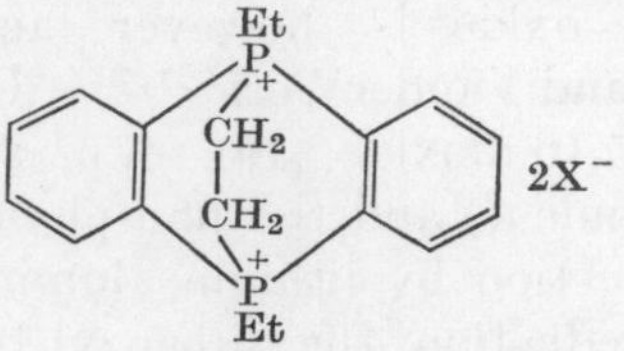   (2)

A solution of this semi-solid product in hot 1-propanol, when boiled under reflux and cooled, deposited 5,10-diethyl-5,10-dihydro-5,10-ethanophosphanthronium dibromide (**2**; X = Br), m.p. 325–326° (dipicrate, m.p. 272°); both salts were recrystallized from ethanolic dimethyl sulfoxide (Davis and Mann[143]).

In view of the ease with which the diethylphosphanthrene combines with 1,2-dibromoethane and with *o*-xylylene dibromide, the failure to obtain a similar combination with 1,3-dibromopropane, with or without a solvent, is unexpected, particularly as 5,10-dihydro-5,10-dimethyl-arsanthrene combines very readily with 1,3-dibromopropane (p. 477).

5,12-*o*-Benzenodibenzo[*b,f*][1,4]diphosphocin

This unknown compound has the structure and numbering shown in (**1**) [RIS —]. The two phosphorus atoms are clearly part of a six-membered ring and of an eight-membered ring, and the suffix -ocin indicates that the compound is named systematically on the basis of the eight-membered ring. It is described here, however, because of its close connection with foregoing phosphanthrene.

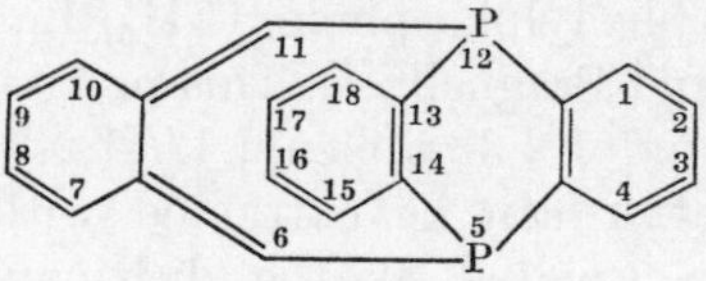

(1) 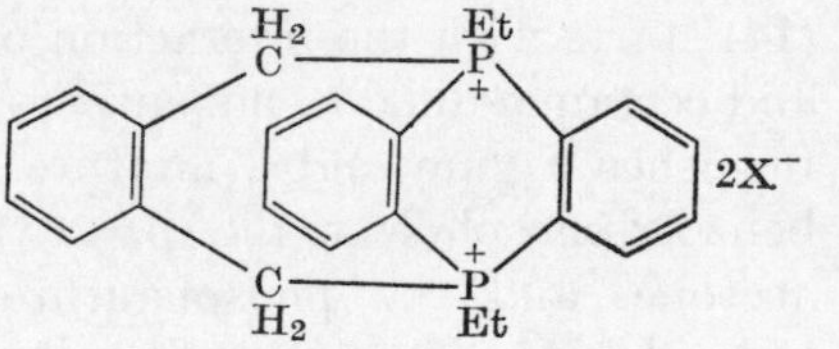   (2)

The only known derivative of (**1**) was obtained by boiling an equimolecular mixture of 5,10-diethyl-5,10-dihydrophosphanthrene and *o*-xylylene dibromide in ethanol, when the crystalline 5,12-diethyl-6,11-dihydro-5,12-*o*-benzenodibenzo[*b,f*][1,4]diphosphocinium dibromide (**2**; X = Br) rapidly separated. This salt was insoluble in most organic solvents but readily soluble in water. Its aqueous solution, treated with aqueous sodium iodide, gave the diiodide (**2**; X = I), m.p. 305–306°, after recrystallization from water; the dipicrate, recrystallized from ethanolic dimethyl sulfoxide, had m.p. 278°. The dibromide was initially termed 5,10-diethyl-5,10-dihydro-5,10-o-xylylenephosphanthronium dibromide (Davis and Mann[143]).

The diiodide, when heated with aqueous sodium hydroxide under reflux, gave o-xylene and the diethyl-phosphanthrene 5,10-dioxide, m.p. 252–253°; a mixture with the authentic dioxide, m.p. 257° (p. 185), melted between these values.

There is strong evidence (p. 481) that the diarsenic analogue of the above dibromide is formed by a 1:1 union of the two components, and it is reasonably certain that this composition also applies to these diphosphorus derivatives. Quaternary phosphonium halides undergo thermal dissociation far less readily than the arsonium halides, and in the preparation of (2; X = Br) there was no indication of the various other derivatives analogous to those obtained in the arsanthrene series.

Seven-membered Ring System containing only Carbon and Two Phosphorus Atoms

5H-Dibenzo[d,f][1,3]diphosphepin

The numbering and presentation of this parent compound, [RIS —], is shown in (1). Apparently the only known members of this class are

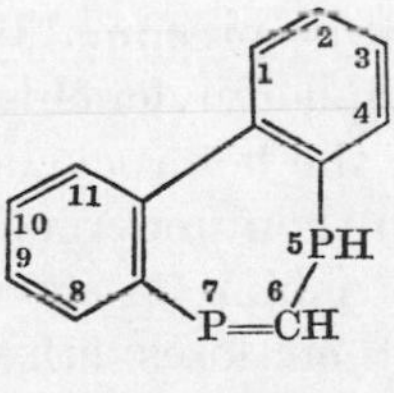

(1) 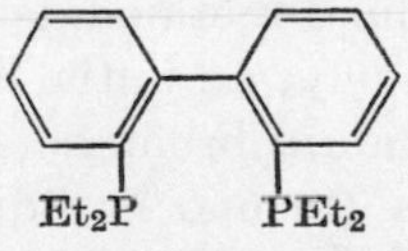   (2) 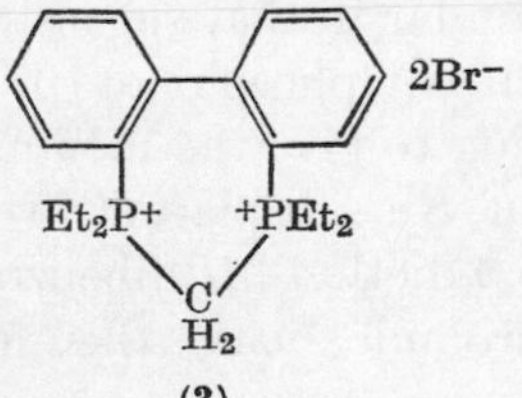   (3)

the diquaternary salts obtained by the combination of 2,2′-biphenylylenebis(diethylphosphine) (2) (p. 78) with the lower αω-dibromoalkanes and with o-xylylene dibromide (p. 201). The diphosphine (2) is structurally very suitable for this purpose, for the rotation about the P—aryl bonds, combined with that about the aryl—aryl bond (although limited), allows the diphosphine to adjust itself to the stereochemical demands of the diquaternizing dibromides.

A mixture of the diphosphine (2) and dibromomethane (one molar equivalent) containing a small proportion of methanol, when heated in a sealed tube at 100° for 5 hours, gives the hygroscopic 5,5,7,7-tetraethyl-6,7-dihydro-5H-dibenzo[d,f][1,3]diphosphepinium dibromide (**3**), m.p. 228–230° (dipicrate, m.p. 221–222°) (Allen, Mann, and Millar[122, 123]). This dibromide was initially termed 5,5:7,7-tetraethyl-5,7-diphosphonia-1,2:3,4-dibenzocycloheptadiene.[123]

Scale models of the cation (**3**) and its homologues (pp. 198, 200, and 201) show that the cations cannot be planar, and that they are twisted about the central biphenyl bond to accommodate the methylene bridge; in this respect, they are, of course, similar to their diarsine counterparts (pp. 486 *et seq.*). The dibromide (**3**) should, therefore, be resolvable into optically active forms which, however, like the diarsine analogues, might readily undergo racemization.

The dibromide (**3**) when heated up to 250° at 0.15 mm in a small bulb-tube distillation apparatus, gave a semi-solid distillate of tertiary phosphine and its hydrobromide. Treatment with aqueous sodium hydrogen carbonate liberated the phosphine component which was extracted with ether. Vapour-phase chromatographic examination of the dried ether extract showed the presence of 9-methyl- and 9-ethyl-9-phosphafluorene (5-methyl- and 5-ethyl-5H-dibenzophosphole) in the proportions of 1:2. Similar examination at 176° indicated the absence of any cyclic diphosphines that might have been formed by simple loss of alkyl groups from the initial dibromide, a process which does occur in the thermal decomposition of the tetramethyldiarsonium dibromide analogue of the dibromide (**3**) (p. 486).

The main mechanism of this ring contraction is presumably similar to that suggested for the dimethiodide of 2,2′-biphenylylenebis-(diethylphosphine) (p. 79), namely, an initial fission of the heterocyclic ring to give the monophosphonium bromide (**4**), which then undergoes an $S_N i$ displacement of the (bromoalkyl)diethylphosphine to give 5,5-diethyl-5H-dibenzophospholium bromide, which then loses ethyl bromide with the formation of 5-ethyl-5H-dibenzophosphole (**5**;

$R = C_2H_5$). An extra factor must enter here, however, for this mechanism should give solely the ethyl-5H-dibenzophosphole (**5**; $R = C_2H_5$). It is noteworthy that the di-, tri-, and tetra-methylene homologues of the dibromide (**3**) on thermal decomposition do give only this ethyl product (pp. 198, 200, and 201). The decomposition of the monomethylene dibromide (**3**) must presumably be also accompanied by the transient formation of the CH_2 radical, which provides the 5-methyl product (**5**; $R = CH_3$). This is not improbable, because Collie[275] reported that tetramethylphosphonium bromide on thermal decomposition gave ethylene as one of the products, indicating (as Collie suggested) that 'methylene' was involved.

Eight-membered Ring Systems containing Carbon and Two Phosphorus Atoms

Three different ring systems come under this heading, and the first two have been recorded only in brief notes.

1,5-Diphosphocin and 1,5-Diphosphocane

1,5-Diphosphocin has the structure and presentation (**1**), [RIS 11648], and its octahydro derivative, conveniently depicted as (**2**), is named 1,5-diphosphocane. Grim and Schaaff[157] have reported that a

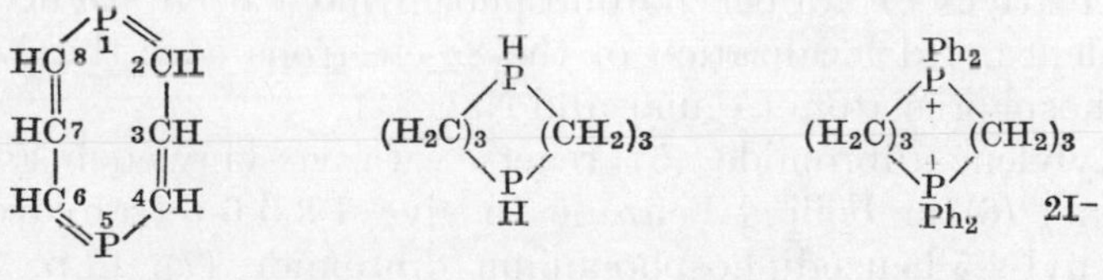

mixture of diphenylphosphine and 1,3-diiodopropane, in boiling acetonitrile under nitrogen, gives the crystalline 1,1,5,5-tetraphenyl-1,5-diphosphocanium diiodide (**3**), m.p. 204–205°, in small yield; its molecular weight in nitrobenzene supports this triionic structure. It is suggested that the intermediate compound is 3-iodopropyldiphenyl-phosphine, $(C_6H_5)_2P(CH_2)_3I$, which undergoes bimolecular quaterniza-tion to (**3**), just as (2-bromoethyl)ethylphenylphosphine forms 1,4-diethyl-1,4-diphenyl-1,4-diphosphorinanium dibromide (p. 170).

It is noteworthy that under the above conditions diphenylphosphine reacts with 1,5-diiodopentane to form 1,1-diphenylphosphorinanium iodide, the intermediate (5-iodopentyl)diphenylphosphine apparently undergoing an intramolecular quaternization (p. 101).[157]

2,5-Benzodiphosphocin

2,5-Benzodiphosphocin (**1**), [RIS —], has the structure and number-ing shown. If the heterocyclic ring were fully reduced, *i.e.*, converted into the octahydro derivative, the latter would be termed 2,5-benzo-diphosphocan; this process would however involve hydrogenation at the

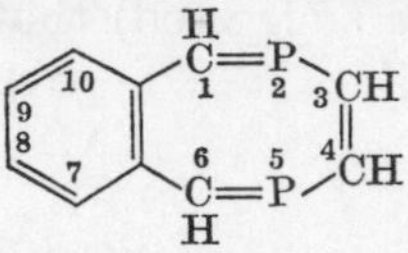

(1) 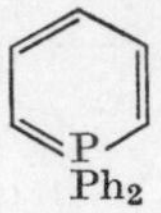   (2) 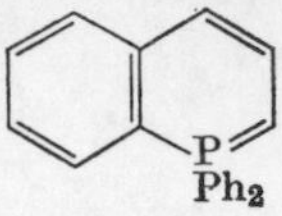   (3) 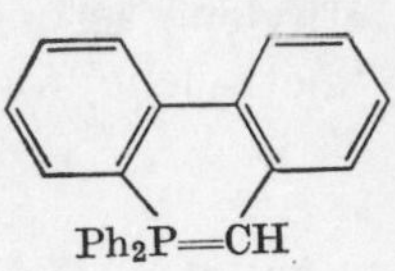   (4)

junction of the benzene ring which in compounds of this general type is not frequently encountered. The intermediate derivatives, in which one or more of the non-benzenoid double bonds have become saturated, must be named as derivatives of di-, tetra-, or hexa-hydro derivatives of (**1**).

A number of six-membered cyclic phosphorus ylides are now known, which apparently show a certain degree of enhanced stability, which may be evidence of delocalization of the 6π-electrons; examples are 1,1-diphenylphosphorin (**2**) (p. 112), 1,1-diphenylphosphinoline (**3**) (p. 126) and 5,5-diphenyldibenzo[*b,d*]phosphorin (**4**) (p. 151); the last two compounds do not show the Wittig reaction with benzaldehyde.

Derivatives of 2,5-benzodiphosphocin have been studied to find any evidence of delocalization of the 8π-electrons over the $3d$-orbitals of the phosphorus atom (Aguiar and Nair[276]).

o-Xylylene dibromide (**5**) reacts with *cis*-vinylenebis(diphenyl-phosphine) (**6**) in boiling benzene to give 1,2,5,6-tetrahydro-2,2,5,5-tetraphenyl-2,5-benzodiphosphocinium dibromide (**7**), m.p. 331–334°

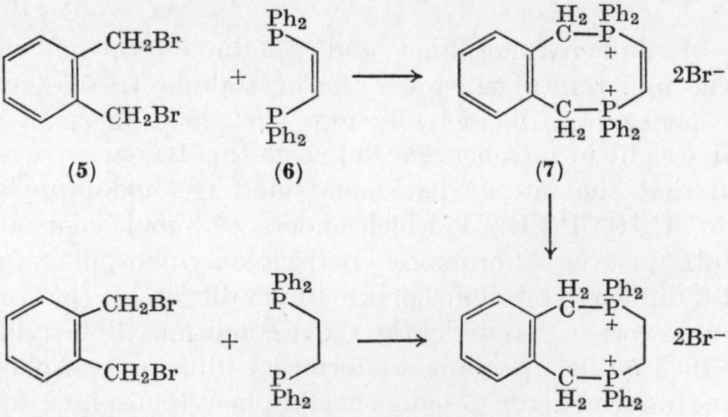

(5)   (6)   (7)

(5)   (8)   (9)

(96%) (dipicrate, m.p. 239–241°). The dibromide (5) similarly reacts with ethylenebis(diphenylphosphine) (8) to give 1,2,3,4,5,6-hexahydro-2,2,5,5-tetraphenyl-2,5-benzodiphosphocinium dibromide (9), m.p. 345–347° (93%). The dibromide (7) may be converted into the dibromide (9) by hydrogenation at room temperature in a methanolic solution containing a rhodium–alumina catalyst. The nmr spectra of (7) and (9) accord with these structures.

Attempts made to remove hydrogen bromide from (7) and thus obtain the diylide, i.e., 2,2,5,5-tetraphenyl-2,5-benzodiphosphocin (10)

(10)

proved fruitless, however. (a) A dry methanolic solution of (7) when treated with collidine deposited collidine hydrobromide, and the red solution on treatment with hydrobromic acid gave an unidentified phosphonium salt. (b) The salt (7) when treated with aqueous sodium hydroxide gave a red amorphous solid which could not be recrystallized, and with hydrobromic acid did not regenerate (7). (c) A pentane solution of butyllithium was added to a suspension of (7) in benzene, which was boiled under nitrogen. The cold deep red solution, when extracted with dilute hydrobromic acid, gave an unidentified phosphonium salt. There is at present, therefore, no decisive evidence that the diylide (10) can be obtained.

The alkaline hydrolysis of the two dibromides (7) and (9) follows different paths. When boiled with 10% aqueous sodium hydroxide for 1 hour, the dibromide (9) gives ethylenebis(diphenylphosphine) dioxide (11), m.p. 265°, with presumably the liberation of o-xylene; this would be the normal action of an alkali on a phosphonium salt.

The dibromide (7), however, when similarly treated undergoes cleavage of the vinyl bridge with the formation of o-xylylenebis(diphenylphosphine) dioxide (12), m.p. 265°. This was identified by treating o-xylylene dichloride with lithiumdiphenylphosphine (2 equivalents) in tetrahydrofuran to form o-xylylenebis(diphenylphosphine) (13), m.p. 282°, which underwent ready atmospheric oxidation to the dioxide (12). It is probable that the mechanism of this hydrolysis is similar to that of 1,2,3,4-tetrahydro-1,1,4,4-tetraphenyl-1,4-diphosphorinium dibromide (p. 177) and that the true hydrolysis product is the monooxide of (13), which is rapidly converted into the dioxide (12).

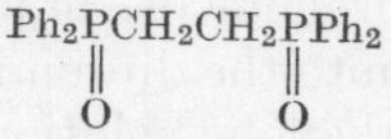

$$Ph_2PCH_2CH_2PPh_2$$

(structures)

(11) 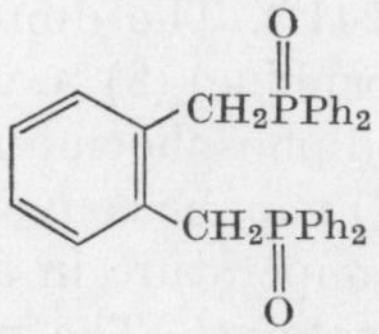   (12) 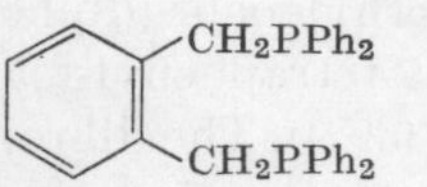   (13)

It should be noted that the following names, which differ only in their terminations, were initially given to the compounds indicated: 1,1,4,4-tetraphenyl-1,4-diphosphonia-6,7-benzocyclo-octadiene-(2,6) dibromide (7); -octene-6 dibromide (9); -octatetarene-2,4,6,8 (10).[276]

Dibenzo[e,g][1,4]diphosphocin

The fundamental compound of this name, [RIS —], is numbered and depicted as (1). The only known members are salts of the cation (2).

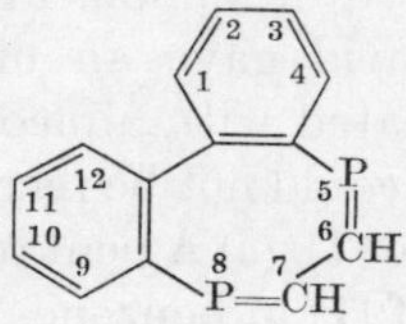

(structures)

$$2Br^-$$

(1) 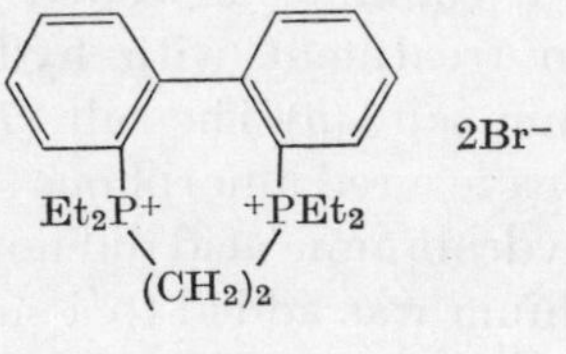   (2)

The diquaternization of 2,2′-biphenylylenebis(diethylphosphine) by 1,2-dibromoethane (1 equivalent), under the conditions described earlier (p. 193) but with 12 hours' heating, gives 5,5,8,8-tetraethyl-5,6,7,8-tetrahydrodibenzo[e,g][1,4]diphosphocinium dibromide (2), m.p. 268–272° (dipicrate monoethanolate, m.p. 142°, resolidifying and remelting at 164–165°). The dibromide was initially termed 5,5:8,8-tetraethyl-5,8-diphosphonia-1,2:3,4-dibenzocyclo-octadiene dibromide (Allen, Mann, and Millar[122, 123]).

Thermal decomposition of the dibromide (2) under the conditions already specified (p. 194) again gave a distillate of tertiary phosphine with some hydrobromide, which was worked up as before. Examination of the ether extract by vapor-phase chromatography showed the presence of 9-ethyl-9-phosphafluorene (5-ethyl-5H-dibenzophosphole) and the absence of 9-methyl-9-phosphafluorene and of diphosphines. The residue from evaporation of the ether extract formed a crude methiodide, readily converted into the pure methopicrate of 9-ethyl-9-phosphafluorene.

The mechanism of this thermal decomposition is apparently identical with the main mechanism for the lower (monomethylene) homologue (p. 194), without further complication by free radicals.

Dibenzo[b,f][1,5]diphosphocin

This parent compound (1), [RIS 12779], is (as expected) unknown, and its stable representative is a derivative of the 5,6,11,12-tetrahydro compound (2).

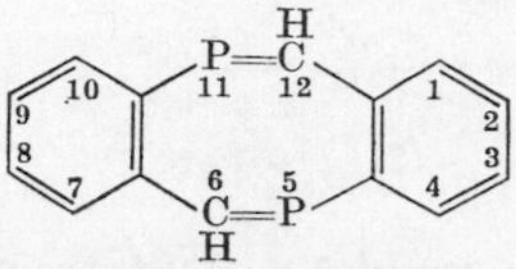

(1) 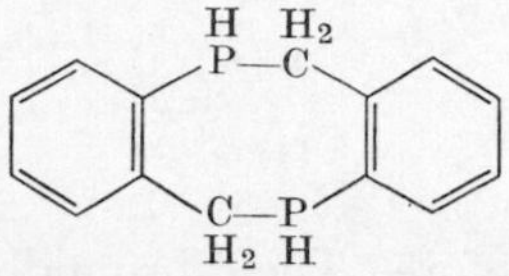 (2)

Märkl[104] has recorded that o-[(methoxymethyl)phenyl]diphenyl-phosphine (3; R = OCH₃) when treated with hydrogen bromide gives

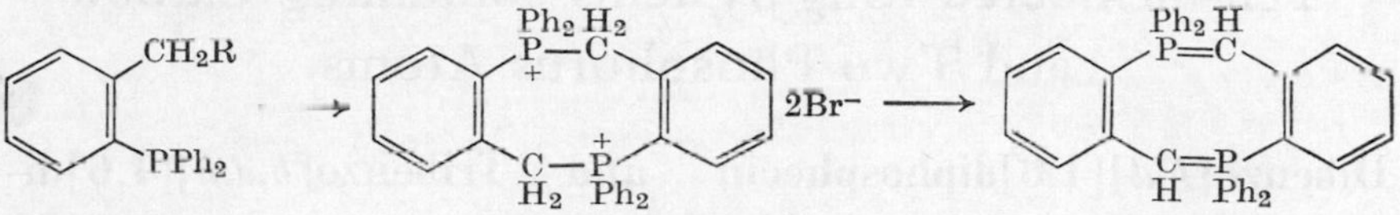

(3) (4) (5)

the o-bromomethyl phosphine (3; R = Br), m.p. 85–87°, which on heating undergoes an intermolecular quaternization to form 5,6,11,12-tetrahydro-5,5,11,11-tetraphenyldibenzo[b,f][1,5]diphosphocinium dibromide (4), m.p. above 360° (p. 67).

It is briefly stated that this dibromide, when treated with aqueous sodium hydroxide, gives 5,5,11,11-tetraphenyldibenzo[b,f][1,5]diphosphocin (5), an interesting example of a cyclic diphosphorane.[277]

Nine-membered Ring System containing Carbon and Two Phosphorus Atoms

5H-Dibenzo[f,h][1,5]diphosphonin

This compound, [RIS —], is numbered and depicted as (1). Its only known derivatives are salts of the cation (2); their chemistry can be very briefly described, as it is closely similar to that of the homologous dimethylene dibromide (p. 198).

2,2′-Biphenylylenebis(diethylphosphine) and 1,3-dibromo-propane (1 equivalent) when heated together at 100° for 16 hours give 5,5,9,9-tetraethyl-6,7,8,9-tetrahydro-5H-dibenzo[f,h][1,5]diphosphoninium dibromide (2), m.p. 335–336° (dipicrate, m.p. 185–186°). Thermal decomposition of the dibromide affords 5-ethyl-5H-dibenzophosphole, identi-

(1)   (2)

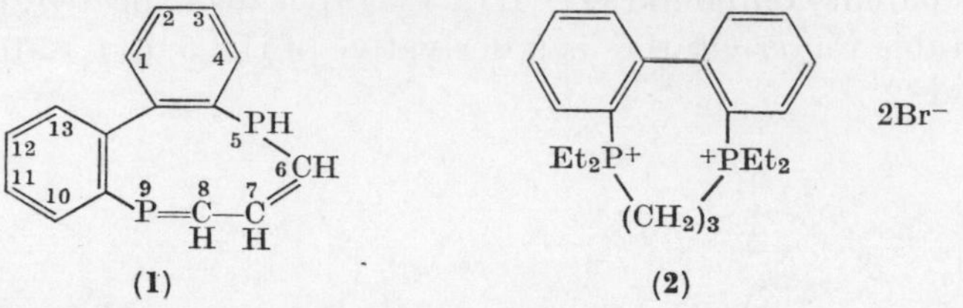

fied as before. The dibromide was earlier named 5,5:9,9-tetraethyl-5,9-diphosphonia-1,2:3,4-dibenzocyclononadiene dibromide (Allen, Mann, and Millar[122, 123]).

Ten-membered Ring Systems containing Carbon and Two Phosphorus Atoms

Dibenzo[b,d][1,6]diphosphecin   and   Tribenzo[b,d,h][1,6]di-phosphecin

The first of these compounds is numbered and depicted as (1), [RIS —], and the second as (2), [RIS —]. Their known chemistry is so similar that they can conveniently be discussed together.

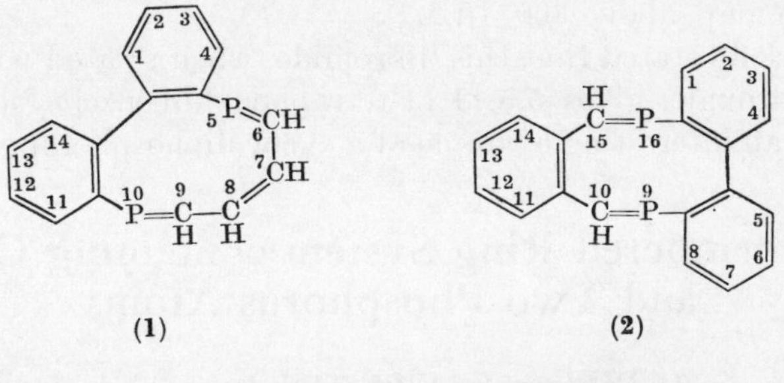

(1)   (2)

2,2′-Biphenylylenebis(diethylphosphine), when heated with 1,4-dibromobutane at 100° for 20 hours, yields 5,5,10,10-tetraethyl-5,6,7,8,9,10-hexahydrodibenzo[b,d][1,6]diphosphecinium dibromide (3), m.p. 152–155°; when heated at 250°/0.15 mm, this salt gives 5-ethyl-5H-dibenzophosphole, identified as before.

The bis(diethylphosphine), when similarly heated with o-xylylene dibromide, gives a solid product which after trituration with ether and extraction with boiling ethanol leaves a residue of the pure 9,9,16,16-tetraethyl - 9,10,15,16 - tetrahydrotribenzo[b,d,h][1,6]diphosphecinium

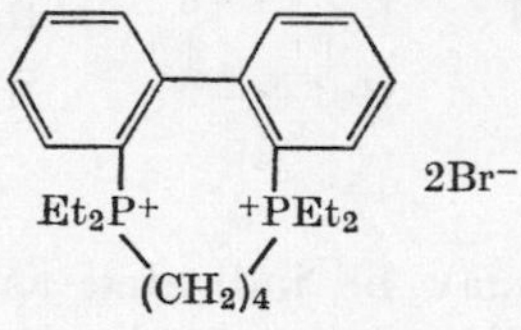

(3) 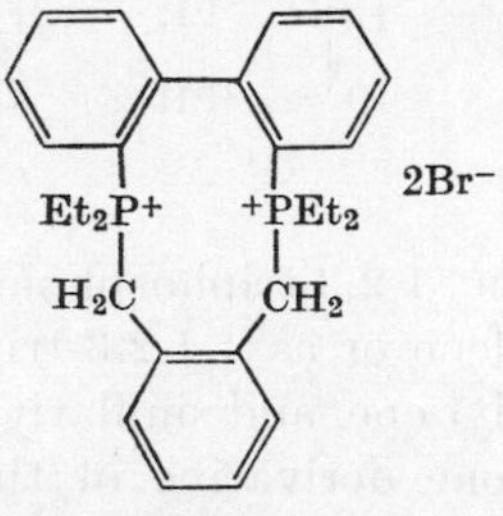   (4)

dibromide (4), melting above 360°, and only very slightly soluble in most organic solvents (dipicrate, m.p. 77–78°).

The dibromide (4), when heated to 300° at 0.15 mm, gave a viscous semi-crystalline distillate, which was basified and extracted with ether. Examination of the dried extract by vapor-phase chromatography at 130° and at 182° showed that the main constituent was 5-ethyl-5H-dibenzophosphole, with a little 2,2′-biphenylylenebis(diethyl-phosphine). The residue from evaporation of the ether extract, treated with methyl iodide, gave a crude product, from which the pure dimeth-iodide, m.p. 254–256°, of the bis(diethylphosphine) was isolated.

Alkaline hydrolysis of the dibromide (4) gave o-xylene and 2,2′-biphenylylenebis(diethylphosphine) dioxide, m.p. 193–195°, the expected preferential cleavage of the benzyl groups having occurred (Allen, Mann, and Millar[122, 123]).

The dibromides (3) and (4) were initially named 5,5:10,10-tetra-ethyl - 5,10 - diphosphonia - 1,2 : 3,4 - dibenzodecadiene dibromide and 5,5:10,10-tetraethyl-5,10-diphosphonia-1,2:3,4:7,8-tribenzocyclodeca-triene dibromide, respectively.[123]

Five-membered Ring Systems containing only Carbon and Three Phosphorus Atoms

1H-1,2,3-Triphosphole

The ring system 1H-1,2,3-triphosphole (1), [RIS —], can theoretically give rise to three isomeric dihydro derivatives, namely, the 2,3-,

8

2,5-, and 4,5-dihydro-1H-1,2,3-triphospholes [(2), (3), and (4), respectively], collectively named 1,2,3-triphospholenes, and to the tetrahydro derivative (5), named 1,2,3-triphospholane. The position of the double

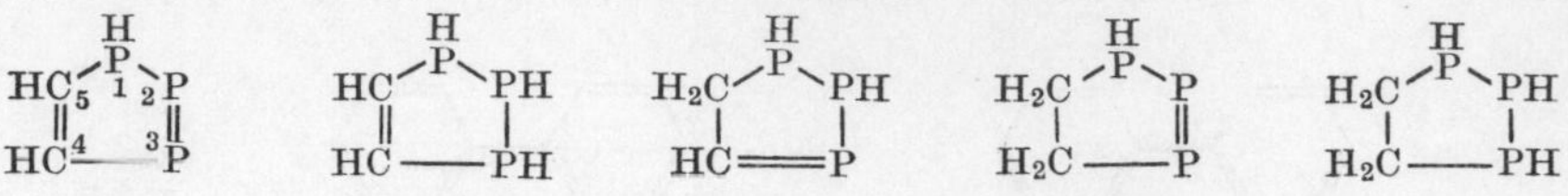

bond in the 1,2,3-triphospholene (2) may be indicated as Δ^4-1,2,3-triphospholene or as 4-1,2,3-triphospholene or (most clearly) as 1,2,3-triphosphol-4-ene, and similarly for the isomers (3) and (4).

Only one derivative of the above monocyclic compounds is at present known. It has already been noted (p. 163) that, when tetrakis-(trifluoromethyl)cyclotetraphosphine $(F_3C—P)_4$, or pentakis(trifluoromethyl)cyclopentaphosphine $(F_3C—P)_5$, is heated with an excess of bis(trifluoromethyl)acetylene, $F_3C—C\equiv C—CF_3$, at 170° for 7 hours, pentakis(trifluoromethyl)-2,3-dihydro-1H-1,2,3-triphosphole [$i.e.$, 1,2,3,4,5-pentakis(trifluoromethyl)-1,2,3-triphospholene] (6) is formed in 31% yield and tetrakis(trifluoromethyl)-1,2-diphosphetin in 55% yield (Mahler[230]).

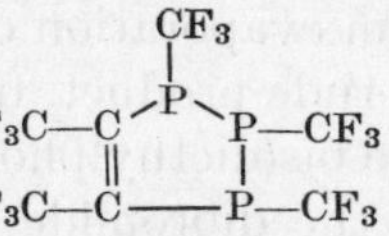

The two products have some closely similar properties. The triphospholene (6) has been identified by analysis for carbon, fluorine, and phosphorus and by a mass-spectrometric molecular-weight determination; its infrared spectrum, which shows a C=C band at 1560 cm^{-1}, and its nmr spectra were fully consistent with this structure. It is a colorless liquid, which has b.p. ca. 160° (calculated from its vapor pressure), and which does not freeze at −120°. It is spontaneously inflammable in air.

Its reaction when heated with iodine has been described on p. 163.

1H-1,2,3-Benzotriphosphole

This compound, [RIS —], has the structure (1); the dihydro derivative (2) is therefore 2,3-dihydro-1H-1,2,3-benzotriphosphole, but is more conveniently termed 1H-1,2,3-benzotriphospholene.

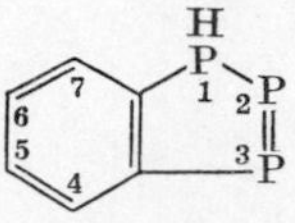

(1) 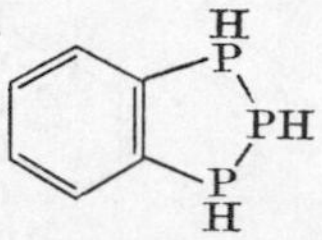   (2) 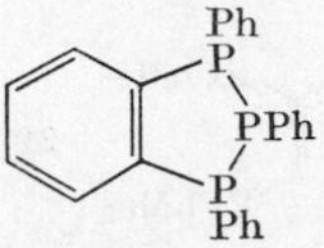   (3) 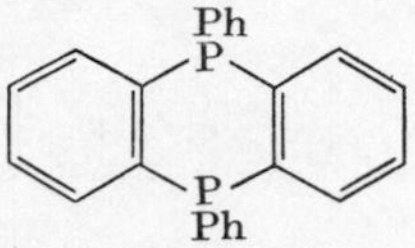   (4)

At present only the triphenyl derivative of (2), *i.e.*, 1,2,3-triphenyl-$1H$-1,2,3-benzotriphospholene (3), has been investigated in detail, although one derivative of the 1,2,3-triethyl member has also been isolated (Mann and Pragnell[274]). The compound (3) was initially termed 1,2,3-triphenyl-1,2,3-triphosphaindane.

Dilithiophenylphosphine, $C_6H_5PLi_2$, can be prepared by the direct action of lithium on phenylphosphonous dichloride (Bloomfield and Parvin[278]) or—more conveniently—by the action of butyllithium on phenylphosphine in petroleum–tetrahydrofuran. The dropwise addition of o-bromoiodobenzene or o-diiodobenzene in tetrahydrofuran to the stirred solution of dilithiophenylphosphine at −40° gives first a dark red and finally a black reaction mixture. Removal of the solvent after a period of boiling leaves a residue which when chilled and shaken with ether–water gives two layers and much insoluble gum. The dried ethereal layer on evaporation gives a residue containing 1,2,3-triphenyl-$1H$-1,2,3-benzotriphospholene (3) and 5,10-dihydro-5,10-diphenylphosphanthrene (4) (p. 189). This residue when stirred with acetone deposits the crystalline compound (3); the filtrate from these crystals when boiled with benzyl bromide deposits the insoluble di(benzylobromide) salt of the phosphanthrene (4); to obtain cleaner products the residue can be distilled at 0.1–0.5 mm and the distillate then treated as before with acetone.

1,2,3-Triphenyl-$1H$-1,2,3-benzotriphospholene (3), m.p. 184–186°, is only slightly soluble in cold acetone, and crystallizes very readily from solution or from the molten state. After analysis and molecular-weight determinations, its structure was deduced from the following reactions:

(i) The triphosphine in cold toluene solution with methyl iodide deposited an unstable monomethiodide. In solution in methyl iodide, heating at 100° caused cleavage of the heterocyclic ring giving o-phenylenebis(dimethylphenylphosphonium iodide) (5), m.p. 311–313° (dipicrate, m.p. 253–255°), accompanied presumably the more soluble trimethylphenylphosphonium iodide.

(ii) A solution of the triphosphine, treated with potassium permanganate, both in cold acetone, after two days gave o-phenylenebis-(phenylphosphinic acid) (6), m.p. 197–198°, almost certainly accompanied similarly by the more soluble phenylphosphonic acid.

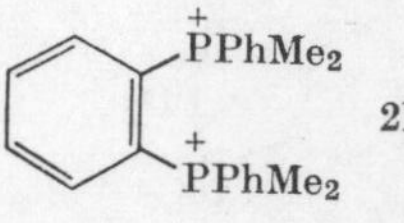

(5) 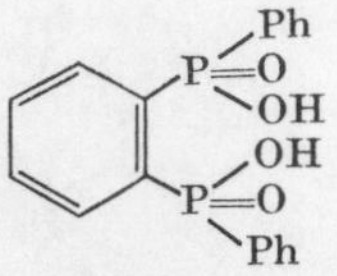   (6)

(iii) The triphosphine, when treated in boiling toluene with sulfur under appropriate conditions, gave a monosulfide, m.p. 122–123°, and two isomeric disulfides, having m.p. 182–183° and 203–204°, respectively. No trisulfide could be prepared, probably because of steric hindrance; the 1,2,3-triethyl analogue of (3), prepared in very small yield and not isolated, readily gave a trisulfide, m.p. 120–122°.

The structure (3) was subsequently confirmed by an X-ray crystal analysis (Daly[279]). This shows that the fused bicyclic system in (3) is almost planar, the heterocyclic ring being very slightly distorted. The pyramidal disposition of the phosphorus atoms causes the phenyl groups to lie outside this plane; the phenyl groups attached to the P(1) and P(3) atoms are *cis* to one another, and that attached to the P(2) atom is *trans* to both. There is thus a plane of symmetry at right angles to the planar bicyclic system, and running through the P(2) atom and its attached phenyl group. It is noteworthy that the intervalency angles within the ring are 99.7° at the P(1) and P(3) atoms, and 97.3° at the P(2) atom. The average is thus very close to the 'natural' angle of *ca.* 98° in symmetric non-hindered phosphines.

The relative disposition of the phenyl groups provides some evidence for the configuration of the two isomeric disulfides, for addition of sulfur to a tertiary phosphine occurs with retention of configuration (Horner and Winkler[280]). The sulfur atoms in one disulfide might

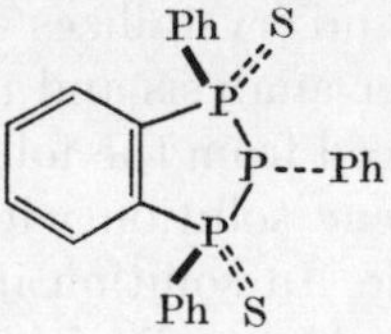

(7) 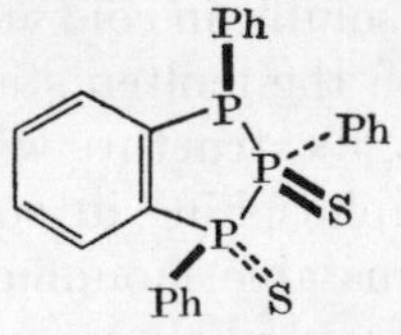   (8)

therefore be in the *cis*-1,3-position (7), and in the other disulfide in the *trans*-1,2-position (8). [In both formulas thick bonds indicate phenyl groups or sulfur atoms projecting above the bicyclic plane, and broken bonds indicate those projecting below this plane.] There is, of course, no evidence to show which of the isomers has the configuration (7) and which has (8).

Numerous experiments have been carried out in which the identity of the o-dihalogenobenzene, its molecular ratio to the dilithiophenylphosphine, and the time, order, and temperature of their addition, have been varied, and the yields of the triphosphine (3) and the phosphanthrene (4) recorded. These results do not, however, give any decisive evidence for the mechanism of the formation of these products, for the numbers of possible factors and intermediates are considerable.

It has been suggested that the initial reaction may be a halogen–lithium interchange between the o-dihalogenobenzene (X = halogen)

(9)   (10)

and the dilithiophosphine to give the reactive intermediates (9) and (10).

The product (9) would readily lose lithium halide to form benzyne, particularly in the presence of an acceptor. A noteworthy example of this process is o-chlorophenyllithium, which at −70° in the presence of tetraethyldiphosphine, $(C_2H_5)_2P$—$P(C_2H_5)_2$, decomposes to benzyne, which reacts with the diphosphine to form o-phenylenebisdiethylphosphine (11) (Hart[260]).

Arylmonohalogenophosphines, RPHX, and arylhalogenolithiophosphines (10) are unstable and, in the absence of other reactive compounds, lose hydrogen halide and lithium halide, respectively, to form the crystalline cyclic polyphosphine, which has usually been considered to be tetraphenylcyclotetraphosphine (12). The highly reactive intermediate in the formation of such polyphosphines is very probably the

(11)   (12)

monomeric phosphinidene, C_6H_5P:, which has been obtained in a transient form by the reduction of phenylphosphonous dichloride by zinc at 25°, by the oxidation of phenylphosphine with iodine in the presence of triethylamine, and by the thermal decomposition of compounds such as (12), and which has been 'trapped' by its reaction with, for example, diethyl disulfide, $(C_2H_5)S$—$S(C_2H_5)$, to form diethyl phenyldithiophosphonite, $C_6H_5P(SC_2H_5)_2$ (Schmidt and Osterroht[281]).

A point of uncertainty arises regarding the compound (12), which occurs in two forms, having m.p. 150° and 190°, respectively, hitherto considered, largely on account of molecular-weight determinations in various solvents,[282] to be isomeric forms of the cyclic tetraphosphine (**12**). Daly has recently shown by X-ray analysis that the first of these compounds has the formula $(C_6H_5P)_5$ with a planar ring of five phosphorus atoms,[235] and the second, m.p. 190°, has a trigonal form of formula $(C_6H_5P)_6$, the six phosphorus atoms being in a ring of chair conformation with the phenyl groups in the equatorial positions[236]; the second compound also exists in a monoclinic and a rhombohedral form (Daly and Maier[283]), the former being very similar in conformation to the trigonal form.[284] It is possible that these compounds undergo a certain degree of dissociation in organic solutions; indeed Fluck and Issleib[285] have adduced evidence, based on ^{31}P-nuclear resonance studies, that the above cyclic phosphines, when molten or in tetrahydrofuran solution, generate a significant proportion of $(C_6H_5P)_1$ and $(C_6H_5P)_2$ units.

Benzyne is known to cleave P—P bonds, and it may possibly also cleave the polyphosphine ring. Hence it might unite with the (C_6H_5P) units to form a diphosphine of which (**13A**) and (**13B**) are contributory forms; these could react with the phosphinidene to form the triphosphine (**3**) and with benzyne to form the diphenylphosphanthrene (**4**).

Alternatively, the benzyne and the phosphinidene might give the product (**14A**), having a canonical form (**14B**): this could unite with more phosphinidene to give the triphosphine (**3**), or by dimerization give the phosphanthrene (**4**).

The addition of phosphinidene units to (**13A** ↔ **13B**) or to (**14A** ↔ **14B**) would stop at the triphospholene stage, because this gives maximum stability by virtue of the 'natural' intervalency phosphorus angle within the ring (p. 186).

The highest recorded yield of the triphosphine (**3**) in the above synthesis from the phosphonous dichloride, calculated on the basis $3C_6H_5PCl_2 \to 1$ triphosphine, is 15%, and that from phenylphosphine, similarly calculated, is 16% (highest recent yield[503] 30%. In all preparations except one, the syrupy polymer was the major product.

The highest yield of the di(benzylobromide) of the phosphanthrene (4) from the phosphonous dichloride, calculated on the basis $2C_6H_5PCl_2 \rightarrow 1$ dibromide, is 3.4%, and that from phenylphosphine is 2.7%. In the earlier synthesis (p. 189), the highest yield of the di(benzylobromide), similarly calculated from the phosphonous dichloride, was 6.2% (Davis and Mann[143]). The present synthesis gives a lower yield, but does not involve isolation and purification of intermediates and may prove to be the quicker and more practical route.

Eight-membered Ring Systems containing Carbon and Four Phosphorus Atoms

1,3,5,7-Tetraphosphocin

1,3,5,7-Tetraphosphocin (1), [RRI —], has the structure and numbering shown in (1), and its 1,2,3,4,5,6,7,8-octahydro derivative (2) is called 1,3,5,7-tetraphosphocane.

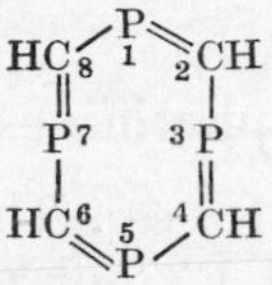

(1) 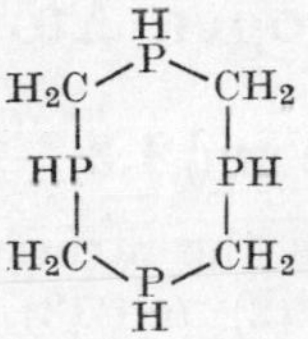   (2) 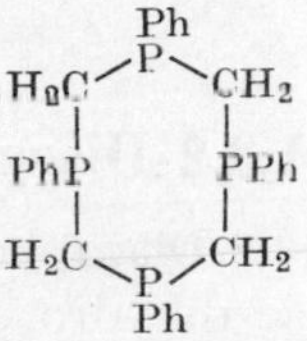   (3)

The only known derivative is 1,3,5,7-tetraphenyl-1,3,5,7-tetraphosphocane (3), which has been prepared by Maier[286, 287] using a novel synthetic route. Formaldehyde in 35% aqueous solution is treated with diethylamine (1 equivalent) under nitrogen with stirring and ice-cooling, and phenylphosphine (0.5 equivalent) is rapidly added; an exothermic reaction occurs. Extraction with petroleum gives bis[(diethylamino)-methyl]phenylphosphine, $C_6H_5P[CH_2N(C_2H_5)_2]_2$, b.p. 136°/0.1 mm (85%). Equimolecular quantities of this compound and of phenyl-phosphine are heated under nitrogen at 140° for 4 hours, diethylamine distilling off in almost theoretical yield. The residue solidifies when set aside for 3 days, and is then extracted with hot ethanol. The extract deposits the tetraphosphine (3), m.p. 125–127°, which has been identified by analysis and molecular-weight determinations; the peaks in the infrared spectrum are noted in full by Maier. The phosphine readily gives a tetraoxide.

The insoluble residue from the ethanolic extraction has m.p. 137–139°, and a composition approximating to that of the phosphine (3); it is considered to be a higher polymeric product.

The compound (3) was initially termed 1,3,5,7-tetraphenyl-1,3,5,7-tetraphosphacyclooctane.[286]

Note. In monocyclic systems containing carbon, nitrogen, and phosphorus, the numbering starts at the nitrogen atom. If the ring is saturated, the normal terminations or suffixes to the names of rings containing phosphorus do not apply: the names of rings containing 3, 4, or 5 atoms end in -iridine, -ctidine, and -olidine, respectively, and those of larger rings retain the names of the corresponding unsaturated rings preceded by the position and number of hydrogen atoms, *e.g.*, '2,3,4,5-tetrahydro'. This applies also to similar rings containing both nitrogen and arsenic (p. 502), antimony, or bismuth.

Four-membered Ring System containing only Carbon and One Phosphorus Atom and Two Nitrogen Atoms

1,3,2-Diazaphosphete and 1,3,2-Diazaphosphetidines

1,3,2-Diazaphosphete has the structure (1), [RIS —]. The two isomeric dihydro derivatives (2) and (3) are termed 1,3,2-diazaphosphetines, and the tetrahydro derivative (4) is 1,3,2-diazaphosphetidine.

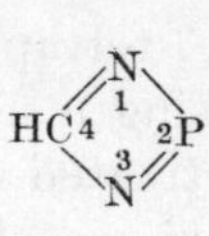

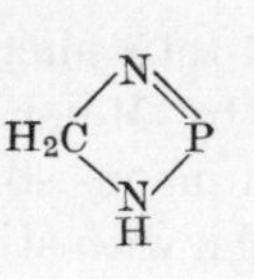

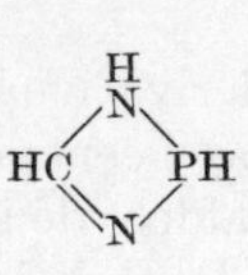

All known compounds having this ring system are derivatives of (4).

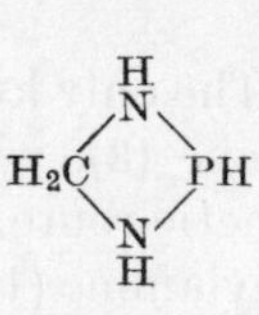

Ulrich and Sayigh [288, 289] have shown that the nature of the product from the interaction of a substituted urea and phosphorus pentachloride depends primarily on the nature of the substituents and to a much smaller extent on the polarity of the solvent. If in a 1,3-disubstituted urea, RNHCOHNR′, both R and R′ are primary alkyl groups, or if R is an aryl group and R′ a primary alkyl group, then the urea in cold carbon tetrachloride will react with PCl_5 with the vigorous evolution of hydrogen chloride and formation of the corresponding 1,3-disubstituted derivative (5). The use of N,N'-dimethylurea thus gave

$$
OC \underset{\underset{R'}{\overset{N}{\big|}}}{\overset{\overset{R}{\big|}}{N}} PCl_3 \longrightarrow RN=PCl_3 + R'NCO \longrightarrow RN=C=NR' + POCl_3
$$

(5)   (6)   (7)

2,2,2-trichloro-1,3-dimethyl-1,3,2-diazaphosphetidin-4-one (5; R = R' = CH_3), b.p. 78–79°/1.5 mm (72%); the 1,3-di-n-butyl derivative, b.p. 105–108°/0.3 mm (71%), and the 1-butyl-3-phenyl derivative (5; R = C_4H_9, R' = C_6H_5), an undistillable oil, were similarly prepared.

The temperature of the reaction mixture must be kept reasonably low, for compounds of type (5) when heated in an inert solvent break down into a trichlorophosphazine (6) and an isocyanate, which then interact to form a disubstituted carbodiimide (7) and phosphorus oxychloride.

The use of a urea having 1,3-secondary alkyl substituents in the above treatment with PCl_5 gives the corresponding N,N'-dialkylchloroformamidine hydrochloride, $RN^+H=C(Cl)NHR'$ Cl^-.

Derkach and Narbut[290] have obtained the same type of reaction by treating 1,3-diarylureas with phosphorus trichloride in phosphorus oxychloride; they thus isolated the diphenyl derivative (5; R = R' = C_6H_5), m.p. 88–90° and the di-1-naphthyl derivative (5; R = R' = $C_{10}H_7$), m.p. 110–112°. These compounds slowly hydrolyze on exposure to air, and when heated also give the corresponding carbodiimide (7).

Five-membered Ring Systems containing Carbon, One Phosphorus, and One Nitrogen Atom

1H-1,2-Azaphosphole and 1,2-Azaphospholidine

Both the 1H-1,2-azaphosphole (1), [RIS 11562], and its 2,3,4,5-tetrahydro derivative, termed 1,2-azaphospholidine (2), are unknown.

$$
\underset{HC{\scriptstyle 4}\text{———}\overset{}{CH}}{HC_5\overset{\overset{H}{\underset{|}{N}}}{\underset{3}{{\scriptstyle 1}}}{}_{2}P} \qquad
\underset{H_2C\text{———}CH_2}{H_2C\overset{\overset{H}{\underset{|}{N}}}{}PH} \qquad
\underset{H_2C\text{———}CH_2}{H_2C\overset{\overset{R}{\underset{|}{N}}}{}\underset{\overset{|}{OH}}{\overset{O}{P}}} \qquad
\underset{H_2C\text{———}CH_2}{H_2C\overset{\overset{R}{\underset{|}{N}}}{}\underset{\overset{|}{NHR'}}{\overset{O}{P}}}
$$

(1)   (2)   (3)   (4)

The 2-hydroxy-1,2-azaphospholidine 2-oxide (**3**; R = H) is sometimes termed a phostamic acid, the corresponding oxa compound in which the —NR— group in (**3**) is replaced by an —O— atom being a phostonic acid (p. 259); consequently the compound (**4**; R = H) is a phostamic acid amide. These general terms are also applied to the corresponding six-membered homologues having the —$(CH_2)_4$— sequence in the ring system.

Members of the class (**4**) have recently been prepared by Helferich and Curtius,[291] who by treating tri-n-butyl phosphite with finely divided sodium obtained dibutyl sodiophosphonate (**5**), which readily condensed with 1-bromo-3-chloropropane to form dibutyl (3-chloropropyl)phosphonate (**6**), b.p. 115–117°/0.01 mm. Boiling hydrochloric acid hydrolyzed this ester to the phosphonic acid (**7**), m.p. 101–103°, which with

$$NaPO(OC_4H_9)_2 \rightarrow Cl(CH_2)_3PO(OC_4H_9)_2 \rightarrow Cl(CH_2)_3PO(OH)_2 \rightarrow$$
$$\text{(5)} \qquad\qquad\qquad \text{(6)} \qquad\qquad\qquad \text{(7)}$$

$$Cl(CH_2)_3POCl_2 \rightarrow Cl(CH_2)_3PO(NHC_6H_5)_2 \rightarrow \text{(4; R = R′ = } C_6H_5\text{)}$$
$$\text{(8)} \qquad\qquad\qquad \text{(9)}$$

phosphorus pentachloride in chloroform readily gave the phosphonic dichloride (**8**), b.p. 65–67°/0.01 mm. A mixture of aniline and triethylamine, when added in anhydrous conditions to a solution of (**8**) in tetrahydrofuran, furnished (3-chloropropyl)phosphonodianilide (**9**), m.p. 161–162°, which when treated in boiling methanol with sodium hydroxide (1 equivalent) underwent cyclization to 2-anilino-1-phenyl-1,2-azaphospholidine 2-oxide (**4**; R = R′ = C_6H_5), m.p. 193–196°; this

(**10**)

compound is sometimes depicted as (**10**) and termed N,N'-diphenyltrimethylenephostamic amide (cf. p. 259). Others of series (**4**), similarly prepared, were those in which R = R′ = $C_6H_4CH_3$-p, m.p. 160–165°; $C_6H_4OCH_3$-p, m.p. 177.5–178°; and $C_6H_4CO_2H$-p, decomp. 180°.

These compounds (**4**) are stable in alkaline solution, but in acids at room temperature they are readily hydrolyzed with ring-opening.

It is noteworthy that the pure diethyl analogue of (**5**) can be prepared expeditiously by the action of sodium hydride on triethyl phosphite (Harvey *et al.*[292]). The product, $NaP(O)(OC_2H_5)_2$, reacts with 1-chloro-3-bromopropane or with 3-chloropropyl toluene-p-sulphonate to give the diethyl analogue of (6), which is similarly hydrolyzed to the phosphonic acid (7).

$1H$-1,3-Azaphosphole and 1,3-Azaphospholidine-2-spirocyclopentane

$1H$-1,3-Azaphosphole (1), [RIS —], has the structure and numbering shown; its tetrahydro derivative (2) is 1,3-azaphospholidine. Both these

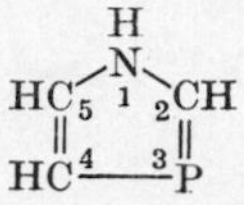

(1) 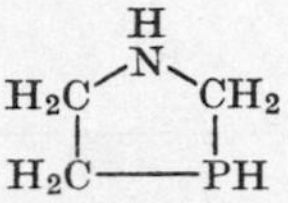   (2) 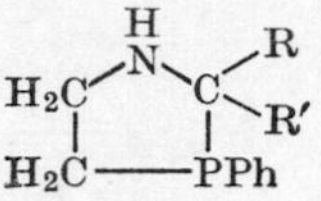   (3)

compounds, like their 1,2-analogues, are unknown in the above unsubstituted condition, but Issleib and Oehme [293, 294] have synthesized a number of derivatives of 3-phenyl-1,3-azaphospholidine (3).

This synthesis can readily be adapted to give the spirocyclic derivative (4), *i.e.*, 3-phenyl-1,3-azaphospholidine-2-spirocyclopentane

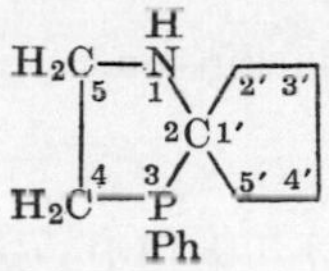

(4) 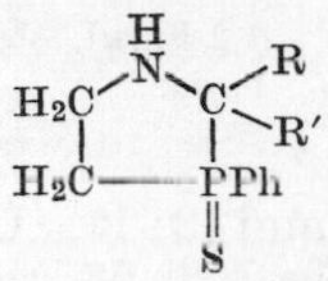   (8)

(4), [RIS —], and its cyclohexane and cycloheptane analogues, which are conveniently described in this Section.

For the synthesis of the 1,3-azaphospholidines (3), phenylphosphine is added to a solution of sodium (one equivalent) in liquid ammonia to

$$C_6H_5PHNa + ClCH_2CH_2NH_2 \rightarrow C_6H_5P(H)CH_2CH_2NH_2$$
$$(5) \qquad\qquad (6) \qquad\qquad\qquad (7)$$

give phenylsodiophosphine (5), followed by an ethereal solution of (2-chloroethyl)amine (6). Distillation gives (2-aminoethyl)phenylphosphine (7), b.p. 115°/0.7 mm (83%).

In the absence of a solvent, this phosphine reacts with many aldehydes and ketones, usually exothermally, with the formation of the corresponding derivative of 3-phenyl-1,3-azaphospholidine (3); the crude product is heated at 100° first at atmospheric and then reduced pressure to complete the reaction and remove the liberated water, and it is finally distilled at 0.2–0.3 mm.

Thus polyformaldehyde gives 3-phenyl-1,3-azaphospholidine (**3**; R = R′ = H), and acetone gives 2,2-dimethyl-3-phenyl-1,3-azaphospholidine (**3**; R = R′ = CH₃).

The use of the appropriate aldehyde or ketone gave the derivatives of (**3**) listed in Table 13.

Table 13. 2,2-Disubstituted derivatives
of 3-phenyl-1,3-azaphospholidine (**3**)

R	R′	Yield (%)
H	H	71
CH₃	H	48
C₂H₅	H	65
C₆H₅	H	83
CH₃	CH₃	79
C₂H₅	C₂H₅	67
C₆H₅	CH₃	59
C₆H₅	C₂H₅	60
C₄H₃O*	H	58

* 2-Furyl, obtained by use of furfuraldehyde.

The compound (**3**; R = C₆H₅, R′ = H) has m.p. 78.5–79.5°; all the other compounds in Table 13 are colorless liquids, readily distilling at 0.2–0.3 mm pressure, and dissolving in most organic solvents. The infrared spectra show a characteristic N—H band at 3350–3310 cm⁻¹ and no indication of a P—H band.

Most of these compounds form stable hydrochlorides by protonation of the ═NH group; these salts can be recrystallized from ethanol or ethanol–ether, and regenerate the base (**3**) on treatment with aqueous sodium hydroxide.

All the compounds in Table 13 combine with sulfur in benzene solution to form the *P*-sulfide, but only two have been obtained crystalline, namely, 2,3-diphenyl-1,3-azaphospholidine 3-sulfide (**8**; R = C₆H₅, R′ = H), m.p. 158–159°, and the 2,2-dimethyl-3-phenyl analogue (**8**; R = R′ = CH₃), m.p. 75°. Both these sulfides form crystalline hydrochlorides.

Methyl iodide reacts with all the compounds (**3**) to form intractable syrupy phosphonium salts.

Compounds of structure (**3**) were initially termed phosphazolidines.[293]

The aminophosphine (7) will also react readily with cyclic ketones to form spirocyclic compounds; with cyclopentanone it gives the compound (4), b.p. 146–147°/0.3 mm, which in turn readily forms a crystalline P-methylphosphonium iodide, m.p. 202–204°.

(9)　　　　　(10)

The use of cyclohexanone in this reaction gives 3-phenyl-1,3-azaphospholidine-2-spirocyclohexane (9), b.p. 159–161°, [RIS —], and cycloheptanone similarly gives 3-phenyl-1,3-azaphospholidine-2-spirocycloheptane (10), b.p. 165–167°, [RIS —]; both these products are colorless liquids.

Five-membered Ring Systems containing Carbon and One Phosphorus and Two Nitrogen Atoms

4H-1,2,4-Diazaphosphole

This ring system (1), [RIS —], is at present known only as the 3,4,5-triphenyl-4H-1,2,4-diazaphosphole (2).

(1)　　　　　(2)

When benzaldazine (3) in chilled carbon tetrachloride solution is treated first with hydrogen chloride, and then with chlorine gas until

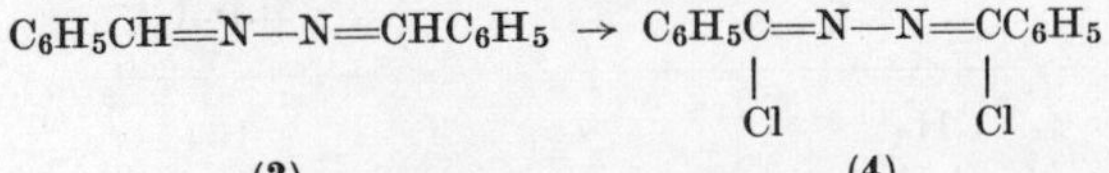

$$C_6H_5CH{=}N{-}N{=}CHC_6H_5 \rightarrow C_6H_5C{=}N{-}N{=}CC_6H_5$$
$$\underset{Cl}{|} \quad \underset{Cl}{|}$$

(3)　　　　　(4)

the precipitated hydrochloride of (3) redissolves, concentration of the solution after 1–2 days gives N,N'-bis-(α-chlorobenzylidene)hydrazine (4), m.p. 121–122° after recrystallization from dioxane.

A suspension of dilithiophenylphosphine, $C_6H_5PLi_2$, in tetrahydro-furan is added slowly to a solution of (4) also in tetrahydrofuran. The solvent is then removed under reduced pressure; after purification the residue gives the yellow 3,4,5-triphenyl-$4H$-1,2,4-diazaphosphole (2), m.p. 195–196°. Benzonitrile is a by-product in this reaction.

The phosphine (2) is soluble in most organic solvents but decomposes in solvents containing water (Issleib and Balszuweit [295]).

A similar cyclization does not occur when (4) is similarly treated with cyclohexyldilithiophosphine even at −60°; the products are tetracyclohexylcyclotetraphosphine, $(C_6H_{11}—P)_4$, m.p. 215–219°, and benzonitrile.

1,2-Diphenyl-1,2-dipotassiodiphosphine, $C_6H_5P(K)—P(K)C_6H_5$, reacts with (4) to give tetraphenylcyclotetraphosphine, $(C_6H_5—P)_4$, m.p. 198° (cf. p. 206).

$1H$-1,4,2-Diazaphosphole and 1,4,2-Diazaphospholidines

$1H$-1,4,2-Diazaphosphole (1) [RIS 11553], is (as expected) unknown and substitution products of the 2,3,4,5-tetrahydro derivative (2), termed 1,4,2-diazaphospholidine, have been very little investigated.

A patent specification (Farben. Bayer A-G. [296]; by G. Schrader [297]) records that the addition of the O-ethyl (chloromethyl)thiophos-phoro(thiocyanatate) acid (3), b.p. 63°/0.01 mm, to stirred 25%

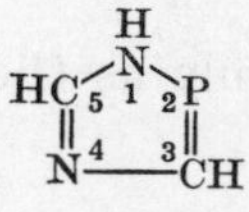

(1) 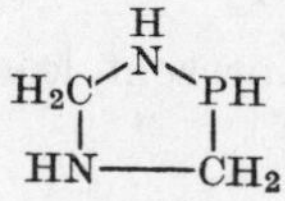   (2) 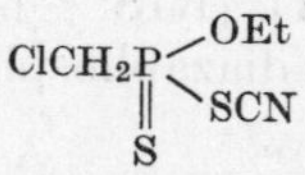   (3) 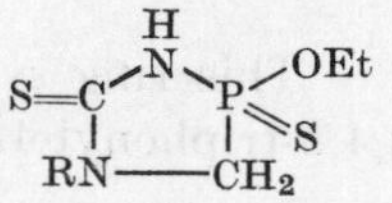   (4)

aqueous ammonia causes the deposition of 2-ethoxy-1,4,2-diazaphos-pholidine-5-thione 2-sulfide (4; R = H), m.p. 155°; the use of hydrazine

Table 14. 4-Substituted 1-ethoxy-1,4,2-diazaphospholidine-5-thione 2-sulfides (4)

R	M.p. (°)
CH_3	104
C_2H_5	52
C_6H_5	144
p-ClC_6H_4	135
p-$C_2H_5OC_6H_4$	143

gives the 4-amino derivative (**4**; $R = NH_2$), m.p. 91°, and use of ammonia gives the unsubstituted parent (**4**; $R = H$), m.p. 155°.

The use of the appropriate primary amines similarly gives the 4-substituted derivatives shown in Table 14.

1H-1,3,2-Diazaphosphole and 1,3,2-Diazaphospholidines

1H-1,3,2-Diazaphosphole, [RRI 86], has the structure and numbering shown in (**1**); an isomeric form of this compound would be 2H-1,3,2-diazaphosphole (**2**). These two compounds are known only as substituted

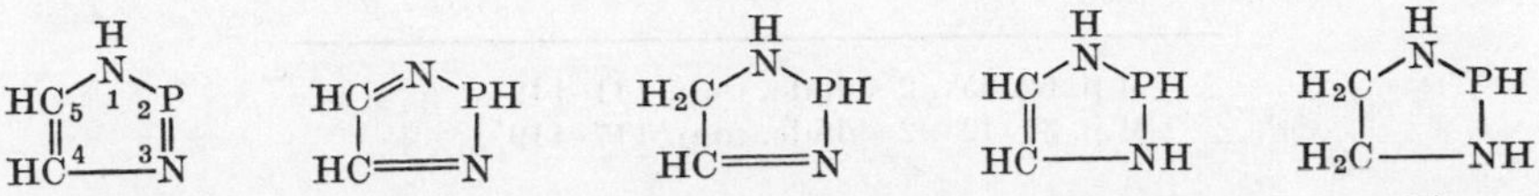

derivatives, in which the number and position of the substituent groups determine the position of the double bonds in the ring. The differences disappear in the dihydro derivatives: the compound (**3**) is termed Δ^3-1,3,2-diazaphospholene, or more conveniently 1,3,2-diazaphosphol-3-ene, and (**4**) is similarly 1,3,2-diazaphosphol-4-ene. The tetrahydro compound (**5**), termed 1,3,2-diazaphospholidine, is by far the most important in the number of its derivatives and the extent of its known chemistry.

An apparently simple route to derivatives of (**2**) is the addition of a phosphonous dichloride, $RPCl_2$, to a chilled ethereal solution of triethylamine and various oxalodiiminic esters of type (**6**) (Dregval and

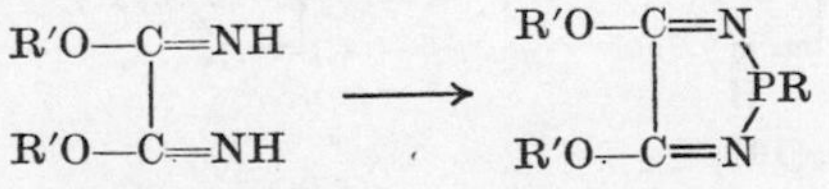

Derkach[298]); thus the action of phenylphosphonous dichloride on the dimethyl ester (**6**; $R' = CH_3$) gives 4,5-dimethoxy-2-phenyl-2H-1,3,2-diazaphosphole (**7**; $R = C_6H_5$, $R' = CH_3$), b.p. 145–146°/0.2 mm, m.p. 61–65°. All the other recorded compounds of type (**7**) have $R' = C_2H_5$, and differ only in the nature of the group R (see Table 15).

The production of compounds of type (**1**) is apparently more complex. It has been claimed (Shevchenko, Standnik, and Kirsanov[299]) that, when a mixture of oxamide (1 equivalent) and phosphorus pentachloride (3 equivalents) in phosphorus oxychloride is maintained at

Table 15. 4,5-Diethoxy-$2H$-1,3,2-diazaphos-
phole derivatives (7)

R	B.p. (°/mm)
C_6H_5*	149–150/0.25
p-$CH_3C_6H_4$†	158–160/0.3
p-ClC_6H_4	152–153/0.25
n-C_4H_9	101–102/0.2
C_6H_5O	137–139/0.2
p-$CH_3C_6H_4O$	151–152/0.2
p-ClC_6H_4O	164–166/0.3

* M.p. 62–65°; 2-sulfide, m.p. 141–142°.
† M.p. 37–42°; 2-sulfide, m.p. 117–119°.

15–17° for 35–40 hours, 2,2,2,4,4,5,5-heptachloro-1,3,2-diazaphospho-
lidine (8) (80–90%) is formed, but, when the mixture is maintained at
15–17° for 2–3 days, the main product is 2,2,4,4,5,5-hexachloro-1,3,2-
diazaphospholidinium hexachlorophosphate (9) (75–80%). The latter

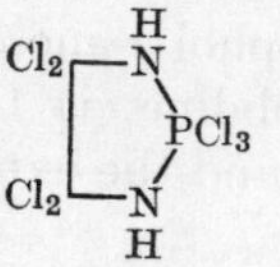

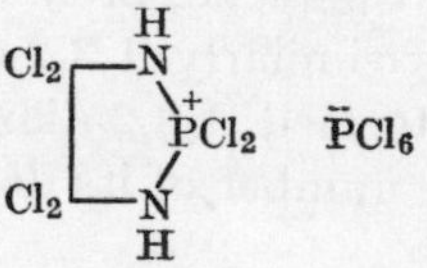

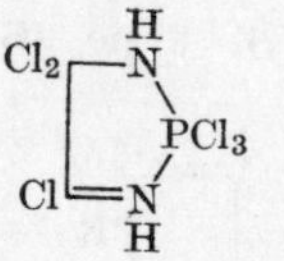

product, when heated with PCl_5 in $POCl_3$ at 40° for 3 hours, gives
2,2,2,4,5,5-hexachloro-1,3,2-diazaphosphol-3-ine (10) (60–70%), which
when similarly heated at 80–85° for 1 hour and then at 100° for 20 minutes
gives 2,2,2,4,5-pentachloro-$1H$-1,3,2-diazaphosphole (11) (50–55%).

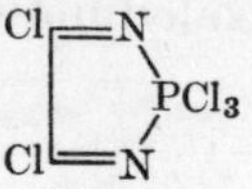

The interest of these compounds would warrant a decisive identification of their structures.

Syntheses of the 1,3,2-diazaphospholidines usually involve solely
the condensation of ethylenediamine and the appropriate dihalogeno-
phosphorus compound.

Phenyl phosphorodichloridate, $C_6H_5OP(O)Cl_2$, when added to a

stirred aqueous solution of ethylenediamine at 0°, gives 2-phenoxy-
1,3,2-diazaphospholidine 2-oxide (12), m.p. 196°: the analogous p-tolyl-
oxy compound has m.p. 204°. The use of phenylphosphoramidic

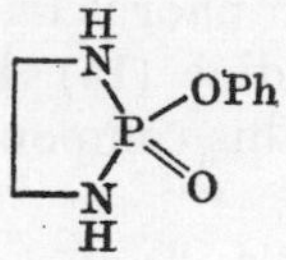

(12) 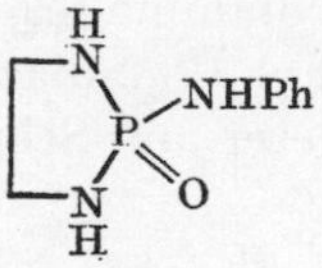   (13) 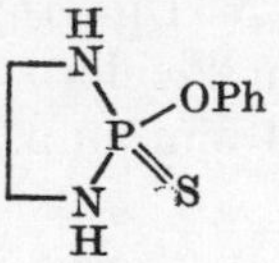   (14)

dichloride, $C_6H_5NHP(O)Cl_2$, similarly gives 2-anilino-1,3,2-diazaphos-
pholidine 2-oxide (13), m.p. 232°, and phenyl thiophosphorodichloridate,
$C_6H_5OP(S)Cl_2$, gives 2-phenoxy-1,3,2-diazaphospholidine 2-sulfide
(14), m.p. 189° (Autenrieth and Bölli[300]; Autenrieth and Meyer[301]).
These three compounds are stable to water, aqueous alkalis, and dilute
acids.

More recently, compounds of type (12) with the phenyl group
halogenated, e.g., 2,4,5- and 2,4,6-trichlorophenyl, 2,4,6-tribromophenyl,
and pentachlorophenyl, have been recorded; it is claimed that they
impart flame resistance to 'cellulosic materials' (Tolkmith and
Britton[302]).

The main point of interest in the compounds (12), (13), and (14) is
that they are all insoluble in water, in all the common organic solvents,
and also in molten naphthalene, and their molecular weights have
apparently not been determined. This insolubility is in striking contrast
to the reasonably ready solubility of the corresponding benzo derivatives (p. 221) and probably implies some type of considerable and stable
molecular association; it is reminiscent of that of oxamide and the much
greater solubility of, for example, succinamide and phthalamide.

It is noteworthy that the above three phosphoric dichlorides react
with aqueous hydrazine under suitable conditions to give the compounds

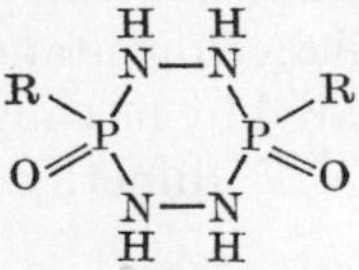

(15a) 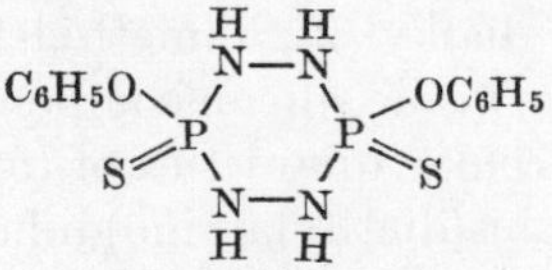   (16a)    (17a)

(15a; $R = C_6H_5O$), m.p. 132°, (15a; $R = C_6H_5NH$), m.p. 208–210°, and
(16a), m.p. 183°, respectively; these products are freely soluble in organic
solvents, in which their molecular weights have been determined. They

are named after the parent inorganic ring system 1,2,4,5,3,6-tetrazadi-phosphorine (**17a**), [RRI 300], and the compound (**15a**; R = C_6H_5O) is therefore hexahydro-3,6-diphenoxy-1,2,4,5,3,6-tetrazadiphosphorine 3,6-dioxide, and the other two compounds are named analogously.

N,N'-Dimethylethylenediamine reacts with phosphorus trichloride to form 2-chloro-1,3-dimethyl-1,3,2-diazaphospholidine (**15**), b.p. 98–100°/14 mm, m.p. 1–3° (Scherer and Schmidt[303]). This compound with

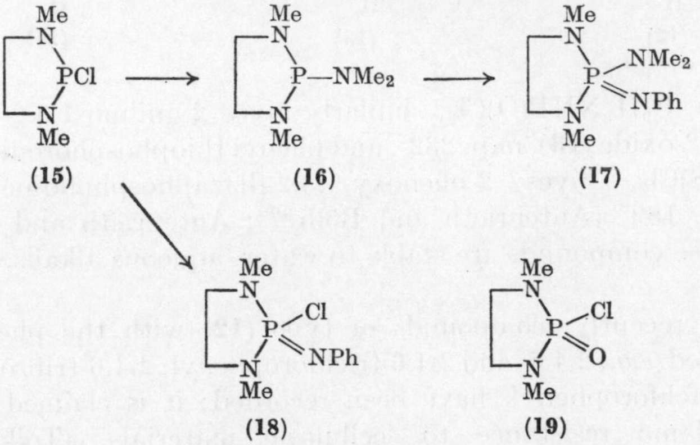

dimethylamine gives the 2-dimethylamino derivative (**16**), b.p. 58–59°, which in turn reacts with phenyl azide in ether to form (2-dimethyl-amino)-1,3-dimethyl-2-(phenylimino)-1,3,2-diazaphospholidine (**17**), m.p. 30–31°. The 2-chloro compound similarly reacts with the azide to form the 2-phenylimino derivative (**18**), m.p. 34–36°. The infrared spectra of (**17**) and (**18**) show intense bands at 1190 and 1220 cm^{-1}, respectively, due to the P—N frequency (Utvary, Gutmann, and Kemenater[304]). Phosphorus oxychloride reacts with N,N'-dimethyl-ethylenediamine (as expected) to give the 2-chloro 2-oxide (**19**), m.p. 213–215°.

Although the compounds (**15**) and (**19**) are thus very readily ob-tained, the corresponding 2-phenyl derivative and its 2-oxide have been prepared from the 1,3-dialkyl-2,2-dimethyl-1,3-diaza-2-silacyclopentane (**20**), [RIS 9742]. These cyclic silicon compounds are prepared by heating a mixture of equimolecular quantities of (for example) N,N'-dimethyl-ethylenediamine and bis(diethylamino)dimethylsilane,

$$[(C_2H_5)_2N]_2Si(CH_3)_2,$$

containing a trace of ammonium sulphate: diethylamine is rapidly evolved and distillation gives 1,2,2,3-tetramethyl-1,3-diaza-2-silacyclo-pentane (**20**; R = CH_3), b.p. 131–132° (82%). The use of N,N'-diethylethylenediamine similarly gives the 1,3-diethyl-2,2-dimethyl analogue (**20**; R = C_2H_5), b.p. 170° (84%).

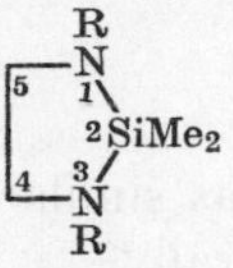

(20)

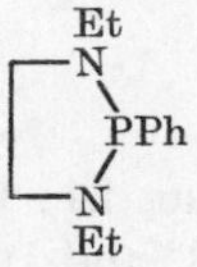

(21)

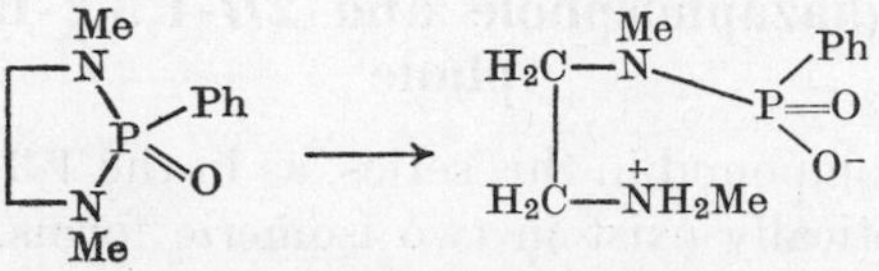

(22)

(23)

Dropwise addition of phenylphosphonous dichloride (1 equivalent) to (**20**; R = C_2H_5) at 0°, followed by heating, gives an early release of dichlorodimethylsilane, $Cl_2Si(CH_3)_2$, followed on distillation by 1,3-diethyl-2-phenyl-1,3,2-diazaphospholidine (**21**), b.p. 98°/0.1 mm (Abel and Bush[305]).

An equimolecular mixture of phenylphosphonic dichloride, $C_6H_5P(O)Cl_2$, and (**20**; R = CH_3), when heated at 150° for 3 hours and then distilled, gives dichlorodimethylsilane and then 1,3-dimethyl-2-phenyl-1,3,2-diazaphospholidine 2-oxide (**22**), b.p. 171°/20 mm, m.p. *ca.* 43°, a compound which can readily be obtained by condensing the phosphonic dichloride with *N,N'*-diethylethylenediamine in the presence of triethylamine.

A solution of (**22**) in dilute aqueous ammonium sulfate, when allowed to evaporate at room temperature, leaves a crystalline residue, m.p. 163–167° after recrystallization from water; this is considered to be the zwitter-ion (**23**) on the basis of its composition and its high solubility in water and alcohols and its insolubility in most organic solvents. When this product is treated with 'strong acids' complete hydrolysis to the initial diamine and phenylphosphonic acid occurs (Yoder and Zuckerman[306]).

The stereochemistry of the ring system (**5**) has been investigated Fel'dman and Berlin,[307] who condensed *meso*-stilbenediamine with by phosphor the amidic dichloride, $(ClCH_2CH_2)_2NP(O)Cl_2$, to form 2-[bis(2'-chloroethyl)-amino]-4,5-diphenyl-1,3,2-diazaphospholidine 2-oxide (**24**; R = CH_2CH_2Cl), which was separated into a less soluble isomer, m.p. 180–181°, and a more soluble isomer, m.p. 150–152°. The

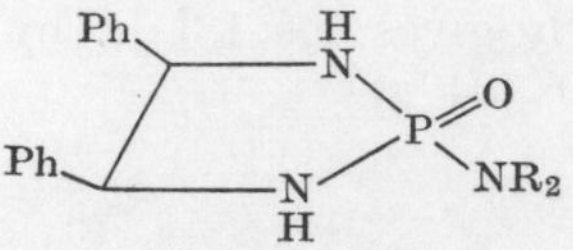

(24)

2-diethylamino analogue (**24**; $R = C_2H_5$) was similarly separated into isomers, m.p. 201.5–203° and 162–164°, severally.

1*H*-1,3,2-Benzodiazaphosphole and 2*H*-1,3,2-Benzodiazaphosphole

The parent compound in this series, as in the 1,3,2-diazaphosphole series, can theoretically exist in two isomeric forms. 1*H*-1,3,2-Benzodiazaphosphole, [RIS 11820], has the structure (**1**), and 2*H*-1,3,2-benzodiazaphosphole, [RRI 1061], has the structure (**2**); the difference

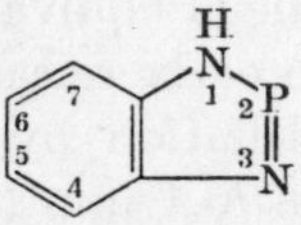

(1) 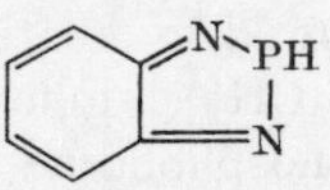   (2) 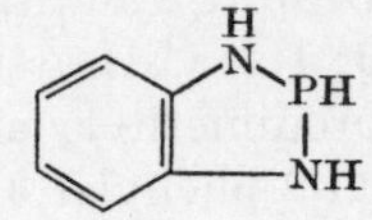   (3)

disappears in the dihydro derivative (**3**), which should be regarded as a derivative of (**1**) and termed 2,3-dihydro-1*H*-1,3,2-benzodiazaphosphole (although the earlier chemists called it 'phenphosphazine'.) Almost all compounds having the benzodiazaphosphole ring system are derivatives of (**3**).

Apparently only one monosubstituted derivative of (**1**) has been prepared. *N*-Phenyl-*o*-phenylenediamine (2-aminodiphenylamine) (**4**)

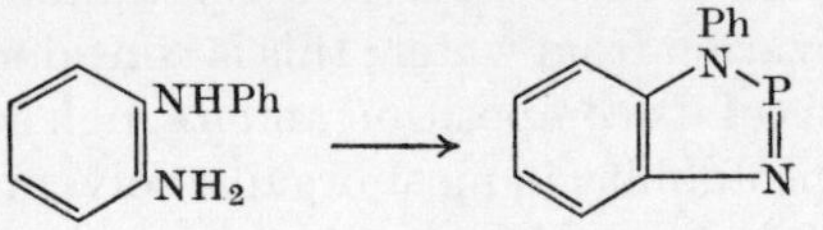

(4)   (5) 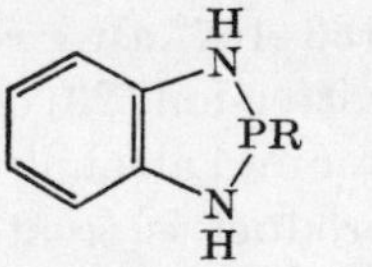   (6)

is added to cooled triphenyl phosphite, $(C_6H_5O)_3P$, and the mixture heated to 150°/1 mm until distillation of phenol ceases, and finally heated at 180°. The residue, crystallized from xylene, gives 1-phenyl-1*H*-1,3,2-benzodiazaphosphole (**5**), m.p. 350°. No analogous reaction occurs (not unexpectedly) when triphenyl phosphate is employed (Pilgram and Korte[308]).

When an equimolecular mixture of o-phenylenediamine and diethyl p-tolylphosphinite, $CH_3C_6H_4P(OC_2H_5)_2$, is heated at 140° for 30 minutes, it yields 2,3-dihydro-2-p-tolyl-1,3,2-benzodiazaphosphole (**6**; $R = C_6H_4CH_3$), m.p. 245–248°.

The preparation of three types of derivative of (**3**) by Autenrieth and his co-workers follows the similar preparations in the 1,3,2-diazaphospholidine series. An ethereal solution of o-phenylenediamine (2 moles) and O-phenyl thiophosphorodichloridate, $C_6H_5OP(S)Cl_2$ (1 mole), when heated first to remove the solvent and then at 150–170°, gives 2,3-dihydro-2-phenoxy-1,3,2-benzodiazaphosphole 2-sulfide (**7**),

(7)   (8)   (9)

m.p. 185° (Autenrieth and Hildebrand[309]). A similar mixture of the diamine and $C_6H_5OP(O)Cl_2$ in boiling benzene gives the corresponding 2-oxide (**8**), m.p. 185° (p-tolyl analogue, m.p. 158°), and the use of $C_6H_5NHP(O)Cl_2$ in boiling benzene gives the 2-anilino 2-oxide (**9**), m.p. 214° (Autenrieth and Bölli[300]). These three compounds differ from their 1,3,2-diazaphospholidine counterparts in being soluble in several organic solvents, in which they show molecular weights which accord with the above formulation.

They are neutral, stable compounds usually unaffected by aqueous alkalis or mineral acids; prolonged boiling with ethanolic potassium hydroxide or concentrated hydrochloric acid does, however, cause decomposition.

'Nitrogen mustard' groups have been introduced as substituents into many heterocyclic systems in order to investigate the therapeutic properties of the products, more particularly any carcinolytic activity which the products might exert on the (presumed) selective action of phosphoramidase in tumor cells (cf. p. 324). When a suspension of bis-(2-chloroethyl)amine hydrochloride, $(ClC_2H_4)_2NH,HCl$, in phosphorus oxychloride is boiled under reflux for 12 hours, distillation gives N,N-bis-(2-chloroethyl)phosphoramidic dichloride, $(ClC_2H_4)_2NP(O)Cl_2$, b.p. 123–125°/0.6 mm, m.p. 54–56° (Friedman and Seligman[310]). This compound gives the usual condensation with o-phenylenediamine to form 2-[bis-(2'-chloroethyl)amino]-2,3-dihydro-1,3,2-benzodiazaphosphole 2-oxide, m.p. 162–164° (Friedman, Papanastassiou et al.[311]); this compound lacked significant therapeutic action.

Six-membered Ring Systems containing Carbon and One Phosphorus and One Nitrogen Atom

Azaphosphorines

There should be three isomeric systems in this class, namely, the 1,2-azaphosphorine (**1**), the 1,3-azaphosphorine (**2**), and the 1,4-azaphosphorine (**3**). None of these actual compounds is known, but substituted hydrogenated derivatives of (**1**) and (**3**) have been prepared: derivatives of (**2**) are at present unknown.

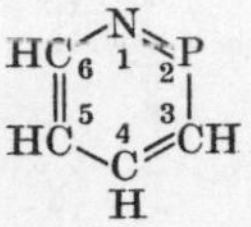

(1) 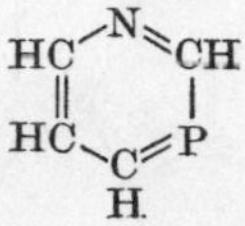   (2) 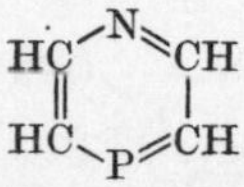   (3)

The benzo derivatives of the azaphosphorines with which we shall be concerned here are the dibenzo compound (**4**) of 1,2-azaphosphorine

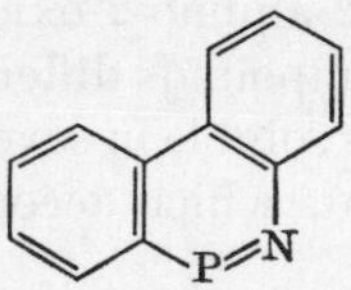

(4) 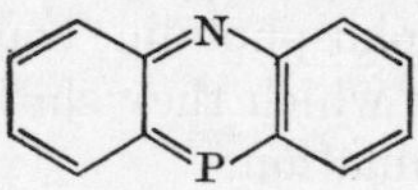   (5)

(p. 224) and the dibenzo compound (**5**) of 1,4-azaphosphorine (p. 232): each is known solely as dihydro derivatives. They are discussed after the corresponding azaphosphorines.

1,2-Azaphosphorine

The fully hydrogenated derivative of 1,2-azaphosphorine (**1**), [RIS 11612], is termed hexahydro-1,2-azaphosphorine (**2**): the term 1,2-azaphosphorane is *not* employed for (**2**) (cf. *Note*, p. 208).

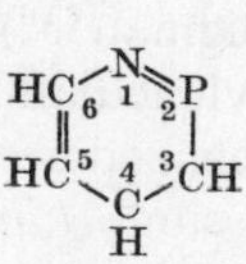

(1) 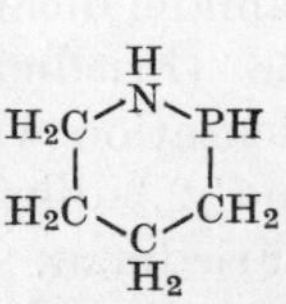   (2) 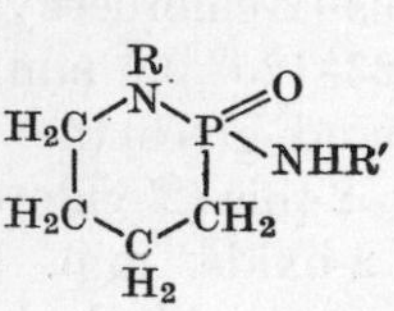   (3)

The known chemistry of (2) is almost solely that of its 1-substituted 2-amino 2-oxide derivatives (3), which are prepared almost precisely as the corresponding lower homologous 2-imino-1,2-azaphospholidine 2-oxide derivatives (p. 210).

4-Chlorobutylphosphonic acid (4) is converted into the phosphonic dichloride (5), which when treated with, for example, aniline (2 equivalents) and triethylamine gives the phosphonic dianilide (6). This

$$Cl(CH_2)_4PO(OH)_2 \rightarrow Cl(CH_2)_4POCl_2 \rightarrow Cl(CH_2)_4PO(NHC_6H_5)_2$$
$$\quad\quad (4) \quad\quad\quad\quad\quad\quad (5) \quad\quad\quad\quad\quad\quad\quad (6)$$

compound in boiling methanol with sodium hydroxide (1 equivalent) gives 2-anilino-hexahydro-1-phenyl-1,2-azaphosphorine 2-oxide (3; R = R' = C_6H_5), m.p. 156–159° (Helferich and Curtius,[291] Helferich and Schroeder[312]).

Other compounds prepared in this series are shown in Table 16.

Table 16. Derivatives of 2-aminohexahydro-1,2-aza-
phosphorine 2-oxide (3)

Compounds	M.p. (°)
(3; R = R' = CH_2C_6H_5)	55–57
(3; R = R' = C_6H_4Cl-p)	207–210
(3; R = R' = C_6H_4SCH_3-p)	187–188
(3; R = R' = C_5H_4N)*	168–170

* The last compound is hexahydro-1-(2'-pyridyl)-2-(2'-pyridylamino)-1,2-azaphosphorine 2-oxide.

These compounds are also called phostamic acid amides, and (like their five-membered analogues) are readily hydrolyzed by acids. The preparation of the free phostamic acids, by cyclization of the monoamides, could not be achieved.

It is noteworthy that when a phosphonic dichloride (5) is treated with one equivalent of a primary amine and triethylamine, a tetrasubstituted derivative of 1,3,2,4-diazadiphosphetidine (7), [RRI 51], is formed; for example, the phosphonic dichloride (5) when thus treated

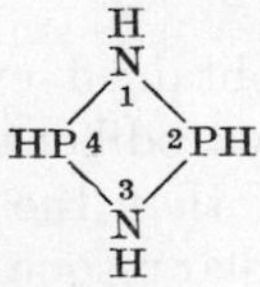

(7) 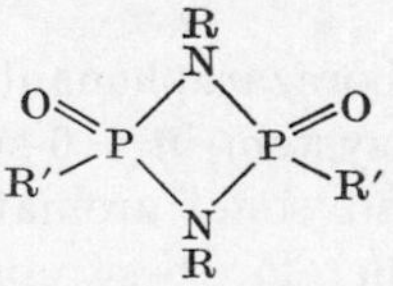   (8)

with p-chloroaniline gives 2,4-bis-(4'-chlorobutyl)-1,3-bis-(p-chloro-phenyl)-1,3,2,4-diazadiphosphetidine 2,4-dioxide [8; R = C_6H_4Cl-p, R' = $(CH_2)_4Cl$], m.p. 183–184°, and n-butylphosphonic dichloride with benzylamine gives 1,3-dibenzyl-2,4-di-n-butyl-1,3,2,4-diazadiphospheti-dine 2,4-dioxide [8; R = $CH_2C_6H_5$, R' = $(CH_2)_3CH_3$], m.p. 126–130°. These compounds are the 'dimeric alkanephosphonic acid imides' of Helferich and Schroeder.[312]

Dibenz[c,e][1,2]azaphosphorine

Dibenz[c,e][1,2]azaphosphorine, [RIS 10449], is now presented and numbered as shown in (1). It is—as one would expect—unknown and

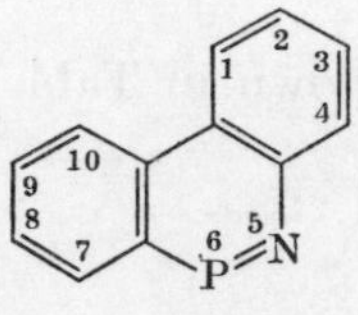

(1) 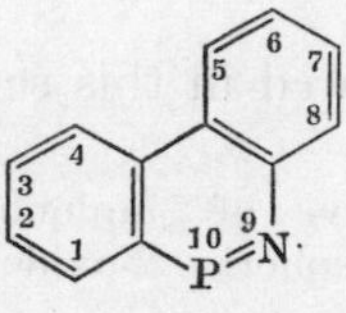   (1A) 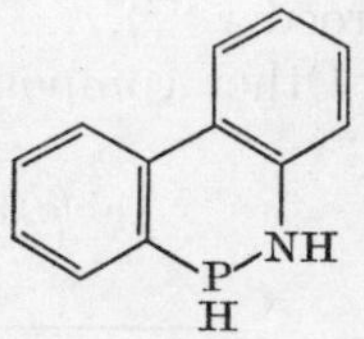   (2)

all derivatives are substitution products of 5,6-dihydrodibenz[c,e][1,2]-azaphosphorine (2).

Compounds having the skeleton (1) were not recorded until the almost simultaneous publication of two major contributions in 1960 (Dewar and Kubba[313]; Campbell and Way[314]): in both publications the ring system (1) was termed 9,10-azaphosphaphenanthrene and numbered as in (1A).

Dewar and his collaborators had earlier investigated certain derivatives of dibenz[c,e][1,2]azaborine (3), [RIS 8536], which they

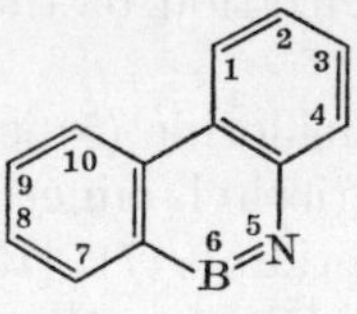

(3) 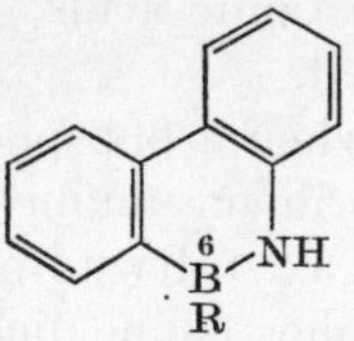   (4) 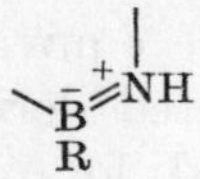   (5)

named 9,10-borazarophenanthrene, and had obtained evidence that the tricyclic system in a 6-substituted 5,6-dihydrodibenz[c,e][1,2]aza-borine (4) had some aromatic character, and that the heterocyclic portion was therefore best represented as the dipolar resonance structure (5). Analogous derivatives of the dihydroazaphosphorine (2) were therefore investigated for possible evidence of a similar —$\overset{-}{P}H{=}\overset{+}{N}H$— structure in which the P=N bond would be a $d\pi$–$p\pi$ bond.

2-Aminobiphenyl in benzene, when treated with phosphorus trichloride, gave a crystalline product which was not identified but was

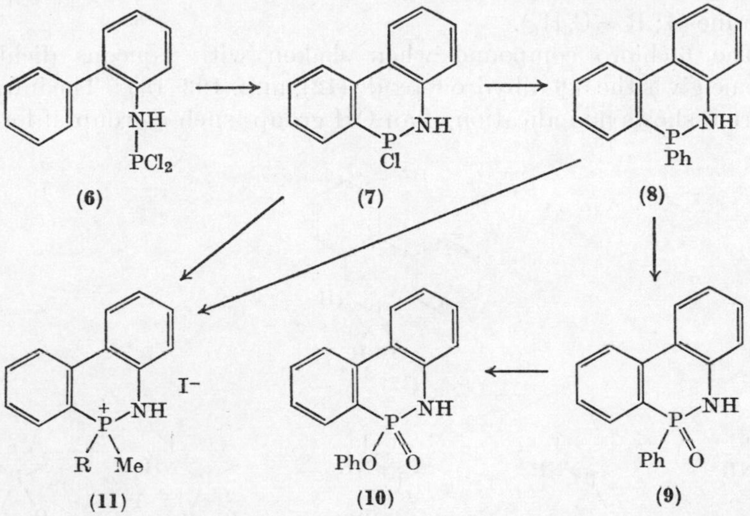

undoubtedly N-(2-biphenylyl)phosphoramidous dichloride (6); when this compound was heated with a 'catalytic' quantity of aluminum chloride at 210–220° for 6 hours it underwent cyclization to 6-chloro-5,6-dihydrodibenz[c,e][1,2]azaphosphorine (7), m.p. 132–134° (42%) after sublimation at 180–190°/0.05 mm. Treatment of (7) with phenyl-magnesium bromide gave the 5,6-dihydro-6-phenyl derivative (8), m.p. 178–179° (58%). This compound was heated with bromobenzene and aluminum chloride in an attempt to prepare the diphenylphosphonium salt (cf. Chatt and Mann[315]; Mann and Millar[316]) but only the 6-oxide (9), m.p. 283° (40%), was obtained. When this compound was exposed to air at room temperature for 6 weeks, it was apparently converted into the 6-phenoxy 6-oxide (10), m.p. 288–289°; this suggested identity was based on (a) the composition, and (b) its ultraviolet spectrum, which was almost identical with that of the 6-phenyl 6-oxide (9) and resembled that of 5,6-dihydro-6-hydroxydibenz[c,e][1,2]azaborine (4; R = OH). A mixture of (9) and (10) melted at 244–250°.

The chloro derivative (7) when treated with an excess of methyl-magnesium iodide gave 5,10-dihydro-6,6-dimethyldibenz[c,e][1,2]aza-phosphorinium iodide (11; R = CH₃), m.p. 230–233°: the 6-phenyl compound (8) in boiling methyl iodide–benzene gave the corresponding

6-methyl-6-phenyl iodide (11; $R = C_6H_5$), m.p. 214°, both salts forming pale yellow crystals. These two methiodides have ultraviolet spectra very similar to that of the 6-phenoxy 6-oxide (10).

The ultraviolet spectrum of the 6-phenyl compound (8) also has a general resemblance to that of 5,10-dihydro-6-phenyldibenz[c,e][1,2]-azaborine (4; $R = C_6H_5$).

The 6-chloro compound when shaken with aqueous dichloromethane gives the 5,6-dihydro 6-oxide (12), m.p. 193–194°. The infrared spectrum shows no indication of an OH group; such a group, if formed

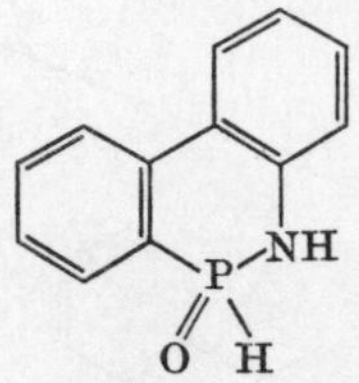

$$P \underset{O}{\overset{NH}{<}} H$$

(12)

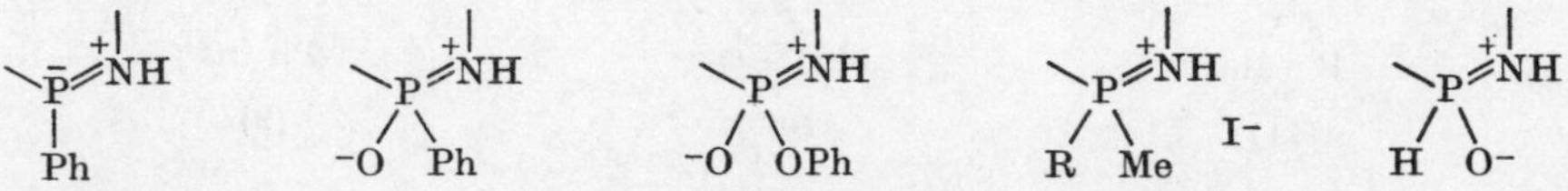

(8A)   (9A)   (10A)   (11A)   (12A)

during this hydrolysis, would be expected to change into the form (12), comparable to a secondary phosphine oxide. The ultraviolet spectrum of (12) in ethanol is very different from that in 10% aqueous sodium hydroxide.

Dewar and Kubba[313] conclude 'The similarity of the U.V. spectra (of these phosphorus compounds) to those of analogous boron compounds which are known to be aromatic suggests that they may be aromatic too.' Acceptance of this conclusion must surely await more decisive evidence.

In accordance with this conclusion, however, they formulate the N–P portion of the compounds (8), (9), (10), (11) and (12) as (8A), (9A), (10A), (11A), and (12A), respectively.

The object of Campbell and Way's investigation[314] was primarily stereochemical, namely, to see if a suitably substituted 5,6-dihydro-6-phenyldibenz[c,e][1,2]azaphosphorine could be resolved into optically active forms by virtue of the asymmetric 3-covalent phosphorus atom. At this time no compound of general type $PR^1R^2R^3$ had been so resolved.

The initial stages were essentially similar to those of Dewar and Kubba,[313] 2-aminobiphenyl or a 4-substituted 2-aminobiphenyl being converted into the 5,6-dihydro-6-chlorodibenz[c,e][1,2]azaphosphorine,

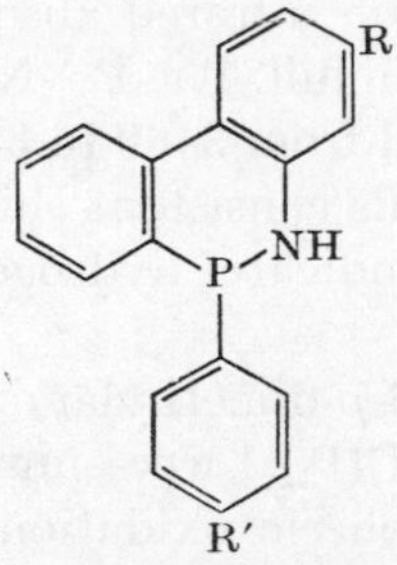

(13) 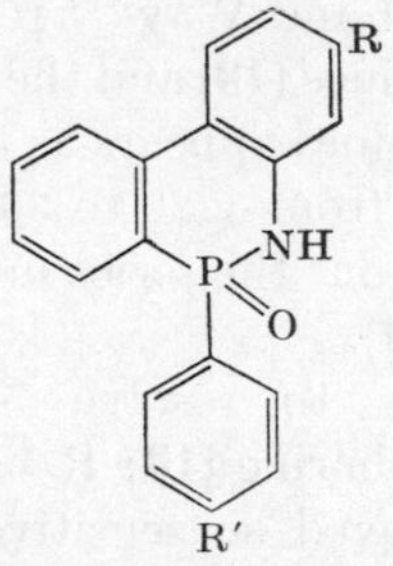   (14)

which was then converted by the appropriate Grignard reagent into the 6-phenyl or the *para*-substituted 6-phenyl derivative (13). The corresponding 6-oxides (14) were also readily prepared at room temperature by hydrogen peroxide oxidation in ethanol, thus giving the following five phosphines (13) and their 6-oxides (14), listed in Table 17.

Table 17.   6-Aryl-5,6-dihydrodibenzo[c,e][1,2]azaphos-
              phorines (13) and their 6-oxides (14)

Substituents	Phosphine (13), m.p. (°)	6-Oxide (14), m.p. (°)
R = R′ = H	181	289–290
R = CH₃, R′ = H	125–128	246–248
R = H, R′ = CH₃	135–138	232–235
R = H, R′ = Br	145–147	223–226
R = H, R′ = N(CH₃)₂	148–153*	258–260

* 158–160° in sealed tube under nitrogen.

This unique set of the 6-phenyl and the 6-phenyl 6-oxide members allowed a wider observation of their chemical and physical properties. The crystalline phosphines (13) were stable in air, but in solution were often converted into the oxides (14). The N—P bond in the oxides was stable in hot dilute alkalis, but was split in boiling 5N-hydrochloric acid, the oxide (14; R = R′ = H) giving the hydrochloride of (2′-amino-2-biphenylyl)phenylphosphinic acid, $NH_2C_6H_4$—C_6H_4—$P(C_6H_5)(O)OH$; this indicates that these 6-oxides are structurally internal amides of a phosphinic acid. Both the ease of ring closure and the stability of the ring system were demonstrated by heating this acid, which melted at 244° with loss of water and hydrogen chloride and solidification to the 6-oxide, which then melted at 290°.

Campbell and Way[314] record the main infrared absorption bands of the phosphines (13) and the oxides (14) in full. The P—NH absorption in the phosphines appears as a broad band from 3460 to 3370 cm⁻¹ and in the oxides from 3220 to 3070 cm⁻¹. This consistent shift of 270–300 cm⁻¹ throughout the series undoubtedly indicates hydrogen bonding in the oxides (14).

Attempts to resolve 5,6-dihydro-6-p-dimethylaminodibenz[c,e]-[1,2]azaphosphorine [13; R = H, R' = N(CH$_3$)$_2$] were unsuccessful, for this base proved so sensitive to atmospheric oxidation that it was entirely converted into the 6-oxide during attempted recrystallization of its salts with optically active acids. This 6-oxide was, however, readily resolved by recrystallization of its salts with (+)- and (−)-camphor-10-sulfonic acid, the less soluble salt in each case having [α]$_D$ −67.9° and +67.4°, respectively. Treatment with aqueous ammonia regenerated the active monohydrated 6-oxides, m.p. 135–136°, [α]$_D$ ±152°.

Reduction of the (+)-oxide in di-n-butyl ether–benzene at 80° with lithium aluminum hydride gave the phosphine, m.p. 143–148°, with an initial rotation of [α]$_D$ −128° (in ethyl acetate), which fell to [α]$_D^{21}$ −97° in 4 hours, in the course of which the solution became cloudy and had to be filtered; the rotation of the filtrate then remained constant. Similar reduction of the (−)-oxide gave the phosphine, m.p. 154–156°, with [α]$_D^{21}$ +141°, which also fell to [α]$_D^{21}$ +114° in the course of 5 hours. A detailed investigation showed that this change in rotation was caused solely by atmospheric conversion into the 6-oxide and that the phosphines showed considerable optical stability in the absence of oxygen.

A model of the compound (13; R = R' = H) indicates that the N–C ring is slightly puckered; if any such puckering were rigid, the molecule would possess molecular dissymmetry as well as an asymmetric phosphorus atom. This factor, and more particularly the high sensitivity of the phosphine to oxidation, caused these workers to investigate other cyclic phosphorus systems (cf. p. 85).

1,4-Azaphosphorine

1,4-Azaphosphorine (1) (formerly termed 1-aza-1-phosphabenzene), [RRI 240], is known only as substituted derivatives of hexahydro-1,4-azaphosphorine (2).

HC$\underset{6}{\overset{N}{\underset{5}{\|}}}\overset{1}{\underset{4}{\underset{P}{\|}}}\overset{2}{\underset{3}{}}$CH
HC$\overset{}{\underset{}{}}$CH

(1)

H$_2$C$\overset{H}{\underset{}{N}}CH_2$
H$_2$C$\overset{}{\underset{H}{P}}CH_2$

(2)

$\overset{Ph}{\underset{Ph}{N}}$ ... $\overset{}{\underset{P}{}}$

(3)

Hexahydro-1,4-diphenyl-1,4-azaphosphorine (**3**) can be readily prepared by the interaction of N,N-di-(2-bromoethyl)aniline,

$$C_6H_5N(CH_2CH_2Br)_2,$$

and phenylphosphinebis(magnesium bromide), $C_6H_5P(MgBr)_2$. The latter was first prepared (Job and Dusollier [317]) by the action of ethereal ethylmagnesium bromide on phenylphosphine; better results are obtained in the preparation of this reagent and its arsenic analogue (p. 369) if phenylmagnesium bromide is used in ether–benzene, from which the ether can be distilled away, although the final product may contain traces of biphenyl.

The reaction with N,N-di-(2-bromoethyl)aniline furnishes the azaphosphorine (**3**) as colorless crystals, m.p. 89–90° (55%) (Mann and Millar [318]). It forms a dihydrochloride, m.p. 165–166°, which is stable in dry air, a monohydriodide, m.p. 190–191°, which becomes brown on exposure to light, and a monopicrate, m.p. 131–132°.

The azaphosphorine (**3**) coordinates with potassium tetrabromo palladate(II) by the phosphine group alone, to give dibromobis(hexahydro-1,4-diphenyl-1,4-azaphosphorine)palladium, orange-yellow crystals, m.p. 184–185°.

The quaternary salts of the azaphosphorine (**3**) are of some interest, particularly in comparison with those of 1,4-diphenylpiperazine (**4**) and

$\overset{Ph}{\underset{Ph}{N}}\cdots\overset{}{\underset{N}{}}$

(4)

$\overset{Ph}{\underset{Ph}{N}}\cdots\overset{}{\underset{As}{}}$

(5)

$\overset{Ph}{\underset{Ph\ R}{N}}\cdots\overset{}{\underset{\overset{+}{P}}{}}$ X$^-$

(6)

$\overset{Ph\ R'}{\underset{Ph\ R}{N}}\cdots\overset{\overset{+}{}}{\underset{\overset{+}{P}}{}}$ 2X$^-$

(7)

of hexahydro-1,4-diphenylazarsenine (**5**), for these three compounds form a eutropic series of which the azaphosphorine is the middle member.

The azaphosphorine with cold methyl bromide forms a monomethobromide (**6**; R = CH$_3$, X = Br), m.p. 152°; with methanolic methyl bromide at 100° it forms a monomethobromide hydrobromide (**7**; R = CH$_3$, R′ = H, X = Br), m.p. 279–283°; both salts with aqueous sodium picrate give the monomethopicrate (6; $R = CH_3$, $X = C_6H_2N_3O_7$), m.p. 118°. It gives similarly a monoethobromide (6; $R = C_2H_5$, $X = Br$), m.p. 175–176°, and a monoethobromide hydrobromide (7; $R = C_2H_5$, $R' = H$, $X = Br$), m.p. 278–284°.

The azaphosphorine (3) reacts vigorously with cold methyl iodide to form a monomethiodide (6; $R = CH_3$, $X = I$), m.p. 155–156°, and in boiling methyl iodide to give a hygroscopic dimethiodide (7; $R = R' = CH_3$, $X = I$), m.p. 118° (dimethopicrate, m.p. 172–174°). This dipicrate can also be obtained by heating the azaphosphorine with methyl toluene-p-sulphonate at 100° and then treating the deliquescent dimethotoluene-p-sulphonate in solution with sodium picrate. The assignment of structure (6) to the monoquaternary salts is based (a) on the generally greater reactivity of tertiary phosphines than of analogous amines, and (b) on confirmation from the infrared spectra.

In contrast, 1,4-diphenylpiperazine (4) (Dunlop and Jones [319]) and hexahydro-1,4-diphenyl-1,4-azarsenine (5) (p. 504) give only monomethiodides even under forcing conditions, confirming the greater reactivity towards quaternization of the P—aryl group compared with similar N—aryl and As—aryl groups. In this eutropic series, as in several others, the nitrogen and arsenic members have closely similar properties, whereas many of those of the phosphorus member are different (cf. p. 106).

Since the azaphosphorine (3) is the only member of the series (3), (4), and (5) to form a diquaternary salt, but 1,3-dimethylpiperazine combines readily with 1,2-dibromoethane to form 1,4-diazabicyclo-[2.2.2]octane dimethobromide (8) (Mann and Baker [255]), attempts were

made to combine 1,2-dibromoethane by similar cyclic diquaternization with the azaphosphorine. These gave, however, only ethylenebis(hexahydro-1,4-diphenyl-1,4-azaphosphorinium) dibromide (9), m.p. 273–274° (dipicrate, m.p. 190–191°); the tertiary amine groups in this compound are sufficiently deactivated to prevent further quaternization even with boiling methyl iodide. The azarsenine (5) behaves similarly (p. 504).

An acetone solution of the azaphosphorine, when heated with 30% hydrogen peroxide at 70–80°, gives the 4-oxide (10), m.p. 143–144°;

the infrared spectrum shows a P$=$O band at 1271 cm^{-1}. When, however, an acetic acid solution of the azaphosphorine is similarly treated, it furnishes crystals of composition $C_{16}H_{18}NO_2P,H_2O_2,H_2O$, *i.e.*, the compound is a monohydrated 1,4-dioxide (11), combined with a molecule of hydrogen peroxide; the compound is stable even at 50°/0.5 mm. It is

(10)   (11)   (12)

noteworthy that Bennett and Glynn [320] showed that 1,4-diphenylpiper-azine (4) when similarly oxidized gave crystals of analogous composition, $C_{16}H_{18}N_2O_2,H_2O_2,2H_2O$. They concluded that the hydrogen peroxide unit formed a hydrogen-bonded bridge between the two oxygen atoms of the dioxide; consideration of the interatomic distances indicated that to accommodate this bridge the piperazine ring must have the chair conformation with *cis*-oxygen atoms. If this is correct, the aza-phosphorine 1,4-dioxide (11) should be similarly linked to the hydrogen peroxide to give the structure and conformation (12). This compound, m.p. 149° with vigorous effervescence, gives a neutral aqueous solution which produces an immediate blue color with potassium iodide–starch.

The azaphosphorine (3) is unaffected when mixed with hydriodic acid of constant b.p. and boiled under reflux for 9 hours, and on cooling crystallizes out as the monohydriodide. This high stability is in marked contrast to that of the analogous azarsenine (5), which when similarly treated undergoes replacement of the 4-phenyl group by iodine (p. 505). Several saturated cyclic arsines having the As—C_6H_5 group are thus converted into the corresponding As—I compound (p. 505) but a similar reaction of the phosphorus analogues is apparently unknown.

When hot benzene solutions of the azaphosphorine and of iodine are mixed, chocolate-brown crystals of a triiodide, m.p. 255–257°, separate and can be recrystallized from ethanol. This triiodide when treated with aqueous sodium hydroxide gives the 4-oxide (10), and when treated with an aqueous solution of sulfur dioxide is rapidly converted into the azaphosphorine 4-oxide hydriodide, m.p. 257–260°, which is also formed by the direct union of the 4-oxide and hydriodic acid; treatment of this salt with boiling water or with cold dilute alkalis liberates the 4-oxide. The structure of the triiodide has not been decisively established.

Hexahydro-4-phenyl-1,4-azaphosphorine (**16**) has recently (1967) been prepared by Issleib and Oehme.[294] A solution of (2-aminoethyl)-phenylphosphine (**13**) (p. 232) in liquid ammonia was treated with

$$C_6H_5P(H)CH_2CH_2NH_2 \rightarrow C_6H_5P(Na)CH_2CH_2NH_2 \rightarrow C_6H_5P(CH_2CH_2NH_2)_2$$
$$\text{(13)} \qquad\qquad\qquad \text{(14)} \qquad\qquad\qquad \text{(15)}$$

sodium (one equivalent) to form the sodiophosphine (**14**), which was converted by the addition of 2-chloroethylamine (one equivalent) into bis-(2-aminoethyl)-phenylphosphine (**15**). After evaporation of the

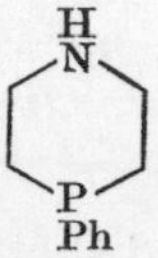

(16) 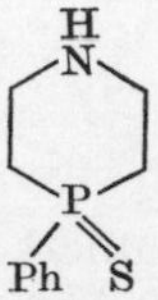   (17)

ammonia the residue underwent cyclization on attempted distillation, with loss of ammonia to give hexahydro-4-phenyl-1,4-azaphosphorine (**16**), b.p. 145–147°/0.3 mm (43%). It was identified by a full analysis and by the addition of sulfur in boiling benzene to form the 4-sulfide (**17**), m.p. 114–115° (70%).

Since the compounds (**14**) and (**15**) were not isolated from the ammonia solution, the presence of the diamine (**15**) was confirmed by evaporating a sample of the ammonia solution and treating the residue with hydrochloric acid to obtain the crystalline dihydrochloride, $[C_{10}H_{19}N_2P]Cl_2$, m.p. 255° (dec.).

Phenophosphazine

Phenophosphazine, [RRI 3305], is the correct name for the parent member (**1**), and the systematic name dibenz[b,e][1,4]azaphosphorine is

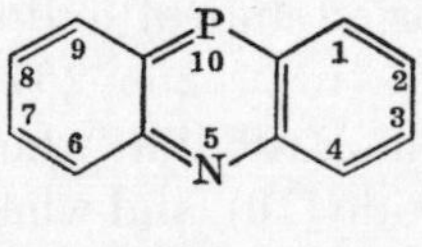

(1) 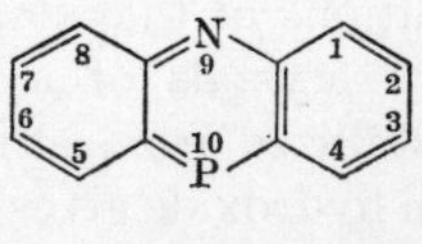   (1A) 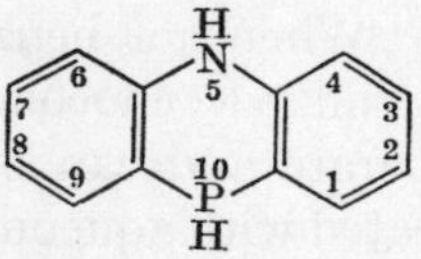   (2)

not used. The numbering in (**1A**) was originally employed and is now obsolete.

This ring system is known solely as substituted derivatives of the 5,10-dihydro compound: since almost all the reactions of this compound occur at or around the phosphorus atom, it is convenient to invert the formula (1) and represent the 5,10-dihydro compound as (2). This compound has been termed 5-aza-10-phospha-5,10-dihydroanthracene.

Michaelis and Schenk [321] stated that diphenylamine and phosphorus trichloride did not interact in the cold, but when heated with zinc chloride in a sealed tube at 250° gave a marked evolution of hydrogen chloride and a viscous oil. The latter, when treated with water, gave a white powdery substance of composition $C_{12}H_{10}NOP$, which they suggested was 5,10-dihydro-10-hydroxyphenophosphazine (3; R = OH). This ring system possessed considerable stability because the compound could be nitrated without rupture of the system.

The formation of this ring system was investigated many years later by Sergeev and Kudryashov [322] who stated that when diphenylamine and phosphorus trichloride had been heated at 200° for 6 hours

(3)   (4)

(no zinc chloride being used), extraction of the cold product with water gave the above 10-hydroxy compound, which shrinks (without melting) at 215–216°. Treatment with thionyl chloride converted it into the 10-chloro compound (3; R = Cl) as a heavy oil which was not purified, but which when treated with sodium ethoxide in ethanol gave 10-ethoxy-5,10-dihydrophenophosphazine (3; R = OC_2H_5), m.p. 151.5–152°. The 10-hydroxy compound in boiling tetralin solution underwent atmospheric oxidation to yield phenophosphazinic acid (10-hydroxy-5,10-dihydrophenophosphazine 10-oxide) (4). This acid gave a silver salt, which in turn gave a methyl ester, m.p. 112–114°, and an ethyl ester, m.p. 99°; a nitro derivative was also described.

Our knowledge of phenophosphazine chemistry was considerably increased by the investigations of Häring. [323] His preparation of the 10-hydroxy compound (3; R = OH), described in detail, [323] consisted essentially in the portionwise addition of diphenylamine (1 mole) to stirred phosphorus trichloride (1.05 mole), the temperature of the complete mixture rising to 50°. This mixture was heated under reflux; at 70° evolution of hydrogen chloride began, and at 140–150° the product

9

formed a clear liquid, which was heated at 200–210° until evolution of HCl ceased. The product was allowed to cool to 100–120° and was then mixed with water, when occasionally crude phosphine was liberated with spontaneous but harmless ignition. The hard reddish-brown product was broken up, filtered off and recrystallized from dimethylformamide (charcoal), giving the monohydrated compound (3; R = OH), m.p. 205–208° (40%). Recrystallization from glacial acetic acid gave the pure anhydrous product (27.3%), m.p. 214.5–216.5° on rather rapid heating; on slow heating it sinters at *ca.* 215° without melting (cf. p. 233).

At room temperature, this 10-hydroxy compound is very slightly soluble in all solvents except formic acid, but it dissolves in hot strongly polar solvents such as ethylene glycol, acetic acid, dimethylformamide, and dimethyl sulfoxide. A solution in hot acetic acid, when treated with a 33% solution of hydrogen bromide also in acetic acid, deposits the hydrobromide, m.p. 156.5°, and the hydrochloride, m.p. 146–147°, is similarly prepared; both salts readily dissociate on attempted recrystallization.

Häring points out that this 10-hydroxy compound (5)(a) might well be expected to have the isomeric 10-oxide structure (5)(b), although the nature of a reagent might determine whether it effectively acted as

(a)   (5)   (b)   (6)

(5)(a) or (5)(b). The infrared spectrum of the anhydrous compound in a mull shows the —NH— portion by bands at 3270, 3185, and 3090 cm^{-1}, a P—H band at 2350 cm^{-1}, and a P=O band at 1355 cm^{-1}; the mono-hydrate shows an additional band at 3390 cm^{-1} attributed to the OH group. This evidence indicates that the solid anhydrous compound has the structure (5)(b), *i.e.*, that of a secondary phosphine oxide; this is expected, for in general the =P—OH group has only a transient existence, passing into the more stable =P(O)H form.

Of the many derivatives of 5,6-dihydrophenophosphazine recorded by Häring, only one involves the =NH group, namely, the reaction of the compound (5) with phenyl isocyanate to form 5(10*H*)-phenophospha-zinecarboxanilide 10-oxide (6), m.p. 227°.

All other reactions concern groups attached to the phosphorus atom, and therefore only this portion of the molecule is depicted in the following reaction sequences.

Häring passed oxygen through a boiling tetralin solution of (5) to obtain phenophosphazinic acid (4), m.p. 277° after crystallization from ethanol; this forms very stable crystals with 0.5 mole of formic acid or

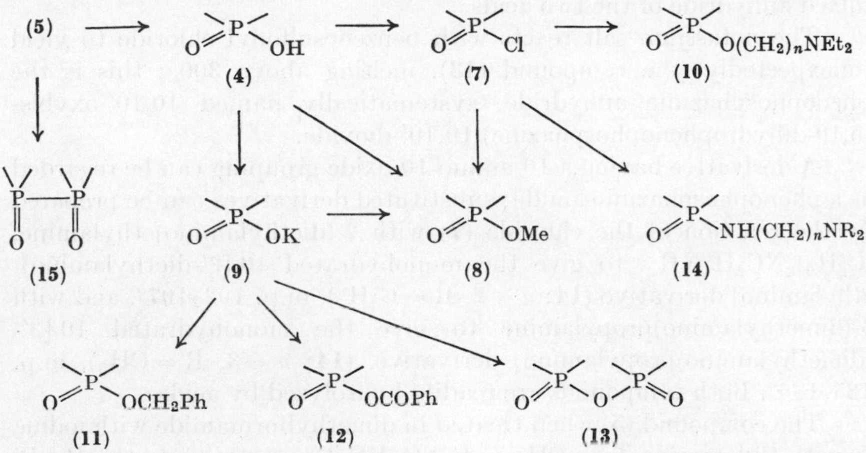

$$(5) \longrightarrow O{=}P{\diagup}^{\diagup}_{OH} \ (4) \longrightarrow O{=}P{\diagup}^{\diagup}_{Cl} \ (7) \longrightarrow O{=}P{\diagup}^{\diagup}_{O(CH_2)_nNEt_2} \ (10)$$

$$\begin{matrix} \overset{\diagup}{P}{-}\overset{\diagup}{P} \\ \parallel \quad \parallel \\ O \quad\ O \end{matrix}\ (15) \qquad O{=}P{\diagup}^{\diagup}_{OK}\ (9) \qquad O{=}P{\diagup}^{\diagup}_{OMe}\ (8) \qquad O{=}P{\diagup}^{\diagup}_{NH(CH_2)_nNR_2}\ (14)$$

$$O{=}P{\diagup}^{\diagup}_{OCH_2Ph}\ (11) \qquad O{=}P{\diagup}^{\diagup}_{OCOPh}\ (12) \qquad O{=}P{\diagup}^{\diagup}_{O}{\diagdown}P{=}O\ (13)$$

ethylene glycol. It is a reasonably strong acid having pK_a 3 (cf. p. 242). When treated with a mixture of phosphorus pentachloride and oxy-chloride, or with thionyl chloride, it readily forms 10-chloro-5,10-dihydrophenophosphazine 10-oxide (7), m.p. 184–185.5°.

The preparation of esters of the acid (4) presents features of some interest. The methyl ester (8), i.e., 5,10-dihydro-10-methoxyphenophosphazine 10-oxide, m.p. 223–224°, can be prepared by (a) the action of diazomethane on the acid (4), (b) the action of sodium methoxide in methanol on the acid chloride (7), and (c) the action of dimethyl sulfate on the potassium salt (9) of the acid in methanolic solution. All three methods gave the same ester, with no indication of a methyl ester, m.p. 112–114°, which the Russian workers [322] obtained by the action of methyl iodide on the silver salt. The corresponding thio derivative occurs in two isomeric forms (see below) but a similar isomerism cannot occur in the dioxygen ester (8).

The n-butyl ester, m.p. 139.5–141°, was prepared by the action of sodium n-butoxide in butanol on the chloride (7), but the 2-(diethyl-amino)ethyl ester (10; $n = 2$) and the 3-(diethylamino)propyl ester (10; $n = 3$) were prepared by the direct action of the corresponding alcohol on the chloride (7): both these esters when anhydrous were oils, but

(10; $n = 2$) readily forms a crystalline hydrate (2.5H_2O), m.p. 80–83.5°, and (10; $n = 3$) forms a dihydrate, m.p. 42–43.5°.

The potassium salt (9) reacts with benzyl bromide in 2-propanol to form the benzyl ester (11), m.p. 179–180°. It also reacts with benzoyl chloride (5 minutes at 100°) to give the 10-benzoyloxy derivative (12), melting above 340°; this compound can be regarded systematically as a mixed anhydride of the two acids.

The potassium salt reacts with benzenesulfonyl chloride to yield (unexpectedly) the compound (13), melting above 300°; this is the phenophosphazinic anhydride, systematically named 10,10′-oxybis-(5,10-dihydrophenophosphazine) 10,10′-dioxide.

A derivative having a 10-amino 10-oxide grouping can be regarded as a phenophosphazinic amide; substituted derivatives can be prepared by the reaction of the chloride (7) with 2-(diethylamino)ethylamine, $(C_2H_5)_2NC_2H_4NH_2$, to give the monohydrated 10-[2′-diethylamino)-ethylamino] derivative (14; $n = 2$, R = C_2H_5), m.p. 192–197°, and with 3-(dimethylamino)propylamine to give the monohydrated 10-[3′-(dimethylamino)propylamino] derivative, (14; $n = 3$, R = CH_3), m.p. 185–186°. Both compounds are readily hydrolyzed by acids.

The compound (5) when treated in dimethylformamide with iodine or with Chloramine-T (p-$CH_3 \cdot C_6H_4SO_2NNaCl$; p. 91) gives the P—P′ derivative (15), named 10,10′-(5H,5′H)-biphenophosphazine 10,10′-dioxide; this crystallizes as a dimethanolate and also as a very stable mono(hydrogen formate), which is unaffected by heating at 180°/0.1 mm; neither melts below 340°.

The compound (5) when treated with sulphur in hot acetic acid solution forms 5,10-dihydro-10-hydroxyphenophosphazine 10-sulfide

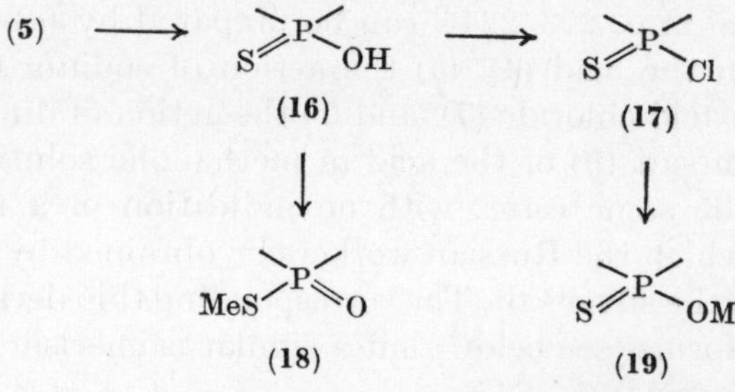

(16), alternatively named thiophenophosphazinic acid, m.p. 213°. This compound could have the structure shown (16) or the isomeric structure =P(O)SH. The infrared spectrum gives strong evidence for (16), for it shows no indication of the P=O band which occurs in the spectrum of (4) at 1340 cm^{-1} but does show a new band at 1235 cm^{-1} attributed to the P=S portion. The acid (16) reacts with phosphorus pentachloride in POCl$_3$ to form the 10-chloro 10-sulfide derivative (17), m.p. 229.5–237.5°. It is noteworthy that methylation of the acid (16) with diazomethane gives 5,10-dihydro-10-(methylthio)phenophosphazine 10-oxide (18), m.p. 266°, whereas treatment of the chloride (17) with sodium methoxide in methanol gives the isomeric 5,10-dihydro-10-methoxyphenophosphazine 10-sulfide (19), m.p. 184–185.5°. The identification of these two esters is based primarily on their infrared spectra: that of (18) shows the P=O band at 1350 cm^{-1}, but the above P=S band at 1235 cm^{-1} is absent, whereas that of (19) shows no P=O at 1340 cm^{-1} but does show the P=S band at 1235 cm^{-1}. Häring[323] suggests that the higher melting point of (18) than of (19) is a result of the more highly polar nature of the P=O link than of the P=S link. The ester (19) can be hydrolyzed by boiling 10% aqueous-ethanolic potassium hydroxide, regenerating the acid (16).

Häring has shown that the initial compound (5), when treated in formic acid solution with bromine (*ca.* 3 moles), gives a dibromophenophosphazinic acid, but the positions of the bromine atoms were unknown.

The synthesis of certain nuclear substituted phenophosphazines has been described in a patent specification (McHattie,[324] to Imperial Chemical Industries). Diphenylamine and certain substituted diphenylamines, when heated with thiophosphoryl chloride, SPCl$_3$, preferably

(20)

under pressure at 175–200°, or in a solvent such as *o*-dichlorobenzene, yield the corresponding 10-chloro-5,10-dihydrophenophosphazine 10-sulfides (20), alternatively named 10-chloro-5,10-dihydro-10-thionophenophosphazines. The specific examples are listed in Table 18.

Several such phenophosphazines possess anthelmintic properties, *i.e.*, remove intestinal parasites, particularly from ruminants.

The following amplification of the chemistry of 10-chloro-5,10-dihydrophenophosphazine 10-sulfide (17) is based on material selected from an unpublished report by Dr. G. V. McHattie,[325] with the kind permission of the author and of the Pharmaceuticals Division of Imperial Chemical Industries Limited.

Table 18. Derivatives of 10-chloro-5,10-dihydrophosphazine
10-sulfides (20)

Diphenylamine	Substituents in the product (20)	M.p. (°)
Unsubstituted	$R^1 = R^2 = R^3 = H$	252–254
3-Chloro	$R^1 = Cl, R^2 = R^3 = H$	270–272
2-Chloro	$R^2 = Cl, R^1 = R^3 = H$	—
2-Methyl	$R^2 = CH_3, R^1 = R^3 = H$	—
3,7-Dimethyl	$R^1 = R^3 = CH_3, R^2 = H$	283–285

This report deals in particular with the 10-sulfide derivatives, whereas Häring's work was mainly on 10-oxide derivatives, and the overlap is consequently very small.

The phosphazine (17) is best prepared by heating a mixture of diphenylamine and thiophosphoryl chloride in an autoclave at 200° for 6 hours, thus forming the crude hydrochloride of (17). This material is pulverized, and then added to chlorobenzene and boiled under reflux, hydrogen chloride being evolved; after filtration the solution deposits the pale yellow crystalline phosphazine (17), m.p. 252–254° (Häring, 229.5–237.5°). Alternatively, the mixture of amine and thiophosphoryl chloride is added to o-dichlorobenzene, and the temperature raised during 3 hours to 175° and maintained thereat for 16 hours; filtration gives a solution which deposits the phosphazine (17), m.p. 250–252°, but in much lower yield.

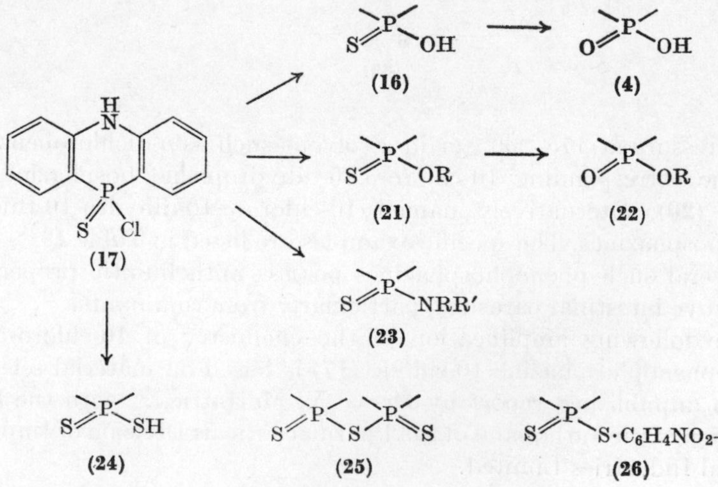

It should be noted that the melting points recorded by McHattie and by Häring for certain compounds are not consistent: in such cases Häring's values are given in parenthesis after McHattie's. The extensive investigation by McHattie and his co-workers often involved repeated preparations of various compounds, for each of which consistent melting points were recorded.

The phosphazine (17) reacts in many ways as an acid chloride; for example, it reacts with water to give the corresponding acid, with alcohols and phenols to give the esters, and with amines to give the amides, i.e., 10-amino derivatives:

Hydrolysis

Hydrolysis is conveniently achieved by adding aqueous sodium sulfide to an acetone solution of (17), which is then boiled under reflux for 1 hour, concentrated, filtered, and acidified. The precipitated product is dissolved in aqueous potassium carbonate, which is filtered and acidified, giving the 5,10-dihydro-10-hydroxyphenophosphazine 10-sulfide (16), m.p. 320–322° (213°). It gives a triethylamine salt, m.p. 238–239°, and a monopiperazine salt, m.p. 304–306°. When the phosphazine (17) is added to a mixture of triethylamine and moist pyridine, which is heated at 100° for 1½ hours, the solution on cooling deposits this triethylamine salt of the acid (16).

It is noteworthy that, when a suspension of (16) in glacial acetic acid at 70° is treated dropwise with hydrogen peroxide ('100 vols.'), a clear purple solution is obtained during 1 hour at 70°. This solution, when cooled and added to concentrated hydrochloric acid and ice, deposits the phenophosphazinic acid (4), which after purification has m.p. 284–286° (277°). (For pK_a value, see pp. 235, 242).

The Esters (21)

Conversion of the phosphazine (17) into, for example, the methyl ester (21; R = CH₃) readily occurs when a suspension of (17) in methanol is boiled until very little remains undissolved: the filtered, cooled solution deposits the methyl ester, m.p. 194–195° (184–185°). To prepare, for example, the p-nitrophenyl ester, sodium p-nitrophenoxide is used in an inert solvent.

The lower alkyl esters do not react with diazotized p-nitroaniline; in boiling acetic anhydride they give their 5-acetyl derivatives (p. 240), the 10-alkoxy group being unaffected. After a solution of the 10-ethoxy compound (21; R = C₂H₅) in absolute ethanol containing Raney nickel had been boiled under reflux in nitrogen for 24 hours, evaporation to dryness and recrystallization gave the 10-ethoxy 10-oxide (**22**; R = C$_2$H$_5$). The desired desulphurized compound (**3**; R = OC$_2$H$_5$) had presumably been formed but had undergone oxidation to the 10-oxide. A list of esters obtained is given in Table 19.

Table 19. Esters: 10-Alkoxy and 10-aryloxy derivatives (**21**)

R	M.p. (°)
CH$_3$*	194–195†
C$_2$H$_5$‡	207–208
n-C$_3$H$_7$	178–180
iso-C$_3$H$_7$	242–244
n-C$_4$H$_9$	130–131
Br$_3$CCH$_2$	193–194
HOCH$_2$CH$_2$	208–209
(CH$_3$)$_2$NCH$_2$CH$_2$§	167–169
C$_6$H$_5$	222–224
p-ClC$_6$H$_4$	204–206
p-NO$_2$C$_6$H$_4$	250–252
o-CH$_3$C$_6$H$_4$	164–166

* 5-Acetyl derivative, m.p. 132–134°.
† Häring, 184–185°.
‡ 5-Acetyl derivative, m.p. 101–103°.
§ Methiodide, m.p. 199–200°.

The Amides (23)

Conversion of the phosphazine (**17**) into the corresponding amides (10-amino derivatives) (**23**) usually occurs when a benzene solution of the phosphazine and the primary or secondary amine is boiled for about 3 hours, the product crystallizing from the cooled solution. The 10-amino derivative (**23**; R = R′ = H) decomposes on attempted recrystallization, but the 10-hydrazyl derivative (**23**; R = NH$_2$, R′ = H) is stable. The known 10-amino compounds are listed in Table 20.

The 10-Mercapto Derivatives (24, 25, 26)

When hydrogen sulfide is passed into a solution of the phosphazine (**17**) in pyridine at 5° and, after 30 minutes, triethylamine is added, the solution becomes bright green. The passage of the gas is continued for a further 1 hour; the solution is left undisturbed for 3 hours and is then poured into an excess of 10% aqueous sodium hydroxide. The solution, when made just acid, deposits pale yellow needles of the pyridine salt (m.p. 314–318°) of 5,10-dihydro-10-mercaptophenophosphazine 10-sulfide (**24**). This salt when treated with an excess of acid yields the free colorless crystalline 10-mercapto derivative (**24**), m.p. 322–326°. When methanolic solutions of this acid and of piperazine are mixed, the monopiperazine salt, m.p. above 380°, is at once precipitated.

Table 20. Amides: 10-Amino derivatives (**23**)

R	R′	M.p. (°)
H	H	233–235
n-C$_4$H$_9$	H	206–207
C$_6$H$_5$CH$_2$	H	184
C$_6$H$_5$	H	247–248
NH$_2$	H	240–241
(CH$_3$)$_2$N(CH$_2$)$_2$	H	142–144
(CH$_3$)$_2$N(CH$_2$)$_3$	H	122–124
(C$_2$H$_5$)$_2$N(CH$_2$)$_3$CH(CH$_3$)*	H	135–136
(C$_4$H$_3$O)CH$_2$†	H	199–200
CH$_3$	CH$_3$	276–278

* 1-Methyl-4-(diethylamino)butyl.
† 2-Furfuryl.

When the above preparation is modified by passing the hydrogen sulfide into a solution of (**17**) in a pyridine–triethylamine mixture, a deposit slowly accumulates. After 2½ hours, the gas is discontinued, and the reaction mixture filtered. The deposit is digested with 20% aqueous potassium carbonate at 100°, and the undissolved material is then thoroughly extracted with hot 2-ethoxyethanol (Cellosolve). The residual pale yellow solid is 10,10′-thiobis-(5,10-dihydrophenophosphazine) 10,10′-disulfide (**25**), m.p. 320–322°.

When the phosphazine (**17**), sodium p-nitrothiophenoxide, and benzene are heated under reflux for 7 hours, the precipitated solid on recrystallization yields 5,10-dihydro-10-[p-(nitrophenyl)thio]phenophosphazine 10-sulfide (**26**), m.p. 259–260°.

The phosphazine (**17**), when treated in ether with phenylmagnesium bromide, gives the 10-phenyl 10-sulfide, m.p. 268–269°.

A mixture of the methyl ester (**21**; R = CH$_3$) and 1:1 aqueous nitric acid, when boiled under reflux for 30 minutes, gives a yellow deposit, which is collected and extracted with boiling acetic acid. The filtered extract deposits 5,10-dihydro-10-methoxy-2,8-dinitrophenophosphazine 10-oxide (**27**), yellow crystals, m.p. 344–345°. Decisive evidence for the

O₂N ... NO₂ ⇌ O₂N ... N–OH

(27)   (28)

position of the nitro groups in (27) is lacking: allocation to the 2,8-positions is based mainly on the deep purple color of a solution of the compound in dilute aqueous potassium hydroxide, indicating a potassium salt of the *aci*-form (28).

With regard to the compounds noted in the patent specification (Table 18), the 3,10-dichloro-5,10-dihydrophenophosphazine 10-sulfide (20; R^1 = Cl, $R^2 = R^3$ = H), m.p. 270–272°, gives a 3-chloro-10-ethoxy derivative, m.p. 190–192°; the isomeric 2,10-dichloro compound (20; R^2 = Cl, $R^1 = R^3$ = H) was converted directly into the 10-ethoxy derivative, m.p. 178–180°; the 10-chloro-2-methyl compound (20; $R^2 = CH_3$, $R^1 = R^3$ = H) was similarly converted into the 10-methoxy derivative, m.p. 250–260° (crude); and the 3,7-dimethyl compound (20; $R^1 = R^3 = CH_3$, R^2 = H), m.p. 283–285°, was converted into the 10-methoxy and the 10-ethoxy derivatives, each having m.p. 228–230°.

With regard to physical properties, Häring has assigned pK_a 3 for phenophosphazinic acid (4) without stating the solvent used (p. 235). The low solubility of this acid in water, however, prevents accurate determinations. In 70% (v/v) aqueous ethanol this acid gives pK_a 4.7 and an equivalent weight of 225 (theory, 231), and the corresponding 10-hydroxy-10-sulfide (16) in 95% (v/v) aqueous ethanol gives pK_a 5.1 and an equivalent weight of 242 (257). It is estimated that the 'true' pK_a values in aqueous solution would be approximately 3.1 and 3.5, respectively.

The infrared spectra (KBr disc) of many esters (Table 19) and amides (Table 20) show that the following characteristic absorptions (cm⁻¹) are always present: 1611 ± 4(vs), 1578 ± 2(s), 1556 ± 5 (weak inflection), 1534 ± 3(vw), 1515 ± 3(s), 1469 ± 4(vs), 1443 ± 5(s-m). In chloroform solution the following shifts appear 1611 → 1602; 1515 → 1504 cm⁻¹.

The typical absorptions of the 5-acetyl derivatives are: 1588(s), 1575 ± 3(s), 1560 (weak inflection), 1537 ± 1(vw), 1471 ± 2(s), 1450 ± 3(vs); the previous bands at 1611 and 1515 cm⁻¹ have disappeared.

A mixture of *N*-methyldiphenylamine and thiophosphoryl chloride was heated in a sealed tube at 190° for 6 hours. The cold crude solid product had a strong mercaptan-like odor, and when digested with boiling ethanol yielded 10-ethoxy-5,10-dihydrophenophosphazine 10-sulfide (**21**; R = C_2H_5). This eviction of the N-methyl group is closely similar to that which occurs when N-methyldiphenylamine is heated with arsenic trichloride, giving 10-chloro-5,10-dihydrophenarsazine (pp. 514, 515).

N-Substituted derivatives of 5,10-dihydrophenophosphazine have recently been recorded (Baum, Lloyd, and Tamborski[326]). 2,2′-Dibromo-diphenyl-N-methylamine (**29**; R = CH_3) was converted into the 2,2′-dilithio derivative, which reacted with phenylphosphonous dichloride

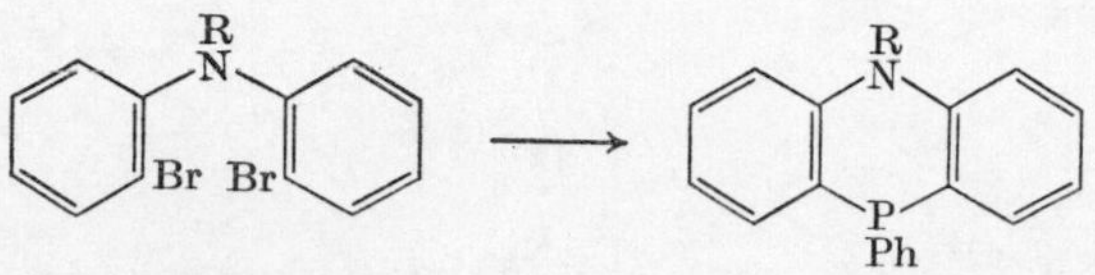

(**29**)　　　　　　(**30**)

to give the pale yellow 5,10-dihydro-5-methyl-10-phenylphenophosphazine (**30**; R = CH_3), m.p. 159–160°; it gave a monomethiodide, namely, 5,10-dihydro-5,10-dimethyl-10-phenylphenophosphazinium iodide, m.p. 295–297°. 2,2′-Dibromodiphenyl-N-ethylamine (**29**; R = C_2H_5) was converted into the 2,2′-di(magnesium bromide), which similarly gave the 5-ethyl-10-phenyl derivative (**30**; R = C_2H_5), m.p. 96.5–98.0°.

Six-membered Ring Systems containing only Carbon and One Phosphorus Atom and Two Nitrogen Atoms

1,2,4-Diazaphosphorine

This system (**1**), [RIS —], is mentioned because of the novel preparation of its only known representative.

Märkl[247] has very briefly recorded that the 1,3-dipolar nitrilimine (**2**) reacts with diphenylvinylphosphine (**3**) in the presence of acid to

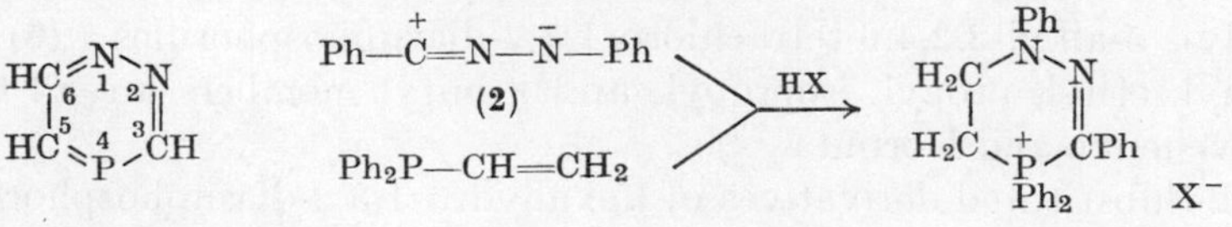

(**1**)　　　　　　(**3**)　　　　　　(**4**)

form the 1,4,5,6-tetrahydro-1,3,4,4-tetraphenyl-1,2,4-diazaphosphor-inium cation (**4**).

1,3,2-Diazaphosphorine

Compounds of type (**1**), [RIS 9755], *i.e.*, six-membered rings having formally a complete series of alternate single and double bonds are encountered usually only as their hydrogenated derivatives.

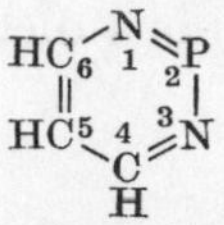

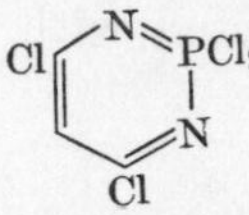

It has been claimed, however, that malonodinitrile, $CH_2(CN)_2$, when treated in benzene solution at 20–25° with an equimolar quantity of phosphorus pentachloride gives 2,2,4,6-tetrachloro-1,3,2-diazaphos-phorine (**2**), b.p. 108–109°/13 mm, m.p. 40–45° (75–80%), and with a bimolar quantity gives the 2,2,4,5,6-pentachloro derivative (**3**), m.p. 150–152°. A benzene solution of the compound (**3**) containing acetic or formic acid (one equivalent), after 24 hours at room temperature, yields the partially hydrolyzed 2,4,5,6-tetrachloro-2-hydroxy-1,3,2-diazaphos-phorine (**4**), m.p. 182–183° (70%). The compound (**2**), when treated with chlorine with ice-cooling, gives the compound (**3**) (60–65%), but it is stated that, when chlorinated in benzene solution at room temperature, it gives 2,2,3,4,4,5,5,6,6-nonachloro-3,4,5,6-tetrahydro-1,3,2-diazaphos-phorine (**5**), b.p. 108–109°/0.015 mm. (Kirsanov *et al.*[327]; Shevchenko *et al.*[328]).

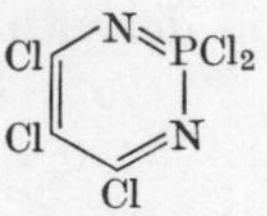

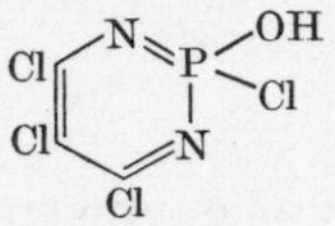

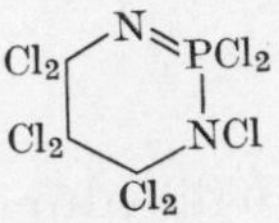

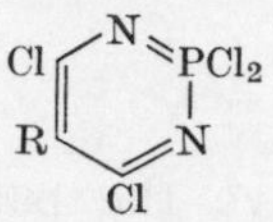

Monoalkylmalonodinitriles, $RCH(CN)_2$, when similarly treated in benzene solution at 30–35° with phosphorus pentachloride (five equiva-lents), give intermediate compounds of type $RC(CN)=CClN=PCl_3$ and finally 5-alkyl-2,2,4,6-tetrachloro-1,3,2-diazaphosphorines (**6**); the methyl, ethyl, propyl, isopropyl, and *n*-butyl members are all liquids (Shevchenko and Kornuta[329]).

2-Substituted derivatives of hexahydro-1,3,2-diazaphosphorine (**7**) are prepared by condensing trimethylenediamine with a compound of

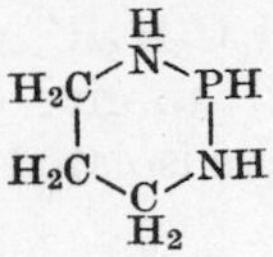

H
H₂C‑N‑PH
H₂C‑C‑NH
  H₂

(7)

H
N‑P=O
   N(CH₂CH₂Cl)₂
NH

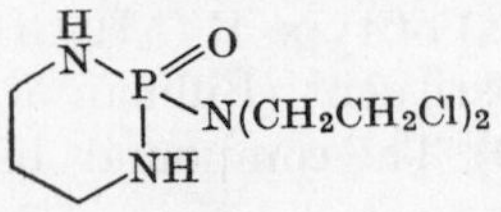

(8)

type $RP(O)Cl_2$ or $RP(S)Cl_2$, a method analogous to that used for preparing similar 1,3,2-diazaphospholidines (p. 217). A number of derivatives bearing various substituents have been recorded by Arnold, Bourseaux, and Brock.[330]

A simpler member was prepared by adding a dioxane solution of N,N-bis-(2-chloroethyl)phosphoramidic dichloride, $(ClC_2H_4)_2NP(O)Cl_2$ (p. 321), to a solution of trimethylenediamine and triethylamine also in dioxane, thus forming 2-[bis-(2'-chloroethyl)amino]hexahydro-1,3,2-diazaphosphorine 2-oxide (8), m.p. 106–107°, which had no antitumor activity (Friedman, Boger, et al.[331]) (cf. p. 320).

A similar condensation of trimethylenediamine with various N-arylphosphoramidic dichlorides, $RNHP(O)Cl_2$ (where R is an aryl group), in benzene solution gave the corresponding 2-anilino 2-oxide

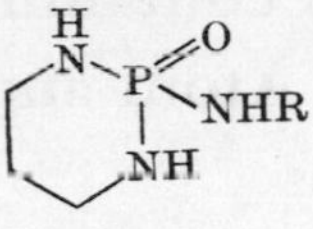

H
N‑P=O
   NHR
NH

(9)

CH₂C₆H₄R′
N‑P=O
   NHR
N‑CH₂C₆H₄R′

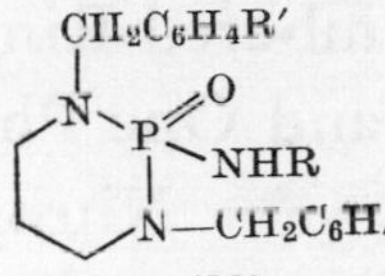

(10)

derivatives (9) (Billman, Meisenheimer, and Awl[332]). The compounds isolated are shown in Table 21.

Table 21. N-Derivatives (9) of 2-amino-
hexahydro-1,3,2-diazaphosphorine 2-oxide

R	M.p. (°)
C_6H_5	217–219
$p\text{-}CH_3C_6H_4$	224–225
$m\text{-}CH_3C_6H_4$	233–234
$o\text{-}ClC_6H_4$	165–166
$p\text{-}ClC_6H_4$	214–215
$p\text{-}CH_3OC_6H_4$	190–190.5

In order to obtain similar compounds having substituents in the 1,3-positions, the above dichlorides, $RNHP(O)Cl_2$ (1 equivalent), were condensed in benzene with N,N'-disubstituted trimethylenediamines (2 equivalents) of type $R'C_6H_4CH_2NH(CH_2)_3NHCH_2C_6H_4R'$, where R' is a *para*-substituent (Billman and Meisenheimer[333]) to give compounds of type (10). The compounds listed in Table 22 were isolated.

Table 22. Mixed *N*-derivatives (10)

R	R'	M.p. (°)
C_6H_5	Cl	128.5–130
C_6H_5	CH_3O	154.5–155.5
m-$CH_3C_6H_4$	Cl	150–150.5
p-$CH_3C_6H_4$	Cl	164.5–166
o-ClC_6H_4	Cl	86.5–87.5
p-ClC_6H_4	Cl	166.5–168
p-ClC_6H_4	CH_3O	188–189
p-$CH_3OC_6H_4$	Cl	143–144
p-$CH_3OC_6H_4$	CH_3O	154.5–155.5

None of the compounds in series (9) and (10) showed cytotoxic activity.

Eight-membered Ring Systems containing only Carbon and One Phosphorus Atom and Two Nitrogen Atoms

1,3,2-Diazaphosphocine

1,3,2-Diazaphosphocine (1), [RIS —], is encountered only as its octahydro derivative (2), named 1,3,2-diazaphosphocane. This is briefly mentioned here for the synthesis of one derivative. DL-Lysine

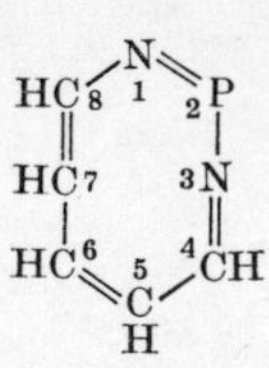

(1)

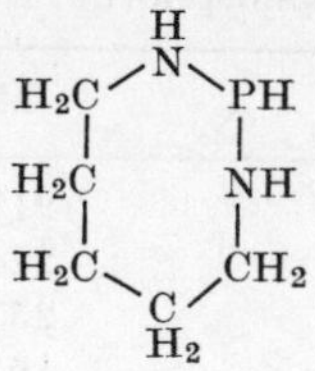

$NH_2(CH_2)_4CH(NH_2)CO_2C_2H_5$
(3)

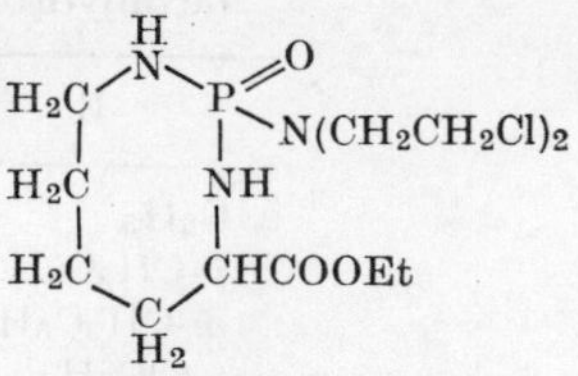

(2)

(5)

$(ClCH_2CH_2)_2NPCl_2$
(4)

ethyl ester (3) condenses readily in dry chloroform with N,N-bis-(2-chloroethyl)phosphoramidic dichloride (4) to give ethyl 2-bis-[(2'-chloroethyl)amino]-1,3,2-diazaphosphocane-4-carboxylate 2-oxide

(5). This was one of a series of cyclic phosphorus compounds having the $\text{>P(O)N(CH}_2\text{CH}_2\text{Cl)}_2$ grouping prepared in order to study the effect of the 'carrier molecules' on the biological activity of the same cytotoxic group (Szerkerke, Kajtar, and Bruckner[334]). Compounds containing this group in the 1,3,2-oxazaphospholidine series and in the tetrahydro-2H-1,3,2-oxazaphosphorine series (p. 320) have been similarly studied in great detail.

Dibenzo[d,g][1,3,2]diazaphosphocine

The compound (1) is 1,2,3,6-tetrahydro-1,3,2-diazaphosphocine (of. p. 246); its dibenzo derivative (2) [RIS —], is presented and numbered

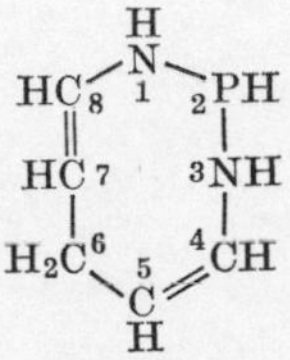

(1) 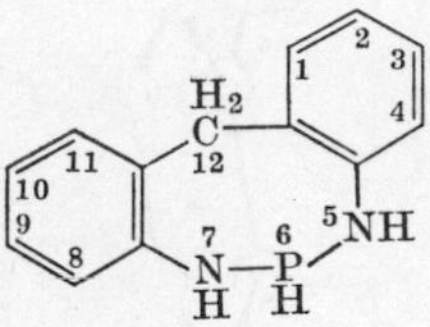   (2) 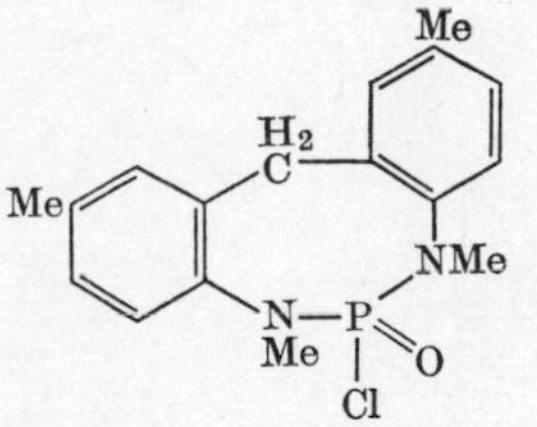   (3)

as shown, and is therefore 5,6,7,12-tetrahydrodibenzo[d,g][1,3,2]diazaphosphocine.

Phosphorus oxychloride reacts with N,N-dimethylaniline in the presence of pyridine at 140° to give tri-(p-N,N-dimethylaminophenyl)-phosphine oxide, [p-$(\text{CH}_3)_2\text{NC}_6\text{H}_4$]$_3$PO, m.p. 290° (cf. Koenigs and Friedrich[335]). It has recently been found (Cheng, Shaw, Cameron, and Prout[336]) that if the *para*-position in N,N-dimethylaniline is blocked by a methyl group, *i.e.*, if N,N-dimethyl-p-toluidine is used in the above reaction without a solvent or a base, the major product is 6-chloro-5,6,7,12 - tetrahydro - 2,5,7,10 - tetramethyldibenzo[d,g][1,3,2]diaza - phosphocine 2-oxide (3), m.p. 170–171°.

Hydrolysis of (3) causes cleavage of both P—N bonds, with the formation of bis-[5-methyl-2-(methylamino)phenyl]methane,

$$\text{CH}_2[\text{C}_6\text{H}_3(\text{CH}_3)(\text{NHCH}_3)]_2,$$

m.p. 84–85°.

The ^{1}H-nmr spectra of the compound (3) supported the above structure, which was fully confirmed by an X-ray analysis of the triclinic crystals. This novel structure, with certain of the intervalency angles and the interatomic distances (in Å units), is shown in Figure 6, the benzene ring on the left being in the plane of the paper. The structure reveals some interesting features. The nitrogen atoms deviate by only

0.01 Å (N-5) and 0.08 Å (N-7) from the plane of the three atoms to which each is attached, suggesting a contribution from a π-bond involving the lone pair of electrons on the nitrogen bonds. More particularly, the

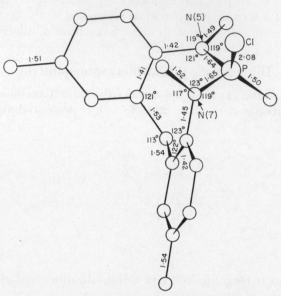

Figure 6. Structure of the compound (**3**), projected on to the least-squares best plane through the benzene ring linked to N(5). (Reproduced, by permission, from C. Y. Chang, R. A. Shaw, T. S. Cameron, and C. K. Prout, *Chem. Commun.*, **1968**, 616)

eight-membered ring containing the phosphorus atom has a distorted chair conformation, and the two *N*-methyl groups are seen to be non-equivalent, owing to this conformation of the ring. The nmr spectra show that this non-equivalence of the methyl groups persists in solution.

Four-membered Ring Systems containing only Carbon, One Phosphorus and One Oxygen Atom

1,2-Oxaphosphete and 1,2-Oxaphosphetane

This ring system in its lowest state of hydrogenation could be represented by two isomers, 2*H*-1,2-oxaphosphete (**1**) and 4*H*-1,2-oxaphosphete (**2**); the fully hydrogenated form is 1,2-oxaphosphetane

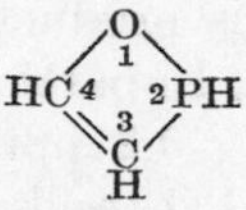

(1) 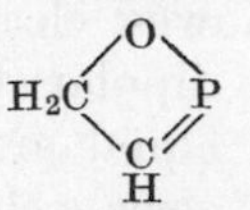   **(2)** 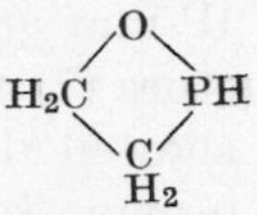   **(3)**

(3), [RRI 36], of which one stable derivative has been recorded. An early name for **(3)** was 1-oxaphosphacyclobutane.

The Wittig olefin synthesis, by which a phosphorane reacts with a carbonyl compound, entails the formation of an intermediate zwitterion

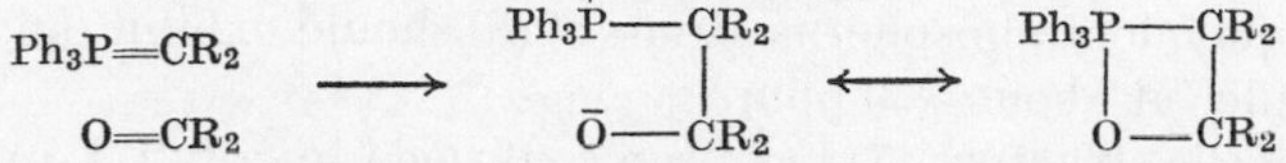

$$Ph_3P{=}CR_2 \longrightarrow \overset{+}{Ph_3P}{-}CR_2 \longleftrightarrow Ph_3P{-}CR_2$$
$$O{=}CR_2 \qquad\qquad \bar{O}{-}CR_2 \qquad\qquad O{-}CR_2$$

(4A)   **(4B)**

(4A), of which the 1,2-oxaphosphetane **(4B)** could be a contributory form. Stabilized dipolar intermediates of type **(4A)** have been isolated, but the cyclic forms of type **(4B)** are usually too unstable for isolation.

Birum and Matthews[337, 338] have shown, however, that when gaseous hexafluoroacetone **(5)** is dispersed in a stirred mixture of hexaphenylcarbodiphosphorane **(6)** and 1,2-diethoxyethane (diglyme) at 40–50° until the yellow color disappears, and the reaction mixture then cooled, colorless crystals, m.p. 155–157°, of the adduct **(7)**, *i.e.*, 4,4-bis(trifluoromethyl)-2,2,2-triphenyl-3-(triphenylphosphoranylidene)-1,2-oxaphosphetane, are deposited.

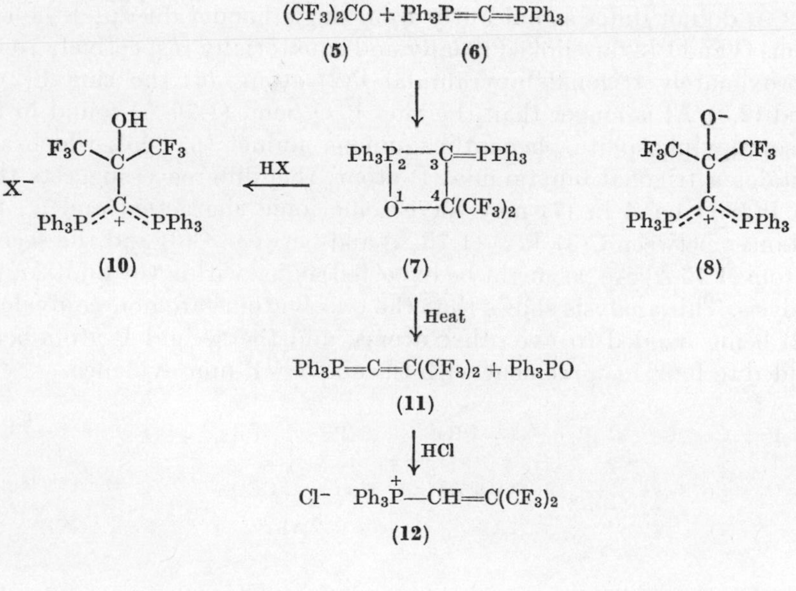

$$(CF_3)_2CO + Ph_3P{=}C{=}PPh_3$$

(5)   **(6)**

$$F_3C{-}\underset{\underset{Ph_3P{\cdots}\overset{+}{}{\cdots}PPh_3}{C}}{\overset{OH}{\underset{|}{C}}}{-}CF_3 \quad\overset{HX}{\longleftarrow}\quad \underset{O{-}C(CF_3)_2}{Ph_3\overset{2}{P}{-}\overset{3}{C}{=}PPh_3} \quad F_3C{-}\underset{\underset{Ph_3P{\cdots}\overset{+}{}{\cdots}PPh_3}{C}}{\overset{O-}{\underset{|}{C}}}{-}CF_3$$

X⁻

(10)   **(7)**   **(8)**

Heat

$$Ph_3P{=}C{=}C(CF_3)_2 + Ph_3PO$$

(11)

HCl

$$Cl^- \quad \overset{+}{Ph_3P}{-}CH{=}C(CF_3)_2$$

(12)

The ^{31}P-nmr spectrum showed clearly that this product had the cyclic structure (7) and not the dipolar structure (8). In particular, this spectrum showed doublets of equal area at -7.3 ± 0.2 ppm and at $+54.0 \pm 1.0$ ppm, $J_{PP} = 47 \pm 7$ c/sec (relative to H_3PO_4 and measured in a saturated CH_2Cl solution at both 24.3 and 40.5 Mc/sec). Spin coupling of the two non-equivalent phosphorus atoms of (7) should produce two doublets, that at -7.3 ppm being within the phosphorus ylide range and that at $+54.0$ ppm being characteristic of cyclic structures having phosphorus covalently linked to five atoms or groups (unless linked to five carbon atoms, when a higher value results[95]). The two equivalent phosphorus atoms of (8) should exhibit only a single ^{31}P resonance at about -20 ppm.

The cyclic structure (7) has been confirmed by a full X-ray crystal analysis (Chioccola and Daly[339]). This shows that: (a) the heterocyclic ring is planar but has considerable angular distortion (9; diagrammatic

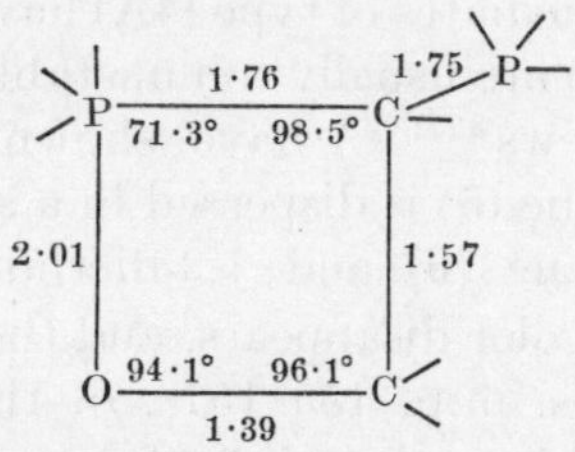

(9)

representation); (b) the ring-bond lengths C(3)–C(4), 1.57; C(4)–O, 1.39 Å, do not differ significantly from single-bond values; (c) the ring atoms O and C(3) are linked axially and equatorially respectively to the approximately trigonal bipyramidal P(2) atom; (d) the ring P(2)–O bond (2.01 Å) is longer than the ring P–O bond (1.76 Å) found in the triisopropylphosphite–phenanthraquinone adduct (p. 265), which also contains a trigonal bipyramidal P atom; this difference suggests that the P(2)–O bond in (7) may have some ionic character; and (e) the distances between C(3)–P(2) (1.76 Å) and between C(3) and the second P atom (1.75 Å) are (as might be expected) equal within the limits of the analysis. This analysis shows that the two P atoms are non-equivalent, P(2) being bonded to five other atoms, and the second P atom being bonded to four, in agreement with the above ^{31}P-nmr evidence.

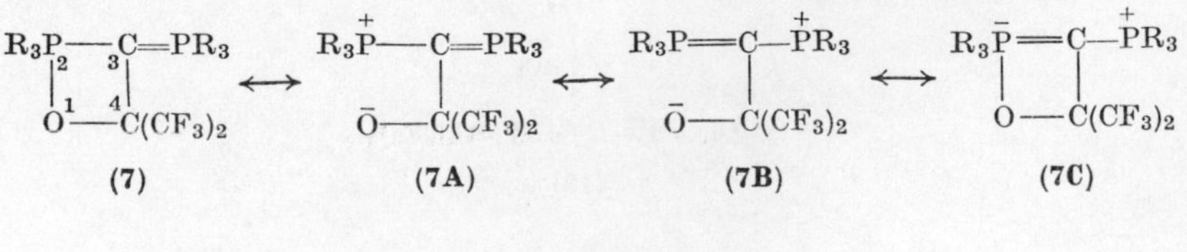

(7)   (7A)   (7B)   (7C)

On the basis of this analysis, the authors suggest that the compound has the 1,2-oxaphosphetane structure (**7**; $R = C_6H_5$) but that it may receive some contribution from forms such as (**7A**), (**7B**), and (**7C**).

On the chemical side, the compound (**7**) was unaffected by methyl iodide at room temperature, or by methanolic iodide at 40–45° for 20 hours. The adducts formed from (**6**) with carbon dioxide or with carbon disulfide are, however, readily methylated by methyl iodide.

Hexafluorophosphoric acid, HPF_6, caused ring-opening of (**7**) with the formation of the symmetrical mesomeric phosphonium salt (**10**; $X = PF_6$), m.p. 213–213.5° (83%); its structure was confirmed by the ^{31}P, the ^1H, and the ^{19}F nmr spectra.

When the compound (**7**) was heated above 110° in inert solvents, the normal reaction sequence occurred with the formation of the vinylidenephosphorane (**11**) and triphenylphosphine oxide; the former proved too reactive for separation from the oxide, but treatment of the mixture in benzene solution with hydrogen chloride gave the pale yellow 2,2-bis(trifluoromethyl)vinyltriphenylphosphonium chloride (**12**), m.p. 153–154°.

Five-membered Ring Systems containing only Carbon, One Phosphorus, and One Oxygen Atom

1,2-Oxaphosphole, 1,2-Oxaphosphol-4-enes, and 1,2-Oxaphospholanes

The compound (**1**) is the unknown 1,2-oxaphosphole, [RRI 130], its 2,3-dihydro derivative (**2**) is 1,2-oxaphosphol-4-ene, and its tetrahydro derivative (**3**) is 1,2-oxaphospholane.

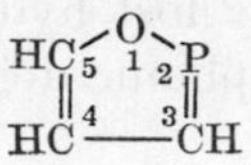

$$
\begin{array}{ccc}
\text{(1)} & \text{(2)} & \text{(3)}
\end{array}
$$

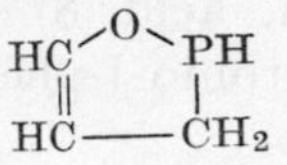

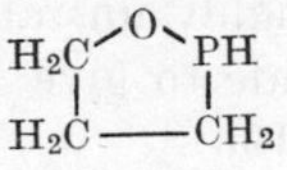

The syntheses of compounds of types (**2**) and (**3**) will be considered separately.

1,2-Oxaphosphol-4-ene (2)

The various syntheses of these compounds are largely variants (numbered below) of the addition of phosphorus derivatives to αβ-unsaturated ketones.

(I) In one of the earliest records, Conant and Cook[340] showed that phosphorus trichloride reacted readily at room temperature with benzylideneacetophenone, $C_6H_5CH=CHCOC_6H_5$, in acetic anhydride

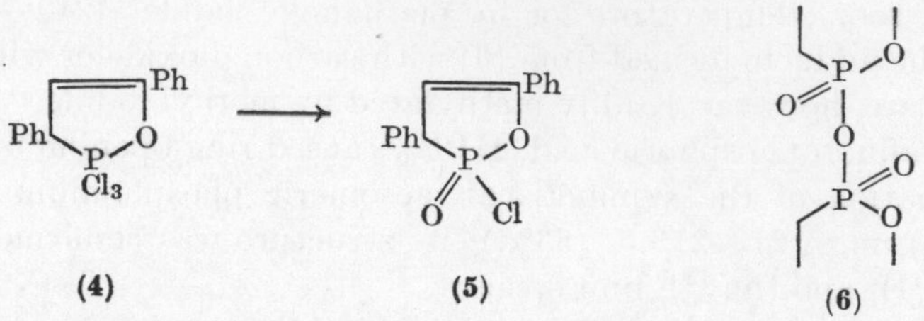

(4)   (5)   (6)

(1 mole) to form acetyl chloride and 2-chloro-3,5-diphenyl-1,2-oxaphosphol-4-ene 2-oxide (5): when an excess of the anhydride was employed, the compound (5) was contaminated with the anhydride (6) of the corresponding acid. They considered that the initial reaction was the direct 1,4-addition of the PCl_3 to the conjugated ketone to give the 2,2,2-trichloro compound (4), which reacted with the acetic anhydride to give (5) and acetyl chloride.

The position of the double bond in (5) was shown by the addition of bromine to give the 4,5-dibromo compound (7), which when treated

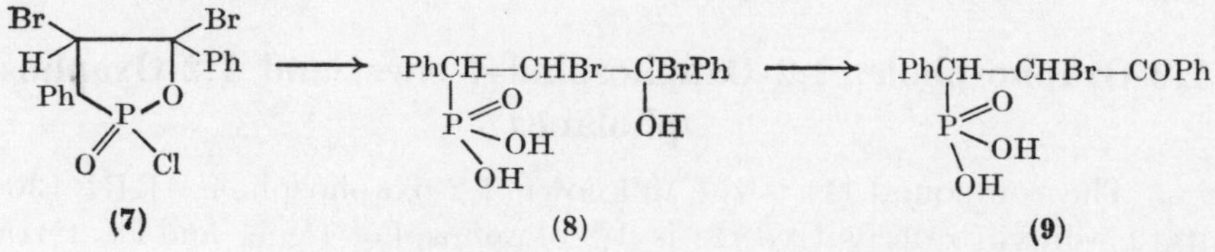

(7)   (8)   (9)

with water (2 moles) underwent hydrolysis and ring cleavage to give the highly unstable phosphonic acid (8), which rapidly lost hydrogen bromide to give 2-benzoyl-2-bromo-1-phenylethylphosphonic acid (9), m.p. 196°.

The compound (5) reacted with phenol to form 2-phenoxy-3,5-diphenyl-1,2-oxaphosphol-4-ene 2-oxide (10); the hydrogen chloride liberated in this reaction converted a portion of (10) into the unstable phenoxy-phosphonyl chloride (11), which underwent ready hydrolysis to phenyl 2-benzoyl-1-phenylethylphosphonate (13), m.p. 146°. The position of the double bond in (10) was determined as before, giving the phenyl ester (12), m.p. 179°, of the acid (9).

(II) Anschutz, Klein, and Cermak[341] much later clarified early work by Michaelis,[342] who by the interaction of acetone and phosphorus

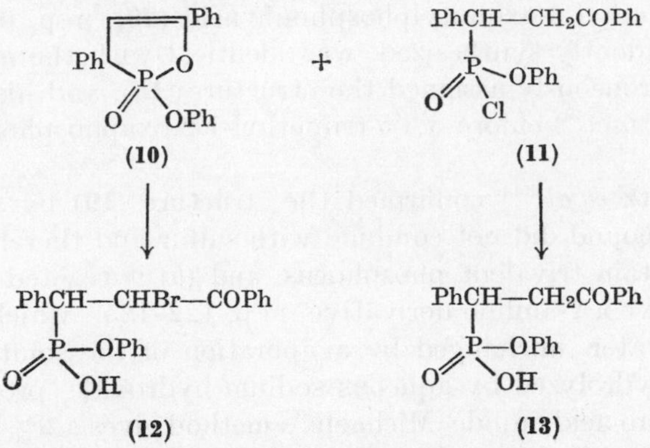

(10)   +   **(11)**

PhCH—CHBr—COPh   PhCH—CH$_2$COPh

(12)   **(13)**

trichloride in the presence of aluminum chloride had isolated 'acetone-
phosphonous chloride', to which he assigned the structure (14), largely
on the basis that with water it apparently gave the phosphonic acid (15),

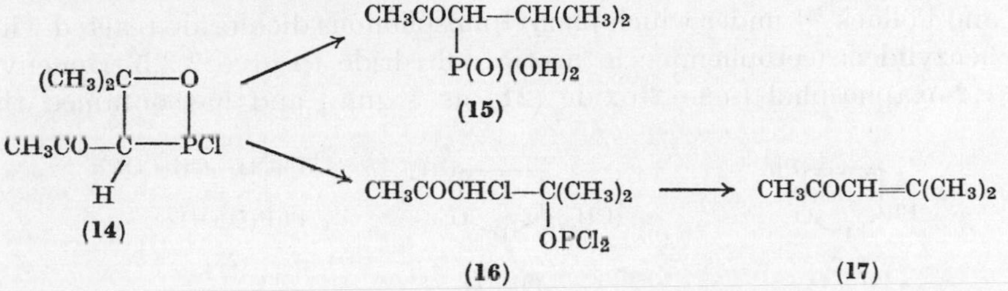

$$CH_3COCH—CH(CH_3)_2$$
$$P(O)(OH)_2$$

(15)

$$(CH_3)_2C—O$$
$$CH_3CO—C—PCl$$
$$H$$

(14)

$$CH_3COCHCl—C(CH_3)_2 \longrightarrow CH_3COCH=C(CH_3)_2$$
$$OPCl_2$$

(16)   **(17)**

while with chlorine it gave the trichloro compound (16), which with
water gave mesityl oxide (17), phosphoric acid, and hydrochloric acid.

Drake and Marvel[343] had later prepared the phosphonic acid (20)
by the action of phosphorus trichloride on mesityl oxide (17) in acetic
anhydride, followed by hydrolysis of the product, and considered the
reaction proceeded as illustrated:

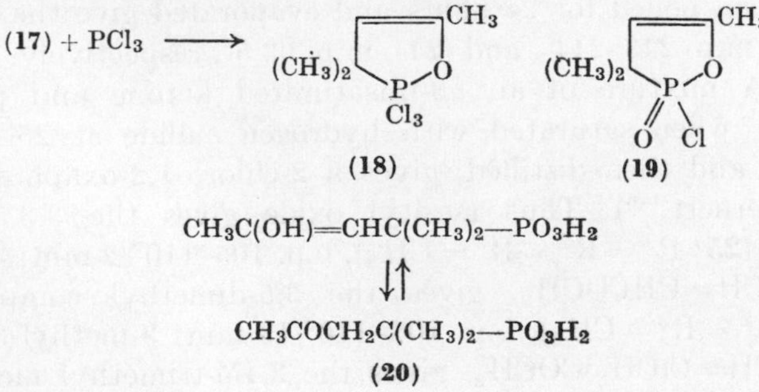

(17) + PCl$_3$ $\longrightarrow$

(18)   **(19)**

$$CH_3C(OH)=CHC(CH_3)_2—PO_3H_2$$
$$\updownarrow$$
$$CH_3COCH_2C(CH_3)_2—PO_3H_2$$

(20)

The 1,1-dimethyl-3-oxobutylphosphonic acid (**20**), m.p. 63–64°, which was independently synthesized, was identical with the acid to which Michaelis erroneously assigned the structure (**15**), and his initial compound is in fact 2-chloro-3,3,5-trimethyl-1,2-oxaphosphol-4-ene oxide (**19**).

Anschutz *et al.*[341] confirmed the structure (**19**) by showing that (*a*) the compound did not combine with sulfur and therefore probably did not contain trivalent phosphorus, and (*b*) it reacted readily with aniline to give a 2-anilino derivative, m.p. 122–125°, which was readily soluble in water, unchanged by evaporation of the solution, and immediately hydrolyzed by aqueous sodium hydroxide, properties to be expected of an acid anilide. Michaelis's method gives a 9% yield of (**19**); if, however, phosphorus trichloride is added dropwise to 'acetone-alcohol', $CH_3COCH_2C(CH_3)_2OH$, with occasional cooling, the solution when kept at 40° for 3 days and then distilled gives a 38% yield of (**19**), b.p. 74°/0.01 mm, m.p. 31–35°.

(III) Bergessen[344] has modified the conditions used by Conant and Pollack[345] under which phenylphosphonous dichloride reacted with benzylideneacetophenone in acetic anhydride to give 2,3,5-triphenyl-1,2-oxaphosphol-4-ene 2-oxide (**21**) as a gum, and has obtained the

crystalline (**21**), m.p. 162–163°; the dichloride reacts similarly with mesityl oxide to give 3,3,5-trimethyl-2-phenyl-1,2-oxaphosphol-4-ene 2-oxide (**22**), b.p. 170–172°/10 mm. Aqueous solutions of these compounds when boiled for 24 hours and evaporated give the phosphinic acids (**23**), m.p. 243–244°, and (**24**), m.p. 93.5°, respectively.

(IV) A mixture of an $\alpha\beta$-unsaturated ketone and phosphorus trichloride, when saturated with hydrogen sulfide at 25°, set aside overnight, and then distilled, gives a 2-chloro-1,2-oxaphosphol-4-ene sulfide (Pernert[346]). Thus mesityl oxide gives the 3,3,5-trimethyl derivative (**25**; $R^1 = R^2 = R^3 = CH_3$), b.p. 105–110°/2 mm; 3-penten-2-one, $CH_3CH=CHCOCH_3$, gives the 3,5-dimethyl compound (**25**; $R^1 = H$, $R^2 = R^3 = CH_3$), b.p. 108–110°/13 mm; 3-methyl-3-penten-2-one, $CH_3CH=C(CH_3)COCH_3$, gives the 3,4,5-trimethyl member (**26**),

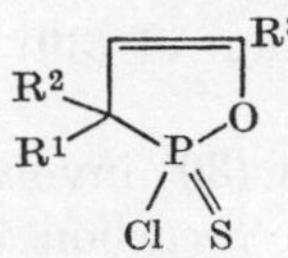

(25) 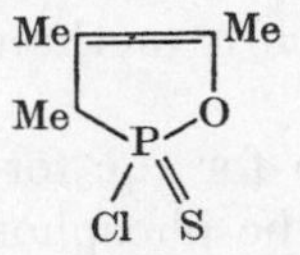   (26) 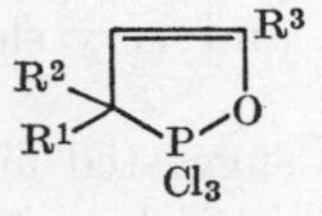   (27)

b.p. 125–127°/14 mm; and dypnone, $C_6H_5(CH_3)C=CHCOC_6H_5$, gives the 3-methyl-3,5-diphenyl member (25; $R^1 = CH_3$, $R^2 = R^3 = C_6H_5$), m.p. 138.0–138.5°.

It is probable that in this reaction the phosphorus trichloride makes the usual 1,4-addition to the unsaturated ketone to form the 2,2,2-trichloro compound (27), which is then converted into the 2-chloro 2-sulfide by the hydrogen sulfide.

(V) An interesting synthetic application of an $\alpha\beta$-unsaturated ketone in this series (Ramirez, Madan, and Heller [347]) occurs when equimolar amounts of 3-benzylidene-2,4-pentanedione (28) and trimethyl phosphite (29) in dichloromethane solution under nitrogen are allowed to react for 24 hours at 20° and then 5 hours at 40°, with the quantitative formation of 4-acetyl-2,2,2-trimethoxy-5-methyl-3-phenyl-1,2-oxaphosphol-4-ene (30), m.p. 49–51°. Considerable nmr spectral evidence is adduced for the structure (30).

The oxaphosphol-4-ene (30) does not react with a second molecule of (28) or with aliphatic aldehydes, but when treated in benzene solution with water (1 mole) it undergoes hydrolysis with ring cleavage to form

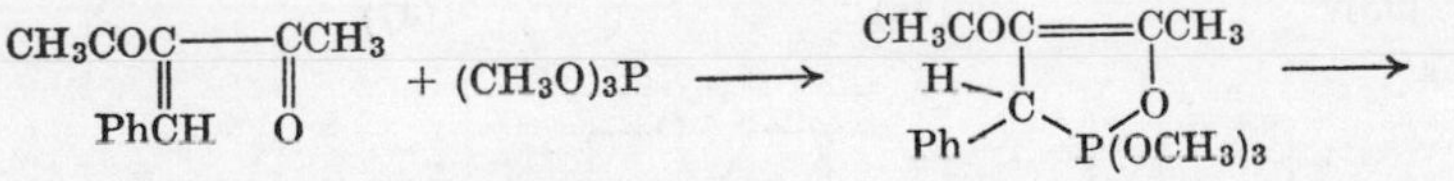

(28)   (29)   (30)

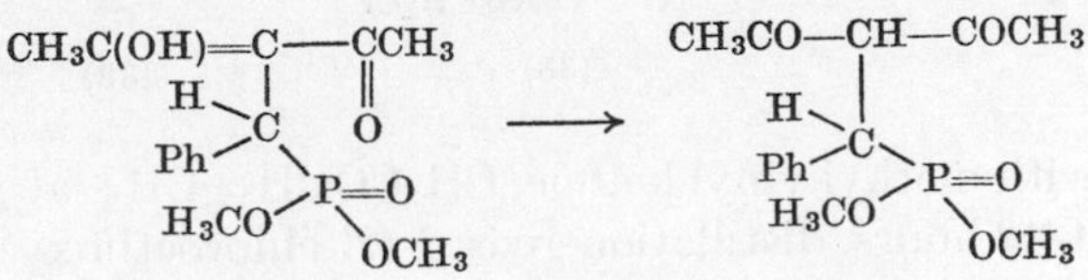

(31)   (32)

the metastable product (31), m.p. 77–81°, the assigned structure of which is based on the infrared spectrum, which shows 'conjugate chelation' between the OH and the CO groups, and by the nmr spectra. The compound (31) undergoes a ready tautomeric change to the 1,2-diketone (32), m.p. 109–110°. The conversion (30) → (33) occurs readily when an ethereal solution of (30) at 0° is treated with dry hydrogen chloride, and very slowly when a methanolic solution of (30) is kept at 20°.

The suggested mechanism for the formation of (30) involves two stages. The first is attack by the phosphorus on the β-carbon atom to give a 1:1 dipolar adduct (33). In the next stage this adduct may undergo

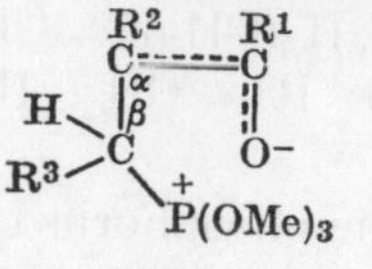

(33) 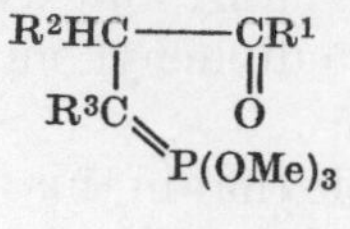   (34)

cyclization to the comparatively stable oxaphosphol-4-ene (30), or it may undergo rapid proton transfer to an alkylidenephosphorane (34). The nature of the substituents R^1, R^2, and R^3 and those on the phosphorus atom will determine the relative stabilities of the cyclic product (30) and of (33) and (34).

An entirely different synthesis has been recorded (Razumova and Petrov[348]) using 2-chloro-1,3,2-dioxaphospholane (35; R = H) (p. 270) and its 4-methyl derivative (35; R = CH₃). If the former is heated

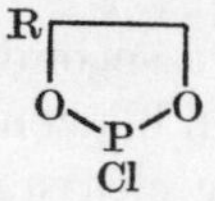

(35) 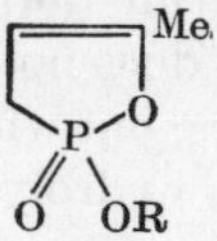   (36) 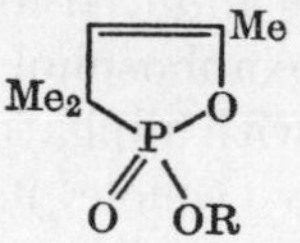   (37)

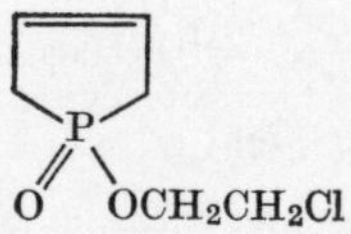

(38) 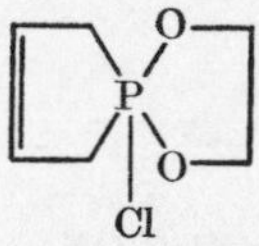   (39)

with methyl vinyl ketone, $CH_3COCH{=}CH_2$, at 100° in a sealed tube for 4–12 hours, distillation gives 2-(2'-chloroethoxy)-5-methyl-1,2-oxaphosphol-4-ene 2-oxide (36; R = ClCH₂CH₂), b.p. 78–79°/0.08 mm; a similar experiment with mesityl oxide gives the 3,3,5-trimethyl analogue (37; R = ClCH₂CH₂), b.p. 159–161°/10 mm. The use of the dioxaphospholane (35; R = CH₃) with methyl vinyl ketone gives the compound [36; R = ClCH₂CH(CH₃)], b.p. 64–66°/0.06 mm, and with mesityl oxide gives [37; R = ClCH₂CH(CH₃)], b.p. 72–73°/0.2 mm.

The reaction is markedly similar to that of (35; R = H) and butadiene (with a trace of zinc chloride), which when heated at 100° for

20 hours gives the 1-(2'-chlorethoxy)-2-phospholene 1-oxide (**38**) (cf. p. 46), apparently by the intermediate formation of a spirocyclic phosphorane (**39**), which undergoes ring-opening of the 1,3,2-dioxaphospholane ring with migration of the chlorine atom. A similar mechanism may well apply to the reaction of (**35**; R = CH$_3$) with methyl vinyl ketone to give (**36**; R = ClCH$_2$CH$_2$).

The compound (**37**; R = ClCH$_2$CH$_2$), when treated with phosphorus pentachloride, gives the corresponding 2-chloro compound (**19**).

1,2-*Oxaphospholanes*

When suitably heated, a phosphinic acid containing a 3-hydroxypropyl group cyclizes to an intramolecular ester (Smith[349]), *i.e.*, a 2-substituted 1,2-oxaphospholane 2-oxide. Thus an equimolar mixture of dimethyl phenylphosphonite (**40**) and 3-bromo-1-propanol, when heated,

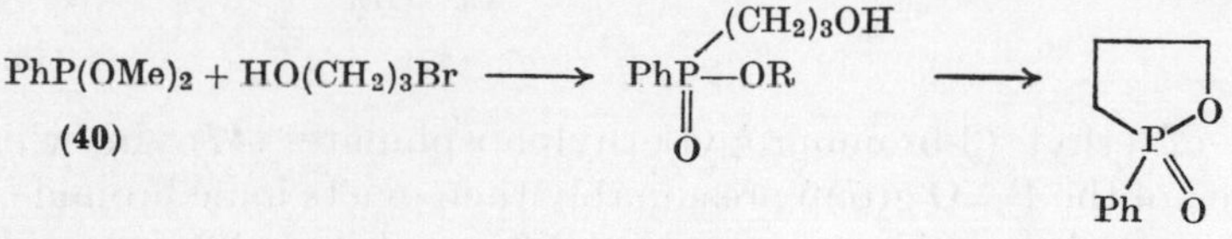

PhP(OMe)$_2$ + HO(CH$_2$)$_3$Br $\longrightarrow$ PhP—OR $\longrightarrow$

(**40**)

(**41**)   (**42**)

undergoes an Arbuzov reaction to give methyl (3-hydroxypropyl)-phenylphosphinate (**41**; R = CH$_3$), which on hydrolysis with hydrochloric acid yields the phosphinic acid (**41**; R = H): this acid, heated at 100–150°, readily cyclizes to form 2-phenyl-1,2-oxaphospholane 2-oxide (**42**), a reaction which emphasizes the 'internal phosphinic acid ester' nature of such 2-oxides.

To obtain a compound of type (**42**) having the unsubstituted $\geq$P(H)O group, the initial compound must contain a PH group, and sodium hypophosphite, NaH$_2$PO$_2$, is conveniently used: a catalyst such as *sec*-butylidenebis-(*tert*-butyl peroxide), CH$_3$(C$_2$H$_5$)C[OOC(CH$_3$)$_3$]$_2$, must now be added. For example, sodium hypophosphite and 2-methylallyl alcohol (**43**), (1 mole), when heated in methanol with the catalyst

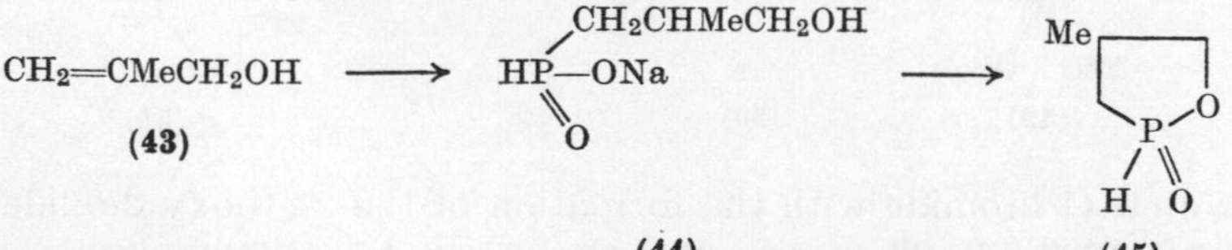

CH$_2$=CMeCH$_2$OH $\longrightarrow$ HP—ONa $\longrightarrow$

(**43**)

(**44**)   (**45**)

gives the sodium salt (**44**); acidification gives the free acid, which when heated at 100–150° yields 4-methyl 1,2-oxaphospholane 2-oxide (**45**), b.p. 100–104°/0.2 mm. Compounds of type (42) and (45) are stated in the specification to be useful as 'surface-active agents, plasticizers, and lubricating-oil additives'. This synthetic route can also be used, *e.g.*, with 3-buten-1-ol, to obtain the analogous six-membered 1,2-oxaphosphorinane (p. 285).

In a related patented synthesis (Garner[350, 351]), an organic phosphonite such as diethyl ethylphosphonite (46) is heated with 1,3-dibromopropane at 150–170° for 2.5 hours, ethyl bromide being continuously distilled off, with the formation by an Arbuzov rearrangement,

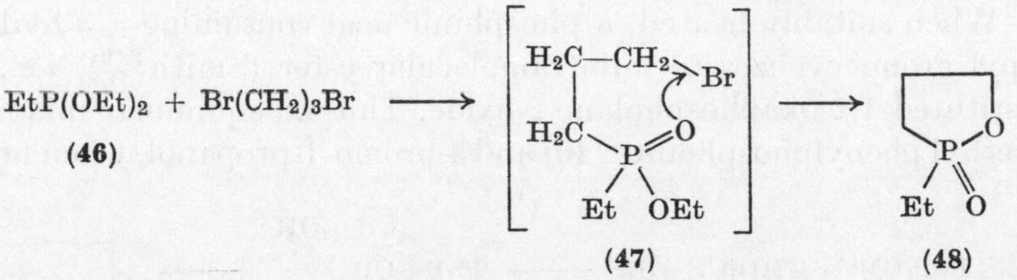

first, of ethyl (3-bromopropyl)ethylphosphinate (47), in which the oxygen of the P=O group presumably then evicts ionic bromide, giving finally ethyl bromide and 2-ethyl-1,2-oxaphospholane 2-oxide (48), b.p. 83°/2 mm (18%).

A mixture of triethyl phosphite and 1,3-dibromopropane, when similarly heated and distilled, gives 2-ethoxy-1,2-oxaphospholane 2-oxide (49), b.p. 106°/2 mm (20%).[351]

In these patent specifications, compounds such as (48) and (49) and their six- and seven-membered ring analogues, are claimed to be 'chemically stable, free-flowing compounds, used for hydraulic fluids and as fire retardants for plastics'.

In a third synthetic route (Eberhard and Westheimer[352]), diethyl 3-bromopropylphosphonate (50) is heated under nitrogen at 180–200° for 2 hours: under these conditions the ester (50) undergoes a similar

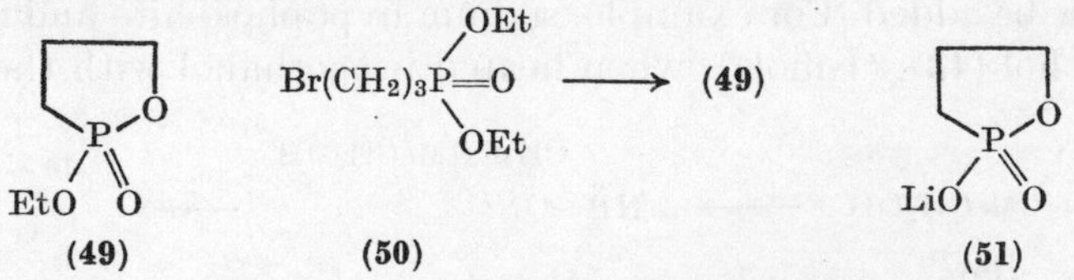

loss of ethyl bromide with the formation of the 2-ethoxy 2-oxide (49), b.p. 74°/0.55 mm. The ester (49) when heated with 48% hydrobromic acid yields the crystalline 3-bromopropylphosphonic acid,

$$Br(CH_2)_3P(O)(OH)_2,$$

m.p. 112–112.8°. The lithium derivative (51) was prepared by treating an aqueous solution of this acid with 1M-lithium hydroxide to attain pH 10, and evaporating to dryness under reduced pressure; extraction with acetone removed lithium bromide from the residue, and when this was dissolved in methanol and the solution filtered and evaporated it gave the pure salt (51).

The ester (49) was prepared to investigate the reasons why such esters and their 2-alkoxy-1,3,2-dioxaphospholane analogues should undergo hydrolysis 10^5–10^8 times faster than the corresponding open-chain esters (cf. pp. 276–278).

A point of nomenclature arises here. The two homologous compounds (52) and (53) have for convenience been called phostonic acids, and the esters (49) and (54), for example, have been termed ethyl

(52)   (53)   (54)

propylphostonate and ethyl butylphostonate, respectively. These ester names are clearly unsatisfactory, for the ester (49) does not contain a propyl group (or, strictly speaking, even a substituted propyl group) and (54) does not contain a butyl group. If the phostonic acid nomenclature is retained, the acids (52) and (53) should be termed trimethylenephostonic acid and tetramethylenephostonic acid respectively, and their esters named accordingly.

The same point arises in the 'phostamic acids' so named by Helferich and Curtius[291] because they differed from the phostonic acids only in having the cyclic —O— atom replaced by the cyclic —NH— group (p. 210).

Hands and Mercer[353] have shown the (3-hydroxypropyl)triphenylphosphonium iodide (55), when treated in tetrahydrofuran with sodium

$[HO(CH_2)_3\overset{+}{P}Ph_3]\,I^-$ ⟶ $\bar{O}(CH_2)_3\overset{+}{P}Ph_3$

(55)   (57)

(56)

hydride, gives hydrogen and 2,2,2-triphenyl-1,2-oxaphospholane (56), m.p. 116–117°. The evidence for the structure of this product, in particular that it is not the intermediate zwitterion (57), is the nmr spectrum, which consists of a double triplet, centered at τ 6.70 ($J_{4,5} = 6.5$ c/sec;

$J_{P,5-H} = 11$ c/sec; $J_{P,3-H} = 11.5$ c/sec) and a complex multiplet at *ca.* τ 8.4 (each multiplet 2H).

The phospholane (**56**), treated with hydriodic acid at 20°, regenerated the iodide (**55**) (90%). When heated under nitrogen at 300° for 5 minutes, the phospholane underwent rearrangement to 3-phenoxypropyldiphenylphosphine (**58**), probably by an intramolecular mechanism similar to that of the normal pyrolysis of phosphonium alkoxides (cf. Eyles and Trippett[354]). The identity of (**58**) was confirmed by its

$$C_6H_5O(CH_2)_3P(C_6H_5)_2 \qquad\qquad C_6H_5O(CH_2)_3P(O)(C_6H_5)_2$$
$$\text{(58)} \qquad\qquad\qquad\qquad \text{(59)}$$

$$C_6H_5O(CH_2)_3P^+(CH_3)(C_6H_5)_2\ I^- \ \rightarrow\ C_6H_5O(CH_2)_3P(O)(CH_3)C_6H_5$$
$$\text{(60)}$$

conversion into the oxide (**59**), m.p. 114–115°, which was independently synthesized, and into the methiodide (**60**): this salt was a glass, but with hot aqueous potassium hydroxide it gave benzene and methyl-3-phenoxypropylphenylphosphine oxide, m.p. 79–80° (75%).

Grayson and Farley[355] have recorded a novel synthesis, in which a secondary phosphine having a 3-hydroxyalkyl group, for example, 3-hydroxypropylphenylphosphine (**61**) reacts in benzene solution at 22° with diphenyl disulfide (1 mole) to give 2-phenyl-1,2-oxaphospholane (**62**), b.p. 112°/0.5 mm (34%), and a resinous residue. The infrared

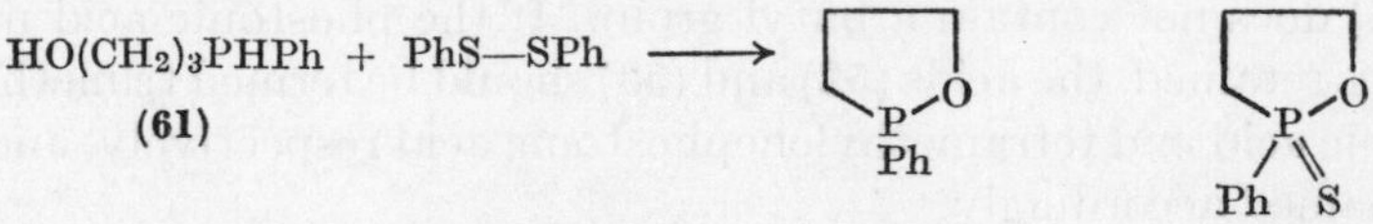

HO(CH₂)₃PHPh + PhS—SPh ⟶ (**61**), (**62**), (**63**)

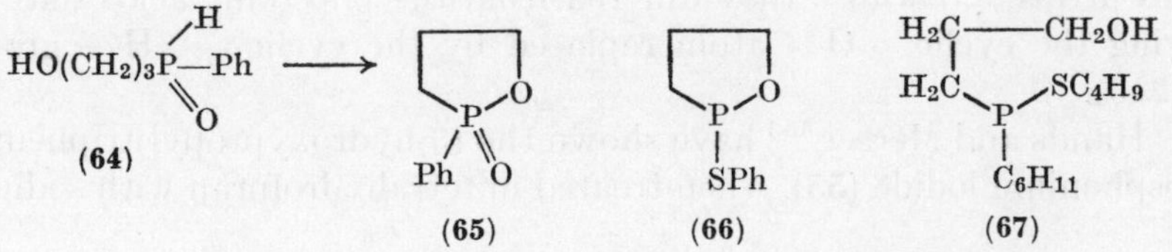

HO(CH₂)₃P(H)(O)—Ph (**64**) ⟶ (**65**), (**66**), (**67**)

spectrum of (**62**) shows an absence of PH and OH bands, but the presence of P—O—C, P—C₆H₅ and P—CH₂ units by bands at 960, 1435, and 1418 cm⁻¹, respectively: the proton count by nmr gave the aryl:OCH₂: CH₂ ratio as 5:2:4. The compound (**62**) readily gave a 2-sulfide (**63**), b.p. 152°/0.2 mm.

3-Hydroxypropylcyclohexylphosphine, $HO(CH_2)_3PH(C_6H_{11})$, reacts similarly with the disulfide to give the 2-cyclohexyl-1,2-oxaphospholane, b.p. 127°/0.15 mm.

Secondary phosphine oxides also give this reaction. (3-Hydroxy-propyl)phenylphosphine oxide (64) reacts similarly to form 2-phenyl-1,2-oxaphospholane 2-oxide (65), b.p. 142°/0.2 mm (84%), identical with the compound (42) prepared by Smith.[349] The analogous 2-n-butyl 2-oxide, b.p. 94°/0.25 mm, is similarly prepared.

The phosphine (61), when treated with diphenyl disulfide (2 moles) under the above conditions, gives 2-phenylthio-1,2-oxaphospholane (66), b.p. 116°/0.2 mm.

The probable mechanism of the initial reaction between (61) and the disulfide to give (62) was shown by treating cyclohexyl-(3-hydroxy-propyl)phosphine with di-n-butyl disulfide (1 mole) in benzene solution at 20–25°. Careful distillation under reduced pressure to remove solvent and butanethiol left a residue the elemental analysis and infrared spectrum of which were consistent with the intermediate compound (67); distillation of this product gave 2-cyclohexyl-1,2-oxaphospholane and a residual glassy resin.

Stereochemistry. Investigations have apparently been limited to *cis–trans*-isomers of suitably substituted 1,2-oxaphospholanes. The α,α-disubstituted lactone (68), when treated in ethanolic solution with potassium hydroxide (2 moles), undergoes ring-opening with the formation of the dipotassium salt (69), which on acidification cyclizes

(68)   (69)   (70)

to form 3-acetyl-2-ethoxy-1,2-oxaphospholane-3-carboxylic acid 2-oxide (70), which has been separated into two isomers, having m.p. 140–142° and 159–162°, respectively (Korte and Röchling[356]; Büchel, Röchling, and Korte[357]). The compound (70) when heated at 170° loses one molecule of CO_2, confirming the presence of a β-keto-acid. The isolation of the dipotassium salt (69) is not essential, for the ethanolic solution of this salt when acidified with HCl–CHCl$_3$ readily gives the product (70).

1,2-Benzoxaphosphole and 1,2-Benzoxaphospholene

The compound (1) is 1,2-benzoxaphosphole, [RRI 1219], and its 2,3-dihydro derivative (2) is 1,2-benzoxaphospholene.

Very little work on this system has been carried out compared with that on the preceding 1,2-oxaphosphol-4-ene and 1,2-oxaphospholane.

Kabachnik and Shepeleva[358] have investigated the action of phosphorus trichloride on aldehydes and ketones at reasonably high temperatures,

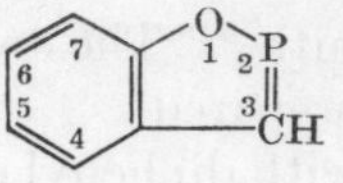

(structures (1) and (2))

(1) 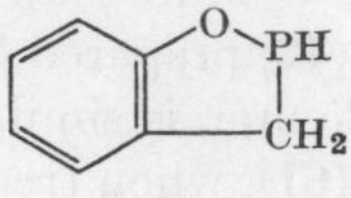   (2)

and find that most aromatic aldehydes give 1-chloro-phosphonyl chlorides, benzaldehyde for example giving α-chlorobenzylphosphonyl chloride, $C_6H_5CHClP(O)Cl_2$, m.p. 60–61° (62%): the overall reaction is an example of the familiar migration of oxygen from carbon to phosphorus. Aliphatic aldehydes, however, give only low yields of the corresponding derivatives.

When salicylaldehyde is heated with PCl_3 at 185–200° for 2.5 hours, there is considerable evolution of hydrogen chloride and a tarry residue is left, which on distillation affords 2,3-dichloro-1,2-benzoxaphospholene

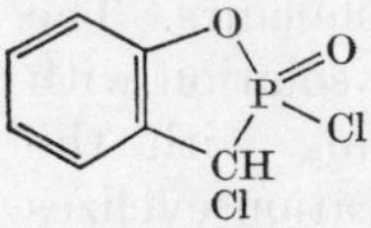

(structures (3), (4), and (5))

(3) 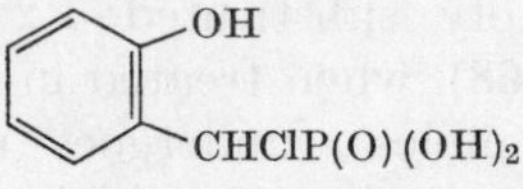   (4) 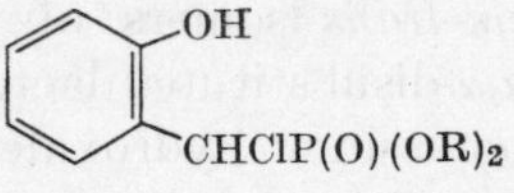   (5)

2-oxide (3), b.p. 138–140°/2.5 mm (40%), the normal reaction having occurred and been followed by cyclization at the phenolic group. The ring system in (3) is readily opened by water or alcohols, the former giving (α-chloro-o-hydroxybenzyl)phosphonic acid (4), m.p. 100–102.5°, and alcohols giving the corresponding dialkyl esters (5).

Five-membered Ring Systems containing only Carbon, One Phosphorus, and Two Oxygen Atoms

1,3,2-Dioxaphospholes and 1,3,2-Dioxaphospholanes

The ring system (1), [RRI 104], is known as 1,3,2-dioxaphosphole and the 4,5-dihydro derivative (2) as 1,3,2-dioxaphospholane. [The presence of the double bond in (1) often causes this parent compound

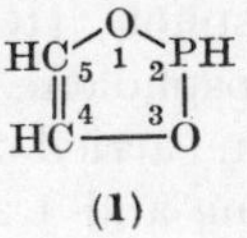

(structures (1) and (2))

(1) 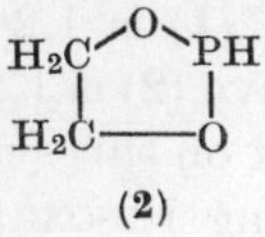   (2)

to be called 1,3,2-dioxaphospholene; the 'phosphole' name is correct, however, because the ring system in (1) is in the lowest state of hydrogenation.] It is often convenient to depict (1) and (2) with the phosphorus atom at the bottom of a pentagon.

Derivatives of both (1) and (2) almost invariably carry substituents in the 2-position. These two series differ considerably in their preparation and properties and will therefore be discussed separately.

1,3,2-Dioxaphospholes

The parent compound (1) is unknown and the system attracted little attention until in 1957 Ramirez and Dershowitz[359] found that trimethyl phosphite attacked the oxygen atom of several *para*-quinones, the methyl group in the product then undergoing migration, but the attack of the phosphite on the oxygen of *ortho*-quinones gave a product which did not show this migration. In 1958 it was recorded both by Birum and Dever[360] and by Kukhtin[361] that tertiary phosphites and 1,2-diketones readily combined to give 1:1 adducts; these were stable if protected from oxygen and water, and those of low molecular weight could be distilled.

The chemistry of this reaction of phosphites with 1,2-diketones and certain other types of compound was then studied in great detail by Ramirez and his school, but only a brief account of some of the main results can be given here. An excellent account of the work up to 1964 has been given by Ramirez,[362, 363] and a later review gives a condensed account[364] of almost all the branches of the work up to 1965. A much briefer account up to 1964 has been given by Quin.[51]

Trimethyl phosphite combines with biacetyl (2,3-butanedione), with benzil, and with 9,10-phenanthraquinone in benzene at room temperature to give the compounds (3; R = CH$_3$), a liquid, b.p. 45–47°/0.5 mm, (4; R = CH$_3$), m.p. 49°, and (5; R = CH$_3$), m.p. 74°. These

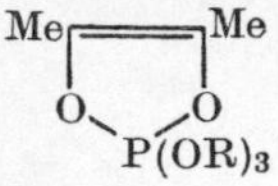

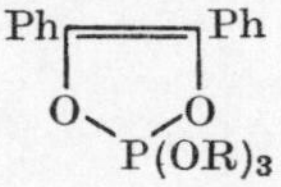

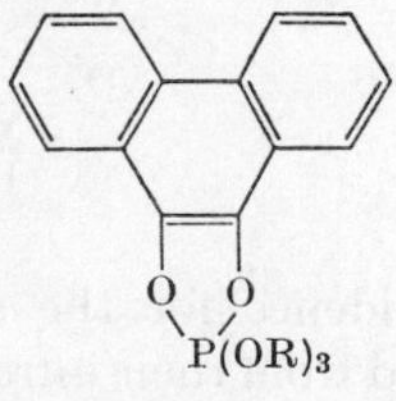

compounds can be regarded as oxyphosphoranes, but systematically they can be termed 2,2,2-trimethoxy-4,5-dimethyl-1,3,2-dioxaphosphole,

2,2,2-trimethoxy-4,5-diphenyl-1,3,2-dioxaphosphole, and as 2,2,2-tri-methoxy-4,5-(9′,10′-phenanthrylene)-1,3,2-dioxaphosphole, respectively. Triphenyl phosphite combines with benzil at 100° to give the compound (4; R = C_6H_5), m.p. 99°, and with phenanthraquinone to give (5; R = C_6H_5), m.p. 147°.

Note. The compound (5; R = CH_3) has a new ring system and the parent compound [RIS 10761] is numbered and presented as shown (6),

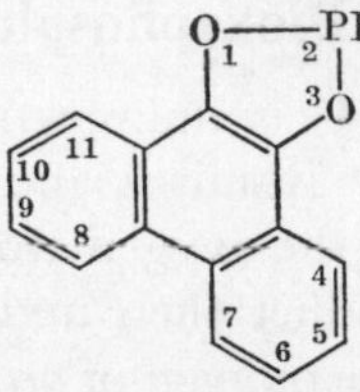

(6)

and named phenanthro[9,10-*d*]-1,3,2-dioxaphosphole. The compound (5; R = CH_3) should accordingly have 2,2,2-trimethoxy prefixed to this name.

It is noteworthy that Kukhtin and his colleagues[365, 366] recorded that interaction of 1,2-diketones and trialkyl phosphites at higher temperatures is accompanied by some molecular rearrangement and may give by-products such as trialkyl phosphate. To avoid such reactions, Ramirez has worked with highly purified compounds under the mildest conditions and (when possible) without a solvent; he has often obtained virtually quantitative yields of the oxyphosphorane compounds and has not detected such alkyl migration.

The probable mechanism of the addition of the trialkyl phosphite to the 1,2-diketone is:

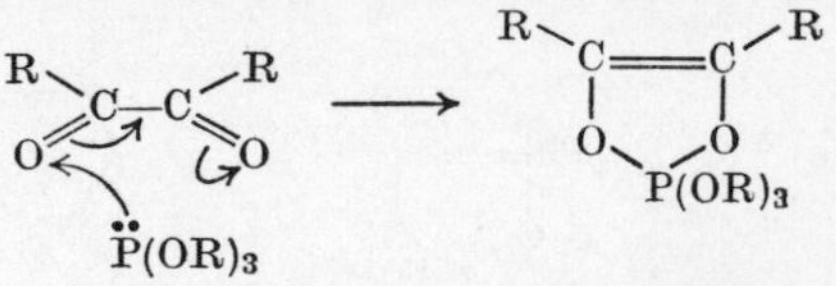

Evidence for the structure of the compounds (3)–(5) has been obtained from their infrared and Raman spectra (Ramirez and Desai[367]) and in particular from a full consideration of their nmr spectra. The ^{1}H-nmr spectrum for (3; R = CH_3), as the pure liquid or in CCl_4 solution, shows one doublet at τ 6.56 (on tetramethylsilane reference) for the three CH_3O groups, indicating that these CH_3O groups are almost certainly equivalent, and moreover that the CH_3O protons are more electronically shielded than those in trimethyl phosphate, $(CH_3O)_3PO$ (τ 6.30), evidence which is consistent with the pentaoxyphosphorane structure but not with a tetraoxyphosphonium structure (cf. p. 266). A singlet (τ 8.25) indicates the two equivalent methyl groups, probably joined to olefinic carbon atoms (Ramirez and Desai[368]). The ^{31}P-nmr spectrum should have ten lines because the P nucleus is coupled to nine CH_3O protons: eight of these have been observed, the others being too faint for observation.

The ^{31}P-nmr spectra for certain 1:1 adducts were first studied by Birum and Dever,[360] who reported positive shifts of +47 to +68 ppm (reference 85% phosphoric acid). Ramirez et al.[367] have recorded positive shifts of +46 to +62 ppm for nine adducts studied, and consider that these relatively large positive shifts [cf. $(CH_3O)_3PO$, +17 ppm] can be used as diagnostic evidence for the pentaoxyphosphorane structure.

Decisive evidence for this oxyphosphorane structure has been obtained (Hamilton, LaPlaca, Ramirez, and Smith[369, 370]) by an X-ray crystal analysis of the orthorhombic form of 2,2,2-triisopropoxy-phenanthro[9,10-d]-1,3,2-dioxaphosphole (5; $R = C_3H_7$), which shows that the molecule has the trigonal bipyramidal structure with the P atom at the center, the phenanthrene ring being linked to one apical and one equatorial oxygen atom, and the isopropyl groups to the remaining two equatorial and one apical oxygen atoms. The X-ray analysis of the monoclinic form of (5; $R = C_3H_7$), carried out with even greater accuracy, should be consulted for interatomic distances, inter-valency angles, and molecular structure (Sprately, Hamilton, and Ladell[371]); no significant differences between the two forms were observed.

The chemical reactivity of the pentoxyphosphoranes has been mentioned. The products formed by the following reagents may depend on the substituent groups in the phosphorane but almost always involve cleavage of the heterocyclic ring. The main reactions can be summarized as follows.

(a) Water (Birum et al.[372]) and (more rapidly) 2% hydrochloric acid[361] convert the compound (7) into a keto phosphate (8) and an alcohol; dry hydrogen chloride[368] and carboxylic acids[365] react similarly when the three R' groups in (7) are alkyl.

(b) Treatment of (7) in benzene solution with 1 mole of water gives, however, the 2-alkoxy 2-oxide (9).[373]

(c) Oxygen gives the trialkyl phosphate (10), the original 1,2-diketone (11), and the corresponding acyl anhydride (12) (Ramirez et al.[374]).

10

(d) Bromine in CCl_4 reacts in two ways. The biacetyl and benzil adducts with trimethyl phosphite give the bromo-keto phosphate (13); the benzil–triphenyl phosphite adduct gives the dibromo ketone (14).[368]

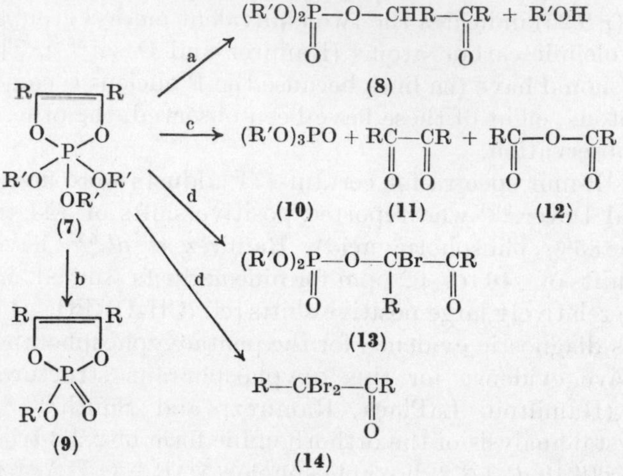

$$(R'O)_2P-O-CHR-CR + R'OH \qquad (8)$$
$$\underset{\parallel}{\overset{}{O}} \qquad \underset{\parallel}{\overset{}{O}}$$

$$(R'O)_3PO + RC-CR + RC-O-CR \qquad (10) \quad (11) \quad (12)$$

$$(R'O)_2P-O-CBr-CR \qquad (13)$$

$$R-CBr_2-CR \qquad (14)$$

(7)

(9)

These reactions are considered to involve the intermediate formation (probably in very small amount) of the reactive polar forms of the oxyphosphorane[368]:

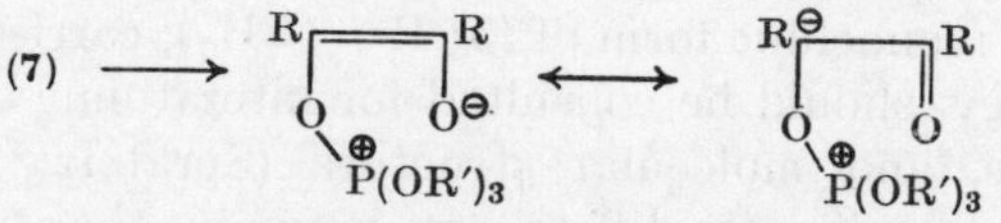

$$(7) \longrightarrow$$

Phenylhydrazine also breaks the ring system with the formation of the bis(phenylhydrazone) of the original 1,2-diketone.[365]

A very interesting extension of this work arose from the observation that whereas, as noted above, biacetyl reacts rapidly and exothermally with trimethyl phosphite to give the 1:1 adduct (3; R = CH_3), a second molecule of biacetyl will then slowly add on to give a 2:1 adduct, which was shown to be (15), i.e., 4,5-diacetyl-2,2,2-trimethoxy-4,5-dimethyl-1,3,2-dioxaphospholane, the overall effect being a saturation of the ring

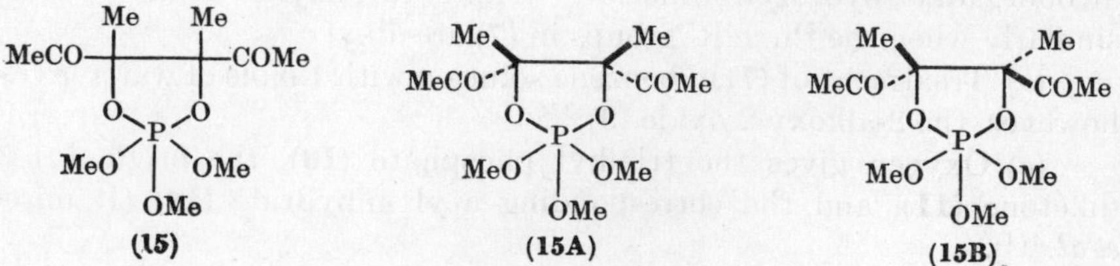

(15)

(15A)

(15B)

system (Ramirez et al.[375-379]). This second addition occurs at 20° and is stereoselective: two isomers are formed, one being the meso-form (15A), m.p. 31° (80%), and the other the racemic form (15B), a liquid at room temperature (20%). Very considerable evidence, based mainly on the nmr spectra, has been adduced for these structures.

This conversion of the original oxyphosphorane shown as (16; $R^1 = R^2 = CH_3$) into the 4,5-diacetyl derivative (15) can be interpreted as the former being able to open the ring to give (possibly to a minute extent) the polar form (17; $R^1 = R^2 = CH_3$). The latter then reacts as shown with a second molecule of diketone (18; in which the methyl groups are numbered afresh for identification), with the formation of a new polar form (19): this process links the original C-5 atom of (16) to

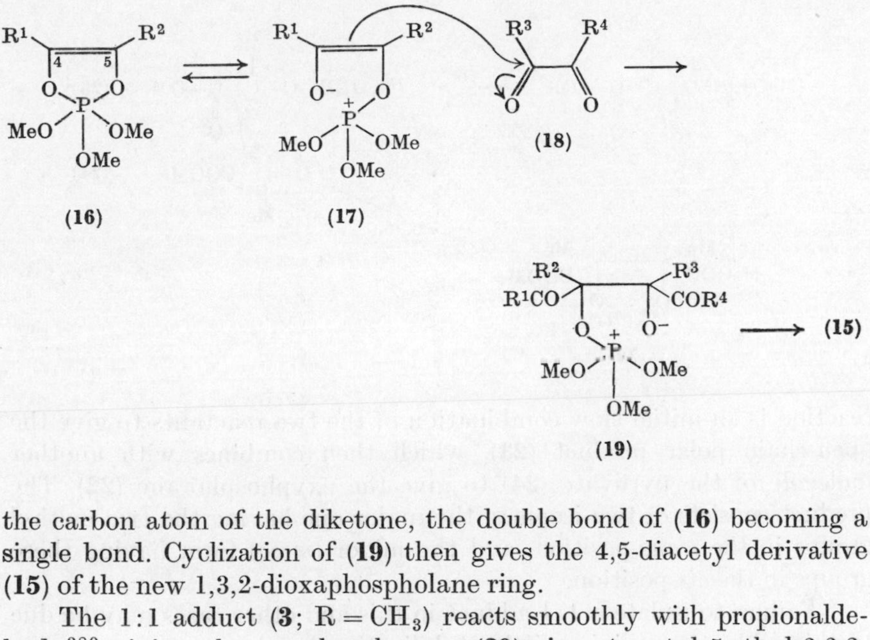

the carbon atom of the diketone, the double bond of (16) becoming a single bond. Cyclization of (19) then gives the 4,5-diacetyl derivative (15) of the new 1,3,2-dioxaphospholane ring.

The 1:1 adduct (3; $R = CH_3$) reacts smoothly with propionaldehyde [380] giving the oxyphosphorane (20), i.e., 4-acetyl-5-ethyl-2,2,2-trimethoxy-4-methyl-1,3,2-dioxaphospholane, in only one isomeric form, in which the 4-methyl and 5-ethyl groups are almost certainly trans to one another: the ^{31}P shift is +51.3 ppm. Butyraldehyde and heptanal react similarly, each giving only one isomeric form, which have ^{31}P shifts of +51.2 and +51.3 ppm, respectively.

The adduct (3; $R = CH_3$) reacts with benzaldehyde very slowly at room temperature [381]: the major isomer (21) has the acetyl group and the phenyl group cis to one another; the second isomer, having the methyl and the phenyl groups in the *cis*-position, is formed in very small amount.

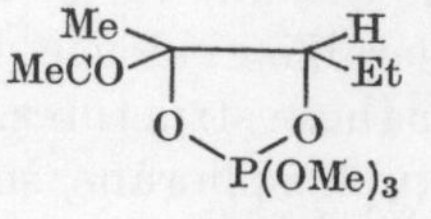

$$
\begin{array}{cc}
\underset{(20)}{\overset{\displaystyle \text{Me} \diagdown \qquad \diagup \text{H}}{\underset{\displaystyle \underset{\text{P(OMe)}_3}{O \diagdown \;\; \diagup O}}{\text{MeCO} \diagup \qquad \diagdown \text{Et}}}}
&
\underset{(21)}{\overset{\displaystyle \text{Me} \diagdown \qquad \diagup \text{H}}{\underset{\displaystyle \underset{\text{P(OMe)}_3}{O \diagdown \;\; \diagup O}}{\text{MeCO} \diagup \qquad \diagdown \text{Ph}}}}
\end{array}
$$

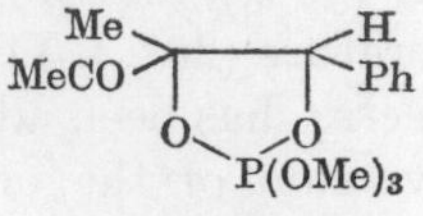

The reaction of trimethyl phosphite with certain other types of compound must be briefly mentioned. Methyl pyruvate, as an example of an α-keto-ester, reacts with the phosphite to give solely a crystalline 2:1 adduct (22), *i.e.*, dimethyl 2,2,2-trimethoxy-4,5-dimethyl-1,3,2-dioxaphospholane-4,5-dicarboxylate. The probable mechanism of this

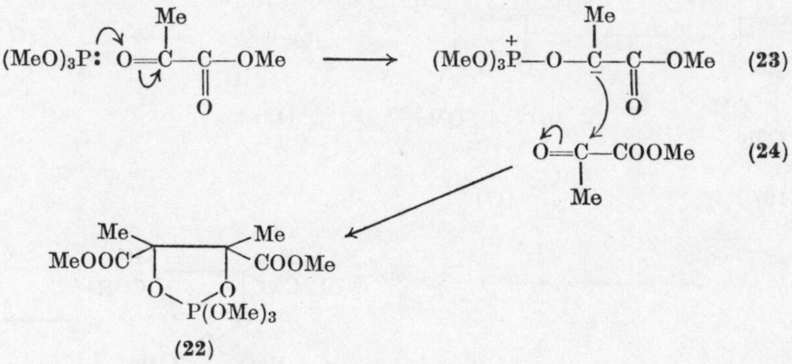

reaction is an initial slow combination of the two reactants to give the open-chain polar product (23), which then combines with another molecule of the pyruvate (24) to give the oxyphosphorane (22). The product consists of two isomers, the major one having the two methyl groups in the *trans*-position, and the minor (*meso*) form having these groups in the *cis*-position.

Failure to isolate a 1:1 adduct in this and other cases may be due to this adduct's receiving little stabilization by the unsaturated oxyphosphorane formation: the adduct is therefore highly reactive and reacts in the polar form with a second molecule of the carbonyl compound as above.

The reaction of trimethyl phosphite with 3-benzylidene-2,4-pentanedione to give a 1,2-oxaphosphol-4-ene derivative (25) has already been discussed (p. 255).

The reaction of trimethyl phosphite with anhydrous propionaldehyde to give 3,5-diethyl-1,4,2-dioxaphospholane (26), [RIS 11556], occurs however very slowly at room temperature, *e.g.*, after 14 days only 60% conversion had occurred: higher temperatures cause side-reactions. The mechanism of the formation of (26) is similar to that of

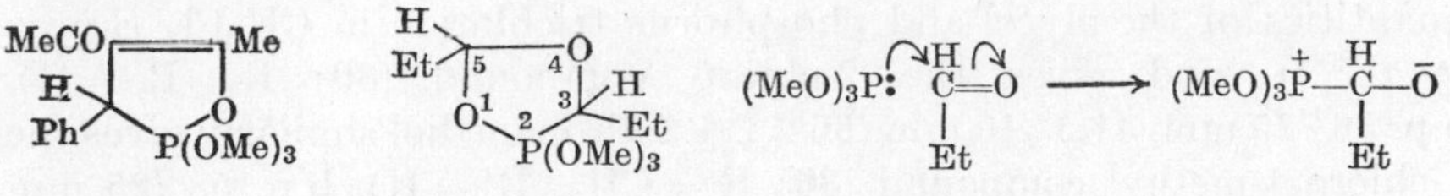

MeCO⌐====⌐Me
H·└────┘·
Ph⁄ ⟍O⁄
     P(OMe)₃

(25)

H⟍
Et⁄⟍⌐₅ ₄⌐⟍H
   │¹   ³│
   O⌐₂   ⌐Et
     P(OMe)₃

(26)

⌢H⌢
(MeO)₃P: C=O ⟶
        │
        Et

(27)

            ⁺ H
(MeO)₃P—C—Ō
         │
         Et

(28)

the compound (22), namely, an initial slow reaction of the phosphite and the aldehyde (27) to give the polar intermediate (28), which then reacts rapidly with a second molecule of propionaldehyde to give the cyclic (26).

The wide and various extensions of this field of work by Ramirez and his co-workers, and the very thorough investigation of the structure of their products, have produced a massive volume of very interesting and valuable work. The original papers should certainly be consulted for the full details. A bibliography of papers from various centers up to 1964 has appeared (Ramirez *et al.*[382]).

1,3,2-Dioxaphospholanes

Members of this class have often been named after the 1,2-glycol of which they can be regarded as esters: thus the compound (29; R = Cl) has been termed glycol chlorophosphonite or phosphorochloridite; the compound (29; R = OCH₃) has been termed glycol methoxyphosphonite or ethylene methyl phosphite; and the compound (29; R = NEt₂) has been termed glycol *N*,*N*-diethylamidophosphonite or phosphoroamidite.

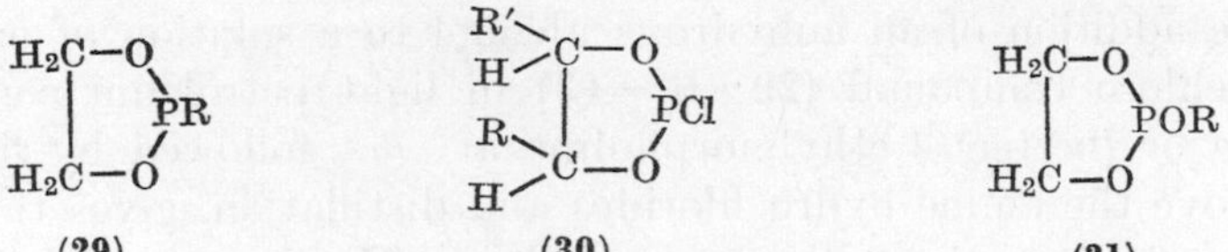

H₂C—O⟍
│   ⟍PR
H₂C—O⁄

(29)

R′⟍
 ⟍C—O⟍
H⁄│   ⟍PCl
R⟍│   ⁄
 ⟍C—O⁄
H⁄

(30)

H₂C—O⟍
│   ⟍POR
H₂C—O⁄

(31)

In one of the earliest syntheses of this system (Rossiĭskaya and Kabachnik[383]), ethylene glycol was added to an excess of stirred, ice-cooled phosphorus trichloride; the reaction mixture when heated at 60–70° for 3 hours and then distilled gave (*a*) 2-chloro-1,3,2-dioxa-phospholane (29; R = Cl), a colorless liquid, b.p. 66–68°/47 mm, which fumed in air, and (*b*) ethylene 1,2-bis(phosphorodichloridite),

$$Cl_2P—O(CH_2)_2O—PCl_2,$$

b.p. 93–100°/4.5 mm, but no 2-chloroethyl phosphorodichloridite, $Cl(CH_2)_2O—PCl_2$. These results conflict with those of Carré (p. 319), whose reactions were performed in ether.

The above preparation is greatly improved by the use of equimolar quantities of the glycol and phosphorus trichloride in CH_2Cl_2 (Lucas *et al.*[384]), and gives the 2-chloro compound (**30**; R = R′ = H), b.p. 56°/25 mm, 41.5°/10 mm (66%). 1,2-Propanediol similarly gives the 2-chloro-4-methyl compound (**30**; R = CH_3, R′ = H), b.p. 58°/25 mm (86%), and *meso*-2,3-butanediol gives the 2-chloro-4,5-dimethyl compound (**30**; R = R′ = CH_3), b.p. 66.0–66.2° (72%).

Table 23. 2-Alkoxy-1,3,2-dioxaphospholanes
(**31**)

R	B.p. (°/mm)
Methyl	60.7–60.9
Ethyl	70.4–70.7
n-Propyl	85.6–85.8
Isopropyl	70.3
n-Butyl	100–100.2
Isobutyl	91.1–91.3
sec-Butyl	87.9–88.3
tert-Butyl	77.9–78.6
n-Pentyl	89/12
n-Hexyl	82.5/6

The unsubstituted chloro compound (**29**; R = Cl) is very readily hydrolyzed; consequently it should always be manipulated under anhydrous conditions.

The addition of an anhydrous alcohol to a solution of one mole of the chloro compound (**29**; R = Cl) in light petroleum containing pyridine or (better) 4-ethylmorpholine at −5°, followed by filtration (to remove the amine hydrochloride) and distillation, gives the corresponding 2-alkoxy-1,3,2-dioxaphospholane (**31**) (Lucas *et al.*[384]). The b.p.s in Table 23 are at 25 mm, except the last two (Foxton *et al.*[385]).

If more than one mole of an alcohol is used, the reaction goes a stage further with the formation of a 2-hydroxyethyl phosphite:

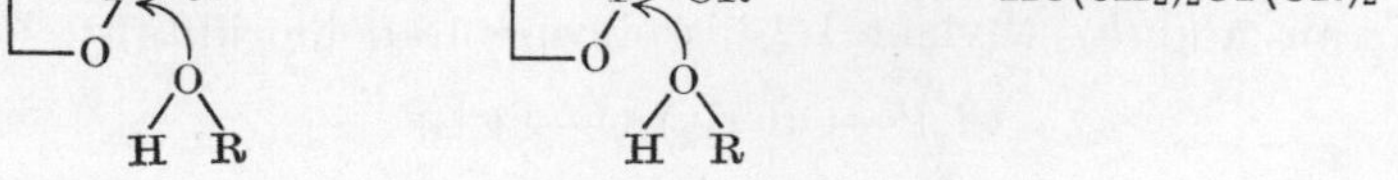

The 2-phenoxy compound (**31**; R = C_6H_5) can be readily prepared by heating triphenyl phosphite at 100° with ethylene glycol (1 mole) in which a small quantity of sodium has been dissolved. Trans-esterification occurs, with liberation of phenol and formation of (**31**; R = C_6H_5) (60%). 1,2-Propanediol similarly gives the 4-methyl-2-phenoxy derivative (52%). The reaction probably proceeds:

$$(C_6H_5O)_3P + HO(CH_2)_2OH \rightarrow C_6H_5OH + (C_6H_5O)_2P-O-(CH_2)_2OH \rightarrow$$
$$(\textbf{31}; R = C_6H_5)$$

The reaction with the lower 1,2-diols will proceed (in lower yield) without the catalytic action of the sodium glycolate, but the latter is essential when more complex diols are used (Ayres and Rydon[386]).

Three other methods for the preparation of 2-alkoxy derivatives (**31**) should be mentioned.

(*a*) Methyl phosphorodichloridite, CH_3OPCl_2, when added to an ethereal solution of glycol and diethylaniline, gives the 2-methoxy compound (**31**; R = CH_3). The 2-ethoxy compound is similarly prepared but only in low yield.[383]

(*b*) Ethylene oxide is added to the 2-chloro compound (**29**; R = Cl) at 15–20°, and the reaction mixture, after 1 hour at 50°, on distillation gives the 2-(2'-chloroethoxy) compound (**31**; R – $ClCH_2CH_2$), b.p. 57°/1 mm (Nagy[387]); many other examples of this reaction have been cited.

(*c*) It is claimed that, in a novel reaction, *n*-hexyl *N,N,N',N'*-tetraethylphosphorodiamidite, $C_6H_{13}O-P[N(C_2H_5)_2]_2$, reacts with ethylene glycol in acetic anhydride to form the 2-*n*-hexyloxy compound (**31**; R = C_6H_{13}), b.p. 52–53°/0.5–1.0 mm (74%). The 2-*n*-propoxy compound, similarly prepared, unites with sulfur to give the 2-*n*-propoxy-1,3,2-dioxaphospholane 2-sulfide, b.p. 81–83°/1 mm (Evdakov, Shlenkova, and Bilevich[388]).

The carefully controlled hydrolysis of the 2-chloro derivatives gives rise to compounds formulated as 2-hydroxy derivatives (**32**), but they are more probably the 1,3,2-dioxaphospholane 2-oxides (**33**). These are

(**32**)   (**33**)   (**34**)

prepared by adding a solution of water (1 mole) in initially anhydrous dioxane in portions with cooling to a solution of the 2-chloro compound (1 mole) also in dioxan; the reaction mixture is kept at room temperature and 40 mm pressure to remove the HCl as thoroughly as possible; the dioxane is then distilled at this pressure, and the residual (**33**) rapidly distilled at very low pressure. The unsubstituted compound (**33**; R = H) is unstable, but the 4-methyl compound (**33**; R = CH₃) is obtained in 96% yield as a colorless viscous liquid, b.p. 76–82°/0.6 mm; it has the curious property that after 5 days at room temperature the viscosity steadily increases until the liquid will hardly flow, yet redistillation at 76.5°/0.65 mm gives a mobile distillate (with very little residue) which again after 1 week becomes immobile.[384]

The 2-chloro derivative when treated with primary amines gives intractable products, but when treated with diethylamine (2 moles) in petroleum and worked up as the 2-alkoxy compounds, gives the 2-diethylamino compound (**34**), b.p. 98.7–99.1°/25 mm; piperidine similarly gives the 2-piperidino compound,[384] b.p. 132°/25 mm.

It is noteworthy that the optically pure D(−)-2,3-butanediol, $[\alpha]_D$ −13.17°, gave the 2-chloro compound (**30**; R = R′ = CH₃), $[\alpha]_D$ +97.12°, b.p. 49.1–49.2°/10 mm (66%), which in turn gave the 2-methoxy compound, $[\alpha]_D$ +53.63°, b.p. 46.0–46.2°/10 mm (47%), and the 2-diethylamino compound, $[\alpha]_D$ −11.18°, b.p. 85.1–85.2°/10 mm (61%); all rotations measured at 25° (Garner and Lucas[389]).

If the above hydrolysis of the 2-chloro compound is modified by the slow addition of water (1 mole) and a base, such as pyridine, dimethylaniline, or trimethylamine, to the ethereal 2-chloro compound (2 moles) at −5°, followed by stirring at room temperature for 3 hours, filtration, and distillation, the anhydride (**35**; R = H), or oxobis-2,2′-(1,3,2-dioxaphospholane), b.p. 100–101°/4 mm (40%), is obtained. The 4,4′-dimethyl analogue (**35**; R = CH₃), b.p. 118–120°/2 mm (35%), is similarly obtained (Arbuzov, Kikonorov, *et al.*[390]).

(**35**)   (**36**)

The interaction of equimolecular quantities of the 2-chloro compound and a sodium dialkyl phosphite, (RO)₂PONa, in chilled ether gives a mixed anhydride; the diethyl and the di-*n*-propyl members (**36**; R = C₂H₅ and C₃H₇) have b.p. 84–85°/2 mm and 93–94°/2 mm, respectively.[390]

The reaction of the 2-chloro compound with 1,3-dienes gives phospholenes (p. 46), and with 2-keto-3-olefins gives 1,2-oxaphosphol-4-enes (p. 256).

As an example of the preparation of 2-aryl derivatives, an equi-molecular mixture of p-tolylphosphonous dichloride, $CH_3C_6H_4PCl_2$, and ethylene glycol in ethereal solution, when treated with pyridine at $-10°$, gives the 2-p-tolyl derivative (**37**; $R = CH_3C_6H_4$), b.p. $128–131°/12$ mm,

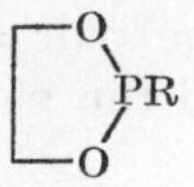

(**37**) 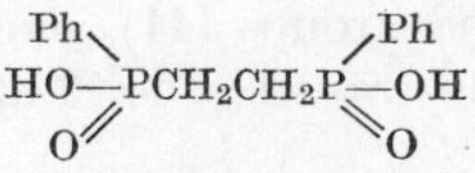   (**38**) 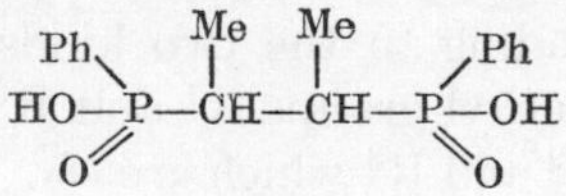   (**39**)

but much polymeric material is formed (Kamai and Ismagilov[391]). Better results are apparently obtained when anhydrous solutions of phenylphosphonous dichloride in benzene and of glycol in pyridine are slowly and simultaneously added to a pyridine–benzene mixture at $10–20°$. Distillation gives the 2-phenyl derivative (**37**; $R = C_6H_5$), b.p. $80.5–82°/1.8$ mm, a compound which is very sensitive to water (Harwood[392]).

When this phenyl derivative, mixed with a very small proportion of aluminum chloride, is heated under nitrogen in a sealed tube at 160° for 65 hours, it gives a hard colorless polymer, soluble in chloroform and insoluble in benzene. This material when immersed in water swells considerably, giving an acidic oil, which on purification yields 1,2-ethylenebis(phenylphosphinic acid) (**38**), m.p. $266–266.7°$. When ethylphosphonous dichloride, 1,2-butanediol, and aluminum chloride are similarly heated together, hydrolysis of the polymeric product with water yields the corresponding 1,2-dimethylethylenebis(ethyl-phosphinic acid) (**39**).[392]

Conformation

The conformation of 1,3,2-dioxaphospholane derivatives having trivalent phosphorus has been investigated by Goldwhite.[393] The ^{1}H-nmr spectra of the 2-chloro-4,4,5,5-tetramethyl compound (**40**) give

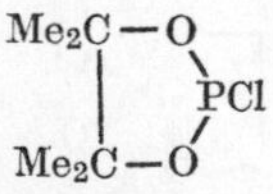

(**40**) 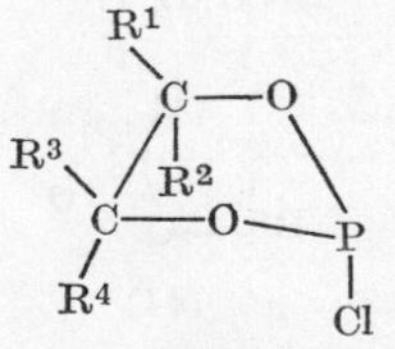   (**41**) 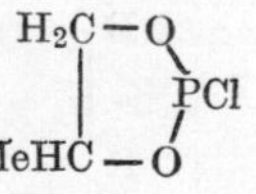   (**42**)

strong evidence that at room temperature this compound has two kinds of methyl group in magnetically different environments. It is considered unlikely that these environments arise from a stable non-planar conformation of the ring, for there is evidence that in similar compounds in which the $>$PCl group is replaced by $>$CH$_2$ or by $>$SO the five-membered ring undergoes rapid flexing. The results for (**40**) are therefore attributed to a stable pyramidal configuration of the phosphorus atom, leading to the two kinds of methyl groups (**41**), namely, the pair of methyl groups R^1 and R^3 which are *trans* to the Cl atom and the pair R^2 and R^4 which are *cis*.

The spectra of the 2-chloro-4-methyl compound (**42**) also indicate the presence of two different methyl groups, which suggests that (**42**) is a mixture of two geometric isomers, one being (**41**; $R^3 = CH_3$, $R^1 = R^2 = R^4 = H$) and the other (**41**; $R^4 = CH_3$, $R^1 = R^2 = R^3 = H$). Similar results were obtained with 2-methoxy-4-methyl-1,3,2-dioxaphospholane.

A similar investigation of 2-methoxy-, 2-ethoxy-, and 2-diethyl-amino-1,3,2-dioxaphospholane (Foster and Fyfe [394]) showed in each case a multiplet splitting of the CH$_2$ protons, which are further split by the nuclear spin of the phosphorus. The phosphorus atom must have a tetragonal (*i.e.*, pyramidal) configuration, but analysis of these multiplets was not attempted.

The following types of 1,3,2-dioxaphospholane derivatives have formally 5-covalent phosphorus.

When a mixture of phenylphosphonyl chloride, C$_6$H$_5$POCl$_2$, and ethylene glycol (1.1 moles) is warmed at 25° under reduced pressure to remove hydrogen chloride, the residue on distillation gives 2-phenyl-1,3,2-dioxaphospholane 2-oxide (**43**; R = H), b.p. 210°/6–7 mm (75%), and 2,3-butanediol similarly affords the 4,5-dimethyl derivative (**43**; R = CH$_3$), b.p. 210–215°/15 mm (79%) (Toy [395]). Historically this is apparently the first recorded synthesis of the 1,3,2-dioxaphosphole ring system.

In a different synthesis, Petrov *et al.*[396] record that (chloromethyl)-phosphonyl dichloride, ClCH$_2$POCl$_2$, when treated cautiously at 20° with water (1 mole) and then heated at 100° for 3 hours and finally under

(43)   (44)   (45)

reduced pressure, gives (chloromethyl)phosphonic anhydride, ClCH$_2$PO$_2$, m.p. 65–72°. The analogous C$_2$H$_5$PO$_2$, m.p. 115–123°, is similarly prepared. When a mixture of an anhydride of this type and an olefin oxide (4 moles) in chloroform is heated in a sealed tube at 80–100° for 3–6 hours, distillation of the product gives the 2,4-disubstituted-1,3,2-dioxaphospholane 2-oxides (44) listed in Table 24.

Table 24. Substituted 1,3,2-dioxaphospholane 2-oxides (44)

R	R'	B.p. (°/mm)
CH_3	H	63–64/0.2
CH_3	CH_3	78–80/0.2
C_2H_5	H	83–84/0.2
C_2H_5	CH_3	87–88/0.2
CH_3	$ClCH_2$	125–130/0.02
C_6H_5	CH_3	130–135/0.03
$ClCH_2$	H	105–115/0.07
$ClCH_2$	CH_3	100–110/0.007

The preparation of 2-chloro-1,3,2-dioxaphospholane 2-oxides (45) by the interaction of a glycol and phosphorus oxychloride in the presence of an amine works well only if the compound (45) is stable at distillation temperatures to traces of amine or amine hydrochloride; ethylene glycol, in particular, reacts with $POCl_3$ under these conditions to give insoluble oils which rapidly polymerize. The 2-oxides (45) can, however, be readily prepared by passing dry oxygen into a solution of the 2-chloro compound (30) in dry benzene until the exothermic reaction is complete, and then isolating the required product by distillation. Starting with the appropriate (30), Edmundson[307] has prepared the 2-chloro oxide (45; R = H), b.p. 79°/0.4 mm, and the 2-chloro-4-methyl oxide (45; R = CH_3), b.p. 74°/0.4 mm, each in 80% yield.

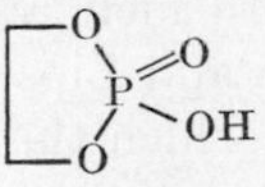

(46) 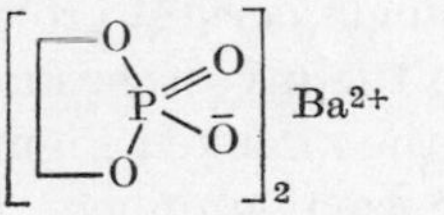   (47) 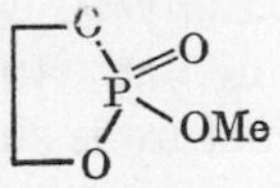   (48)

The 2-chloro 2-oxide (45; R = H) reacts with water to form the 2-hydroxy 2-oxide (46) ('ethylene phosphate') which, however, readily hydrolyzes further to 2-hydroxyethyl phosphate,

$$HOC_2H_4O—P(O)(OH)_2,$$

and the internal ester-acid (46) is isolated usually as its metallic salts. The barium salt has been prepared by adding water (1 mole) slowly to phosphorus oxychloride and, after complete reaction, cooling to 0° and adding ethylene bromohydrin. The reaction mixture is warmed under reduced pressure to remove hydrogen chloride, and then suitably treated with barium hydroxide to give barium 2-bromoethyl phosphate. An aqueous solution of this phosphate is warmed to 75° for 15 minutes, and then treated again with barium hydroxide; rapid evaporation under reduced pressure gives a mixture of barium bromide and the barium salt (47), which can be isolated by suitable aqueous-ethanolic extraction (Kumamoto, Cox, and Westheimer[398]).

Esters of the acid (46) are obtained by oxidation of the corresponding 2-alkoxy-1,3,2-dioxaphospholane: thus the 2-methoxy compound (31; R = CH$_3$) when treated in dichloromethane solution at

Figure 7. Structure of 2-methoxy-1,3,2-dioxaphospholane 2-oxide (48).
Interatomic Distances (Å): P—O(3) = P—O(6) = P—O(7) = 1.57; P—O(6) = 1.44; O(3)—C(4) = 1.41; O(1)—C(5) = 1.45; O(7)—C(8) = 1.44; C(4)—C(5) = 1.52. Intervalency Angles: P—O(3)—C(4) = P—O(1)—C(5) = 112.0°; P—O(7)—C(8) = 118.0°; O(1)—P—O(3) = 99.1°; O(1)—P—O(7) = 109.2°; O(3)—P—O(7) = 105.7°; O(6)—P—O(1) = 117.3°; O(6)—P—O(3) = 116.0°; O(6)—P—O(7) = 108.7°; O(3)—C(4)—C(5) = 107.8°; C(4)—C(5)—O(1) = 106.8°. [Adapted, by permission, from T. A. Steitz and W. N. Lipscomb, *J. Amer. Chem. Soc.*, **87**, 2488 (1965)]

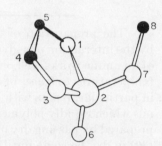

−10° to −15° with dinitrogen tetraoxide gives the crystalline 2-methoxy 2-oxide (48), commonly known as methyl ethylene phosphate, m.p. −6° to −5° (75%) (Covitz and Westheimer[399]).

The structure of the compound (48) has been determined by X-ray crystal analysis, using a single crystal grown from a melt at −5° and maintained at −40° during the investigation (Steitz and Lipscomb[400]). The structure is shown in the annexed Figure 7, in which the atoms in the ring are given their normal numbers, the remaining two oxygen and one carbon atoms being numbered 6, 7, and 8.

Certain points in this structure are of particular interest.

(a) The five-membered ring is puckered, so that the normal of the O(1)—P—O(3) plane is about 11° from the normal of the O(3)—C(4)—O(1) plane.

(b) The angle O(1)—P—O(3) is 99°, i.e., 10° less than the tetrahedral angle. The fact that this angle (99°) is virtually identical with the 'natural' angle (p. 186) of a symmetric phosphine, PR$_3$, may be fortuitous.

(c) The P—O(7)—C(8) angle (118.8°) in the unstrained part of the molecule is reduced to 112° within the 'strained' ring; cf. P—O(3)—C(4), P—O(1)—C(5).

(d) The equality of the three 'esterified' P—O bonds is apparently independent of angle strain; this value (1.57 Å) is significantly shorter than that of the ring P—O bonds (1.60 and 1.76 Å) found in the penta-oxyphosphorane (5; R = iso-C_3H_7) (p. 265).

An X-ray structure analysis of 2-methoxy-4,4,5,5-tetramethyl-1,3,2-dioxaphospholane 2-oxide, termed 'methyl pinacol phosphate', shows a very similar ring arrangement to that of (48), although the C atom of the CH_3O group has changed position; the corresponding bond lengths and intervalency angles of the two compounds are tabulated (Newton, Cox, and Bertrand [401]).

The conditions and the mechanism of the hydrolysis of esters such as (48) are of considerable theoretical and biochemical importance, as such phosphate esters are intermediates in the hydrolysis of ribonucleic acids. The subject has been studied in great detail by Westheimer and his co-workers.

The particular qualities of a five-membered ring containing a phosphorus atom, namely, that ring closure, ring opening, and displacement of exocyclic substituents all occur far more rapidly than in larger analogous ring systems, have become increasingly recognized. The considerable readiness with which five-membered rings containing an arsenic atom are formed was recognized much earlier (pp. 536–539).

The effect of ring size is not limited to the hydrolysis rate of esters. The rate of conversion of the methylphenylphospholanium cation (49) by hydroxyl ions into 1-methylphospholane 2-oxide and benzene is

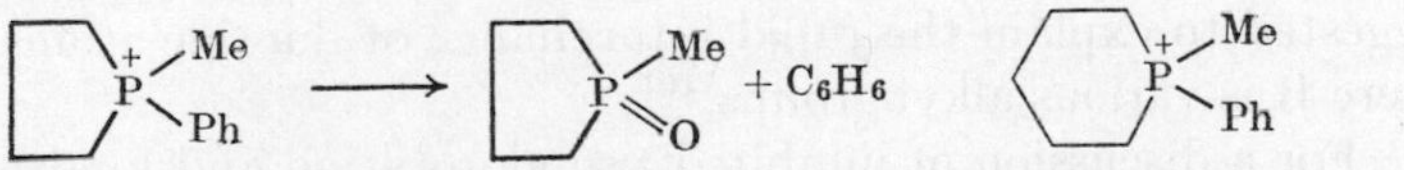

ca. 1300 times more rapid than the analogous reaction of the methyl-phenylphosphorinanium cation (50) (Aksnes and Bergesen [402]).

More strikingly, five-membered cyclic esters of phosphoric acid such as (48), and of phosphonic acid such as (51), undergo acid and alkaline hydrolysis at rates 10^5–10^8 times as fast as their acyclic analogues, and some of these reactions, in both acid and base hydrolysis, occur at these high rates without ring cleavage (Dennis and West-heimer,[403, 404] and references given therein). The rate of alkaline hydrolysis of (51; R = C_2H_5) drops rapidly in turn in the six-membered and seven-membered homologous ethyl esters (Aksnes and Bergesen[405]).

Consideration of these and earlier results, and of the fact that the esters (52), (53), and (54) undergo acid and alkaline hydrolysis more slowly than their open-chain analogues, led Dennis and Westheimer[403, 404] to postulate a trigonal bipyramid as an intermediate in the hydrolysis of esters of phosphorus acids. It is considered that an alkyl group is energetically less favorably situated in an apical position of the bi-pyramid than in an equatorial position (or alternatively that oxygen atoms are most favorably situated at the apical positions). This general situation is analogous to that of the stable alkylfluorophosphoranes, R_nPF_{5-n}, where alkyl groups occupy equatorial positions (Schmutzler[406]).

The formation of a trigonal bipyramidal intermediate would relieve the normal strain of a five-membered ring system, but this relief is apparently insufficient in the esters (52), (53), and (54) to overcome the barrier to the presence of an alkyl group in an apical position, while angle strain prevents the ring adopting a diequatorial position.

The hydrolysis of the methyl ester (51; R = CH_3) occurs at the very high rate but almost exclusively by ring cleavage, only traces of methanol being detectable, yet the acid-catalyzed hydrolysis of the methyl ester (48) occurs with 30% of the product arising from hydrolysis of the methyl group without ring opening. The rates of both hydrolyses—with and without ring opening—are about 1,000,000 times those of their acyclic analogues. These results can be interpreted on the basis of a 'pseudo-rotation' of the trigonal bipyramidal intermediates, similar to that suggested to explain the rapid interchange of fluorine atoms in R_2PF_3, where R is various alkyl groups.[406]

For a discussion of inhibited pseudorotation and kindred matters, see Gorenstein and Westheimer.[407]

A brief account cannot possibly do justice to Westheimer's very considerable work in this field—work which will assuredly be regarded as a classical elucidation of reaction mechanisms in a very difficult field. For a fuller account see Kirby and Warren.[408]

For the preparation of 2-[bis(2'-chloroethyl)amino]-1,3,2-dioxa-phospholane 2-oxide, see p. 324.

1,4,6,9-Tetraoxa-5-phospha(v)spiro[4.4]nona-2,7-diene

This is a correct name for the spirocyclic system (1), [RIS —]; it can however more conveniently but less accurately be numbered as in

(1A) and termed P-spirobi[1,3,2-dioxaphosphole]; hence its inclusion at this stage.

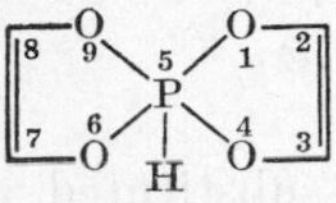

(1) 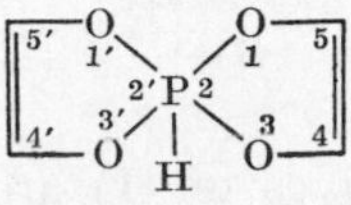   (1A)

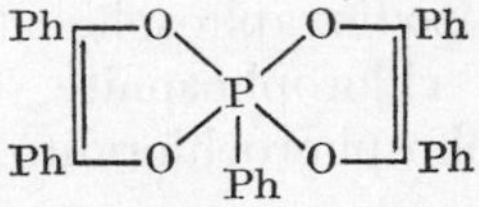

(2)

A pentaphenyl derivative (2) of this spirocyclic system has been prepared by Schmidt and Osterroht[281] in their investigations on the monomeric carbene-like phenylphosphinidene, $C_6H_5P:$, which they detected in a transient state during (a) the reduction of phenylphosphonous dichloride at 25°, (b) the oxidation of phenylphosphine with iodine in the presence of triethylamine, and (c) the thermal disintegration of the compound which was considered at that time to be tetraphenylcyclotetraphosphine (cf. p. 205).

When phenylphosphonous dichloride in tetrahydrofuran was stirred with zinc dust and benzil at 25° for 2 hours, the reaction mixture yielded 2,3,5,7,8-pentaphenyl-1,4,6,9-tetraoxa-5-phospha(v)spiro[4.4]-nona-2,7-diene or P-phenyl-P-spirobi[4,5-diphenyl-1,3,2-dioxaphosphole] (2).

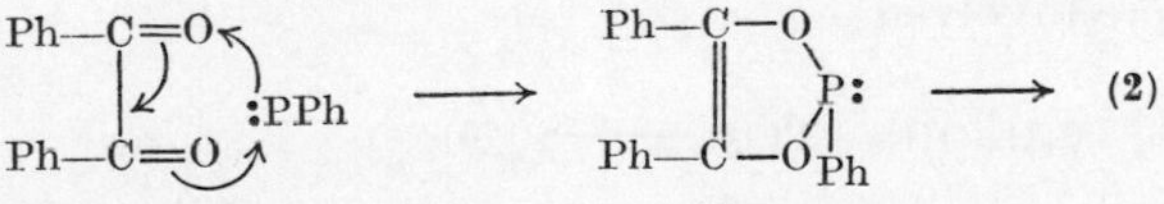

(3)   (4)   (2)

The mechanism of this reaction is probably the initial 1,4-addition of the phenylphosphinidene to the benzil (3) to give the cyclic ester (4) of phenylphosphonous acid, which is then capable of a similar addition to a second molecule of benzil to form (2).

1,3,2-Benzodioxaphosphole and 2,2′-Spirobi[1,3,2-benzo-dioxaphosphole]

1,3,2-Benzodioxaphosphole, [RRI 1100], has the structure (1) and 2,2′-spirobi-[1,3,2-benzodioxaphosphole], [RIS 10683], has the structure (2). The two systems are conveniently discussed in one section. A very early name of (1) was 1,3-dioxa-2-phosphaindan.

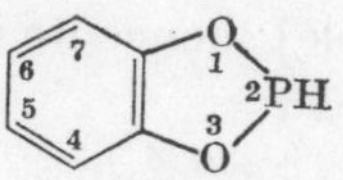

(1)

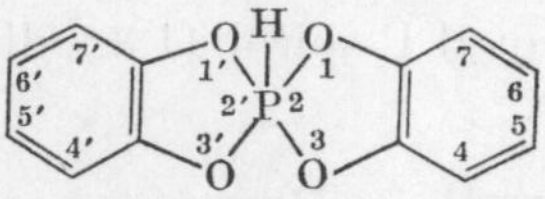

(2)

Derivatives of (1) are almost always 2-substituted products. A variety of names have been given to these derivatives. One of the best known is 2-chloro-1,3,2-benzodioxaphosphole (3), which in the past has been termed o-phenylene chlorophosphite, o-phenylene chlorophosphonite, and o-phenylene phosphorochloridite. This compound can now

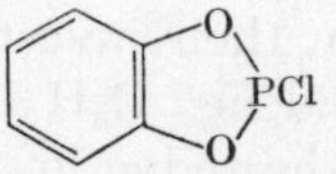

(3)

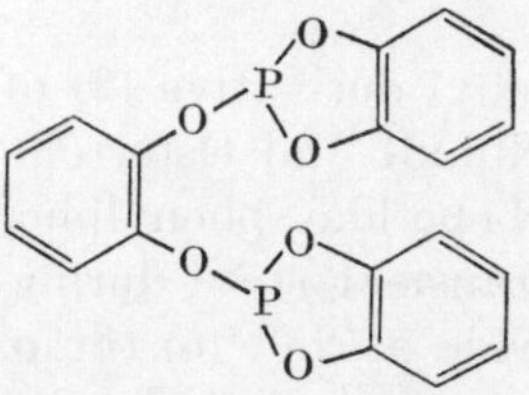

(4)

be readily prepared, but presented some difficulty to the earlier chemists. Knauer[409] showed that a mixture of pyrocatechol and phosphorus trichloride, when boiled for 12 hours and then distilled, gave the 2-chloro compound (3) and 2,2'-(o-phenylenedioxy)di-(1,3,2-benzodioxaphosphole) (4). This process was investigated in detail by L. Anschütz et al.,[410, 411] who considered that under these conditions three main reactions were involved:

(a) $C_6H_4(OH)_2 + PCl_3 \longrightarrow$ (3)

(b) $C_6H_4(OH)_2 + (3) \longrightarrow$

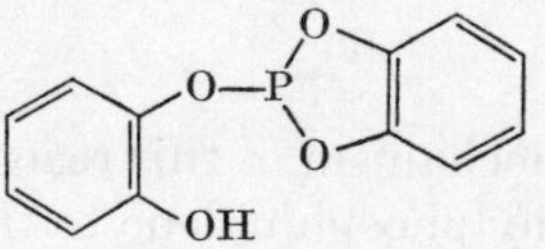

(5)

(c)   (3) + (5) $\longrightarrow$ (4)

The 2-(o-hydroxyphenoxy)-1,3,2-benzodioxaphosphole (5), b.p. 155°/12 mm, m.p. 112–113°, has been isolated as an intermediate; when pyrocatechol and phosphorus trichloride had been heated in benzene solution distillation of the crude product gave the three compounds (3), (5), and (4) in this order. A mixture of the compounds (3) and (5) when heated at 120° gave the compound (4), which in turn when heated with phosphorus trichloride in a sealed tube at 160° regenerated the compound (3).

Much later, L. Anschütz et al.[412] simplified the matter by discovering the activating effect of traces of water. Equimolecular quantities of pyrocatechol and phosphorus trichloride when dissolved in moist ether at 20–25° gave an initial deposit of the crystalline (5), which dissolved when the mixture was boiled; distillation ultimately gave the 2-chloro compound (3), b.p. 81–82°/10 mm, in 94% yield. A solution of pyrocatechol (2 equivs.) and PCl_3 (1 equiv.) in anhydrous ether at 20–25° deposited the compound (5) in 85% yield; the compound (5) when boiled with PCl_3 in moist ether for 5 hours gave the 2-chloro compound (3) in 86% yield. It was considered therefore that the PCl_3 when activated by a trace of water could convert the compound (5) into (3).

The use of benzene or xylene as a solvent, with the addition of water (0.05 mole) gave similar results, but in the absence of water the main product was the compound (4), a viscous liquid, b.p. 242–248°/14 mm, which was characterized by its ready union with sulfur to form the PP'-disulfide, m.p. 114–115°.

In the quickest and best preparation of (3), the two reactants are heated at 100° in the presence of the water but without a solvent; distillation gives the pure product (94%) (Crofts, Markes, and Rydon[413]).

The 2-alkoxy derivatives (6) were initially obtained by the action of the appropriate alkyl phosphorodichlorite, $ROPCl_2$, on pyrocatechol: the compounds (6), where $R = CH_3$, C_2H_5, n-C_3H_7, and n-C_4H_9, were thus prepared (L. Anschütz, Walbrecht, and Broeker[411]).

Arbuzov and Valitova used a laborious method for the preparation of the 2-chloro compounds (3),[414] but by treating this with the appropriate sodium alkoxide they obtained a smooth production of the above four 2-alkoxy derivatives (6) and also the members (6; $R = $ iso-C_3H_7 and iso-C_4H_9).[415] These esters form 1:1 complexes with cuprous chloride, bromide and iodide; these complexes have sharp m.p.s and are probably tetramers, having a structure similar to that of the compounds $[R_3P,CuX]_4$, where R is an alkyl group (Mann, Purdie, and Wells[416]).

(6)   (7)

The modern and simplest method for the preparation of these 2-alkoxy derivatives is exemplified by the reaction of the 2-chloro compound in ether (a) with ethanol–diethylaniline to give the 2-ethoxy derivative (**6**; $R = C_2H_5$), b.p. $97°/14$ mm (89%), and (*b*) with benzyl alcohol–diethylaniline to give the 2-benzyloxy derivative (**6**; $R = CH_2C_6H_5$), b.p. $97°/0.13$ mm (71%).[413]

The 2-phenoxy compound (**6**; $R = C_6H_5$), b.p. $150°/12$ mm, can be readily obtained by the action of phenol on the 2-chloro compound (**3**) at $100°$[411]: an earlier method, employed before (**3**) was available, was the reaction of pyrocatechol with diphenyl phosphorochloridite, $(C_6H_5O)_2PCl$, in warm benzene, which gave (**6**; $R = C_6H_5$), phenol, and hydrogen chloride (L. Anschütz and Broeker[417]). The compound (**6**; $R = C_6H_5$) readily gives a sulfide[411] (**7**), m.p. $70°$.

Interest in the chemistry of oxy-2,2'-bis(1,3,2-benzodioxaphosphole) (**8**), sometimes termed bis-*o*-phenylene pyrophosphite, was aroused by the introduction of tetraethyl pyrophosphite (**9**) as a reagent

(**8**)   (**9**)

$$(3) + (EtO)_2POH \longrightarrow \quad\quad\quad \longrightarrow (8) + (9)$$

(**10**)   (**11**)

in peptide synthesis (Anderson, Blodinger, and Welcher[418]) and its subsequent extensive use for this purpose by du Vigneaud and his school.[419] The low yield and difficult purification of (**9**) led to attempts to obtain comparable but more suitable compounds. Thus a solution of the 2-chloro compound (**3**) in petroleum was added to a similar solution of diethyl phosphite (**10**) containing triethylamine: the product on working up gave, however, the oxo compound (**8**) and the pyrophosphite (**9**), presumably by dismutation of the required intermediate compound (**11**). A similar dismutation had been earlier noted in the interaction of (**3**) with sodium diethyl phosphite, which also gave the compound (**8**) in addition to (**11**) (Arbusov and Valitova[420]). An excellent method for the preparation of (**8**) on a reasonably large scale was subsequently devised, entailing the controlled hydrolysis of the 2-chloro compound, and giving the oxo-compound (**8**), b.p. $156°/0.25$ mm, m.p. $72°$ (87%) (see Crofts, Markes, and Rydon[413] for details and apparatus).

Isomeric hexahydro compounds (**12A–B**) have been isolated by ester-interchange between triphenyl phosphite and the appropriate diol containing a catalytic quantity of the sodium derivative (p. 271)

(12A)   (12B)

(Ayres and Rydon[386]). *cis*-1,2-Cyclohexanediol (1 mole) in dioxan is treated with sodium (0.1 mole), the solvent removed from the clear solution, and the residue stirred with triphenyl phosphite (1 mole). Phenol is distilled off at 80°/0.2 mm, and the residue extracted with cold petroleum, leaving insoluble polymeric material. Distillation of the petroleum extract gives *cis*-3a,4,5,6,7,7a-hexahydro-2-phenoxy-1,3,2-benzodioxaphosphole (12A), b.p. 97°/0.004 mm, m.p. 43–45° (33%).

trans-1,2-Cyclohexanediol similarly gives the *trans*-isomer (12B), b.p. 84°/0.0045 mm. Hydrolysis of (12A) and (12B) with boiling 4N-hydrochloric acid regenerated the *cis*-diol (88%) and the *trans*-diol (91%), respectively.

Molecular models show that the size of the phosphorus atom allows this ready formation of the heterocyclic ring from the *cis*-diol, having one axial and one equatorial hydroxyl group, and from the *trans*-diol, having two equatorial hydroxyl groups, and that the cyclohexane ring in (12A) and (12B) undergoes very little distortion from the normal chair conformation. It is noteworthy that *cis*-1,2-cyclopentanediol, with favorably placed OH groups, undergoes a similar condensation with triphenyl phosphite, but *trans*-1,2-cyclopentanediol does not.

Certain compounds having the ring system (1) but with quinque-valent phosphorus can be briefly discussed.

Pyrocatechol when heated with phosphorus oxychloride gives the *PP'*-dioxide of (4) as a viscous syrup, which when heated in turn with an excess of the oxychloride gives 2-chloro-1,3,2-benzodioxaphosphole 2-oxide[409] (13), b.p. 162°/55 mm, m.p. 35°. The corresponding 2-sulfide,[410] m.p. 49–50°, is readily prepared by the addition of sulfur to (3) at 195°.

L. Anschütz,[417] in his earliest work on this ring system, showed that, when pyrocatechol (0.8 mole) was slowly added to a suspension of phosphorus pentachloride in dry benzene, the product on working up gave the 2,2,2-trichloro compound (14), b.p. 132–135°/13 mm, m.p.

(13)   (14)   (15)

61–62° (55%), and 2-chloro-2,2'-spirobi[1,3,2-benzodioxaphosphole] (15), b.p. 194°/11 mm, m.p. 166–168° (28%). This compound undoubtedly arose by the action of (14) on unchanged pyrocatechol, for he also prepared (15) directly from these two compounds.

Gross and Gloede [421, 422] have noted three practical methods for the preparation of the trichloro compound (14): (a) the action of phosphorus pentachloride on pyrocatechol (1:1 moles) in boiling benzene (69%); (b) the similar action of the pentachloride on the 2-chloro 2-oxide (13) (88%); and (c) the action of chlorine on the 2-chloro compound (3) in chilled carbon tetrachloride solution followed by distillation (94%).

The trichloro compound (14) is a useful reagent. When heated with anhydrous oxalic acid in boiling benzene, CO, CO_2, and HCl are evolved, and the residue on distillation gives the 2-chloro 2-oxide (13). When methanol (3 moles) is added to (14), both in ethereal solution, hydrogen chloride and methyl chloride are liberated, and distillation then gives

(16)   (17)

the 2-methoxy 2-oxide (16), alternatively named methyl o-phenylene phosphate, a viscous liquid, b.p. 148°/11 mm; the 2-ethoxy analogue is similarly prepared.[417]

A mixture of (14) and the 2-chloro 2-oxide (13) when heated at high temperature gives phosphorus oxychloride and the spirocyclic compound (15).[422] The latter is, however, best prepared by warming the trichloro compound (14) and 2-ethoxy-1,3-benzodioxole (17), [RRI 1236], in petroleum solution at 30–40° for 3 hours; distillation gives dichloromethyl ethyl ether (55%) and the compound (15) (87%).[421]

This spirocyclic compound (15) is dimeric: it usually forms needles, m.p. 166–168°, but a second form, melting with decomposition at 180–210°, can be obtained; both forms have been fully analyzed and their molecular weights determined.[417]

Six-membered Ring Systems containing only Carbon, One Phosphorus Atom, and One Oxygen Atom

2H-1,2-Oxaphosphorin and 1,2-Oxaphosphorinanes

2H-1,2-Oxaphosphorin, [RRI 251], has the structure and numbering shown in (1), but the known chemistry of this ring system is almost entirely that of the 3,4,5,6-tetrahydro derivative, named 1,2-oxaphosphorinane (2).

$$HC_6^{\;\;O_1\;\;_2}PH$$
$$HC^5{\;}_{4\;}^{\;3}CH$$
$$\underset{H}{C}$$
(1)

$$H_2C^{\;O}PH$$
$$H_2C{\;\;\;\;}CH_2$$
$$\underset{H_2}{C}$$
(2)

The syntheses of derivatives of (2) are almost identical with those of the corresponding derivatives of 1,2-oxaphospholanes (p. 257) and need therefore be only briefly enumerated.

(a) Sodium hypophosphite, NaH_2PO_2, condenses with 3-buten-1-ol (3) in methanolic solution containing a catalytic quantity of sec-butylidenebis(tert-butyl peroxide) (p. 257), when heated at 130° for 1 hour,

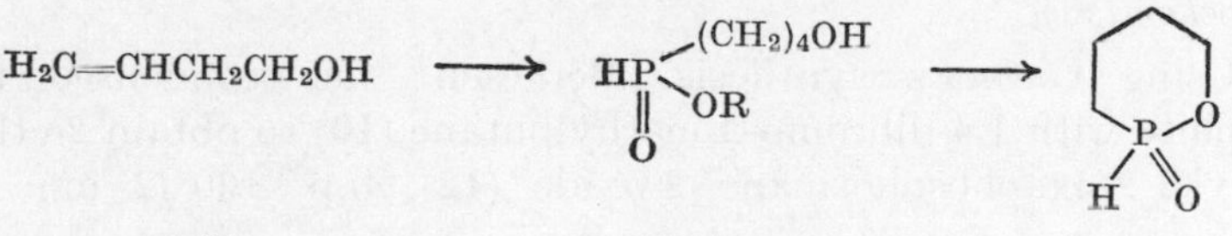

$$H_2C\!\!=\!\!CHCH_2CH_2OH \longrightarrow HP\!\!\begin{array}{c}(CH_2)_4OH\\\\OR\end{array}\!\!\underset{O}{\overset{\parallel}{}} \longrightarrow$$

(3)   (4)   (5)

to give the salt (4; R = Na); acidification gives the acid (4; R = H), which when heated at 100–150° undergoes cyclic dehydration to 1,2-oxaphosphorinane 2-oxide (5) (Smith[349]).

(b) The diethyl ester of an alkylphosphonous acid when heated with 1,4-dibromobutane should give finally the 2-ethyl-1,2-oxaphosphorinane 2-oxide (cf. p. 258). The example cited by Garner[350] is the condensation of silver ethyl isobutylphosphonite with 1,4-dichlorobutane to form 2-isobutyl-1,2-oxaphosphorinane 2-oxide (6), b.p. 120°/1 mm.

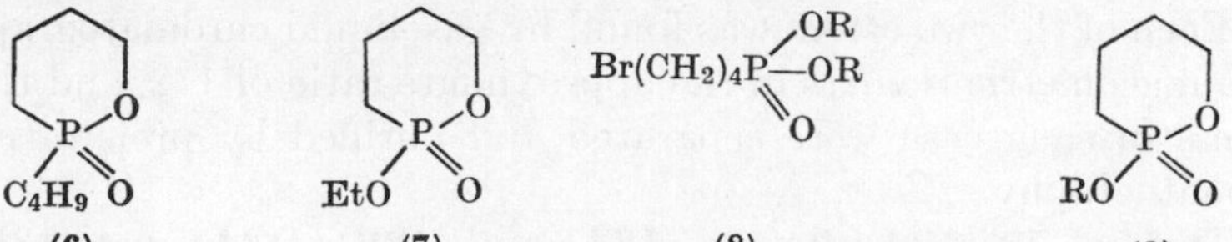

$$\underset{C_4H_9}{}\overset{}{}P\overset{O}{\underset{O}{}}$$
(6)

$$\underset{EtO}{}P\overset{O}{\underset{O}{}}$$
(7)

$$Br(CH_2)_4P\!\!\begin{array}{c}OR\\\\OR\end{array}\!\!\underset{O}{\overset{}{}}$$
(8)

$$\underset{RO}{}P\overset{O}{\underset{O}{}}$$
(9)

Triethyl phosphite and 1,4-dibromobutane when similarly heated loses ethyl bromide with the final formation of 2-ethoxy-1,2-oxaphosphorinane 2-oxide (7), b.p. ca. 100°/1 mm (Garner[351]). This ester was earlier termed ethyl butyl phostonate, a correct version of which is ethyl tetramethylenephostonate (cf. p. 259).

(c) Eberhard and Westheimer[352] have shown that diethyl 4-bromobutylphosphonate (8; $R = C_2H_5$), when heated at 105–110°/0.7 mm, gives the ethoxy ester (7), b.p. 56°/0.08 mm. The ester (8; $R = C_2H_5$) when hydrolyzed with 48% hydrobromic acid gives the phosphonic acid (8; $R = H$), m.p. 126.5–127.8°, the lithium salt of which readily cyclizes to the lithium salt (9; $R = Li$) of 2-hydroxy-1,2-oxaphosphorinane 2-oxide, which in turn furnishes the free acid (9; $R = H$), m.p. 104.5–106°.

It is noteworthy that the compound (6) can be regarded as the internal ester of (4-hydroxybutyl)isobutylphosphinic acid, and the compound (9; $R = H$) as a similar ester of (4-hydroxybutyl)phosphonic acid.

The hydrolysis of the ester (7) proceeds far more slowly than that of its five-membered analogue (cf. p. 278).

Stereochemistry

Using Garner's synthesis, Bergesen[5] has condensed triethyl phosphite with 1,4-dibromo-2-methylbutane (10) to obtain 2-ethoxy-5-methyl-1,2-oxaphosphorinane 2-oxide (12), b.p. 130°/2 mm (40%),

$$BrCH_2CH(CH_3)CH_2CH_2Br$$

(10)

$$BrCH(CH_3)CH_2CH_2CH_2Br$$

(11)

(12)   (13)

and with 1,4-dibromopentane (11) to obtain the 6-methyl isomer (13), b.p. 95°/0.5 mm (30%). It was assumed that the least hindered bromine atom in (10) and (11) would make the initial union with the phosphorus atom; the identities of the two esters were confirmed by infrared and pmr spectra.

Each of the two esters was found by gas–liquid chromatography to contain geometric isomers in the approximate ratio of 1:2, and the two isomers in each case were separated and purified by preparative gas chromatography.

The two pairs of isomers, (12A) and (12B), (13A) and (13B) (cf. Table 25) have probably the chair conformation, and on the basis of the uncertain application of the Auwers–Skita rule to such compounds, the *cis*-isomer in the 1,4-series, and the *trans*-isomer in the 1,3-series, should have the higher refractive index and the higher density. The probable conformations are shown in the Table. *cis*-Isomers usually have a shorter retention time than *trans*-isomers, and these values agree

Table 25. Properties of the pairs of stereoisomeric esters (12) and (13)

Ester	n_D^{20}	d_4^{20}	P=O (cm^{-1})	Retention time (min) at 185°	Probable conformation
(12A)	1.4597	1.148	1240	31.5	cis
(12B)	1.4541	1.134	1250, 1275	42.5	trans
(13A)	1.4492	1.095	1240, 1255	27.5	cis
(13B)	1.4525	1.106	1250	46.0	trans

with the allocated conformations. The structural relationship between the two isomers of either ester, i.e., whether for example the compound termed cis has the methyl group cis to the OC$_2$H$_5$ group or to the =O atom, has not been established.

An unusual approach to the 1,2-oxaphosphorinane system has recently (1967) been very briefly recorded (Cummins[423]). When air is passed through a benzene solution of styrene and phosphorus at 30° for 160–165 hours, three products are deposited and can be filtered off. One of these products is termed a 'phosphorate', of undetermined structure, whereas the two others are cyclic phosphorus derivatives. When the mixture is dissolved in aqueous sodium hydroxide and the

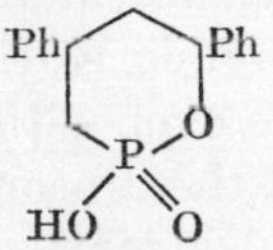

(14) 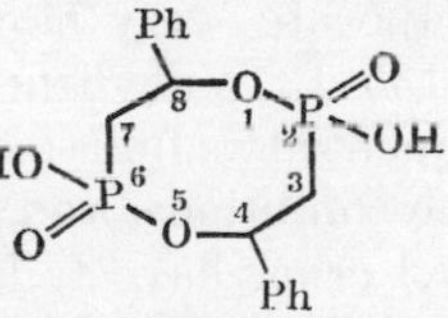   (15)

solution then acidified, the hydrolytic products of the phosphorate remain in solution, and the cyclic derivatives are precipitated. It is claimed that one of these is 2-hydroxy-4,6-diphenyl-1,2-oxaphosphorinane 2-oxide (14), and that the other has the structure (15), i.e., it is 2,6-dihydroxy-4,8-diphenyl-1,5,2,6-dioxadiphosphocan 2,6-dioxide, [RIS —]. Compound (14), m.p. 131–133°, can be sublimed at 220°/2 mm, whereas (15) decomposes under these conditions. These two products have been characterized by elemental analysis, molecular-weight determination, potentiometric titration, and by infrared and nmr analysis, but no details are recorded.

The relative proportions of (14) and (15) are determined by the initial ratio of styrene/phosphorus; the higher this proportion, the higher is that of the compound (14).

Similar pairs of compounds are obtained when cyclohexene, α-pinene, or 1-octene is used instead of styrene.

The interest of these compounds and their analogues calls for a fuller investigation of their properties and structure.

4H-1,4-Oxaphosphorin and 1,4-Oxaphosphorinanes

This compound (1), [RRI 252], is unknown: its tetrahydro derivative (2), named 1,4-oxaphosphorinane, is recorded only as the 4-phenyl-1,4-oxaphosphorinane (3).

```
     O                     O                      O
 HC6 1 2CH           H2C      CH2            H2C      CH2
  ‖  4 3 ‖            |        |             |        |
 HC5    CH          H2C      CH2            H2C      CH2
     P                     P                      P
     H                     H                      Ph

    (1)                   (2)                    (3)
```

Lecoq[424] has claimed to have prepared the compound (3) by the interaction of phenylphosphinylidenebis(magnesium bromide), usually termed phenylphosphinebis(magnesium bromide), $C_6H_5P(MgBr)_2$, and 2,2'-diiodoethyl ether in boiling benzene solution. The crude black product was extracted with ethanol to remove unchanged diiodoethyl ether; the undissolved residue was dissolved in ether, and the solution allowed to evaporate 'very slowly' in the open air, depositing white crystals, m.p. 135–137°, identified as the 4-phenyl derivative (3).

All subsequent attempts to repeat this work, even with considerable variation of the conditions, and with the use of the dibromo- as well as the diiodo-ethyl ether have failed (Mann and Millar[316]). The reaction gave a pale amber resin, which could be distilled under nitrogen at ca. 300°/0.5 mm, but the distillate rapidly re-formed the resin. All attempts to crystallize the resin, or to obtain crystalline derivatives, failed.

A similar attempt to prepare the analogous 4-phenyl-1,4-oxarsenane (p. 554), using phenylarsinylidenebis(magnesium bromide), gave a viscous residue which however when heated in nitrogen at 0.1 mm underwent progressive decomposition with the formation of the required oxarsenane, a mobile liquid, b.p. 149–151°/18 mm, which gave no indication of crystallization.[316] In both types of experiment the viscous major product was probably formed by extensive linear condensation of the main reactants.

A compound such as (3) would very probably be a liquid at room temperature, and being essentially a dialkylarylphosphine would oxidize readily on exposure to air, particularly on slow evaporation of its ethereal solution. It is possible that Lecoq's compound, in spite of the analytical evidence, was 4-phenyl-1,4-oxaphosphorinane 4-oxide, but the interpretation of his results remains uncertain.

The incorrect identification of the addition product of phenacylidenetriphenylphosphorane and dimethyl acetylenedicarboxylate as dimethyl 4,4,4,6-tetraphenyl-4H-1,4-oxaphosphorin-2,3-dicarboxylate has already been noted (p. 58).

Phenoxaphosphine

Phenoxaphosphine, [RRI 3403], is the recognized name for the compound (1), and the systematic name, 10H-dibenz[1,4]oxaphosphorin,

(1)   (2)   (3)

is rarely used. As in so many cyclic phosphorus systems, most reactions occur at the phosphorus atom, and it is convenient to invert the formula (1), as in (3).

In the first synthetic route to the phenoxaphosphine system (Mann and Millar[316]), 2,2'-dibromodiphenyl ether (2; R = Br) was treated in benzene–petroleum solution with n-butyllithium to form the 2,2'-dilithio derivative (2; R = Li), which was treated with phenylphosphonous dichloride. The reaction mixture was boiled, cooled, and hydrolyzed with water; the dried organic layer on distillation gave 10-phenylphenoxaphosphine (3), b.p. 191–193°/0.5 mm, m.p. 94.5–95° (63%), and a resinous solid residue. The initial stages of this synthesis have been shortened by the subsequent demonstration (Oita and Gilman[425]) that diphenyl ether undergoes direct lithiation to (2; R = Li).

The phosphine (3) is (as expected) a very weak base, dissolving in hot concentrated hydrochloric acid but crystallizing unchanged on cooling. It reacts as a normal tertiary phosphine, giving a methiodide, m.p. 236–237°, systematically termed 10-methyl-10-phenylphenoxaphosphonium iodide, with hydrogen peroxide in aqueous acetone forming a 10-oxide (4), m.p. 173–174°, and with anhydrous 'Chloramine-T', p-CH$_3$C$_6$H$_4$NSO$_2$NNaCl, in boiling ethanol giving the 10-(p-toluenesulphonyl)imino derivative (5), m.p. 164–164.5°.

The heterocyclic ring in (3) has great stability, for the compound is unaffected when its solutions in concentrated hydrochloric, hydrobromic, and hydriodic acids are boiled, or when it is heated with hydrobromic acid in a sealed tube at 150–155° for 4 hours. Moreover, a mixture

Ph O   Ph NSO₂C₆H₄CH₃   Ph Ph
(4)   (5)   (6)

of the phosphine, bromobenzene, and aluminum chloride, when heated at 210° for 2 hours, extracted with boiling water, and treated with potassium iodide, deposits 10,10-diphenylphenoxaphosphonium iodide (6), m.p. 195–197°. These are the conditions under which triphenyl-phosphine is converted into tetraphenylphosphonium iodide (Chatt and Mann [315]) and their rather vigorous nature has apparently no disruptive effect on the phenoxaphosphine system.

The phosphine (3) readily forms metallic co-ordination compounds, such as the colorless iodo(phosphine)gold, $C_{18}H_{13}OP,AuI$, m.p. 200°, the yellow dibromobis(phosphine)palladium, $(C_{18}H_{13}OP)_2PdBr_2$, m.p. 298–302°, and the reddish-brown bridged dibromobis-(phosphine)-$\mu\mu$-di-bromodipalladium, $(C_{18}H_{13}OP)_2(PdBr)_2$, m.p. 279–281° (cf. also p. 295).

The ultraviolet spectrum of the phenoxaphosphine (3) has λ_{max} 294 mμ (ϵ 4300), and λ_{min} 275 mμ (ϵ 3000), and that of the corresponding methiodide has λ_{max} 303 and 294 mμ (ϵ 7700, 7100) and λ_{min} 297 and 258 mμ (ϵ 6980, 1800). It is noteworthy that the spectrum of (3) bears a strong general resemblance to those of triphenylphosphine and 10,11-dihydro-5-phenyl-5H-dibenzo[b,f]phosphepin (7) (discussed with those of the arsenic analogue, p. 436), and that of the methiodide bears a similar resemblance to those of the corresponding salts of the last two phosphines, the chief difference being that the principal band of the phenoxaphosphine (3) and its methiodide shows a marked displacement to higher wavelengths compared with the corresponding bands of triphenylphosphine, the phosphine (7) (p.153), and their methiodides.

It should be noted that if the phenoxaphosphine possessed three independently absorbing rings, in accordance with the formal structure (3), several absorption bands, with the main absorption at about

Ph   Ph   Ph
(7)   (8)   (9)

255 mμ, would be expected. The structure of the phosphine (**3**) may, however, involve resonance with numerous polar forms of type (**8**), precisely as triphenylphosphine and the phosphine (**7**) may, but in addition it may show further resonance with similar forms of type (**9**). This increased resonance may be responsible, not only for the absence of benzenoid bands, but also for the general shift to higher wavelengths. In the methiodide of the phenoxaphosphine (**3**) the polar forms of type (**8**) are necessarily suppressed, but those of type (**9**) may still operate. Possibly it is this factor which allows the emergence of two bands at 294 and 303 mμ (which may represent the normal absorption of the benzene rings considerably displaced) but which are merged into one broad band in the more highly resonating parent phosphine (**3**).

A second synthetic route to the phenoxaphosphine system was sought in the production of 10-chlorophenoxaphosphine (**10**; R = R′ = H) and, by hydrolysis and oxidation, of phenoxaphosphinic acid, alternatively named 10-hydroxyphenoxaphosphine 10-oxide (**11**; R = R′ = H).

(**10**)   (**11**)   (**12**)

The direct approach is blocked by a striking difference in the chemistry of phosphorus and arsenic, namely, that arsenic trichloride reacts readily with diphenyl ether in the presence of aluminum chloride to give 10-chlorophenoxarsine (p. 556), whereas phosphorus trichloride reacts almost exclusively at the 4-position of the ether under these conditions, the reaction mixture on hydrolysis and oxidation giving p-phenoxyphenylphosphonic acid, $C_6H_5O—C_6H_4P(O)(OH)_2$ (Davies and Morris[426]).

It has recently been shown, however, that when a mixture of diphenyl ether and phosphorus trichloride containing a high proportion of aluminum chloride is boiled for 4 hours and the excess of trichloride is removed, the residue on being mixed with pyridine and distilled gives a crude fraction, b.p. 135–150°/0.005 mm, which on hydrolysis and oxidation gives the acid (**11**; R = R′ = H) in 2% yield (Levy, Freedman, and Doak[427]).

All attempts to achieve the cyclic dehydration of o-phenoxyphenylphosphonic acid (**12**) failed (Freedman, Doak, and Edmisten[428]). These workers therefore treated di-p-tolyl ether with aluminum chloride and phosphorus trichloride: attack at the 4-position was thus prevented, and cyclization occurred to form 10-chloro-2,8-dimethyl-phenoxaphosphine (**10**; $R = R' = CH_3$), which however underwent oxidation during the reaction or the subsequent hydrolysis, giving 2,8-dimethylphenoxaphosphinic acid (**11**; $R = R' = CH_3$), m.p. above 300° (73%).

The phenoxaphosphine system in the acid (**11**; $R = R' = CH_3$) also showed great stability, and permanganate oxidation gave 2,8-dicarboxy-phenoxaphosphinic acid (**11**; $R = R' = CO_2H$), m.p. above 300° (56%): attempts to decarboxylate this acid left it unaffected.

The acid (**11**; $R = R' = CH_3$), when heated with diphenylsilane, $(C_6H_5)_2SiH_2$ (1.6 moles) at 250° for 3 hours under nitrogen, is however reduced to the secondary phosphine, i.e., 2,8-dimethylphenoxaphosphine, b.p. 128°/0.25 mm, m.p. 56° (85%) (Fritzsche, Hasserodt, and Korte[69]).

The acid (**11**; $R = R' = CH_3$) when treated with fuming nitric acid at room temperature gave the 2,8-dimethyl-4,6-dinitro acid; the allocation of the nitro groups was based on the factors (a) the 4,6-positions are those most activated by the ring oxygen atoms and least deactivated by the $=P(O)OH$ group, (b) diphenyl ether undergoes nitration solely in the positions ortho to the oxygen atom, and (c) the spectral confirmation.

The acid (**11**; $R = R' = CO_2H$) underwent nitration similarly to give the 2,8-dicarboxy-4,6-dinitro acid. Both the nitrated acids melted above 300°.

The third synthetic approach was the application of the Doak–Freedman[132, 133] reaction (pp. 83, 84) to o-phenoxybenzenediazonium fluoroborate (**13**; $R = H$). A suspension of this salt in ethyl acetate

(13)   (14)

containing a small quantity of cuprous bromide was treated with phosphorus trichloride; after completion of the reaction, powdered aluminum (the reducing agent of Quin and Montgomery[137]) was added, and the mixture was boiled under reflux and then distilled. The highest-boiling fraction was 10-chlorophenoxaphosphine (**10**; $R = R' = H$), m.p. 62–64° (24%), which on hydrogen peroxide oxidation in alkaline solution gave the phenoxaphosphinic acid (11; R = R′ = H), m.p. 231–234° (Doak, Freedman and Levy[138]).

The ultraviolet spectra of this acid, of phenoxarsinic acid (p. 557), of 5,10-dihydrodibenzo[b,e]phosphorin 5-oxide (p. 153), and of diphenylphosphinic acid have been recorded and discussed[138]; the spectra of the first two acids are virtually identical.

The above synthesis was repeated, but using phenylphosphonous dichloride instead of phosphorus trichloride in the reaction with (13; R = H). Distillation gave 10-phenylphenoxaphosphine (3), b.p. 150°/ca. 0.005 mm, m.p. 97.5–98.0° (26%); oxidation as before gave the 10-oxide (4), m.p. 176–179°. The use of p-tolylphosphonous dichloride similarly gave 10-p-tolylphenoxaphosphine (14), m.p. 59–62° (19%), which gave the 10-oxide, m.p. 195–197°; both these 10-oxides showed strong absorption at 1200 cm^{-1} characteristic of the P=O group. The low yield of (14) was caused by incomplete cyclization, for the mother-liquors from its crystallization yielded after oxidation (o-phenoxyphenyl)-p-tolylphosphinic acid (Levy, Doak, and Freedman[141]).

To obtain nuclear-substituted derivatives, 5-chloro-2-phenoxy-benzenediazonium tetrafluoroborate (13; R = Cl) was treated as above with phosphorus trichloride to give 2,10-dichlorophenoxaphosphine (10; R = Cl, R′ = H), b.p. 135°/ca. 0.005 mm, m.p. 120–122°, which in turn gave 2-chlorophenoxaphosphinic acid (11; R = Cl, R′ = H), m.p. 240–242°. The interaction of phenylphosphonous dichloride and (13; R = Cl) similarly gave 2-chloro-10-phenylphenoxaphosphine, b.p. 160°/ca. 0.005 mm, m.p. 41.5–42.0° (methiodide, m.p. 238.0–240.5°; benzyl chloride quaternary salt, m.p. above 300°), which gave the 10-oxide, m.p. 150–151°.

The ethiodide, m.p. 255–257°, and the benzyl chloride quaternary salt, m.p. above 300°, of the 10-phenyl compound (3) are also recorded.[427]

Infrared Spectra

Levy, Freedman, and Doak[427] record that the spectra of twelve phenoxaphosphine derivatives having 'quadruply connected phosphorus' (*i.e.*, phenoxaphosphinic acids, 10-oxides, and quaternary salts) all show three characteristic bands of medium (m.) to very strong (v.s.) intensity. The first band (m.–s.) appears at 1330–1320 cm^{-1}, the second (s.–v.s.) at 1277–1268 cm^{-1}; and the third (m.–s.) at 1223–1215 cm^{-1}. This set of bands has not been observed in closely related non-heterocyclic organophosphorus compounds or in trivalent phenoxaphosphine derivatives.

The conformation of a 10-substituted phenoxaphosphine such as (3) is at present unknown, but it is almost certainly closely similar to that of a 10-substituted phenoxarsine, the considerable evidence for which is discussed later (pp. 562 *et seq.*). In brief, the normal intervalency angle of a phosphorus atom in a symmetrical compound of type PR_3 is *ca.* 98°; that of the oxygen atom in cyclic compounds such as dioxan and trioxymethylene is *ca.* 112°. These angles, if retained in the phenoxaphosphine ring system, might cause the molecule to be folded about the O–P axis, although the degree of folding would presumably be less than that in the 5,10-dihydrophosphanthrene ring system (p. 186) and the corresponding arsanthrene system (p. 466). If this folding did occur, the compound (3) could exist in two geometrically isomeric forms, in one of which the 10-phenyl group would be projecting within the angle subtended by the two *o*-phenylene groups, and the second in which this group would project behind this angle (*cf.* p. 564, diagram 29, for the arsenic equivalent). No evidence of such isomerism has been obtained, possibly because one form is markedly more stable than the other, or because a rapid oscillation between the two forms may occur, or because the conformation has at present received only meagre experimental investigation.

It is noteworthy that 8-chloro-10-phenylphenoxaphosphine-2-carboxylic acid 10-oxide, m.p. 328–330°, and 2-chloro-8-methylphenoxaphosphine-10-propionic acid 10-oxide, m.p. 211°, have been synthesized (Campbell[429]), but neither acid gave evidence of optical resolution by fractional recrystallization of its salts with various optically active amines.

Recent work (Allen and Millar[221]) has shown that 2,2′-dilithiodiphenyl ether (2; R = Li) reacts with methylphosphonous dichloride to give 10-methylphenoxaphosphine (15), b.p. 95°/0.05 mm, which

(15)   (16)   (17)

forms a methiodide (16; R = CH$_3$), m.p. 298–301°, and with diiodomethane forms the iodomethiodide (16; R = CH$_2$I), m.p. 241°. Nmr studies of the methiodide dissolved in deuteriochloroform or trifluoroacetic acid show the equivalence of the two methyl groups, even (in DCCl$_3$) at −55°. This indicates that the tricyclic system in (16) is effectively planar, as would be expected in a simple quaternary salt of this type.

Alkaline hydrolysis of the iodomethiodide (**16**; $R = CH_2I$) served solely to remove the iodomethyl group with the formation of the 10-oxide (**17**), m.p. 32–33°, identical with that obtained by direct oxidation of the phosphine (**15**). No indication of expansion of the heterocyclic ring, comparable to that undergone by 5-iodomethyl-5-phenyl-5H-dibenzophospholium iodide under similar conditions (p. 150), was observed.

The infrared spectrum of the oxide (**17**) shows a set of bands at 1320, 1275, and 1222 cm^{-1}, in harmony with the observations of Levy et al.[427]

10-Methylphenoxaphosphine (**15**) reacts with nickel(II) chloride to give the 5-co-ordinate complex, $(C_{13}H_{11}OP)_3NiCl_2$, whereas the 10-phenyl analogue (**3**) gives only the normal 4-co-ordinate complex, $(C_{18}H_{13}OP)_2NiCl_2$ (Allen, Millar and Mann[128]); in this respect these phosphines react similarly to the 5-methyl- and 5-phenyl-5H-dibenzophospholes (p. 80).

Six-membered Ring Systems containing Carbon, One Phosphorus, and Two Oxygen Atoms

4H-1,3,2-Dioxaphosphorin and 1,3,2-Dioxaphosphorinanes

4H-1,3,2-Dioxaphosphorin (**1**), [RRI 218], is encountered almost invariably as its 5,6-dihydro derivative (**2**), termed 1,3,2-dioxaphosphorinane: this compound was formerly called 1,3-dioxa-2-phosphacyclohexane.

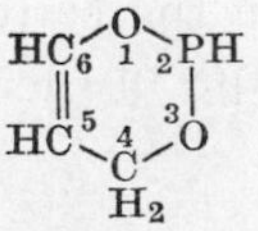

(**1**) 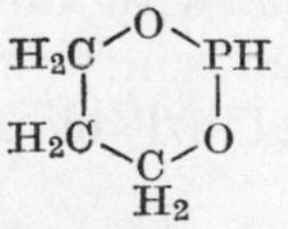   (**2**)

The synthesis of several types of derivatives of (**2**) consists essentially in the condensation of simple phosphorus compounds with 1,3-propanediol or (more frequently) with 2,2-dimethyl-1,3-propanediol, alternatively named neopentanediol, $(CH_3)_2C(CH_2OH)_2$. These syntheses are therefore often very similar to those of 1,3,2-dioxaphospholane (p. 269).

1,3-Propanediol reacts with phosphorus trichloride in dichloromethane solution to give 2-chloro-1,3,2-dioxaphosphorinane (**3**), b.p. 67°/15 mm, which with absolute ethanol gives the 2-ethoxy compound (**4**; $R = C_2H_5$), b.p. 77°/25 mm, and with diethylamine (2 equivalents) gives the 2-diethylamino compound (5), b.p. 102°/25 mm (Lucas, Mitchell, and Scully[384]). The 2-phenoxy compound (4; $R = C_6H_5$), b.p. 88°/0.4 mm, m.p. 44–46°, can be readily obtained by transesterification between the diol and triphenyl phosphite (Ayres and Rydon[386]) (cf. p. 271). The corresponding 2-phenoxy 2-oxide, m.p. 75°, can be obtained by the action of $C_6H_5OP(O)Cl_2$ on the diol dissolved in dry acetone containing pyridine (Meston,[430] who termed it phenyl trimethylene phosphate). For a discussion of the ^{31}P-nmr spectra of (4; $R = C_2H_5$) and the corresponding 2-oxide, of the 4-methyl derivative of each compound, and of the above 2-phenoxy 2-oxide see Blackburn, Cohen, and Todd.[431]

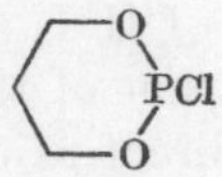

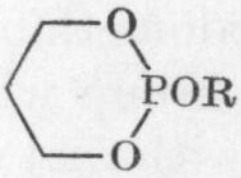

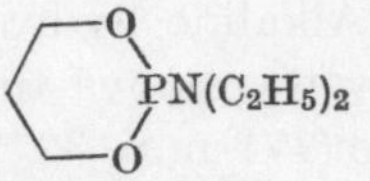

Neopentanediol similarly condenses with phosphorus trichloride to give 2-chloro-5,5-dimethyl-1,3,2-dioxaphosphorinane (6), b.p. 66°/13

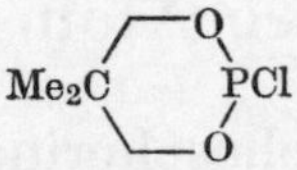

mm. The compound (6) when treated in cold ether with 1 equivalent each of an alcohol and pyridine gives the corresponding 2-alkoxy derivative (7); these are liquids, but the lower members are characterized by their ready union with sulfur to give the corresponding 2-alkoxy 2-sulfides (8) (Edmundson[397, 432]). The b.p.s of the compounds (7) and m.p.s of (8) are listed in Table 26.

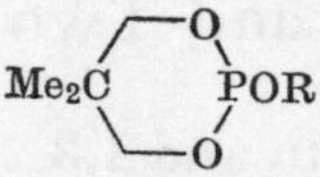

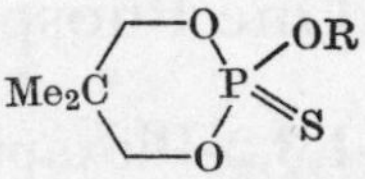

Table 26. 2-Alkoxy-5,5-dimethyl-1,3,2-dioxaphosphorinanes (7) and their sulfides (8)

R =	CH_3	C_2H_5	$n\text{-}C_3H_7$	iso-C_3H_7	$n\text{-}C_4H_9$
(7) B.p. (°/mm)	66/18	76–77/16	85/15	74/15	98/14
(8) M.p. (°)	93.5–94.5	62–63	(b.p. 90–94°/0.25 mm)	—	—

It is noteworthy that when the 2-ethoxy compound (7; $R = C_2H_5$) is treated at room temperature with methyl iodide, an Arbuzov rearrangement occurs with eviction of the ethoxy group and the formation of the 2-methyl 2-oxide (9), m.p. 119–121°.[432] 2-Ethoxy-1,3,2-dioxa-phospholane when similarly treated undergoes cleavage of the ring system.

2-Chloro-5,5-dimethyl-1,3,2-dioxaphosphorinane 2-oxide (10), m.p. 105–106°, can be prepared by passing dry oxygen through a benzene solution of (6),[397] or by the reaction of the diol with phosphorus oxychloride in benzene containing pyridine (McConnell and Coover[433]); the yields are 20% and 62%, respectively. It is a compound of rather low

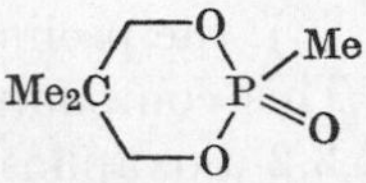

(9) 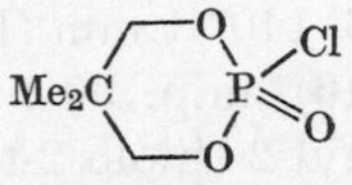   (10) 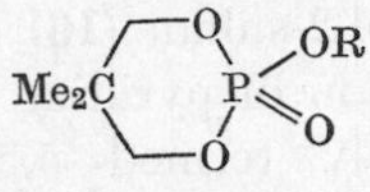   (11)

reactivity, but when treated in benzene solution with sodium methoxide it yields the 2-methoxy 2-oxide (11; R = CH₃), m.p. 94°, and with sodium isopropoxide yields the 2-isopropoxy compound (11; R = C₃H₇), m.p. 57–58°. The 2-ethoxy compound (11; R = C₂H₅) has been prepared by the reaction of ethyl phosphorodichloridate, C₂H₅OP(O)Cl₂, on the diol in pyridine–ether, and the n-butoxy compound is similarly prepared.[432]

In addition to PCl₃, POCl₃, and ROP(O)Cl₂, the following phosphorus compounds react with neopentanediol to give the 5,5-dimethyl-1,3,2-dioxaphosphorinane system:

(a) PCl₃ + C₂H₅OH. When PCl₃ is added to a cooled equimolar mixture of the diol and ethanol, the reaction mixture on distillation gives 5,5-dimethyl-1,3,2-dioxaphosphorinane 2-oxide (12), m.p. 48–50°: this probably arises from the intermediary 2-ethoxy compound (7; R – C₂H₅), which reacts with the hydrogen chloride to form ethyl chloride and (12). The use of isobutyl alcohol in place of ethanol also gives the compound (12).[433] A benzene solution of (12) containing triethylamine, when treated with sulfur, deposits the 2-hydroxy 2-sulfide (13), as the triethylamine salt, m.p. 70.5–72.5° (Edmundson[434]).

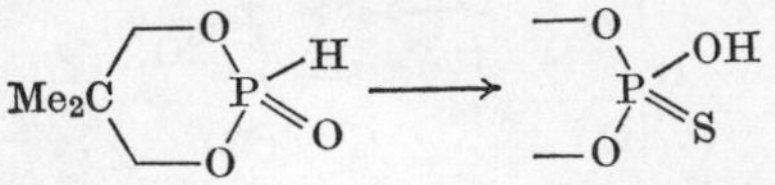

(12)   (13) 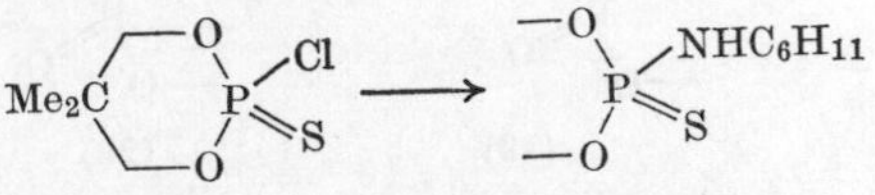   (14)   (15)

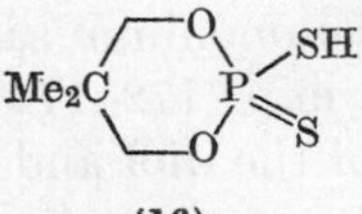

(16) 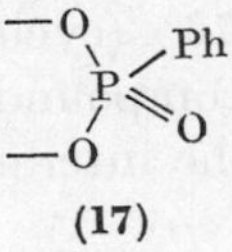   (17) 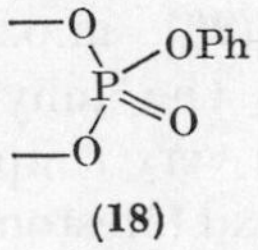   (18)

(*b*) SPCl$_3$. A solution of the diol in benzene containing pyridine, when treated with phosphorus thiochloride and warmed to 50–70°, gives 2-chloro-5,5-dimethyl-1,3,2-dioxaphosphorinane 2-sulfide (**14**), m.p. 90–91.5°, termed by Edmundson[432] 2-chloro-5,5-dimethyl-2-thiono-1,3,2-dioxaphosphorinane. When warmed with cyclohexylamine, it gives the 2-cyclohexylamino 2-sulfide (**15**), m.p. 98–99°.[434]

(*c*) P$_2$S$_5$. When phosphorus pentasulfide is added over a period of 5 minutes to a cold benzene solution of the diol, which is then heated at 70–85° for 1.25 hours, distillation of the reaction mixture gives the 2-thiol 2-sulfide (**16**), b.p. 126–140°/1 mm (Bartlett[435]); the preliminary addition of pyridine gives (**16**), m.p. 90–91.5°.[434] This compound was initially termed 5,5-dimethyl-2-thiolo-2-thiono-1,3,2-dioxaphosphorinane.[434]

(*d*) C$_6$H$_5$P(O)Cl$_2$. When an equimolar mixture of the diol and phenylphosphonic chloride is warmed at 25° under reduced pressure to remove all hydrogen chloride produced, the residue on distillation gives the 2-phenyl 2-oxide (**17**), b.p. 212–214°/7.5 mm (Toy[395]). The reaction when carried out in dry dioxane at 25° gives (**17**), m.p. 103–105°,[433] 108.5–110°.[432]

(*e*) C$_6$H$_5$OP(O)Cl$_2$. A mixture of phenol, phosphorus oxychloride, and anhydrous magnesium chloride is heated to form the crude phenyl phosphorodichloridate, to which the diol is added. After 5 hours at room temperature and 3 hours at 100°, the product yields the 2-phenoxy 2-oxide (**18**), m.p. 136–137°.[433] The reaction of phenol on the 2-chloro 2-oxide (**10**) in the presence of pyridine–toluene gave a low yield of (**18**) with by-product formation.

Two further types of derivative prepared from the 2-chloro 2-oxide (**10**) should be mentioned. A warm benzene solution of (**10**) containing triethylamine, when treated with cyclohexylamine, gives the 2-cyclohexylamino 2-oxide (**19**), m.p. 236–239°. The compound (**10**) undergoes

(**19**)   (**20**)   (**21**)   (**22**)

rapid hydrolysis in boiling 50% aqueous acetone to the 2-hydroxy 2-oxide (**20**), m.p. 174–176° (cyclohexylammonium salt, m.p. 240–245°).[434] The anhydrous compound (**20**), m.p. 172–174°, can also be prepared very simply by the interaction of the diol and hot polyphosphoric acid (Meston[430]).

The infrared spectra of almost all the above neopentyl derivatives have been recorded by Edmundson, who has discussed them in detail.[432]

The so-called pyrophosphites and pyrophosphates containing two 1,3,2-dioxaphosphorinane units have some interesting features. Arbuzov, Kikonorov, *et al.*[390] have shown that the careful addition of water (1 mole) with a base such as pyridine to 2-chloro-1,3,2-dioxaphosphorinane (**3**) (2 moles) in ether at −5° gives the compound (**23**), b.p. 118–120°/2 mm, now systematically called 2,2'-oxybis-(1,3,2-dioxaphosphorinane); the

(**23**)   (**24**)

readiness with which this compound unites with sulfur and also forms complexes with cuprous halides is strong evidence for the trivalent state of the phosphorus atoms. Khorana *et al.*[436] have prepared the corresponding 2,2'-oxybis-(1,3,2-dioxaphosphorinane) 2,2'-dioxide (**24**; R = R' = H), m.p. 137–137.5°, by treating the 2-hydroxy 2-oxide (**20**) in acetonitrile with dicyclohexylcarbodiimide, C_6H_{11}—N=C=N—C_6H_{11}; the latter forms the anhydride of the 2-hydroxy 2-oxide, being itself converted into *N,N'* dicyclohexylurea, C_6H_{11}—NH—CO—NH—C_6H_{11}. These authors term this anhydride (**24**; R = R' = H) 'bis(propane-1,3-diol cyclic)-pyrophosphate'.

A series of nuclear-substituted compounds of type (**24**) have been prepared by Lanham.[437] As an example, 2-chloro-5,5-diethyl-1,3,2-dioxaphosphorinane 2-oxide is slowly added to a mixture of anhydrous sodium acetate and toluene at 100°; after 30 minutes at this temperature, the reaction mixture is chilled and extracted with water and the organic layer is dried and distilled, giving 2,2'-oxybis-(5,5-diethyl-1,3,2-dioxaphosphorinane) 2,2'-dioxide (**24**; R = R' = C_2H_5), m.p. 182–190° (81%). The 5-ethyl-5-methyl member (**24**; R = C_2H_5, R' = CH_3), m.p. 155–159°; the 5-butyl-5-ethyl member (**24**; R = C_4H_9, R' = C_2H_5); the 4-methyl; and the 4,6-dimethyl member, m.p. 160–166°; these are some of the compounds recorded in the patent specification.[437] Lanham terms the above compound (**24**; R = R' = C_2H_5) '5,5,5',5'-tetraethyl-2,2'-dioxopyro-1,3,2-dioxaphosphorinane'.

It is claimed that these compounds impart flame resistance to polyurethane foams.

Edmundson, in a study of sterically hindered pyrophosphates, has prepared a series of compounds of type (**24**; R = R′ = CH₃) and of their thio analogues.[434, 438]

2,2′-Oxybis-(5,5-dimethyl-1,3,2-dioxaphosphorinane) 2,2′-dioxide (**25**), m.p. 193–195°, was prepared by the action of dicyclohexylcarbodiimide on the 2-hydroxy 2-oxide (**20**): Edmundson terms this compound

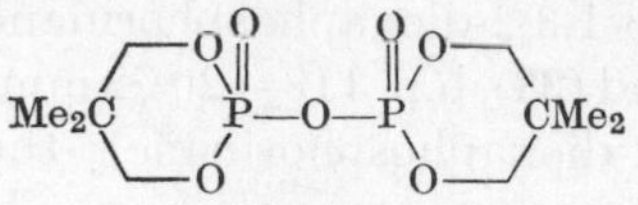

(**25**) 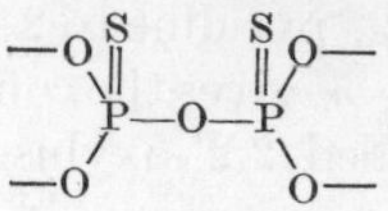   (**26**) 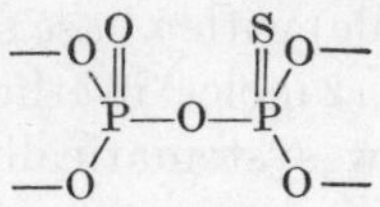   (**27**)

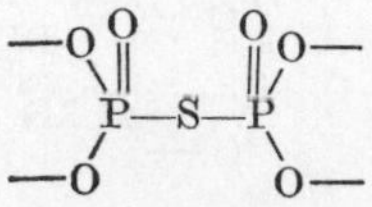

(**28**) 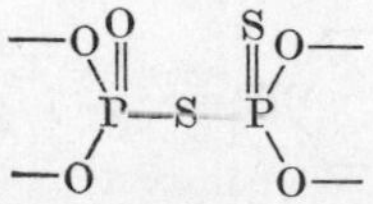   (**29**) 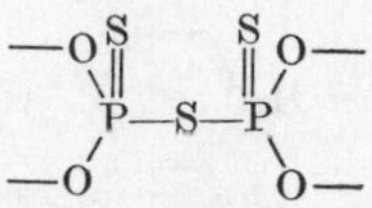   (**30**)

bis(5,5-dimethyl-2-oxo-1,3,2-dioxaphosphorinanyl) oxide.[434] The other members of the series differ from (**25**) only in the central O and/or S atoms joined to the phosphorus atoms, and only this portion is represented here in their formulae. The corresponding 2,2′-disulfide (**26**), m.p. 233°, was best obtained (66%) by the interaction of the 2-chloro 2-sulfide (**14**) and mercaptoacetamide, $HSCH_2CONH_2$, in chloroform containing pyridine.[434]

The interaction of the 2-chloro 2-oxide (**10**) and the triethylamine salt of the 2-thiol 2-oxide in boiling benzene gave the 2-oxide 2′-sulfide (**27**), m.p. 155°. This triethylamine salt reacted with the 2-chloro compound (**6**) to form an oil which when freshly prepared reacted with hydrogen peroxide to give the 2,2′-thiobis-(5,5-dimethyl-1,3,2-dioxaphosphorinane) 2,2′-dioxide (**28**), m.p. 183–185°. The infrared spectrum of (**27**) showed strong absorption at 733 cm⁻¹ (P=S) and at 1311 and 1318 cm⁻¹ (P=O), and very strong absorption at 942 cm⁻¹ assigned to the P—O—P group; the spectrum of (**28**) showed no absorption at 630–770 cm⁻¹, but strong P=O absorption at 1282 and 1287 cm⁻¹ (P=O) and very weak absorption at 940 cm⁻¹ caused by traces of (**26**).[438]

The remaining two members were prepared from 2,2′-dithiobis-(1,3,2-dioxaphosphorinane) 2,2′-disulfide (**31**), m.p. 133.5–134°, obtained

(**31**)

(86%) by treating an aqueous solution of the ammonium salt of the 2-thiol 2-sulfide (16) with potassium iodide. A benzene solution of (31) and of the 2-oxide (12), when boiled for 5 hours, gave the 2,2'-thio 2-oxide 2'-sulfide (29), m.p. 171–173°. When the compound (31) in benzene solution was treated with triphenylphosphine, a gentle exothermic reaction occurred with deposition of the 2,2'-thio 2,2'-disulfide (30), m.p. 223°; evaporation of the filtrate afforded triphenylphosphine sulfide.[434]

The structural identity of the compounds (25)–(30) can often be confirmed by the cleavage products resulting from the action of cyclohexylamine in boiling toluene. Thus (25) when so treated gives the 2-cyclohexylamino compound (19) (75%) and the cyclohexylamine salt of the compound (20) (81%). The compound (27) similarly gives (19) and the cyclohexylamine salt of the 2-hydroxy 2-sulfide.

Edmundson[434] has made a thorough investigation of the infrared spectra of the compounds (25)–(30) and has adduced much evidence, both chemical and spectroscopic, for the conformation of these molecules; his report also contains a considerable number of references to more general work in this field.

For a later study of alternative syntheses of (31) and of its reactions, see Edmundson.[439]

For the preparation of 2-[bis(2'-chloroethyl)amino]-1,3,2-dioxaphosphorinane 2-oxide, see p. 324.

The various names, cited above, which have been applied in particular to the compounds (24)–(30), emphasize the value of the modern unambiguous nomenclature, in which all such compounds are named as substitution or addition derivatives of 1,3,2-dioxaphosphorinane.

4H-1,3,2-Benzodioxaphosphorin

This compound, [RRI 1528], has the structure and numbering shown in (1); hydrogenation of the heterocyclic ring clearly cannot occur, apart from the exceptional case of hydrogenation at the carbon atoms

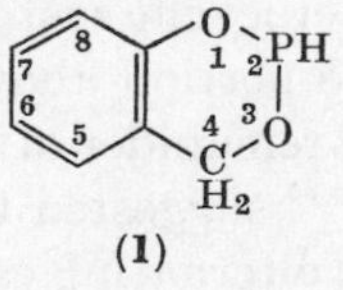

(1)

common to both rings. Oxidation derivatives which carry the $>$POH and the $>$P(O)OH groups can be regarded as cyclic derivatives of phosphorous acid and phosphoric acid, respectively, and members of the latter class have in the past been frequently named as phosphates.

Interest in compound (1) has centered mainly around two different types of derivative, and the study of the two types has occupied chemists for periods of widely different duration. The first arose from Couper's investigation of the action of PCl_5 on salicylic acid in 1858 and has only recently approached finality: the second arose during the investigations which followed the recognition in 1930 of the highly toxic character of tri-o-tolyl phosphate (often erroneously termed tri-o-cresyl phosphate).

Couper's early papers (1858)[440-442] on the action of PCl_5 on salicylic acid to form suggested heterocyclic products must rank collectively as one of the most strikingly original contributions to our understanding of chemical structure. In assessing their quality, it should be borne in mind that Couper's papers preceded by seven years Kekulé's conception of the 'benzene ring'[443]: moreover, Couper considered that the carbon atom could be bivalent or tetravalent, whilst Kekulé maintained strongly the uniform tetravalency of carbon.

Couper used the commonly accepted atomic weight of eight for oxygen, and his formulae therefore contained twice as many oxygen atoms as their modern equivalents. He represented salicylic acid as (A), and the product of its interaction with PCl_5 as (B): this compound in warm moist air was transformed into product (C).

$$
C \begin{cases} C\text{---}H^2 \\ C\text{---}H \end{cases}
C \begin{cases} C\text{---}H \\ C\text{---}O\text{---}OH \end{cases}
C \begin{cases} O^2 \\ O\text{-----}OH \end{cases}
\qquad
C \begin{cases} C\text{---}H^2 \\ C\text{---}H \end{cases}
C \begin{cases} C\text{---}H \\ C\text{---}O\text{---}O \end{cases}
C \begin{cases} O^2 \\ O\text{-----}O \end{cases} PCl^3
\qquad
C \begin{cases} C\text{---}H^2 \\ C\text{---}H \end{cases}
C \begin{cases} C\text{---}H \\ C\text{---}O\text{---}O \end{cases}
C \begin{cases} O^2 \\ O\text{-----}O \end{cases} P \begin{cases} O^2 \\ Cl \end{cases}
$$

      (A)   (B)   (C)

When we allow for the duplication of oxygen, the compounds (B) and (C) contain the same heterocyclic system as that in the more modern formulae (2A) and (4A): the portion above the heterocyclic system in (B) and (C) accounts for the remainder of the unknown benzene ring.

One year later Couper[444] suggested that cyanuric acid had a six-membered ring consisting of alternating carbon and nitrogen atoms.

It is still not widely realized that the first two cyclic systems to be suggested were heterocyclic, and that they were published so far in advance of Kekulé's simple homocyclic system. In 1858 Couper, aged

27, had had only a few years of chemical experience, and the brilliant insight of his work and its historical importance have received little recognition.

Later studies by various workers of the action of PCl_3, PCl_5, and $POCl_3$ on salicylic acid and its esters produced a very tangled skein. This arose largely from the very reactive and often unstable nature of the products, the constitution of which was often not revealed clearly by their reactions but has now been considerably elucidated by modern spectroscopic techniques. An excellent discussion of work in this field up to 1957 has been given by Atherton,[445] but only the main points of this early work can be given here.

R. Anschütz[446] in 1885 gave the 'open' structure (2B) to Couper's compound (B), an opinion which after long investigation in this field he still held in 1906.[446a] Meanwhile R. Anschütz and Emery[447] (1887) had

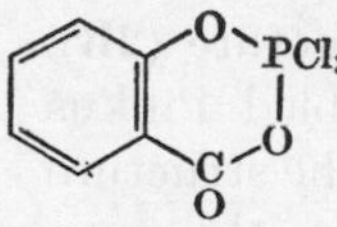

(2A) 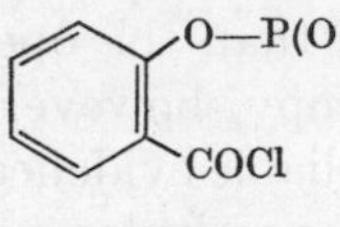   (2B) 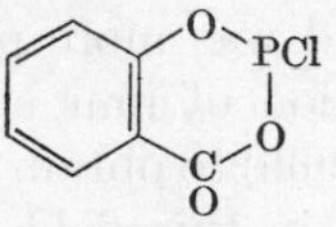   (3A)   (3D)

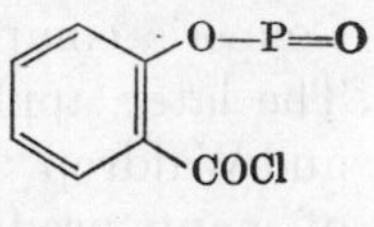

isolated by the interaction of PCl_3 and salicylic acid a compound, $C_7H_4ClO_3P$, termed salicylchlorophosphite, but decisive evidence to differentiate between the structures (3A) and (3B) was lacking. Couper's compound (C), of formula $C_7H_4ClO_4P$, was a well-defined product, but again its structure, (4A) or (4B), was uncertain.

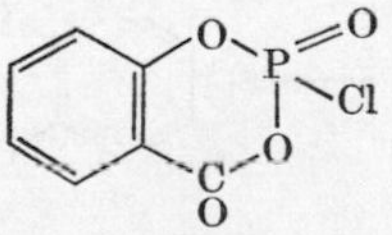

(4A) 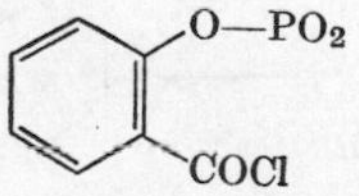   (4B)

Even as late as 1924, L. Anschütz[448] considered that the PCl_3 derivative had the structure (3B), because it reacted with p-toluidine to give a p-toluidide, apparently (5A), for on hydrolysis it gave only salicyloyl-p-toluidide (6; R = H, R' = $C_6H_4CH_3$-p). Young,[449] however, showed that the initial product was (5B), because it reacted with aniline

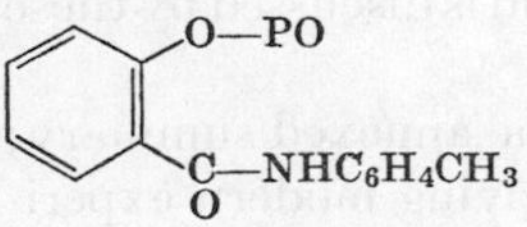

(5A) 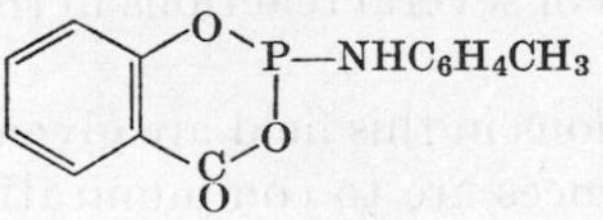   (5B) 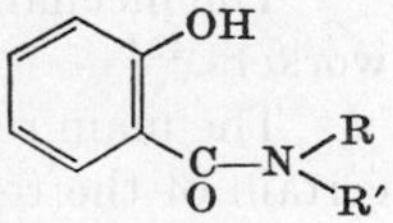   (6)

to give a product which on hydrolysis gave only salicyloylanilide (6; R = H, R′ = C$_6$H$_5$) and *p*-toluidine. Similarly the PCl$_3$ derivative reacted with diethylamine (1 mole) in the presence of triethylamine (1 mole) to give the diethylamino analogue of (5B), which when treated with *p*-toluidine gave a product which on hydrolysis gave only salicyloyl-diethylamide (6; R = R′ = C$_2$H$_5$). It follows that the initial product, *e.g.*, (5B), must be an acylating reagent towards primary and secondary amines.

A considerable number of papers by various workers appeared on this general subject, but in more recent times Atherton[445] produced much evidence for the 'open' structure (2B), and very strong evidence for the cyclic structure (4A), which can now be termed 2-chloro-4*H*-1,3,2-benzodioxaphosphorin-4-one 2-oxide.

Pinkus, Waldrep, and Collier,[450] combining a study of reactions with that of infrared spectra, decided that the cyclic structure (2A) for Couper's compound was more probable than the open structure (2B). The later application of nmr spectroscopy, however, enabled Pinkus and Waldrep[451] (1966) to obtain more reliable evidence for the structure of many products in this field, and in particular to assign the open structure (2B) to Couper's original compound. This assignment was supported chemically by two syntheses. Salicylic acid was converted by thionyl chloride with a trace of aluminum chloride into salicyloyl chloride, C$_6$H$_5$(OH)COCl, which when added in benzene solution at room temperature to a solution of POCl$_3$ (1 equivalent) and pyridine (1 equivalent) in benzene gave (2B), identical with Couper's original

compound. Secondly, phenyl salicylate (7) when treated with POCl$_3$ in pyridine gave the phosphoryl compound (8), which in turn with phenol (2 moles) and triethylamine gave *o*-phenoxycarbonyltriphenyl phosphate (9), m.p. 75.5–76.3° (90%). This compound, however, was also obtained (91%) when Couper's compound (2B) was similarly treated with phenol (3 moles) and triethylamine (3 moles).

The mechanism of several reactions in this field is discussed by these workers.[451]

The main reactions in this field are given in the annexed summary; certain of the references are to communications giving modern experimental details of the earlier preparations.

Salicylic acid (10) reacts with PCl_3 (a) in boiling toluene to give 2-chloro-4H-1,3,2-benzodioxaphosphorin-4-one (3A), b.p. 127–128°/11 mm, m.p. 36–37° (70%),[449] (b) in ether–pyridine at −10° to give (3A)

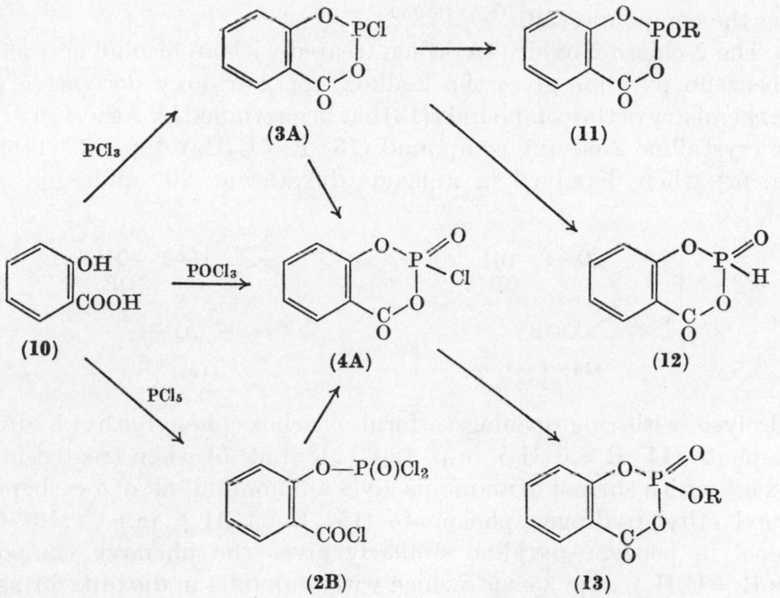

(3A)   (11)

PCl_3

(10) $POCl_3$ (4A)   (12)

OH
COOH

PCl_5

O—P(O)Cl$_2$

COCl

(2B)   (13)

(85%) (Cade and Gerrard[452, 452a]), and (c) with no solvent, using PCl_3 (1.6 moles) at 85° to give (3A) (97%)[450]; the corresponding bromo compound, b.p. 143°/9 mm, is also prepared by method (b).[452a]

The 2-n-butoxy derivative (11; $R = C_4H_9$), b.p. 97–99°/0.02 mm, can be obtained by treating the compound (3A) with 1-butanol in the presence of a tertiary base,[450] or by the direct interaction of the acid (10) in pyridine with butyl phosphorodichloridite, $C_4H_9OPCl_2$ (1 equivalent) in ether at −10°.[452, 452a]

The chloro compound (3A) in ether, when added to a solution of propionic acid in pyridine–ether at −10°, gives propionic anhydride and 4H-1,3,2-benzodioxaphosphorin-4-one 2-oxide (12), m.p. 97–100°; both this compound and (3A) are recorded as breaking down to salicylic acid and PCl_3 when treated with hydrogen chloride.[452a]

The acid (10) reacts with $POCl_3$ to give the 2-chloro 2-oxide (4A), m.p. 95°, and with PCl_5 to give Couper's compound, i.e., o-(chloro-carbonyl)phenyl phosphorodichloridate[450] (2B), b.p. 169–176°/13 mm.

The compound (4A) can be prepared in three other ways: (a) by passing oxygen through a solution of (3A) in pure dry benzene[445, 450]; (b) by treating an ethereal solution of (2B) with water (1 equivalent) in ethanol–pyridine[450]; (c) by heating the compound (2B) with anhydrous oxalic acid at 75–80° (R. Anschütz and Moore[453]), a method now solely of historic interest.

The addition of chlorine (1 mole) to the chloro compound (3A) gives the compound (2B).[447, 445, 450]

The 2-chloro 2-oxide (4A) when treated with an alcohol or a phenol in benzene–pyridine gives the 2-alkoxy or 2-aryloxy derivative (13). The chemistry of the compounds (13) has been studied by Atherton.[454,455] The crystalline 2-ethoxy compound (13; R = C$_2$H$_5$), b.p. 142–146°/0.3 mm, (a) when dissolved in aqueous dioxane at 20° undergoes slow

(14)   (15)

hydrolysis with ring opening to form o-carboxyphenyl ethyl hydrogen phosphate (14; R = C$_2$H$_5$), m.p. 120–121°, but (b) when treated in dry dioxane with a current of ammonia gives ammonium salt of o-carbamoyl-phenyl ethyl hydrogen phosphate (15; R = C$_2$H$_5$), m.p. 121.5–123°. Phenol in benzene–pyridine similarly gives the phenoxy compound (13; R = C$_6$H$_5$), m.p. 85–86°, which with ammonia in dioxane forms the ammonium salt of o-carbamoylphenyl phenyl hydrogen phosphate (15; R = C$_6$H$_5$) (90%).

Esters of type (13) can alternatively be prepared[455] directly from salicylic acid by the action of an aryl phosphorodichloridate, ROP(O)Cl$_2$. Thus the acid (10) in toluene–pyridine reacts with o-tolyl phosphoro-dichloridate to give (13; R = C$_6$H$_4$CH$_3$-o), b.p. 164–174°/0.3–0.4 mm, m.p. 78–82°. This compound undergoes hydrolysis to o-carboxyphenyl o-tolyl hydrogen phosphate (14; R = C$_6$H$_4$CH$_3$-o), and also reacts with ammonia (as above) to give the ammonium salt of o-carbamoylphenyl o-tolyl hydrogen phosphate (15; R = C$_6$H$_4$CH$_3$-o), m.p. 154.5–155.5°. The two Patent Specifications (Atherton[454, 455]) give a number of compounds of type (13), (14), and (15) investigated for possible use in rheumatic disorders and in salicylic acid therapy generally.

For the action of PCl$_5$ on phenyl salicylate, see Pinkus, Waldrep, and Ma,[456] and on methyl salicylate, see reference 450; for the action of POCl$_3$ on phenyl salicylate, see reference 451.

The toxic properties of tri-o-tolyl phosphate (16) probably first became evident in 1930 when an estimated number of 15,000 persons in the U.S.A. became affected (some fatally) by consuming 'Jamaica

Ginger', an alcoholic drink which had been adulterated with 2% of the phosphate (16) (Sax [457]). Technical chemists in Germany also became aware of its toxicity when it was used as a plasticizer. A second severe outbreak of 'T.C.P.' poisoning occurred in Morocco in 1958 when edible oil sold for cooking purposes had been dishonestly mixed with lubricating oil which in turn contained the phosphate.

In a preliminary note (Casida, M. Eto, and Baron [458]) on the cause of these toxic properties, it was stated that the phosphate was metabolized *in vivo* (in rats) to form probably three similar compounds which were potent esterase inhibitors. These were formed by the hydroxylation of an *o*-methyl group, which then underwent cyclization with the phosphorus atom with ejection of an *o*-tolyl group. The principal compound thus formed was 2-*o*-tolyloxy-4*H*-1,3,2-benzodioxaphosphorin 2-oxide (17); this was synthesized by the interaction of 2-hydroxybenzyl alcohol (saligenin) and *o*-tolyl phosphorodichloridate, *o*-$CH_3C_6H_4O$—$P(O)Cl_2$, and had infrared spectra and chromatographic characteristics identical with those of the *in vivo* product.

(16)   (17)   (18)

Further details of the synthesis of (17), b.p. 159–161°/0.1 mm, were given by M. Eto, Casida, and T. Eto, [459] who obtained indirect evidence from experiments with rats that the other two metabolites were probably the 2-*o*-(hydroxymethyl)phenyl and the 2-(2'-hydroxybenzyl) analogues of (17), *i.e.*, (18; R = $C_6H_4CH_2OH$-*o* and $CH_2C_6H_4OH$-*o*, respectively). The yields of these products *in vivo* are low, owing possibly to excretion or to enzymically catalyzed hydrolysis to acyclic esters.

It is noteworthy that when ethyl di-*o*-tolyl phosphate and phenyl di-*o*-tolyl phosphate were given to house flies (*Muscara domestica*), a similar process occurred, with hydroxylation and cyclization of one *o*-tolyl group and ejection of the other, with the formation of (18; R = C_2H_5 and C_6H_5, respectively) (Eto, Matsuo, and Oshima [460]).

The fact that the toxic metabolites formed from tri-*o*-tolyl phosphate have the ring system (18) harmonizes with relatively innocuous properties of the isomeric *m*- and *p*-tolyl phosphates.

Further investigation showed that although compounds of type (18; R = aryl) and in particular (17) were highly toxic to mammals, their

insecticidal properties—based on experiments on house flies—were very
weak (M. Eto, T. Eto, and Oshima[461]). Replacement of the aryl group
by a methyl group, to give 2-methoxy-4H-1,3,2-benzodioxaphosphorin
2-oxide (18; R = CH$_3$), commonly known as Salioxon, resulted in strong
insecticidal activity, which the corresponding 2-sulfide (19; R = CH$_3$),
known as Salithion, also possessed. In this respect they were comparable

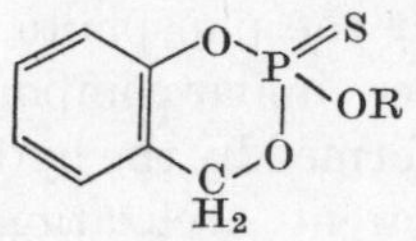

(19) 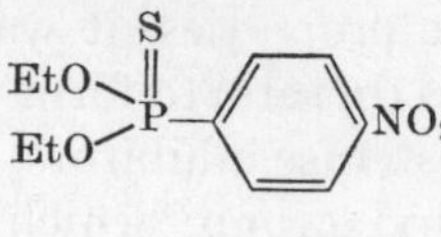   (20)

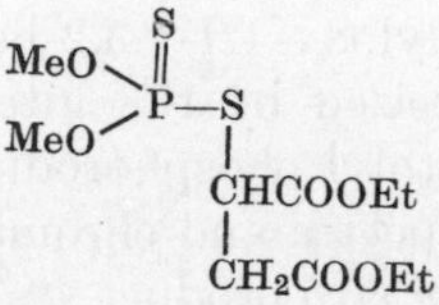

(21) 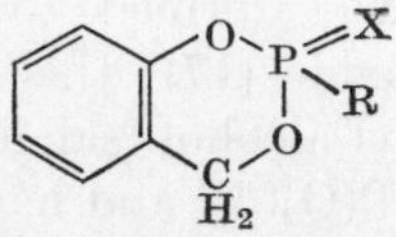   (22)

to parathion [$OO'O''$-diethyl p-(nitrophenyl)thiophosphonate] (20) and
to malathion S-[1,2-bis(ethoxycarbonyl)ethyl] O,O'-dimethyl dithio-
phosphate (21), but were less toxic to animals (M. Eto, Kinoshita,
Kato, and Oshima[462, 463]).

Salioxon, a colorless oil, b.p. 110–112°/0.05 mm, can be readily
prepared by the interaction of o-hydroxybenzyl alcohol and methyl
phosphorodichloridate in the presence of pyridine. The preparation of
Salithion, m.p. 51–53°, requires more vigorous conditions, a mixture of
o-hydroxybenzyl alcohol and methyl phosphorodichloridothionate,
CH$_3$OP(S)Cl$_2$, in toluene containing potassium carbonate and copper
powder being heated at 80–90° for 15–20 hours.

Compounds of type (18) are clearly cyclic phosphates, but those of
type (22) are cyclic phosphonates (X = O) or phosphonothionates (X = S).

Table 27. 2-Alkyl-1,3,2-benzodioxaphosphorin
2-oxides and 2-sulfides (22)

R	X	B.p. (°/mm)	M.p. (°)
CH$_3$	O	140/0.5	35
C$_2$H$_5$	O	143–149/0.3	25
CH$_3$	S	130/0.6	—
C$_2$H$_5$	S	120/0.6	—

These can be readily prepared by the action of the appropriate phosphonic dichlorides, $RP(O)Cl_2$, or phosphonothoic dichlorides, $RP(S)Cl_2$, on 2-hydroxybenzyl alcohol in pyridine. The members in Table 27 have been prepared by M. Eto et al.[464]

It is noteworthy that these compounds also show insecticidal activity to house-flies, 2-ethyl-4H-1,3,2-benzodioxaphosphorin 2-sulfide (**22**; $R = C_2H_5$, $X = S$) having the greatest toxicity of the above members.

4H-1,3,5-Dioxaphosphorin and 1,3,5-Dioxaphosphorinanes

4H-1,3,5-Dioxaphosphorin, [RIS 7749], has the structure (**1**) and its 5,6-dihydro derivative is 1,3,5-dioxa-phosphorinane (**2**).

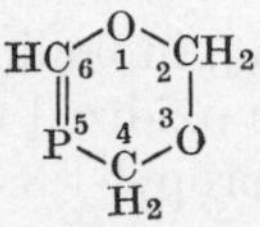

It has already been noted (p. 61) that straight-chain aliphatic aldehydes react with phosphine in an ethereal solution of hydrogen chloride to give tetrakis(1-hydroxyalkyl)phosphonium chlorides:

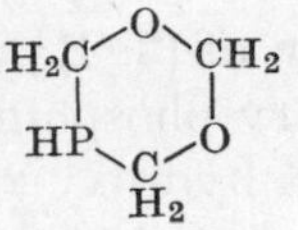

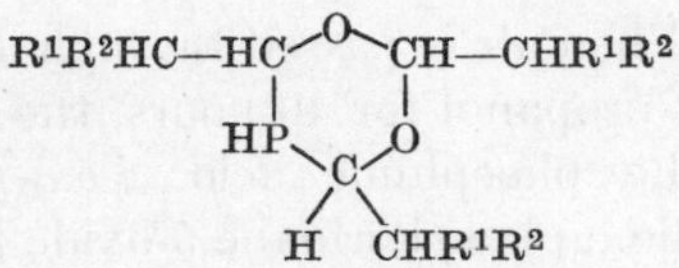

$$4RCHO + PH_3 + HCl \rightarrow [RCH(OH)]_4P^+ \, Cl^-$$

By using tetrahydrofuran as a solvent, with concentrated hydrochloric acid, Buckler and Wystrach[98, 99] have extended this reaction up to heptanal, n-$C_6H_{13}CHO$, and dodecanal, n-$C_{11}H_{23}CHO$, and obtained almost theoretical yields of the corresponding hydroxyphosphonium chlorides.

When, however, an α-branched aldehyde is employed, the reaction takes a different course. Thus the interaction of isobutyraldehyde, $(CH_3)_2CHCHO$, with phosphine in the presence of concentrated hydrochloric acid at 20–25° gives 2,4,6-triisopropyl-1,3,5-dioxaphosphorinane (**3**; $R^1 = R^2 = CH_3$), b.p. 110°/14 mm (75%); the similar use of 2-ethylhexaldehyde $(C_2H_5)(n$-$C_4H_9)CHCHO$, gives 2,4,6-tris-(2-ethylpentyl)-1,3,5-dioxaphosphorinane (**3**; $R^1 = C_2H_5$, $R^2 = n$-C_4H_9), b.p. 149–150°/0.02 mm (90%).

These cyclic secondary phosphines are liquids which show a typical P—H infrared absorption at 2300 cm^{-1}; they have a 'not unpleasant' odor, are reasonably stable to atmospheric oxygen, and undergo no appreciable hydrolysis when boiled with concentrated hydrochloric acid for 2 hours.

When the 2,4,6-triisopropyl derivative is treated with p-chlorophenyl isocyanate, the secondary phosphine group is converted into the 5-p-chlorophenylcarbamoyl-phosphine group, p-ClC$_6$H$_4$NHCOP$\mathopen{<}$.

(**3**; R^1 = R^2 = CH$_3$)

(CH$_3$)$_2$HC—HC$\underset{}{\overset{}{\bigvee}}$CH—CH(CH$_3$)$_2$

HO—P—C—O

H   CH(CH$_3$)$_2$

(**4**)

$\longrightarrow$

(CH$_3$)$_2$CHCHO
+
(CH$_3$)$_2$CHCH(OH)
$\underset{\text{P}}{\overset{\text{O}}{\diagup}}$OH
(CH$_3$)$_2$CHCH(OH)

(**5**)

When air is passed through a solution of (**3**; R^1 = R^2 = CH$_3$) in boiling 2-propanol for 40 hours, the secondary phosphine group is oxidized to the phosphinic acid, *i.e.*, giving 5-hydroxy-2,4,6-triisopropyl-1,3,5-dioxaphosphorinane 5-oxide (**4**), m.p. 159–160°, which undergoes ready hydrolysis with hot dilute hydrochloric acid to give isobutyraldehyde and bis-(1-hydroxy-2-methylpropyl)phosphinic acid (**5**), m.p. 168–169°.

It is suggested [99] that this marked difference between the behavior of straight-chain and α-branched-chain aldehydes towards phosphine plus hydrochloric acid arises because the increasing steric hindrance prevents further substitution in the disubstituted intermediate [R^{1}R^{2}CH(OH)]$_2$PH, which therefore undergoes cyclic acetal formation with another molecule of aldehyde to give (**3**). This, however, implies that the OH groups in the intermediate are not also hindered in reaction with the aldehyde. Yet there are examples known in other types where bulky substituents seem to accelerate cyclization and to retard hydrolytic fission of the ring system so formed (Bordwell, Osborne, and Chapman [465]), and the strong resistance of compounds of type (**3**) to such fission may be significant.

Compounds of type (**3**) were initially termed 2,4,6-trialkyl-1,3-dioxa-5-phosphacyclohexane. [98, 99]

The same ring system, but with substitution on the phosphorus atom, is obtained by a similar acid-catalyzed condensation of primary phosphines and certain aromatic aldehydes. The addition of benzaldehyde (3 equivalents) at room temperature to a solution of phenylphosphine (1 equivalent) in acetonitrile containing concentrated hydrochloric acid causes the rapid crystallization of 2,4,5,6-tetraphenyl-1,3,5-dioxaphosphorinane (**6**), m.p. 195–198° (53%); this compound, when added to a solution of hydrogen peroxide in hot methanol, gives the corresponding 5-oxide, m.p. 275–278° (Buckler and Epstein [466]).

In a neighboring investigation, Buckler[467] saturated a solution of benzaldehyde in anhydrous ether with dry hydrogen chloride, and then

$$PhHC \underset{PhP}{\overset{O}{\diagdown}} \underset{CHPh}{\overset{CHPh}{\diagdown}} O$$

(6)

$$PhHC \underset{PhCH_2}{\overset{O}{\diagdown}} \underset{O}{\overset{CHPh}{\diagdown}} \underset{CHPh}{\overset{O}{\diagdown}}$$

(9)

treated the solution with phosphine. The crystalline product, m.p. 152–153°, was expected to be tris-(α-hydroxybenzyl)phosphine (7); its chemical properties and its infrared and nmr spectra showed decisively,

$(C_6H_5CHOH)_3P$

(7)

$(C_6H_5CHOH)_2P(O)CH_2C_6H_5$

(8)

however, that it was the isomeric benzylbis-(α-hydroxybenzyl)phosphine oxide (8). This isomerization is another example of the familiar process of transference of oxygen from a carbon to a phosphorus atom, a process which almost invariably proceeds by an intermolecular mechanism. When compound (8) was added to a benzene solution of benzaldehyde containing a trace of p-toluenesulfonic acid, the reaction mixture being then boiled for 20 hours with removal of the liberated water, the solution yielded 5-benzyl-2,4,6-triphenyl-1,3,5-dioxaphosphorinane 5-oxide (9) (66%). This product was initially a mixture of isomers, of m.p. 176–195°, but two of these isomers, of m.p. 220–222° and 199–201°, respectively, were isolated by fractional crystallization.

The compounds (6) and (9) were originally termed 2,4,5,6-tetra-phenyl-1,3-dioxa-5-phosphacyclohexane[466] and 5-benzyl-2,4,6-tri-phenyl-1,3-dioxa-5-phosphacyclohexane 5-oxide,[467] respectively.

3,5,8-Trioxa-1-phosphabicyclo[2.2.2]octane

This compound, shown formally as (1), [RIS —], or alternatively as (2; R = H), is clearly isomeric with 2,6,7-trioxa-1-phosphabicyclo-[2.2.2]octane (p. 312).

$$\begin{array}{c} H_2C{-}_6\underset{}{\overset{}{\rule{0pt}{0pt}}}P_1{-}_2CH_2 \\ \quad\quad _7CH_2 \\ \quad\quad _8O \\ O{-}_5\underset{H}{\overset{}{C}}_4{-}_3O \end{array}$$

(1)

$$\begin{array}{c} H_2C \overset{P}{\diagdown} CH_2 \\ \quad CH_2 \\ O \quad O \quad O \\ \overset{C}{\underset{R}{}} \end{array}$$

(2)

The parent compound (**1**) is prepared by the interaction of tris-(hydroxymethyl)phosphine, $P(CH_2OH)_3$, with trimethyl orthoformate, $HC(OCH_3)_3$, in the presence of 'two drops' of triethylamine: as the temperature reaches *ca.* 90°, nitrogen is passed through to remove methanol. The residue after sublimation gives the pure colorless phosphine (**1**), m.p. 88–89°. When oxygen is passed through an ethereal solution of the phosphine irradiated with ultraviolet light, the 1-oxide, m.p. 151–155°, is formed: the infrared spectrum of the oxide shows a $P=O$ band at 1198 cm^{-1} (Boros, Coskran, King, and Verkade[468]).

The phosphine (**1**) when heated with sulfur gives the 1-sulfide (28%).

Repetition of the above preparation using, however, trimethyl orthoacetate, $CH_3C(OCH_3)_3$, gives 4-methyl-3,5,8-trioxa-1-phosphabicyclo[2.2.2]octane (**2**; R = CH$_3$). This phosphine (*a*) when treated with ethanolic hydrogen peroxide gives the 2-oxide (20%), having $\nu(P=O)$ 1200 cm^{-1}, and (*b*) when heated with sulfur at 110° for 1 hour gives the 1-sulfide, purified by sublimation at 60° under reduced pressure (33%); it shows $\nu(P=S)$ 744 cm^{-1} (Boros, Compton, and Verkade[469]). [The m.p.s of these three compounds are apparently not cited.] The interpretation of the proton- and the ^{31}P-nmr spectra of the compounds (**2**; R = H) and (**2**; R = CH$_3$) and of the isomeric 2,6,7-trioxa compounds and their derivatives is discussed in some detail by Boros *et al.*[468, 469]

2,6,7-Trioxa-1-phosphabicyclo[2.2.2]octane

This bicyclic system (**1**), also depicted as (**1A**), [RRI 1525], bears an obvious relationship to 2,6,7-trioxa-1,4-diphosphabicyclo[2.2.2]octane

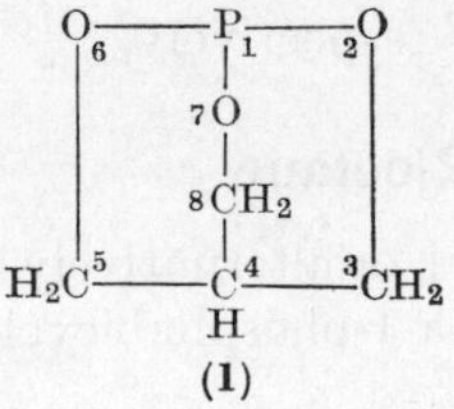

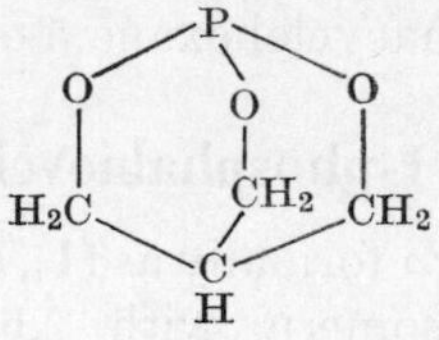

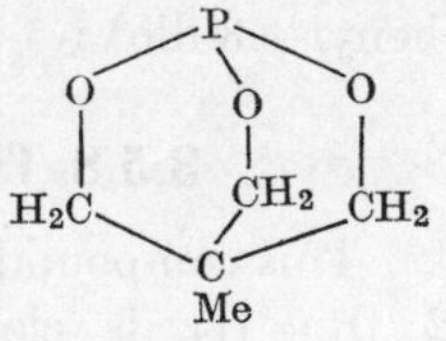

$NO_2C(CH_2OH)_3$

(3)

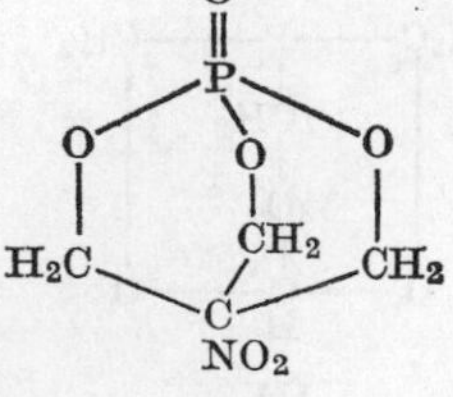

(p. 317), in which the CH unit at position 4 of (1) is replaced by a phosphorus atom, and to 1,4-diphosphabicyclo[2.2.2]octane (p. 318), in which additionally the three oxygen atoms of (1) are replaced by CH_2 groups.

4-Methyl-2,6,7-trioxa-1-phosphabicyclo[2.2.2]octane (2) can be readily prepared by the interaction of PCl_3 and 2-hydroxymethyl-2-methyl-1,3-propanediol, $CH_3C(CH_2OH)_3$ (Verkade and Reynolds[470]). To reduce the proportion of polymer formation, the reaction was carried out at high dilution in anhydrous solvents. For this purpose, two solutions, one of PCl_3 in tetrahydrofuran (THF) and one of the diol in tetrahydrofuran containing pyridine, were added dropwise and simultaneously to stirred THF under nitrogen at room temperature. After filtration to remove pyridine hydrochloride, the solvent was distilled off, and the white residue was sublimed at 1 mm with heating from room temperature to 80°. This sublimation, when performed three times at 50°/1 mm, gave the 4-methyl compound (2), m.p. 97–98° (40%).

This compound (a) when treated with ethanol–hydrogen peroxide gave the 1-oxide, m.p. 249–250° (92%), showing in the infrared a strong P=O band at 1325 cm^{-1}, and (b) when heated with sulfur at 140° for 5 minutes gave the 1-sulfide, m.p. 224–225° (89%), showing a P=S band of medium intensity at 800 cm^{-1}. The structure of the 1-oxide is discussed later.

It is noteworthy that the tertiary phosphine (2), its 1-oxide, and its 1-sulfide were each sublimed three times to obtain a pure product. This recalls the similar treatment which 1,4-diphosphabicyclo[2.2.2]-octane and its derivatives (p. 181) required to eliminate traces of solvent.

The phosphine (2) is hygroscopic, but it is chemically unaffected on exposure to air for several months. This is unexpected, as the phosphorus atom is recorded as being an excellent donor, presumably because the virtual absence of steric hindrance leaves fully available the lone pair of electrons on this atom. The considerable stability of the phosphine (2) contrasts markedly with the low stability of its arsenic analogue (p. 571).

The donor properties of the phosphine (2) are well shown in its reaction with nickel carbonyl, in which each CO group is replaced in turn by one molecule of the phosphine (Verkade, McCarley, Hendricker, and King[471]).

The nmr spectra of the phosphine (2) show the following ^{31}P shifts (in ppm from external 85% H_3PO_4): the phosphine (2), −91.5; the 1-oxide, +7.97; the 1-sulfide, −57.4 (Verkade and King[472]). The value for the 1-sulfide provides some confirmation for the position of the sulfur in the 2,6,7-trioxa-1,4-diphosphabicyclo[2.2.2]octane sulfide (p. 318).

The parent phosphine (**1**), m.p. 126–127°, was later prepared by the interaction of 2-hydroxymethyl-1,3-propanediol and trimethyl phosphite with a small quantity of triethylamine. It gives a 1-oxide, m.p. 245–247°, which shows a P=O band at 1308 cm^{-1}: the two compounds readily sublimed at *ca.* 20°/0.01 mm and 100°/0.01 mm, respectively (Boros, Coskran, King, and Verkade[468]). The ^{1}H- and ^{31}P-nmr spectra of both (**1**) and (**2**), and of similar compounds, have been discussed.[468]

In an earlier synthesis of the ring system (**1**), a solution of POCl$_3$ (1 mole) in dry pyridine was added dropwise with shaking to a solution of tris(hydroxymethyl)nitromethane (**3**) also in dry pyridine at −10°,

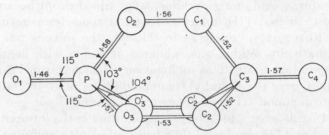

Figure 8. Structure of 4-methyl-2,6,7-trioxa-1-phosphabicyclo[2.2.2]octane. Interatomic distances in Å. [Reproduced, by permission, from D. M. Nimrod, D. R. Fitzwater, and J. G. Verkade, *J. Amer. Chem. Soc.*, **90**, 2780 (1968)]

the reaction mixture being poured next day into chilled dilute sulfuric acid. The crystalline material which separated was recrystallized from hot water, giving 4-nitro-2,6,7-trioxa-1-phosphabicyclo[2.2.2]octane 1-oxide (**4**), m.p. 243° (62%). This compound is remarkably stable to acids, and can be recrystallized unchanged from 10% sulfuric acid; on the other hand, cold aqueous alkali—even sodium carbonate—causes rapid hydrolysis with the production of a yellow color (Zetzsche and Zurbrügg[473]).

The compounds (**3**) and (**4**) were initially termed nitroisobutyl glycerin and nitroisobutyl-glycerin-bicyclo-phosphate, respectively.[473] The compound (**2**) was formerly numbered from the bridgehead C atom and thus termed 1-methyl-3,5,8-trioxa-4-phosphabicyclo[2.2.2]octane.[470]

The structure and conformation of the bicyclic system (**1**) has been clearly shown by an X-ray crystal analysis of 4-methyl-2,6,7-trioxa-1-phosphabicyclo[2.2.2]octane (**2**) (Nimrod, Fitzwater, and Verkade[474]). The intervalency angles and the interatomic distances (Å) are given in Figure 8 in which the atoms are numbered for crystallographic convenience and not as in (**1**). The symmetry of the central bicyclic 'cage' and the colinear disposition of the carbon atom of the methyl group and the oxygen of the oxide group are manifest. The original paper[474] should be consulted for a detailed discussion of this structure.

2,8,9-Trioxa-1-phosphaadamantane

The formula of this interesting compound (1), [RRI 3252], is presented and numbered as shown.

Stetter and Steinacker[475] have recorded the hydrogenation of phloroglucinol in ethanolic solution in the presence of Raney nickel to give two isomeric 1,3,5-cyclohexanetriols, the α-form (60%), m.p.

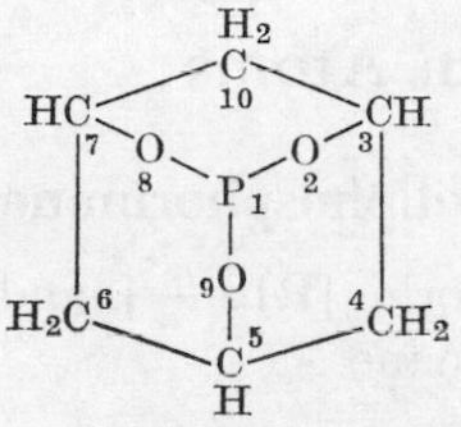

(1) 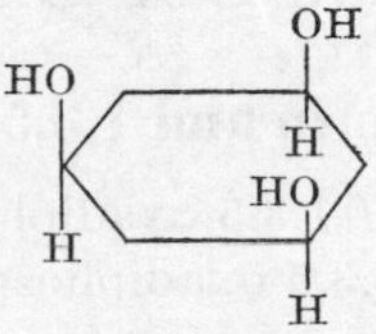   (2) 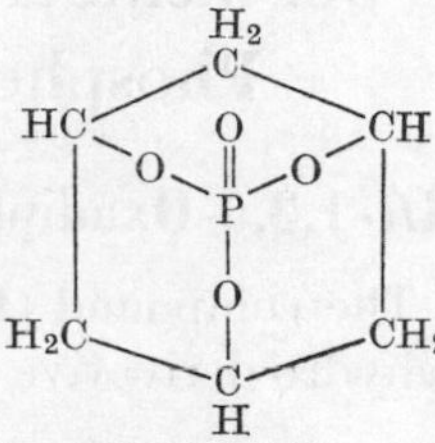   (3)

184°, and the β-form (10%), m.p. 145°. The fact that the α-form is the *cis-cis*-isomer (2) is confirmed by its reaction in pyridine with phosphorus trichloride to give 2,8,9-trioxa-1-phosphaadamantane (1), m.p. 207° (20%), and with phosphorus oxychloride to give the 1-oxide (3), m.p. 267–268° (40%).

These preparations, for which the reagents, solvents, and apparatus must be thoroughly dry, follow what is now a familiar pattern: for example, in the preparation of (3) a solution of (2) in pyridine, and a solution of $POCl_3$ (1 mole) in carbon tetrachloride, are added from separate dropping-funnels dropwise and simultaneously into pyridine which is vigorously stirred and chilled in ice-salt. The complete mixture is stirred for 1 hour at room temperature and then at 80°, cooled, and decanted from pyridine hydrochloride. Removal of the solvents under reduced pressure gives a syrupy residue of the 1-oxide which slowly crystallizes and can then be recrystallized from dioxane or butanol. The oxide is moderately soluble in these solvents, easily soluble in water, methanol, ethanol, and acetic acid and almost insoluble in benzene, ether, acetone, and chloroform. When heated, the oxide undergoes considerable sublimation at *ca.* 230°.

The tertiary phosphine (1) is similarly prepared, and is purified by recrystallization from butanol or by sublimation under reduced pressure at *ca.* 140°. It is soluble in most solvents except ether. The phosphine is converted into the 1-oxide (**3**) slowly on exposure to air, and rapidly and quantitatively by dilute hydrogen peroxide.

The above preparation, when carried out using thiophosphoryl chloride, $SPCl_3$, gives the 1-sulfide, colorless needles (from methanol), m.p. 250–251° (30%). This compound is also formed when a mixture of the phosphine (**1**) and sulfur is heated to 180–190°. It is insoluble in water and ether, moderately soluble in methanol, ethanol, and dioxane, and readily soluble in acetone, chloroform, and benzene.

Six-membered Ring System containing Two Phosphorus and One Oxygen Atoms

4*H*-1,3,5-Oxadiphosphorin and 1,3,5-Oxadiphosphorinane

The compound (**1**) is 4*H*-1,3,5-oxadiphosphorin, [RIS —], and its tetrahydro derivative (**2**) is 1,3,5-oxadiphosphorinane.

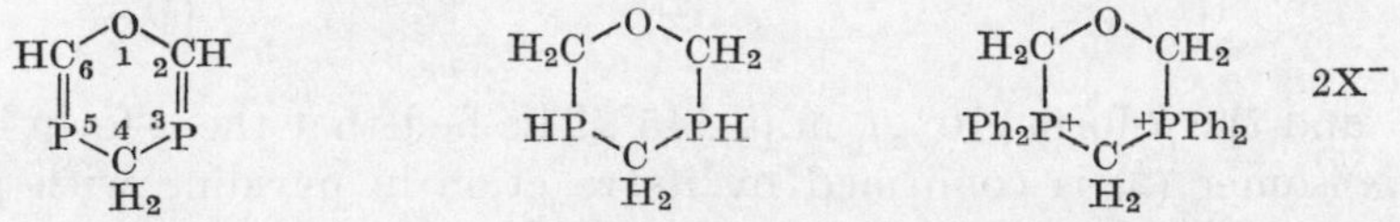

A benzene solution of methylenebis(diphenylphosphine), $CH_2[P(C_6H_5)_2]$ (1 mole) and bis(chloromethyl) ether, $(ClCH_2)_2O$, (2.5 mole), when boiled under reflux for 24 hours, deposits 3,3,5,5-tetraphenyl-1,3,5-oxadiphosphorinanium dichloride (**3**; X = Cl) (61%), a hygroscopic salt which gives a crystalline bis(tetraphenylborate) [**3**; X = $B(C_6H_5)_4$], m.p. 135–137° (Aguiar, Hansen, and Mague[476]).

In an alternative synthesis, lithium diphenylphosphide in tetrahydrofuran is treated with bis(chloromethyl) ether, giving an immediate exothermic reaction with the formation of bis[(diphenylphosphino)-methyl] ether, $[(C_6H_5)_2PCH_2]_2O$, m.p. 86–88° (80%). A mixture of this ether and an excess of dibromomethane, when boiled under reflux for 24 hours without a solvent, gives the non-hygroscopic dibromide (**3**; X = Br), m.p. 235–241° (91%); this salt gives the same bis(tetraphenyl-borate), m.p. 135–137° alone and when mixed with the earlier sample.

The dichloride and dibromide have essentially identical nmr spectra, which are in accordance with the structure (**3**). Reliable molecular-weight determinations of the dibromide in solution are difficult to obtain, but a graph of conductance measurements in dilute methanolic solutions at 25° is very similar to those of cobalt dichloride and calcium dichloride at equivalent concentration, indicating a 1:2 ionic salt.

The above simple diquaternization of methylenebis(diphenylphosphine) by the bis(chloromethyl) ether contrasts markedly with the complex reaction of the latter with *cis*-vinylenebis(diphenylphosphine), whereby 1,4-dihydro-1,1,4,4-tetraphenyl-1,4-diphosphorinium dichloride is formed (p. 179).

The dibromide (3; X = Br) was initially termed 3,3,5,5-tetraphenyl-3,5-diphosphonia-1-oxacyclohexane dibromide.[476]

Six-membered Ring Systems containing Two Phosphorus and Two Oxygen Atoms in Each Ring

2,6,7-Trioxa-1,4-diphosphabicyclo[2.2.2]octane

This compound (1), alternatively depicted as (1A), [RIS —], can be readily prepared by the dropwise addition of trimethyl phosphite

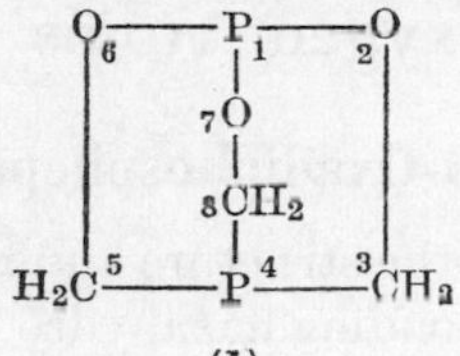

(1) 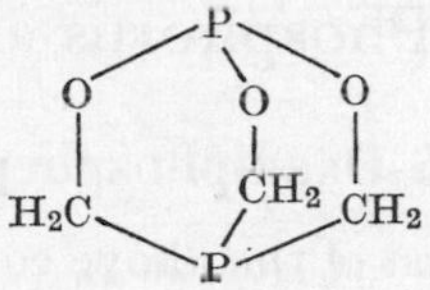   (1A)

to an equimolecular quantity of tris(hydroxymethyl)phosphine, $P(CH_2OH)_3$, dissolved in dry, vigorously stirred tetrahydrofuran, which is then boiled under reflux for 12 hours. After distillation of the solvent, the residue is sublimed at 50°/0.1 mm, giving the colorless diphosphine (1), m.p. 75–76° (35%) (Coskran and Verkade[477]).

(If the above two reagents are mixed in the absence of a solvent, a violent exothermic reaction occurs.)

The diphosphine (1) with ethanol–hydrogen peroxide gives the 1,4-dioxide, m.p. 210–213° (95%). A mixture of (1) and sulfur (1 mole), when heated in a sealed tube at 110° for 20 minutes, gives the 1-sulfide, $SP(OCH_2)_3P$, m.p. 235–237°; the same monosulfide is formed even when an excess of sulfur is present. A solution of this 1-sulfide in ethanol–hydrogen peroxide, when briefly boiled and cooled, gives the colorless 1-sulfide 4-oxide, $SP(OCH_2)_3PO$, which decomposes at *ca*. 210°.

The structure of (1) is confirmed by the proton nmr spectrum which shows a straightforward A_2XY pattern for the six equivalent hydrogen atoms. The curious fact that only the P-1 or 'phosphite' atom will form a sulfide is confirmed by the infrared evidence. The spectrum of this monosulfide shows a strong bond at 807 cm^{-1} almost certainly due to the P=S stretching frequency; that of the 1-sulfide 4-oxide shows the P=S band at 807 and the 'phosphine' P=O band at 1215 cm^{-1}; that of the 1,4-dioxide shows a strong phosphate P=O band at 1325 cm^{-1} and a strong phosphine P=O band at 1220 cm^{-1}.

The nmr spectra give the following ^{31}P shifts (in ppm against 85% H_3PO_4): the tertiary diphosphine (1) −90.0; the 1,4-dioxide, +18.1; the monosulfide, −51.8. These values for the diphosphine and the 'phosphite' sulfide agree reasonably well with those for 4-methyl-2,6,7-trioxa-1-phosphabicyclo[2.2.2]octane (p. 313) and its sulfide; the latter must be a 'phosphite' sulfide.

The tris(hydroxymethyl)phosphine used in this work was prepared by the action of sodium hydroxide on tetrakis(hydroxymethyl)phosphonium chloride in absolute ethanol (Grayson[478]).

Seven-membered Ring Systems containing Two Phosphorus and One Oxygen Atoms

1,3,6-Oxadiphosphepin and 1,3,6-Oxadiphosphepane

The first of the above compounds has the structure and numbering (1), [RIS —], and the second is the corresponding hexahydro derivative (2).

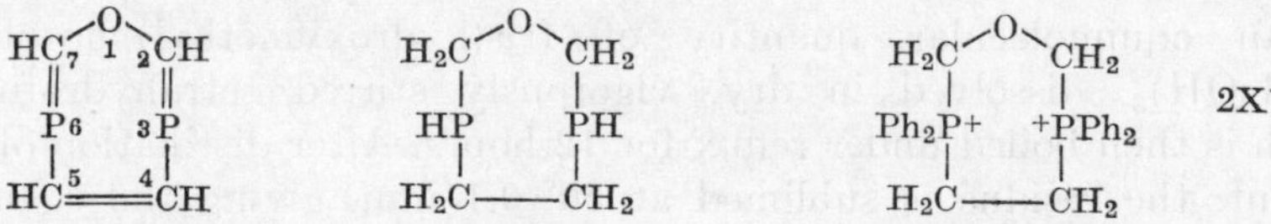

(1)   (2)   (3)

An equimolecular mixture of bis[(diphenylphosphino)methyl] ether (p. 316) and 1,2-dibromoethane in benzene solution, when boiled under reflux for 24 hours, afforded 3,3,6,6-tetraphenyl-1,3,6-oxadiphosphepanium dibromide (3; X = Br), m.p. 292–293° (52.5%). It gives an orange dipicrate (3; X = $C_6H_2N_3O_7$), m.p. 223–225°, thus confirming the complete quaternization in the formation of the dibromide (Aguiar, Hansen, and Mague[476]).

As in the case of the lower homologue (p. 316), the graph of conductance measurements of the above dibromide in dilute methanolic solutions is closely similar to those of cobalt dibromide and of calcium dichloride at equivalent concentrations, thus supporting the above 1 : 2 formulation of the dibromide.

The dibromide (**3**; X = Br) was initially termed 3,3,6,6-tetraphenyl-1-oxa-3,6-diphosphoniacycloheptane dibromide.[476]

Ten-membered Ring Systems containing Two Phosphorus and Four Oxygen Atoms

1,3,6,8,2,7-Tetraoxadiphosphecin

The ring-system (**1**) [RRI 448] is termed 1,3,6,8,2,7-tetraoxadiphosphecin, and its tetrahydro derivative (**2**; R = H) is 1,3,6,8,2,7-tetraoxadiphosphecane; an earlier name for the latter compound was 1,3,6,8-tetraoxa-2,7-diphosphacyclodecane.

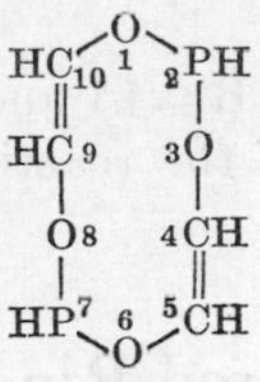

$$
\begin{array}{ll}
\text{HC}^{10} \quad ^1 \quad _2\text{PH} & \\
\text{HC}_9 \qquad _3\text{O} & \\
\text{O}_8 \qquad _4\text{CH} & \\
\text{HP}^7 \quad _6 \quad ^5\text{CH} &
\end{array}
$$

(1) 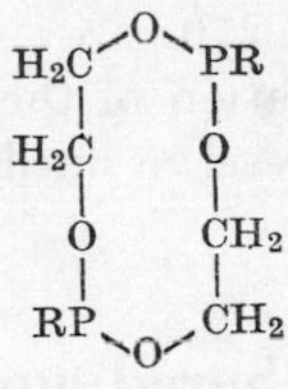   (2)

It has been stated by Carré[470] that, if phosphorus trichloride is added to a solution of ethylene glycol in anhydrous ether, a reaction occurs with the formation of two products, the major product being soluble in ether and the minor product insoluble. Evaporation of the ether from the solution of the major product leaves a liquid residue that cannot be distilled without decomposition, even in a vacuum. Carré claims, however, that analysis of this residue shows that it has the composition corresponding to the structure (**2**; R = Cl), and that this is confirmed by its molecular weight in freezing benzene (experimental value, 244; theoretical value, 253).

Carré states that when the compound (**2**; R = Cl) is hydrolyzed with water using external cooling, and the solution is subsequently treated with calcium carbonate, two calcium salts can be isolated, analysis indicating that one has the composition $P_2(OCH_2CH_2O)_2O_2Ca$ and the other the composition $P_2(OCH_2CH_2O)(OH)_2O_2Ca$. He therefore argues that the dichloro compound (**2**; R = Cl) undergoes hydrolysis initially to give the corresponding dihydroxy compound (**2**; R = OH), and that this compound is very unstable and rapidly undergoes further hydrolysis with liberation of ethylene glycol and the formation of the linear derivative (**3**).

The minor product which is formed in the initial reaction and is insoluble is ether has been identified as the compound (**4**) by Carré.

$(HO)_2P—OCH_2CH_2O—P(OH)_2$ $ClCH_2CH_2O—P(OH)Cl$ $ClCH_2CH_2O—PCl_2$
         (**3**)   (**4**)   (**5**)

He adds that, if phosphorus trichloride is added to ethylene glycol without ether as a solvent, the compound (**5**) is formed and this can be distilled in a vacuum, but no boiling point is quoted.

These results are in conflict with the much later work of Rossiĭskaya et al.,[383] who found that phosphorus trichloride reacts with the glycol in the absence of a solvent to give the volatile 2-chloro-1,3,2-dioxaphospholane and without the formation of the compound (**5**) (p. 270), and with the work of Lucas et al.[384] who also found that these reactants in dichloromethane solution give 2-chloro-1,3,2-dioxaphospholane in high yield (cf. p. 270).

Carré's isolation of the compounds (**2**; R = Cl and OH) must be accepted with reserve until the products of the reaction in ether are confirmed.

Five-membered and Six-membered Ring Systems containing only Carbon and One Phosphorus, One Nitrogen and One Oxygen Atom

1,3,2-Oxazaphosphole and 2H-1,3,2-Oxazaphosphorine

These two systems differ mainly in having a five- and a six-membered ring, respectively. The study of the synthesis and properties of their derivatives has so much in common that the two systems are conveniently discussed in one section

1,3,2-Oxazaphosphole, [RRI 73], has the structure and numbering shown in (**1**), but the ring system is almost invariably encountered in

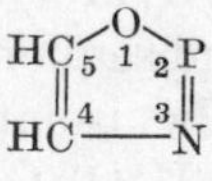

         (**1**) 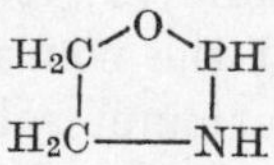   (**2**) 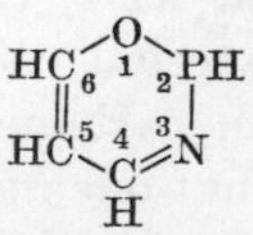   (**3**) 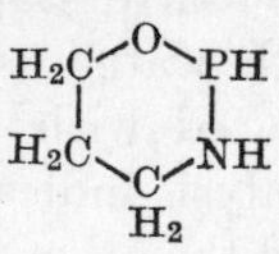   (**4**)

derivatives of the tetrahydro compound, 1,3,2-oxazaphospholidine (**2**). Similarly 2H-1,3,2-oxazaphosphorine (**3**), [RIS 11963], is encountered as derivatives of tetrahydro-2H-1,3,2-oxazaphosphorine (4); in this compound the position of the additional four hydrogen atoms is unambiguous and their 3,4,5,6 position is not usually specified in the name.

The fundamental type of reaction for the synthesis of derivatives of (2) and (4) is shown by two examples. A mixture of 2-aminoethanol hydrochloride, $NH_2(CH_2)_2OH,HCl$, and phosphoryl chloride is heated for 2 hours at 110–140°, and unchanged $POCl_3$ then removed by vacuum-distillation. The residue on purification gives the crystalline 2-chloro-1,3,2-oxazaphospholidine 2-oxide (5) (Bersin et al.[480]).

(5)   (6)

A dichloromethane solution of $POCl_3$ is added dropwise to a mixed solution of 3-aminopropanol and triethylamine also in dichloromethane, at 5–10°, and the reaction mixture is stirred for 16 hours. The solution, after filtration to remove triethylamine hydrochloride, is evaporated under reduced pressure, and the residual syrup taken up in dichloromethane; this solution is filtered and again evaporated. The final syrup is extracted with cold benzene: evaporation of the benzene extract gives the crystalline 2-chlorotetrahydro-2H-1,3,2-oxazaphosphorine 2-oxide (6) (48%). A short-path distillation of a sample at 85–90°/0.015 mm gave the pure product (6), m.p. 80–83° (5%) with considerable residue (Iwamoto et al.[481]).

Almost all the work on these two classes has involved the preparation of derivatives of (5) and (6) having a bis-(2-chloroethyl)amino or other bis(chloroalkyl)amino groups linked to the phosphorus atom, in the hope that the introduction of the 'nitrogen-mustard' unit might produce valuable pharmacological properties. This work initially appeared in three patent specifications having numerous examples (Aste-Werke Akt.-Ges. Chem. Fab.[482–484]), and the salient points have been discussed by Arnold and Bourseaux.[485] The pharmaceutical promise of certain of these compounds has evoked a very full investigation of their synthesis, their potential therapeutic action, and their clinical examination.

A reagent used in much of this work is N,N-bis-(2-chloroethyl)phosphoramidic dichloride (7), which is prepared by boiling a mixture of

$(ClCH_2CH_2)_2N—POCl_2$ $(ClCH_2CH_2)_2N—PSCl_2$ $(ClCH_2CH_2)N—PO(OC_6H_5)Cl$

(7)   (8)   (9)

di-(2-chloroethyl)amine hydrochloride and phosphoryl chloride under reflux for 12 hours; fractional distillation then gives the dichloride (7), b.p. 123–125°/0.6 mm, m.p. 54–56°. Of less importance is the corresponding thio derivative (8) which is prepared by boiling the above hydrochloride with thiophosphoryl chloride, $SPCl_3$. for 45 hours. After concentration under reduced pressure, the residue is extracted with cold ether; evaporation of the extract gives the N,N-bis-(2-chloroethyl)thiophosphoramidic dichloride [483] (8), m.p. 34°.

A third reagent is phenyl phosphoramidochloridate (9), obtained when a benzene solution of dry triethylamine is added to a mixed solution of the dichloride (7) and phenol in boiling benzene, which is then boiled for 3 hours and cooled. Deposited triethylamine hydrochloride is removed and the filtrate when fractionally distilled gives the chloridate (9), a pale yellow oil, b.p. 167–169°/0.2 mm (Friedman and Seligman [310]).

The synthesis of compounds of the required type is illustrated by the following examples.

A dioxane solution of 2-aminoethanol ('ethanolamine') (1 mole) and triethylamine (2 moles) is added dropwise at room temperature to a

stirred dioxane solution of the dichloride (7); after filtration to remove the triethylamine hydrochloride, the solution is evaporated under reduced pressure at 40–45°, leaving the residual 2-[bis-(2′-chloroethyl)-amino]-1,3,2-oxazaphospholidine 2-oxide (10), m.p. 99.4°. It is noteworthy that if an N-acyl derivative of 2-aminoethanol, e.g.,

$$HOCH_2CH_2NHCOCH_3,$$

is used in this preparation, the same product (10) is obtained, the acyl group being evicted during the reaction. The use of 2-methylaminoethanol, $(HOCH_2CH_2)CH_3NH$, gives, as expected, the 3-methyl derivative of (10), an almost colorless oil, and 1-amino-2-propanol $CH_3CH(OH)CH_2NH_2$ gives the 5-methyl derivative of (10), m.p. 81°.

Under the same conditions, a dry dioxane solution of 3-amino-1-propanol, $HO(CH_2)_3NH_2$, and triethylamine reacts with the dichloride (7) to give 2-[bis-(2′-chloroethyl)amino]tetrahydro-2H-1,3,2-oxazaphosphorine 2-oxide (11), an oil which forms a monohydrate, m.p. 48–49°.

The therapeutic properties of this compound, briefly discussed later, make it the most important member of the double series (10), (11), and their derivatives. All the above compounds are soluble in water.

This condensation of $\alpha\omega$-alkanolamines, $HO(CH_2)_nNH_2$, with dichloride (7) also occurs with similar alkanolamines having $n = 4$, 5, or 6. The reaction of 2-aminoethanol in particular is solely that of cyclization; even when an excess of this amine is used, no indication of the formation of a compound such as (12) has been obtained.

A wide variety of compounds of type (10) and (11) can clearly be prepared by using suitably substituted derivatives of 2-aminoethanol and 3-amino-1-propanol. Thus condensation of the dichloride (7) with 'diethanolamine', $(HOC_2H_4)_2NH$, gives the 3-(2'-hydroxyethyl) derivative of (10); condensation with ($\pm$)-serine methyl ester hydrochloride gives the methyl ester of the ($\pm$)-4-carboxy derivative of (10).

The dichloride (7) has been similarly condensed with the ethyl ester of both *threo*- and *erythro*-($\pm$)-β-phenylserine, to give the corresponding isomers (13), and with the diethyl ester of ($\pm$)-*threo*-β-hydroxyglutamic acid to give (14) (Szekerke, Kajtav, and Bruckner[334]). For the preparation of similar compounds in various stereoisomeric forms by the condensation of the dichloride (7) with (+)-, (−)- and ($\pm$)-ψ-ephedrine and with (+)-, (−)- and ($\pm$)-ephedrine, see Fel'dman and Berlin.[486]

(13) (14)

A different type of derivative of (11) has been obtained by condensing the chloride (6) with phenylalanine to give the compound (15).

The use of the thio-dichloride (8) in place of the dichloride (7) gives the corresponding 2-sulfides: thus condensation of (8) with 2-amino-ethanol and 3-amino-1-propanol gives respectively the 2-sulfides (16),

(15) (16) (17)

m.p. 66–67°, and (17), a yellow viscous oil; both are insoluble in water.[483]

The 2-chloroethyl groups in the dichloride (7)—and therefore in the corresponding derivatives (10) and (11)—have also been changed,[484] the $(ClCH_2CH_2)_2N$ group having been replaced by the following (and other analogous) groups:

$(CH_3CHClCH_2)_2N$ $(ClCH_2CH_2CH_2)_2N$

ClCH$_2$CH$_2$ \   CH$_3$CHClCH$_2$ \   ClCH$_2$CH$_2$ \
                    N   N   N
CH$_3$CHCl—CH$_2$ /   ClCH$_2$CH$_2$CH$_2$ /   CH$_3$CH$_2$CHClCH$_2$ /

In an alternative synthesis[482] of the oxazaphospholidine (**10**), 2-aminoethanol in dioxane is added dropwise to a boiling dioxane solution of phenyl phosphoramidochloridate (**9**); the mixture when worked

$$\begin{array}{c} PhO \diagdown \quad P \diagup\!\!=\!\! O \\ \qquad \Big| \quad N(C_2H_4Cl)_2 \\ HOH_2C \diagdown \\ \qquad H_2C—NH \end{array}$$

(**18**)

up in the usual way gives the compound (**10**). This reaction presumably involves intermediate formation of the product (**18**), which undergoes rapid cyclization to the compound (**10**) with ejection of phenol.[482, 485]

The hydrolysis of the compounds (**10**) and (**11**), both by pure water at 37° and by phosphatase–water at 37°, has been investigated in some detail:[485] in these circumstances the heterocyclic ring of (**10**) undergoes cleavage far more readily than that of (**11**).

It is convenient to mention here that a 1,2- or a 1,3-glycol in dioxane also condenses readily at room temperature with the dichloride

(**19**)   (**20**)

(**7**). Thus ethylene glycol gives 2-[bis-(2′-chloroethyl)amino]-1,3,2-dioxaphospholane 2-oxide (**19**), a pale yellow oil, sparingly soluble in water, and trimethylene glycol gives 2-[bis-(2′-chloroethyl)amino]-1,3,2-dioxaphosphorinane 2-oxide[482] (**20**), m.p. 49–50°.

Of the many hundreds of compounds containing the 'nitrogen-mustard' unit, 2-[bis-(2′-chloroethyl)amino]tetrahydro-2H-1,3,2-oxaza-phosphorine 2-oxide (**11**)—known commonly as Endoxan, Endoxana, Cytoxan, and Cyclophosphamide—has probably been of greatest use for delaying, or even arresting, the development of certain types of malignant tissue. It has been suggested that this comparatively powerful cytostatic action arises from the compound's being an active 'transport' form of the $N(C_2H_4Cl)_2$ portion, and that this alkylating group is selectively liberated in the tumor tissue by phosphoramidase enzymes, which are present in greater amounts in such tissue than in normal tissue;[481] consequently the C_2H_4Cl group may be the active cytostatic group. The compound has a greater margin of safety than other cytostatic agents, and although it does affect the bone-marrow causing a fall in the leucocyte (white cell) level, the normal leucocyte level is regained if the treatment is reduced or completely interrupted. The compound is devoid of the vesicant action of nitrogen-mustards, and its side effects, such as nausea and vomiting, are much less marked than those caused by nitrogen-mustards.

This cytostatic action, although to a lesser degree, is claimed for all derivatives of (10) and (11) cited in the specification.[482]

It is noteworthy that the replacement of one or both chlorine atoms in the compound (11) by fluorine atoms destroys the anti-tumor activity (determined against rodent tumors) (Papanastassiou et al.[487]).

For a discussion of the pharmacology of (10) and (11), see Brock,[488] also Arnold, Bourseaux, and Brock;[489] for a more recent and detailed discussion of (11), see 'Cyclophosphamide', edited by G. H. Fairley and J. M. Simister,[490] an account of the Proceedings of a Symposium held at the Royal College of Surgeons of England on 4th October, 1963. In this volume it is stated that by 1964 over 800 publications on various aspects of Endoxan had appeared.

Note. The compounds (10) and (11) were earlier termed NN-bis-(β-chloroethyl)-N^1,O-ethylene phosphoric acid ester diamide and NN-bis-(β-chloroethyl)-N^1,O-propylene phosphoric acid ester diamide, respectively[485]; the same names, but often with the omission of 'ester', were used in the specification.[482]

The compounds (19) and (20) were termed, respectively, NN-bis-(β-chloroethyl)-O,O^1-ethylene (and propylene) phosphoric acid amide.[482]

Five-membered Ring Systems containing only Carbon and One Phosphorus Atom and Two Sulfur Atoms

1,3,2-Dithiaphosphole and 1,3,2-Dithiaphospholanes

1,3,2-Dithiaphosphole (1), [RRI 112], is encountered only as substitution products of its 4,5-dihydro derivative (2; R = H), termed 1,3,2-dithiaphospholane, but originally termed 1,3-dithia-2-phospha-cyclopentane.

Compounds of type (**2**) were first prepared (A. E. Arbuzov and Zoroastrova[491]) by the interaction of 1,2-ethanedithiol, $HSCH_2CH_2SH$, and phosphorus trichloride in ether; the solution when heated at 50°

(**1**)   (**2**)   (**3**)

for 20 minutes and then distilled gave 2-chloro-1,3,2-dithiaphospholane (**2**; $R = Cl$), b.p. 102–103°/10 mm (67%), and a more complex material, considered (correctly) to be 2,2′-(ethylenedithio)bis-1,3,2-dithiaphospholane (**3**), m.p. 130°.

The chloro compound (**2**; $R = Cl$) when added to a dry methanol–pyridine mixture with ice-cooling gave the 2-methoxy derivative (**2**; $R = OCH_3$), b.p. 97–98.5°/7.5 mm; the dithiol, when mixed in ether–pyridine with ethyl phosphorodichloridite, $C_2H_5OPCl_2$, gave the 2-ethoxy derivative (**2**; $R = OC_2H_5$), a liquid, b.p. 98–99°/5 mm, having an unpleasant odor and readily forming crystalline cuprous chloride and iodide complexes, m.p. 122–124° and 134°, respectively. The formation of these complexes indicates that the organic portion very probably contains trivalent phosphorus. When this compound (**2**; $R = OC_2H_5$) was heated with ethyl iodide at 80–90° in a sealed tube for 3 hours, the product contained the crystalline 1,4-dithian, $S(CH_2CH_2)_2S$, and a liquid portion, b.p. 172–174°/4 mm, which had the same composition as (**2**; $R = OC_2H_5$) and was probably the isomeric 2-ethyl-1,3,2-dithiaphospholane 2-oxide (**4**).

(**4**)   (**5**)   (**6**)

The 2-methyl derivative (**2**; $R = CH_3$), b.p. 90°/5 mm, m.p. −5°, can be prepared by addition of the dithiol to a solution of methylphosphonous dichloride in ether–trimethylamine (Wieber, Otto, and Schmidt[492]); it readily gives by direct addition a 2-sulfide, b.p. 132–134°/2 mm, m.p. 74°, and a 2-selenide, b.p. 148°/2 mm, m.p. 68°.

A novel synthetic route in this field involves the use of 2,2-dimethyl-1,3-dithia-2-silacyclopentane (**5**), a liquid, b.p. 188°, which can be readily prepared by the action of 1,2-ethanedithiol on bis(diethylamino)dimethylsilane, $[(C_2H_5)_2N]_2Si(CH_3)_2$, or (less readily) on tri-N-ethylhexamethylcyclotrisilazane, also termed 1,3,5-triethyl-2,2,4,4,6,6-hexa-methyl-1,3,5-triaza-2,4,6-trisilacyclohexane, $[C_2H_5N \cdot Si(CH_3)_2]_3$, (Abel, Armitage, and Bush[493]). When a mixture of equimolecular quantities of the silane (5) and phosphorus trichloride is first warmed to 50° to effect an exchange reaction and to eliminate dichlorodimethylsilane, $Cl_2Si(CH_3)_2$, distillation of the residue gives the 2-chloro derivative (2; R = 3l), b.p. 67°/1 mm (75%). When a mixture of the silane (5) and phosphorus trichloride in the molecular ratio of 3:2 is similarly treated, the compound (3), m.p. 130° (79% crude) is formed, thus confirming the identity of the product obtained earlier.[491]

Equimolar quantities of the silane (5) and phenylphosphonous dichloride give the 2-phenyl derivative (2; R = C_6H_5), b.p. 136–137°/0.2 mm (58%), with a small amount of 2-phenyl-1,3,2-dithiaphospholane 2-oxide, b.p. 180°/0.001 mm, m.p. 70°.

1,2-Ethanedithiol reacts with N,N-bis-(2-chloroethyl)phosphoramidic dichloride, $(ClCH_2CH_2)_2NP(O)Cl_2$ (p. 321), to form 2-[bis-(2'-chloroethyl)amino]-1,3,2-dithiaphospholane 2-oxide (6), m.p. 115–117°, one of the many cyclic phosphorus compounds having the $\!>\!PN(C_2H_4Cl)_2$ side-chain, which have been prepared in order to investigate their possible action as neoplasm inhibitors (Arnold et al.[330]) (cf. p. 324).

1,3,2-Benzodithiaphosphole

Derivatives of 1,3,2-benzodithiaphosphole (1), [RIS 9922], initially termed 1,3-dithia-2-phosphaindan, were first prepared by Campbell and

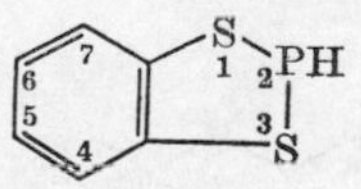

(1)

Way[314] in a stereochemical investigation of the optical stability of compounds in which the activity was due solely to a 3-covalent phosphorus atom. For this purpose, a dry stirred ethereal solution of

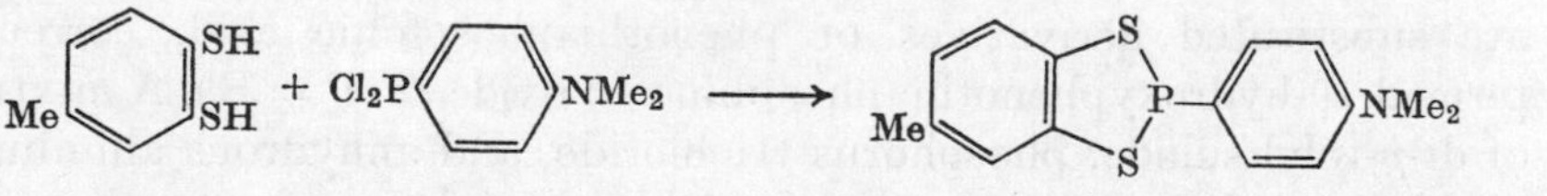

(2)   **(3)**   **(4)**

3,4-toluenedithiol (2) containing pyridine was treated with p-dimethyl-aminophenylphosphonous dichloride (3); the precipitated pyridine hydrochloride was rapidly removed and the solution deposited the pale yellow crystalline 2-[p-(dimethylamino)phenyl]-5-methyl-1,3,2-benzodithiaphosphole (4), m.p. 125–126° (monomethiodide, m.p. 125–127°, slowly decomposing at room temperature).

Satisfactory salts of (4) with tartaric acid, dibenzoyltartaric acid, and camphor-10-sulfonic acid could not, however, be obtained and the investigation on these lines was abandoned.

It is noteworthy that attempts to condense the dithiol (2) under the same conditions with phenylphosphonous dichloride gave phenylphosphonic acid, m.p. 158–159°, and with p-bromophenylphosphonous dichloride gave p-bromophenylphosphonous acid, m.p. 143°, no oxidation having occurred in the latter case.

The simpler 2,5-dimethyl-1,3,2-benzodithiaphosphole, b.p. 110–114°/2 mm, has been prepared by the similar condensation of the dithiol (2) with CH_3PCl_2; it gave a 2-sulfide and a 2-selenide, which could not be distilled (Wieber et al.[492]).

Six-membered Ring Systems containing only Carbon and One Phosphorus and One Sulfur Atom

Phenothiaphosphine

Phenothiaphosphine, [RIS —], is the accepted name for the compound (1), and the systematic name, 10H-dibenzo[1,4]thiaphosphorin, is not employed (cf. phenoxphosphine, p. 289).

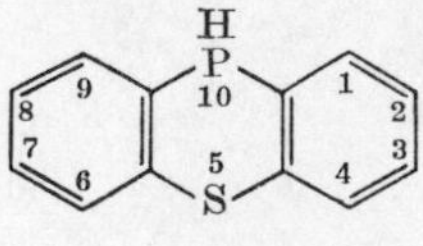

(1) 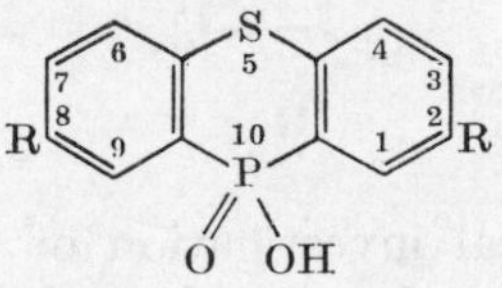   (2) 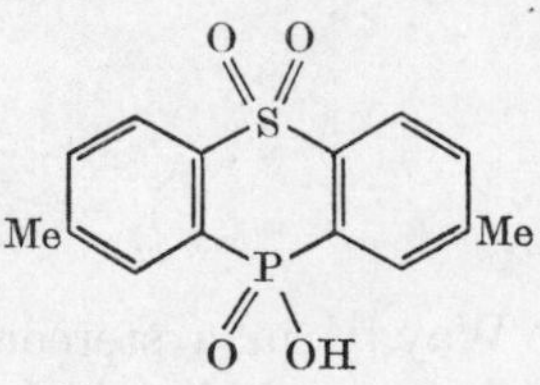   (3)

At present, the only known compounds having the ring system (1) are substituted derivatives of phenothiaphosphinic acid, correctly termed 10-hydroxyphenothiaphosphine 10-oxide (2; R = H). A mixture of di-p-tolyl sulfide, phosphorus trichloride, and anhydrous aluminum trichloride is boiled under reflux for 7 hours, cooled, and poured on ice; the heavy oil which separates is dissolved in boiling 5% aqueous sodium hydroxide (charcoal), and the filtered solution treated with an excess of concentrated hydrochloric acid. The white precipitate, when thoroughly extracted with boiling ether and then recrystallized from ethanol, yields 2,8-dimethylphenothiaphosphinic acid, or 10-hydroxy-2,8-dimethyl-phenothiaphosphine 10-oxide (**2**; R = CH$_3$), m.p. above 300° (25%) (Freedman and Doak[494]). The mechanism of this formation of the ring system (**1**) is unknown: the fact that traces of p-tolylphosphonic acid and of di-p-tolyl disulfide were isolated from the above reaction mixture shows that some cleavage of the aryl—S bond must occur. The aluminum trichloride is clearly essential, for in its absence the above reaction mixture of the sulfide and phosphorus trichloride can be heated under reflux for 24 hours without interaction.

The acid (**2**; R = CH$_3$), when suspended in boiling acetic acid and treated with 30% hydrogen peroxide, is converted into 10-hydroxy-2,8-dimethylphenothiaphosphine 5,5,10-trioxide (**3**), m.p. above 300°. The infrared spectrum of (**3**) shows strong bands at 1158 and 1315 cm^{-1}, assigned to the symmetric and the asymmetric stretching modes, respectively, of the SO$_2$ group; these bands do not occur in the spectrum of the parent acid.

Six-membered Ring Systems containing only Carbon and One Phosphorus and Two Sulfur Atoms

4H-1,3,2-Dithiaphosphorin

This compound, [RIS 11606], has the structure and numbering shown in (**1**), and its 5,6-dihydro derivative (**2**; R = H) is termed 1,3,2-dithiaphosphorinane.

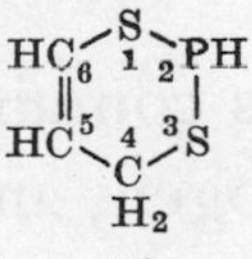

(**1**) 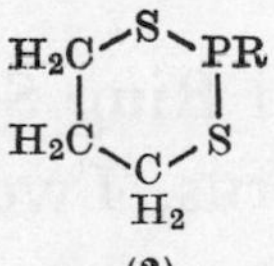   (**2**)

The synthesis of the dithiaphosphorinane system follows that of the five-membered 1,3,2-dithiaphospholanes (Wieber *et al.*[492]). When 1,3-propanedithiol, HS(CH$_2$)$_3$SH, is added to an ethereal solution of methylphosphonous dichloride containing triethylamine (2 moles), a smooth condensation occurs with the formation of 2-methyl-1,3,2-dithiaphosphorinane (**2**; R = CH$_3$), b.p. 102°/5 mm. Treatment in solution with sulfur gives the corresponding 2-sulfide, m.p. 121°.

A large number of derivatives could clearly be prepared by ring closure on the above lines.

12

The compound (**2**; R = CH$_3$) was initially[492] termed 2-methyl-1,3,2-dithiaphosphane.

Five-membered Ring Systems containing Carbon and One Phosphorus, One Nitrogen, and One Sulfur Atom

1,3,2-Thiazaphosphole

1,3,2-Thiazaphosphole, [RIS 9741], has the structure (**1**), and its tetrahydro derivative (**2**) is 1,3,2-thiazaphospholidine.

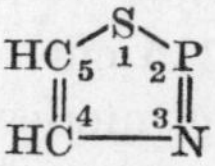

(1) 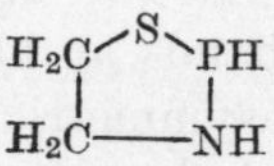   (2) 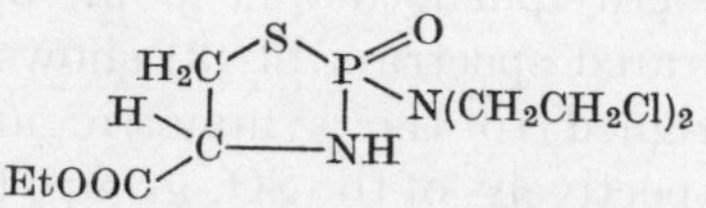   (3)

The latter is one of the many heterocyclic systems into which a 'nitrogen-mustard' unit has been inserted in the search for neoplasm inhibitors, *i.e.*, drugs which will slow up or arrest the growth of tumor tissue.

The ethyl ester, HSCH$_2$CH(NH$_2$)CO$_2$C$_2$H$_5$, of L-cysteine has been condensed with N,N-bis-(2-chloroethyl)phosphoramidic dichloride, (ClCH$_2$CH$_2$)$_2$NP(O)Cl$_2$ (p. 321), to form the ethyl ester (**3**) of 2-[bis-(2'-chloroethyl)amino]-1,3,2-thiazaphospholidine-4-carboxylic acid 2-oxide (Szekerke, Kajtar, and Bruckner[334]).

Six-membered Ring Systems containing Carbon, Two Phosphorus, Two Nitrogen, and One Sulfur Atom

4*H*-1,2,6,3,5-Thiadiazadiphosphorine

An interesting series of compounds, described in a preliminary note by Appel and Hänssgen,[495] can be regarded as derivatives of a novel ring system, 4*H*-1,2,6,3,5-thiadiazadiphosphorine (**1**) [RIS —].

Dimethyl sulfide bis(bromoimine) (**2**) combines with methylene-bis(diphenylphosphine) (**3**) in benzene–acetonitrile to give the salt formulated as (**4**), *i.e.*, 1,1-dimethyl-3,3,5,5-tetraphenyl-4*H*-1,2,6,3,5-thia(VI)diazadiphosphorinium dibromide (**4**). This salt, when heated at

110–120° in a high vacuum, loses one molecule of methyl bromide to form 1 - methyl - 3,3,5,5 - tetraphenyl - 1H - 1,2,6,3,5 - thia(IV)diazaphospha - (V)phosphorinium bromide (5). Liquid ammonia abstracts a proton from

(1) 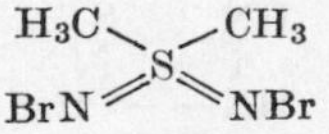   (2) 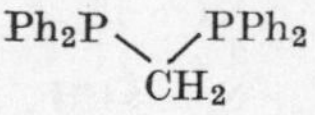   (3)

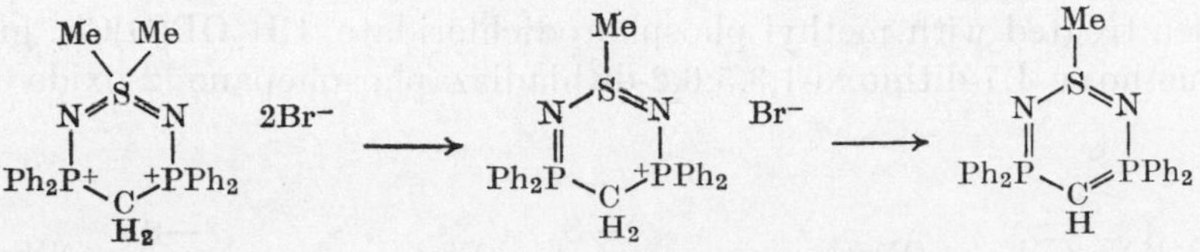

(4)   (5)   (6)

the methylene group of (5), with the formation of 1-methyl-3,3,5,5-tetraphenyl-1,2,6,3,5-thia(IV)diazadiphosph(V)orine (6).

The structures assigned to the compounds (4), (5), and (6) are based on elemental analysis, the ionic equivalents of (4) and (5), the determination of the molecular weight of (6), and on the different types of proton shown in the pmr spectra. The ionic nature of (4) and (5) is also confirmed by their conductivities, and by the formation of the corresponding nitrates and picrates by simple double decomposition.

The salts (4) and (5) are soluble in polar solvents such as alcohols and dimethyl sulfoxide, but are insoluble in ether and hydrocarbons. The compound (6) dissolves readily in benzene and chloroform.

The authors[495] term the compound (6) 1-methyl-3,3,5,5-tetraphenyl-1-thia-3,5-diphospha(V)-2,6-diazine.

Seven-membered Ring System containing only Carbon and One Phosphorus Atom, Two Nitrogen Atoms, and Two Sulfur Atoms

1,3,5,6,2-Dithiadiazaphosphepine

This parent compound (1), [RIS 11627], is encountered as substitution derivatives of the 4,5,6,7-tetrahydro compound (2).

Brief discussion of this ring system is warranted by its unusual type of synthesis.

Hydrazine combines with carbon disulfide to give bis(dithio-carbonyl)hydrazine, which forms a stable dipotassium salt (**3**). It is

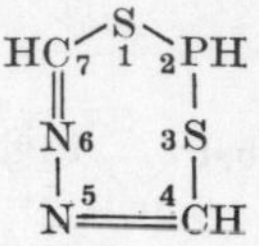

(1) 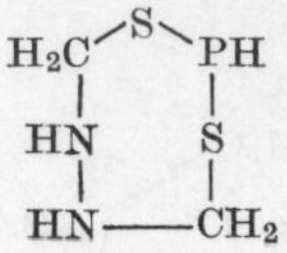   (2) 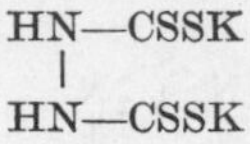   (3)

claimed in a patent specification [496] that a suspension of (**3**) in acetone, when treated with methyl phosphorodichloridate, $CH_3OP(O)Cl_2$, gives 2-methoxy-4,7-dithioxo-1,3,5,6,2-dithiadiazaphosphepane 2-oxide (**4**).

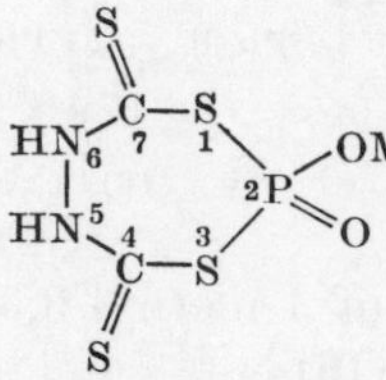

(4) 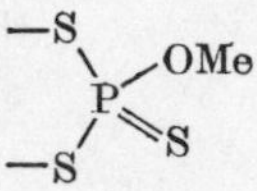   (5) 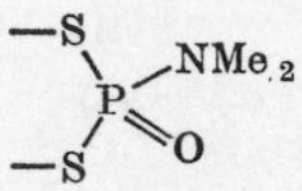   (6)

The use of *O*-methyl thiophosphorodichloridate, $CH_3OP(S)Cl_2$, gives the corresponding 2-sulfide (**5**), and *N,N*-dimethylphosphoramidic dichloride, $(CH_3)_2NP(O)Cl_2$, gives the 2-dimethylamino 2-oxide (**6**).

The similar preparation of the 2-ethoxy and the 2-*n*-propoxy analogues of (**4**), the 2-ethoxy analogue of (**5**), and the 2-di-*n*-propyl-amino analogue of (**6**), is also mentioned (Nagasawa, Imamiya, and Sugiyama [496]). It is stated that these compounds are useful as insecticides and bactericides.

Six-membered Ring Systems containing Carbon and One Phosphorus and One Arsenic Atom

1,4-Benzophospharsenin

Phospharsenin (**1**), and its hexahydro derivative phospharsenan (**2**), alone or as substituted products, are at present unknown, but the ring system 1,4-benzophospharsenin (**3**), [RRI 1536], is known as derivatives of 1,2,3,4-tetrahydro-1,4-benzophospharsenin (**4**).

o-Bromophenylarsonous dichloride (**5**; R = Cl, R′ = Br) is readily converted by methylmagnesium bromide into *o*-bromophenyldimethyl-arsine (**5**; R = CH$_3$, R′ = Br) (76%), which when treated in petroleum

(1)   (2)   (3)   (4)

with n-butyllithium gives the lithio derivative (**5**; $R = CH_3$, $R' = Li$); this in turn with diethylphosphinous chloride gives [o-(diethylphosphino)phenyl]dimethylarsine (**6**; $R = CH_3$), b.p. $105–106°/0.6$ mm, identified as the P-methiodide, m.p. $162–163°$.

(5)   (6)   (7)   (8)

Repetition of these reactions using o-bromophenyldiethylarsine (**5**; $R = C_2H_5$, $R' = Br$) gives [o-(diethylphosphino)phenyl]diethylarsine (**6**; $R = C_2H_5$), b.p. $136°/1.0$ mm. This compound readily gives a P-monomethiodide, m.p. $148–149°$, and a dimethiodide, m.p. $165°$: the allocation of the first methyl group is based solely on the greater reactivity of tertiary phosphine groups than of analogous arsine groups to quaternizing agents.

The phosphine-arsine (**6**; $R = CH_3$) reacts vigorously with 1,2-dibromoethane to give the crystalline 1,1-diethyl-1,2,3,4-tetrahydro-4,4-dimethylbenzophospharseninium dibromide (**7**; $R = CH_3$), m.p. $235–245°$ (dipicrate, m.p. $196–197°$). The phosphine-arsine (**6**; $R = C_2H_5$) similarly gives 1,1,4,4-tetraethyl-1,2,3,4-tetrahydrobenzophospharseninium dibromide (**7**; $R = C_2H_5$), m.p. $240°$ (dipicrate, m.p. $221°$) (Jones and Mann[497]).

The thermal decomposition of diquaternary salts such as (**7**) is of considerable interest. Quaternary arsonium halides containing one or more alkyl groups usually readily undergo thermal loss of an alkyl halide with formation of the tertiary arsine; similar quaternary phosphonium halides usually gives the tertiary phosphine far less readily (pp. 68, 69) and almost certainly by a different mechanism (p. 125). The dibromide (**7**; $R = C_2H_5$), when heated at $270°/15$ mm for 15 minutes, effervesced with loss of ethyl bromide, giving a residue of 1,1,4-triethyl-1,2,3,4-tetrahydrobenzophospharseninium bromide (**8**; $R = C_2H_5$), m.p. $167–169°$; further heating, up to $350°/2$ mm, of this monobromide, either as the crude residue or as the purified crystals, gave a mild secondary effervescence, with general decomposition and deposition of arsenic, and the minute distillate gave no crystalline derivatives. The pure bromide (8; $R = C_2H_5$) when heated in ethanolic ethyl bromide regenerated the dibromide (7; $R = C_2H_5$).[497]

The salt (7; $R = CH_3$) was originally[497] named 4,4-diethyl-1,1-dimethylethylene-o-phenylene-1-arsonium-4-phosphonium dibromide.

Therapeutic investigation showed that the water-soluble dibromide (7; $R = C_2H_5$) had no significant value as a possible schistosomicide; the monobromide (8; $R = C_2H_5$) had some such action combined, however, with relatively high toxicity.

The various types of complex metallic derivatives which the above intermediate phosphine-arsine (6; $R = C_2H_5$) forms in particular with Pd(II) and with Cu(I), Ag(I), Au(I) and Pt(II) (Cochran, Hart, and Mann[498]) have been investigated in some detail. The isomorphous Cu(I) and Au(I) derivatives (9; $M = $ Cu or Au) are of particular interest

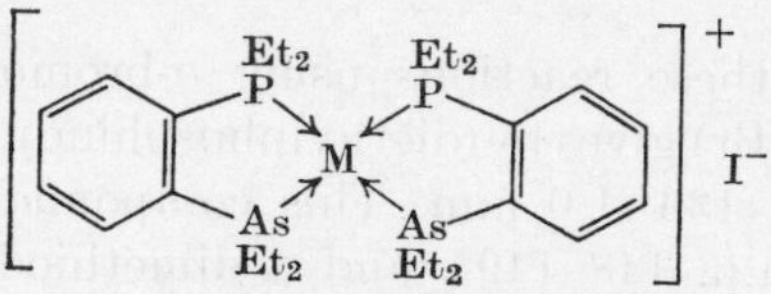

$$\left[\begin{array}{c} \underset{As}{\overset{P}{}} \quad M \quad \underset{As}{\overset{P}{}} \\ Et_2 \qquad Et_2 \end{array} \right]^{+} \quad I^{-}$$

(9)

because the crystal structure shows that although the positions of the P and As atoms are clearly defined, the atoms of these two elements cannot be differentiated owing to an apparently random distribution, i.e., sites conveniently labeled A and B may at one place in the crystal be occupied by P and As atoms, but crystallographically equivalent sites A′ and B′ may be occupied by P and As atoms, or by As and P atoms; this random distribution is quite distinct from the regular arrangement to be expected in a racemate. This structure confirms the tetrahedral configuration of the Cu(I) and Au(I) atoms in the above salts and probably explains the failure to resolve into optically active forms the aurous salt (9; $M = $ Au). The insertion of substituents into the o-phenylene groups of this salt in the hope that the increased dissymmetry of the cation might suppress this random distribution gave aurous salts of type (9) which, however, in spite of the use of various optically active anions, could not be resolved (Davis and Mann[499]). It has been suggested that the much higher solubility of the iodides of these cations (9; $M = $ Au) in cold methanol and ethanol than in many other solvents may indicate coordination with the former solvents to form a symmetric cation, the salts crystallizing however in the solvent-free condition.[499]

Dibenzo[1,4]phospharsenin

This ring system (1), [RIS 12646], is found, as one would expect, only as 5,10-disubstituted derivatives (2) of the 5,10-dihydrodibenzo-[1,4]phospharsenin.

The synthesis of compounds of type (2) (Davis and Mann [264, 143]) is identical with that of similar phosphanthrene derivatives (p. 184) except

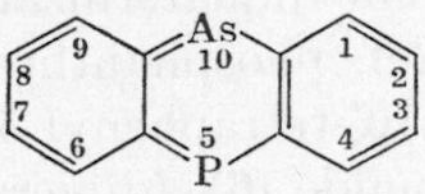

(1) 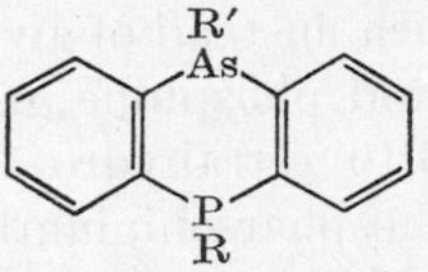   (2) 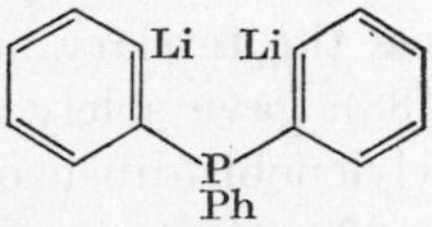   (3)

for the penultimate step. 2,2′-Dilithiotriphenylphosphine (3) in tetra-hydrofuran at −60° was treated with phenylarsonous dichloride, $C_6H_5AsCl_2$, also in tetrahydrofuran; the crude product, when distilled at 0.009 mm, gave a low-boiling fraction containing mainly triphenyl-arsine and 9-phosphafluorene (5H-dibenzophosphole), and a higher fraction (mainly of b.p. 216°) which when stirred with acetone deposited the colorless 5,10-dihydro-5,10-diphenyldibenzo[1,4]phospharsenin (2; $R = R' = C_6H_5$), m.p. 189–190°. This reacted with potassium palladobromide to give insoluble orange crystals of composition $C_{24}H_{18}AsP,PdBr_2$ but of unknown molecular weight.

The compound (2; $R = R' = C_6H_5$), which like the phosphanthrene analogue showed no indication of isomeric forms, readily gave the phos-phonium methiodide (4; $R = CH_3$, $X = I$), m.p. 300–302° (picrate, m.p. 171–172°); the identity of the methiodide was confirmed by the action

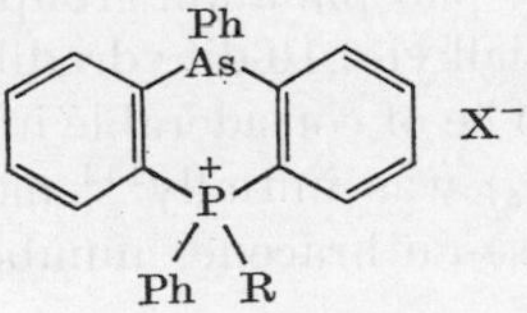

(4) 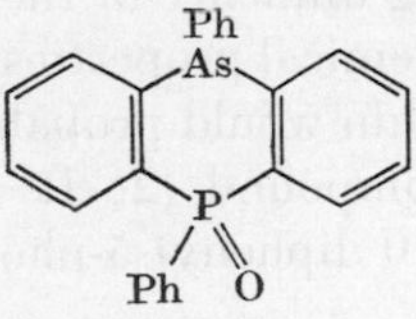   (5)

of hot aqueous sodium hydroxide, which converted it into the 5-oxide (5), m.p. 214–216°, having a strong P—O band in its infrared spectrum. Oxidation of the 5-oxide with hydrogen peroxide gave the 5,10-dihydro-5,10-diphenyldibenzo[1,4]phospharsenin 5,10-dioxide, m.p. 360°.

The methiodide, even when heated with methyl p-toluenesulfonate at 180° for 4 hours, did not undergo further quaternization. Similarly the compound (2; $R = R' = C_6H_5$) readily gave a quaternary 5-benzyl bromide (**4**; R = CH$_2$C$_6$H$_5$, X = Br), m.p. 318–322° (picrate, m.p. 170–171°), but this bromide, when treated again with benzyl bromide under forcing conditions, gave a very small yield of a quaternary 5,10-di(benzyl bromide), identified as its dipicrate, m.p. 212–213°.

This very low activity of the tertiary arsine in salts such as (**4**) would be expected in view of the strong inductive effect of the positive pole on the phosphorus atom. It was demonstrated further by the action of *o*-xylylene dibromide which, instead of giving cyclic diquaternization across the tertiary arsine and phosphine groups (cf. phosphanthrene, p. 188), gave solely 5,5′,10,10′-tetrahydro-5,5′,10,10′-tetraphenyl-5,5′-*o*-xylylenebis(dibenzo[1,4]phospharseninium) dibromide (**6**), (dipicrate, m.p. 239–240°).

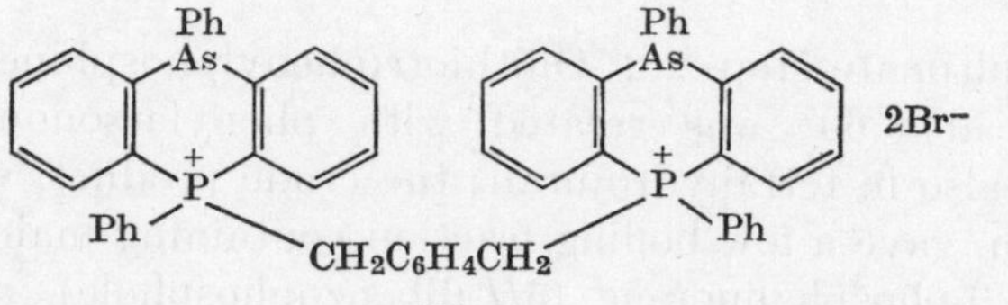

(**6**)

10-Ethyl-5,10-dihydro-5-phenyldibenzo[1,4]phospharsenin (**2**; R = C$_6$H$_5$, R′ = C$_2$H$_5$) was similarly prepared by the action of ethylarsonous dichloride on the dilithio compound (**3**), and when heated with benzyl bromide at 100° for 5 hours gave a mixture of the 5-benzyl bromide, m.p. 295–296° (picrate, m.p. 180–182°) and the 5,10-dibenzyl dibromide, m.p. 363–365° (dipicrate, m.p. 206°). The ethylarsine group in the parent compound would have greater reactivity than the phenylarsine group in (**2**; R = R′ = C$_6$H$_5$) and hence could more successfully resist the deactivating influence of the quaternary phosphonium group.

The chemical properties of a 5,10-dialkyl-5,10-dihydrodibenzo[1,4]-phospharsenin would probably prove to be of considerable interest.

The compound (**2**; R = R′ = C$_6$H$_5$) was initially[143] named 5,10-dihydro-5,10-diphenyl-5-phospha-10-arsa-anthracene, numbered as in (**1**).

Seven-membered Ring Systems containing Carbon and One Phosphorus and One Arsenic Atom

1*H*-1,5-Benzophospharsepin

Our knowledge of 1*H*-1,5-benzophospharsepin (**1**), [RRI 1816], is limited to substituted products of 2,3,4,5-tetrahydro-1*H*-1,5-benzophospharsepin (**2**), and follows the same lines as that of the hydrogenated 1,4-benzophospharsenin (p. 332).

[o-(Diethylphosphino)phenyl]dimethylarsine (p. 333) reacts readily with 1,3-dibromopropane by cyclic diquaternization to form 1,1-diethyl-2,3,4,5 - tetrahydro - 5,5 - dimethyl - 1H - 1,5 - benzophospharsepinium dibromide (**3**; R = CH_3), m.p. 245–247° (dipicrate, m.p. 201–202°). The

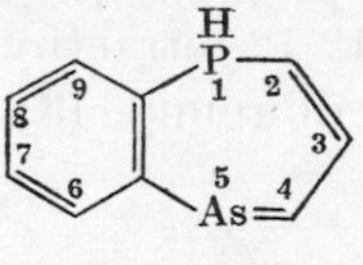

(1)

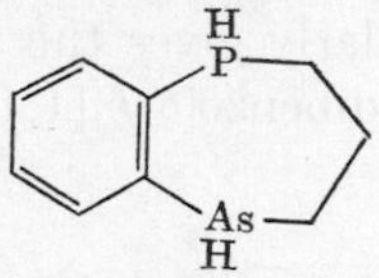

(2)

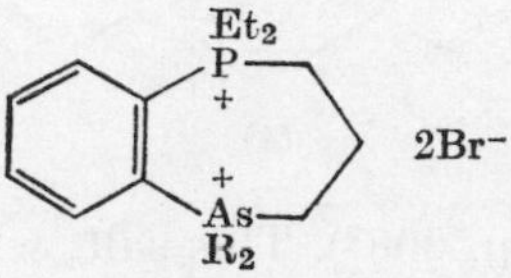

2Br⁻

(3)

Br⁻

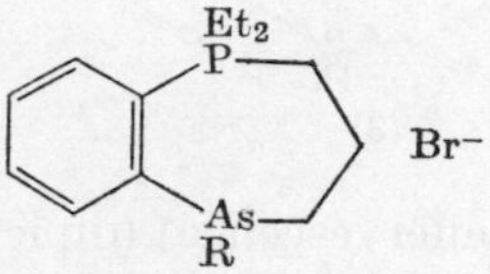

(4)

use of [o-(diethylphosphino)phenyl]diethylarsine similarly gives the 1,1,5,5-tetraethyl salt (**3**; R = C_2H_5), m.p. 240° (dipicrate, m.p. 221°) (Jones and Mann[497]). The thermal decomposition of this salt has not been investigated: it would almost certainly give 1,1,5-triethyl-2,3,4,5-tetrahydro-1H-1,5-benzophospharsepinium bromide (**4**; R = C_2H_5).

The salt (**3**; R = CH_3) was originally termed 4,4-diethyl-1,1-dimethyltrimethylene - o - phenylene - 1 - arsonium - 4 - phosphonium dibromide.[497]

Eight-membered Ring Systems containing Carbon and One Phosphorus and One Arsenic Atom

Dibenzo[b,f][1,4]phospharsocin

This system (**1**), [RRI 3736], is known only as substituted derivative of 5,6,11,12-tetrahydrodibenzo[b,f][1,4]phospharsocin (**2**), and its known

(1)

(2)

chemistry follows closely that of 1,4-benzophospharsenin (p. 332) and 1H-1,5-benzophospharsepin (p. 336).

[o-(Diethylphosphino)phenyl]dimethylarsine (p. 333) reacts readily with o-xylylene dibromide, $C_6H_4(CH_2Br)_2$, in warm methanolic solution, depositing 5,5-diethyl-5,6,11,12-tetrahydro-12,12-dimethyldibenzo[b,f]-[1,4]phospharsocinium dibromide (**3**; R = CH_3), m.p. 208–209° (effervescence) (dipicrate, m.p. 188°). [o-(Diethylphosphino)phenyl]diethylarsine similarly gives the corresponding 5,5,12,12-tetraethyl-5,6,11,12-tetrahydrodibenzo[b,f][1,4]phospharsocinium dibromide (**3**; R = C_2H_5),

(**3**)   (**4**)

m.p. 234° (effervescence) (dipicrate, m.p. 208°). This salt, when heated at 270°/15 mm under nitrogen, lost ethyl bromide and gave a residue of the glassy bromide (**4**; X = Br), which when treated in aqueous solution with sodium iodide deposited the crystalline 5,5,12-triethyl-5,6,11,12-tetrahydrodibenzo[b,f][1,4]phospharsocinium iodide (**4**; X = I), m.p. 222–224°; the residue from the thermal dissociation, when further heated to 330°/2 mm, underwent extensive decomposition with liberation of arsenic and gave no distillate (Jones and Mann[497]).

There is a possibility that the cyclic diquaternization of [o-(diethylphosphino)phenyl]diethylarsine with o-xylylene dibromide might involve two molecules of each reagent, giving a 16-membered ring cation as the tetrabromide. This possibility can almost certainly be dismissed, for the strictly analogous union of o-phenylenebis(dimethylarsine) and o-xylylene dibromide gave a crystalline product, shown by X-ray crystal analysis to be the true 'monomeric' 5,6,11,12-tetrahydro-5,5,12,12-tetramethyldibenzo[b,f][1,4]diarsocinium dibromide, $i.e.$, the As–As analogue of the As–P salt (**3**). In both these ring systems the two CH_2 groups in the ring force the two o-phenylene rings out of coplanarity, so that each cation could exist in cis- and $trans$-forms (see pp. 457, 458 and Figures 12 and 13 for a detailed discussion). The X-ray crystal analysis afforded strong evidence that the diarsocinium dibromide has the $trans$-structure, which very probably applies also to the highly crystalline homogeneous dibromides (**3**; R = CH_3 and C_2H_5). In the case of these phospharsocinium dibromides, the presence of the phosphorus and arsenic atoms would cause both the cis- and the $trans$-form to be dissymmetric. The dibromide (**3**; R = C_2H_5) was therefore converted into the di-(+)-10-camphorsulfonate and the di(+)-3-bromo-8-cam-phorsulfonate, but on fractional recrystallization neither of these salts gave evidence of optical resolution.

The compound (**3**; R = CH$_3$) was originally [497] called 4,4-diethyl-1,1-dimethyl-o-phenylene-o-xylylene-1-arsonium-4-phosphonium dibromide.

Six-membered Ring System containing Carbon and Two Phosphorus Atoms and One Magnesium Atom

1,3,2-Diphosphamagnesiane

Isslieb and Deylig [500] have shown that when a xylene solution of P,P-diphenyltrimethylenediphosphine, $C_6H_5PH(CH_2)_3PHC_6H_5$, and a

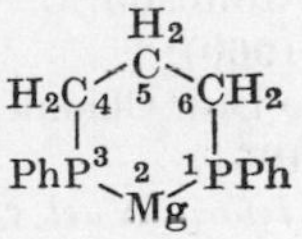

$$
\begin{array}{c}
\overset{H_2}{\underset{}{C}} \\
H_2C\overset{4}{\underset{}{}}\ {}^{5}\ {}_{6}\ CH_2 \\
PhP\overset{3}{\underset{}{}}\ {}_{2}\ {}^{1}PPh \\
\underset{Mg}{}
\end{array}
$$

(1)

n-butyl ether solution of diethylmagnesium are simultaneously added to boiling xylene, an unusual type of compound, 1,3-diphenyl-1,3,2-diphosphamagnesiane (**1**), [RIS —], crystallizes out in 72% yield. In addition to analysis for phosphorus and magnesium, its molecular weight in boiling tetrahydrofuran has been determined. Interaction of the corresponding tetramethylenediphosphine and diethylmagnesium under these conditions gave polymeric material.

The compound (**1**) can also be prepared by the interaction of magnesium bromide in tetrahydrofuran and the dilithio derivative $C_6H_5(Li)P$—$(CH_2)_3$—$P(Li)C_6H_5$. It is, as would be expected, readily affected by air and by hydrolyzing agents.

The compound (**1**) was termed PP'-trimethylene-magnesium-bisphenylphosphide by the above workers.

References (Part I: Phosphorus)

1 R. I. Wagner, U.S. Pat. 3,086,053; 3,086,056 (to American Potash and Chemical Corporation), April 13, 1963; *Chem. Abstr.*, **59**, 10124 (1963): these specifications are almost identical.

2 E. Jungermann and J. J. McBride, Jr, *J. Org. Chem.*, **26**, 4182 (1961) (preliminary note).

3 E. Jungermann, J. J. McBride, Jr, R. Clutter, and A. Mais, *J. Org. Chem.*, **27**, 606 (1962).

4 E. Jungermann, J. J. McBride, Jr, J. V. Killheffer, and R. J. Clutter, *J. Org. Chem.*, **27**, 1833 (1962).

5 K. Bergesen, *Acta. Chem. Scand.*, **21**, 1587 (1967).

6 W. Hawes and S. Trippett, *Chem. Commun.*, **1968**, 577.

6a D. D. Swank and C. N. Caughlan, *Chem. Commun.*, **1968**, 1051.

7 S. E. Fishwick, J. Flint, W. Hawes, and S. Trippett, *Chem. Commun.*, **1967**, 1113.

8 S. E. Fishwick and J. A. Flint, *Chem. Commun.*, **1968**, 182.

8a W. Hawes and S. Trippett, *Chem. Commun.*, **1968**, 295.

9 S. E. Cremer and R. J. Chorvat, *Tetrahedron Lett.*, **1968**, 413.

9a S. E. Cremer, *Chem. Commun.*, **1968**, 1132. *Note.* In the title of this paper, '2,2,3,4,4-Pentamethyl' should be replaced by '1,2,2,3,4,4-Hexamethyl'.

10 G. M. Kosolapoff and R. F. Struck, *J. Chem. Soc.*, **1957**, 3739.

11 L. I. Smith and H. H. Hoehn, *J. Amer. Chem. Soc.*, **63**, 1184 (1941).

12 F. C. Leavitt, T. A. Manuel, and F. Johnson, *J. Amer. Chem. Soc.*, **81**, 3163 (1959).

13 F. C. Leavitt, T. A. Manuel, F. Johnson, L. V. Matternas, and D. S. Lehman, *J. Amer. Chem. Soc.*, **82**, 5099 (1960).

14 F. C. Leavitt and F. Johnson (to Dow Chemical Co.), U.S.P. 3,116,307 (Dec. 31, 1963); *Chem. Abstr.*, **60**, 6872 (1964).

15 W. Hübel and E. H. Braye, *J. Inorg. Nucl. Chem.*, **10**, 250 (1959).

16 W. Hübel and E. H. Braye, International Conference on Co-ordination Chemistry, London, April 7th, 1959.

17 E. H. Braye and W. Hübel, *Chem. & Ind. (London)*, **1959**, 1250.

18 E. H. Braye, W. Hübel, and I. Caplier, *J. Amer. Chem. Soc.*, **83**, 4406 (1961).

19 W. Hübel, E. H. Braye, and I. H. Caplier (to Union Carbide Corp.), U.S.P. 3,151,140 (Sept. 29, 1964); *Chem. Abstr.*, **61**, 16097 (1964).

20 W. Reppe and W. J. Schweckendiek, *Annalen*, **560**, 104 (1948).

21 E. Weiss and W. Hübel, *J. Inorg. Nucl. Chem.*, **11**, 42 (1959).

22 I. G. M. Campbell, R. C. Cookson, and M. B. Hocking, *Chem. & Ind. (London)*, **1962**, 359.

23 I. G. M. Campbell, R. C. Cookson, M. B. Hocking, and A. N. Hughes, *J. Chem. Soc.*, **1965**, 2184.

24 G. Märkl and R. Potthast, *Angew. Chem.*, **79**, 58 (1967); *Int. Ed. Engl.*, **6**, 86 (1967).

25 E. Howard and R. E. Donadio, Abstracts, 136th National Meeting of the American Chemical Society, Atlantic City, N.J., Sept. 1959, p. 100P.

26 E. H. Braye, IUPAC Symposium on Organo-Phosphorus Compounds, Heidelberg, 1964.

27 E. H. Braye, personal communication.

28 E. H. Braye and K. K. Joshi, personal communication.

29 K. Bergesen, *Acta Chem. Scand.*, **20**, 899 (1966).

30 D. A. Brown, *J. Chem. Soc.*, **1962**, 929.

31 A. F. Bedford, D. M. Heinekey, I. T. Millar, and C. T. Mortimer, *J. Chem. Soc.*, **1962**, 2932.

32 R. A. Walton, *J. Chem. Soc.*, A, **1966**, 365.

33 R. C. Cookson, G. W. A. Fowles, and D. K. Jenkins, *J. Chem. Soc.*, **1965**, 6406.

34 G. S. Reddy and C. D. Weis, *J. Org. Chem.*, **28**, 1822 (1963).

35 M. A. Shaw, J. C. Tebby, R. S. Ward, and D. H. Williams, *J. Chem. Soc., C*, **1968**, 1609.

36 J. B. Hendrickson, R. E. Spenger, and J. J. Sims, *Tetrahedron*, **19**, 707 (1963).

37 A. W. Johnson and J. C. Tebby, *J. Chem. Soc.*, **1961**, 2126.

38 M. A. Shaw, J. C. Tebby, J. Ronayne, and D. H. Williams, *J. Chem. Soc., C*, **1967**, 944.

39 L. D. Quin, J. A. Peters, C. E. Griffin, and M. Gordon, *Tetrahedron Lett.*, **1964**, 3689.

40 H. Fritzsche, U. Hasserodt, and F. Korte, *Chem. Ber.*, **98**, 171 (1965).

41 L. Quin and J. G. Bryson, *J. Amer. Chem. Soc.*, **89**, 5984 (1967).

42 W. A. Anderson, R. Freeman, and C. A. Reilly, *J. Chem. Phys.*, **39**, 1518 (1963).

43 L. D. Quin and T. P. Barket, *Chem. Commun.*, **1967**, 914.

44 T. F. Page, Jr, T. Alger, and D. M. Grant, *J. Amer. Chem. Soc.*, **87**, 5333 (1965).

45 D. A. Usher and F. H. Westheimer, *J. Amer. Chem. Soc.*, **86**, 4732 (1964).

46 R. Kluger, F. Kerst, D. G. Lee, and F. H. Westheimer, *J. Amer. Chem. Soc.*, **89**, 3919 (1967).

47 W. B. McCormack (to E. I. du pont de Nemours & Co.), U.S. Pat. 2,663,736 (Dec. 22, 1953); *Chem. Abstr.*, **49**, 7602 (1955).

48 W. B. McCormack (to E. I. du Pont de Nemours & Co.), U.S. Pat. 2,663,737 (Dec. 22, 1953); *Chem. Abstr.*, **49**, 7601 (1955).

49 W. B. McCormack, *Org. Syn.*, **43**, 73 (1963).

50 L. D. Quin and D. A. Mathewes, *J. Org. Chem.*, **29**, 836 (1964).

51 L. D. Quin, 'Trivalent Phosphorus Compounds as Dienophiles', being Chapter 3 of 1,4-*Cyclo-addition Reactions*, J. Hamer, Ed., Academic Press Inc., New York, 1967.

52 W. B. McCormack (to E. I. du Pont de Nemours & Co.), U.S. Pat. 2,663,738 (Dec. 22, 1953); *Chem. Abstr.*, **49**, 7602 (1955).

53 W. B. McCormack (to E. I. du Pont de Nemours & Co.), U.S. Pat. 2,663,739 (Dec. 22, 1953); *Chem. Abstr.*, **49**, 7602 (1955).

54 W. J. Balon (to E. I. du Pont de Nemours & Co.), U.S. Pat. 2,853,518 (Sept. 23, 1958); *Chem. Abstr.*, **53**, 5202 (1959).

55 T. W. Campbell, J. J. Monagle, and V. S. Foldi, *J. Amer. Chem. Soc.*, **84**, 3673 (1962).

56 U. Hasserodt, K. Hunger, and F. Korte, *Tetrahedron*, **19**, 1563 (1963).

57 K. Hunger, U. Hasserodt, and F. Korte, *Tetrahedron*, **20**, 1593 (1964).

58 K. Hunger and F. Korte, *Tetrahedron Lett.*, **1964**, 2855.

59 H. Weitkamp and F. Korte, *Z. Anal. Chem.*, **204**, 245 (1964).

60 G. A. Wiley and W. R. Stine, *Tetrahedron Lett.*, **1967**, 2321.

61 L. D. Quin, J. P. Gratz, and T. P. Barket, *J. Org. Chem.*, **33**, 1034 (1968).

62 L. D. Quin, J. P. Gratz, and R. E. Montgomery, *Tetrahedron Lett.*, **1965**, 2187.

63 N. A. Razumova and A. A. Petrov, *Zhur. Obshch. Khim.*, **31**, 3144 (1961); *Chem. Abstr.*, **56**, 12720 (1962).

64 B. A. Arbuzov, L. A. Shapshinskaya, and V. M. Erokhina, *Izv. Akad. Nauk SSSR, Otd. Khim. Nauk*, **1962**, 2074; *Chem. Abstr.*, **58**, 11396 (1963).

65 N. A. Razumova and A. A. Petrov, *Zhur. Obshch. Khim.*, **33**, 783 (1963); *Chem. Abstr.*, **59**, 8783 (1963).

66 G. M. Bogolyubov, N. A. Razumova, and A. A. Petrov, *Zhur. Obshch. Khim.*, **33**, 2419 (1963); *Chem. Abstr.*, **59**, 14018 (1963).

[67] B. A. Arbuzov and L. A. Shapshinskaya, *Izv. Akad. Nauk SSSR, Otd. Khim. Nauk*, **1962**, 65; *Chem. Abs.*, **57**, 13791 (1962).

[68] H. Fritzsche, U. Hasserodt, and F. Korte, *Chem. Ber.*, **97**, 1988 (1964).

[69] H. Fritzsche, U. Hasserodt, and F. Korte, *Chem. Ber.*, **98**, 1681 (1965).

[70] R. O. Sauer, W. J. Scheiber, and S. D. Brewer, *J. Amer. Chem. Soc.*, **68**, 962 (1946).

[71] G. Grüttner and E. Krause, *Ber.*, **49**, 437 (1916).

[72] R. C. Evans, F. G. Mann, H. S. Peiser, and D. Purdie, *J. Chem. Soc.*, **1940**, 1209.

[73] K. Issleib and S. Häusler, *Chem. Ber.*, **94**, 113 (1961).

[74] K. Issleib, K. Krech, and K. Gruber, *Chem. Ber.*, **96**, 2186 (1963).

[75] G. Märkl, *Angew. Chem.*, **75**, 669 (1963); *Int. Ed. Engl.*, **2**, 479 (1963).

[76] G. Märkl, *Angew. Chem.*, **75**, 859 (1963); *Int. Ed. Engl.*, **2**, 620 (1963).

[77] J. H. Davies, J. D. Downer, and P. Kirby, *J. Chem. Soc.*, *C*, **1966**, 245.

[78] W. B. McCormack, Lecture delivered at Gordon Conference, New Hampton, N.H., U.S.A., during week of 15th July 1957.

[79] L. Maier, *Helv. Chim. Acta*, **48**, 133 (1965).

[80] R. Schmutzler, *Inorg. Chem.*, **3**, 421 (1964).

[81] B. Helferich and E. Aufderhaar, *Annalen*, **658**, 100 (1962).

[82] G. Hilgetag, H. G. Henning, and D. Gloyna, *Z. Chem.*, **4**, 347 (1964).

[83] A. B. Burg and P. J. Slota, *J. Amer. Chem. Soc.*, **82**, 2148 (1960).

[84] R. Schmutzler, *Inorg. Chem.*, **3**, 410 (1964).

[85] E. L. Muertterties, W. Mahler, and R. Schmutzler, *Inorg. Chem.*, **2**, 613 (1963).

[86] J. F. Nixon and R. Schmutzler, *Spectrochim. Acta*, **20**, 1835 (1964).

[87] C. G. Krespan (to E. I. du Pont de Nemours and Co.), U.S. Pat. 2,931,803 (April 5, 1960); *Chem. Abstr.*, **55**, 12436 (1961).

[88] C. G. Krespan and C. M. Langkammerer, *J. Org. Chem.*, **27**, 3584 (1962).

[89] F. W. Bennett, H. J. Emeléus, and R. N. Haszeldine, *J. Chem. Soc.*, **1954**, 3598.

[90] S. A. Buckler and M. Epstein, *J. Org. Chem.*, **27**, 1090 (1962).

[91] S. A. Buckler and M. Epstein, Fr. Pat., 1,348,669 (to American Cyanamid Co.) (Jan. 10, 1964): *Chem. Abstr.*, **60**, 15912 (1964); U.S. Pat. 3,142,685 (July 28, 1964).

[92] J. B. Hendrickson, *J. Amer. Chem. Soc.*, **83**, 2018 (1963).

[93] S. T. D. Gough and S. Trippett, *Proc. Chem. Soc.*, **1961**, 302.

[94] J. B. Hendrickson, C. Hall, R. Rees, and J. F. Templeton, *J. Org. Chem.*, **30**, 3312 (1965).

[95] D. Hellwinkel, *Chem. Ber.*, **98**, 576 (1965).

[95a] L. D. Quin and J. A. Caputo, *Chem. Commun.*, **1968**, 1463.

[95b] K. L. Marsi, *Chem. Commun.*, **1968**, 846.

[96] J. Messinger and C. Engels, *Ber.*, **21**, 326, 2919 (1888).

[97] A. Hoffmann, *J. Amer. Chem. Soc.*, **43**, 1684 (1921).

[98] S. A. Buckler and V. P. Wystrach, *J. Amer. Chem. Soc.*, **80**, 6454 (1958).

[99] S. A. Buckler and V. P. Wystrach, *J. Amer. Chem. Soc.*, **83**, 168 (1961).

[100] G. E. McCasland and S. Proskow, *J. Amer. Chem. Soc.*, **77**, 4688 (1955).

[101] E. H. Braye, personal communication.

[102] A. N. Hughes and S. Uaboonkul, *Tetrahedron*, **24**, 3437 (1968).

[103] F. G. Mann and I. T. Millar, *J. Chem. Soc.*, **1951**, 2205.

[104] G. Märkl, *Z. Naturforsch.*, **18b**, 84 (1963).

[105] F. G. Mann, I. T. Millar, and F. H. C. Stewart, *J. Chem. Soc.*, **1954**, 2832.

[106] J. Meisenheimer, J. Casper, M. Höring, W. Lauter, L. Lichtenstadt, and W. Samuel, *Annalen*, **449**, 213 (1926).

[107] F. G. Mann and F. H. C. Stewart, *J. Chem. Soc.*, **1954**, 2819.

[108] F. G. Mann, I. T. Millar, and H. R. Watson, *J. Chem. Soc.*, **1958**, 2516.

[109] F. G. Mann and H. R. Watson, *Chem. & Ind.* (*London*), **1958**, 1264.

[110] J. W. Collier, F. G. Mann, D. G. Watson, and H. R. Watson, *J. Chem. Soc.*, **1964**, 1803.

[111] J. W. Collier and F. G. Mann, *J. Chem. Soc.*, **1964**, 1815.

[112] J. W. Collier, A. R. Fox, I. G. Hinton, and F. G. Mann, *J. Chem. Soc.*, **1964**, 1819.

[113] G. Wittig and G. Geissler, *Annalen*, **580**, 44 (1953).

[114] G. A. Razuvaev and N. A. Osanova, *Dokl. Akad. Nauk SSSR*, **104**, 552 (1955); *Chem. Abstr.*, **50**, 11268 (1956).

[115] G. A. Razuvaev and N. A. Osanova, *Zh. obshchei Khim.*, **26**, 2531 (1956); *Chem. Abstr.*, **51**, 1875 (1957).

[116] G. A. Razuvaev, N. A. Osanova, and I. A. Shlyapnikova, *Zh. obshchei Khim.*, **27**, 1466 (1957); *Chem. Abstr.*, **52**, 3715 (1958).

[117] G. Wittig and A. Maercker, *Chem. Ber.*, **97**, 747 (1964).

[118] D. Seyferth, M. A. Eisert, and J. K. Heeren, *J. Organometal. Chem.*, **2**, 101 (1964).

[119] D. Seyferth, W. B. Hughes, and J. K. Heeren, *J. Amer. Chem. Soc.*, **87**, 3467 (1965).

[120] G. Wittig and E. Benz, *Chem. Ber.*, **92**, 1999 (1959).

[121] E. Zbiral, *Tetrahedron Lett.*, **1964**, 1649.

[122] D. W. Allen, F. G. Mann, and I. T. Millar, *Chem. & Ind.* (*London*), **1966**, 196.

[123] D. W. Allen, I. T. Millar, and F. G. Mann, *J. Chem. Soc.*, *C*, **1967**, 1869.

[124] L. Mascarelli, *Atti Accad. naz. Lincei, Rend. Cl. Sci. Fis. Mat. Natur.*, **16**, II, 582 (1907).

[125] R. B. Sandin and A. S. Hay, *J. Amer. Chem. Soc.*, **74**, 274 (1952).

[126] A. S. Nesmeyanov, *Bull. Soc. Chim. France*, **1965**, 897, and references quoted therein.

[127] D. W. Allen, F. G. Mann, and I. T. Millar, *Chem. & Ind.* (*London*), **1966**, 2096.

[128] D. W. Allen, I. T. Millar, and F. G. Mann, *J. Chem. Soc.*, *A*, **1969**, 1101.

[129] Shell Int. Res. Maatschappij N.V., Belg. Pat. 631,416 (Nov. 18, 1963); *Chem. Abstr.*, **61**, 689 (1964).

[130] E. H. Braye, personal communication.

[131] L. D. Freedman and G. O. Doak, *J. Org. Chem.*, **21**, 238 (1956).

[132] G. O. Doak and L. D. Freedman, *J. Amer. Chem. Soc.*, **73**, 5658 (1951).

[133] L. D. Freedman and G. O. Doak, *J. Amer. Chem. Soc.*, **74**, 2884 (1952).

[134] M. Busch and W. Weber, *J. Prakt. Chem.*, **146**, 1 (1936).

[135] P. W. Morgan and B. C. Herr, *J. Amer. Chem. Soc.*, **74**, 4526 (1952).

[136] L. D. Quin and G. O. Doak, *J. Org. Chem.*, **24**, 638 (1959).

[137] L. D. Quin and R. E. Montgomery, *J. Org. Chem.*, **27**, 4120 (1962).

[138] G. O. Doak, L. D. Freedman, and J. B. Levy, *J. Org. Chem.*, **29**, 2382 (1964).

[139] I. G. M. Campbell and J. K. Way, *Proc. Chem. Soc.*, **1959**, 231.

[140] I. G. M. Campbell and J. K. Way, *J. Chem. Soc.*, **1961**, 2133.

[141] J. B. Levy, G. O. Doak, and L. D. Freedman, *J. Org. Chem.*, **30**, 660 (1965).

[142] E. R. Lynch, *J. Chem. Soc.*, **1962**, 3729.

[143] M. Davis and F. G. Mann, *J. Chem. Soc.*, **1964**, 3770.

[144] K. Issleib and H. Völker, *Chem. Ber.*, **94**, 392 (1961).

[145] A. D. Britt and E. T. Kaiser, *J. Phys. Chem.*, **69**, 2775 (1965).

[146] G. Wittig and E. Kochendörfer, *Angew. Chem.*, **70**, 506 (1958).

[147] H. Standinger and E. Hauser, *Helv. Chim. Acta*, **4**, 861 (1921).
[148] H. Standinger and W. T. K. Braunholtz, *Helv. Chim. Acta*, **4**, 897 (1921).
[149] G. Wittig and E. Kochendörfer, *Chem. Ber.*, **97**, 741 (1964).
[150] G. Wittig and D. Hellwinkel, *Angew. Chem.*, **74**, 76 (1962); *Int. Ed. Engl.*, **1**, 53 (1962).
[151] F. G. Mann and E. J. Chaplin, *J. Chem. Soc.*, **1937**, 527.
[152] D. Hellwinkel, *Angew. Chem.*, **77**, 378 (1965); *Int. Ed. Engl.*, **4**, 356 (1965).
[153] G. Grüttner and M. Wiernik, *Ber.*, **48**, 1473 (1915).
[154] P. J. Slota, Jr., Dissertation, Temple University, Philadelphia, U.S.A., Jan. 1954, 'Studies in the Synthesis of Organophosphorus Heterocycles'.
[155] R. B. Fox, *J. Amer. Chem. Soc.*, **72**, 4147 (1950).
[156] R. Rabinowitz and R. Marcus, *J. Amer. Chem. Soc.*, **84**, 1312 (1962).
[157] S. O. Grim and R. Schaaff, *Angew. Chem.*, **75**, 669 (1963); *Int. Ed. Engl.*, **2**, 486 (1963).
[158] G. M. Kosolapoff, *J. Amer. Chem. Soc.*, **77**, 6658 (1955).
[159] E. Howard, Jr., and M. Braid, Abstracts of National Amer. Chem. Soc. Meeting, Chicago, Ill., Sept. 1961, p. 40Q.
[160] M. Braid, Dissertation, Temple University, Philadelphia, U.S.A., 1962, 'Studies in Heterocyclic Phosphorus Chemistry. Pentamethylenephosphine and Related Compounds'.
[161] R. P. Welcher, G. A. Johnson, and V. P. Wystrach, *J. Amer. Chem. Soc.*, **82**, 4437 (1960).
[162] R. P. Welcher, G. A. Johnson, and V. P. Wystrach, Ger. Pat., 1,419,004 (to Amer. Cyanamid Co.) (May 22nd, 1963): *Chem. Abstr.*, **59**, 14023 (1963); U.S. Pat. 3,094,549 (June 18th, 1963).
[163] F. G. Mann and I. T. Millar, *J. Chem. Soc.*, **1952**, 4453.
[164] M. M. Rauhut, I. Hechenbleikner, H. A. Currier, F. G. Schaefer, and V. P. Wystrach, *J. Amer. Chem. Soc.*, **81**, 1103 (1959).
[165] M. J. Gallagher and F. G. Mann, *J. Chem. Soc.*, **1962**, 5110.
[166] R. P. Welcher and N. E. Day, *J. Org. Chem.*, **27**, 1824 (1962).
[167] R. P. Welcher, U.S. Pat. 3,105,096 (to Amer. Cyanamid Co.) (Sept. 24, 1963); *Chem. Abstr.*, **60**, 5553 (1964); Ger. Pat., 1,162,840 (Feb. 13, 1964).
[168] R. P. Welcher (to Amer. Cyanamid Co.), Brit. Pat. 969,129 (Sept. 9, 1964); *Chem. Abstr.*, **60**, 16095 (1964); U.S. Pat. 3,218,358 (Nov. 16, 1965).
[169] L. D. Quin and H. E. Shook, Jr, *Tetrahedron Lett.*, **1965**, 2193.
[170] R. A. Pickering and C. C. Price, *J. Amer. Chem. Soc.*, **80**, 4931 (1958).
[171] H. E. Shook, Jr, and L. D. Quin, *J. Amer. Chem. Soc.*, **89**, 1841 (1967).
[172] L. D. Quin and D. A. Mathewes, *Chem. & Ind. (London)*, **1963**, 210.
[173] L. D. Quin and H. E. Shook, Jr, *J. Org. Chem.*, **32**, 1604 (1967).
[174] S. A. Buckler and M. Epstein, *Tetrahedron*, **18**, 1221 (1962).
[175] G. Märkl and H. Olbrich, *Angew. Chem.*, **78**, 598 (1966); *Int. Ed. Engl.*, **5**, 588 (1966).
[176] G. Märkl and H. Olbrich, *Angew. Chem.*, **78**, 598 (1966); *Int. Ed. Engl.*, **5**, 589 (1966).
[177] C. C. Price, *Chem. & Ind. (Japan)*, **16**, 109 (1963).
[178] C. C. Price, T. Parasaran, and T. V. Lakshminarayan, *J. Amer. Chem. Soc.*, **88**, 1034 (1966).
[179] G. Märkl, *Angew. Chem.*, **78**, 907 (1966); *Int. Ed. Engl.*, **5**, 846 (1966).
[180] G. Märkl, F. Lieb, and A. Merz, *Angew. Chem.*, **79**, 59 (1967); *Int. Ed. Engl.*, **6**, 87 (1967).

[181] G. Märkl, F. Lieb, and A. Merz, *Angew. Chem.*, **79**, 475 (1967); *Int. Ed. Engl.*, **6**, 458 (1967).

[182] K. Dimroth, F. Kalk, and G. Neubauer, *Chem. Ber.*, **90**, 2058 (1957).

[183] K. Dimroth, A. Berndt, and R. Volland, *Chem. Ber.*, **99**, 3040 (1966).

[184] K. Dimroth, A. Berndt, F. Bär, A. Schweig, and R. Volland, *Angew. Chem.*, **79**, 69 (1967); *Int. Ed. Engl.*, **6**, 34 (1967).

[185] K. Dimroth, N. Greif, H. Perst, and F. W. Steuber, *Angew. Chem.*, **79**, 58 (1967); *Int. Ed. Engl.*, **6**, 85 (1967).

[186] K. Dimroth, N. Greif, W. Städe, and F. W. Steuber, *Angew. Chem.*, **79**, 725 (1967); *Int. Ed. Engl.*, **6**, 711 (1967).

[187] K. Dimroth and N. Mach, *Angew. Chem.*, **80**, 489 (1968); *Int. Ed. Engl.*, **7**, 460 (1968).

[188] K. Dimroth and F. W. Steuber, *Angew. Chem.*, **79**, 410 (1967); *Int. Ed. Engl.*, **6**, 445 (1967).

[189] R. C. Cookson and F. G. Mann, *J. Chem. Soc.*, **1949**, 67.

[190] F. G. Mann, *J. Chem. Soc.*, **1949**, 2816.

[191] J. T. Braunholtz and F. G. Mann, *J. Chem. Soc.*, **1955**, 381.

[192] M. H. Beeby and F. G. Mann, *J. Chem. Soc.*, **1951**, 411.

[193] G. Märkl, *Angew. Chem.*, **75**, 168 (1963); *Int. Ed. Engl.*, **2**, 153 (1963).

[194] G. Märkl, personal communication.

[195] M. J. Gallagher, E. C. Kirby, and F. G. Mann, *J. Chem. Soc.*, **1963**, 4846.

[196] L. D. Quin and J. S. Humphrey, Jr, *J. Amer. Chem. Soc.*, **83**, 4124 (1961).

[197] L. D. Quin and R. E. Montgomery, *J. Org. Chem.*, **27**, 4120 (1962).

[198] F. G. Mann and F. A. Hart, *Nature* (*London*), **175**, 952 (1955).

[199] F. A. Hart and F. G. Mann, *J. Chem. Soc.*, **1955**, 4107.

[200] M. J. Gallagher and F. G. Mann, *J. Chem. Soc.*, **1963**, 4855.

[201] F. G. Mann and F. G. Holliman, *Nature* (*London*), **159**, 438 (1947).

[202] F. G. Holliman and F. G. Mann, *J. Chem. Soc.*, **1947**, 1634.

[203] A. Michaelis and L. Gleichmann, *Ber.*, **15**, 801 (1882).

[204] A. Michaelis and A. Reese, *Ber.*, **15**, 2876 (1882).

[205] W. Cule Davies and W. P. G. Lewis, *J. Chem. Soc.*, **1934**, 1599.

[206] L. Horner, H. Winkler, A. Rapp, A. Mentrup, H. Hoffmann, and P. Beck, *Tetrahedron Lett.*, **1961**, 161.

[207] P. de Koe and F. Bickelhaupt, *Angew. Chem.*, **79**, 533 (1967); *Int. Ed. Engl.*, **6**, 567 (1967).

[208] H. Oediger, H. J. Kabbe, Fr. Möller, and K. Eiter, *Chem. Ber.*, **99**, 2012 (1966).

[209] H. Oediger, K. Eiter, Fr. Möller, and H. J. Kabbe, Ger. Pat., 1,186,063 (15th May, 1962); *Chem. Abstr.*, **62**, 14498 (1965).

[210] H. Oediger and Fr. Möller, *Angew. Chem.*, **79**, 53 (1967); *Int. Ed. Engl.*, **6**, 76 (1967).

[211] E. R. Lynch, *J. Chem. Soc.*, **1962**, 3729.

[212] E. R. Lynch (to Monsanto Chemicals Ltd.) Brit. Pat. 933,800 (Aug. 14, 1963); *Chem. Abstr.*, **60**, 1796 (1964).

[213] P. J. Wheatley, *J. Chem. Soc.*, **1962**, 3733.

[214] P. de Koe, R. van Veen, and F. Bickelhaupt, *Angew. Chem.*, **80**, 486 (1968); *Int. Ed. Engl.*, **7**, 465 (1968).

[215] W. A. Henderson, S. A. Buckler, N. E. Day, and M. Grayson, *J. Org. Chem.*, **26**, 4770 (1961).

[216] M. Schlosser, *Angew. Chem.*, **74**, 291 (1962); *Int. Ed. Engl.*, **1**, 266 (1962).

[217] D. W. Allen, J. C. Tebby, and D. H. Williams, *Tetrahedron Lett.*, **1965**, 2361.

[218] D. W. Allen and J. C. Tebby, *Tetrahedron*, **23**, 2795 (1967).

[219] E. M. Richards and J. C. Tebby, *Chem. Commun.*, **1967**, 957.

[220] D. W. Allen and I. T. Millar, *Chem. & Ind.* (*London*), **1967**, 2178.

[221] D. W. Allen and I. T. Millar, *J. Chem. Soc.*, *C*, **1969**, 252.

[222] E. A. Cookson and P. C. Crofts, *J. Chem. Soc.*, *C*, **1966**, 2003.

[222a] E. A. Cookson and P. C. Crofts, I.U.P.A.C. Symposium on Organo-Phosphorus Compounds, Heidelberg, 1964.

[223] F. G. Mann, I. T. Millar, and B. B. Smith, *J. Chem. Soc.*, **1953**, 1130.

[224] J. Kenner and J. Wilson, *J. Chem. Soc.*, **1927**, 1108.

[225] M. Green, *Proc. Chem. Soc.*, **1963**, 177.

[226] M. Green, *J. Chem. Soc.*, **1965**, 541.

[227] T. J. Katz, *J. Amer. Chem. Soc.*, **82**, 3784 (1960).

[228] T. J. Katz and P. J. Garratt, *J. Amer. Chem. Soc.*, **85**, 2852 (1963); **86**, 4876, 5194 (1964).

[229] T. J. Katz, C. R. Nicholson, and C. A. Reilly, *J. Amer. Chem. Soc.*, **88**, 3832 (1966).

[230] W. Mahler, *J. Amer. Chem. Soc.*, **86**, 2306 (1964).

[231] W. Mahler and A. B. Burg, *J. Amer. Chem. Soc.*, **80**, 6161 (1958).

[232] K. Issleib and F. Krech, *Chem. Ber.*, **94**, 2656 (1961).

[233] K. Issleib, *Z. Chem.*, **2**, 163 (1962).

[234] K. Issleib and K. Krech, *Chem. Ber.*, **99**, 1310 (1966).

[235] J. J. Daly, *J. Chem. Soc.*, **1964**, 6147.

[236] J. J. Daly, *J. Chem. Soc.*, **1965**, 4789.

[237] L. Horner, J. P. Bercz, and C. V. Bercz, *Tetrahedron Lett.*, **1966**, 5783.

[238] K. Issleib and K. Krech, *Chem. Ber.*, **98**, 2545 (1965).

[239] G. Märkl, *Z. Naturforsch.*, **18b**, 1136 (1963).

[240] M. Davis and F. G. Mann, *J. Chem. Soc.*, **1964**, 3786.

[241] C. H. S. Hitchcock and F. G. Mann, *J. Chem. Soc.*, **1958**, 2081.

[242] R. C. Hinton and F. G. Mann, *J. Chem. Soc.*, **1959**, 2835.

[243] W. J. Bailey and S. A. Buckler, *J. Amer. Chem. Soc.*, **79**, 3567 (1957).

[244] K. Issleib and K. Standtke, *Chem. Ber.*, **96**, 279 (1963).

[245] K. Issleib and G. Döll, *Chem. Ber.*, **96**, 1544 (1963).

[246] M. M. Rauhut, G. B. Borowitz, and H. C. Gillham, *J. Org. Chem.*, **28**, 2565 (1963).

[247] G. Märkl, *Angew. Chem.*, **77**, 1109 (1965); *Int. Ed. Engl.*, **4**, 1023 (1965).

[248] M. Zanger, C. A. VanderWerf, and W. E. McEwen, *J. Amer. Chem. Soc.*, **81**, 3806 (1959).

[249] A. M. Aguiar, H. Aguiar, and D. Daigle, *J. Amer. Chem. Soc.*, **87**, 671 (1965).

[250] A. M. Aguiar and H. Aguiar, *J. Amer. Chem. Soc.*, **88**, 4090 (1966).

[251] A. M. Aguiar and D. Daigle, *J. Amer. Chem. Soc.*, **86**, 2299 (1964).

[252] A. M. Aguiar, K. C. Hansen, and G. S. Reddy, *J. Amer. Chem. Soc.*, **89**, 3067 (1967).

[253] A. M. Aguiar and K. C. Hansen, *J. Amer. Chem. Soc.*, **89**, 4235 (1967).

[254] A. M. Aguiar, personal communication.

[255] F. G. Mann and F. C. Baker, *J. Chem. Soc.*, **1957**, 1881.

[256] J. Meisenheimer, *Annalen*, **420**, 190 (1920).

[257] C. G. Krespan, B. C. McKusick, and T. L. Cairns, *J. Amer. Chem. Soc.*, **82**, 1515 (1960).

[258] C. G. Krespan, *J. Amer. Chem. Soc.*, **83**, 3432 (1961).

259 C. G. Krespan, B. C. McKusick, and T. L. Cairns, *J. Amer. Chem. Soc.*, **83**, 3428 (1961).
260 F. A. Hart, *J. Chem. Soc.*, **1960**, 3324.
261 F. A. Hart and F. G. Mann, *Chem. & Ind. (London)*, **1956**, 574.
262 F. A. Hart and F. G. Mann, *J. Chem. Soc.*, **1957**, 3939.
263 L. Kalb, *Annalen*, **423**, 39 (1921).
264 M. Davis and F. G. Mann, *Chem. & Ind. (London)*, **1962**, 1539.
265 F. G. Mann and F. C. Baker, *J. Chem. Soc.*, **1961**, 3845.
266 J. Chatt and F. G. Mann, *J. Chem. Soc.*, **1940**, 1184.
267 K. Mislow, A. Zimmerman, and J. T. Melillo, *J. Amer. Chem. Soc.*, **85**, 594 (1963).
268 Emrys R. H. Jones and F. G. Mann, *J. Chem. Soc.*, **1955**, 411.
269 O. Kennard, F. G. Mann, D. G. Watson, J. K. Fawcett, and K. A. Kerr, *Chem. Commun.*, **1968**, 269.
270 P. M. de Wolff and O. Kennard (incorporated in ref. 269).
271 K. F. Kumli, W. E. McEwen, and C. A. VanderWerf, *J. Amer. Chem. Soc.*, **81**, 248 (1959).
272 K. F. Kumli, C. A. VanderWerf, and W. E. McEwen, *J. Amer. Chem. Soc.*, **81**, 3805 (1959).
273 L. Horner, H. Fuchs, H. Winkler, and A. Rapp, *Tetrahedron Lett.*, **1963**, 965.
274 F. G. Mann and M. J. Pragnell, *J. Chem. Soc.*, *C*, **1966**, 916.
275 N. Collie, *J. Chem. Soc.*, **53**, 636, 714 (1888).
276 A. M. Aguiar and M. G. R. Nair, *J. Org. Chem.*, **33**, 579 (1968).
277 G. Märkl, *Angew. Chem.*, **75**, 1121 (1963); *Int. Ed. Engl.*, **3**, 147 (1964).
278 P. R. Bloomfield and K. Parvin, *Chem. & Ind. (London)*, **1959**, 541.
279 J. J. Daly, *J. Chem. Soc.*, *A*, **1966**, 1020.
280 L. Horner and H. Winkler, *Tetrahedron Lett.*, **1964**, 175.
281 U. Schmidt and C. Osterroht, *Angew. Chem.*, **77**, 455 (1965); *Int. Ed. Engl.*, **4**, 437 (1965).
282 W. A. Henderson, Jr, M. Epstein, and F. S. Seichter, *J. Amer. Chem. Soc.*, **85**, 2462 (1963), and references cited therein.
283 J. J. Daly and L. Maier, *Nature (London)*, **203**, 1167 (1964).
284 J. J. Daly, *J. Chem. Soc.*, *A*, **1966**, 428.
285 E. Fluck and K. Issleib, *Z. Naturforsch.*, **21b**, 736 (1966).
286 L. Maier, *Helv. Chim. Acta*, **48**, 1034 (1965).
287 L. Maier, U.S.P. 3,253,033 (to Monsanto Co.) (May 24, 1966); *Chem. Abstr.*, **65**, 5488 (1966).
288 H. Ulrich and A. A. R. Sayigh, *Angew. Chem.*, **76**, 647 (1964); *Int. Ed. Engl.*, **3**, 585 (1964).
289 H. Ulrich and A. A. R. Sayigh, *J. Org. Chem.*, **30**, 2779 (1965).
290 G. I. Derkach and A. V. Narbut, *Zh. Obshch. Khim.*, **35**, 932 (1965); *Chem. Abs.*, **63**, 6984 (1965).
291 B. Helferich and U. Curtius, *Annalen*, **655**, 59 (1962).
292 R. G. Harvey, T. C. Myers, H. I. Jacobsen, and E. V. Jensen, *J. Amer. Chem. Soc.*, **79**, 2612 (1957).
293 K. Issleib and H. Oehme, *Tetrahedron Lett.*, **1967**, 1489.
294 K. Issleib and H. Oehme, *Chem. Ber.*, **100**, 2685 (1967).
295 K. Issleib and A. Balszuweit, *Chem. Ber.*, **99**, 1316 (1966).
296 Farben. Bayer A.-G. (by G. Schrader), Ger. Pat. 1,111,196 (Mar. 26th, 1960); *Chem. Abstr.*, **56**, 8750 (1962).
297 G. Schrader: see Farben. Bayer A.-G.296

298 G. F. Dregval and G. I. Derkach, *Zh. Obshch. Khim.*, **33**, 2952 (1963); *Chem. Abstr.*, **60**, 1792 (1964).

299 V. I. Shevchenko, V. I. Standnik, and A. V. Kirsanov, *Probl. Organ. Sinteza, Akad. Nauk SSSR, Otd. Obshch. i Tekhn. Khim.*, **1965**, 272; *Chem. Abstr.*, **64**, 5071 (1966).

300 W. Autenrieth and E. Bölli, *Ber.*, **58**, 2144 (1925).

301 W. Autenrieth and W. Meyer, *Ber.*, **58**, 848 (1925).

302 H. Tolkmith and E. C. Britton (to Dow Chemical Co.), U.S. Pat. 2,805,256 (Sept. 3, 1957); *Chem. Abstr.*, **52**, 2086 (1958).

303 O. J. Scherer and M. Schmidt, *Angew. Chem., Int. Ed.*, **3**, 702 (1964).

304 K. Utvary, V. Gutmann, and C. Kemenater, *Inorg. Nucl. Chem. Lett.*, **1**, 75 (1965).

305 E. W. Abel and R. P. Bush, *J. Organometal. Chem.*, **3**, 245 (1965).

306 C. H. Yoder and J. J. Zuckerman, *J. Amer. Chem. Soc.*, **88**, 2170 (1966).

307 I. K. Fel'dman and A. I. Berlin, *Zh. Obshch. Khim.*, **32**, 1604 (1962); *Chem. Abstr.*, **58**, 12563 (1963).

308 K. Pilgram and F. Korte, *Tetrahedron*, **19**, 137 (1963).

309 W. Autenrieth and O. Hildebrand, *Ber.*, **31**, 1111 (1898).

310 O. M. Friedman and A. M. Seligman, *J. Amer. Chem. Soc.*, **76**, 655 (1954).

311 O. M. Friedman, Z. B. Papanastassiou, R. S. Levi, H. R. Till, Jr, and W. M. Whaley, *J. Med. Chem.*, **6**, 82 (1963).

312 B. Helferich and L. Schröder, *Annalen*, **670**, 48 (1963).

313 M. J. S. Dewar and V. P. Kubba, *J. Amer. Chem. Soc.*, **82**, 5685 (1960).

314 I. G. M. Campbell and J. K. Way, *J. Chem. Soc.*, **1960**, 5034.

315 J. Chatt and F. G. Mann, *J. Chem. Soc.*, **1940**, 1192.

316 F. G. Mann and I. T. Millar, *J. Chem. Soc.*, **1953**, 3746.

317 A. Job and G. Dusollier, *Compt. rend.*, **184**, 1454 (1927).

318 F. G. Mann and I. T. Millar, *J. Chem. Soc.*, **1952**, 3039.

319 J. G. M. Dunlop and H. O. Jones, *J. Chem. Soc.*, **95**, 419 (1909).

320 G. M. Bennett and E. Glynn, *J. Chem. Soc.*, **1950**, 211.

321 A. Michaelis and A. Schenk, *Annalen*, **260**, 1 (1890).

322 P. G. Sergeev and D. G. Kudryashov, *J. Gen. Chem. USSR.*, **8**, 266 (1938); *Chem. Abstr.*, **32**, 5403 (1938).

323 M. Häring, *Helv. Chim. Acta*, **43**, 1826 (1960).

324 G. V. McHattie (to Imperial Chemical Industries), Brit. Pat. 860,629 (Feb. 8, 1961); *Chem. Abstr.*, **57**, 2256 (1962).

325 G. V. McHattie, personal communication.

326 G. Baum, H. A. Lloyd, and C. Tamborski, *J. Org. Chem.*, **29**, 3410 (1964).

327 O. V. Kirsanov, M. D. Bodnarchuk, and V. I. Shevchenko, *Dopovidi Akad. Nauk Ukr. RSR*, **1963**, 221; *Chem. Abstr.*, **59**, 12666 (1963).

328 V. I. Shevchenko, P. P. Kornuta, M. D. Bodnarchuk, and A. V. Kirsanov, *Zh. Obshch. Khim.*, **36**, 730 (1966); *Chem. Abstr.*, **65**, 8912 (1966).

329 V. I. Shevchenko and P. P. Kornuta, *Zh. Obshch. Khim.*, **36**, 1254 (1966); *Chem. Abstr.*, **65**, 15381 (1966).

330 H. Arnold, F. Bourseaux, and N. Brock, *Arzneimittel-Forsch.*, **11**, 143 (1961); *Chem. Abstr.*, **55**, 16816 (1961).

331 O. M. Friedman, E. Boger, V. Grubliauskas, and H. Sommer, *J. Med. Chem.*, **6**, 50 (1963).

332 J. H. Billman, J. L. Meisenheimer, and R. A. Awl, *J. Med. Chem.*, **7**, 366 (1964).

333 J. H. Billman and J. L. Meisenheimer, *J. Med. Chem.*, **8**, 264 (1965).

334 M. Szekerke, M. T. Kajtar, and V. Bruckner, *Acta Chim. Acad. Sci. Hung.*, **47**, 231 (1966); *Chem. Abstr.*, **64**, 17706 (1966).

335 E. Koenigs and H. Friedrich, *Annalen*, **509**, 138 (1934).

336 C. Y. Cheng, R. A. Shaw, T. S. Cameron, and C. K. Prout, *Chem. Commun.*, **1968**, 616.

337 G. H. Birum and C. N. Matthews, *Chem. Commun.*, **1967**, 137.

338 G. H. Birum and C. N. Matthews, *J. Org. Chem.*, **32**, 3554 (1967).

339 G. Chioccola and J. J. Daly, *J. Chem. Soc.*, (*A*), **1968**, 568.

340 J. B. Conant and A. A. Cook, *J. Amer. Chem. Soc.*, **42**, 830 (1920).

341 L. Anschutz, E. Klein, and G. Cermak, *Chem. Ber.*, **77**, 726 (1944).

342 A. Michaelis, *Ber.*, **17**, 1273 (1884); **18**, 898 (1885).

343 L. R. Drake and C. S. Marvel, *J. Org. Chem.*, **2**, 387 (1938).

344 K. Bergesen, *Acta Chem. Scand.*, **19**, 1784 (1965).

345 J. B. Conant and S. M. Pollack, *J. Amer. Chem. Soc.*, **43**, 1665 (1921).

346 J. C. Pernert (to Albright and Wilson (Mfg.) Ltd.), Brit. Pat. 801,568 (Sept. 17, 1958); *Chem. Abstr.*, **53**, 9251 (1959).

347 F. Ramirez, O. P. Madan, and S. R. Heller, *J. Amer. Chem. Soc.*, **87**, 731 (1965).

348 N. A. Razumova and A. A. Petrov, *Dokl. Akad. Nauk SSSR*, **158**, 907 (1964); *Chem. Abstr.*, **62**, 2790 (1965).

349 C. W. Smith (to Shell Development Co.), U.S. Pat. 2,648,695 (Aug. 11, 1953); *Chem. Abstr.*, **48**, 8252 (1954).

350 A. Y. Garner (to Monsanto Chemical Co.), U.S. Pat. 2,916,510 (Dec. 8, 1959); *Chem. Abstr.*, **54**, 5571 (1960).

351 A. Y. Garner (to Monsanto Chemical Co.), U.S. Pat. 2,953,591 (Sept. 20, 1900); *Chem. Abstr.*, **55**, 5346 (1961).

352 A. Eberhard and F. H. Westheimer, *J. Amer. Chem. Soc.*, **87**, 253 (1965).

353 A. R. Hands and A. J. H. Mercer, *J. Chem. Soc.*, (*C*), **1967**, 1099.

354 C. T. Eyles and S. Trippett, *J. Chem. Soc.*, (*C*), **1966**, 67.

355 M. Grayson and C. E. Farley, *Chem. Commun.*, **1967**, 830.

356 F. Korte and H. Röchling, *Tetrahedron Lett.*, **1964**, 2099.

357 K. H. Buechel, H. Röchling, and F. Korte, *Annalen*, **685**, 10 (1965).

358 M. I. Kabachnik and E. S. Shepeleva, *Dokl. Akad. Nauk SSSR*, **75**, 219 (1950); *Chem. Abstr.*, **45**, 6569 (1951).

359 F. Ramirez and S. Dershowitz, *J. Org. Chem.*, **22**, 856, 1282 (1957); **23**, 778 (1958).

360 G. H. Birum and J. L. Dever, *Abstr. Amer. Chem. Soc.*, 134th *Natl. Meeting, Chicago, Illinois*, 1958, p. 101-P.

361 V. A. Kukhtin, *Dokl. Akad. Nauk SSSR*, **121**, 466 (1958); *Chem. Abstr.*, **53**, 1105 (1959).

362 F. Ramirez, 'Condensations of Carbonyl Compounds with Phosphite Esters', being a Chapter in *Organo-Phosphorus Compounds*, p. 337 (I.U.P.A.C., International Symposium, Heidelberg, 1964), Butterworth and Co. Ltd., London, 1964.

363 F. Ramirez, *Pure and Applied Chemistry*, **9**, 337 (1964) (identical with reference 362).

364 F. Ramirez, *Bull. Soc. Chim. France*, **1966**, 2443.

365 V. A. Kukhtin and K. M. Orekhova, *Zh. Obshch. Khim.*, **30**, 1208 (1960); *Chem. Abstr.*, **55**, 358 (1961).

366 V. A. Kukhtin, T. N. Voskoboeva, and K. M. Kirillova, *Zh. Obshch. Khim.*, **32**, 2333 (1962); *Chem. Abstr.*, **58**, 9127 (1963).

[367] F. Ramirez and N. B. Desai, *J. Amer. Chem. Soc.*, **85**, 3252 (1963).

[368] F. Ramirez and N. B. Desai, *J. Amer. Chem. Soc.*, **82**, 2652 (1960).

[369] W. C. Hamilton, S. J. LaPlaca, and F. Ramirez, *J. Amer. Chem. Soc.*, **87**, 127 (1965).

[370] W. C. Hamilton, S. J. LaPlaca, F. Ramirez, and C. P. Smith, *J. Amer. Chem. Soc.*, **89**, 2268 (1967).

[371] R. D. Sprately, W. C. Hamilton, and J. Ladell, *J. Amer. Chem. Soc.*, **89**, 2272 (1967).

[372] G. H. Birum and J. L. Dever, U.S. Pat. 3,014,949 (1961); *Chem. Abstr.*, **56**, 10191 (1961).

[373] F. Ramirez, O. P. Madan, N. B. Desai, S. Meyerson, and E. M. Banas, *J. Amer. Chem. Soc.*, **85**, 2681 (1963).

[374] F. Ramirez, R. B. Mitra, and N. B. Desai, *J. Amer. Chem. Soc.*, **82**, 2651, 5763 (1960).

[375] F. Ramirez and N. Ramanathan, *J. Org. Chem.*, **26**, 3041 (1961).

[376] F. Ramirez, N. Ramanathan, and N. B. Desai, *J. Amer. Chem. Soc.*, **84**, 1317 (1962).

[377] F. Ramirez, N. B. Desai, and N. Ramanathan, *J. Amer. Chem. Soc.*, **85**, 1874 (1963).

[378] F. Ramirez, A. V. Patwardhan, N. B. Desai, N. Ramanathan, and C. V. Greco, *J. Amer. Chem. Soc.*, **85**, 3056 (1963).

[379] F. Ramirez, N. Ramanathan, and N. B. Desai, *J. Amer. Chem. Soc.*, **85**, 3465 (1963).

[380] F. Ramirez, A. V. Patwardhan, N. Ramanathan, N. B. Desai, C. V. Greco, and S. R. Heller, *J. Amer. Chem. Soc.*, **87**, 543 (1965).

[381] F. Ramirez, A. V. Patwardhan, and C. P. Smith, *J. Org. Chem.*, **31**, 474 (1966).

[382] F. Ramirez, S. B. Bhatia, R. B. Mitra, Z. Hamlet, and N. B. Desai, *J. Amer. Chem. Soc.*, **86**, 4394 (1964).

[383] P. A. Rossiĭskaya and M. I. Kabachnik, *Bull. Acad. Sci. URSS, Cl. Sci. Chim.*, **1947**, 509; *Chem. Abstr.*, **42**, 2924 (1948).

[384] H. J. Lucas, F. W. Mitchell, and C. N. Scully, *J. Amer. Chem. Soc.*, **72**, 5491 (1950).

[385] A. A. Foxton, G. H. Jeffery, and A. I. Vogel, *J. Chem. Soc.*, A, **1966**, 249.

[386] D. C. Ayres and H. N. Rydon, *J. Chem. Soc.*, **1957**, 1109.

[387] G. Nagy (to Establissements Kuhlmann), French Pat. 1,384,809 (Jan. 8, 1965); *Chem. Abstr.*, **62**, 9061 (1965).

[388] V. P. Evdakov, E. K. Shlenkova, and K. A. Bilevich, *Zh. Obshch. Khim.*, **35**, 728 (1965); *Chem. Abstr.*, **63**, 2889 (1965).

[389] H. K. Garner and H. J. Lucas, *J. Amer. Chem. Soc.*, **72**, 5497 (1950).

[390] B. A. Arbuzov, K. V. Kikonorov, O. N. Fedorova, G. M. Vinokurova, and Z. G. Shishova, *Dokl. Akad. Nauk SSSR*, **91**, 817 (1953); *Chem. Abstr.*, **48**, 10539 (1954).

[391] G. Kamai and R. K. Ismagilov, *Zh. Obshch. Khim.*, **34**, 439 (1964); *Chem. Abstr.*, **60**, 13267 (1964).

[392] H. J. Harwood (to Monsanto Co.), U.S. Pat. 3,157,694 (Nov. 17, 1964); *Chem. Abstr.*, **62**, 4053 (1965).

[393] H. Goldwhite, *Chem. & Ind.* (*London*), **1964**, 494.

[394] R. Foster and C. A. Fyfe, *Spectrochem. Acta*, **21**, 1785 (1965).

[395] A. D. F. Toy (to Victor Chemical Works), U.S. Pat. 2,382,622 (Aug. 14, 1945); *Chem. Abstr.*, **40**, 604 (1946).

[396] K. A. Petrov, R. A. Baksova, and L. V. Khorkhoyanu, *Zh. Obshch. Khim.*, **35**, 732 (1965); *Chem. Abstr.*, **63**, 4328 (1965).

[397] R. S. Edmundson, *Chem. & Ind. (London)*, **1962**, 1828.

[398] K. Kumamoto, J. R. Cox, Jr, and F. H. Westheimer, *J. Amer. Chem. Soc.*, **78**, 4858 (1956).

[399] F. Covitz and F. H. Westheimer, *J. Amer. Chem. Soc.*, **85**, 1773 (1963).

[400] T. A. Steitz and W. N. Lipscomb, *J. Amer. Chem. Soc.*, **87**, 2488 (1965).

[401] M. G. Newton, J. R. Cox, Jr, and J. A. Bertrand, *J. Amer. Chem. Soc.*, **88**, 1503 (1966).

[402] G. Aksnes and K. Bergesen, *Acta Chem. Scand.*, **19**, 931 (1965).

[403] E. A. Dennis and F. H. Westheimer, *J. Amer. Chem. Soc.*, **88**, 3431 (1966) and references given therein.

[404] E. A. Dennis and F. H. Westheimer, *J. Amer. Chem. Soc.*, **88**, 3432 (1966) and references given therein.

[405] G. Aksnes and K. Bergesen, *Acta Chem. Scand.*, **20**, 2508 (1966).

[406] R. Schmutzler, *Angew. Chem.*, **77**, 530 (1965); *Int. Ed. Engl.*, **4**, 496 (1965).

[407] D. G. Gorenstein and F. H. Westheimer, *J. Amer. Chem. Soc.*, **89**, 2762 (1967).

[408] A. J. Kirby and S. G. Warren, *The Organic Chemistry of Phosphorus*, Monograph No. 5 in 'Reaction Mechanisms in Organic Chemistry', ed. C. Eaborn and N. B. Chapman, Elsevier, Amsterdam/London/New York, 1967, pp. 343–352.

[409] W. Knauer, *Ber.*, **27**, 2569 (1894).

[410] L. Anschütz and W. Broeker, *Ber.*, **61**, 1264 (1928).

[411] L. Anschütz and H. Walbrecht (with W. Broeker), *J. Prakt. Chem.*, **133**, 65 (1932).

[412] L. Anschütz, W. Broeker, R. Neher, and A. Ohnheiser, *Chem. Ber.*, **76**, 223 (1943).

[413] P. C. Crofts, J. H. H. Markes, and H. N. Rydon, *J. Chem. Soc.*, **1958**, 4250.

[414] A. E. Arbuzov and F. G. Valitova, *Trans. Kirov Inst. Chem. Tech. Kazan*, **8**, 12 (1940); *Chem. Abstr.*, **35**, 2485 (1941).

[415] A. E. Arbuzov and F. G. Valitova, *Bull. Acad. Sci. USSR, Cl. Sci. Chim.*, **1940**, 529; *Chem. Abstr.*, **35**, 3990 (1041).

[416] F. G. Mann, D. Purdie, and A. F. Wells, *J. Chem. Soc.*, **1936**, 1503.

[417] L. Anschütz and W. Broeker, *Annalen*, **454**, 71 (1927).

[418] G. W. Anderson, J. Blodinger, and A. D. Welcher, *J. Amer. Chem. Soc.*, **74**, 5309 (1952).

[419] V. du Vigneaud, C. Ressler, J. M. Swan, C. W. Roberts, P. G. Katsoyannis, and S. Gordon, *J. Amer. Chem. Soc.*, **75**, 4879 (1953) and subsequent papers in that Journal.

[420] A. E. Arbusov and F. G. Valitova, *Izvest. Akad. Nauk SSSR, Otdel. Khim. Nauk* **1956**, 681; *Chem. Abstr.*, **51**, 1877 (1957).

[421] H. Gross and J. Gloede, *Chem. Ber.*, **96**, 1387 (1963).

[422] H. Gross and J. Gloede, *Z. Chem.*, **5**, 178 (1965); *Chem. Abstr.*, **63**, 5622 (1965).

[423] R. W. Cummins, *Chem. and Ind. (London)*, **1967**, 918.

[424] H. Lecoq, *Bull. Soc. Chim. Belges*, **42**, 199 (1933).

[425] K. Oita and H. Gilman, *J. Amer. Chem. Soc.*, **79**, 339 (1957).

[426] W. Cule Davies and C. J. O. R. Morris, *J. Chem. Soc.*, **1932**, 2880.

[427] J. B. Levy, L. D. Freedman, and G. O. Doak, *J. Org. Chem.*, **33**, 474 (1968).

[428] L. D. Freedman, G. O. Doak and J. R. Edmisten, *J. Org. Chem.*, **26**, 284 (1961).

[429] I. G. M. Campbell, *J. Chem. Soc.*, *C*, **1968**, 3026.

[430] A. M. Meston, *J. Chem. Soc.*, **1963**, 6059.

431 G. M. Blackburn, J. S. Cohen, and A. R. (Lord) Todd, *Tetrahedron Lett.*, **1964**, 2873.

432 R. S. Edmundson, *Tetrahedron*, **20**, 2781 (1964).

433 R. L. McConnell and H. W. Coover, Jr, *J. Org. Chem.*, **24**, 630 (1959).

434 R. S. Edmundson, *Tetrahedron*, **21**, 2379 (1965).

435 J. H. Bartlett (to Esso Research and Engineering Co.), U.S. Pat. 3,159,664 (Dec. 1, 1964); *Chem. Abstr.*, **62**, 7639 (1965).

436 H. G. Khorana, G. M. Tener, R. S. Wright, and J. G. Moffatt, *J. Amer. Chem. Soc.*, **79**, 430 (1957).

437 W. M. Lanham (to Union Carbide Co.), U.S. Pat. 3,159,591 (Dec. 1, 1964); *Chem. Abstr.*, **62**, 7639 (1965).

438 R. S. Edmundson, *Chem. & Ind. (London)*, **1963**, 784.

439 R. S. Edmundson, *J. Chem. Soc., C*, **1967**, 1635.

440 A. S. Couper, *Phil. Mag.* [4], **16**, 104 (1858).

441 A. S. Couper, *Edinburgh New Philosophical Journal, New Series*, **8**, 213 (1858), republished in Alembic Club Reprints No. 21, '*On a New Chemical Theory and Researches on Salicylic Acid*', Edinburgh, 1933.

442 A. S. Couper, *Compt. rend.*, **46**, 1157 (1858).

443 F. A. Kekulé, *Bull. Soc. chim. France*, **3**, 98 (1865).

444 A. S. Couper, *Ann. chim. phys.* (3), **53**, 489 (1859).

445 F. R. Atherton, '*Some Aspects of the Chemistry of Phosphorus Compounds derived from Salicylic Acid*', being a Chapter in '*Phosphoric Esters and Related Compounds*', Special Publication No. 8 of the Chemical Society, London, 1957.

446 R. Anschütz, *Annalen*, **228**, 314 (1885).

446a R. Anschütz, *Annalen*, **346**, 286 (1906).

447 R. Anschütz and W. O. Emery, *Annalen*, **239**, 301 (1887).

448 L. Anschütz, *Annalen*, **439**, 265 (1924).

449 R. W. Young, *J. Amer. Chem. Soc.*, **74**, 1672 (1952).

450 A. G. Pinkus, P. G. Waldrep, and W. J. Collier, *J. Org. Chem.*, **26**, 682 (1961).

451 A. G. Pinkus and P. G. Waldrep, *J. Org. Chem.*, **31**, 575 (1966).

452 J. A. Cade and W. Gerrard, *Chem. & Ind. (London)*, **1954**, 402.

452a J. A. Cade and W. Gerrard, *J. Chem. Soc.*, **1960**, 1249.

453 R. Anschütz and G. D. Moore, *Annalen*, **239**, 314 (1887).

454 F. R. Atherton (to Roche Products Ltd.), Brit. Pat. 793,722 (April 23, 1958); *Chem. Abstr.*, **52**, 20063 (1958).

455 F. R. Atherton (to Roche Products Ltd.), Brit. Pat. 806,879 (Jan. 7, 1959); *Chem. Abstr.*, **53**, 14125 (1959).

456 A. G. Pinkus, P. G. Waldrep, and S. Y. Ma, *J. Heterocyclic Chem.*, **2**, 357 (1965).

457 N. I. Sax, 'Dangerous Properties of Industrial Materials', Reinhold Publ. Corp., New York (Chapman and Hall, Ltd., London), 1957, p. 1207.

458 J. E. Casida, M. Eto, and R. L. Baron, *Nature (London)*, **191**, 1396 (1961).

459 M. Eto, J. E. Casida, and T. Eto, *Biochem. Pharmacol.*, **11**, 337 (1962).

460 M. Eto, S. Matsuo, and Y. Oshima, *Agr. Biol. Chem. (Tokyo)*, **27**, 870 (1963); *Chem. Abstr.*, **60**, 11311 (1964).

461 M. Eto, T. Eto, and Y. Oshima, *Agr. Biol. Chem. (Tokyo)*, **26**, 630 (1962); *Chem. Abstr.*, **59**, 3269 (1963).

462 M. Eto, Y. Kinoshita, T. Kato, and Y. Oshima, *Nature (London)*, **200**, 171 (1963).

463 M. Eto, Y. Kinoshita, T. Kato, and Y. Oshima, *Agr. Biol. Chem. (Tokyo)*, **27**, 789 (1963); *Chem. Abstr.*, **60**, 7944 (1964).

[464] M. Eto, K. Kishimoto, K. Matsumura, N. Oshita, and Y. Oshima, *Agr. Biol. Chem. (Tokyo)*, **30**, 181 (1966); *Chem. Abstr.*, **64**, 17627 (1966).

[465] F. G. Bordwell, C. E. Osborne, and R. D. Chapman, *J. Amer. Chem. Soc.*, **81**, 2698 (1959).

[466] M. Epstein and S. A. Buckler, *Tetrahedron*, **18**, 1231 (1962).

[467] S. A. Buckler, *J. Amer. Chem. Soc.*, **82**, 4215 (1960).

[468] E. J. Boros, K. J. Coskran, R. W. King, and J. G. Verkade, *J. Amer. Chem. Soc.*, **88**, 1140 (1966).

[469] E. J. Boros, R. D. Compton, and J. G. Verkade, *Inorg. Chem.*, **7**, 165 (1968).

[470] J. G. Verkade and L. T. Reynolds, *J. Org. Chem.*, **25**, 663 (1960).

[471] J. G. Verkade, R. E. McCarley, D. G. Hendricker, and R. W. King, *Inorg. Chem.*, **4**, 228 (1965).

[472] J. G. Verkade and R. W. King, *Inorg. Chem.*, **1**, 948 (1962).

[473] F. Zetzsche and E. Zurbrügg, *Helv. Chem. Acta*, **9**, 298 (1926).

[474] D. M. Nimrod, D. R. Fitzwater, and J. G. Verkade, *J. Amer. Chem. Soc.*, **90**, 2780 (1968).

[475] H. Stetter and K. H. Steinacker, *Chem. Ber.*, **85**, 451 (1952).

[476] A. M. Aguiar, K. C. Hansen, and J. T. Mague, *J. Org. Chem.*, **32**, 2383 (1967).

[477] K. J. Coskran and J. G. Verkade, *Inorg. Chem.*, **4**, 1655 (1965).

[478] M. Grayson (to American Cyanamid Co.), Ger. Pat. 1,151,255 (July 11, 1963); *Chem. Abstr.*, **60**, 554 (1964).

[479] P. Carré, *Comp. rend.*, **136**, 756 (1903).

[480] T. Bersin, H. G. Moldtmann, H. Nafziger, B. Marchand, and W. Leopold, *Z. physiol. Chem.*, **269**, 241 (1941).

[481] R. H. Iwamoto, E. M. Acton, L. Goodman, and B. R. Baker, *J. Org. Chem.*, **26**, 4743 (1961).

[482] Asta-Werke A.G. Chemische Fabrik, Brit. Pat. 812,651 (April 29, 1959); *Chem. Abstr.*, **54**, 5472 (1960).

[483] Asta-Werke A.G. Chemische Fabrik, Brit. Pat. 822,110 (Oct. 21, 1959); *Chem. Abstr.*, **55**, 3431 (1961).

[484] Asta-Werke A.G. Chemische Fabrik, Brit. Pat. 853,044 (Nov. 2, 1960); *Chem. Abstr.*, **55**, 19793 (1961).

[485] H. Arnold and F. Bourseaux, *Angew. Chem.*, **70**, 539 (1958).

[486] I. K. Fel'dman and A. I. Berlin, *Zh. Obshch. Khim.*, **32**, 575 (1962); *Chem. Abstr.*, **58**, 6820 (1963).

[487] Z. B. Papanastassiou, R. J. Bruni, F. P. Fernandes, and P. L. Levins, *J. Med. Chem.*, **9**, 357 (1966).

[488] N. Brock, *Arzneimittel-Forsch.*, **8**, 1 (1958); *Chem. Abstr.*, **52**, 9439 (1958).

[489] H. Arnold, F. Bourseaux, and N. Brock, *Naturwiss.*, **45**, 64 (1958); *Chem. Abstr.*, **52**, 12229 (1958).

[490] G. H. Fairley and J. M. Simister (Editors), 'Cyclophosphamide', J. Wright and Sons, Ltd., Bristol, 1964.

[491] A. E. Arbuzov and V. M. Zoroastrova, *Izvest. Akad. Nauk SSSR, Otdel. Khim. Nauk*, **1952**, 453; *Chem. Abstr.*, **47**, 4833 (1953).

[492] M. Wieber, J. Otto, and M. Schmidt, *Angew. Chem.*, **76**, 648 (1964); *Int. Ed. Engl.*, **3**, 586 (1964).

[493] E. W. Abel, D. A. Armitage, and R. P. Bush, *J. Chem. Soc.*, **1965**, 7098.

[494] L. D. Freedman and G. O. Doak, *J. Org. Chem.*, **29**, 1983 (1964).

[495] R. Appel and D. Hänssgen, *Angew. Chem.*, **79**, 577 (1967); *Int. Ed. Engl.*, **6**, 560 (1967).

496 M. Nagasawa, Y. Imamiya, and H. Sugiyama (to Ihara Agricultural Chemical
    Co.), Jap. Pat. 17,211 (Oct. 23, 1962); *Chem. Abstr.*, **59**, 11254 (1963).
497 Emrys R. H. Jones and F. G. Mann, *J. Chem. Soc.*, **1955**, 4472.
498 W. Cochran, F. A. Hart, and F. G. Mann, *J. Chem. Soc.*, **1957**, 2816.
499 M. Davis and F. G. Mann, *J. Chem. Soc.*, **1964**, 3791.
500 K. Issleib and H.-J. Deylig, *Chem. Ber.*, **97**, 946 (1964).
501 W. Hawes and S. Trippett, *J. Chem. Soc.*, *C*, **1969**, 1465.
502 L. D. Quin, J. D. Bryson, and C. G. Moreland, *J. Amer. Chem. Soc.*, **91**, 3308
    (1969).
503 F. G. Mann and A. J. H. Mercer, unpublished results.

PART II

Heterocyclic Derivatives of Arsenic

Five-membered Ring Systems containing only Carbon and One Arsenic Atom

Arsoles and Arsolanes

Arsoles

The ring system (1) (RRI 140) is known as arsole*: the two isomeric dihydro-derivatives (2) and (3) would be 2- and 3-arsolenes (following

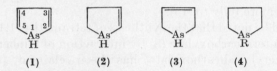

(1)   (2)   (3)   (4)

the phosphorus nomenclature) and the tetrahydro derivative (4; R = H) is arsolane. [The formula (1) is given as shown for convenience, but the Ring Index presents it with the arsenic atom uppermost.] Arsolane is one of the earliest known cyclic arsenic systems, and was long known as cyclotetramethylenearsine, and sometimes (more recently) as arsolidene.

Arsole (1) is unknown, and the chemistry of its stable substituted derivatives closely follows that of the corresponding phosphole compounds, but the arsole derivatives have not been studied in such detail.

Leavitt, Manuel, *et al.*[1, 2, 3] have shown that 1,4-dilithio-1,2,3,4-tetraphenylbutadiene (5) (p. 13) reacts with phenylarsonous dichloride, $C_6H_5AsCl_2$, in ether to give 1,2,3,4,5-pentaphenylarsole (6),

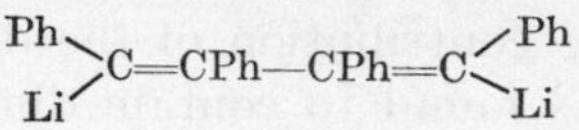

(5) 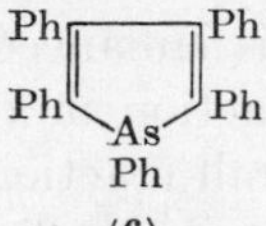   (6) 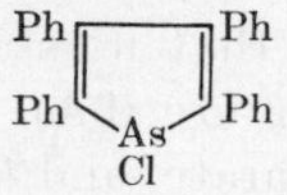   (7)

yellow needles, m.p. 215–216°, in 93% yield. The dilithio compound (5) reacts similarly with arsenic trichloride to form 1-chloro-2,3,4,5-tetraphenylarsole (7), yellow needles, m.p. 182–184°; this was not obtained pure, but its identity was confirmed by its conversion by phenyllithium or by phenylmagnesium bromide into the pure arsole (6).

---

* In English this ring system has frequently been named arsenole 'for euphony'.

357

This synthesis has been carried out independently by Braye *et al.*,[4, 5] who by the action of phenylarsonous dichloride on the dilithio derivative in ether obtained the pentaphenylarsole (**6**), greenish-yellow needles, m.p. 213–214.5°, in 56% yield. There is a very close resemblance between the ultraviolet spectra in cyclohexane of pentaphenylphosphole and pentaphenylarsole (**6**).[4]

Braye *et al.*[4] show that the arsole (**6**), when treated with hydrogen peroxide, gives an oxide as yellow needles, m.p. 252°, and when heated with iron pentacarbonyl, $Fe(CO)_5$, in isooctane at 150° for 15 hours in a sealed tube gives pale yellow crystals of composition $C_{34}H_{25}As,Fe(CO)_3$, m.p. 155–170° (dec.). The infrared spectrum of the latter shows the terminal CO groups by bands at 2053, 1996, and 1980 cm^{-1}. The chemistry of similar iron carbonyl derivatives of pentaphenylphosphole has been studied more fully (p. 15).

It should be noted that the synthesis of tetramethyl 1,1,1-triphenyl-arsole-2,3,4,5-tetracarboxylate (**8**) by interaction of triphenylarsine and dimethyl acetylenedicarboxylate has been claimed (Hendrickson, Spenger, and Sims[6]); this product has considerable stability (unlike its apparent phosphorus analogue, p. 26) and could be sublimed unchanged at 300° under reduced pressure. Further investigation by these workers[7] has, however, provided considerable evidence that this adduct has the structure (**9**), i.e., dimethyl 2-oxo-3-(triphenylarsoranylidene)succinate,

and that its stability is enhanced by the contribution of the dipolar structure (**9A**). The reaction mixture was found to contain dimethyl fumarate, and the overall reaction may be regarded as the interaction of the triphenylarsine and the dimethyl acetylenedicarboxylate in the presence of water, whereby one molecule of the ester is reduced to the fumarate and the other oxidized with formation of (**9**). The probable mechanism has been discussed in detail.[7]

The electronic structures of the phosphole and arsenole ring systems in terms of molecular-orbital theory have been discussed by Brown,[8] who suggests that the planar configuration of the ring system has considerable conjugation energy by the use of the np_π-orbitals of the phosphorus or arsenic atom. The interested reader should consult the original paper.

Arsolenes

Apparently neither of the two isomeric arsolenes (**2**) and (**3**) has been prepared. The McCormack reaction for preparation of the corresponding phospholenes by the interaction of a 1,3-diene and an aryl-phosphonous dichloride (p. 31) has no parallel when an arylarsinous dichloride is employed.[9]

Arsolanes

Grüttner and Krause[10] have shown that phenylarsonous dichloride, $C_6H_5AsCl_2$, reacts with the di-Grignard reagent obtained from 1,4-dibromobutane, *i.e.*, with $(CH_2CH_2MgBr)_2$ in cold ether to give 1-phenyl-arsolane (also called *As*-phenylarsolane) (**4**; $R = C_6H_5$), a colorless liquid, b.p. 128.5°/15–16 mm, having a faint odor. Modern experimental details for this preparation have been given by Monagle.[11] The arsine is not readily oxidized in air. It has the normal properties of a tertiary arsine; for example, it combines with chlorine to give a colorless crystalline dichloride, $C_{10}H_{13}AsCl_2$, m.p. 120.5°, and with mercuric chloride to form a mercurichloride, m.p. 160–162°, $C_{10}H_{13}As \cdot HgCl_2$, which probably has the 'double' bridged structure, dichlorobis-1-phenylarsolane-$\mu\mu'$-dichlorodimercury (Evans, Mann, Peiser, and Purdie[12]). It also forms the following colorless crystalline quaternary salts: methiodide, m.p. 135–136°; ethiodide, m.p. 85–86°; propyl iodide, m.p. 123–124°; isopropyl iodide, m.p. 113–114°. 1-Phenylarsolane 1-oxide was found to be inert as a catalyst for the conversion of isocyanates into carbodiimides, although triphenylarsine oxide has high activity.[11]

Steinkopf, Schubart, and Roch[13] have similarly prepared 1-methyl-arsolane (**4**; $R = CH_3$). This arsine also combines with chlorine to give the arsine dichloride, m.p. 112–115°, which when heated under reduced pressure loses methyl chloride with the formation of 1-chloroarsolane (**4**; $R = Cl$), b.p. 77°/18 mm. This method of replacing an alkyl group directly joined to the arsenic atom of a tertiary arsine has been frequently employed, as in the arsenanes (p. 392), the tetrahydroarsinolines (p. 401) and the tetrahydroisoarsinolines (p. 416).

1-Chloroarsolane is of synthetic value because treatment with an alkyl or aryl Grignard reagent or lithium reagent will readily furnish the corresponding 1-alkyl(or aryl)arsolane.

It has been stated by Das-Gupta[14] that succinoyl chloride $(CH_2COCl)_2$, phenylarsonous dichloride, and sodium react together in benzene containing some ethyl acetate to give 'succinoylphenylarsine' or 2,5-dioxo-1-phenylarsolane (**10**), a colorless liquid, b.p. 119–120°/10 mm. This compound gives a quaternary methiodide, m.p. 176°, and an ethiodide, m.p. 165–167°. It is also stated to give a picrate, m.p. 117°

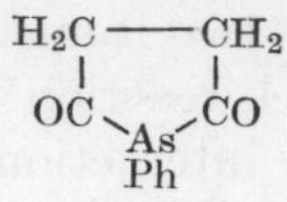

$$H_2C\text{---}CH_2$$
$$OC\diagdown_{\underset{Ph}{As}}\diagup CO$$

(**10**)

(although tertiary arsines do not usually give picrates), and a mercuri-chloride, m.p. 245°. When reduced with sodium in an ethanol–toluene solution, it gives 1-phenylarsolane, b.p. 125–130°/15 mm.

Wiberg and Mödritzer[15] have reduced 1-chloroarsolane (**4**; R = Cl) with lithium borohydride in ether at −60° to obtain arsolane (**4**; R = H), a colorless liquid, m.p. −29° to −28°, in 55% yield. It is described as having 'a phosphine odor'.

Two classes of compound that can be regarded systematically as arsoles to which a further five-membered ring-system is attached can be conveniently recorded here.

7-Arsabicyclo[2.2.1]heptane

The compound (**1**) is termed 7-arsabicyclo[2.2.1]heptane, [RRI 972], and the 1,2,3,4,5,6-hexadehydro form (**2**) as 7-arsabicyclo[2.2.1]hepta-1,3,5-triene.

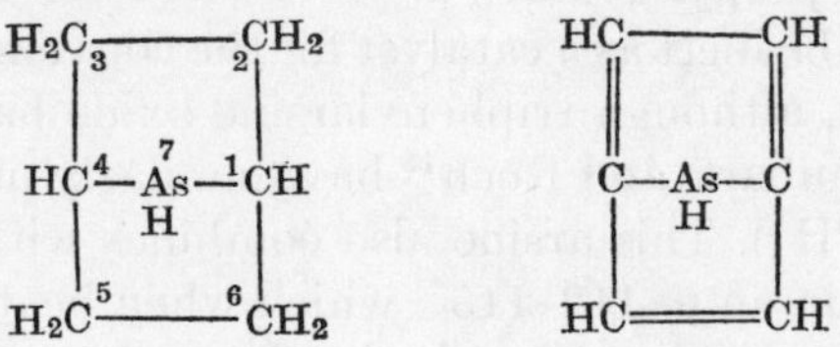

(**1**)   (**2**)

Schmidt and Hoffmann[15a] have claimed that p-aminophenyl-arsonic acid (**3**), when diazotized in 5N-hydrochloric acid with sodium

$$p\text{-}H_2NC_6H_4AsO(OH)_2 \rightarrow p\text{-}ClN_2C_6H_4AsO(OH)_2 \rightarrow p\text{-}ClN_2C_6H_4AsCl_2$$
(**3**)   (**4**)   (**5**)

nitrite, gives p-chlorodiazoniophenylarsonic acid (**4**) which was not isolated; reduction in the reaction mixture with sulfur dioxide and a trace of potassium iodide gave p-dichloroarsinophenyldiazonium chloride (**5**) in quantitative yield as a yellow powder. This material, when suspended in ice-cold aqueous sodium hydrogen carbonate, evolved nitrogen, and the filtered solution, when neutralized to Congo Red with hydrochloric acid, gave (in 70% yield) the 'clear brown' arsinic acid (6), *i.e.*, 7-hydroxy-7-arsabicyclo[2.2.1]hepta-1,3,5-triene 7-oxide, as a monohydrate, which in a vacuum-desiccator gave the anhydrous acid. This acid, when added to ice-cold ethanolic hydrochloric acid and treated with sulfur dioxide, yielded the 7-chloro compound (7), as a reddish-brown powder of indefinite melting point; its molecular weight in chloroform indicated the monomeric form (7).

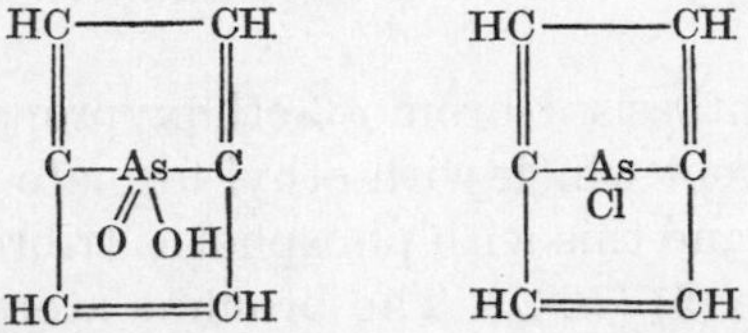

HC———CH
C—As—C
O OH
HC====CH
(6)

HC———CH
C—As—C
Cl
HC====CH
(7)

This series of reactions has some curious features. The conversion of the compound (5) into (6) involves the formation of a link through an arsenic atom across the *para*-positions of a benzene ring, which is surprising; it also involves the oxidation of the arsenic atom from the three- to the five-valent state, although if the cyclization occurred by a modified Bart reaction this increase in valency could result. The alternative formation of a dimeric diarsonic acid (8) is discounted by the authors

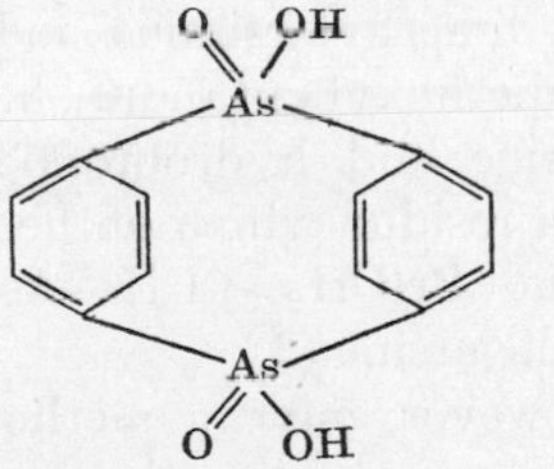

O   OH
As
As
O   OH
(8)

because the compound (7) is apparently monomeric. The compounds (5) and (7) have been analyzed only for arsenic and chlorine, and the acid (6), dihydrated and anhydrous, only for arsenic. A fuller investigation of these reactions and of the properties of the products, with carbon and hydrogen analyses for additional confirmation, has apparently not been made.

1*H*-Arsolo[1,2-*a*]arsole

The saturated bridge-head compound (2) was prepared by Heinekey, Millar, and Mann,[16] who named it 1-arsabicyclo[3.3.0]octane and

13

numbered it as shown in (**2**). The parent compound (**1**) was subsequently named 1H-arsolo[1,2-a]arsole, [RIS 11804], and the compound (**2**) is

$$
\begin{array}{cc}
(\mathbf{1}) & (\mathbf{2})
\end{array}
$$

therefore now hexahydro-1H-arsolo[1,2-a]arsole, with the numbering as indicated in (**1**).

In the above synthesis, 1-bromo-3-ethoxypropane (**3**) was converted into a Grignard reagent which with ethyl formate gave 1,7-diethoxy-4-heptanol (**4**) (57%), and this with phosphorus tribromide gave 4-bromo-1,7-diethoxyheptane (**5**) (76%). The bromine atom in (**5**) did not react

$$C_2H_5O(CH_2)_3Br \;\rightarrow\; [C_2H_5O(CH_2)_3]_2CHOH \;\rightarrow\; [C_2H_5O \cdot (CH_2)_3]_2CHBr \;\rightarrow$$
$$\qquad (\mathbf{3}) \qquad\qquad\qquad (\mathbf{4}) \qquad\qquad\qquad\qquad (\mathbf{5})$$

$$[C_2H_5O(CH_2)_3]_2CH\!-\!As(CH_3)_2 \;\rightarrow\; [C_2H_5O(CH_2)_3]_2CH\!-\!\overset{+}{A}s(CH_3)_3\,Br^- \;\rightarrow$$
$$\qquad (\mathbf{6}) \qquad\qquad\qquad\qquad\qquad (\mathbf{7})$$

$$[Br(CH_2)_3]_2CH\!-\!\overset{+}{A}s(CH_3)_3\,Br^- \;\rightarrow\; (\mathbf{1})$$
$$(\mathbf{8})$$

readily with magnesium, but the use of the 'entrainment' method with an excess of magnesium and some methyl iodide gave the Grignard reagent, which with dimethylarsinous iodide, $(CH_3)_2AsI$, yielded 4-(dimethylarsino)-1,7-diethoxyheptane (**6**), b.p. 144–146°/13 mm (48%). A mixture of this arsine and hydrobromic acid, when boiled and evaporated, furnished a residue which on heating gave 1,7-dibromo-4-(dimethylarsino)heptane, $[Br(CH_2)_3]_2CH\!-\!As(CH_3)_2$, without apparent cyclization to the dicyclic arsine (**2**).

The arsine (**6**), however, gave a methobromide (**7**) which when heated with hydrobromic acid gave the arsonium bromide (**8**). This compound when heated to 200° underwent loss of methyl bromide followed by cyclic quaternization; repetition of these processes occurred during the decomposition and gave the arsine (**2**), b.p. 100–102°/17 mm, and a residue of the corresponding methobromide. The very readily oxidized arsine (**2**) was characterized as its red crystalline pallado-bromide derivative, $(C_7H_{13}As)_2(PdBr_2)_2$, *i.e.*, dibromobis(hexahydro-1H-arsolo[1,2-a]arsole)-$\mu\mu'$-dibromodipalladium, m.p. 150°, and the methobromide as the methopicrate, hexahydro-4-methyl-1H-arsolo-[1,2-a]arsolium picrate, yellow crystals, m.p. 232°.

The very low yields in this last stage were not improved by variation of the conditions of the thermal decomposition. Many arsenic compounds show a strong tendency to form five-membered rings (cf. pp. 536–539), and the low yield may be due to the event that on thermal decomposition the intermediate cyclized bromides undergo considerable ring disruption in addition to simple loss of methyl bromide.

Benzo Derivatives of Arsoles

The fusion of one cyclic benzo ring to an arsole molecule can give rise to two isomeric series of compounds, the arsindoles and the iso-arsindoles, and the fusion of two such rings to an arsole molecule gives rise to the dibenzarsoles, alternatively named the 9-arsafluorenes. These classes will be considered in turn.

Arsindoles and Arsindolines

The system (1; R = H) is named arsindole (RRI 1249), and its 2,3-dihydro derivative (2; R = H) is arsindoline. These names indicate that these compounds are the arsenic analogues of indole and indoline,

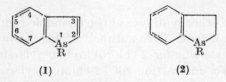

(1)   (2)

respectively, but the chemistry of the two arsenic systems differs markedly from that of their nitrogen analogues; in particular, the comparatively low reactivity of the NH group of indole is not shown by the AsR group (where R is alkyl or aryl) of the arsindoles, for this group—on the basis of recorded work—has the normal high reactivity of tertiary arsines.

As in the case of so many heterocyclic arsenic derivatives, the parent compounds (1; R = H) and (2; R = H) have not been isolated.

Arsindoles

The first derivative of arsindole was synthesized by Mannich,[17] who showed that N,N-diethyl-3-phenyl-2-propynylamine,

$$(C_2H_5)_2NCH_2—C\equiv C—C_6H_5,$$

reacted with arsenic trichloride at 150–170° to give 1,3-dichloro-2-[(diethylamino)methyl]arsindole (3). This compound, by virtue of its tertiary amine group, forms a hydrochloride, m.p. 199°. The chloro group in the 1-position is (as expected) more reactive than that in the

3-position; the compound when treated with the appropriate acid therefore forms the 1-bromo-3-chloro hydrobromide, decomp. 205°,

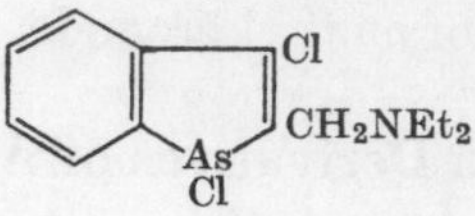

(3)

and the 3-chloro-1-iodo hydriodide, decomp. 194.5°; it also gives a 3-chloro-1-hydroxy hydrochloride, m.p. 135°. Aqueous potassium cyanide reacts with the original hydrochloride to give 3-chloro-1-cyano-2-[(diethylamino)methyl]arsindole, m.p. 65°.

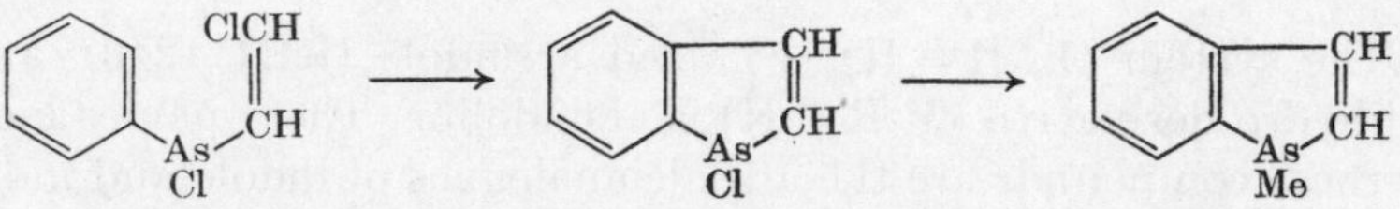

(4)   (5)   (6)

A number of other syntheses of the arsindole ring system have been claimed by Das-Gupta. This author[18, 19] has shown that chloro-(2-chlorovinyl)phenylarsine (or 2-chlorovinylphenylarsinous chloride) (4), obtained by the interaction of 2-chlorovinylarsonous dichloride, $ClCH{=}CHAsCl_2$, and benzene, undergoes cyclization when boiled in carbon disulfide solution with aluminum chloride for 7 hours, with the formation of 1-chloroarsindole (5). It is claimed that the latter can also be obtained by cyclization of o-vinylphenylarsonous dichloride, $CH_2{=}CH{-}C_6H_4{-}AsCl_2$. 1-Chloroarsindole, when treated with methylmagnesium iodide, gives 1-methylarsindole (6), a colorless liquid, b.p. 142–145°/6 mm. The reactivity of the arsenic atom in this compound is shown by the fact that it gives a quaternary methiodide, decomp. 216–218°, a mercurichloride, m.p. 150–151°, and a picrate, m.p. 106–107°. 1-Ethylarsindole was similarly obtained as a liquid, b.p. 138–145°/6 mm; it gives a picrate, m.p. 100–102°. Both 1-methyl- and 1-ethyl-arsindole are described as having offensive odors and vesicant action.

In an entirely different type of synthesis,[20] dimethylstyrylarsine, $C_6H_5CH{=}CH{-}As(CH_3)_2$, was treated with chlorine in carbon tetrachloride solution, and the product when evaporated in a vacuum and heated at 185–190° yielded 1-methylarsindole. It is clear that in these circumstances the tertiary arsine first formed the dichloride (7), which when heated lost methyl chloride, and also hydrogen chloride on cyclization, to give the 1-methylarsindole; the synthesis is thus a direct application of Turner and Bury's much earlier synthesis [21] of 1-methyl-arsindoline (see p. 416). In another synthesis, [22] β-bromostyrene,

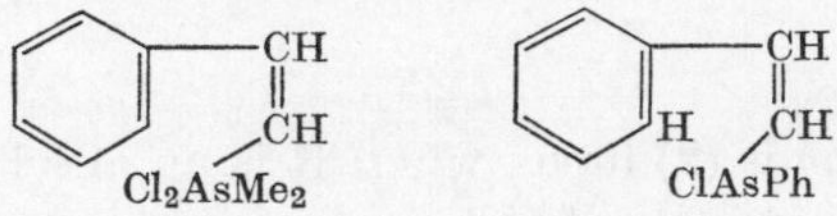

(7)   (8)

C_6H_5CH=CHBr, was heated with arsenic trichloride and sodium in benzene containing ethyl acetate to give tristyrylarsine, $(C_6H_5CH$=$CH)_3As$, m.p. 82°; this compound when heated with more arsenic chloride was supposed to give an unstable styrylarsonous dichloride, C_6H_5CH=CH—$AsCl_2$, which on further heating lost hydrogen chloride during cyclization to form 1-chloroarsindole (5). [22]

It is also claimed [20] that when a mixture of β-bromostyrene and phenylarsonous dichloride is dissolved in ethanolic 33% sodium hydroxide solution, kept, first at room temperature for 24 hours and then on a boiling water-bath for 10 hours, and finally reduced by sulfur dioxide in the presence of hydrochloric acid, the ultimate product was 1-phenylarsindole, b.p. 165–170°/3 mm. It is clear that in these circumstances the bromostyrene has reacted with the phenylarsonous dichloride by a Meyer reaction to give phenylstyrylarsinic acid, $C_6H_5(C_6H_5CH$=$CH)AsO(OH)$, and that the latter has then been reduced to phenylstyrylarsinous chloride (8). An unusually ready cyclization with loss of hydrogen chloride must now have occurred, as shown, to form 1-phenylarsindole.

Finally, Das-Gupta [22] has shown that, when phenylacetylene and phenylarsonous dichloride are heated together at 140–150° for 7 hours, condensation occurs with the evolution of hydrogen chloride and the

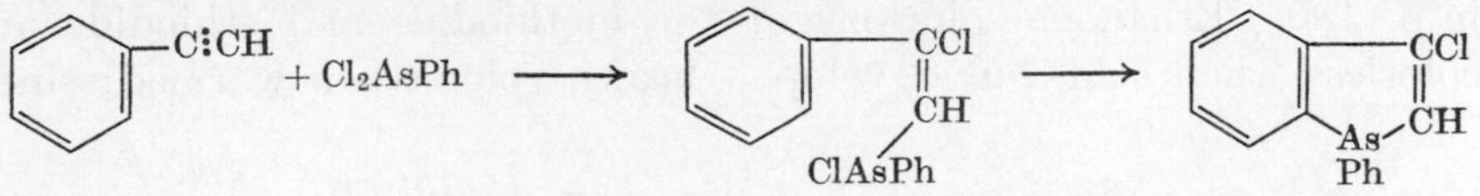

(9)   (10)

formation of 3-chloro-1-phenylarsindole (10), with the probable intermediate formation of (β-chlorostyryl)phenylarsinous chloride (9). The 3-chloro-1-phenylarsindole (10) is a colorless liquid, b.p. 165–175°/10 mm; it forms a picrate, m.p. 115–116°, and a mercurichloride, m.p. 232–235°, and also the quaternary methiodide, m.p. 152–153°, and ethiodide, m.p. 161°. Evidence for the structure (10) is obtained from the fact that oxidation with hydrogen peroxide breaks the heterocyclic ring, with the formation of (o-carboxyphenyl)phenylarsinic acid, $(HOOCC_6H_4)C_6H_5AsO(OH)$, m.p. 166°.

Arsindolines

These compounds (2) have, when R is an alkyl or aryl group, the expected normal properties of tertiary arsines.

Two synthetic routes are available for the preparation of arsindolines. Turner and Bury [21] have shown that the Grignard reagent prepared from phenethyl bromide, $C_6H_5CH_2CH_2MgBr$, when treated with dimethylarsinous iodide, $(CH_3)_2AsI$, gives dimethylphenethylarsine (11).

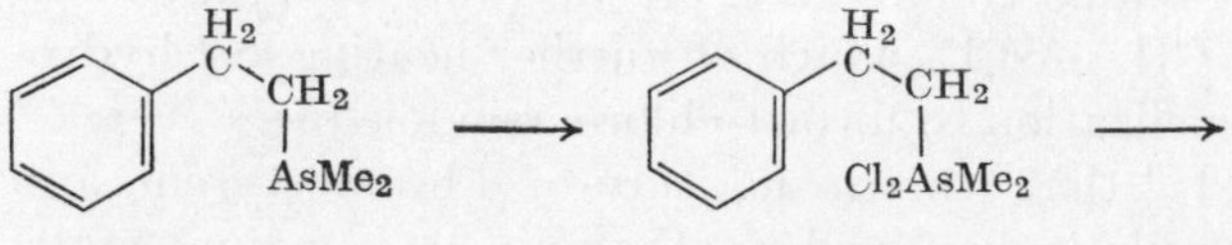

(11)   (12)

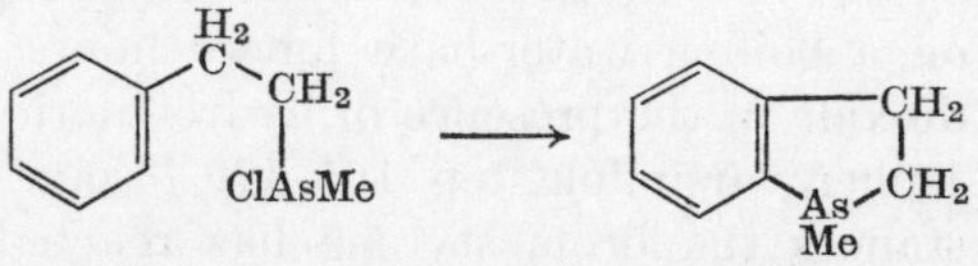

(13)   (14)

This arsine adds chlorine to give the dichloride (12), which on heating loses methyl chloride to give methylphenethylarsinous chloride (13). This compound, when gently warmed in carbon disulfide solution with aluminum chloride, cyclizes with loss of hydrogen chloride to give 1-methylarsindoline (14), a colorless liquid, b.p. 112–113°/15 mm.

This arsindoline gives the colorless crystalline quaternary salts: methiodide, m.p. 250°; ethiodide, m.p. 162–163°; benzyl bromide, m.p. 180°. Ethanolic solutions of the methiodide and ethiodide are colorless when cold, but develop a brown color when warmed; since

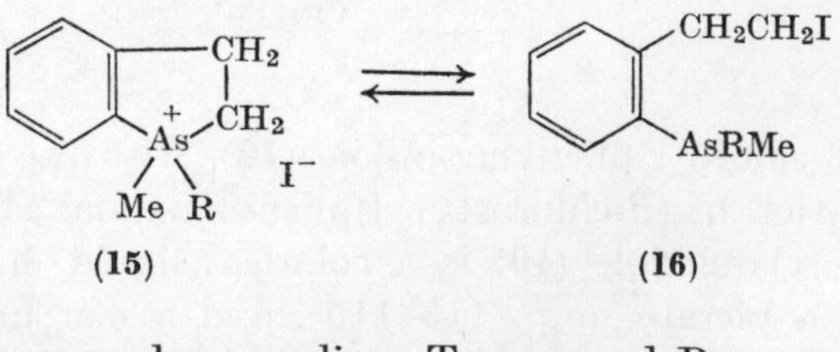

(15)   (16)

this change is reversed on cooling, Turner and Bury suggest that the quaternary salt (15) in solution may be in equilibrium with the tertiary arsine (16). Heat shifts the equilibrium point from left to right, but on cooling this point must move almost entirely back to the left, since decolorization is complete.

This explanation is not entirely satisfactory, for the arsine (16) should itself be colorless, and partial decomposition with liberation of iodine would be required to give a brown color; such a decomposition would not, however, be so readily reversible. If the iodine became covalently linked to the arsenic atom, a brown color might ensue, but such a reaction seems very improbable.

The second synthetic route, due to Jones and Mann,[23] is much shorter. Phenethyl bromide (17) is added to phenylarsonous dichloride,

$$Ph(CH_2)_2Br \longrightarrow \underset{Ph}{Ph(CH_2)_2}\!\!\!\nearrow\!\!\!\underset{OH}{As}\!\!\!\stackrel{O}{\nwarrow} \longrightarrow$$

(17)   (18)   (19)

$C_6H_5AsCl_2$, in boiling aqueous sodium hydroxide (the Meyer reaction) to furnish phenethylphenylarsinic acid (18), m.p. 142–145°, in 33% yield. This acid is readily cyclized, by heating with concentrated sulfuric acid at 100° for 15 minutes, to 1-phenylarsindoline oxide, which without isolation is reduced by sulfur dioxide to the 1-phenylarsindoline (19), b p 126–128°/0.6 mm, in 76% yield from the acid (18). It is not known whether this ready cyclization of arsinic acids of type (18) is limited to those having an aryl group directly linked to the arsenic atom.

When a stream of sulfur dioxide was passed through a suspension of the acid (18) in 1:1 aqueous hydrochloric acid containing a trace of potassium iodide, the acid was reduced to phenethylphenylarsinous chloride, $C_6H_5(CH_2)_2As(C_6H_5)Cl$, which when heated in benzene containing aluminum chloride also underwent cyclization to 1-phenylarsindoline, but only in 17% yield.

1-Phenylarsindoline is readily characterized as its methiodide (1-methyl-1-phenylarsindolinium iodide), m.p. 174–175°, or as its orange palladium dibromide derivative, bis-(1-phenylarsindoline)dibromopalladium, m.p. 229–230°.

It is noteworthy that attempts to dehydrogenate the arsine (19) with tetrachloro-o-benzoquinone in xylene or with palladized charcoal in ethylene glycol failed; the failure with the palladium was not unexpected, as phosphines and arsines usually rapidly 'poison' palladium and platinum, almost certainly by complex formation. Attempts to oxidize the arsine (19) to the *As*-oxide of the 2- or 3-oxo derivative (cf.

p. 421), using hot aqueous permanganate or selenium dioxide in hot ethanol, left the ring system unchanged.

Isoarsindole and Isoarsindolines

The ring system (**1A**), [RRI 1250], is named isoarsindole. This compound, like its nitrogen analogue isoindole, is unknown. Attempts to

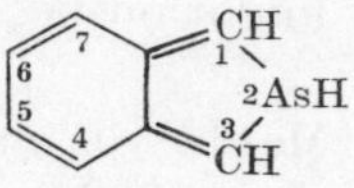

obtain the true 'unhindered' isoindole result in the formation of a true aromatic (as distinct from quinonoid) isomer analogous to (**1B**); in the arsenic series, this would almost certainly not occur, for arsenic shows a very strong reluctance to form part of a ring system having C=As—C bonding. The non-occurrence of isoarsindole, as either (**1A**) or the isomeric (**1B**), is therefore not surprising.

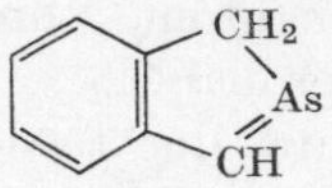

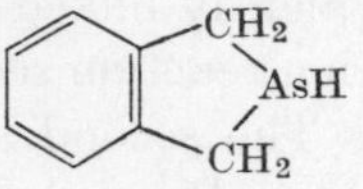

The 1,3-dihydro derivative, isoarsindoline (**2**) has, however, a stable ring system and 2-substituted derivatives in particular are stable compounds, which are synthesized without difficulty.

Isoarsindolines

The preparation and properties of 2-substituted isoarsindolines have been examined in some detail. These compounds may be prepared in the following ways.

(*a*) Lyon and Mann[24] have shown that when *o*-xylylene dibromide (**3**) and, for example, phenylarsonous dichloride, are boiled together in

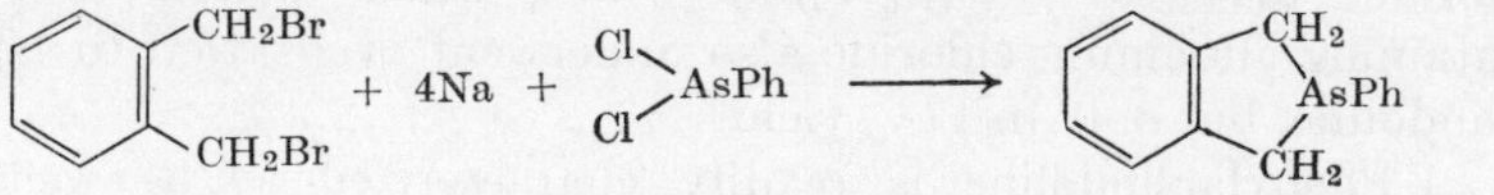

ethereal solution with sodium and a small quantity of ethyl acetate under an atmosphere of nitrogen, direct condensation occurs with the formation of 2-phenylisoarsindoline (**4**) in 19% yield. Several such compounds have been prepared by this method (p. 415).

(*b*) The second method (Lyon, Mann, and G. H. Cookson[25]) is less troublesome, since the long boiling under nitrogen and the distillation of the ether are avoided. In this method, equimolecular quantities of *o*-xylylene dibromide (**3**) and, for example, dimethylphenylarsine are mixed and warmed. Immediate reaction occurs, the chief product being the diquaternary arsonium salt (**5**) (cf. the analogous phosphonium salts, p. 68). When this crude product is heated under reduced pressure,

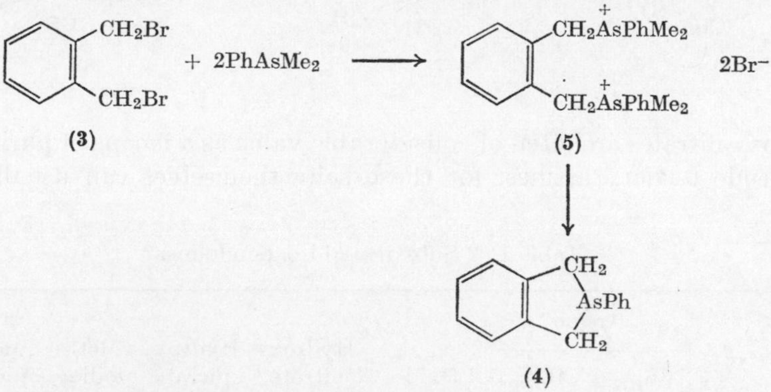

methyl bromide is lost from the quaternary dibromide, cyclization occurs, and the 2-phenylisoarsindoline distils over. The method is rapid and the yield is 60%.

(*c*) When the di-Grignard reagent prepared from phenylarsine, $C_6H_5AsH_2$, *i.e.*, phenylarsinebis(magnesium bromide) $C_6H_5As(MgBr)_2$,* is treated with *o*-xylylene dibromide, a rapid reaction occurs with the formation of 2-phenylisoarsindoline.[26] This method gives the arsine (**4**)

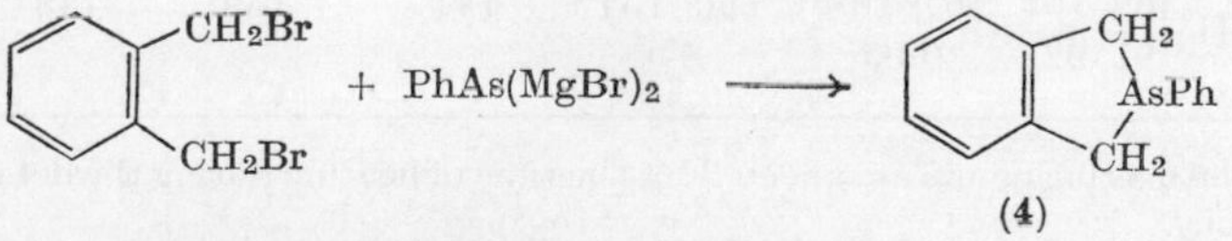

in *ca*. 80% yield. The probable mechanism of the more complex reaction of the di-Grignard reagent with *o*-xylylene dichloride has been discussed by Beeby, G. H. Cookson, and Mann.[26]

(*d*) More complex 2-substituted isoarsindolines may arise in other thermal decompositions (see p. 456).

The 2-alkyl- and the lower 2-aryl-isoarsindoline thus prepared are liquids at room temperature, but the higher 2-aryl derivatives are crystalline. They possess many of the properties of stable tertiary arsines. For example, when the 2-phenylisoarsindoline is oxidized with nitric acid, it forms the oxide (**6**), but the highly polar As—O group causes this compound to form the crystalline hydroxy nitrate (**7**),† which, systematically, should be termed 2-hydroxy-2-phenylisoarsindolinium nitrate.[25]

---

* For preparation of the P and As analogues see p. 229.   † But see p. 372.

These hydroxy nitrates are ionized salts and when treated with, for example, picric acid give the corresponding hydroxy picrates. The

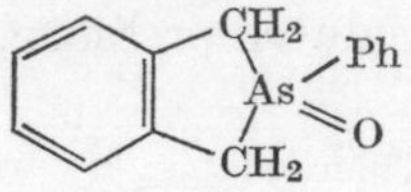

(6) 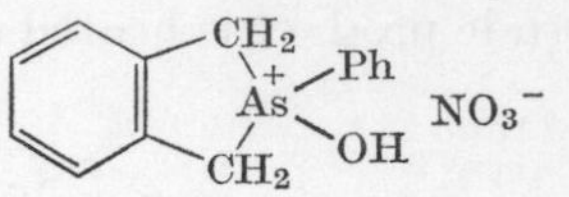   (7) 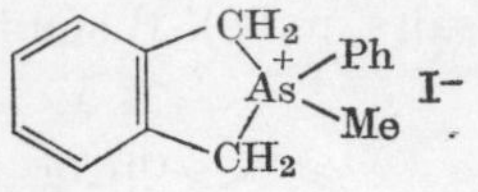   (8)

hydroxy nitrates are often of considerable value as a means of purifying the crude tertiary arsines, for these salts themselves can usually be

Table 1. 2-Substituted isoarsindolines*

2-Substituent	Arsine M.p. (°)	Arsine B.p. (°/mm)	Oxide, m.p. (°)	Hydroxy nitrate, m.p. (°)	Hydroxy picrate, m.p. (°)	Meth-iodide, m.p. (°)	Metho-picrate, m.p. (°)
CH_3	Liquid	115/17	—	—	—	—	—
C_6H_5	Liquid	136–138/ 0.3	130–131	149–150	164–166	—	—
p-$CH_3C_6H_4$	Liquid	138–139/ 0.3	134–136	146	147	156	134–135
p-ClC_6H_4	63–64	153–154/ 0.05	145–146	144–145	147–148	204	162
p-$CH_3OC_6H_4$	{69–70† 91–92	167–168/ 0.03	150–151	141	159	148	134

* Many of the melting points are affected by the time of heating during the determination.
† Dimorphous.

readily purified by recrystallization from water or dilute nitric acid, and then on treatment with alkali give the arsine oxide (6), which in turn can be expeditiously reduced in chloroform solution by sulfur dioxide to the pure tertiary isoarsindoline. The isoarsindolines also readily form quaternary salts with alkyl iodides: 2-phenylisoarsindoline thus gives the methiodide (8), i.e., 2-methyl-2-phenylisoarsindolinium iodide, which in turn can also be converted into the corresponding picrate. The properties of a number of the isoarsindolines and their chief derivatives are summarized [25] in Table 1.

It is noteworthy that 2-phenylisoarsindoline oxide (6), and therefore also the dihydroxide which it apparently forms in presence of water, show marked activity against *Trypanosoma congolense* in mice, but very little activity against *Trypanosoma rhodesiense* and *Trypanosoma cruzi*; replacement of the 2-phenyl group by the other aryl groups listed in Table 1 does not affect these comparative activities.

The heterocyclic ring in the isoarsindolines has remarkable stability. When the 2-aryl derivatives are heated with an excess of constant-boiling aqueous hydriodic acid, the aryl group is split off and the highly crystalline yellow 2-iodoisoarsindoline (9), m.p. 107–108°, is obtained.

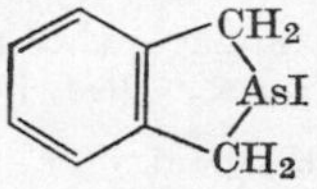

(9)

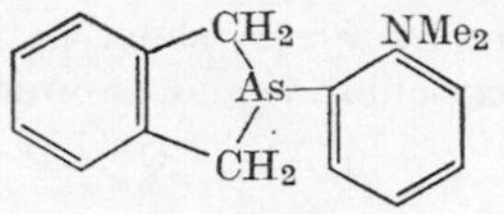

(10)

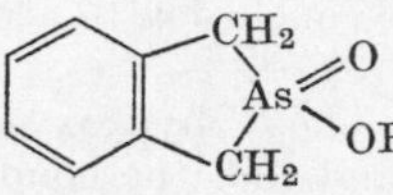

(11)

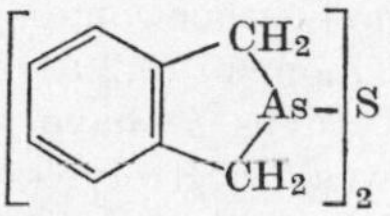

(12)

By the application of a suitable Grignard reagent, the 2-iodo group can now be replaced by any desired alkyl or aryl group, and the iodo compound (9) is therefore a valuable intermediate in converting one 2-substituted isoarsindoline into another.[25] For example, the 2-iodo-arsindoline when treated with *o*-(dimethylamino)phenylmagnesium bromide gives 2-[*o*-(dimethylamino)phenyl]isoarsindoline (or *o*-2-iso-arsindolyl-*N,N*-dimethylaniline) (10), m.p. 75° (Mann and H. R. Watson[27]).

When 2-iodoisoarsindoline is treated first with sodium carbonate solution and then with hydrochloric acid, the 2-chloroisoarsindoline is obtained as colorless crystals, m.p. 73–74°. Oxidation of the 2-iodo compound with nitric acid yields 2-hydroxyisoarsindoline 2-oxide (or *o*-xylylenearsinic acid) (11), which separates from solution as the crystalline dihydroxy nitrate; treatment of the latter with one equivalent of alkali liberates the arsinic acid (11), colorless crystals, m.p. 144°. Furthermore, when the 2-iodo compound (9) is hydrolyzed with sodium hydroxide and the product is treated with hydrogen sulfide, di-2-isoarsindolyl sulfide (12) is obtained as magnificent white crystals, m.p. 146°. It will be seen, therefore, that the heterocyclic ring has remained unchanged both in the formation of the 2-iodo derivative, and in all the subsequent reactions of this compound noted above; in particular, it is unaffected by such vigorous reducing and oxidizing agents as hydriodic acid and nitric acid, respectively.[25]

Note. These highly crystalline tertiary arsine 'hydroxynitrates' (pp. 369 ff.) can be readily obtained by the action of reasonably concentrated nitric acid on tertiary arsines or tertiary arsine oxides. Salts with other acids can be obtained by the action of the appropriate acid on the arsine oxide. It is convenient to refer to them as, *e.g.*, 'hydroxynitrates', implying salt formation of type $[R_3AsOH]^+NO_3^-$. Their structure may not always be so simple. Ferguson and Macaulay[27a] have carried out an X-ray crystal structure of triphenylarsine 'hydroxybromide' and have studied the nmr spectra of this bromide and the corresponding chloride. They find that in the bromide, the arsenic atom is tetrahedrally linked to three phenyl groups and to one oxygen atom, and that the latter is strongly hydrogen-bonded to the halogen by a 'very short' hydrogen bond. They suggest that the structure of the hydroxy-chloride and -bromide can be represented as a hybrid:

$$Ph_3As^+\!\!-\!\!O^-\ldots H\!\!-\!\!Hal \leftrightarrow Ph_3As^+OH\ldots Hal^-$$

or more concisely as

$$Ph_3As^+\!\!-\!\!O^{\frac{1}{2}-}\ldots H\ldots Hal^{\frac{1}{2}-}$$

Jensen[27b] had earlier come to much the same conclusion to account for the very high dipole moment (9.2 D) of the chloride.

Harris and Inglis[27c] have shown that the above hydroxy-chloride and -bromide form weak electrolytes in acetonitrile, and that the conductivity and mode of electrolysis indicate simple ionization. It is probable that salts such as the hydroxy-nitrate and -perchlorate are fully ionized.

Arsonic acids, $RAs(O)(OH)_2$, and arsinic acids, $R_2As(O)(OH)$, usually also form crystalline 'hydroxynitrates', and these acids are therefore amphoteric.

Tertiary phosphine oxides very rarely form compounds of this type (Mann and Chaplin[27d]).

Spiro Derivatives of Isoarsindole

The reactions of the isoarsindolines provided the first synthetic approach to the spiro arsonium salts in which the arsenic atom is linked by all its four covalencies to carbon atoms. Lyon and Mann[24] have shown that 2-methylisoarsindoline combines readily with *o*-xylylene dibromide (one equivalent) to give 2-[2'-(bromomethyl)benzyl]-2-methylisoarsindolinium bromide (**13**), which on being heated under reduced pressure loses methyl bromide to form 2,2'-spirobi[isoarsindolinium] bromide (**14**), [RRI 4222], initially termed *As-spiro*-bis-isoarsindolinium bromide. This salt, and the corresponding iodide and picrate, form extremely beautiful crystals.

(13)   (14)

Since the four-covalent arsonium atom has a tetrahedral configuration, it follows that a spirocyclic arsonium salt similar to (**14**) but having a suitable substituent in the benzene rings would possess molecular dissymmetry and hence should be susceptible to resolution into optically active forms. Lyon, Mann, and G. H. Cookson[25] therefore converted 4-chloro-*o*-xylylene dibromide (15) into 5-chloro-2-methylisoarsindoline (16), which in turn was combined with a second equivalent of the dibromide (15) to form the arsonium salt (17). This bromide, when heated at 160–170°/15 mm, readily lost methyl bromide with the formation of the highly crystalline 5,5'-dichloro-2,2'-spirobi[isoarsindolinium] bromide (18), initially termed *As-spiro*-bis-5-chloroisoarsindolinium bromide. This salt must clearly lack any elements of symmetry, although the arsenic atom is not asymmetric. Yet, although a number of salts with optically active anions were prepared and examined, no decisive

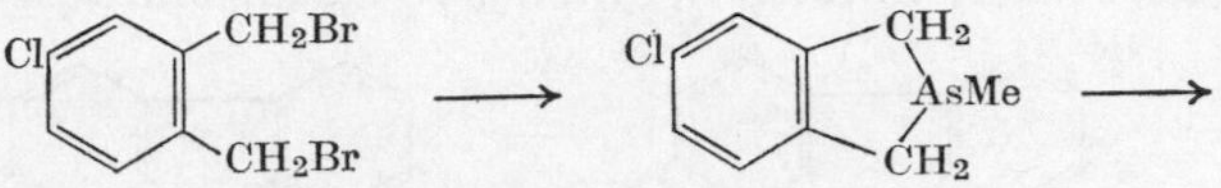

(15)   (16)

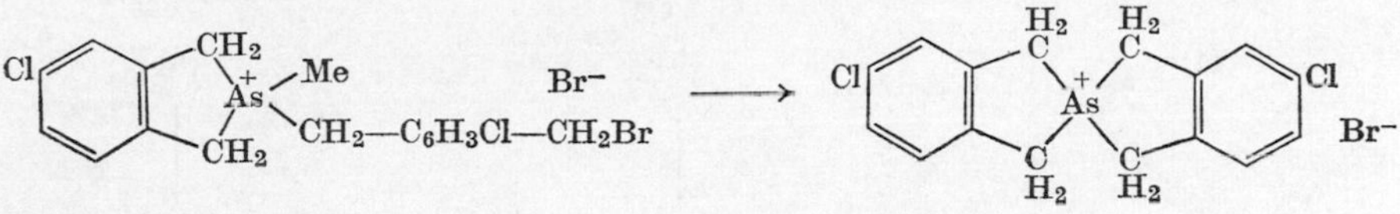

(17)   (18)

evidence of optical resolution was obtained. This result is in striking contrast to the ready resolution of another spirocyclic salt, 3,3',4,4'-tetrahydro-2,2'[1*H*,1'*H*]-spirobi[isoarsinolinium] bromide, earlier termed *As-spiro*-bis-1,2,3,4-tetrahydroisoarsinolinium bromide, by Holliman and Mann[28] (see p. 419).

Dibenzarsoles (9-Arsafluorenes)

The systematic naming and numbering of the structure (1) present the same decisions as those of the corresponding phosphorus compound

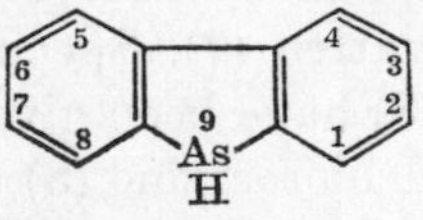

(1) 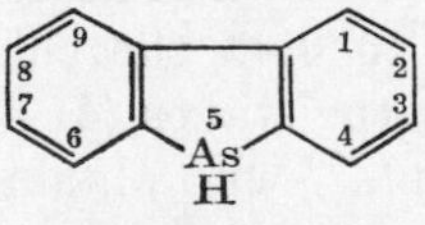   (2) 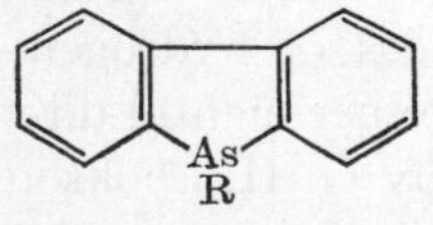   (3)

(p. 72). Aeschlimann *et al.*,[29] who first prepared this ring structure, termed it '*o,o'*-diphenylylenearsine'. For many years, however, it has been termed in English literature 9-arsafluorene and numbered as in (1):

this system retains the numbering of the 'parent' hydrocarbon fluorene, [RRI 3127], and of the nitrogen analogue carbazole, [RRI 2927], and is particularly convenient for naming the arsinic acid having this ring system. The Ring Index name, [RRI 2840], is dibenzarsole* (or dibenz-arsenole) with numbering as in (2). At the present time the system (1) is more deeply embedded in chemical literature, but the authority of the Ring Index will ultimately replace it by (2). Both names are therefore given for the chief compounds in this Section.

The parent compound (1)—as in so many heterocyclic systems— is unknown, and most derivatives (3) have alkyl or aryl groups replacing the arsine hydrogen atom.

Aeschlimann et al.[29] have shown that 2-aminobiphenyl (4), when subjected to the usual Bart reaction, furnishes biphenylyl-2-arsonic acid

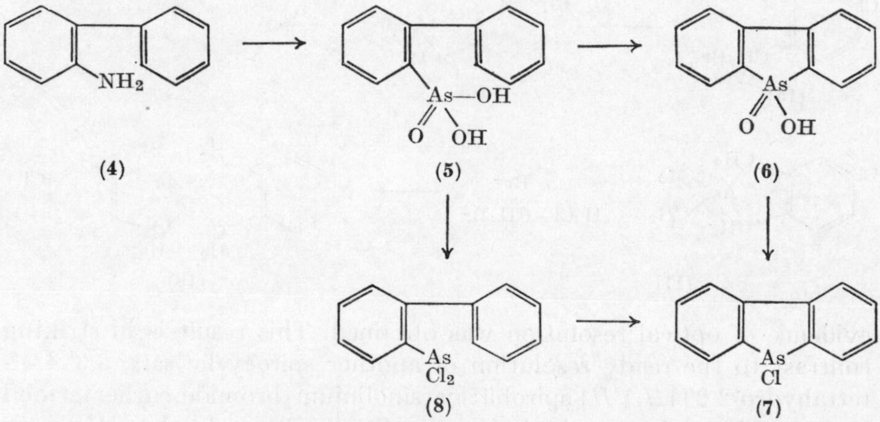

(4) → (5) → (6)

(8) → (7)

(5), which when heated with sulfuric acid at 100° for a few minutes readily cyclizes to the arsinic acid (6), i.e., 9-arsafluoreninic acid or dibenzarsenolic acid. This acid, when reduced with sulfur dioxide in the presence of hydrochloric acid, gives 9-chloroarsafluorene (7), colorless crystals, m.p. 161°, b.p. approx. 230°/25 mm. Alternatively, the arsonic acid (5) can be similarly reduced to 2-biphenylylarsonous dichloride (8), a heavy oil which when heated in a vacuum cyclizes with loss of hydrogen chloride to give the chloro derivative (7). Specific experimental directions for the stages (4) → (5) → (8) have been given by G. H. Cookson and Mann,[30] who obtained the dichloroarsine (8) as crystals, m.p. 47.5–48.5°.

Blicke et al.[31] have shown that the chloro derivative (7) can be obtained by heating biphenyl with a mixture of arsenic trichloride and

_____

* The 5H is omitted from [RRI 2840] and [RRI 3065] (p. 596) but accords with their system.

aluminum trichloride for $4\frac{1}{2}$ hours at 165–185°; the modest yield of the compound (7) must be amply compensated by the rapidity of the preparation. The chloro compound when treated with sodium iodide in acetone solution gives the yellow 9-iodo compound (3; R = I), m.p. 166–167°, which when shaken in benzene over mercury gives 5,5'-bis(dibenzarsole), $C_{12}H_8As$—$AsC_{12}H_8$, m.p. 269–273°. The 9-cyano derivative (3; R = CN) forms colorless needles, m.p. 178°. The chloro compound on mild hydrolysis furnishes the diarsine 'oxide', i.e., 5,5'-oxybis(dibenzarsole), $C_{12}H_{18}As$—O—$AsC_{12}H_{18}$.

The chloro compound (7) can also be obtained by heating a suspension of the arsinic acid (5) in acetic acid to 80°, and then adding phosphorus trichloride until a clear solution is obtained; the similar use of phosphorus tribromide gives the bromo compound (3; R = Br), m.p. 178° (Garascia and Mattei[32]). In these reactions, the phosphorus trihalide is clearly acting as both a reducing and a cyclizing agent.

The chloro derivative (7) is a valuable intermediate for the preparation of the corresponding alkyl or aryl derivatives by the action of appropriate Grignard reagents; for example, methylmagnesium iodide converts it into the methyl derivative[29] (3; R = CH$_3$), m.p. 46°, which gives a methiodide,[33] m.p. 206–207°, systematically termed 9,9-dimethylarsafluorenonium iodide or 5,5-dimethyldibenzoarsolium iodide. Alternatively, 2,2'-dilithiobiphenyl, when treated, for example, with phenylarsonous dichloride, $C_6H_5AsCl_2$, gives 9-phenyl-9-arsafluorene or 5-phenyldibenzarsole (3; R = C$_6$H$_5$), b.p. 182–186°/0.3 mm, m.p. 88°, which with hydrogen peroxide gives the tertiary arsine 9 oxide, m.p. 152–153°. and with methyl bromide gives the methobromide, m.p. 164–165° (Heinekey and Millar[34]).

Quaternary salts can also be prepared by the action of Grignard reagents on the tertiary arsine oxides (Blicke and Cataline[35]); thus 5-phenyldibenzarsole 5-oxide with phenylmagnesium iodide gives 5,5-diphenyldibenzarsolium bromide, m.p. 240–241°.

The stability of the ring system (1) is shown by the production of 9-methyl-9-arsafluorene, i.e., 5-methyldibenzarsole, (3; R = CH$_3$) during the thermal decomposition of 5,6,7,8-tetrahydro-5,5,8,8-tetramethyl-dibenzo[e,g][1,4]diarsocinium dibromide (p. 489) and of 6,7,8,9-tetrahydro-5,5,9,9-tetramethyl-5H-dibenzo[f,h][1,5]diarsoninium dibromide (p. 491).

This stability of the tricyclic ring system (1) under other conditions is shown by the arsinic acid (6) which when treated with sulfuric acid containing one equivalent of nitric acid gives the 3-nitrodibenzarsenolic acid (9), m.p. >345°, and thence by reduction the 3-amino acid, m.p. 240°; more vigorous nitration of the acid (6) gives the 3,7-dinitro acid

(10), m.p. >300°. It is noteworthy that 4-nitro-2-biphenylylarsonic acid **(11)** also undergoes cyclization with sulfuric acid at 65° to give the

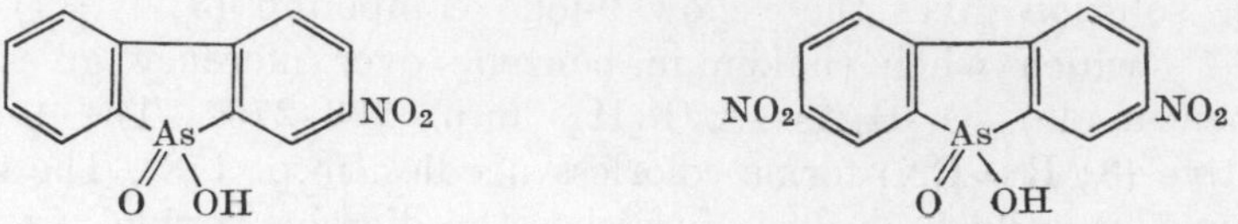

2-nitro acid **(12)**, m.p. >330°, which similarly furnishes the 2-amino acid, m.p. >350° (Feitelson and Petrov[36]).

Garascia, Carr, and Hauser[37] have shown that 2-aminodibenz-arsenolic acid (**13**; R = NH$_2$, R′ = H) gives the following reactions: (a) with acetic anhydride it forms the 2-acetamido derivative (**13**;

R = NHCOCH$_3$, R′ = H), m.p. 245°; (b) with chloroacetamide in alkaline solution it forms the 2-carbamoylmethylamino derivative (**13**; R = NHCH$_2$CONH$_2$, R′ = H); (c) when diazotized in hydrochloric acid and treated with cuprous chloride, it forms the 2-chloro derivative (**13**; R = Cl, R′ = H), decomp. >300°, which reacts with phosphorus trichloride in chloroform to give 2,5-dichlorodibenzarsole (**14**; R = Cl, R′ = H, X = Cl).

Similarly 3-aminodibenzarsenolic acid (**13**; R = H, R′ = NH$_2$), when diazotized and then treated (a) with potassium cyanide–cuprous cyanide gives the 3-cyano derivative (**13**; R = H, R′ = CN), decomp. >360°, and (b) with hydrobromic acid and cuprous bromide gives the 3-bromo acid (**13**; R = H, R′ = Br), m.p. 314–315°, and thence with hydrochloric acid and zinc amalgam under benzene gives 3-bromo-5-chlorodibenzarsole (**14**; R = H, R′ = Br, X = Cl), m.p. 176°.

The 3-chlorodibenzarsenolic acid (**13**; R = H, R′ = Cl), m.p. 313–314°, with phosphorus trichloride in acetic acid gives 3,5-dichlorodibenzarsole (**14**; R = H, R′ = X = Cl), m.p. 154–155°, and with phosphorus tribromide in chloroform gives 5-bromo-3-chlorodibenzarsole (**14**; R = H, R′ = X = Cl), m.p. 153°; these two derivatives were not analyzed.

Bactericidal and fungicidal properties have been claimed for the 5-(dimethylaminodithiocarbamoyl) derivative (**14**; R = R′ = H, X = —S—C(S)=N(CH₃)₂ (Urbschat[38]).

It is noteworthy that Wittig and Benz,[39] in their study of the 'anionization' reactions of sodium diphenyllithium, Na[Li(C₆H₅)₂], have shown that this compound in ethereal solution reacts with triphenylarsine during two weeks at room temperature to give 5-phenyl-9-dibenzarsole in 4% yield, with a 30% recovery of the triphenylarsine.

The stereochemistry of the 9-substituted 9-arsafluorenes (**3**) has been investigated in detail by Campbell and Poller.[40] Their work, quite apart from its intrinsic stereochemical value, ably illustrates synthetic methods in this field. At the time when this work was published (1956) there was no decisive evidence that the tricyclic system (**1**) was planar, but consideration of a number of factors indicated strongly the coplanarity of the system. If the system were planar, and if the trivalent arsenic atom retained its normal pyramidal configuration, derivatives of type (**3**) could be regarded as having the tricyclic system in the plane of the paper, with the substituent R projecting above or below this plane. If other suitable substituents were inserted into one of the *o*-phenylene groups, the molecule would be dissymmetric and, if any oscillation of the group R between the upper and lower positions were slow, the compound should be susceptible to optical resolution.

(15)   (16)

Two compounds were chosen for investigation, namely, 9-*p*-carboxyphenyl-2-methoxy-9-arsafluorene, i.e. [*p*-(3-methoxydibenzarsole-5-yl)benzoic acid] (**15**) and 2-amino-9-phenyl-9-arsafluorene

[3-amino-5-phenyldibenzarsole] (16); the substituents in these compounds ensure dissymmetry and provide an acidic and a basic group, respectively, for optical resolution.

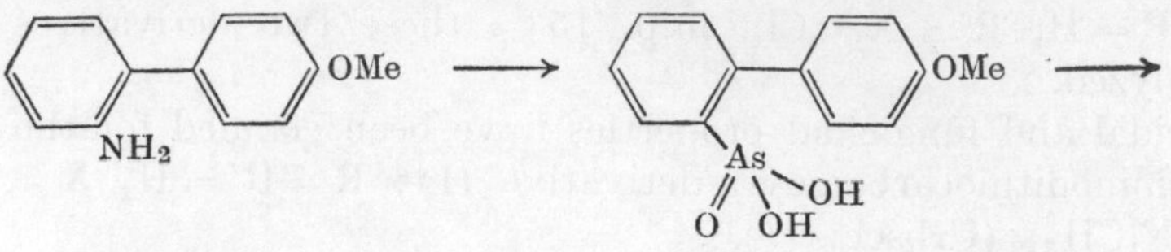

(17)   (18)

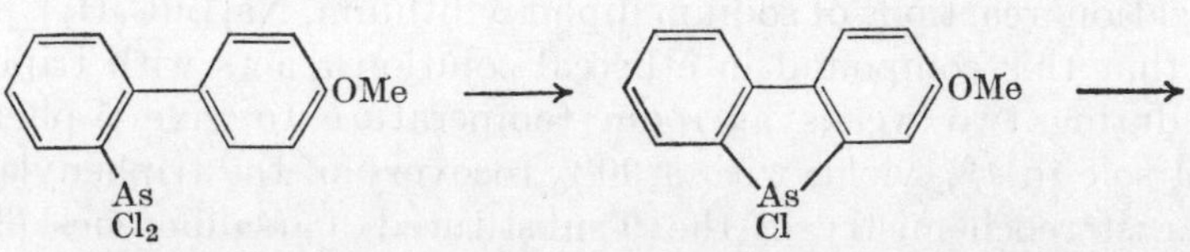

(19)   (20)

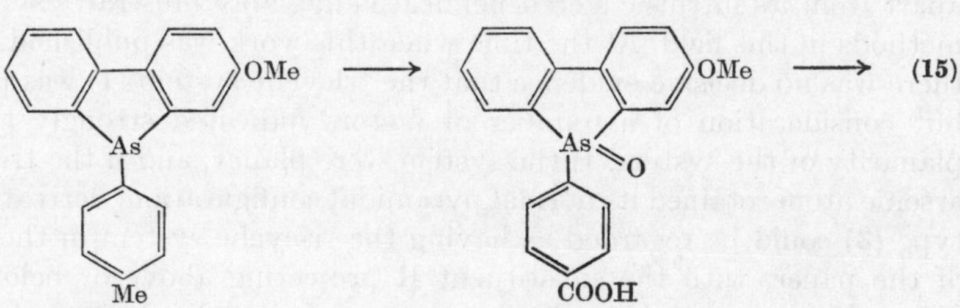

(21)   (22)   (15)

For the synthesis of the acid (15), 4'-methoxybiphenyl-2-amine (17) was diazotized in hydrochloric acid with sodium nitrite, and the solution added to an aqueous solution of sodium arsenite and carbonate containing copper sulfate maintained at 60°. When this Bart reaction was complete, the solution, made just acid to Congo Red, deposited 4'-methoxy-2-biphenylylarsonic acid (18), m.p. 209–222°; this on reduction with sulfur dioxide in hydrochloric acid afforded 4'-methoxy-2-biphenylylarsonous dichloride (19), which when heated at 200° underwent cyclization to 9-chloro-2-methoxy-9-arsafluorene (20), m.p. 133–137°. p-Tolylmagnesium bromide converted (20) into the 9-p-tolyl derivative (21), which with alkaline permanganate gave 9-p-carboxyphenyl-2-methoxy-9-arsafluorene 9-oxide (22), which was readily reduced with sulfur dioxide to the required acid (15), m.p. 223°.

In an alternative synthesis of this acid, the 4'-methoxybiphenyl-2-amine (17) in ethanol saturated with hydrogen chloride was diazotized with pentyl nitrite, and the solution was added to an ethanolic solution of p-carboxyphenylarsonous dichloride, p-$HO_2CC_6H_4AsCl_2$, containing copper-bronze at 40–50°, thus forming p-carboxyphenyl-(4'-methoxy-2-biphenylyl)arsinic acid (23), m.p. 280–281°. This acid, reduced with

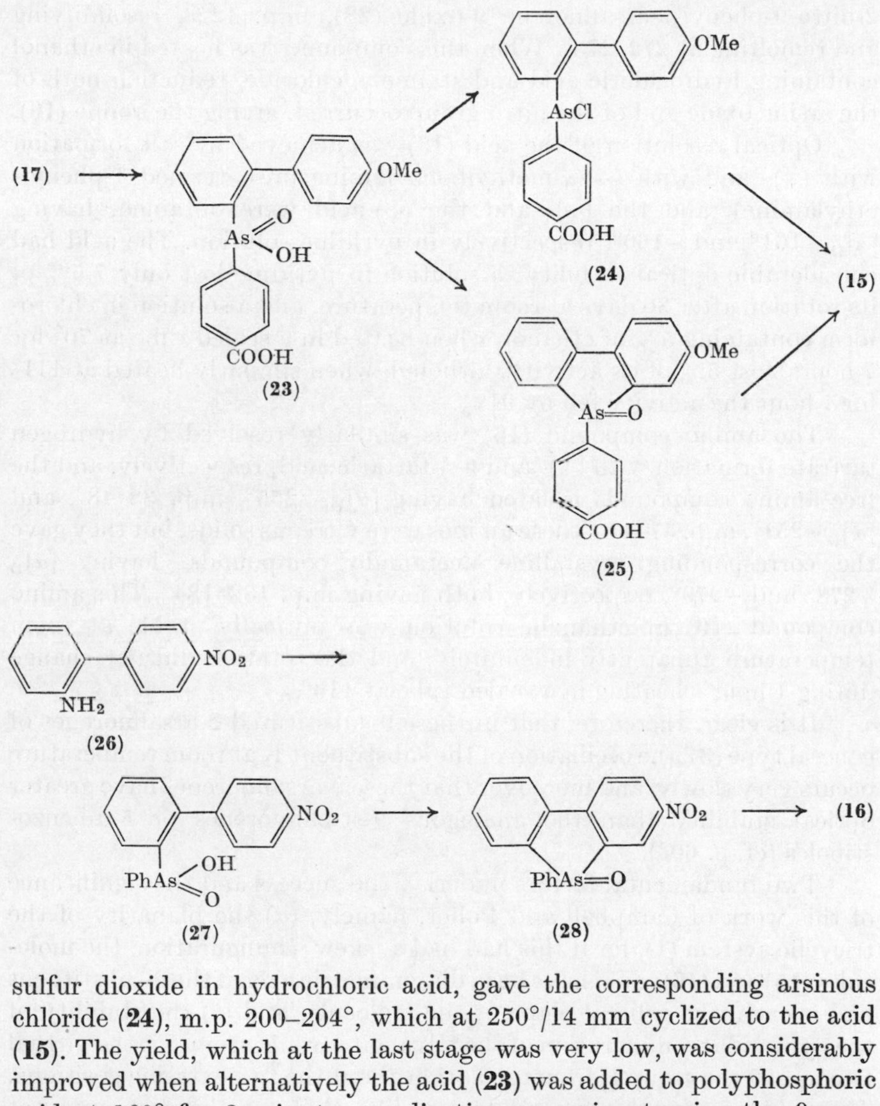

sulfur dioxide in hydrochloric acid, gave the corresponding arsinous chloride (24), m.p. 200–204°, which at 250°/14 mm cyclized to the acid (15). The yield, which at the last stage was very low, was considerably improved when alternatively the acid (23) was added to polyphosphoric acid at 160° for 3 minutes, cyclization occurring to give the 9-arsa-fluorene 9-oxide (25), m.p. 296–302°; reduction with sulfur dioxide then gave the acid (15).

To prepare the amino compound (**16**), 4'-nitrobiphenyl-2-amine (**26**) was diazotized with pentyl nitrite as above and the solution added to ethanolic phenylarsonous dichloride, $C_6H_5AsCl_2$, with copper bronze at 60°. The (4'-nitro-2-biphenylyl)phenylarsinic acid (**27**), m.p. 241–244°, thus obtained was cyclized with sulfuric acid at 140–150° to give 2-nitro-9-phenyl-9-arsafluorene 9-oxide (**28**), m.p. 122°, resolidifying and remelting at 272–273°. When this compound was heated in ethanol containing hydrochloric acid and stannous chloride, reduction both of the arsine oxide and of the nitro group occurred, giving the amine (**16**).

Optical resolution of the acid (**15**) was achieved by salt formation with (+)- and with (−)-α-methylbenzylamine (also termed 1-phenyl-ethylamine), and the (+)- and the (−)-acid were obtained, having $[\alpha]_D$ +161° and −160°, respectively in pyridine solution. The acid had considerable optical stability: a solution in pyridine lost only 7.5% of its rotation after 30 days at room temperature, and a solution in chloroform containing 5% of ethanol, when heated in a sealed tube at 70° for 7 hours, lost 5% of its activity, although when similarly heated at 111° for 1 hour the activity fell by 91%.

The amino compound (**16**) was similarly resolved by hydrogen tartrate formation with (+)- and (−)-tartaric acid, respectively, and the free amino compounds isolated having $[\alpha]_D$ +255°, m.p. 38–48°, and $[\alpha]_D$ −251°, m.p. 37–47°; these amines were vitreous solids, but they gave the corresponding crystalline acetamido compounds, having $[\alpha]_D$ +278° and −279°, respectively, both having m.p. 182–184°. This amino compound (**16**) in ethanolic solution was optically stable at room temperature apparently indefinitely, and the rotation did not change during 1 hour's heating in a sealed tube at 110°.

It is clear, therefore, that in these 9-substituted 9-arsafluorenes of general type (**3**), the oscillation of the substituent R at room temperature occurs very slowly, and moreover that these 9-arsafluorenes have greater optical stability than the analogous 9-stibafluorenes or 5-dibenzo-stiboles (cf. p. 603).

Two fundamental factors underlay the success and the significance of this work of Campbell and Poller, namely, (*a*) the planarity of the tricyclic system (**1**), for if this had had a 'skew' configuration, the molecules (**15**) and (**16**) could have been dissymmetric even if the 9-substituent had been in the general plane of the molecule, and (*b*) the stability of the pyramidal configuration of the trivalent arsenic atom, *i.e.* the virtual freedom of the group R from oscillation about the two extreme positions. The evidence for this second factor, like that for the first, was not decisive at the time of this work. Decisive evidence on both factors has, however, subsequently appeared.

Sartain and Truter[41] have shown that 9-phenyl-9-arsafluorene (3; $R = C_6H_5$) crystallizes in a monoclinic form and an orthorhombic form. An X-ray crystal-structure analysis has shown that the monoclinic form has the planar tricyclic arsafluorene system: the arsenic atom has the pyramidal configuration with the As–C(phenyl) bond projecting upwards at an angle of 80° to the arsafluorene plane. Consequently the molecule is symmetric about a plane which passes through this As–C-(phenyl) bond and is perpendicular to the arsafluorene plane. The essential bond lengths and intervalency angles are shown in Figure 1.

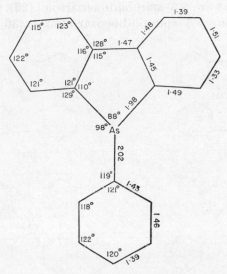

Figure 1. Dimensions of the 9-phenyl-9-arsafluorene molecule. Standard deviations are ±0.03 Å for C–C bond lengths, ±0.02 Å for As–C bond lengths, ±2° for C–C–C angles, and ±1° for As–C–C and C–As–C angles. (Reproduced, by permission, from D. Sartain and M. R. Truter, *J. Chem. Soc.*, **1963**, 4414)

It is noteworthy that the As–C–C angle in the five-membered ring of this compound is the same as the corresponding one in the hydrocarbon fluorene,[42] but the C–C–C bond angle is 7° larger, indicating that there is less strain in the central ring of the arsafluorene molecule than in that of the fluorene molecule.

With regard to the second factor, Horner and Fuchs[43] have shown that cathodic reduction of optically active quaternary arsonium salts gives rise to optically active tertiary arsines, *e.g.*, ethylmethylphenyl-arsine, *n*-butylmethylphenylarsine, which can be distilled at low pressures without racemization. The rate of inversion of the pyramidal arsine molecule, even when the substituents are small, must therefore be very slow.

Arsoranes and Kindred Compounds containing the 5-Dibenz-arsole (9-Arsafluorene) Ring System

The study of these compounds involves three new ring systems: the most important are 5,5′-spirobi[dibenzarsole] (**29**), [RIS 14004] and its derived cation, 5,5′-spirobi[dibenzarsolium] (**30**), [RIS 14005],

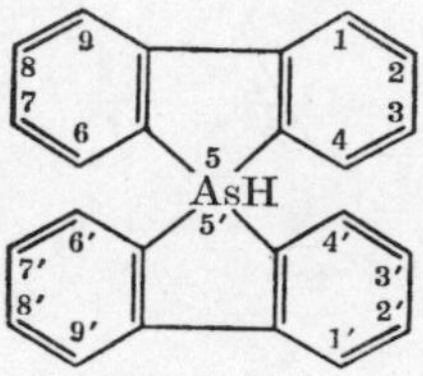

(**29**) 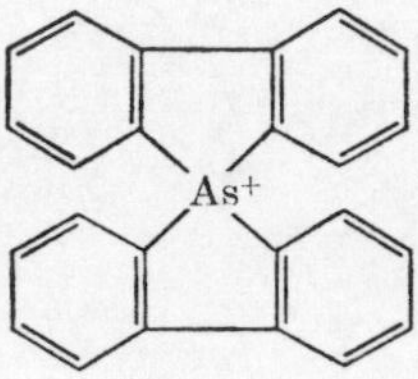   (**30**)

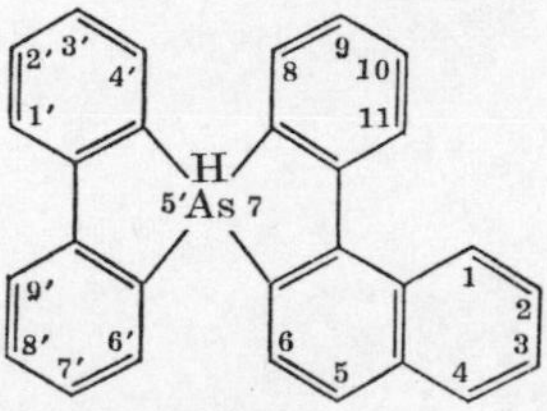

(**31**) 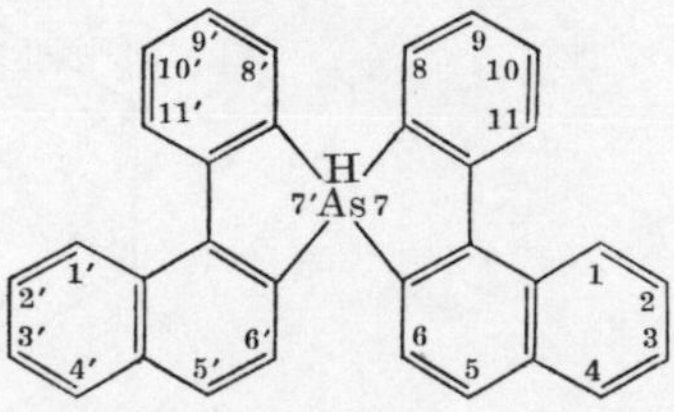   (**32**)

similarly numbered; the others are systematically benzo derivatives of (**29**), namely, spiro[7*H*-benzo[*b*]naphth[1,2-*d*]arsole-7,5′-[5*H*]dibenz-arsole] (**31**), [RIS —], and 7,7′-spirobi[7*H*-benzo[*b*]naphth[1,2-*d*]arsole] (**32**), [RIS —], and their corresponding arsolium cations.

The compounds indicated by the above title have been studied in detail by Wittig and Hellwinkel,[45] on lines closely similar to the work of Wittig and Maercker[44] on the analogous phosphoranes (p. 91), but there are some significant differences. Wittig and Hellwinkel in two preliminary Notes[45] point out that the method of synthesis of pentaarylphosphoranes by the interaction of triarylphosphine-alkyl-aryliminium bromides and two equivalents of aryllithium (Wittig and Kockendörfer[46]) (p. 90) cannot be employed in the arsenic series because, for example, triphenylarsine does not react with phenyl azide to give the required intermediate arsinimine. A synthetic route is available, however, by using the p-tolylsulphonyliminotriarylarsines which Mann and Chaplin[47] had obtained by the action of anhydrous p-tolylsulfonchlorosodioamide ('Chloramine T'), p-$CH_3C_6H_4SO_2NNaCl$, on tertiary arsines in absolute ethanol. These compounds, alternatively called triarylars-N-(p-tolylsulfonimines), react with aryllithium to give

$$R_3As{=}NSO_2C_6H_4CH_3 + 2RLi \rightarrow R_5As + Li_2NSO_2C_6H_4CH_3$$

the pentaarylarsorane in good yield, a reaction which, it is suggested, goes through the intermediate compound:

$$R_4As{-}NLiSO_2C_6H_4CH_3 \leftrightarrow [R_4As]^+[NLiSO_2C_6H_4CH_3]^-$$

This synthetic route is therefore analogous to the alternative route for the preparation of comparable phosphoranes (p. 91).

Another synthetic approach is outlined in the second of these Notes.[45] When 2-biphenylylmagnesium iodide is treated in ether–benzene at 0° with arsenious oxide, and the hydrolyzed, solvent-free product is then boiled with hydrochloric acid, bis-(2-biphenylyl)-chloroarsine [bis-(2-biphenylyl)arsinous chloride], m.p. 112–114°, is obtained. This compound in carbon tetrachloride adds chlorine to form bis-(2-biphenylyl)trichloroarsorane (33), which when heated at 265°

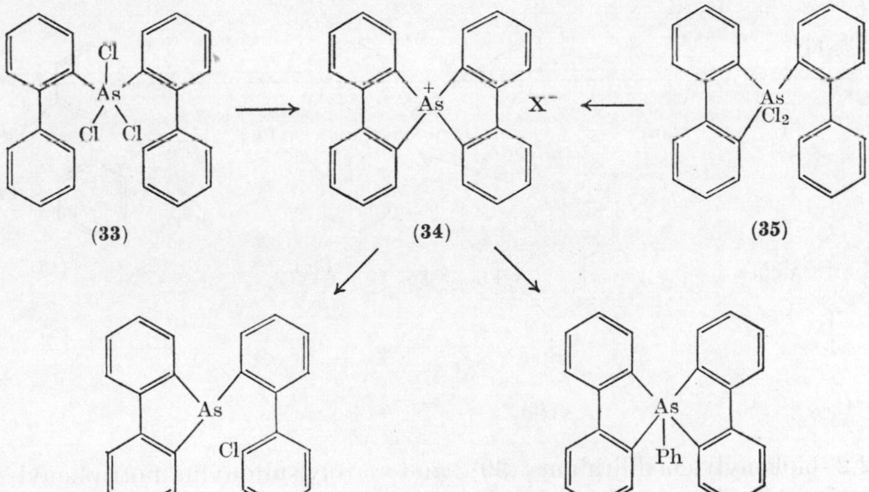

(33)   (34)   (35)

(36)   (37)

under reduced pressure cyclizes to form 5,5'-spirobi[dibenzarsolium]-chloride (34; X = Cl), m.p. 319°. Similarly 5-chlorodibenzarsole (7) when treated with 2-biphenylyllithium gives 5-(2-biphenylyl)-5-dibenzarsole, m.p. 131–133°, which adds chlorine to give 2-biphenylyl-5,5-dichloro-5-dibenzarsole (35), which in turn when heated at 270° under reduced pressure also yields the chloride (34; X = Cl). The cation (34) has been isolated also as the iodide (34; X = I), m.p. 309–311°, the tetraphenylborate [34; X = B(C_6H_5)_4], m.p. 257.5–259.5°, and the pale yellow bis-(2,2′-biphenylylene)borate [34; X = B(C_6H_4C_6H_4)_2], m.p. 290–291°, a salt having two spirocyclic ions.

The chloride (34; X = Cl), when heated at its melting point, undergoes isomerization with cleavage of one of the heterocyclic rings to form 9-(2′-chloro-2-biphenylyl)-9-arsafluorene [5-(2′-chloro-2-biphenylyl)-dibenzarsole] (36), m.p. 179–181°. An aryllithium, e.g., phenyllithium, reacts with the chloride (34; X = Cl) to form 5-phenyl-5,5′-spirobi-[dibenzarsole] (37), m.p. 228–230°.

In their third and detailed paper, Wittig and Hellwinkel [48] illustrate the use of the sulphonimines in the synthesis of arsoranes containing the 5-dibenzarsole ring system. The arsorane, 5,5,5-triphenyl-5-dibenzarsole (38), m.p. 188–189°, may be prepared either by the interaction of

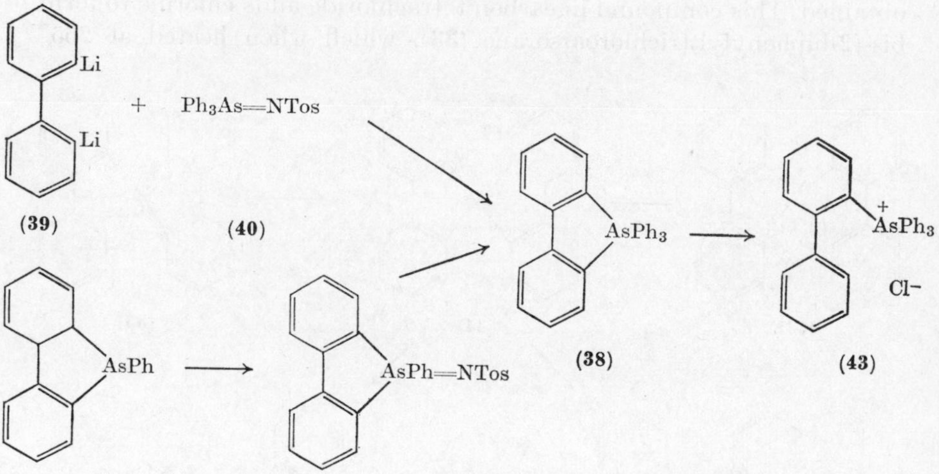

(39)   (40)   (38)   (43)

(41)   (42)

2,2′-biphenylylenedilithium (39) and p-tolylsulfonyliminotriphenylarsine (40; Tos = p-CH_3C_6H_4SO_2), or by converting 5-phenyldibenzarsole (41) into the sulfonyl-arsimine (42), m.p. 189–192°, and then treating the latter with two equivalents of phenyllithium. The arsorane (38), when boiled with methanol containing a small proportion of concentrated hydrochloric acid for 15 minutes, is converted into

2-biphenylyltriphenylarsonium chloride (**43**), characterized as the iodide, m.p. 291–293°.

The following reactions of the arsorane, 5-phenyl-5,5'-spirobi[dibenzarsole] (**37**) provide further illustrations of the ready cleavage of a heterocyclic ring in salts such as (**34**; X = Cl) and arsoranes of type (**37**) and (**38**).

(*a*) The arsorane (**37**) when boiled with hydrochloric acid for 45 minutes gives 5-(2-biphenylyl)-5-phenyldibenzarsolium chloride (**44**; X = Cl), characterized as the iodide (**44**; R = I), m.p. 267–267.5°. This reaction is comparable to the conversion of the arsorane (**38**) into the arsonium salt (**43**).

(*b*) The arsorane when heated with methyl iodide at 100° for 14 days gives 5-(2'-methyl-2-biphenylyl)-5-phenyldibenzarsolium iodide (**45**; R = CH$_3$, X = I), m.p. 263.5–265.5°.

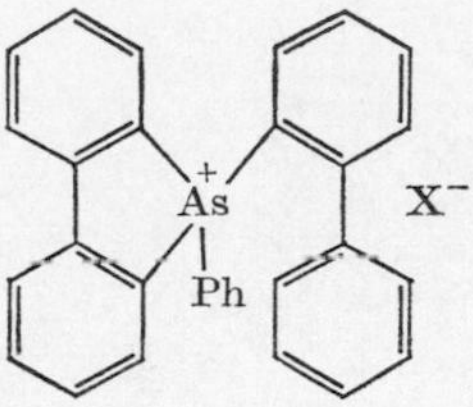

(**44**) 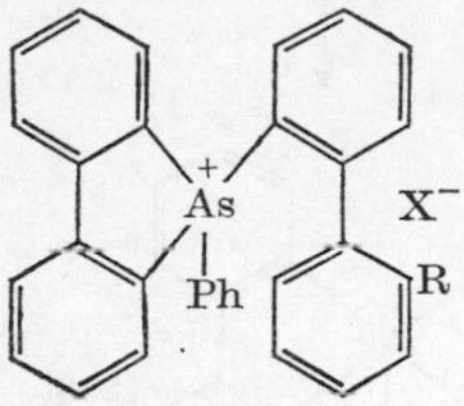   (**45**) 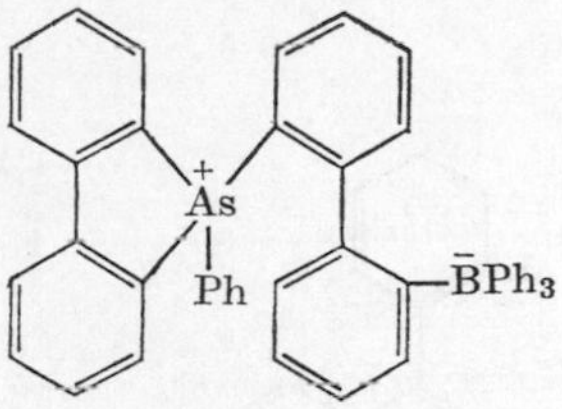   (**46**)

(*c*) The arsorane in methylene dichloride reacts readily with bromine to form the orange crystalline 5-(2'-bromo-2-biphenylyl)-5-phenyldibenzarsolium tribromide (**45**; R = Br, X = Br$_3$), m.p. 99–102°, which in boiling ethanol reacts with potassium tetraphenylborate to give the tetraphenylborate [**45**; R = Br, X = B(C$_6$H$_5$)$_4$], m.p. 181–183°.

(*d*) A solution of the arsorane (**37**) in ether–tetrahydrofuran when heated with triphenylborine for 5 weeks at 100° gives the betaine-like compound (**46**), termed 2'-(5-phenyldibenzarsolio)-2-biphenylyltriphenylborate(1−) or triphenyl-[2'-(5-phenyldibenzarsolio)-2-biphenylyl]borate, m.p. 170–172°.

Other arsoranes, similar in type to (**37**) but having substituents in the phenyl group or in the polycyclic system, have been prepared.

Thus 5-chlorodibenzarsole (**7**) reacts with *p*-(dimethylamino)-phenyllithium to give 5-[*p*-(dimethylamino)phenyl]dibenzarsole (**47**), m.p. 118.5–120°, which gives in turn the sulfonylarsinimine (**48**), m.p. 243–244.5°. Reaction of (**48**) with 2,2'-biphenylylenedilithium yields *N*,*N*-dimethyl-*p*-(5,5'-spirobi[dibenzarsol]-5-yl)aniline (**49**), and with the 2,2'-dilithio derivative (**50**) of 1-phenylnaphthalene yields *N*,*N*-

dimethyl-p-(spiro[7H-benzo[b]naphth[1,2-d]arsole-7,5′-[5H]dibenzo-
arsol]-yl)aniline (**51**), more conveniently but less correctly termed (5,6-
benzo-2,2′-biphenylylene)-2,2′-biphenylylene-p-dimethylaminophenyl-
arsorane, m.p. 190°.

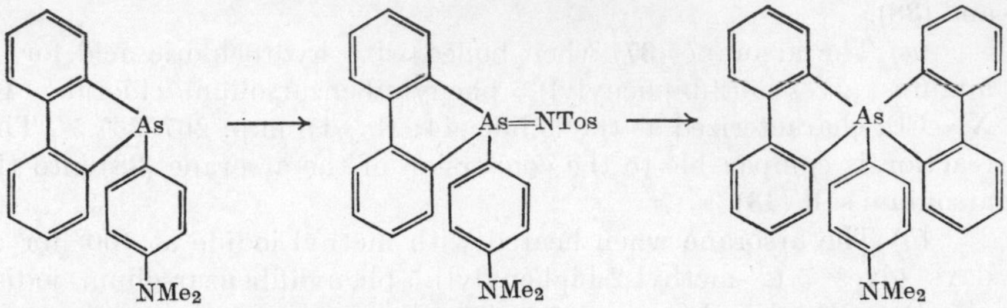

(47)   (48)   (49)

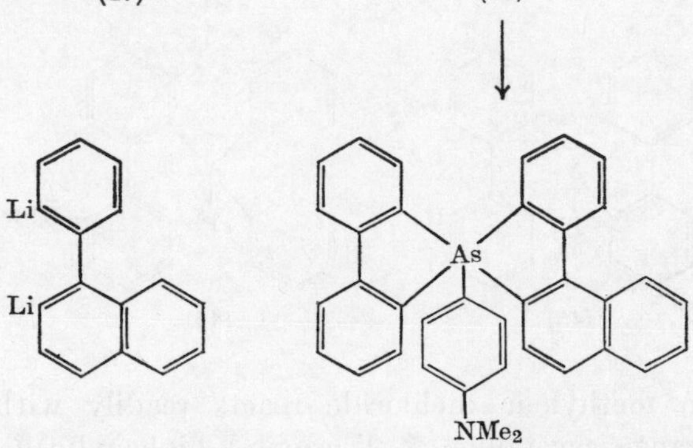

(50)   (51)

The arsorane (**51**) shows the familiar cleavage of one of the arsole
rings: attempted recrystallization from methanol–tetrahydrofuran gives
the covalent 5-(p-dimethylaminophenyl)-5-methoxy-5-(o-1-naphthyl-
phenyl)dibenzarsole (**52**), m.p. 166°. The arsorane in methyl iodide at
room temperature for 2 days undergoes the same cleavage to form

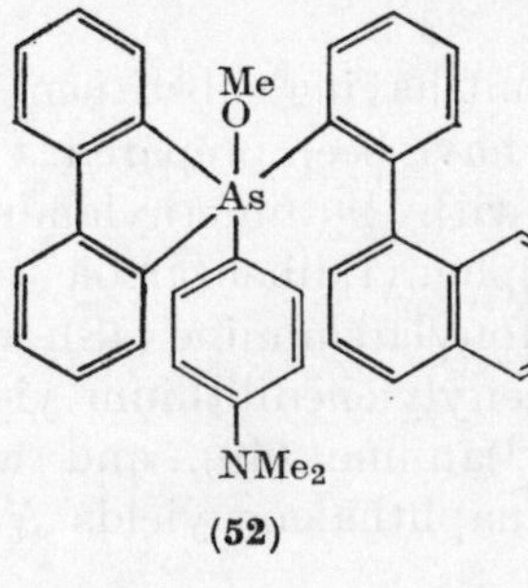

(52) 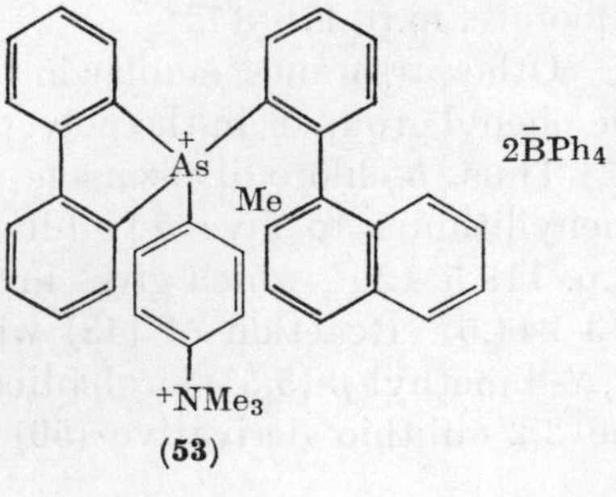   (53)

(probably) the quaternary ammonium-arsonium cation (**53**), isolated as its bistetraphenylborate, named 5-(p-dimethylaminophenyl)-5-[o-(2'-methyl-1'-naphthyl)phenyl]dibenzarsolium As-tetraphenylborate N-methotetraphenylborate.[48]

Thermal Decomposition

These arsoranes having the 9-arsafluorene (5-dibenzoarsole) ring undergo comparatively ready isomerization, similar to that of the corresponding phosphoranes (p. 92). The arsorane (**38**) when heated above its melting point for a few minutes gives the isomeric diphenyl-2-o-terphenylylarsine (**54**), the structure of which is confirmed by its alternative formation by the action of 2-o-terphenylyllithium on diphenylarsinous chloride. The arsorane (**37**) when heated at 233° for 2–3 minutes forms 17-phenyl-17H-tetrabenz[b,d,f,h]arsonin (**55**; R = C_6H_5), m.p. 132.5–133.5°, the proof of the structure of which is given on

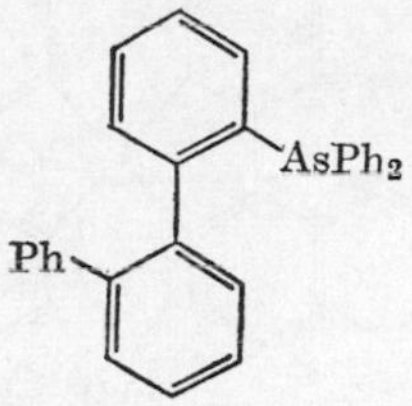

(54) 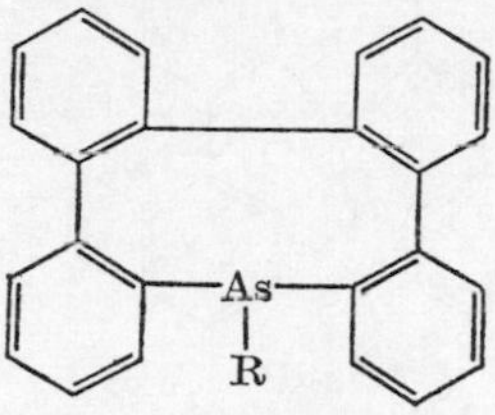   (55)

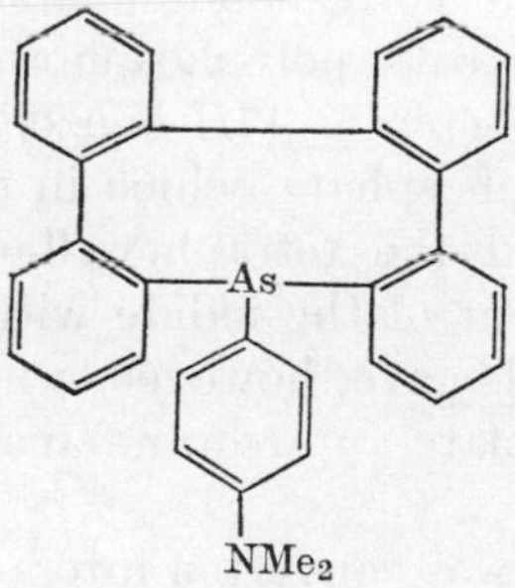

(56) 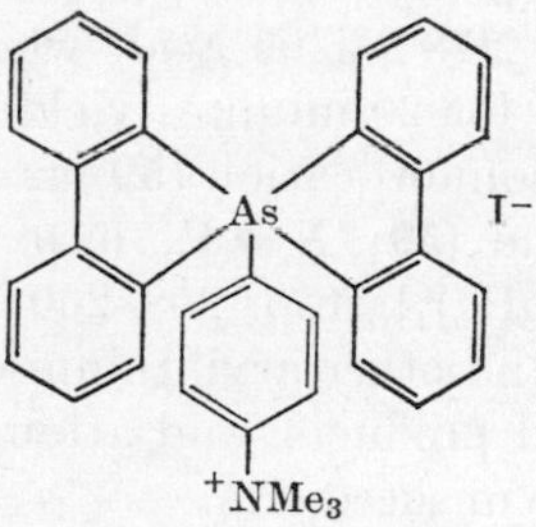   (57)

p. 442. The arsorane (**49**) when quickly heated gives N,N-dimethyl-p-(17H-tetrabenz[b,d,f,h]arsonin-17-yl)aniline (**56**) (cf. p. 95); alternatively, if the arsorane (**49**) is kept in methyl iodide at room temperature for 7 days it undergoes methiodide formation on the tertiary amino group to give the iodide (**57**), which when heated quickly to 290° also gives the tertiary arsine.

The thermal decomposition of the spirocyclic arsonium chloride (**34**; X = Cl) to the tertiary arsine (**36**) has already been noted. In view of the general type of the above thermal decompositions, this product (**36**) might have been expected to be the isomeric chloroarsine (**55**; R = Cl): the fact that the product would not react with phenyllithium indicates strongly, however, that it is the compound (**36**).

Since arsoranes of type (**37**)—like their phosphorane analogues— probably have a trigonal bipyramid structure, a suitably substituted compound might be resolvable into optically active forms. Attempts to resolve the cation (**57**) by fractional crystallization of its (+)-camphor-10-sulfonate or its D-(−)-hydrogen dibenzoyltartrate failed.

In another attempt, the Grignard reagent from 1-(*o*-bromophenyl)-naphthalene was treated with arsenious oxide and the product boiled with hydrochloric acid, giving the arsinous chloride (**58**), m.p. 193–195°,

(**58**)   (**59**)

which with hydrogen peroxide gave the corresponding arsinic acid, m.p. 276–278°. This acid, when heated with polyphosphoric acid at 160–200° for 20 minutes yielded the 7,7′-spirobis-[7*H*-benzo[*b*]naphth-[1,2-*d*]arsolium] cation (**59**) as the metaphosphate, which in turn gave the iodide (**59**; X = I), m.p. 165°, and the tetraphenylborate [**59**; X = B(C$_6$H$_5$)$_4$], m.p. 258–260°. Treatment of the iodide with phenyllithium and other aryllithium compounds gave, however, a mixture of undefined products, and attempts to isolate an arsorane analogous to (**36**) were unsuccessful.

Finally, 2-bromo-4-methylbiphenyl was converted into a Grignard reagent, which on treatment with arsenious oxide and then hydrochloric acid (as before) gave bis-(4-methyl-2-biphenylyl)arsinous chloride (**60**), m.p. 126°. This compound added chlorine to form the trichloroarsorane, which when heated at 230° under reduced pressure formed 3,3′-dimethyl-5,5′-spirobi[dibenzarsolium] chloride (**61**; X = Cl), m.p. 275–276°, from which the iodide, m.p. 277–278°, and the tetraphenylborate, m.p. 283–285°, were also obtained. The iodide (**61**; X = I) with phenyllithium readily gave the arsorane, 5-phenyl-5,5′-spirobi[3,3′-dimethyldibenz-

Me Me Me Me Me Me

As
Cl

As⁺ X⁻

As
Ph

(60)   (61)   (62)

arsole] (62), m.p. 193–194° (solvent-free), 125–127° with 1 molecule of cyclohexane.

For resolution, the iodide (61; X = I) was converted into the D-(−)-hydrogen dibenzoyltartrate, which after fractional crystallization from acetone–methanol had $[\alpha]_{436}$ −150°, m.p. 146–147°. This salt in dichloromethane–methanol was chromatographically converted into the (−)-chloride, and then to the (−)-iodide, having $[\alpha]_{436}$ −14.5° in methanol, m.p. 277–278°. Working up the salt in the initial recrystallization mother-liquors ultimately gave the (+)-iodide, $[\alpha]_{436}$ +15°. The activity of these iodides, when the latter were treated with phenyllithium in ether, rapidly decreased and the solutions yielded solely the inactive arsorane (62).[48]

Tris-(2,2′-biphenylylene)arsenate Anion

Hellwinkel[48a] has developed the study of phosphoranes containing dibenzophosphole units to the preparation of the analogous tris-(2,2′-biphenylylene)phosphate anion (p. 96). This anion has the octahedral configuration and has been resolved into optically active forms.

Hellwinkel and Kilthau[48b] have later extended the study of the analogous arsoranes to synthesize salts having the tris-(2,2′-biphenylylene)arsenate anion (1), [RIS —]. This anion can be represented more simply by the formula (1A), in which the curved portion of each section represents a 2,2′-biphenylylene unit linked to the central arsenic atom.

A mixture of 5,5′-spirobi[dibenzarsolium] iodide (2) (2 moles) and 2,2′-dilithiobiphenyl (3) (1 mole), when stirred in ether for 12 hours deposits the salt, 5,5′-spirobi[dibenzarsolium] tris-(2,2′-biphenylylene)-arsenate (4), yellow crystals (from dichloromethane–methanol), decomposing at 209–210°.

Either of two reactions may be employed for resolution purposes. The salt (4), when treated with a large excess of sodium iodide in acetone, undergoes simple double decomposition with regeneration of the iodide (2) and the formation of sodium tris-(2,2′-biphenylylene)arsenate (5;

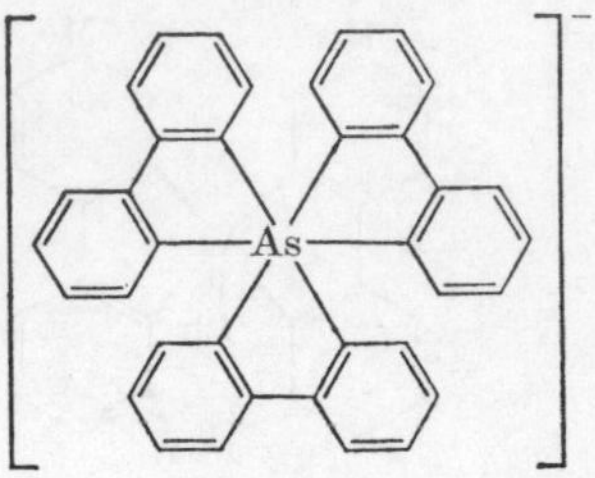

(1) 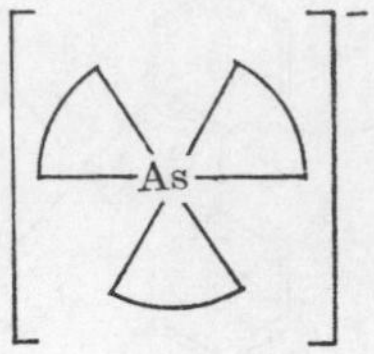   (1A)

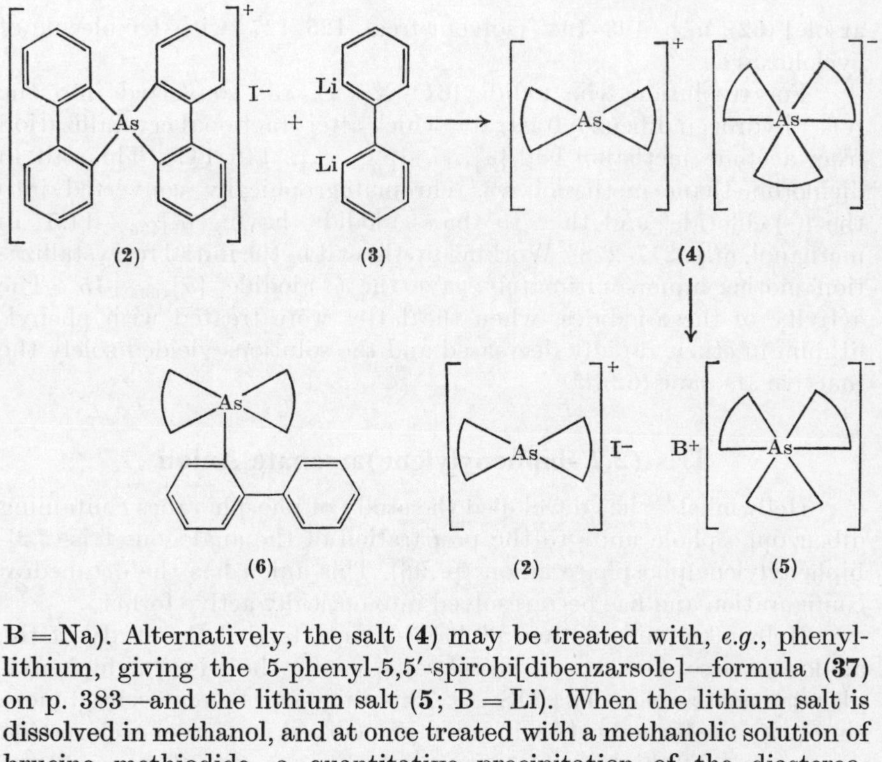

(2)   +   (3)   ⟶   (4)

(6)   (2)   (5)

B = Na). Alternatively, the salt (4) may be treated with, *e.g.*, phenyl-lithium, giving the 5-phenyl-5,5′-spirobi[dibenzarsole]—formula (37) on p. 383—and the lithium salt (5; B = Li). When the lithium salt is dissolved in methanol, and at once treated with a methanolic solution of brucine methiodide, a quantitative precipitation of the diastereo-isomeric methylbrucinium tris-(2,2′-biphenylylene)arsenates occurs. Ten recrystallizations of this mixture from acetone affords the optically pure (−)-methylbrucinium (−)-tris-(2,2′-biphenylylene)arsenate (5; B = $C_{24}H_{29}N_2O_4$), m.p. 207–209° (dec.) (11%), $[\alpha]_{578}$ −1220 ± 20° in dichloro-methane.

This salt reacts with the iodide (2) to give the optically active 5,5′-spirobi[dibenzarsolium] tris-(2,2′-biphenylylene)arsenate (4), which after five recrystallizations from dichloromethane–ether had m.p. 201–209°, $[\alpha]_{578}$ −1180 ± 10° in dimethylformamide.

Salts of the optically pure (+)-anion could not be isolated from the initial acetone mother-liquor.

Treatment of the (−)-methylbrucinium salt (5; B = $C_{24}H_{29}N_2O_4$) in aqueous acetone with hydrochloric acid causes cleavage of one of the dibenzarsole units of the anion, with the formation of the optically inactive 5-(2-biphenylyl)-5,5′-spirobi[dibenzarsole] (6).

Six-membered Ring Systems Containing Carbon and One Arsenic Atom

Arsenins and Arsenanes

The fundamental compound in this series is arsenin (1) [RRI 272, in which the arsenic atom is depicted uppermost], and the hexahydro

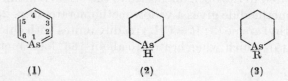

(1)   (2)   (3)

derivative (2) is termed arsenane. The compound (1), like its phosphorus analogue, phosphorin (p. 99), is unknown. This absence of the first or the fundamental member of a heterocyclic arsenic or phosphorus system is common, and in certain cases may well be due to the simple fact that no chemist has had occasion to prepare the compound. In the present case, however, arsenin may be too unstable to exist, for arsenic (like phosphorus and antimony) very rarely forms a component of a true aromatic system.

Various names have been applied in the past to the compounds (1) and (2). Zappi[49] gave the name arsedine to (1) and arsepidine to (2) because they were the arsenic analogues of pyridine and piperidine, respectively; these names have not been widely adopted, but the compound (2) is still occasionally named arsacyclohexane or (cyclo)-pentamethylene-arsine.

1-Substituted derivatives (3) of arsenane are readily prepared, the methods being similar to those for the preparation of 1-substituted arsolanes (p. 359). Grüttner and Wiernik[50] prepared 1-phenylarsenane (3; R = C_6H_5) and 1-p-tolylarsenane by the interaction of the di-Grignard reagent prepared from 1,5-dibromopentane, i.e.,

$$CH_2(CH_2CH_2MgBr)_2,$$

with phenyl- and p-tolyl-arsonous dichloride, $ArAsCl_2$, respectively. Historically these were the first compounds having a ring system of carbon and arsenic to be recorded: the analogous 1-phenylarsolane was recorded one year later by Grüttner and Krause.[10]

Grüttner and Wiernik also prepared 1-phenylarsenane (but less effectively) by the interaction of 1,5-dibromopentane, phenylarsonous dichloride, and sodium in an ether medium.

1-Methylarsenane (**3**; $R = CH_3$) was prepared by the action of the above di-Grignard reagent on methylarsonous dichloride (Zappi[49]) and 1-ethylarsenane was similarly prepared (Steinkopf, Donat, and Jaeger[51]).

1-Methylarsenane is a colorless liquid, b.p. 156°/760 mm, 76°/36 mm, 65°/20–22 mm; it is volatile in steam, smells like oil of mustard and is readily oxidized in air. With methyl iodide it forms a methiodide (**4**; $R = CH_3$, $X = I$), systematically named 1,1-dimethylarsenanium iodide, white crystals which when heated in a sealed tube melt at 290°, but when heated in the open dissociate to the original arsine and methyl iodide. The methiodide gives a yellow methopicrate, m.p. 258°.

The methyl arsine (**3**; $R = CH_3$) readily unites with chlorine to form a dichloride (**5**), which when heated to about 166° loses methyl chloride

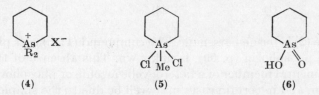

(**4**)   (**5**)   (**6**)

to form 1-chloroarsenane (**3**; $R = Cl$), b.p. 84–86°/13 mm, the constitution of which was confirmed by its reaction with sodium thiocyanate to give the 1-thiocyanato analogue (**3**; $R = SCN$) (Steinkopf, Schubart, and Roch[13]). This production of 1-chloroarsenane (a standard reaction, cf. p. 359) was confirmed by Gorski, Schpanski, and Muljav,[52] who showed that treatment with methanolic sodium methoxide converted it into the diarsine 'oxide', i.e. 1,1'-oxydiarsenane, $(CH_2)_5As—O—As(CH_2)_5$, which in turn, on oxidation with hydrogen peroxide, gave the colorless crystalline 1-hydroxyarsenane 1-oxide, also called cyclopentamethylenearsinic acid (**6**), m.p. 200.5–202°. Zappi's earlier observation[49] that the dichloride (**5**) decomposed to methyl chloride and 1,5-dichloropentane was therefore probably incorrect; he notes that the methyl arsine (**3**; $R = CH_3$) unites with bromine to give a dibromide which on heating decomposes to form methyl bromide and 1-bromoarsenane (**3**; $R = Br$), and unites with iodine to give a yellow diiodide, m.p. 120°.

Much later, Zappi and Degiorgi[53] stated that 1-methyl-arsenane (3; R = CH$_3$) reacts in cold carbon tetrachloride with one atomic proportion of bromine to give a compound formulated as [(CH$_2$)$_5$As(CH$_3$)Br]$_2$, hygroscopic white crystals, m.p. 60°; this compound when exposed to the air forms a peroxide, and when treated with more bromine forms the above dibromide. Further, the red oily 1-bromoarsenane (3; R = Br) combines with a molecular equivalent of bromine to form the colorless crystalline 1,1,1-tribromoarsenane, m.p. 102°; this compound when treated with water forms hydrogen bromide and the arsinic acid (6), but when heated gives arsenic tribromide and gaseous hydrocarbons, and not the 2,3-dihydroarsenin as these workers had hoped.

Wiberg and Mödritzer[15] have shown that 1-chloroarsenane (3; R = Cl) in ether, when reduced with lithium borohydride at −50°, or with lithium aluminum hydride at −60°, gives the colorless arsenane (2), m.p. −13 to −11°, in 77% yield. It has a 'phosphine smell', and is very readily oxidized by air to the arsinic acid (6), termed arsenaninic acid; the arsenane when treated in ethereal solution with iodine gives the red crystalline 1-iodoarsenane (3; R = I), m.p. 27°.

The di-Grignard reagent from 1,5-dibromopentane reacts (a) with ethylarsonous dichloride to give 1-ethylarsenane,[51] b.p. 62–64°/12.5 mm (methiodide, m.p. 276°), and (b) with (2-chlorovinyl)arsonous dichloride to give 1-(2-chlorovinyl)arsenane (3; R = CH=CHCl),[52] b.p. 89–91°/5 mm. The ethylarsenane combines with cyanogen bromide to give a deliquescent addition product, which is readily hydrolyzed to the ethylarsenane hydroxy-bromide, m.p. 71°. This compound when heated gives various products, of which the original ethylarsine, cyanogen bromide, and ethyl bromide have been identified.

The di-Grignard reagent from 1,5-dichloro-n-hexane reacts with methylarsonous dichloride to form 1,2-dimethylarsenane, b.p. 169°/760 mm, 85°/22 mm[54]; this gives a methiodide, a methopicrate, m.p. 231°, and an arsine oxide, which when exposed to the air, and also when treated in petroleum solution with iodine, is stated to form a resin.

1-Phenylarsenane[49] is a slightly viscous oil, b.p. 153–154°/18–20 mm, which has a feeble odor and (unlike its As-alkyl analogues) is not markedly oxidized in air. It forms a crystalline mercurichloride, m.p. 201.5–202°, of composition C$_{11}$H$_{15}$As,HgCl$_2$, but probably having the double bridged structure (p. 50). The arsenane forms a very deliquescent crystalline dichloride which on heating decomposes giving 1,5-dichloropentane, a reaction again unlike that of its As-alkyl analogues. 1-Phenylarsenane gives the quaternary salts: methiodide, m.p. 179.5°; ethiodide, m.p. 185°; n-propyl iodide, m.p. 137–138°; isopropyl iodide, m.p. not recorded; and n-butyl iodide, m.p. 140°. Tzschach and Lange[55]

14

have recently recorded a novel reaction whereby in effect quaternary salts with aryl halides may be prepared: when a dioxan solution of 1,5-dibromopentane is treated at room temperature with one equivalent of diphenylpotassioarsine, $(C_6H_5)_2AsK$, the reaction mixture when filtered and evaporated leaves an oily residue which on treatment with ether deposits the crystalline 1,1-diphenylarsenanium bromide (4; $R = C_6H_5$, $X = Br$), m.p. 232–234° (cf. the P analogue, p. 52).

1-Phenylarsenane gives a white crystalline cyanogen bromide addition product, m.p. 107°, which is readily hydrolyzed to the hydroxy-bromide, m.p. 162.5°, and which when heated under reduced pressure gives a number of products including the original tertiary arsine (Steinkopf and Wolfram [56]).

1-p-Tolylarsenane (3; $R = C_6H_4CH_3$) [50] is an oil, b.p. 162–163°/20 mm, 177–178°/50 mm; it gives a dichloride, m.p. 134°, and a mercuri-chloride, m.p. 175°, of identical type to that noted above. 2-Methyl-1-phenylarsenane has been described in a patent specification. [57]

4-Arsenanones

Phenylarsonic acid, $C_6H_5As(O)(OH)_2$, can be readily reduced to phenylarsine, $C_6H_5AsH_2$, which in turn reacts with acrylonitrile to form bis-(2-cyanoethyl)phenylarsine (7), m.p. 59–60°; alkaline hydrolysis then furnishes bis-(2-carboxyethyl)phenylarsine [As-phenyl-3,3′-arsinyl-idenedipropionic acid] (10; $R = H$), m.p. 83–84.5° (R. C. Cookson and Mann [58]). Welcher, Johnson, and Wystrach, [59] following their synthesis of the analogous 4-phosphorinanone (p. 105), cyclized the dinitrile (7) by treatment with sodium *tert*-butoxide in boiling toluene, and so obtained 4-amino-1,2,5,6-tetrahydro-1-phenylarsenin-3-carbonitrile (8), m.p. 65–67°, but did not record its hydrolysis to 1-phenyl-4-arsenanone (9).

Gallagher and Mann, [60] in a comparative study of the properties of 1-phenyl-4-arsenanone and its phosphorus and nitrogen analogues

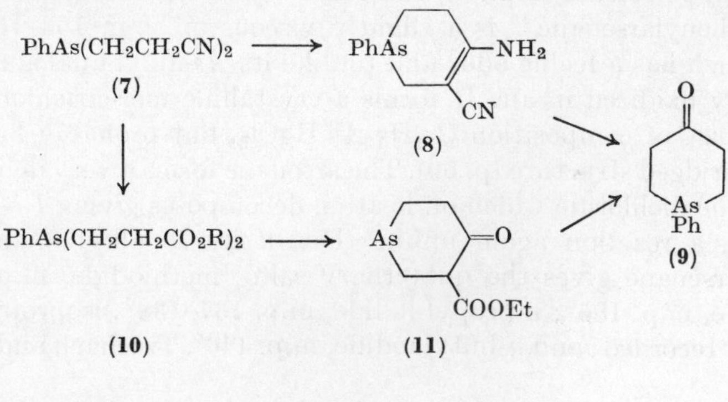

(p. 105), converted the bis-(2-carboxyethyl)phenylarsine (**10**; R = H) into the diethyl ester (**10**; R = C_2H_5), b.p. 153–154°/1.25 mm, which when treated in boiling benzene with sodium methoxide under nitrogen gave 3-ethoxycarbonyl-1-phenyl-4-arsenanone [ethyl 4-oxo-1-phenyl-3-arsenanecarboxylate] (**11**). This compound on acidic hydrolysis gave 1-phenyl-4-arsenanone (**9**), b.p. 114–115°/0.3 mm, m.p. 9–10° (4-phenylsemicarbazone, m.p. 175–176°).

Two interesting heterocyclic arsenic systems were prepared from these intermediates. The diethyl ester (**10**; R = C_2H_5) underwent

(12)   (13)

hydrolysis and oxidation, slowly on exposure to damp air and rapidly in hot acetone–hydrogen peroxide, to form the crystalline 5-phenyl-1,6-dioxa-5-arsaspiro[4.4]nonane (**12**), [RIS 7889], initially named di-(2-carboxyethyl)phenylarsine dihydroxide dilactone and previously prepared by the direct oxidation of the acid (**10**; R = H)[61] (p. 533). The keto ester (**11**), when heated at 100° with phenylhydrazine and a trace of acetic acid, gave 4,5,6,7-tetrahydro-2,5-diphenylarsenino[4,3-*c*]-pyrazol-3(2*H*)-one (**13**), [RIS 11861], originally named 5-oxo-1,1'-diphenyl-1'-arsacyclohexano(4',3'-3,4)-2-pyrazoline (p. 397).

1-Phenyl-4-arsenanone (**9**) gives a series of reactions with methyl iodide which are not shown by 1-phenyl-4-phosphorinanone (p. 106) but which are shown by the analogous 1-phenyl-4-piperidone.[60] The arsenanone (**9**) with boiling methyl iodide behaves normally, giving the methiodide (**14**) [1-methyl-4-oxo-1-phenylarsenanium iodide], m.p. 145–146°; with boiling methanolic methyl iodide, however, it gives a ketal, 4,4-dimethoxy-1-methyl-1-phenylarsenanium iodide (**15**), m.p. 146–147°, which gives a picrate, m.p. 105.5–108.5°. A mixture of the iodides (**14**) and (**15**) melted at 124–125°. The evidence for the structure of the compound (**15**) may be summarized: (i) the analysis of the iodide and picrate; (ii) the iodide is unaffected by heating at 60°/0.1 mm for 1 hour, and methanol of crystallization is thus unlikely; (iii) the infrared spectrum of the iodide shows no C=O or OH absorption, and the compound (which had been crystallized from ethanol) cannot be a monoethanolate of the iodide (**14**). The spectrum also shows a strong doublet at 1107 and 1057 cm^{-1}: Tschamler and Leutner[62, 63] have

recorded a strong doublet in the 1150–1080 cm^{-1} region, characteristic
of the C—O—C—O—C grouping; (iv) the nmr spectra provided strong

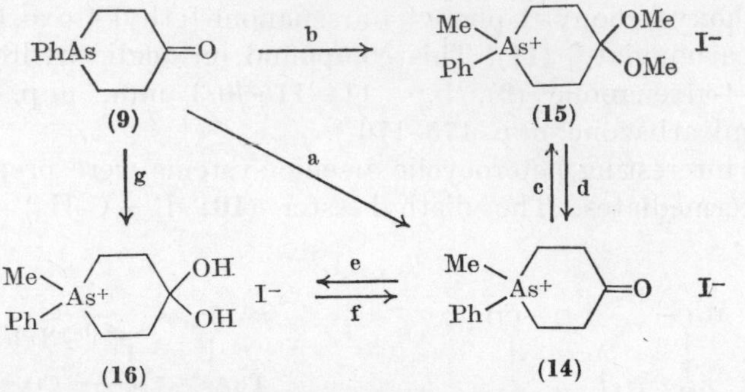

a, CH$_3$I. b, CH$_3$OH–CH$_3$I. c, CH$_3$OH + trace HI.
d, Aq. HI. e, Aq. (CH$_3$)$_2$CO–C$_2$H$_5$OH, or C$_2$H$_5$OH + trace HI.
f, Recryst. from C$_2$H$_5$OH. g, C$_2$H$_5$OH–CH$_3$I.

confirmation, for details of which the original should be consulted;
(v) the dimethoxy-iodide (15) was readily formed when the methiodide
(14) was boiled in methanol containing a trace of hydriodic acid, but the
arsenanone (9) when similarly treated was unaffected. The iodide (15)
conversely was rapidly hydrolyzed by boiling aqueous hydriodic acid,
with the formation of the methiodide (14).[60]

The methiodide (14), on recrystallization from aqueous acetone–
ether, gave 4,4-dihydroxy-1-methyl-1-phenylarsenanium iodide (16),
m.p. 120–121°; this compound was also obtained when the arsenanone
(9) was boiled with ethanolic methyl iodide. The infrared spectrum of
(16) showed only weak C=O absorption but a strong doublet at 1087
and 1055 cm^{-1} and normal OH stretching absorption by a sharp band
at 3340 cm^{-1}; the compound has considerable thermal stability, being
unaffected by heating at 60°/0.1 mm for 36 hours, but repeated recrystallization from absolute ethanol converted it back into the methiodide (14).

The formation of compounds of type (15) may be limited to the
methyl member. This is indicated by the fact that the dihydroxy-iodide
(16) is formed when the ketone (9) is boiled with ethanolic methyl iodide
and also when the methiodide (14) is boiled with ethanol containing a
trace of hydriodic acid; in neither case was the diethoxy analogue of
(15) detected.[60]

The arsenanone (9) with phenylhydrazine formed a viscous product
which gave only tars under the normal conditions of the Fischer indolization reaction. In this respect, the behavior of the arsenanone again

differs strikingly from that of the 1-phenylphosphorinanone (p. 121), but strongly resembles that of 1-phenyl-4-piperidone.

1-Phenylarsenanone (**9**) behaved, however, precisely similarly to 1-phenylphosphoranone and 1-phenyl-4-piperidone when treated with alkaline *o*-aminobenzaldehyde and with isatin in hot alkaline solution. The arsenanone with *o*-aminobenzaldehyde (the Friedländer reaction) gave 1,2,3,4-tetrahydro-2-phenylarsenino[4,3-*b*]quinoline (**17**; R = H)

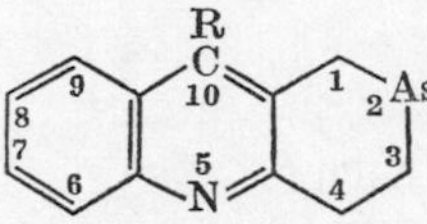

(17) 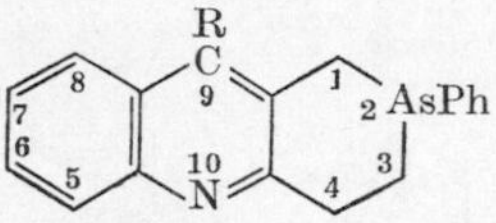   (18)

[RIS —] and with alkaline isatin (the Pfitzinger reaction) gave the corresponding 10-carboxylic acid (**17**; R = CO$_2$H), which are discussed under the main heading Arsenino-[4,3-*b*]quinoline (p. 398).

The compounds (**17**; R = H and CO$_2$H) were initially named as arsa-aza-anthracenes and therefore numbered as anthracenes: they were thus called 1,2,3,4-tetrahydro-2-phenyl-10-aza-2-arsa-anthracene (**18**; R = H) and its 9-carboxylic acid (**18**; R = CO$_2$H), respectively.[60]

3*H*-Arsenino[4,3-*c*]pyrazole

This system (**1**) [RIS 11861] is known only in the hydrogenated substituted form (**2**), which is therefore 4,5,6,7-tetrahydro-2,5-diphenyl-arsenino[4,3-*c*]pyrazol-3(2*H*)-one (**2**) (p. 395). This compound is

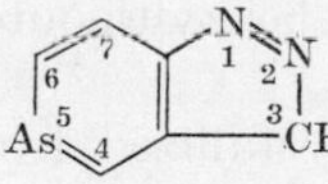

(1) 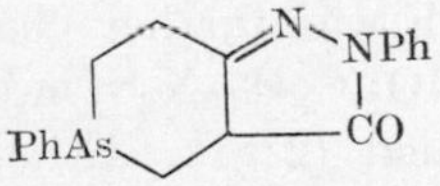   (2)

readily obtained when a mixture of ethyl 4-oxo-1-phenyl-3-arsenane-carboxylate (**3**) and phenylhydrazine is heated with a trace of acetic acid under nitrogen at 100° for 1 hour; it forms colorless crystals, m.p. 173–174°.

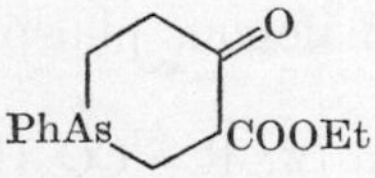

(3) 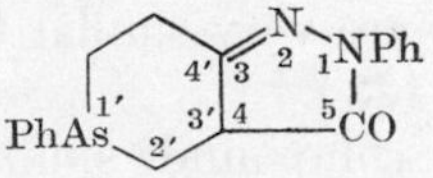   (4)

It was earlier named 5-oxo-1,1'-diphenyl-1'-arsacyclohexano(4',3'-3,4)2-pyrazoline, based on the numbering given in (4) (Gallagher and Mann [60]).

Arsenino[4,3-*b*]quinolines

The parent compound (1) [RIS —] is, as expected, unknown, but its 1,2,3,4-tetrahydro derivatives (2) have been briefly studied (Gallagher and Mann [60]).

(1)   (2)   (3)

The Friedländer reaction, when applied to 1-phenyl-4-arsenanone (3), requires critical conditions and even then does not proceed well. A mixture of the freshly distilled arsenanone and *o*-aminobenzaldehyde, when heated at 100° for 1 hour under nitrogen without a solvent or an alkaline catalyst, gave a crude product which furnished the colorless crystalline 1,2,3,4-tetrahydro-2-phenylarsenino[4,5-*b*]quinoline (2; R = H), m.p. 76–77°, which gave a stable hydrochloride hemihydrate, m.p. 156–157°. This reaction proceeds far less satisfactorily than the corresponding reaction with 1-phenyl-4-phosphorinanone (p. 122).

When the arsenanone (3) was subjected to the Pfitzinger reaction, being added to a boiling solution of isatin in ethanol–aqueous sodium hydroxide, it furnished the cream-colored 1,2,3,4-tetrahydro-2-phenyl-arsenino[4,3-*b*]quinoline-10-carboxylic acid (2; R = CO_2H), m.p. 266–268°, which was further characterized as its benzylthiouronium salt, $C_{19}H_{16}AsNO_2,C_8H_{10}N_2S$, m.p. 150–151°.

The acid (2; R = CO_2H) could be recrystallized from aqueous dimethylformamide and sublimed unchanged at 250°/0.05 mm. Its infrared spectrum showed a broad band of low intensity centered at *ca.* 2000 cm^{-1}, attributed to strongly hydrogen-bonded OH groups, probably arising from intermolecular interaction of the CO_2H groups and the quinoline-nitrogen atoms. The spectrum also showed C=O absorption at 1633 cm^{-1}, the low value being influenced by the above bonding: no evidence of a CO_2^- ion was detected. These physical properties are very similar to those of the analogous phosphorus compound (p. 123).

Many compounds similar in structure to (2; R = CO_2H) but with the arsenic replaced by nitrogen exist as deeply colored zwitterions, which are insoluble in most solvents and usually undergo decarboxylation on sublimation (cf. p. 136).

The compounds (2; R = H) and (2; R = CO$_2$H) were initially[60] termed 1,2,3,4-tetrahydro-2-phenyl-2-arsa-10-aza-anthracene and its 9-carboxylic acid, respectively.

Benzo and 1,8-Naphtho Derivatives of Arsenin

A number of polycyclic derivatives of arsenin can be regarded systematically as being formed by the fusion of a benzo or a 1,8-naphtho nucleus to the arsenin ring. Thus the fusion of a benzo group to the 2,3-position of this ring gives arsinoline (or 1-arsanaphthalene) (below), and to the 3,4-position gives isoarsinoline (or 2-arsanaphthalene) (p. 414). The fusion of two benzo groups symmetrically in the 2,3- and 5,6-positions gives acridarsine (p. 419), and unsymmetrically in the 2,3- and 4,5-positions gives arsanthridin (p. 424). Finally, the fusion of a 1,8-naphtho group to the 3,4,5-positions of arsenin would give 1H-naphth[1,8-cd]arsenin (or 1H-2-arsaphenalene) (p. 428), and its fusion to the 1,2,6-positions would give the bridgehead 1H,7H-benz[ij]-arsinolizine (or arsuline) (p. 430).

None of these parent compounds is known in the free state, but hydrogenated derivatives have been prepared and will be discussed in the above order.

It should be added that one benzo group could theoretically be added to arsenin in a third position, namely to the 1,2-position; this would give arsinolizine, a system which has not yet been synthesized.

Arsinoline and 1,2,3,4-Tetrahydroarsinolines

Arsinoline (1) [RRI 1678] derives its name from being the arsenic analogue of quinoline: it has also been termed 1-arsanaphthalene.

(1)   (2)

1,2,3,4-Tetrahydroarsinoline (2; R = H) is known mainly as 1-substituted derivatives, in which R could be an alkyl or aryl group.

Compounds of this type were first prepared by Burrows and Turner,[64] using a method of synthesis virtually identical with that later employed by Turner and Bury[21] for the preparation of 1-alkylars-indolines (p. 366). For the arsinoline synthesis, 3-phenylpropyl bromide,

$C_6H_5(CH_2)_3Br$, was converted into the corresponding Grignard reagent, which when treated with dimethylarsinous iodide, $(CH_3)_2AsI$, gave dimethyl-(3-phenylpropyl)arsine (3), a colorless liquid, b.p. 133°/14 mm. This tertiary arsine was characterized by the preparation of its quaternary methiodide, m.p. 144°. The tertiary arsine (3), however, showed one unusual property in that it readily gave an additive compound with an equimolecular proportion of dimethylarsinous iodide; this additive compound, which consequently had the composition $C_{11}H_{17}As,(CH_3)_2AsI$, formed colorless prisms, m.p. 78–81°, which could be recrystallized from a number of solvents. Molecular-weight determinations in benzene solution, however, showed that it was completely dissociated into its components in this solvent.

The arsine (3) was converted in carbon tetrachloride solution into the arsine dichloride (4), which when heated at 160–180° lost methyl chloride, giving chloromethyl-(3-phenylpropyl)arsine [or methyl-(3-phenylpropyl)arsinous chloride] (5), a colorless liquid, b.p. 164–167°/14

(3)   (4)

(5)   (6)

mm. The monobromoarsine, b.p. 177–180°/16 mm, corresponding to (5), was prepared similarly. When the monochloroarsine (5) was boiled in carbon disulfide solution with aluminum chloride for 3 hours, cyclization occurred with loss of hydrogen chloride and the formation of 1,2,3,4-tetrahydro-1-methylarsinoline (6), a highly refractive liquid, b.p. 140°/14 mm. This arsine has a faint odor suggestive of quinoline and is oxidized slowly on exposure to the air.

The 1-methylarsinoline (6) readily gives a quaternary methiodide (1,2,3,4-tetrahydro-1,1-dimethylarsinolinium iodide) (7), colorless crystals, m.p. 235°. Burrows and Turner[64] state that this iodide in cold ethanol gives a colorless solution, which becomes yellow on heating; cooling reverses this color change. Moreover, a rapid volumetric estimation of the ionic iodine in the pure salt (7) in cold dilute ethanolic solution gave values about 7% lower than those obtained by the usual analytical methods. They therefore consider that the methiodide (7) in solution is in equilibrium with the o-(3-iodopropylphenyl)dimethylarsine

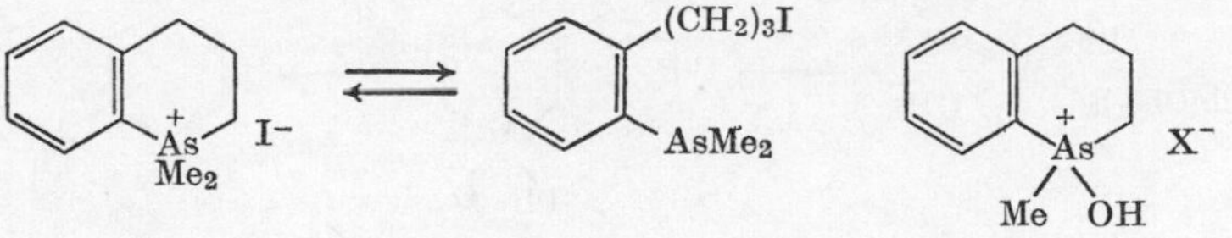

(7)   (8)   (9)

(8). These results are closely similar to those subsequently observed by Turner and Bury[21] for similar salts in the arsindoline series (p. 366).

Holliman and Thornton[65] have more recently prepared the iodide (7) and the corresponding bromide, m.p. 250–251°; both salts give the methopicrate, m.p. 119–120°. They find that ethanolic solutions of the iodide remain colorless when heated, and suggest that the brown color in Burrows and Turner's experiment may have been caused by photochemical decomposition.

Tetrahydro-1-methylarsinoline is oxidized by nitric acid to the hydroxy-nitrate (9; X = NO$_3$) which with picric acid gives the crystalline hydroxy-picrate (9; X = C$_6$H$_2$N$_3$O$_7$).

Roberts, Turner, and Bury[66] have shown that tetrahydro-1-methylarsinoline (6) combines with chlorine to form the arsine dichloride (10). This white solid product, when heated under reduced

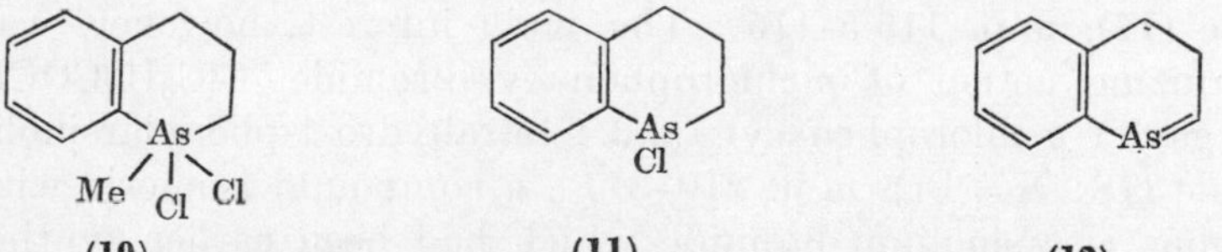

(10)   (11)   (12)

pressure, evolved methyl chloride smoothly with the formation of 1-chloro-1,2,3,4-tetrahydroarsinoline (11). Attempts to split off hydrogen chloride from (11) by boiling it with diethylaniline and so obtain 3,4-dihydroarsinoline (12) failed and the latter compound is still unknown (cf. p. 417).

A second synthesis of the arsinoline ring was developed by Holliman and Thornton,[65] who showed that although 1,2,3,4-tetrahydro-1-phenylarsinoline (15) could be prepared in 75% yield by the action of phenylmagnesium bromide on the 1-chloroarsinoline (2; R = Cl), a more rapid over-all method consisted in a Meyer reaction between phenylarsonous dichloride, C$_6$H$_5$AsCl$_2$, in warm aqueous sodium hydroxide solution, and

3-phenylpropyl bromide, thus giving phenyl-(3-phenylpropyl)arsinic acid (**13**). This acid when heated with concentrated sulfuric acid underwent cyclization to 1,2,3,4-tetrahydro-1-phenylarsinoline 1-oxide (**14**),

(**13**)   (**14**)   (**15**)

which was readily reduced in chloroform under hydrochloric acid by sulfur dioxide, giving the phenylarsinoline (**15**), the yield being 6% based on the initial dichloride employed.

The oily phenylarsinoline (**15**) gave a methiodide, m.p. 165–166°, and a methopicrate, m.p. 105.5–107°. It readily formed a dibromide, which (*a*) when treated with hydrogen sulfide formed the sulfide (**16**),

(**16**)   (**17**)   (**18**)

m.p. 103.5–105°, and (*b*) when treated with ammonia gave the corresponding dihydroxide, characterized by conversion into the hydroxypicrate (**17**), m.p. 113.5–115°. The main interest, however, was the quaternizing action of p-chlorophenacyl bromide, $ClC_6H_4COCH_2Br$, which gave 1-p-chlorophenacyl-1,2,3,4-tetrahydro-1-phenylarsinolinium bromide (**18**; X = Br), m.p. 210–211°, a compound isomeric with the analogous isoarsinolium bromide which had been earlier synthesized and resolved into optically active forms (p. 418).

The cation (**18**) has an asymmetric arsenic atom and should also be susceptible to optical resolution. It was isolated as the (+)-camphorsulfonate, m.p. 178–180°, and as the (−)-menthyloxyacetate, but neither salt on fractional crystallization gave evidence of resolution. It was also converted into the (+)-3-bromo-8-camphorsulfonate ($X = C_{10}H_{14}BrO_4S$); this salt, m.p. 182–195°, was initially recrystallized five times from ethanol, the m.p. and molecular rotation in methanol after each operation being: (i) m.p. 182–185°, $[M]_D$ +321°; (ii) m.p. 187–191°, $[M]$ +342°; (iii) not recorded; (iv) m.p. 261°, $[M]$ +690°; (v) m.p. 188–191°, $[M]$ +404°. Resolution had clearly been proceeding during the recrystallizations but the salt had formed a less soluble partial racemate before or during the fifth recrystallization, and all subsequent repetitions of the preparation of this salt gave a product, m.p. 184–185°, which did not change during six recrystallizations from ethanol.[65]

1-*Substituted* 1,2,3,4-*Tetrahydro-4-arsinolones*

Early attempts to prepare 1,2,3,4-tetrahydro-1-phenyl-4-arsinolone (**19**; $R = C_6H_5$) by the cyclization of 3-(diphenylarsino)propionic acid,

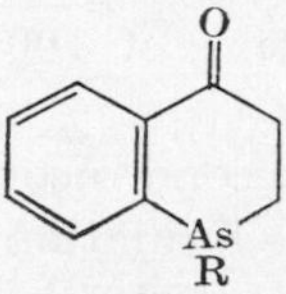

(**19**)

$(C_6H_5)_2As(CH_2)_2CO_2H$, or its chloride, $(C_6H_5)_2As(CH_2)_2CO_2Cl$, under a variety of conditions failed, although the nitrogen analogue of the acid can be cyclized to 1,2,3,4-tetrahydro-1-phenyl-4-quinolone (R. C. Cookson and Mann[67]). Later attempts to cyclize the analogous methyl-phenylarsino acid, $CH_3(C_6H_5)As(CH_2)_2CO_2H$, or its nitrile also failed. 3-[(*m*-Methoxyphenyl)methylarsino]propionic acid (**23**; $R = CO_2H$) was therefore synthesized to obtain appropriate activation of the phenyl group by the methoxyl group and thus facilitate cyclization on to the *ortho*-position (Mann and Wilkinson[68]).

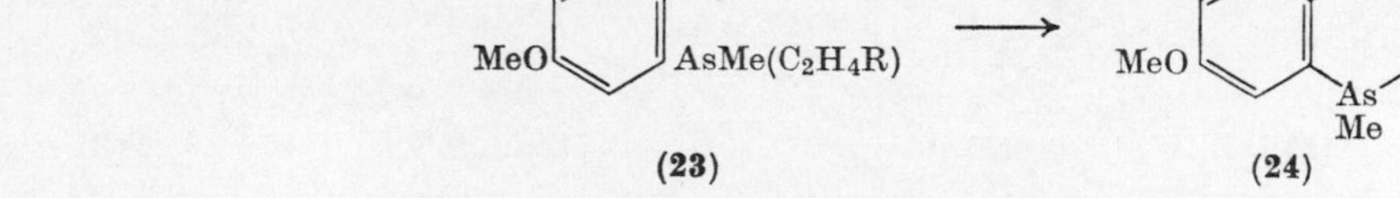

(**20**)   (**21**)   (**22**)

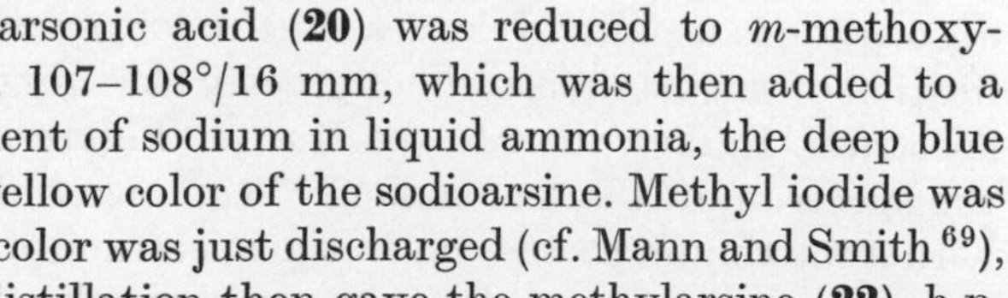

(**23**)   (**24**)

m-Methoxyphenylarsonic acid (**20**) was reduced to *m*-methoxy-phenylarsine (**21**), b.p. 107–108°/16 mm, which was then added to a solution of one equivalent of sodium in liquid ammonia, the deep blue color changing to the yellow color of the sodioarsine. Methyl iodide was added until the yellow color was just discharged (cf. Mann and Smith[69]), and evaporation and distillation then gave the methylarsine (**22**), b.p. 117–119°/15 mm (85%). The arsine when heated with acrylonitrile

under nitrogen gave 3-[(*m*-methoxyphenyl)methylarsino]propionitrile
(**23**; R = CN), b.p. 148–152°/1 mm (90%), which on alkaline hydrolysis
gave the syrupy 3-[(*m*-methoxyphenyl)methylarsino]propionic acid (**23**;
R = CO$_2$H), b.p. 166–172°/0.02 mm (crude, 97%).

The acid (**23**; R = CO$_2$H) when heated in boiling xylene containing
phosphoric anhydride and freshly heated 'Hyflo Supercel' gave the
colorless crystalline 1,2,3,4-tetrahydro-7-methoxy-1-methyl-4-arsinol-
one (**24**), m.p. 52° (35%).[68] The alternative use of a phosphoric anhydride–phosphoric acid mixture at 100° without a solvent gave the
ketone (**24**) in 26% yield.

This cyclization was the second recorded example of the formation
of a cyclic arsine system by the formation of a carbon–carbon bond, the
first being Das Gupta's cyclization of chloro-(2-chlorovinyl)phenylarsine
by an internal Friedel–Crafts reaction to form 1-chloroarsindole[12, 19];
almost all other cyclizations have involved the formation of an arsenic–
carbon bond.

The cyclization of the acid (**23**; R = CO$_2$H) could also have given
the isomeric 4-arsinolone bearing the methoxyl group in the 5-position.
The infrared spectrum of the homogeneous acid however showed a
strong band at 8310 cm^{-1} and none in the 7840–7410 cm^{-1} region;
1,2,4- and 1,2,3-trisubstituted benzenes have a well-defined band at
8197 and 7634 cm^{-1}, respectively.

There is a marked contrast between the colorless 4-arsinolone (**24**)
and the yellow liquid 1,2,3,4-tetrahydro-1-methyl-4-quinolone (**25**)
(Allison, Braunholtz, and Mann[70]). The methoxyl group in the former
prevents a strict comparison of the ultraviolet spectra of the two compounds. However, the effect of a CH$_3$O substituent on benzenoid

(25)   (25A)   (24A)   (24B)

absorption is usually to cause a shift to longer wavelengths with intensi-
fication of the maxima without greatly changing the general form of the
spectrum. In fact, the two spectra (see Figure 2) are entirely different,
the spectrum of the 4-arsinolone[68] completely lacking the broad band
with a maximum at 384 mμ in the spectrum of the 4-quinolone (**25**), a
band which is responsible for the yellow color of the latter compound.
(The spectrum of the arsinolone has however a general resemblance to

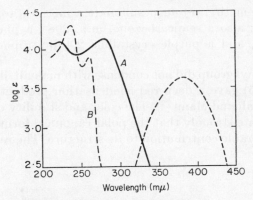

Figure 2. Ultraviolet spectra of: (*A*), 1,2,3,4 - Tetrahydro - 7 - methoxy - 1 - methyl-4-arsinolone (**24**) in ethanol; (*B*), 1,2,3,4 - tetrahydro - 1 - methyl - 4 - quinolone (**25**) in ethanol; there is a minimum at 289 mμ (logϵ 1.74) not shown in the Figure. (Reproduced, by permission, from F. G. Mann and A. J. Wilkinson, *J. Chem. Soc.*, **1957**, 3336)

that of 1,2,3,4-tetrahydro-1-phenyl-4-phosphinolone, p. 129). There is little doubt that, whereas the polar form (**25A**) makes an appreciable contribution to the 4-quinolone, a similar polar form (**21A**) makes no contribution to the 4-arsinolone. The latter fact is to be expected, for it is exceedingly rare for an arsenic atom in a ring to be linked to two flanking ring carbon atoms by a single and a double bond severally. This structure almost certainly exists in the 5,10-disubstituted 5,10-dihydroarsanthrenium dibromides (pp. 265, 268–270) and in the acridarsinones (p. 419), and it has been doubtfully claimed for arsanthrene (p. 462) and phenarsazine (p. 509).

A second possible polar form (**24B**) must make a negligible contribution to the 4-arsinolone, otherwise the latter would be colored.

It is noteworthy that the infrared spectra of the 4-arsinolone (**24**) and the 4-quinolone (**25**) show the CO band at 1653 and 1675 cm^{-1}, respectively; for comparison, *o*- and *p*-methoxyacetophenone show this band at 1653 and 1658 cm^{-1}, respectively.[71] The strong evidence for conjugation of the CO group in the quinolone compound with the absence of such evidence in the arsinolone, shown so clearly by the ultraviolet spectra, is therefore not shown by the infrared spectra.

The arsinolone (**24**) has the normal properties of a tertiary arsine and of a ketone. It forms a methiodide, m.p. 239°, a methopicrate, m.p.

226°, and with methyl p-toluenesulfonate a metho-p-toluenesulfonate, m.p. 180–205°; also a semicarbazone, m.p. 208°, a phenylhydrazone, m.p. 121–122°, and a purple crystalline 2,4-dinitrophenylhydrazone, m.p. 183°.[68]

The carbonyl group did not condense with malonitrile, although the 4-quinolone (25) gave a deep red condensation product (26) with this reagent (Ittyerah and Mann[72]). The color and stability of this product, however, indicate strongly that the polar canonical forms of type (26A) make a considerable contribution to its structure. The reluctance of the

(26)   (26A)   (27)

arsenic atom in the 4-arsinolone to acquire a similar internal positive charge ('internal' as distinct from that acquired by quaternization) is apparently strong enough to prevent the formation of a structure analogous to that of (26A).

The CH_2 group adjacent to the CO group in the 4-arsinolone (24) condensed under alkaline conditions with p-dimethylaminobenzaldehyde and with p-nitrosodimethylaniline to give the yellow 3-(p-dimethylaminobenzylidene)-1,2,3,4-tetrahydro-7-methoxy-1-methyl-4-arsinolone (27) and the corresponding orange-brown 3-(p-dimethylaminophenylimino) compound, respectively.[68]

The 4-arsinolone phenylhydrazone gives the Fischer reaction, and the arsinolone itself gives the Friedländer reaction with o-aminobenzaldehyde and the Pfitzinger reaction with alkaline isatin. Since these reactions give rise to new polycyclic systems, they are considered later (pp. 410, 422).

Spiro Derivatives of 1,2,3,4-Tetrahydroarsinoline

Following the synthesis and optical resolution of 3,3',4,4'-tetrahydro-2,2'(1H,1'H)-spirobi[isoarsinolinium] iodide (p. 419) and 3,3',4,4'-tetrahydro-1,1'(2H,2'H)-spirobiphosphinolinium iodide (p. 131), the synthesis and resolution of 3,3',4,4'-tetrahydro-1,1'(2H,2'H)-spirobisarsinolinium iodide (28; X = I) has been achieved (Holliman, Mann, and Thornton[73]). This compound is the tetrahydro derivative of the parent compound, 1,1'(2H,2'H)-spirobisarsinolinium iodide (29), [RIS 10887],

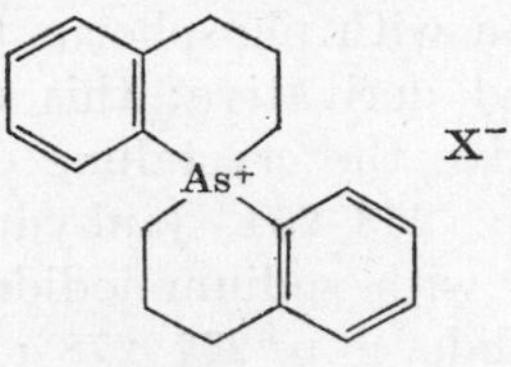

(28) 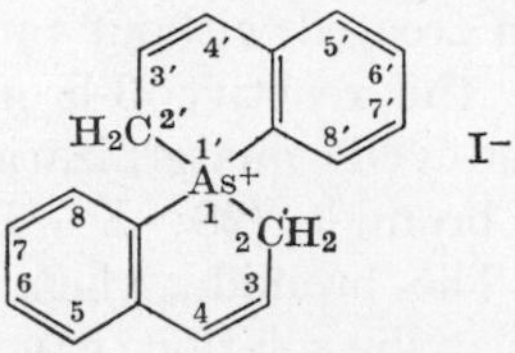   (29)

but was originally termed *As*-spirobis-1,2,3,4-tetrahydroarsinolinium iodide.

In the synthesis of the compound (28), 3-(*o*-bromophenyl)propyl methyl ether (30; R = CH₃) was converted into its lithium derivative, which when treated with 1-chloro-1,2,3,4-tetrahydroarsinoline (2;

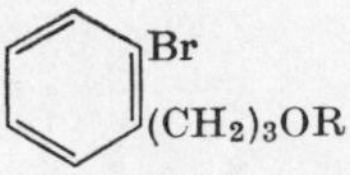

(30) 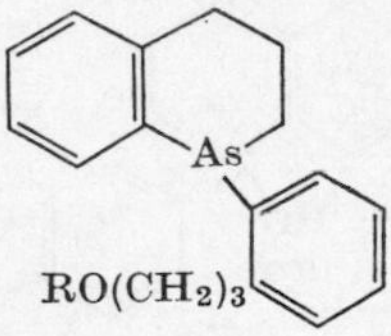   (31) 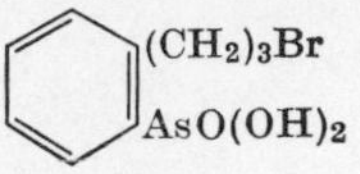   (32)

R = Cl) gave 1,2,3,4-tetrahydro-1-[*o*-(3′-methoxypropyl)phenyl]arsinoline (31; R = CH₃). This compound was treated with hydrogen bromide under various conditions in order to replace the methoxyl group by a bromine atom, which might then undergo cyclic quaternization with the arsenic atom. These attempts were unsuccessful. The compound (31; R = CH₃), in a boiling hydrobromic acid–acetic acid solution through which hydrogen bromide was passed, yielded 3-phenylpropyl bromide, $C_6H_5(CH_2)_3Br$; a similar experiment without the hydrogen bromide yielded 1-bromo-tetrahydroarsinoline (2; R = Br), which was identified by oxidation to *o*-(3′-bromopropyl)phenylarsonic acid (32), m.p. 134–136°.

To obtain a derivative of type (31) which would be more sensitive to acids and thus less likely to undergo the above cleavage, pure crystalline 1-bromotetrahydroarsinoline (2; R = Br), b.p. 174–180°/11 mm, prepared by addition of bromine to the 1-methyl compound (2; R = CH₃) followed by thermal elimination of methyl bromide, was treated with the Grignard reagent formed from 3-(*o*-bromophenyl)propyl triphenylmethyl ether [30; R = C(C₆H₅)₃], thus forming 1,2,3,4-tetrahydro-1-[*o*-(3′-triphenylmethoxypropyl)phenyl]arsinoline [31; R = C(C₆H₅)₃]. This compound in boiling dilute acetic acid readily lost the trityl group with the formation of the hydroxy compound (31; R = H), which when treated without purification with phosphorus tribromide at 0° gave the reactive 3-bromopropyl derivative; this underwent spontaneous cyclic quaternization to give the crystalline dihydrated spirocyclic bromide (**28**; X = Br), m.p. 253–254° (anhydrous, m.p. 270–271°). The bromide when treated with sodium iodide, both in aqueous ethanolic solution, gave the iodide, m.p. 277–278°; both gave the picrate (**28**; X = $C_6H_2N_3O_7$), m.p. 102–103°.

It is noteworthy that an attempt to distil the gummy product [**31**; R = $C(C_6H_5)_3$] under reduced pressure proved unsatisfactory; the residue, when boiled with thionyl chloride, gave the unstable 1-[o-(3′-chloropropyl)phenyl]-1,2,3,4-tetrahydro-1-hydroxyarsinolinium chloride (**33**; X = Cl), m.p. 158–161°, which with dilute nitric acid gave the

(33)

crystalline nitrate (**33**; X = NO_3), m.p. 128.5–129°. Reduction of (**33**; X = Cl) in chloroform with sulfur dioxide was immediately followed by quaternization, giving the highly soluble spirocyclic chloride (**28**; X = Cl), which was converted into the iodide (**28**; X = I). This synthetic route is inferior to the above phosphorus tribromide route.

The cation in the salts (**28**) is clearly dissymmetric by virtue of the tetrahedral configuration of the arsonium atom. The bromide (**28**; X = Br) was therefore treated with an equivalent of silver (−)-menthyl-oxyacetate[74] and the product, after 4 recrystallizations from diethyl ketone, afforded the optically pure (−)-arsonium (−)-menthyloxyacetate dihydrate (**28**; X = $C_{12}H_{21}O_3$), m.p. 88.5–90°. A 0.484% chloroform solution of this salt had $[\alpha]_D^{20}$ −48.6°, $[M]$ −288°; it gave the (−)-arsonium iodide, m.p. 223–223.5°, a 0.593% chloroform solution having $[\alpha]^{22.5}$ −30.0°, $[M]_D$ −131.5°, the rotation being unaffected after 24 hours at room temperature. The (−)-menthyloxyacetate also gave the (−)-picrate, m.p. 95–97°, a 0.757% solution having $[\alpha]^{20}$ −24.4°, $[M]$ −132°.

The diethyl ketone mother-liquor from the first recrystallization of the (−)-menthyloxyacetate was evaporated and the residue converted into the iodide, a mixture of the (±)- and the (+)-form. Careful recrystallization gave the pure (+)-arsonium iodide, m.p. 223–223.5°; a 0.655% chloroform solution had $[\alpha]^{21}$ +30.4°, $[M]$ +133°. A mixture of approximately equal amounts of the (+)- and the (−)-iodide had m.p. 220–274°.[73]

The (±)-arsonium (+)-3-bromo-8-camphorsulfonate (**28**; X = $C_{10}H_{14}BrO_4S$), m.p. 191.5–193°, gave no evidence of resolution after recrystallization from diethyl carbonate.

The significance of the above resolutions is discussed with that of the rotations of the isomeric isoarsinoline salts (pp. 417, 418).

The moderate ease with which the exocyclic As–C bond in compounds of type (**31**) is broken may well have caused the failure of attempts to synthesize a spirocyclic salt isomeric with (**28**) but having one arsinoline and one isoarsinoline ring.[65] The Grignard reagent from o-(2′-chloroethyl)benzyl methyl ether (**34**) (Holliman and Mann[75])

(**34**)   (**35**)   (**36**)

reacted with 1-chlorotetrahydroarsinoline (**2**; R = Cl) to give 1,2,3,4-tetrahydro-1-[2′-o-(methoxymethyl)phenethyl]arsinoline (**35**), b.p. 176–179°/0.19 mm, characterized as its methiodide, m.p. 114–116°. Treatment of the compound (**35**) with hydrobromic acid under various conditions, in the hope that replacement of the methoxyl group by bromine would be followed by cyclic quaternization, gave cleavage of the above As–C bond with formation of the 1-bromo-arsinoline (**2**; R = Br). Similarly, treatment of the 1-methylarsinoline (**2**; R = CH₃) with o-(2′-bromoethyl)benzyl bromide, $C_6H_4(CH_2Br)(CH_2CH_2Br)$ (p. 415), gave the quaternary bromide (**36**) as a glass which when heated at various temperatures and pressures effervesced leaving a gum which could not be identified; no indication of the desired loss of methyl bromide followed by quaternization could be obtained.

These results are in striking contrast to the union of the 1-methyl-arsinoline and o-xylylene dibromide, $C_6H_4(CH_2Br)_2$, to form 1-[o-(bromomethyl)benzyl]tetrahydro-1-methylarsinolinium bromide (**37**), a glass which when heated under nitrogen to 190° effervesced with the formation of the spirocyclic bromide (**38**), which gave the crystalline iodide, m.p. 241–243°, and picrate, m.p. 163–164°.

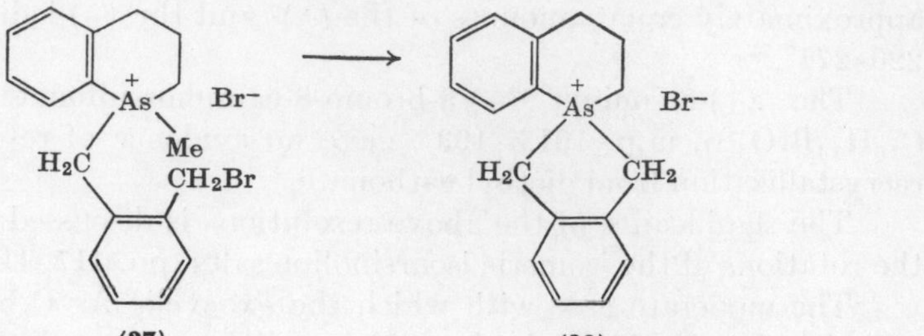

(37)   **(38)**

The 'parent' cation of the bromide (**38**) will presumably be 1,2′(2*H*)-spiro(arsinoline-1,2′-isoarsindolium) (**39**), [RIS —], and the bromide (**38**)

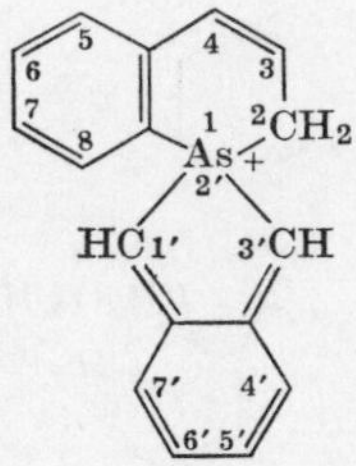

(39)

is thus named 1′,3,3′,4-tetrahydro-1,2′(2*H*)-spiro(arsinoline-1,2′-iso-arsindolium) bromide.

11*H*-Arsinolino[4,3-*b*]indoles

1,2,3,4-Tetrahydro-7-methoxy-1-methyl-4-arsinolone (**1**) gives a phenylhydrazone which when subjected to the Fischer indolization reaction gives the compound (**2**; R = H), originally termed 1,2-dihydro-7-methoxy-1-methylindolo(3′,2′:3,4)arsinoline (Mann and Wilkinson [76]).

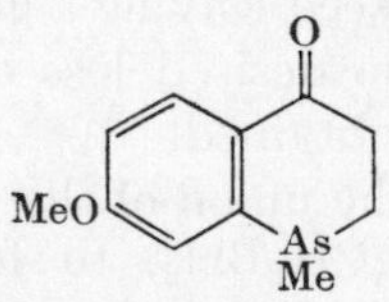

(1) 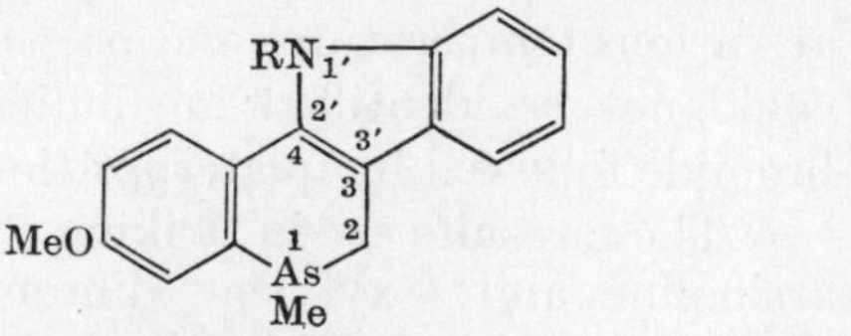   **(2)**

The Ring Index [RIS 8907] presents and numbers the parent ring system as (**3**), named 11*H*-arsinolino[4,3-*b*]indole, and the compound (**2**) should therefore be named 5,6-dihydro-3-methoxy-5-methyl-11*H*-arsinolino[4,3-*b*]indole (**4**; R = H). This cream-colored compound, m.p. 137–139°, was obtained when a solution of the phenylhydrazone of (**1**)

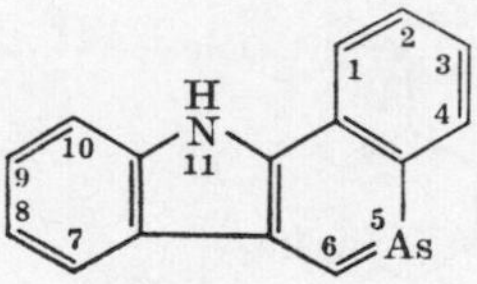

(**3**) 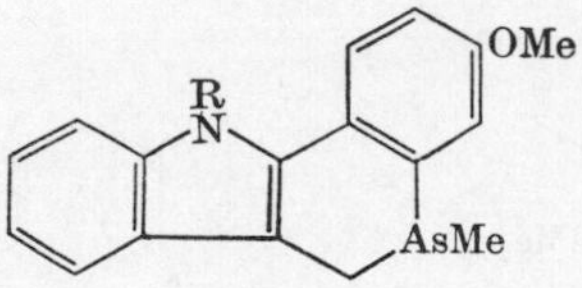   (**4**) 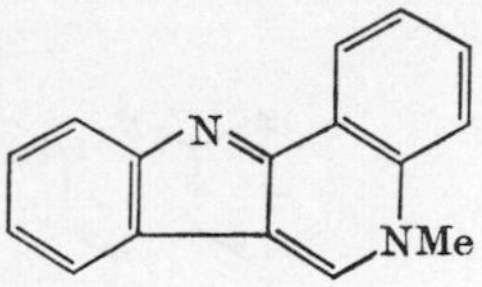   (**5**)

in acetic acid containing zinc chloride was boiled for 15 minutes. The 3-methoxy-5,11-dimethyl analogue (**4**; R = CH₃), m.p. 115°, and the 3-methoxy-5-methyl-11-phenyl analogue (**4**; R = C₆H₅), m.p. 157°, were obtained when an acetic acid solution of the arsinolone (**1**) was boiled with *N*-methyl-*N*-phenylhydrazine, H_2N—$N(CH_3)C_6H_5$, and *N*,*N*-diphenylhydrazine, respectively.

It is noteworthy that the phenylhydrazone of the 4-arsinolone (**1**), like that of the analogous 4-phosphinolone (p. 133), gives the true indole (**4**), whereas the phenylhydrazone of 1,2,3,4-tetrahydro-1-methyl-4-quinolone undergoes dehydrogenation to give a colored, strongly basic ψ-indole or 3*H*-indolenine (**5**) (Braunholtz and Mann[77]). The structure of the indole (**4**; R = H) is shown by its lack of color, its non-basic character, and its infrared spectrum, which shows a marked band at 3378 cm⁻¹ due to the NH group. The point is discussed more fully in connection with the phosphorus analogue (pp. 133, 134).

Arsinolino[4,3-*b*]quinolines

The unknown parent compound, arsinolino[4,3-*b*]quinoline [RIS 9055] has the structure (**1**). Derivatives of this compound are obtained

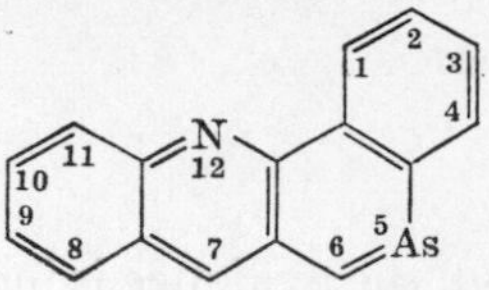

(**1**) 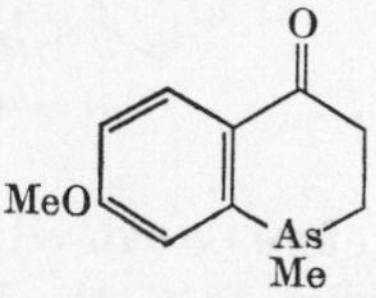   (**2**)

when 1,2,3,4-tetrahydro-7-methoxy-1-methyl-4-arsinolone (**2**) is subjected to the Friedländer and the Pfitzinger reaction (Mann and Wilkinson[76]): the product of the former reaction shows a strong resemblance to the corresponding phosphinolino[4,3-*b*]quinoline derivative (p. 134).

The 4-arsinolone (2), when condensed with o-aminobenzaldehyde in an alkaline medium, gives the colorless 5,6-dihydro-3-methoxy-5-methylarsinolino[4,3-b]quinoline (3), m.p. 136°, which with cold methyl

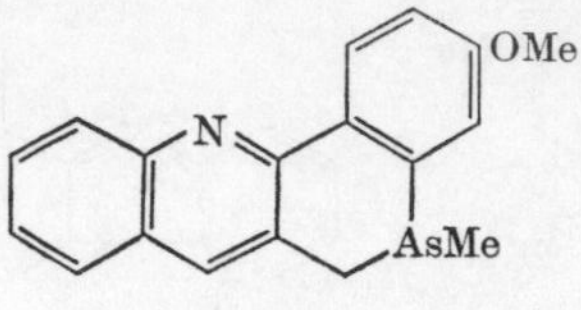

(3) 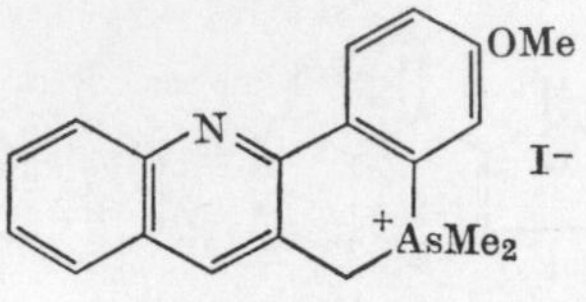   (4)

iodide gives a colorless methiodide, m.p. 225°. Since in most fused-ring systems containing a quinoline unit, the nitrogen atom of the latter accepts a proton readily, but usually undergoes quaternization only under vigorous conditions, this methiodide almost certainly has the structure (4). The compound (3), however, readily forms a yellow crystalline hydrochloride, m.p. 206°, in which protonation has occurred on the nitrogen atom (5), since tertiary arsines are neutral; the yellow

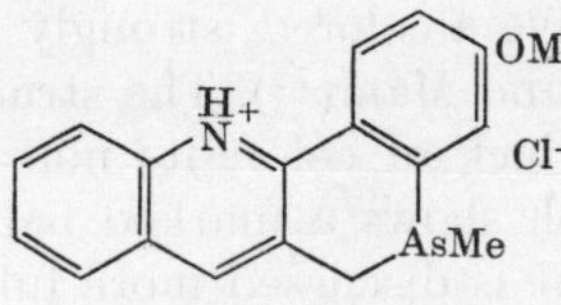

(5) 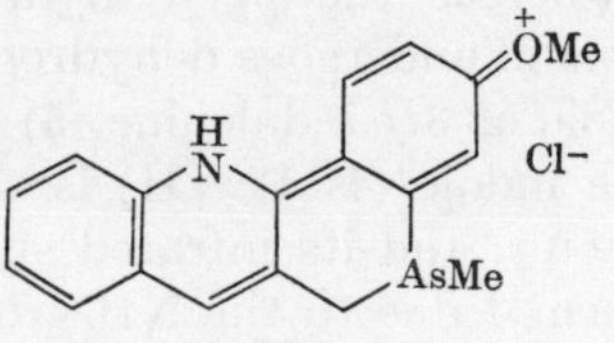   (5B)

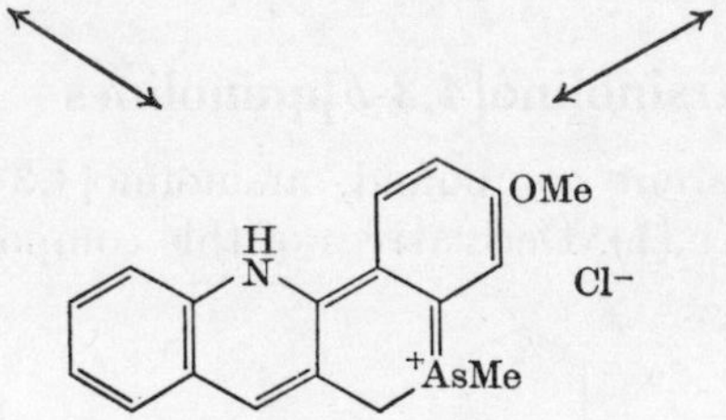

(5A)

color may therefore be caused (although the evidence is inconclusive) by some contribution from the form (5A) and even from (5B). Figure 3 shows the ultraviolet spectra of the base (3) in (A) ethanol and in (B) ethanolic hydrochloric acid. The curve (B) is generally similar to (A) but the main features are more prominent, and there is a general shift to longer wavelengths, the broad band at 385 mμ being responsible for the yellow color. The two curves are closely similar to those of the analogous 5,6-dihydro-5-phenylphosphinolino[4,3-*b*]quinoline in the same solvents: see p. 135 where their significance is more fully discussed.

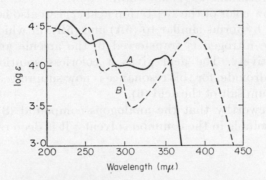

Figure 3. Ultraviolet spectra of the base (**3**) in (*A*) ethanol and in (*B*) ethanol diluted with an equal volume of 0.1N-hydrochloric acid. (Reproduced, by permission, from F. G. Mann and A. J. Wilkinson, *J. Chem. Soc.*, **1957**, 3346)

The base (**3**) when boiled with hydrochloric acid gave no indication of the allylic rearrangement which the corresponding dihydroquinolino-quinoline undergoes (cf. p. 136).

Application of the Pfitzinger reaction, in which the arsinolone (**2**) and isatin in ethanolic-aqueous potassium hydroxide were boiled under nitrogen for 48 hours, gave the highly insoluble crude yellow 5,6-dihydro-3-methoxy-5-methylarsinolino[4,3-*b*]quinoline-7-carboxylic acid (**6**); sublimation at 250°/0.0005 mm gave the pure acid (**6**), m.p.

(6)   (7)

220–250° (dec.). This proved, however, to have the zwitterion structure (**7**): its infrared spectrum showed no NH band in the 3330 cm^{-1} region, but showed a broad band at 2040 cm^{-1} attributed to the NH$^+$ group, and two bands at 1626 and 1603 cm^{-1} characteristic of the CO$_2^-$ ion in amino acids having the zwitterion structure. This volatilization without decomposition of amino acids having this structure is rare and may indicate that the zwitterion (7) reverts to the covalent state (6) in the vapor phase, returning to (7) on solidification.

The yellow color of the zwitterion acid (7) can also be attributed to contributions by forms similar to (5A) and (5B), in which the positive charge on the nitrogen is transferred to the arsenic and the oxygen atom, respectively. The acid gives a colorless solution in aqueous potassium hydroxide, for this resonance is now suppressed by formation of the potassium salt of the acid (6).[76]

It is noteworthy that the analogous compound (8) also forms a zwitterion insoluble in the common solvents; it is deep red in color and

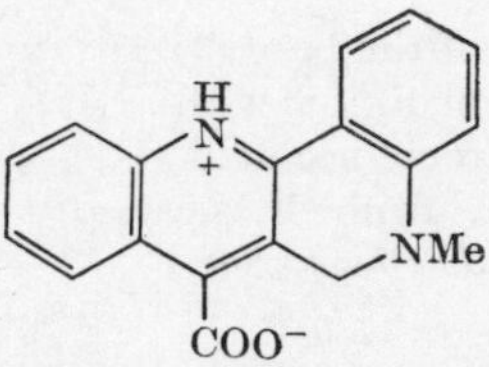

(8) 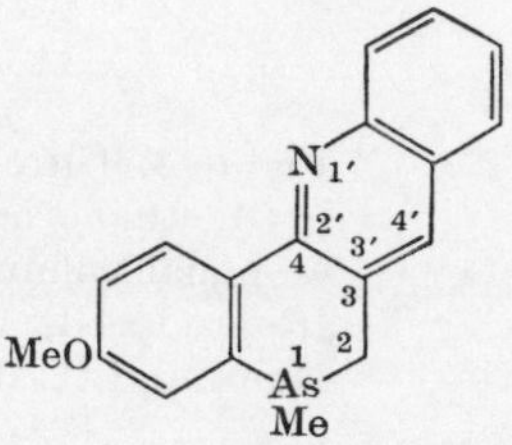   (9)

(like many other nitrogen acids of this type) readily undergoes decarboxylation on attempted sublimation (Braunholtz and Mann[77]). The acid (6) did not undergo decarboxylation even when heated with barium hydroxide or soda-lime at 180°/0.002 mm.

The base (3) was initially named 1,2-dihydro-7-methoxy-1-methyl-quinolino(3',2':3,4)arsinoline in accordance with the numbering (9).[76]

Isoarsinoline and 1,2,3,4-Tetrahydroisoarsinolines

Isoarsinoline (1), [RRI 1679], analogously to arsinoline, derives its name as being the arsenic analogue of isoquinoline. 1,2,3,4-Tetrahydro-isoarsinoline (2; R = H), in the form of its 2-alkyl and 2-aryl derivatives, has been more fully studied than its tetrahydroarsinoline isomer.

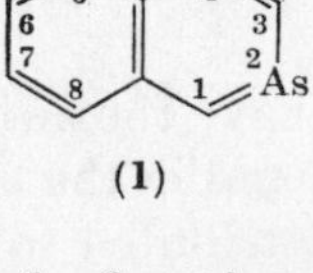

(1) 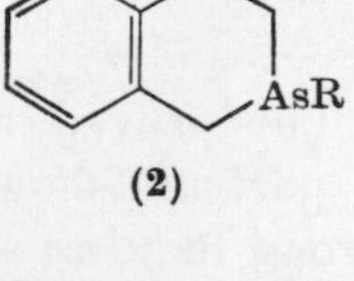   (2)

Two methods for the synthesis of compounds of type (2) are available. The first, by Holliman and Mann,[78] uses o-(bromomethyl)-

phenethyl bromide (= *o*-(2'-bromoethyl)benzyl bromide) (**3**) synthesized earlier by these workers,[79] but for which two other syntheses are now available (Anderson and Holliman[80]; Colonge and Boisde[81]). An equimolecular mixture of the dibromide (**3**) and phenylarsonous dichloride in ether was boiled under nitrogen with an excess of sodium

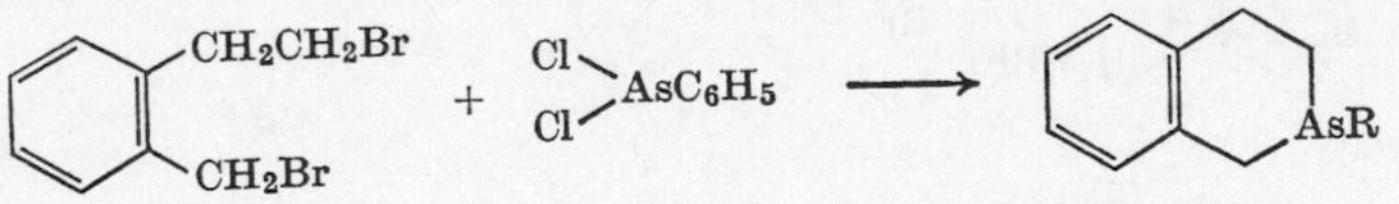

(**3**)   (**4**)

wire for several hours with the occasional addition of small quantities of ethyl acetate, giving 1,2,3,4-tetrahydro-2-phenylisoarsinoline (**4**; R = C_6H_5) (31%), a colorless liquid, b.p. 110–112°/0.01 mm and 128–130°/0.05 mm. The similar use of methylarsonous dichloride gave the 2-methyl derivative (**4**; R = CH_3) (16%), b.p. 131°/18 mm. This method is similar to that used later[24] for the preparation of 2-substituted isoarsindoles (p. 368).

In the second method (Beeby, G. H. Cookson, and Mann[26]), a benzene solution of the dibromide (**3**) was added to an ethereal solution of phenylarsinebis(magnesium bromide), $C_6H_5As(MgBr)_2$, prepared by the action of phenylmagnesium bromide on phenylarsine. This gave the crude 2-phenyl derivative (**4**; R = C_6H_5) in 70% yield, with a small yield of 2-methylphenethyl bromide (**5**) and phenylarsonous dibromide. This indicates that, although the main reaction may follow the stages (i),

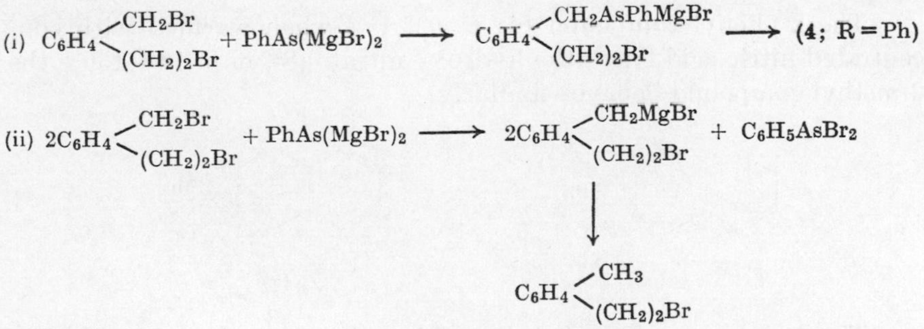

(**5**)

a minor may follow the stages (ii), the bromide (**5**) arising from the hydrolysis of the *o*-(2-bromoethyl)benzylmagnesium bromide. The mechanisms of these reactions and of the comparable action of phenylbis(magnesium bromide) with *o*-xylylene dichloride and dibromide have been discussed.[26]

A possible third method, in which an equimolecular mixture of the dibromide (**3**) and dimethylphenylarsine was heated, in the hope that an initially formed quaternary arsonium bromide (**6**) would lose methyl

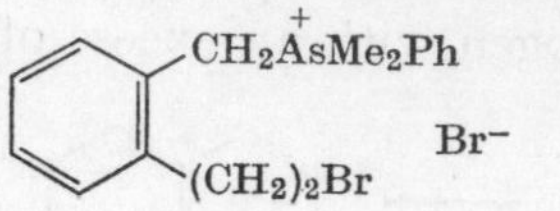

(**6**) 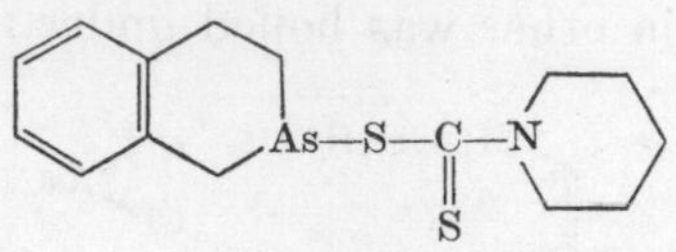   (**7**)

bromide, then undergo cyclic quaternization, and finally again lose methyl bromide to give the phenyl compound (**4**; $R = C_2H_5$), gave the latter in only *ca.* 8% yield accompanied by various by-products.[26] This synthetic method, which is so valuable in the isoarsindoline series (p. 369), is not of practical value in the present field.

The heterocyclic ring in the tetrahydroisoarsinolines also possesses considerable stability. The 2-phenyl compound (**4**; $R = C_6H_5$) when heated in constant-boiling hydriodic acid loses benzene with the formation of 1,2,3,4-tetrahydro-2-iodoarsinoline (**4**; $R = I$), a red oil,[26] and in boiling hydrobromic acid gives the 2-bromo derivative, a colorless oil (36%), b.p. 173–176°/12 mm.[65] The two compounds were characterized by treatment with hot ethanolic piperidine N-pentamethylenedithiocarbamate, which gave the crystalline 1,2,3,4-tetrahydro-2-(N-pentamethylenethiocarbamoylthio)isoarsinoline [alternative name: 1,2,3,4-tetrahydro-2-isoarsinolinyl 1-piperidinedithiocarboxylate] (**7**), m.p. 132–133°.

The 2-phenyl compound (**4**; $R = C_6H_5$) when warmed with concentrated nitric acid forms the hydroxy-nitrate (**8**), m.p. 149–150°: the 2-methyl compound behaves similarly.

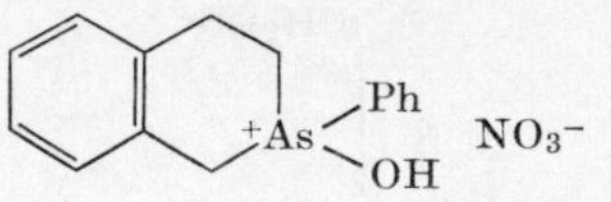

(**8**) 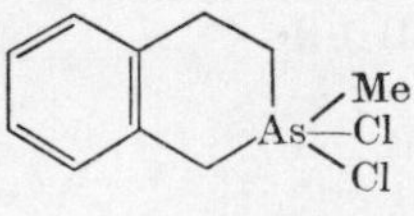   (**9**)

The 2-phenyl compound (**4**; $R = C_6H_5$) gives also a hydroxy-picrate, m.p. 116–118° (from the hydroxy-nitrate), as well as directly a methiodide, m.p. 136–137°, and a colorless sulfide, m.p. 124°; the 2-methyl analogue gives similarly a hydroxy-picrate, m.p. 164–165.5°, a methiodide, m.p. 179–181°, and a methopicrate, m.p. 163–164°.

The 2-methyl compound also behaves normally in uniting with a molecular equivalent of chlorine to give the colorless arsine dichloride

(9), which on heating at 15 mm loses methyl chloride to form 2-chloro-1,2,3,4-tetrahydroisoarsinoline (4; R = Cl), b.p. 157°/14 mm. Attempts to split off hydrogen chloride from the latter compound by boiling it with pyridine and thus obtain a dihydroisoarsinoline failed. It is noteworthy, however, that no fission of the heterocyclic ring was detected either in the preparation of 2-halogeno derivatives by the action of hydrogen halides or by the thermal decomposition of the dichloride (9), or in the above oxidation reaction.

The reason for the failure to split off hydrogen chloride from the 2-chloroisoarsinoline (4; R = Cl) and for the similar earlier failure with 1-chloroarsinoline (p. 401) is probably that the tetrahydro derivatives are much more stable than the dihydro derivatives into which, therefore, they cannot be readily converted. The heterocyclic ring in the tetrahydro compounds is almost certainly 'buckled' in order to avoid strain and so obtain maximum stability. The true arsinoline and isoarsinoline compounds, if they were known, might be found to possess an enhanced stability owing to their 'semi-aromatic' nature and to the consequent resonance. However, the intermediate dihydro derivatives possess neither of these stabilizing factors and hence are difficult to isolate.

The stereochemical investigation of derivatives of the above isoarsinolines led to novel results in two directions.

Many attempts by earlier workers had been made to resolve into optically active forms a quaternary arsonium salt of type [abcdAs]X, where a, b, c, and d represent unlike groups and X is an acid radical. Partial success had been obtained only by Burrows and Turner,[64] and by Kamai,[82] but in each case these workers had succeeded solely in isolating an arsonium iodide of very low rotation, which, moreover, racemized rapidly in solution. Burrows and Turner attributed this failure to the fact that such quaternary arsonium iodides give rise in organic solvents to a 'dissociation-equilibrium' of type [abcdAs]I ⇌ abcAs + dI. This equilibrium was expected to cause rapid racemization and, since moreover all such arsonium salts investigated possessed (for synthetic reasons) at least one alkyl group attached to the arsenic atom, the formation of such a dissociation-equilibrium was always possible.

The formation of a dissociation equilibrium was originally suggested by Pope and Harvey[83] to explain why their optically active quaternary ammonium salts, [abcdN]X, underwent rapid racemization in chloroform solution when X was an iodide ion, but were optically stable when X was a nitrate or sulfate ion. The suggestion received further support in the quaternary arsonium field when Burrows and Turner[64] showed that dimethylphenylarsine when treated with an excess of ethyl iodide furnished not only the expected ethyldimethylphenylarsonium iodide but also trimethylphenylarsonium iodide.

The optical stability of many dissymmetric tertiary arsines and quaternary arsonium salts has recently been demonstrated by Horner and Fuchs,[43] but in 1943 Holliman and Mann,[84] having the structural stability of the arsonium cation in mind, united the very stable 1,2,3,4-tetrahydro-2-phenylisoarsinoline (4; $R = C_6H_5$) with the strongly quaternizing p-chlorophenacyl bromide to form 2-(p-chlorophenacyl)-1,2,3,4-tetrahydro-2-phenylisoarsinolinium bromide (10; X = Br), m.p. 190–191°. This salt, which contains an asymmetric arsenic atom, was converted into the (+)-3-bromo-8-camphorsulfonate (10; X = $C_{10}H_{14}BrO_4S$), which on fractional recrystallization gave the optically

(10)

pure (−)-arsonium (+)-sulfonate: this was then converted into the (−)-2-p-chlorophenacyl-1,2,3,4-tetrahydro-2-phenylisoarsinolinium iodide (10; R = I), $[M]_D$ −354° in chloroform solution, and into the corresponding picrate, $[M]_D$ −450° in chloroform. Repetition of the resolution using the (−)-3-bromo-8-camphorsulfonate similarly gave the (+)-picrate, $[M]_D$ +457°, also in chloroform. A solution of the iodide, $[M]_D$ −354° in chloroform, showed no perceptible racemization at room temperature for several days: hence no dissociation-equilibrium could have occurred, although such equilibria were considered to occur more readily in chloroform than in any other common solvent.

The iodide (10; R = I) was therefore the first known optically stable quaternary arsonium salt.

Spiro Arsonium Salts

In a second direction, notable results were obtained by applying Lyon and Mann's spirocyclic arsonium salt synthesis,[24] to the corresponding isoarsinoline derivatives. For this purpose, Holliman and Mann[28] combined 1,2,3,4-tetrahydro-2-methylisoarsinoline (11) with an equimolecular quantity of the dibromide (3) to form the arsonium salt (12). When this salt was heated at 14 mm pressure, with the temperature ultimately rising to 200°, it lost methyl bromide with the formation of the spirocyclic salt, originally named As-spiro-bis-1,2,3,4-tetrahydroisoarsinolinium bromide (13; X = Br). The cation in this salt should now be named after the 'parent' cation (14), [RRI 5047], which is termed 2,2′[1H,1′H]-spirobi[isoarsinolinium] and depicted as shown; hence the salt (**13**; X = Br) is 3,3′,4,4′-tetrahydro-2,2′[1H,1′H]-spirobi-[isoarsinolinium] bromide.

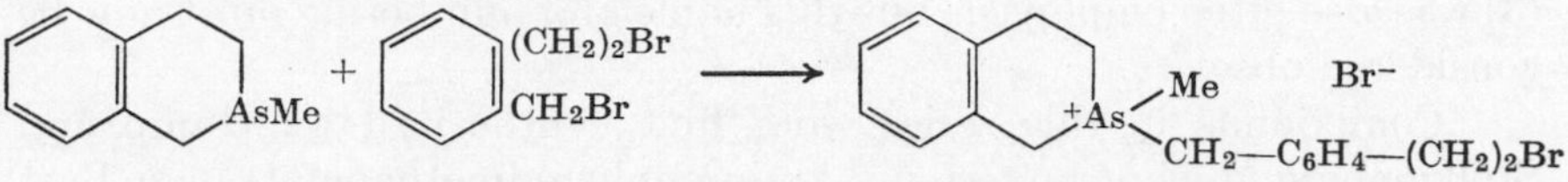

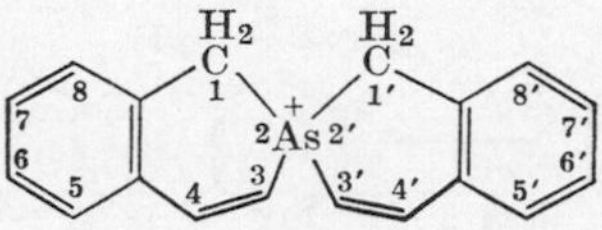

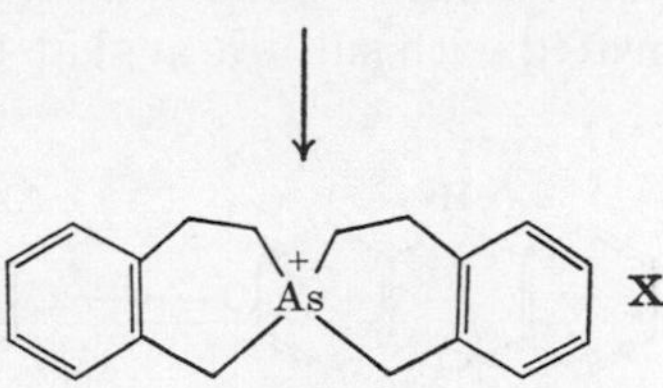

This bromide was converted into the corresponding (+)-3-bromo-8-camphorsulfonate (**13**; X = $C_{10}H_{14}BrO_4S$), which on recrystallization from water ultimately furnished the optically pure (−)-isoarsinolinium (+)-sulfonate. This compound was converted into the (−)-isoarsinolinium iodide (**13**; X = I), which had $[M]_D$ −344° in chloroform solution. The resolution was then repeated using the (−)-bromocamphorsulfonate, and the optically pure (+)-iodide, having $[M]_D$ +342° in chloroform, thus isolated. No racemization of the (+)- and (−)-iodides in chloroform solution at room temperature could be detected over a period of several days.

Historically the bromide (**13**; X = Br) was the first spirocyclic quaternary arsonium salt containing only carbon and the arsenic atom in the cation to be resolved into optically active forms.

Acridarsines

The fundamental compound in this series is the unknown acridarsine (**1**) [RRI 3454], and most compounds are 5-substituted derivatives of 5,10-dihydroacridarsine (**2**; R = H). Although acridarsine was the

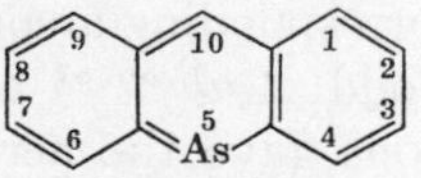

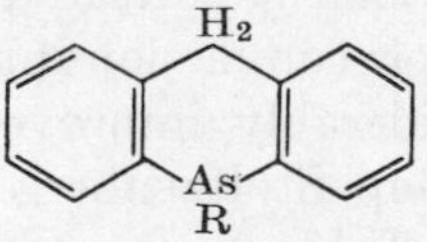

original name for the system (**1**), it was subsequently frequently termed arsacridine to denote the arsenic analogue of acridine, and the numbering

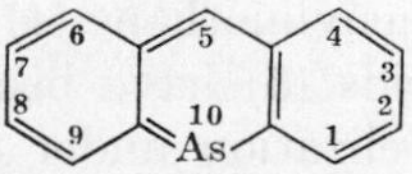

(3) was also often employed; but this name and numbering must now be considered obsolete.

Compounds in this series were first synthesized by Gump and Stoltzenberg,[85] who converted 2-aminodiphenylmethane (4) by a Bart reaction into o-benzylphenylarsonic acid (5), m.p. 161–162°, which when heated with sulfuric acid at 100° underwent a smooth cyclization to give

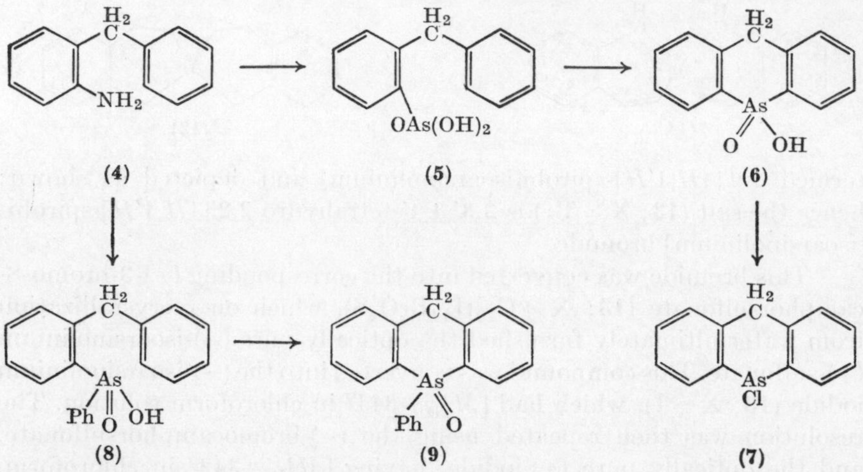

acridarsinic acid (6), m.p. 235–236°. This acid (systematically termed 5,10-dihydro-5-hydroxyacridarsine 5-oxide, but also in the past termed arsacridinic acid), when reduced with sulfur dioxide in the presence of hydrochloric acid, gave 5-chloro-5,10-dihydroacridarsine (7), m.p. 114–115°. The chloro group in the compound (7) can then readily be replaced by an alkyl or aryl group by using a Grignard or lithium reagent.

The above synthesis may be modified by using, for example, oxophenylarsine, C_6H_5AsO, in the Bart reaction, thus converting the amine (4) into the arsinic acid (8; R = C_6H_5), which is similarly cyclized to the tertiary arsine oxide (9; R = C_6H_5); reduction with sulfur dioxide then gives 5,10-dihydro-5-phenylacridarsine.

The synthesis of various substituted derivatives of 2-aminodiphenylmethane (4) and their conversion into the corresponding arsonic acids (5) have been considerably improved by Todd et al.,[86, 87] and cyclization under the Gump–Stoltzenberg conditions gave the acridarsine derivatives listed in Table 2.

1-(p-Carboxyphenyl)-5,10-dihydro-3-methylacridarsine (10) has been prepared by the above modified route and resolved into optically active forms, having $[\alpha]_D$ +84.1° and −83.9° in methanol (Mislow,

Table 2. Some acridarsine derivatives

	M.p. (°)
2-Methylacridarsinic acid	184 (dec.)
2-Chloroacridarsinic acid	210–212 (dec.)
5-Cyano-5,10-dihydroacridarsine	114–115
5-Chloro-5,10-dihydro-3-methylacridarsine	65.5–66.5
5-Chloro-5,10-dihydro-2-methylacridarsine	87
2,5-Dichloro-5,10-dihydroacridarsine	116
2-Chloro-5-cyano-5,10-dihydroacridarsine	113–114

Zimmerman, and Melillo[88]); the dissymmetry of this compound is attributed to the configurational stability of the pyramidal (tertiary) arsenic atom.

In view of the ease with which the 5,10-dihydroacridarsine ring system can be synthesized, the preparation of the corresponding 5-substituted $10(5H)$-acridarsone system (11) resisted earlier attempts.

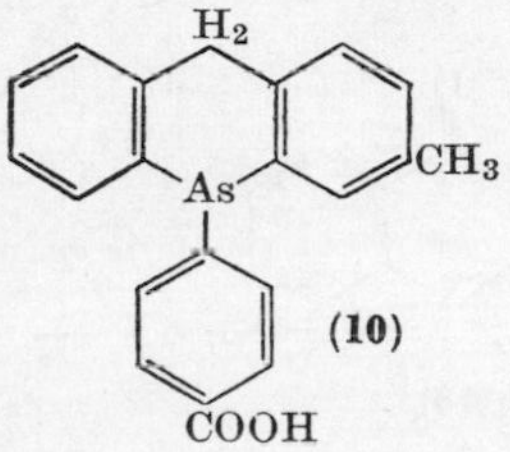

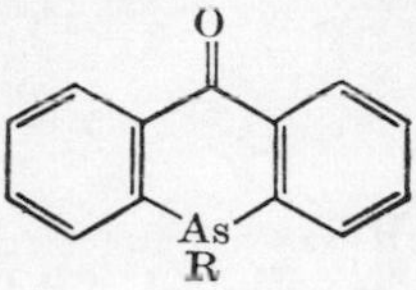

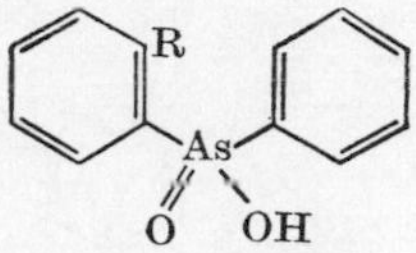

Aeschlimann and McCleland[89] failed to prepare the acridarsone (11; R = Cl), and Sakellarios[90] failed to obtain the corresponding arsinic acid by the required cyclization of the unstable 2-carboxydiphenylarsinic acid (12; R = COOH) (cf. p. 424).

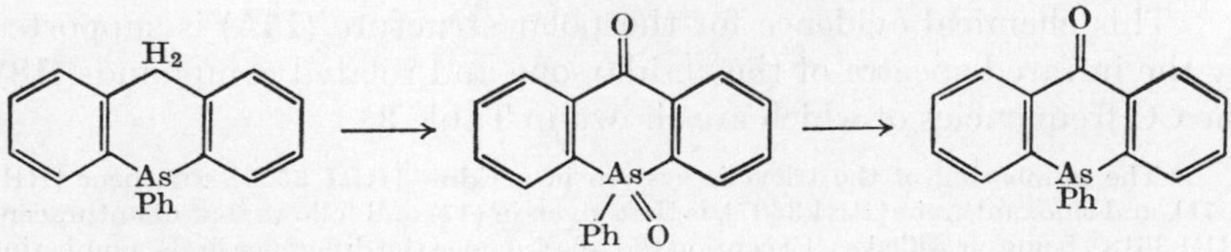

When, however, 5,10-dihydro-5-phenylacridarsine (13) was briefly treated with hot aqueous permanganate, it gave 5-phenyl-10-acridarsone 5-oxide (14), m.p. 238°, which when reduced in hydrochloric acid by sulfur dioxide gave the colorless 5-phenyl-10-acridarsone (**15**), m.p. 138–139°, the yield at each stage being almost theoretical (Jones and Mann[91]).

The structure of the acridarsone (**15**) is of considerable interest. It shows very few of the normal properties of a ketone or of a tertiary arsine. Thus it does not react with hydroxylamine, hydrazine, phenylhydrazine, or 2,4-dinitrophenylhydrazine; the arsine group does not apparently react with boiling methyl iodide, and when heated with methyl *p*-toluene sulfonate at 200° gives only a small yield of the quaternary metho-*p*-toluenesulfonate. Whereas most tertiary arsines undergo a ready exothermic oxidation with cold acetone–hydrogen peroxide, the acridarsone (**15**) requires treatment with the boiling reagent for reconversion into the 5-oxide (**14**). The acridarsone (**15**) does react with

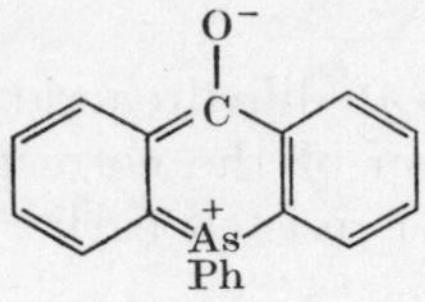

(15A)

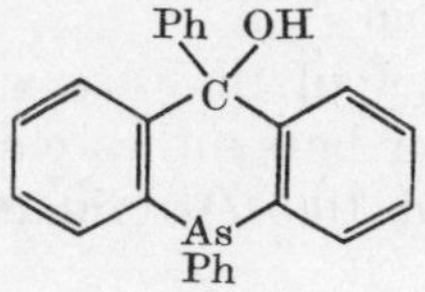

(16)

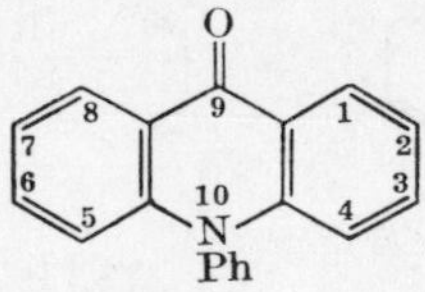

(17)

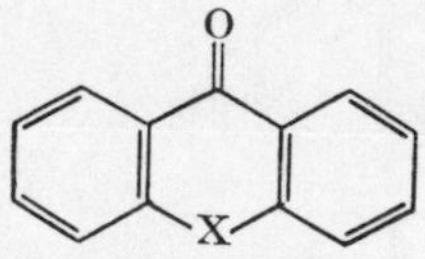

(18)

potassium tetrabromopalladite, $K_2[PdBr_4]$, to give the covalent bis(acridarsone)dibromopalladium, $(C_{19}H_{13}AsO)_2PdBr_2$, but the highly insoluble nature of this compound might promote its formation.

These chemical properties point strongly to a major contribution to the structure of the molecule from the polar form (**15A**), in which it resembles the highly inert analogous 10-phenylacridone (**17**)*. Phenylmagnesium bromide reacts with the acridarsone (**15**) to form the 5,10-dihydro-10-hydroxy-5,10-diphenylacridarsine (**16**), but this attack by the $C_6H_5^-$ ion is shown by various otherwise inert acridones.

This chemical evidence for the polar structure (**15A**) is supported by the infrared spectra of the acridarsone and related compounds (**18**), the CO frequencies of which are shown in Table 3.

* The numbering of the tricyclic system in acridine [RRI 3523], xanthene [RRI 3571], and thioxanthrene [RRI 3607], is that given in (**17**) and follows that of anthracene [RRI 3618], being classified as 'Exception to Rules'; hence the difference in the numbering of acridine and acridarsine.

Table 3. Infrared CO frequencies of some compounds (**18**)

Compound	X in (**18**)	CO frequency (cm^{-1})
10-Phenyl-9-anthrone	$CH(C_6H_5)$	1662
10,10-Diphenyl-9-anthrone	$C(C_6H_5)_2$	1665
Xanthone	O	1660
Thioxanthone	S	1645
10-Phenylacridone (**17**)	NC_6H_5	1632
The acridarsone (**15, 15A**)	AsC_6H_5	1647
The acridarsone oxide (**14**)	$AsO(C_6H_5)$	1662

The CO frequency of the acridarsone (**15, 15A**) is lower by *ca.* 15 cm^{-1} than the normal frequency in comparable anthrones, and that of 10-phenylacridone (**17**) is even lower; in the acridarsone oxide (**14**), in

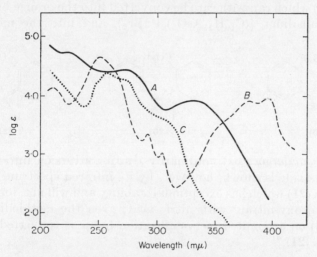

Figure 4. Ultraviolet absorption spectra: (*A*) 5-phenyl-acridarsone (**15, 15A**), (*B*) 10-phenylacridone, and (*C*) anthrone, taken from R. M. Jones.[92] (Reproduced, by permission, from Emrys R. H. Jones and F. G. Mann, *J. Chem. Soc.*, **1958**, 294)

which the above charge separation cannot occur, the CO frequency is restored to its normal value. Moreover, the CO frequency of xanthone is only very slightly lower than that of the two anthrones, but that of thioxanthone, in which greater charge separation might well be expected, falls to almost exactly that of the acridarsone.

The ultraviolet absorption spectra of the acridarsone (**15, 15A**) and the comparable 10-phenylacridone (Figure 4) are not greatly unlike, but

by comparison with that of anthrone[92] the region of large absorption
due to the electronic change introduced by the Group V element has
undergone a greater shift to longer wavelength in the acridone than in
the acridarsone; this indicates a greater degree of conjugation in the
acridone than in the acridarsone and explains the yellow color of the
former and the colorless appearance of the latter.

There is thus considerable chemical and physical evidence that the
heterocyclic ring in 5-phenyl-10-acridarsone has the 'semi-aromatic'
structure (15A). This type of structure almost undoubtedly exists in the
5,10-disubstituted arsanthronium ions (pp. 268–270), and there is some
evidence for this structure in the compounds claimed to be arsanthrene
(p. 462) and phenarsazine (p. 509).

It is noteworthy that the failure to cyclize compounds of the type
(12) may be related to the particular properties of 2-carboxytriphenyl-
arsine (19) which can with care be converted into the orange bis(arsine)-
dibromopalladium, $[(C_{19}H_{15}AsO_2)_2PdBr_2]$, and into the methiodide,

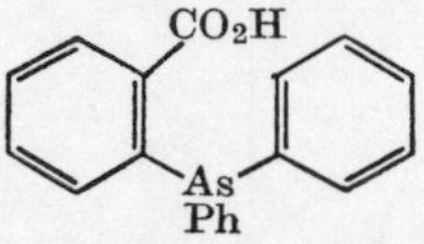

(19) 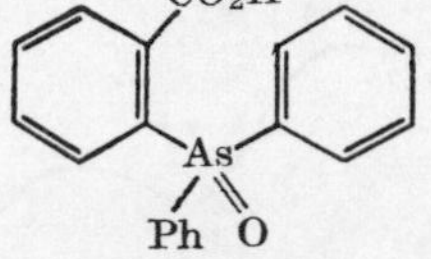   (20) 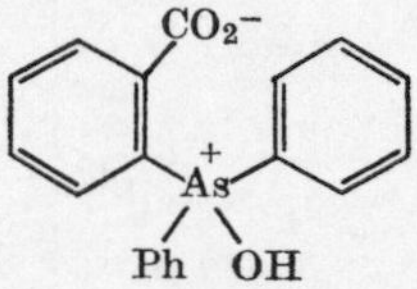   (21)

but which undergoes extraordinarily ready conversion into the oxide
(20). This oxide is shown, however, by its infrared spectrum to be the
zwitterion (21) which has exceptional stability and will not, for example,
give a hydroxy-nitrate with nitric acid; even the methiodide of the
arsine (19) on attempted recrystallization in air is converted into this
zwitterion (21).[91]

Arsanthridins

Just as the acridarsines derive their name from being the arsenic
analogues of the acridines, the arsanthridins—of which (1) is the fundamental compound [RRI 3455]—derive their name from being the similar

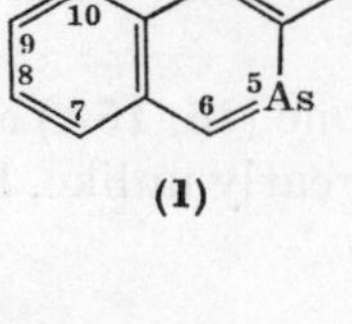

(1) 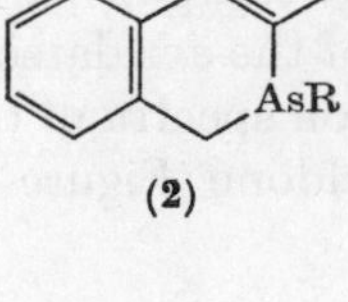   (2) 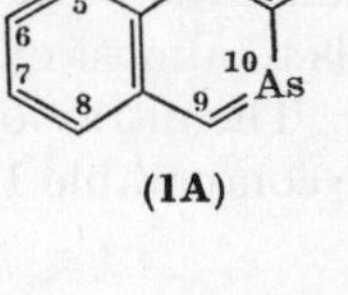   (1A)

analogues of the phenanthridines. The compound (1) has not been isolated, and members of this series are all derivatives of 5,6-dihydro-arsanthridin (2; R = H). The numbering (1A) was initially employed for the arsanthridin system.

Two somewhat similar synthetic routes have been developed to give compounds of type (2) (G. H. Cookson and Mann [93]).

In the first method, o-phenylbenzyl chloride (3) was converted into a Grignard reagent, which reacted with dimethylarsinous iodide, $(CH_3)_2AsI$, to give dimethyl-(o-phenylbenzyl)arsine (4), b.p. 124–126°/ 0.4 mm (60%), characterized by the methiodide, m.p. 232–235°, and the orange bis(arsine)dichloropalladium, $(C_{15}H_{27}As)_2PdCl_2$, m.p. 150°. The arsine (4) was united with chlorine to give the colorless crystalline arsine

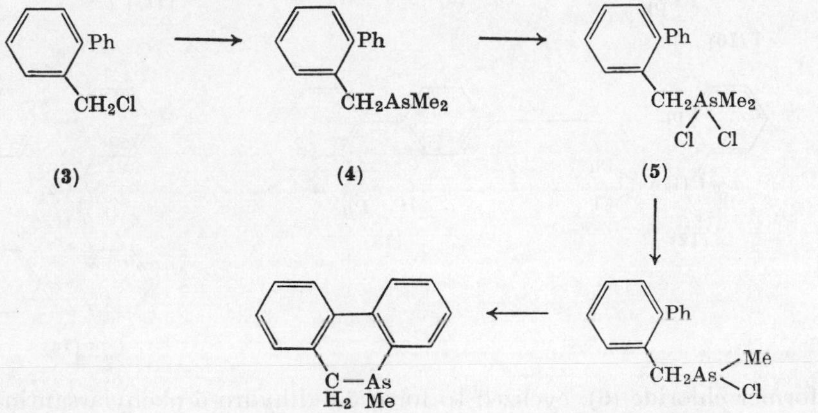

dichloride (5), which when heated at 150°/30 mm gave the methyl-(o-phenylbenzyl)arsinous chloride (6). This crude chloride, when treated in carbon disulfide solution with aluminum chloride, underwent cyclization to furnish 5,6-dihydro-5-methylarsanthridin (7), which was isolated as the methiodide, i.e., 5,6-dihydro-5,5-dimethylarsanthridinium iodide, m.p. 212–215°.

In the second method, o-phenylbenzyl bromide (8) was treated with phenylarsinebis(magnesium bromide), $C_6H_5As(MgBr)_2$ (pp. 229, 415) giving phenylbis-(o-phenylbenzyl)arsine (9), m.p. 70–77°, characterized as the bis(arsine)dichloropalladium, $(C_{32}H_{27}As)_2PdCl_2$, m.p. 188–190°, and the mercurichloride, $(C_{32}H_{27}As)_2(HgCl_2)_2$, m.p. 172–173°. Alternatively, the bromide (8) with one equivalent of dimethylphenylarsine gave the quaternary bromide (10), which when heated gave the arsine (9) and a small proportion of 1,2-di-(2'-biphenylyl)ethane (14).

15

The arsine (9) gave the dichloride (11) which when briefly heated at 150°/0.2 mm gave o-phenylbenzyl chloride and phenyl-(o-phenylbenzyl)arsinous chloride (12). This compound, when treated as the

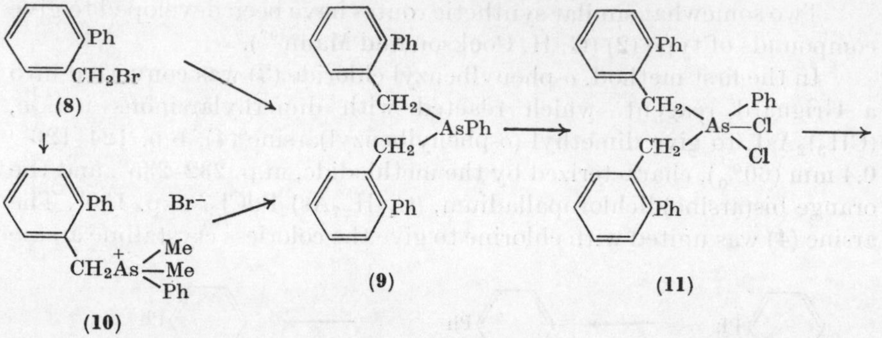

former chloride (6), cyclized to form 5,6-dihydro-5-phenylarsanthridin (13) in 15% yield based on the arsine (9). The arsanthridin (13) formed a bis(arsine)dichloropalladium, m.p. 244–245°, an analogous dibromo derivative, also m.p. 244–245°, and a methiodide, 5,6-dihydro-5-methyl-5-phenylarsanthridinium iodide. This iodide occurred in two forms; the first form, yellow leaflets, when scratched or crushed changing rapidly to a cream-colored form, m.p. 195°, which was the only form subsequently obtained.

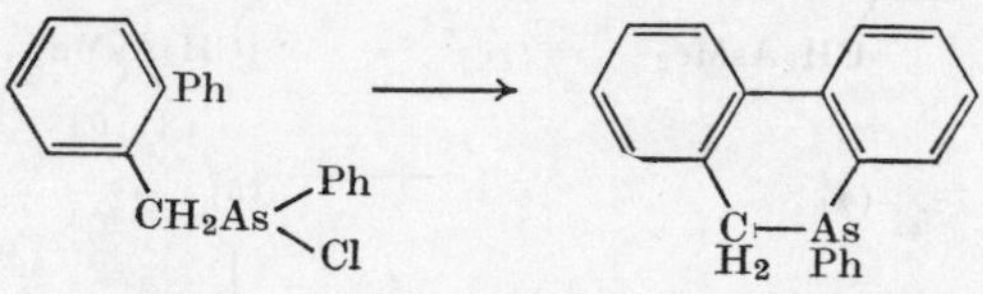

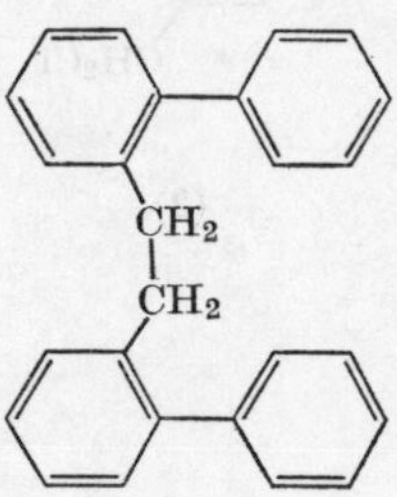

A scale diagram (Figure 5) of the 5,6-dihydroarsanthridin ring system indicates that the tricyclic system cannot be flat, and that the two o-phenylene units of the biphenylene group must be twisted about their common axis to accommodate the methylene and the arsine groups. The angle of twist between these two units, calculated on the accepted bond lengths and atomic radii, is ca. 34°. [It is noteworthy that in the compound (15), which has a structure generally analogous to that of the dihydroarsanthridins, X-ray analysis shows that the angle between the two benzene rings is 22° (p. 147)[94]]. If the dihydroarsanthridin

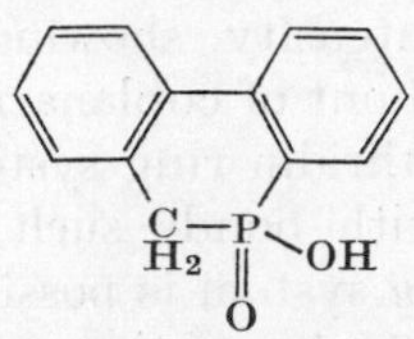

(15)

molecule has this angle of *ca.* 34°, its ultraviolet absorption spectrum should show possibly small but significant differences from the characteristic absorption band of biphenyl at 252 mμ (ϵ_{max} 17,000).

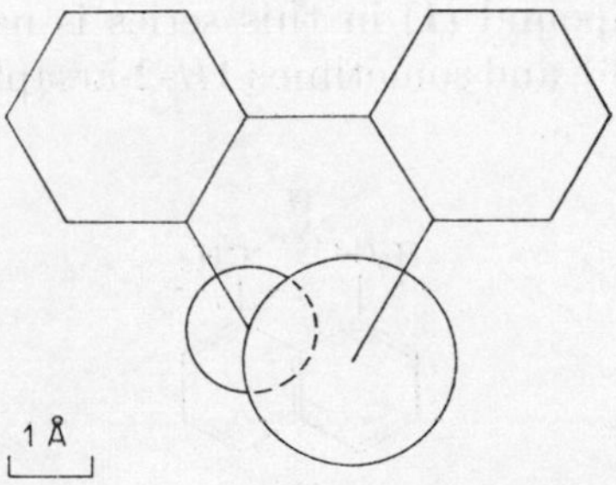

1 Å

Figure 5. Scale diagram of the 5,6-dihydroarsanthridin ring system. (The large and the small circles represent respectively the arsenic atom and the carbon atom of the methylene group.) (Reproduced, by permission, from G. H. Cookson and F. G. Mann, *J. Chem. Soc.*, 1949, 2888)

The absorption spectra of the methiodides of the methylarsanthridin (7) and the phenylarsanthridin (13) are very similar and show only one band, which is at 268 mμ (ϵ_{max} 11,100) and 269 mμ (ϵ_{max} 9,600), respectively. For comparison the spectrum of trimethyl-(*o*-phenylbenzyl)arsonium iodide, *i.e.*, the methiodide of the arsine (4), shows no indication of the biphenyl band, and restricted rotation is presumably complete, whereas the spectrum of dimethylphenyl-(*p*-phenylbenzyl)-arsonium iodide [the *para* cationic isomer of the highly deliquescent *ortho*-bromide (10) and -iodide], in which restricted rotation of the biphenyl unit cannot occur, shows one strong band at 261 mμ (ϵ_{max} 26,700).

The absorption bands at 268 and 269 mμ of the above arsanthridinium salts are undoubtedly biphenyl absorption bands slightly displaced and reduced in intensity, showing that the two *o*-phenylene units are somewhat twisted out of coplanarity. It is noteworthy that a model of the dihydroarsanthridin ring system, constructed with fixed intervalency angles but with bonds such that (when the structure permits) 'flexing' of the ring system is possible, does allow—with some intermediate resistance—a flexing of the system about the two extreme positions through the planar conformation. There is, however, no decisive evidence that this flexing occurs.

1*H*-Naphth[1,8-*cd*]arsenins

The parent compound (**1**) in this series is named 1*H*-naphth[1,8-*cd*]arsenin, [RRI 3456], and sometimes 1*H*-2-arsaphenalene; it is known,

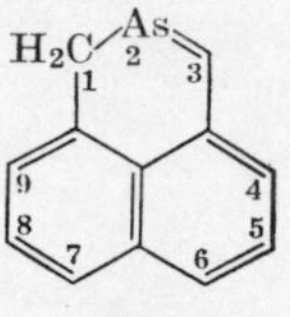

(1) 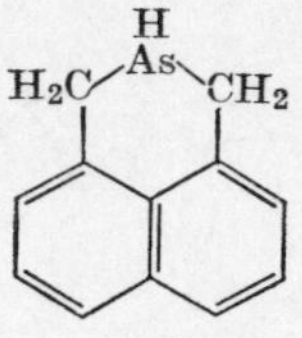   (2) 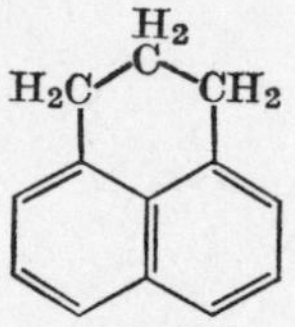   (3)

however, only as the 2-substituted derivatives of 2,3-dihydro-1*H*-naphth[1,8-*cd*]arsenin (**2**), which were earlier known as arsaperinaphthenes after the hydrocarbon perinaphthene (**3**) which has the same numbering.

1,8-Bis(bromomethyl)naphthalene (**4**) reacts with phenylarsinebis-(magnesium bromide), $C_6H_5As(MgBr)_2$, to give 2,3-dihydro-2-phenyl-1*H*-naphth[1,8-*cd*]arsenin (**5**; R = C_6H_5), as an oil (60%) which could

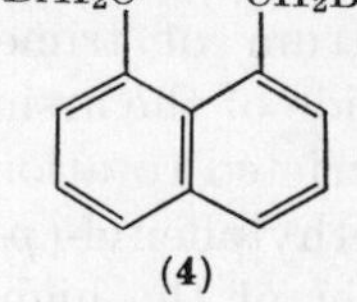

(4) 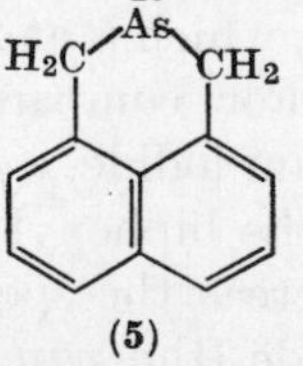   (5)

not be obtained crystalline and did not distil below 180°/0.1 mm (Beeby, G. H. Cookson, and Mann[26]). It was identified (*a*) by treatment with aqueous-ethanolic potassium chloropalladite, which furnished the orange microcrystalline bis(arsine)dichloropalladium, $(C_{18}H_{15}As)_2PdCl_2$, and

(b) by conversion into the methiodide, *i.e.*, 2,3-dihydro-2-methyl-2-phenyl-1*H*-naphth[1,8-*cd*]arseninium iodide, m.p. 196–199°.

The stability of the heterocyclic ring in compound (**5**; R = Ph) is shown by its conversion in boiling hydriodic acid into a pale yellow 2-iodo derivative (**5**; R − I), m.p. 117–119°, from which other 2-alkyl or 2-aryl derivatives could clearly be obtained by the action of appropriate Grignard or lithium derivatives.

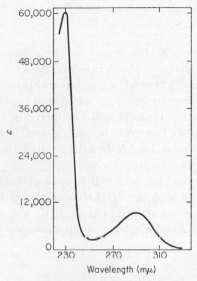

Figure 6. Ultraviolet absorption spectrum of the methiodide of 2,3 - dihydro - 2 - phenyl - 1*H* - naphth[1,8-*cd*]arsenin (**5**; R = C_6H_5). (Reproduced, by permission, from M. H. Beeby, G. H. Cookson, and F. G. Mann, *J. Chem. Soc.*, **1950**, 1917)

The absorption spectrum of the methiodide (Figure 6) is almost identical with that of the 2-iodo derivative, which is therefore not depicted. The virtual identity of the two spectra shows that this absorption is not affected by the valency of the arsenic atom or (in this case) by the nature of the groups attached to this atom. There is in fact very little doubt that the two marked bands in this spectrum are the two typical bands of naphthalene, but displaced to wavelengths longer by 10–12 mμ.

1H,7H-Benz[ij]arsinolizines

The bridgehead arsenic compound (**1**) was originally termed arsuline, its dihydro derivative (**2**) arsuloline, and its tetrahydro

(**1**)   (**2**)   (**3**)

derivative arsulolidine (**3**), these names being based on those of their nitrogen analogues, namely juline, juloline, and julolidine, respectively (Mann and Wilkinson[68]). Since the double bonds in the two heterocyclic rings of (**1**) could move to the 2,3- and the 4,5-positions, there could be two isomers of the arsuline (**1**) and similarly one of the arsuloline (**2**).

These names must, however, be abandoned, for the 'parent' compound (**4**) is now termed 1H,7H-benz[ij]arsinolizine [RIS 8574], with the numbering and presentation shown.

This nomenclature can be readily applied to the isomeric forms, for the compounds (**1**) and (**5**) are 3H,5H- and 1H,5H-benz[ij]arsinolizines respectively, and the compounds (**2**) and (**6**) are 5,6-dihydro-3H,7H- and 5,6-dihydro-1H,7H-benz[ij]arsinolizines respectively.

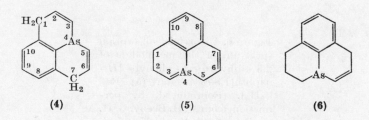

(**4**)   (**5**)   (**6**)

Of these various compounds, the 2,3,5,6-tetrahydro-1H,7H-benz-[ij]arsinolizine (**3**) is the only known member, isolated solely as substituted derivatives. The synthesis of the ring system is similar to that of the 1,2,3,4-tetrahydro-4-arsinolones (p. 403). m-Methoxyphenylarsine (**7**), when heated under nitrogen with an excess of acrylonitrile, gave di-(2-cyanoethyl)-(m-methoxyphenyl)arsine (**8**; R = CN), which on alkaline hydrolysis furnished di-(2-carboxyethyl)-m-methoxyphenyl-arsine (**8**; R = CO_2H), characterized as the methiodide, m.p. 162°, and as the bis-S-benzylthiouronium salt, m.p. 134°. Under the activation of the benzene ring by the methoxyl group, the dicyclization of the acid in maximum yield (11%) was obtained by boiling its solution in toluene with phosphoric anhydride and 'Hyflo-Supercel' for only 15 minutes,

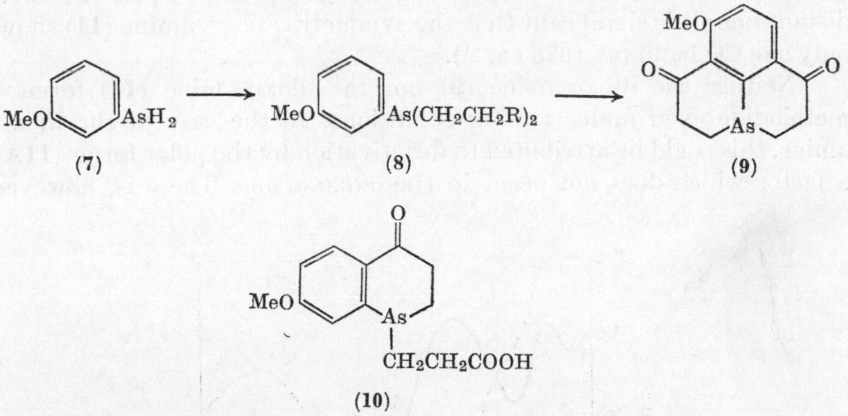

(7)   (8)   (9)

(10)

giving the crystalline colorless 2,3,5,6-tetrahydro-8-methoxy-1,7-dioxo-1H,7H-benz[ij]arsinolizine (9), m.p. 150°, the first 'bridgehead' arsenic–carbon system to be isolated.

The acid (8; R = CO$_2$H), when heated with a phosphoric acid–phosphoric anhydride mixture, gave the compound (9) in lower yield, and a small yield of the crude intermediate 1-(2'-carboxyethyl)-1,2,3,4-tetrahydro-7-methoxy-4-oxo-1-arsinolinpropionic acid (10).

The series of reactions (7) → (8) → (9) is precisely similar to the di-cyanoethylation of aniline and, after hydrolysis, the dicyclization of the acid to the yellow 1,6-dioxojulolidine (11) (Braunholtz and Mann[95,96]), now termed 2,3,6,7-tetrahydro-1,7-dioxo-1H,5H-benzo[ij]quinolizine [RRI 3493]. The marked difference in the color of the compounds

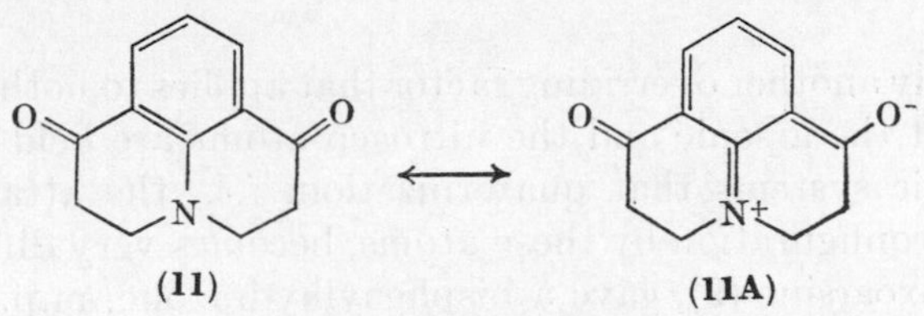

(11)   (11A)

(9) and (11) is undoubtedly due to an appreciable contribution from the polar form (11A) to the structure of (11), whereas there is no evidence of any similar charge separation in the dioxoarsine (9). This essential difference in their structures is indicated by the marked difference in their ultraviolet absorption spectra (Figure 7).

The infrared spectrum of the dioxoarsine (9) shows two CO bands, at 1709 and 1672 cm^{-1}, due undoubtedly to the greater influence of the methoxyl group on the neighboring CO group than upon the more distant one; it is significant that the symmetric dioxo-amine (11) shows only one CO band (at 1675 cm^{-1}).

Neither the dioxo-arsine (9) nor the dioxo-amine (11) forms a methiodide even under forcing conditions. In the case of the dioxo-amine, this could be attributed to deactivation by the polar forms (11A), a factor which does not occur in the dioxo-arsine. There is, however,

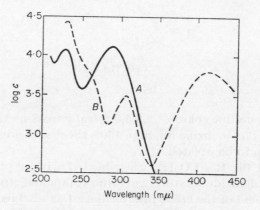

Figure 7. Ultraviolet absorption spectra of (A) 2,3,5,6-tetrahydro-8-methoxy-1,7-dioxo-1H,7H-benz[ij]arsinolizine (9) and (B) 2,3,6,7-tetrahydro-1,7-dioxo-1H,5H-benzo[ij]quinolizine (11), also termed 1,6-dioxojujolidine. (Reproduced, by permission, from F. G. Mann and A. J. Wilkinson, J. Chem. Soc., 1957, 3336)

very probably another overriding factor that applies to both compounds, namely, that the arsenic and the nitrogen atoms are held so rigidly in these dicyclic systems that quaternization, i.e., the attainment of a tetrahedral configuration by these atoms, becomes very difficult.

The dioxoarsine (9), gave a bisphenylhydrazone, m.p. 169–171°, a bis-N-methylphenylhydrazone, m.p. 160°, a gummy bis-N-diphenyl-hydrazone, a mono- and a di-2,4-dinitrophenylhydrazone, m.p.s 125° and 140°, respectively, and a monooxime, m.p. 181°.

No satisfactory product could be obtained by the condensation of p-dimethylaminobenzaldehyde or p-nitrosodimethylaniline with the 2-methylene group of the dioxo-arsine (9). The application of the Fischer indolization and the Friedländer reactions are noted in the following Section.

4H-Benz[1,9]arsinolizino[2,3-b]indole and 4H,8H-Benz[1,9]arsinolizino[2,3-b]quinoline

These two systems are considered together, because they have a common origin and certain similarities, and comparatively little at present is known of them.

The compound (1) is termed 4H-benz[1,9]arsinolizino[2,3-b]indole [RIS 9276], and the compound (2) is 4H,8H-benz[1,9]arsinolizino-[2,3-b]quinoline [RIS 9363].

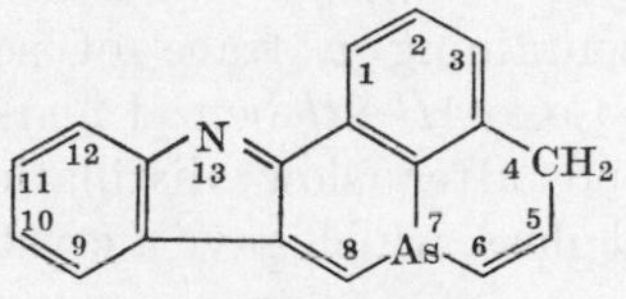

(1) 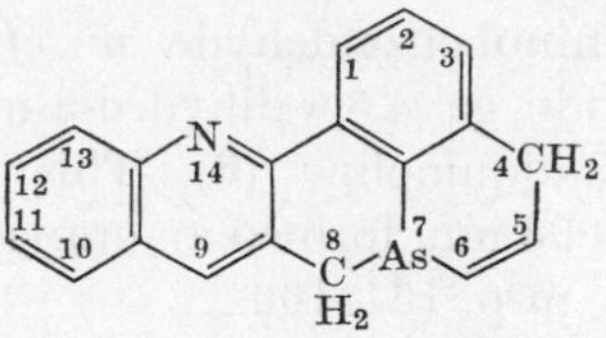   (2)

The bisphenylhydrazone of 2,3,5,6-tetrahydro-8-methoxy-1,7-di-oxo-1H,7H-benz[ij]arsinolizino (3), when heated in acetic acid containing

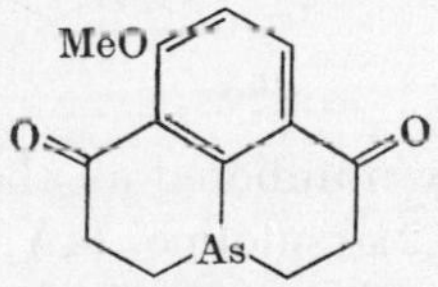

(3) 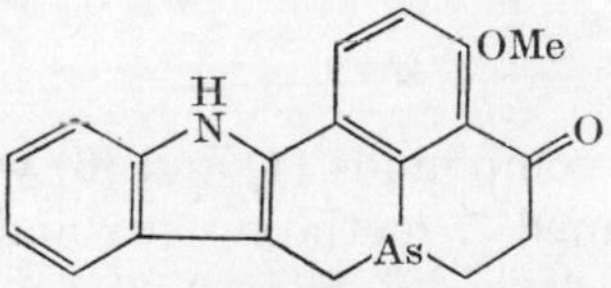   (4)

zinc chloride, furnished the almost colorless 5,6,8,13-tetrahydro-3-methoxy-4-oxo-4H-benz[1,9]arsinolizino[2,3-b]indole (4), one of the phenylhydrazone units having been hydrolyzed off with regeneration of the oxo group (Mann and Wilkinson [76]). There is little doubt that this phenylhydrazone unit was that nearest the methoxyl group, for it is known that in the analogous julolidine or benzo[ij]quinolizine derivative

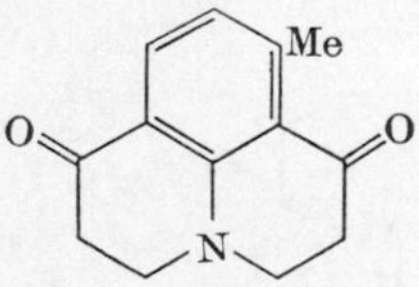

(5) 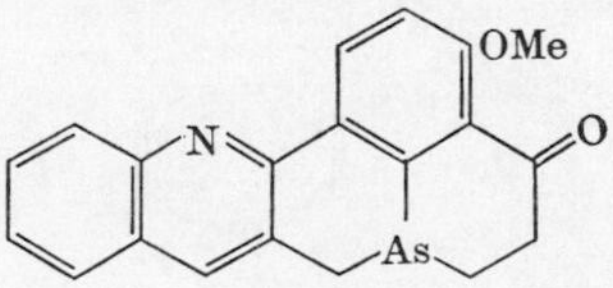   (6)

(5), the methyl group exerts a powerful steric influence on the neighbouring CO group and, whilst allowing bisphenylhydrazone formation, suppresses other reactions involving this group; for example, only the CO group most distant from the methyl group will condense with malononitrile (Ittyerah and Mann [72]).

The structure of the compound (4) is confirmed by its infrared spectrum, which shows the NH group by a band at 3300 cm^{-1}, and the CO group by a band at 1661 cm^{-1}. No question of the formation of a ψ-indole arises, undoubtedly for the same reasons as were discussed for the analogous compound (4) on p. 411 (cf. p. 134).

The compound (3) gave only tarry products with hot aqueous alkaline isatin. The far milder conditions of the Friedländer reaction, when the compound (3) was treated at room temperature with an excess of o-aminobenzaldehyde in ethanol containing a trace of sodium hydroxide, gave 5,6-dihydro-3-methoxy-4-oxo-4H,8H-benz[1,9]arsinolizino[2,3-b]quinoline (6). This compound after slow distillation at 200°/0.003 mm formed an orange colored glass, which gave a crystalline picrate, m.p. 164–166°.

(4A)   (6A)

The compounds (4) and (6) were initially numbered as above, and were named 7-methoxy-6-oxoindolo(2',3':1,2)arsuloline (4A) and 7-methoxy-6-oxoquinolino(2',3':1,2')arsuloline (6A), respectively.[76]

Seven-membered Ring Systems containing Carbon and One Arsenic Atom

Arsepins and Arsepanes

The compound of structure (1) is named 1H-arsepin [RIS —] and the fully reduced system (2) is 1H-arsepane. These simple compounds are known, however, only as their fused benzo derivatives, and in this respect they resemble the analogous nine-membered ring systems (p. 441).

5H-Dibenz[b,f]arsepin

The compound (1), which can be regarded as 1H-arsepin with benzo groups fused across the 2,3- and the 6,7-positions, is termed

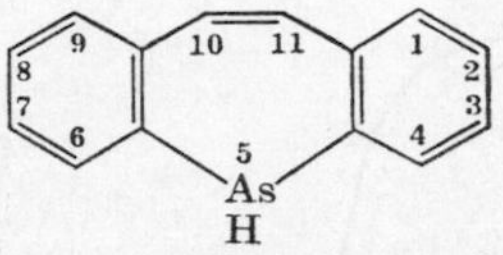

(1) 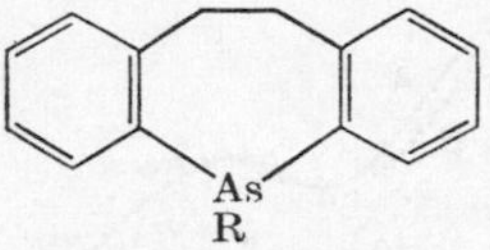   (2)

5H-dibenz[b,f]arsepin [RRI 3686], and the compound (2; R = H), which contains a partly reduced arsepin ring, as 10,11-dihydro-5H-dibenz-[b,f]arsepin.

10,11-Dihydro-5-phenyldibenz[b,f]arsepin (2; R = C_6H_5) was prepared by Mann, Millar, and Smith [97] by the same method as the analogous 10,11-dihydro-5-phenyldibenz[b,f]phosphepin (3) [RRI 3705] (p. 154), 2,2'-dibromobibenzyl, $BrC_6H_4CH_2CH_2C_6H_4Br$, being converted with butyllithium into 2,2'-dilithiobibenzyl, which when condensed with phenylarsonous dichloride, $C_6H_5AsCl_2$, gave the required arsepin (2; R = C_6H_5). This compound, which the above workers called 1-phenyl-1-arsa-2,3:6,7-dibenzocyclohepta-2,6-diene, could be obtained pure either

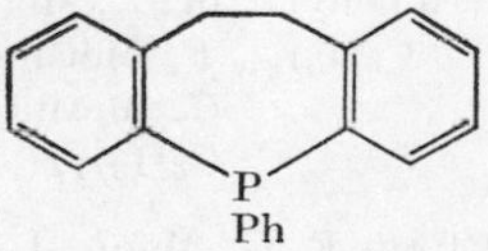

(3)

by recrystallization of the distilled material, or by conversion of the latter into the methiodide, which when heated under nitrogen at 260°/0.001 mm underwent complete dissociation; each method gave the arsine, m.p. 59–59.5°. When however the molten arsine was treated just above its melting point with a trace of the high-melting form of the phosphine (3), it immediately solidified and after recrystallization had m.p. 78.5–79°: this high-melting form of the arsine could now be used similarly to convert the low-melting into the high-melting arsine. The two forms of the arsine are each isomorphous with the corresponding forms of the phosphine, having m.p.s 75–75.5 and 94.5–95°, respectively, and the arsine and the phosphine are therefore isodimorphous.

The arsine gives an oxide, m.p. 188–190°, and a bis(arsine)dichloropalladium, $(C_{20}H_{17}As)_2PdCl_2,\frac{1}{2}C_4H_8O_2$, dark yellow crystals, m.p. 275–277°, having (like the phosphorus analogue) dioxane of crystallization.

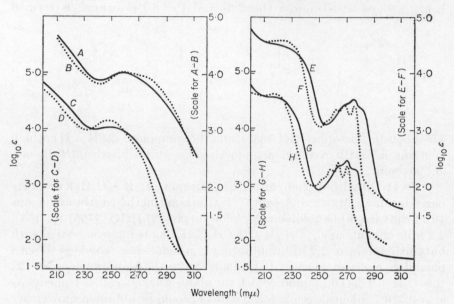

Wavelength (mμ)

Figure 8. Ultraviolet absorption spectra of: A, The phosphine (**3**); B, Triphenylphosphine; C, The arsine (**2**; R = C₆H₅); D, Triphenylarsine

Figure 9. Ultraviolet absorption spectra of: E, Methiodide of the phosphine (**3**); F, Methiodide of triphenylphosphine; G, Methiodide of the arsine (**2**; R = C₆H₅); H, Methiodide of triphenylarsine

(Reproduced, by permission, from F. G. Mann, I. T. Millar, and B. B. Smith, *J. Chem. Soc.*, **1953**, 1130)

A comparison of some of the physical and chemical properties of the arsine (**2**; R = C₆H₅) and of the phosphine (**3**) is of interest.

One pronounced chemical difference is the action of boiling hydriodic acid, which leaves the phosphine chemically unchanged but rapidly breaks down the arsine ring system with the formation of bibenzyl $(C_6H_5CH_2—)_2$ and arsenic triiodide. This behavior differs from that of several other cyclic arsines also having a phenyl group joined to the arsenic atom, from which the phenyl group is split off as benzene by the boiling hydriodic acid, with the formation of the iodoarsine, the ring system being unaffected (see pp. 371, 416 for examples, and p. 448 for a diarsine that is cleaved by boiling hydriodic acid but not by hydrobromic acid, which gives the dibromodiarsine).

The arsine (2; $R = C_6H_5$) differs, however, from the phenylarsines described in the above pages in that it is systematically a triarylarsine whereas the other cyclic arsines more closely approach dialkylarylarsines in type.

The triaryl nature of the phosphine (3) and the arsine (2; $R = C_6H_5$) is indicated by a comparison of the absorption spectra of these compounds with those of the simple triphenyl analogues. (In both Figures 8 and 9, the scale of $\log_{10} \epsilon$ values for the phosphorus compounds has been raised above that of the arsenic compounds to avoid overmuch superposition of the curves.) The spectrum of the phosphine (3) (Figure 8) is closely similar to that of triphenylphosphine, and that of the arsine (2; $R = C_6H_5$) is also similar to that of triphenylarsine. If any one of these compounds had three independently absorbing benzenoid rings, several absorption bands would be expected, with main absorption at ca. 255 mμ, $\log_{10} \epsilon_{max}$ 2.8. The absence of such bands and the greater intensity indicate considerable resonance from the numerous forms such as (4a), (4b), and (4c), where $X = P$ or As; this type of resonance would be

(4a) (4b)

(4c)

very little affected by the —$CH_2 \cdot CH_2$— chain joining the two benzene rings. Hence the members of each pair of compounds would be expected to have very similar spectra.

The spectra of the methiodides of the four compounds under consideration are shown in Figure 9. The positive charges on the phosphorus and arsenic atoms suppress resonance of the above type, and the main absorptions have considerably lower intensities. The interesting feature is the close resemblance between the spectra of the methiodides of the phosphine (3) and the arsine (2; $R = C_6H_5$), and also between those of the methiodides of triphenyl-phosphine and -arsine. Apparently the several bands in the 250–280 mμ region shown by the methiodides of triphenyl-phosphine and -arsine indicate the normal absorption of the benzene groups, which are merged into one main broad absorption in the spectra of the heterocyclic methiodides.

6H-Dibenz[c,e]arsepin

The compound (1) has been prepared by condensing 2,2′-bisbromomethylbiphenyl (2) in ether–benzene with phenylarsinebis(magnesium

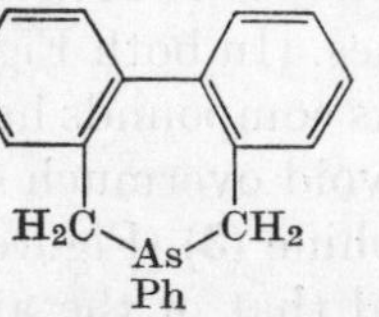

(1) 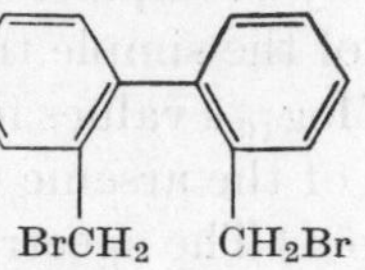   (2)

bromide), $C_6H_5As(MgBr)_2$ (p. 229); Beeby, Mann and Turner,[98] named the compound 6-phenyl-6-arsa-1,2:3,4-dibenzcyclohepta-1,3-diene.

The Ring Index [RRI 3687] bases the numbering and presentation on the hypothetical structure (3), termed 6H-dibenz[c,e]arsepin, and

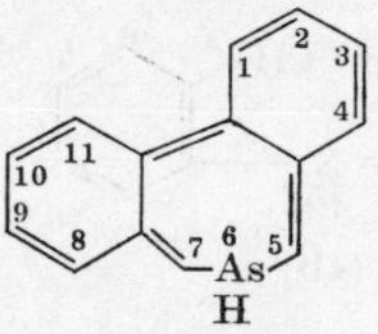

(3) 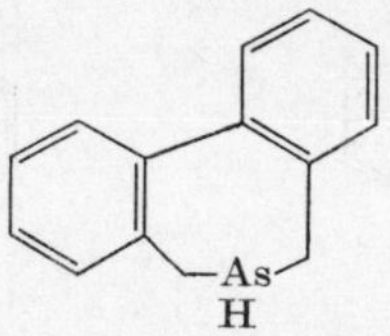   (4) 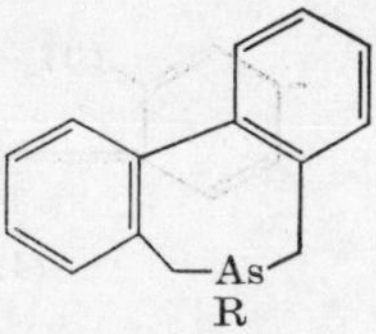   (5)

the dihydro derivative (4) is termed 5,7-dihydrodibenz[c,e]arsepin. Consequently the compound (1) is shown in the form (5; $R = C_6H_5$) and named 5,7-dihydro-6-phenyldibenz[c,e]arsepin; it was the first compound having a ring of six carbon atoms and one arsenic atom to be isolated.

This compound (1) forms colorless crystals, m.p. 118–118.5°; when boiled with methyl iodide it gives the methiodide, 5,7-dihydro-6-methyl-6-phenyldibenz[c,e]arsepinium iodide, m.p. 223–224°, and with phenacyl chloride, $C_6H_5COCH_2Cl$, in boiling benzene it gives the corresponding 6-phenacyl-6-phenyl arsepinium chloride, crystals which tenaciously retain benzene but on being heated at 100°/0.1 mm for 3 hours give the solvent-free salt, m.p. 110–115°.

The heterocyclic ring possesses considerable stability, for the 6-phenyl derivative (5; $R = C_6H_5$), in boiling hydriodic acid for two hours, gave the 6-iodo derivative (5; $R = I$), yellow crystals, m.p.

117–117.5°. In this respect, the heterocyclic ring in the compound
(5; R = C₆H₅) is vastly more stable than that in the isomeric 10,11-
dihydro-5-phenyldibenz[b,f]arsepin (p. 436).

With regard to the stereochemistry of the tricyclic system in
compounds of type (5), if the atomic radii and the bond lengths utilized
by G. H. Cookson and Mann (p. 427) are applied to this system and an
intervalency angle of 100° at the arsenic atom is accepted, calculation

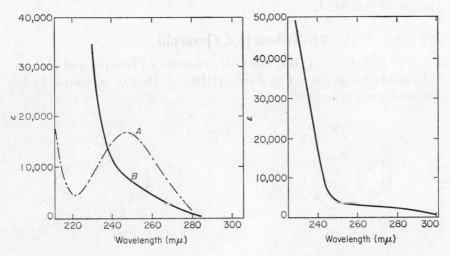

Figure 10. Ultraviolet absorption spectra of
(A) biphenyl and (B) 5,7-dihydro-6-methyl-
6-phenylbenz[c,e]arsepinium iodide

Figure 11. Ultraviolet absorption
spectrum of 5,7-dihydro-6-iodo-
dibenz[c,e]arsepin (5; R = I)

(Reproduced, by permission, from M. H. Beeby, F. G. Mann, and E. E. Turner,
J. Chem. Soc., **1950**, 1923)

indicates that the two linked benzene rings subtend an angle of 63°;
further, the intervalency angle within the ring at carbon atoms 5 and 7
becomes 110°, and the heterocyclic ring is apparently almost strainless.
A model based on these dimensions shows that the central heterocyclic
ring locks the two benzene rings in position, and no rotation about their
common axis is possible. When the compound forms a quaternary salt,
e.g., the methiodide, the intervalency angle within the ring at the As
atom increases to ca. 109°; this will, however, be accompanied by a
shrinkage of the interatomic As–C distance, and the overall process will
probably be without significant effect on the angle between the benzene
rings and on the stability of the heterocyclic ring.

Although the final result of such considerations must be accepted with some reserve, it does receive strong support from the spectroscopic evidence.

The absorption spectra of biphenyl and of the methiodide of the 6-phenyl arsepin (5; $R = C_6H_5$) are shown for comparison in Figure 10. The biphenyl shows the characteristic band at 248 mμ, ϵ_{max} 17,000, which is strongly associated with the coplanar conformation; this band is entirely absent from the spectrum of the methiodide. The spectrum (Figure 11) of the 6-iodo compound (5; $R = I$) also shows a complete absence of this band.

9H-Tribenz[b,d,f]arsepin

The compound (1) is termed 9H-tribenz[b,d,f]arsepin and presumably would be numbered as shown [RIS —]. Clearly no simple hydro derivative can arise here.

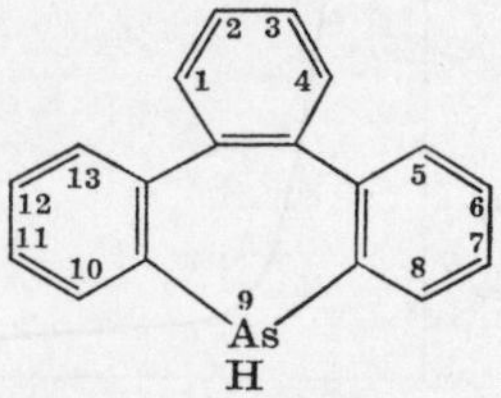

(1) 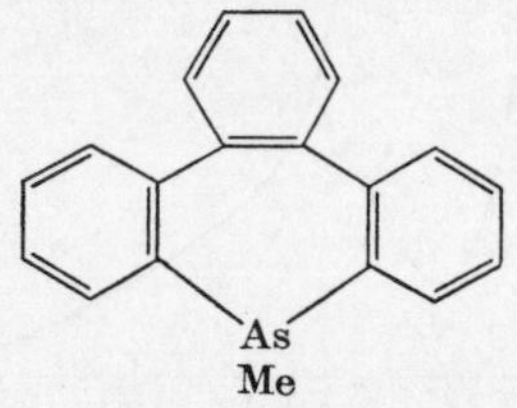　　　　　　　　　　　(2)

The cyclic diarsonium dibromide (3; $n = 2$) occurs in two forms, having m.p. 201–202° and 272–275°, respectively (p. 497): the former when heated under nitrogen at 200° rising to 224°/14 mm, and the latter when similarly heated at 250–260°/14 mm, undergo decomposition with the formation of 9-methyl-9H-tribenz[b,d,f]arsepin (2), m.p. 159–160°. The homologous diarsonium dibromide (3; $n = 3$) (p. 499), when heated

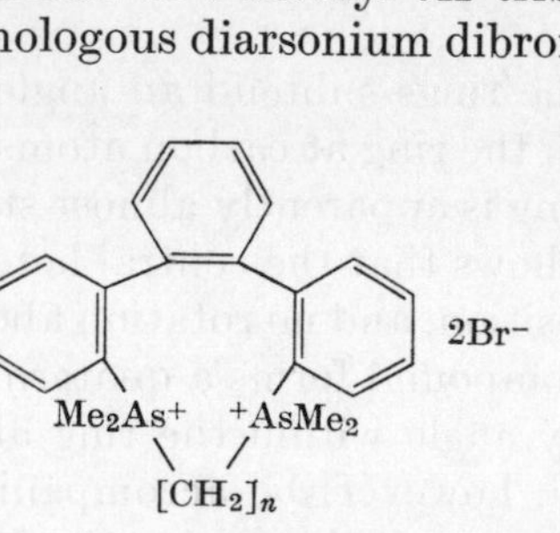

(3) 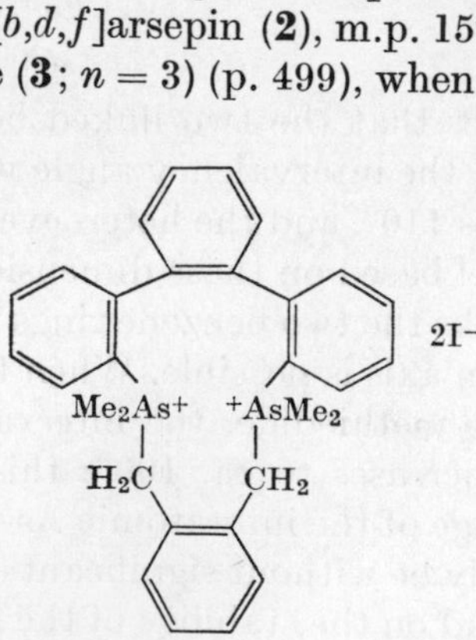　　　　　　　　　　　(4)

at 220°/0.1 mm, also gives the arsepin (2) in low yield. The twelve-membered-ring diarsonium diiodide (4) (p. 500), when heated at 245–250°/14 mm, also gives the arsepin (2) in low yield; the corresponding

(5)   2Br⁻   (6)

dibromide loses methyl bromide without cleavage of the ring system although the conditions are critical (Mann, Millar, and Baker[99]).

The 9-methyl-9H-tribenz[b,d,f]arsepin (2), initially termed 7-methyl-7-arsa-1,2:3,4:5,6-tribenzocycloheptatriene,[99] is apparently the only known member of this system; it forms an unstable monohydrated methiodide, m.p. 218–221°, and a stable yellow methopicrate, m.p. 190–191°.

The thermal decomposition of the diarsonium dibromides (3; $n = 2$ and 3) is thus very similar to that of the diarsonium dibromides (5; $n = 2$ and 3), both of which furnish 9-methyl-9-arsafluorene (5-methyldibenzarsole) (6) (pp. 489, 491).

Nine-membered Ring Systems containing Carbon and One Arsenic Atom

Arsonins and Arsonanes

The compound of structure (1) is termed 1H-arsonin, [RIS —], and the fully reduced compound (2) is arsonane. Both are at present unknown,

but the arsonin system is known as its tetrabenzo derivatives (see below).

17H-Tetrabenz[b,d,f,h]arsonins

This system has the structure and numbering (1), [RIS —], and known derivatives have aryl substituents in the 17-position. These

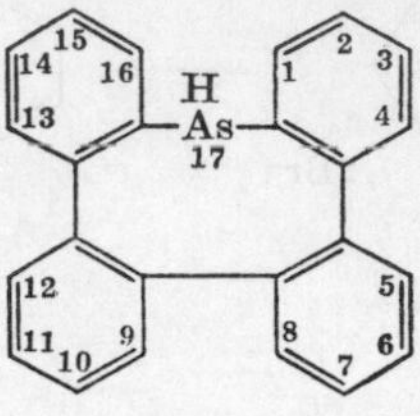

(1)

derivatives have been described (pp. 387 ff.) and are collected here for quick reference. The formula (1) is inverted below for convenience.

Wittig and Hellwinkel[48] have shown that 5-phenyl-5,5'-spirobi-[dibenzarsole] (2; R = H) when briefly heated at 233° gave the isomeric 17-phenyl-17H-tetrabenz[b,d,f,h]arsonin (3; R = H), m.p. 132–135.5°

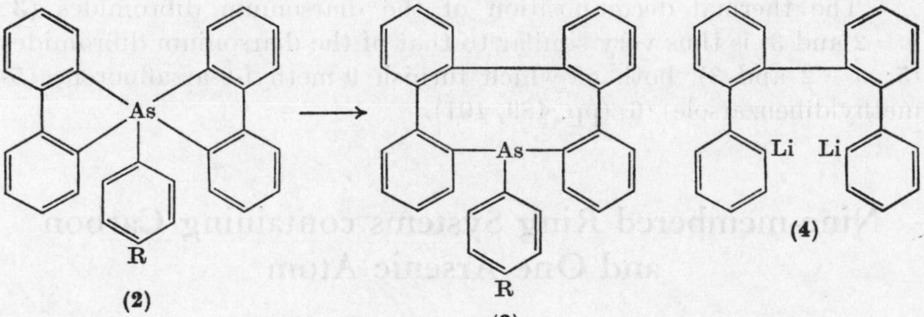

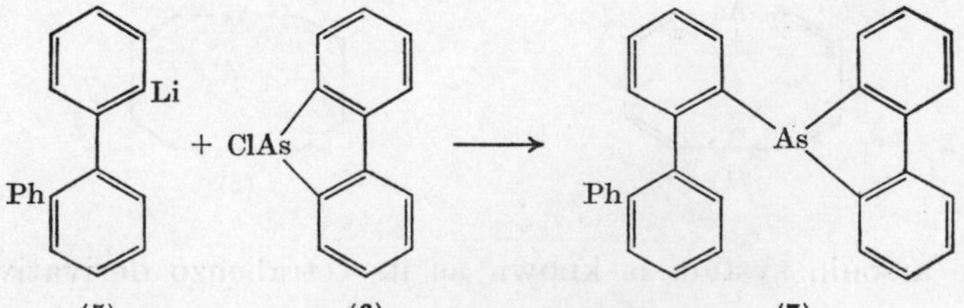

They proved the structure of this compound by an alternative synthesis in which 2,2'''-o-quaterphenylylenedilithium (4) was condensed with phenylarsonous dichloride, $C_6H_5AsCl_2$. The possibility of an isomeric structure (**7**) in place of (**3**; R = H) was disproved by condensing 2-*o*-terphenylyllithium (**5**) with 9-chloro-9-arsafluorene (5-chlorodibenz-arsole) (**6**), thus obtaining the authentic 5-(2-*o*-terphenylyl)dibenzarsole (**7**), m.p. 185–187°.

N,N - Dimethyl - *p* - 5,5'spirobi[dibenzarsol] - 5 - ylaniline [**2**; R = N(CH₃)₂] when quickly heated similarly gave *N,N*-dimethyl-*p*-17*H*-tetrabenz[*b,d,f,h*]arsonin-17-ylaniline [**3**; R = N(CH₃)₂], m.p. 223–225°: the *N*-methiodide of [**2**; R = N(CH₃)₂] when plunged into a bath at 255° and then rapidly heated to 290° also gave the arsonin [**3**; R = N(CH₃)₂].

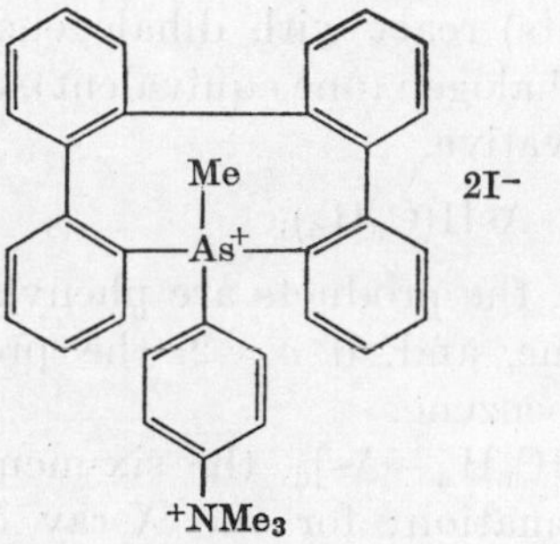

(8) 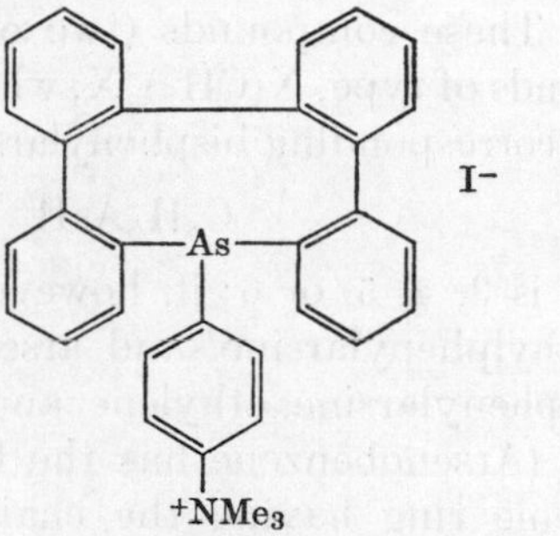   (9)

A solution of the arsonin [**3**; R = N(CH₃)₂] in methyl iodide at room temperature slowly deposited the dimethiodide (**8**), which after recrystallization from acetone–ethanol melted at 194–195.5° with loss of methyl iodide, formation of the *N*-methiodide (**9**) and resolidification, for the *N*-methiodide has m.p. 246°. A solution of methiodide (**9**) in methyl iodide, when boiled for 4 hours, gave the dimethiodide (**8**).

Five-membered Ring Systems containing Carbon and Two Arsenic Atoms

1,2-Diarsole and 1,2-Diarsolanes

There are theoretically two isomeric diarsole systems, namely, 1*H*-1,2-diarsole (**1**), [RIS —], and 1*H*-1,3-diarsole (**2**), [RIS —]; the ring

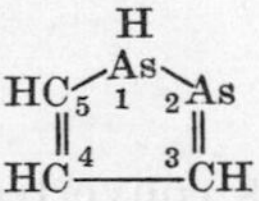

(1) 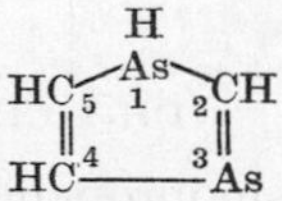   (2) 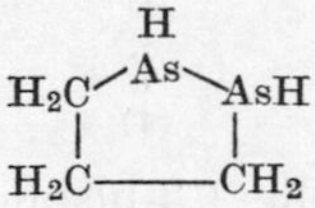   (3) 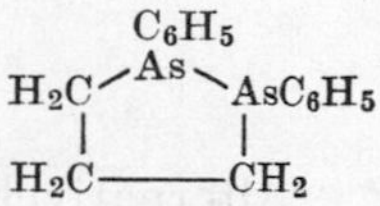   (4)

system (2) is at present unknown, and the system (1) is known only as derivatives of 2,3,4,5-tetrahydro-1,2-diarsole, known as 1,2-diarsolane (3).

Tzschach and Pacholke[100] have shown that when phenylarsine, $C_6H_5AsH_2$, in dioxane or tetrahydrofuran is treated with potassium or sodium it forms the potassium (or sodium) phenylarsine,

$$K(Na)AsH(C_6C_5);$$

each of these compounds forms a crystalline compound with dioxane, of composition $K(Na)AsH(C_6H_5),2C_4H_8O_2$, the potassium compound being yellow and the sodium compound orange-yellow.

These compounds (two equivalents) react with dihalogeno compounds of type, $X(CH_2)_nX$, where $X =$ halogen (one equivalent), to give the corresponding bisphenylarsine derivative,

$$C_6H_5AsH—(CH_2)_n—AsH(C_6H_5),$$

if n is 3, 4, 5, or 6; if, however, $n = 1$, the products are phenylarsine, methylphenylarsine, and arsenobenzene, and, if $n = 2$, the products are phenylarsine, ethylene, and arsenobenzene.

(Arsenobenzene has the formula $[C_6H_5—As]_6$, the six-membered arsenic ring having the chair conformation; for the X-ray crystal structure, see Hedberg, Hughes and Waser.[100a] Its systematic name is hexaphenylhexarsenane [RIS 9728]).

Tzschach and Pacholke[101] have later shown that, when potassium phenylarsine is prepared under argon as above, the subsequent addition of 1,3-dichloropropane gives 1,3-bisphenylarsinopropane (5), some of which undergoes dehydrogenation to form 1,2-diphenyl-1,2-diarsolane (4), b.p. 187–195°/3 mm.

$$(C_6H_5)HAs—(CH_2)_3—AsH(C_6H_5) \longrightarrow (4)$$
$$(5)$$
$$\downarrow$$
$$Li(C_6H_5)As—(CH_2)_3—As(C_6H_5)Li$$
$$(6)$$
$$\downarrow Br(CH_2)_2Br$$
$$Li(C_6H_5)As—(CH_2)_3—As(C_6H_5)Br + Li(CH_2)_2Br$$
$$(7) \qquad\qquad\qquad (8)$$
$$\downarrow \qquad\qquad\qquad\qquad \downarrow$$
$$(4) \qquad\qquad\qquad CH_2{=}CH_2$$

Alternatively, if the 1,3-bisphenylarsinopropane (5) is converted into the dilithio derivative (6), the addition of one equivalent of 1,2-

dibromoethane then gives the bromolithiodiarsine (**7**) and 2-bromo-ethyllithium (**8**). These compounds lose lithium bromide with the formation of the 1,2-diarsolane (**4**), b.p. 193–195°/0.3 mm (56%), and ethylene, respectively. Some undistilled polymeric material is also formed.

The As—As bond in the diarsolane (**4**) is readily cleaved by both nucleophilic and electrophilic reagents. Butyllithium and phenyllithium react with (**4**) in warm ethereal solution to give the open-chain dilithio-diarsine (**6**). Methyl iodide reacts with (**4**) in warm ethanolic solution to give 3-(iodo-phenylarsino)propyldimethylphenylarsonium iodide (**9**),

$$C_6H_5As(I)—(CH_2)_3—\overset{+}{As}(CH_3)_2C_6H_5 \ I^-$$
(9)

$$(C_6H_5)_2As—As(C_6H_5)_2$$
(10)

m.p. 151–152°. These reactions are analogous to the action of butyl(or phenyl)lithium on tetraphenyldiarsine (**10**), giving lithiodiphenylarsine (Issleib and F. Krech[102]), and to the action of methyl iodide on the diarsine (**10**), giving diphenylarsinous iodide, $(C_6H_5)_2AsI$, and dimethyl-diphenylarsonium iodide (Steinkopf and Schwen[103]).

When, however, the diarsole (**4**) is treated with lithium aluminum hydride in boiling dioxane, a more extensive cleavage occurs with the formation of phenylarsine and n-propylphenylarsine, $(C_3H_7)(C_6H_5)AsH$ (Issleib, Tzschach, and Schwarzer[104]).

1H-1,3-Benzodiarsole and 2,2′-Spirobi[1H-1,3-benzodiarsolene]

The ring system (**1**) is 1H-1,3-benzodiarsole, [RIS —], and the 2,3-dihydro derivative (**2**) is 1H-1,3-benzodiarsolene. The corresponding spirocyclic compound (**3**) is therefore 2,2′-spirobi[1,3-benzodiarsolene], [RIS —]; both (**2**) and (**3**) are known only as As-substituted derivatives.

(1)   (2)   (3)

It has been briefly recorded (Collinge, Nyholm, and Tobe[105]) that a solution of o-phenylenebis(dimethylarsine) (**4**) (cf. p. 540) in an excess of dibromomethane, when boiled under reflux for 3 hours, deposits the colorless 1,1,3,3-tetramethyl-1H-1,3-benzodiarsolenium dibromide (**5**; X = Br), m.p. 220° (diiodide, m.p. 239°; diperchlorate, m.p. 270°).

When a mixture of the diarsine (**4**) (2 moles) and carbon tetra-bromide (1 mole) is either kept at room temperature for several weeks or heated in a sealed tube at 100° for 3 hours, it gives 1,1,1′,1′,3,3,3′,3′-octamethyl-2,2′-spirobi-(1,3-benzodiarsolenium) tetrabromide (**6**; X = Br), m.p. 82°, resolidifying to a higher-melting product (tetraiodide, m.p. 231°; tetraperchlorate, m.p. 243°), all colorless salts.

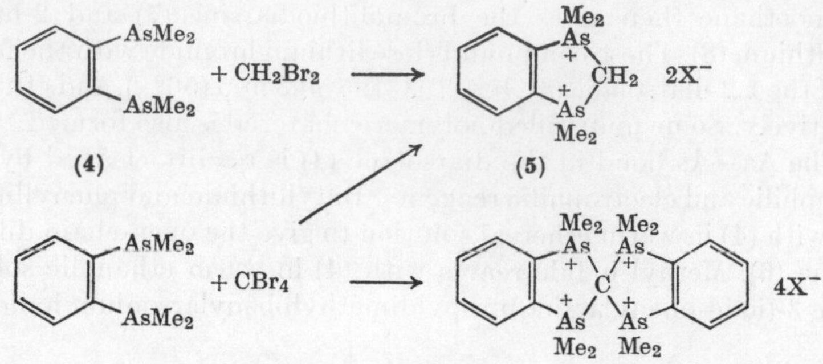

When, however, this mixture of the diarsine (**4**) and carbon tetrabromide in ethanolic solution is boiled for 3 hours, it gives the dibromide (**5**; X = Br). This is an unexpected reaction, for the initial cyclized product is presumably the 2,2-dibromo derivative of (**5**; X = Br), from which the ethanol apparently replaces the nuclear bromine by hydrogen, being itself brominated and/or oxidized; this would imply that the action of the ethanol is more rapid than the cyclic diquaternization of the second diarsine molecule.

The conductivities of the three salts (**5**; X = Br, I, or ClO_4) in aqueous solution are 250, 246, and 202 ohm^{-1} at 10^{-3} M, respectively; the conductivities of the salts (**6**; X = Br, I, or ClO_4) also in aqueous solution indicate a considerable degree of ion-association.

The ready decomposition of the dibromide (**6**; X = Br) at its melting point to a higher-melting product deserves further investigation. It should be possible to evict four molecules of methyl iodide from this salt to form the covalent 1,1′,3,3′-tetramethyl derivative of the compound (**3**).

Six-membered Ring Systems containing Carbon and Two Arsenic Atoms

Diarsenins and Diarsenanes

There is clearly a possibility of three isomeric compounds, 1,2-diarsenin (**1**), [RRI —], 1,3-diarsenin (**2**), [RRI —], and 1,4-diarsenin

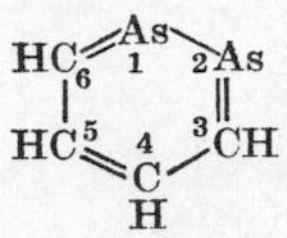

HC≡As As (1)

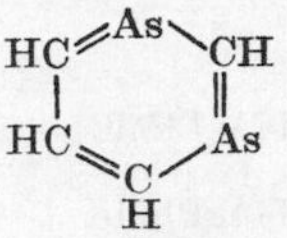

HC≡As CH (2)

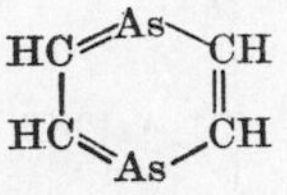

HC≡As CH (3)

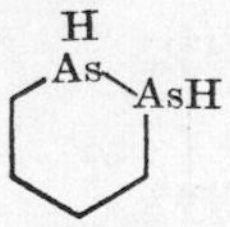

(4)

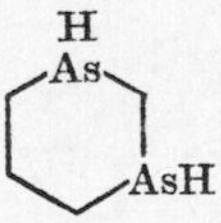

(5)

(6)

(3), [RRI 232]. All three are unknown. The hexahydro derivatives are named diarsenanes, and 1,2- and 1,4-disubstituted derivatives of 1,2-diarsenane (4) and 1,4-diarsenane (6), respectively, are known.

1,2-Diarsenanes

The preparation and properties of 1,2-diphenyl-1,2-diarsenane (7) closely follow those of 1,2-diphenyl-1,2-diarsolane (p. 444) (Tzschach and Pacholke[100, 101]). 1,4-Dichlorobutane when treated with 2 equivalents of potassiophenylarsine in a dioxane–benzene mixture gives

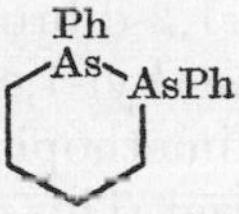

(7)

$PhRAs-(CH_2)_4-AsRPh$

(8)

$[IPhAs-(CH_2)_4-AsMe_2Ph]^+ I^-$

(9)

a mixture of 1,4-bisphenylarsinobutane (8; R = H) and 1,2-diphenyl-1,2-diarsenane (7) (73%), b.p. 196–198°/0.3 mm, m.p. 59°. Similarly the dilithiodiarsine (8; R = Li)—used conveniently as its bisdioxane addition product—when treated in benzene–tetrahydrofuran with 1,2-dibromoethane (one equivalent) gives ethylene and the diarsenane (7) (68%), with some polymeric material.

The diarsenane (7) is very slowly affected by the air. With butyl(or phenyl)lithium the As—As link is cleaved, with the formation of the dilithio derivative (8; R = Li), and warm ethanolic methyl iodide converts the diarsenane into 4-(iodophenylarsinobutyl)dimethylphenylarsonium iodide (9), m.p. 98–99°.

1,4-Diarsenanes

1,4-Diphenyl-1,4-diarsenane (5; R = C_6H_5), known earlier as 1,4-diphenyldiethylenediarsine and as 1,4-diphenyl-1,4-diarsacyclohexane, has been prepared by two methods. In the first method (Jones

$$
\begin{array}{ccccc}
\text{CH}_2\text{AsPhCl} & & \text{CH}_2\text{AsPhMe} & & \\
| & \longrightarrow & | & \longrightarrow & \textbf{(4)} \\
\text{CH}_2\text{AsPhCl} & & \text{CH}_2\text{AsPhMe} & & \\
\textbf{(2)} & & \textbf{(3)} & &
\end{array}
$$

(4): cyclic structure with two As$^+$ centres each bearing Ph and Me, $2X^-$

$$
\begin{array}{ccccc}
\text{CH}_2\text{As(O)(OH)Ph} & & \text{CH}_2\text{AsHPh} & & \\
| & \longrightarrow & | & \longrightarrow & \textbf{(5)} \\
\text{CH}_2\text{As(O)(OH)Ph} & & \text{CH}_2\text{AsHPh} & & \\
\textbf{(1)} & & \textbf{(6)} & &
\end{array}
$$

(5): diarsenane ring with R on each As.

and Mann[106]), ethylenebis(phenylarsinic acid) (1) was reduced in hydrochloric acid with sulfur dioxide to ethylenebis(chlorophenyl-arsine) (2) (*cf.* Chatt and Mann[107]), which with methylmagnesium iodide gave ethylenebis(methylphenylarsine) (3), b.p. 163–165°/0.2 mm (dimethiodide, decomp. *ca.* 250°; dimethopicrate, m.p. 217–218°; dihydroxypicrate, m.p. 171–172°). The diarsine (3), when heated in methanol at 100° for 6 hours with one equivalent of 1,2-dibromoethane, underwent cyclic diquaternization to form 1,4-dimethyl-1,4-diphenyl-1,4-diarsenanium dibromide (4; X = Br) (21%) (dimethopicrate, m.p. 244°), together with much gummy polymeric material. This dibromide, when heated at 0.5 mm, lost methyl bromide, giving 1,4-diphenyl-1,4-diarsenane (5; R = C_6H_5), m.p. 142–144° (monomethiodide, m.p. 211°; dimethiodide, m.p. 221°; bishydroxynitrate, m.p. 188°).

The yields in the last two stages were low: therefore in the second synthesis (Mann and Pragnell[108]), the diarsinic acid (1) in ether was added to zinc dust in hydrochloric acid containing mercuric chloride under nitrogen at room temperature. The reduction gave the ethylene-bis(phenylarsine) (6) (crude, 87%), b.p. 150–153°/2×10^{-6} mm when rapidly distilled in small quantity. An ethereal solution of (6), when treated with butyllithium (2 equivalents) and 1,2-dichloroethane (1 equivalent) gave the diphenyldiarsenane (5; R = C_6H_5) (82%).

The stability towards boiling hydriodic acid shown by many cyclic arsenic systems (cf. pp. 416, 428) was not shown by the diphenyldi-arsenane (5; R = C_6H_5), which under these conditions underwent ring cleavage with the formation of ethylenebis(diiodoarsine), [—CH$_2$AsI$_2$]$_2$, but boiling hydrobromic acid readily gave the crystalline 1,4-dibromo-1,4-diarsenane (5; R = Br), m.p. 168–170°, which with methylmagnesium iodide gave 1,4-dimethyl-1,4-diarsenane[106] (**5**; R = CH$_3$), b.p. 113–114°/24 mm.

The dibromodiarsenane, when reduced in tetrahydrofuran with lithium aluminum hydride, gave 1,4-diarsenane (**5**; R = H), a very reactive liquid having b.p. 87–89°/14 mm when distilled rapidly in small quantities; it was identified[108] by the action of hot methanolic methyl iodide, which gave 1,1,4,4-tetramethyl-1,4-diarsenanium diiodide, and by the similar action of ethyl iodide which gave the 1,1,4,4-tetraethyl-1,4-diarsenanium diiodide, characterized as the dipicrate, m.p. 212–215°.

It is noteworthy that whereas 1,4-dimethylpiperazine (**7**) combines readily with 1,2-dibromoethane to give in high yield the dimethiodide (**8**) of 1,4-diazabicyclo[2.2.2]octane, [RRI 1595] (Mann and Baker[109]),

(7)   (8)

neither 1,4-dimethyl-1,4-diarsenane (**5**; R = CH$_3$) nor the 1,4-diphenyl analogue (**5**; R = C$_6$H$_5$) gives a similar reaction.[106] The dimethyl compound (**5**; R = CH$_3$), which has been investigated in only small amount, gives with 1,2-dibromoethane an amorphous material, apparently by extensive linear condensation. The diphenyl compound (**5**; R = C$_6$H$_5$) also gives amorphous products with 1,2-dibromoethane, 1,3-dibromopropane, and o-xylylene dibromide, o-C$_6$H$_4$(CH$_2$Br)$_2$. The reaction of the dimethylpiperazine (**7**) to form the diquaternary salt (**8**) must involve an easy conversion of the ring system of (**7**) into the boat conformation. It is possible that the 1,4-diarsenane system, irrespective of its substituents, either will not oscillate from the chair to the boat form or alternatively oscillates so slowly that linear condensation of the chair form is virtually the exclusive reaction. An X-ray crystal investigation of 1,4-diphenyl-1,4-diarsenane (**5**; R = C$_6$H$_5$) carried out by two dimensional analysis shows that the molecule is centrosymmetric and therefore has the chair conformation, and the electron-density projection can be interpreted in terms of equatorial (**9**), rather than axial, positions for the phenyl groups (Nyberg and Hilton[110]). This conformation, if it persisted in solution and in the molten condition, would inhibit cyclic diquaternization of the type (**7**) → (**8**). The liquid 1,2,2,4,5,5-hexamethylpiperazine (**10**) also fails to give this diquaternization, and its

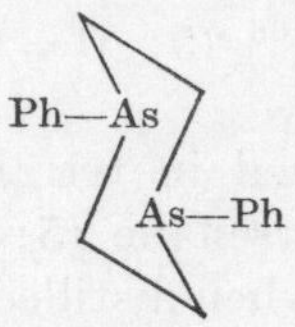

(9) 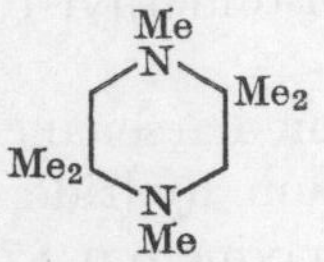   (10)

crystalline dinitrate is also centrosymmetric (W. Cochran, quoted in Mann and Senior[111]); this diamine, which readily forms a mono- and a di-methobromide, may also remain firmly in the chair conformation.

1,4-Diarsabicyclo[2.2.2]octane

1,4-Diarsabicyclo[2.2.2]octane (1), [RIS 11987], is conventionally depicted and numbered as shown; it can be regarded systematically as a derivative of 1,4-diarsenane, in which a —CH_2CH_2— group bridges the 1,4-arsenic atoms. The 2,3,5,6,7,8-hexadehydro form (2) is therefore

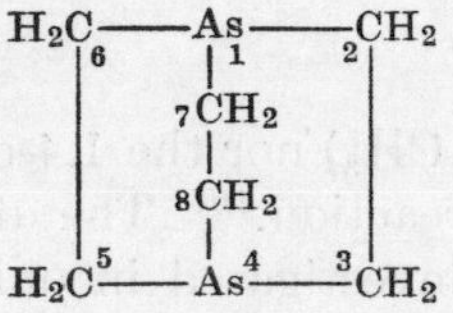

(1) 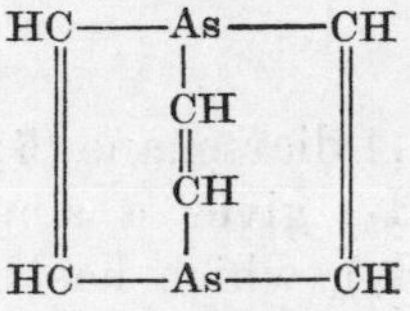   (2) 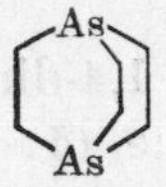   (3)

named 1,4-diarsabicyclo[2.2.2]octa-2,5,7-triene. These formulae may tend to mask the highly symmetrical nature of these compounds; since the unhindered intervalency angle of the 3-covalent arsenic atom is *ca.* 98°, the compound (1) can be depicted as (3).

The compounds (1) and (2) are unknown at present. Krespan[112] has shown, however, that when 1,1,1,4,4,4-hexafluoro-2,3-diiodo-2-butene (4) is heated with arsenic at 200° for 10 hours, the product after

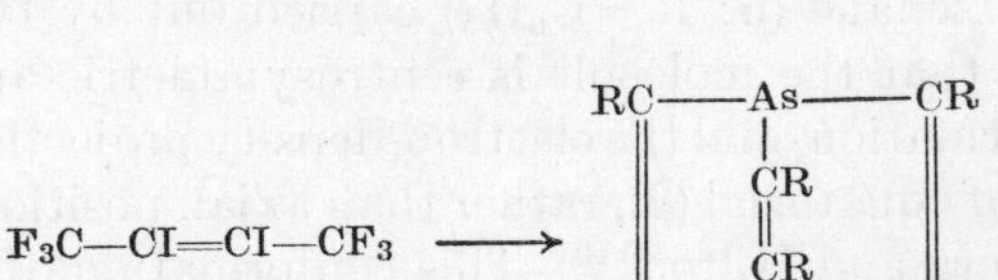

F_3C—CI=CI—CF_3 $\longrightarrow$

(4)

(5) 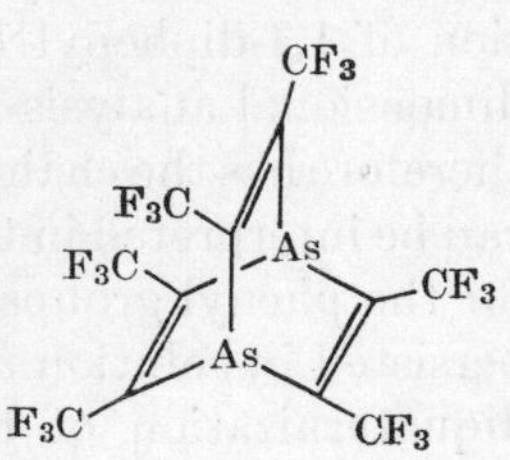   (6)

purification yields 2,3,5,6,7,8-hexakis(trifluoromethyl)-1,4-diarsabi-cyclo[2.2.2]octa-2,5,7-triene (5; R = CF$_3$), alternatively depicted as (6), colorless crystals, m.p. 139–140° [44%, based on the compound (4)]. This reaction is precisely similar to that used by Krespan[112] to obtain the 1,4-diphospha analogue of this compound (p. 182).

The structure of the compound (5; R = CF$_3$) is mainly based on physical evidence very similar to that adduced for the diphospha analogue. In addition to analysis and molecular-weight determination, the infrared spectrum of the diarsa compound shows C=C absorption at 1616 cm^{-1}, the ultraviolet spectrum shows a band at λ 297 mμ (ϵ 960) with a shoulder at 238 mμ (ϵ 900), and the nuclear magnetic resonance spectrum in dimethylformamide shows a single peak for CF$_3$ at −1325 cps. Although the chemical properties are not specified, this compound undoubtedly shows the same resistance to atmospheric oxidation and to quaternization as the diphospha analogue does.

The diarsa compound and its diphospha analogue retain the high symmetry indicated by (3), and both sublime very readily. This association of high symmetry and ready sublimation is in line with quinuclidine, with 1,4-diazabicyclo[2.2.2]octane [RRI 1595], and with 1,4-diphosphabicyclo[2.2.2]octane (p. 180), all of which also sublime readily.

Benzo Derivatives of 1,4-Diarsenin

The structural fusion of benzo groups to the 1,4-diarsenin ring system can give rise to the monobenzo derivative or 1,4-benzodiarsenin (1 on p. 452), [RRI 1537], and to the dibenzo derivative arsanthrene (1 on p. 460), [RRI 3262]. These two systems and certain derivatives with extra rings will be considered in turn.

1,4-Benzodiarsenin

The parent compound, 1,4-benzodiarsenin (1), [RRI 1537], is unknown, but derivatives having the heterocyclic ring hydrogenated are readily prepared (Glauert and Mann[113]). For this purpose, o-phenyl-enebis(dimethylarsine) (2), previously prepared by Chatt and Mann[107] (p. 540), was heated with one molecular equivalent of 1,2-dibromoethane at 125–130°, giving the dimethobromide of 1,2,3,4-tetrahydro-1,4-dimethyl-1,4-benzodiarsenin (4), the dimethobromide itself (3; X = Br) being systematically named 1,2,3,4-tetrahydro-1,1,4,4-tetramethyl-1,4-benzodiarseninium dibromide and having m.p. 254° [dipicrate (3; X = C$_6$H$_2$N$_3$O$_7$), m.p. 236–238°]. The dibromide is freely soluble in cold water, moderately soluble in boiling methanol, and almost insoluble in boiling ethanol.

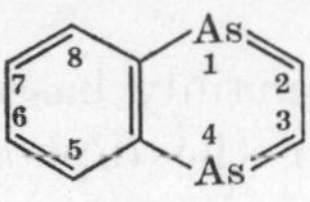

(1) 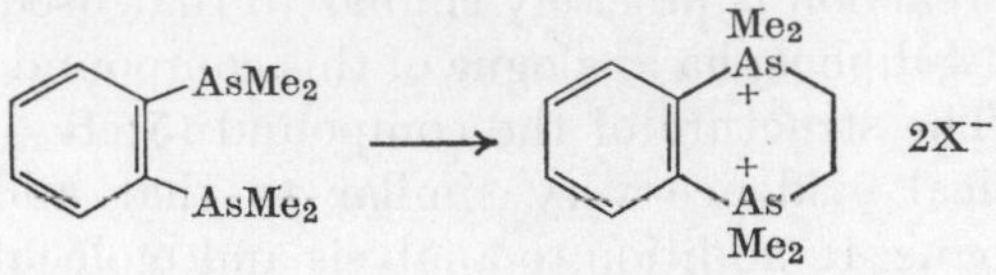   (2)   (3)

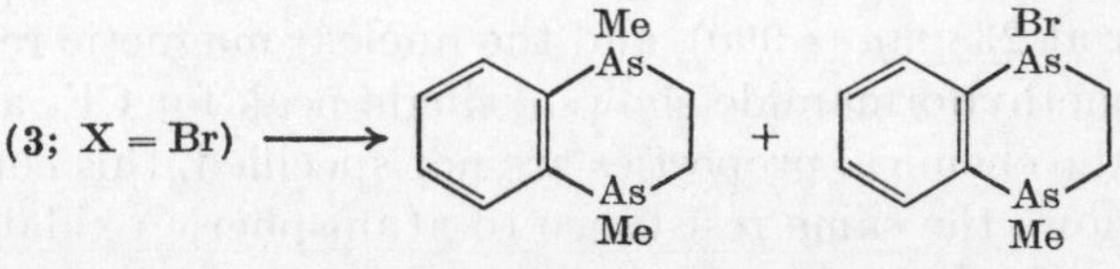

(3; X = Br) ⟶   +

(4)   (5)

It is noteworthy that the diarsine (2) when dissolved in methanolic methyl bromide at room temperature gives the monomethobromide

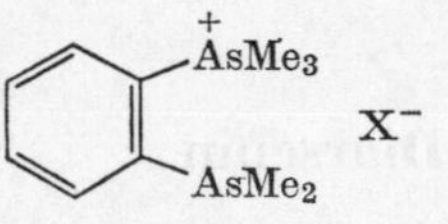

(6) 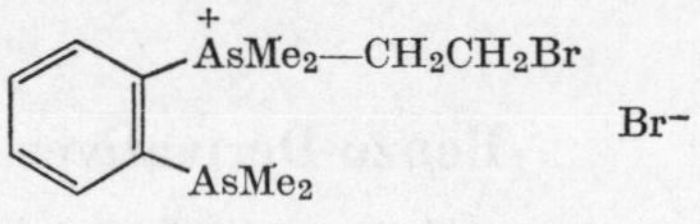   (7)

(6; X = Br), m.p. 225–226°, but the solution when heated at 100° in a sealed tube gives the dimethobromide, m.p. 222–223°. Similarly, a solution of the diarsine (2) in boiling methyl iodide after several hours gives only the monomethiodide (6; X = I), m.p. 228–230°, but when heated at 100° gives the dimethiodide, m.p. 222–224°. This low reactivity of the monomethohalides (6) is to be expected, for the inductive effect of the strong positive pole on the arsonium group would partially deactivate the tertiary arsenic atom. [The melting points of the dimetho-bromide and dimethiodide indicate that each has probably dissociated to the more stable mono salt before melting.] On the other hand, the diarsine (2) unites readily and apparently completely with the 1,2-dibromoethane at 125–130°, for the hard, solid product contains no indication of the intermediate 2-bromoethobromide (7).

The pure dry diquaternary dibromide (3; X = Br) when rapidly heated under nitrogen at 0.03 mm undergoes decomposition with loss of methyl bromide to give 1,2,3,4-tetrahydro-1,4-dimethyl-1,4-benzodiarsenin (**4**), b.p. 94–97°/0.015 mm, with some 1-bromo-1,2,3,4-tetra-hydro-4-methylbenzodiarsenin (**5**) (Mann and Baker[114]).

The diarsenin (**4**), earlier termed 1,4-dimethylethylene-*o*-phenylene-diarsine,[113] forms mono- and di-methobromides and methiodides under the same conditions as the similar derivatives of the diarsine (**2**): mono-methobromide, m.p. 228°, dimethobromide (**3**; X = Br); monometh-iodide, m.p. 250°, dimethiodide, m.p. 245–246°.

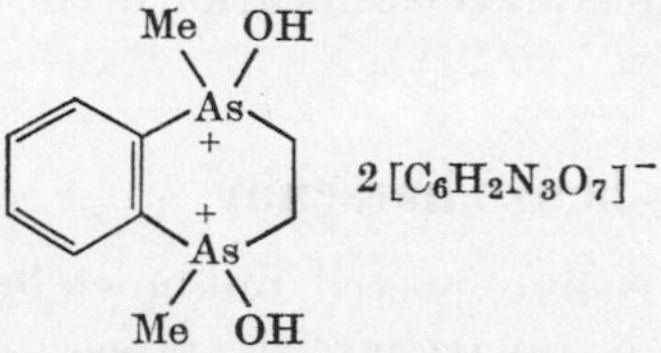

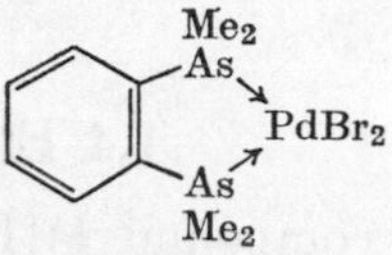

Oxidation of the diarsenin (**4**) gives the very soluble 1,4-dioxide, characterized as the bishydroxypicrate (**8**), m.p. 179–180°. The diarsenin in aqueous ethanol reacts with potassium tetrabromopalladite to give the deep orange crystalline bis(diarsenin)dibromopalladium,

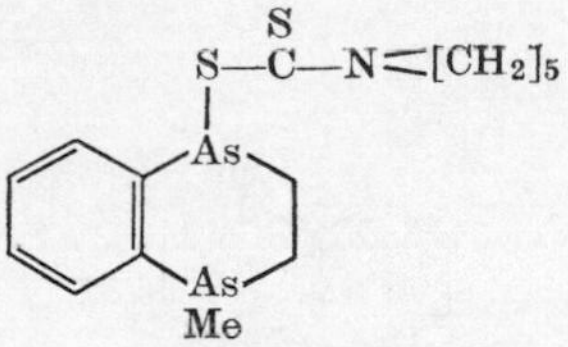

$$(C_{10}H_{14}As_2)_2PdBr_2,$$

m.p. 360°; this compound has an exceptional composition, in that only one arsenic atom in each diarsine molecule has coordinated with the palladium, probably for steric reasons, aided by the slight deactivation of the second arsenic atom by the coordination of the first atom. This behavior is in strong contrast to that of the *o*-phenylenebis(dimethyl-arsine) (**2**), which readily forms the very stable chelated bis(arsine)-dichloropalladium (**9**) (Chatt and Mann[115]); many similarly chelated metallic derivatives of this diarsine have since been prepared.

The 1-bromo-1,2,3,4-tetrahydro-4-methylbenzodiarsenin (**5**) could not be obtained pure by fractional distillation. It was identified by the following reactions. (*a*) With methanolic methyl bromide at 100° it forms the monomethobromide, m.p. 240°, but with boiling methyl iodide it forms the yellow monomethiodide of 1,2,3,4-tetrahydro-1-iodo-4-methylbenzodiarsenin. (*b*) With ethanolic piperidine 1-piperidine-dithiocarboxylate, it gives 1,2,3,4-tetrahydro-4-methyl-1-(1′-piperidine-thiocarbonylthio)-1,4-benzodiarsenin (**10**), m.p. 101–101.5°. (*c*) With methylmagnesium iodide it gives solely the diarsenin (**4**).

The diarsenin (**4**) was united in turn by cyclic diquaternization with compounds such as 1,2-dibromoethane and *o*-xylylene dibromide. Since the products have new heterocyclic systems, they are considered separately under the Ring Index names and classification in the following Sections.

1,4-Ethano-1,4-benzodiarsenin

This compound (**1**) [RRI 3261] is, as expected, unknown but certain derivatives may be prepared by a continuation of the process described in the previous Section.

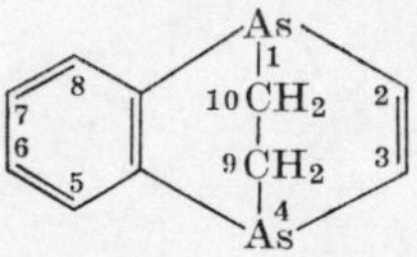

(1) 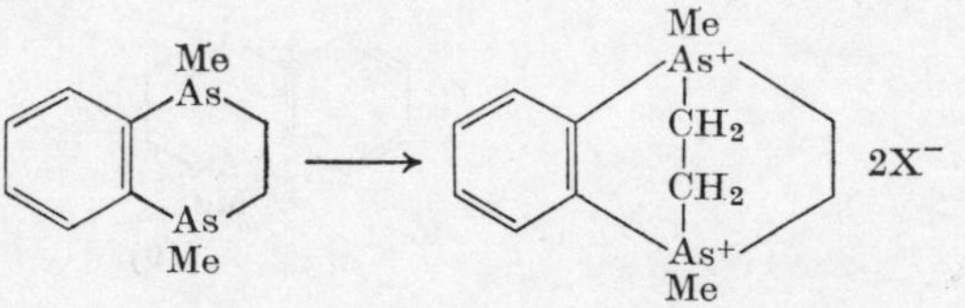   (2)   (3)

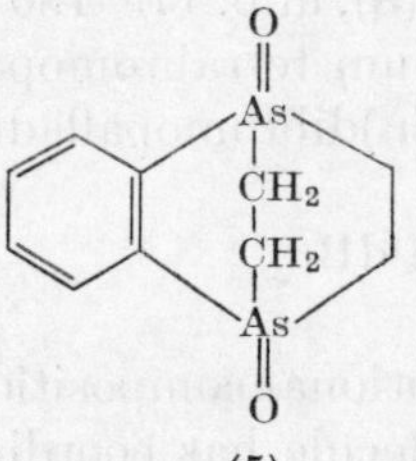

(5)

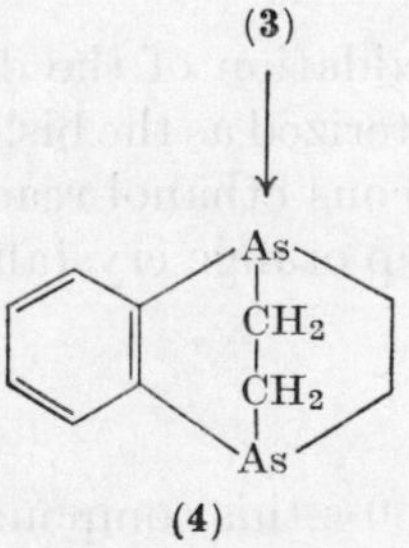

(4)

1,2,3,4 - Tetrahydro - 1,4 - dimethyl - 1,4 - benzodiarsenin (**2**), when heated with one equivalent of 1,2-dibromoethane, undergoes ready cyclic diquaternization to form 1,2,3,4-tetrahydro-1,4-dimethyl-1,4-ethano-1,4-benzodiarseninium dibromide (**3**; X = Br), m.p. 240° [dipicrate (**3**; X = $C_6H_2N_3O_7$), m.p. 230°]. X-Ray powder photographs give independent evidence that this dipicrate is structurally distinct from, although similar to, the dimethopicrate of the diarsenin (**2**), from which it differs in composition by only two hydrogen atoms. The dibromide, when heated under nitrogen at 270°/0.5 mm, decomposed smoothly giving methyl bromide and a distillate which after fractionation gave 1,2,3,4-tetrahydro-1,4-ethano-1,4-benzodiarsenin (**4**), b.p. 95–96°/0.01 mm (Mann and Baker[114]). This compound was earlier known as diethylene-o-phenylenediarsine.

The diarsenin (**4**) closely resembles the diarsenin (**2**) and also o-phenylenebis(dimethylarsine) in the comparative readiness of the formation of its simple mono- and di-quaternary salts. Thus methanolic methyl bromide in the cold gives the monomethobromide, m.p. 244°, and at 100° regenerates the dimethobromide (**3**; X = Br); boiling methyl iodide gives the monomethiodide, m.p. 240°, and at 100° gives the dimethiodide, m.p. 260°.

Oxidation of the diarsenin (**4**) with hydrogen peroxide gives the rather sticky 1,4-dioxide (**5**), which gives a crystalline bis(hydroxy-picrate), m.p. 184°.

The diarsenin (**4**) has a more feeble odor than that of the diarsenin (**2**), and is less susceptible to atmospheric oxidation.

Dibenzo[b,f][1,4]diarsocin and 5,12-Ethanodibenzo[b,f][1,4]diarsocin

Dibenzo[b,f][1,4]diarsocin (**1**), [RRI 3737], and 5,12-ethanodibenzo[b,f][1,4]diarsocin (**2**), [RRI 5315], are considered together because

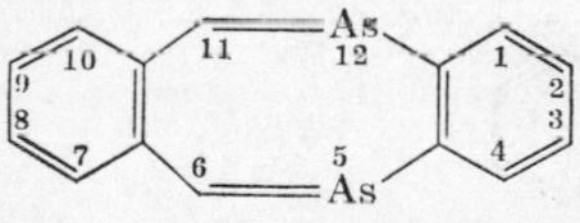

(1) 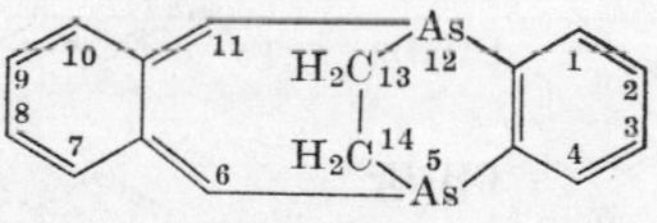   (2)

both the preparation and the properties of their known derivatives are very closely similar. (The names of both compounds indicate an eight-membered ring containing two arsenic atoms, but since compound (**2**) also has a six-membered ring containing two arsenic atoms, the pair of compounds are discussed at this stage.)

When equimolecular quantities of o-phenylenebisdimethylarsine (**3**) and powdered o-xylylene dibromide, $C_6H_4(CH_2Br)_2$, were mixed, they combined exothermally and the solid product, after being heated at 100° and crystallized from methanol, consisted of 5,6,11,12-tetrahydro-5,5,12,12-tetramethyldibenzo[b,f][1,4]diarsocinium dibromide monohydrate (**4**; R = H, X = Br), m.p. 214–216°, initially termed 1,4-dimethyl-o-phenylene-o-xylylenediarsine dimethobromide monohydrate (Jones and Mann[116]). The same dibromide rapidly separated from a warm methanolic solution of the above components. It gave a corresponding anhydrous dipicrate (4; R = H, X = $C_6H_2N_3O_7$), m.p. 219–220°.

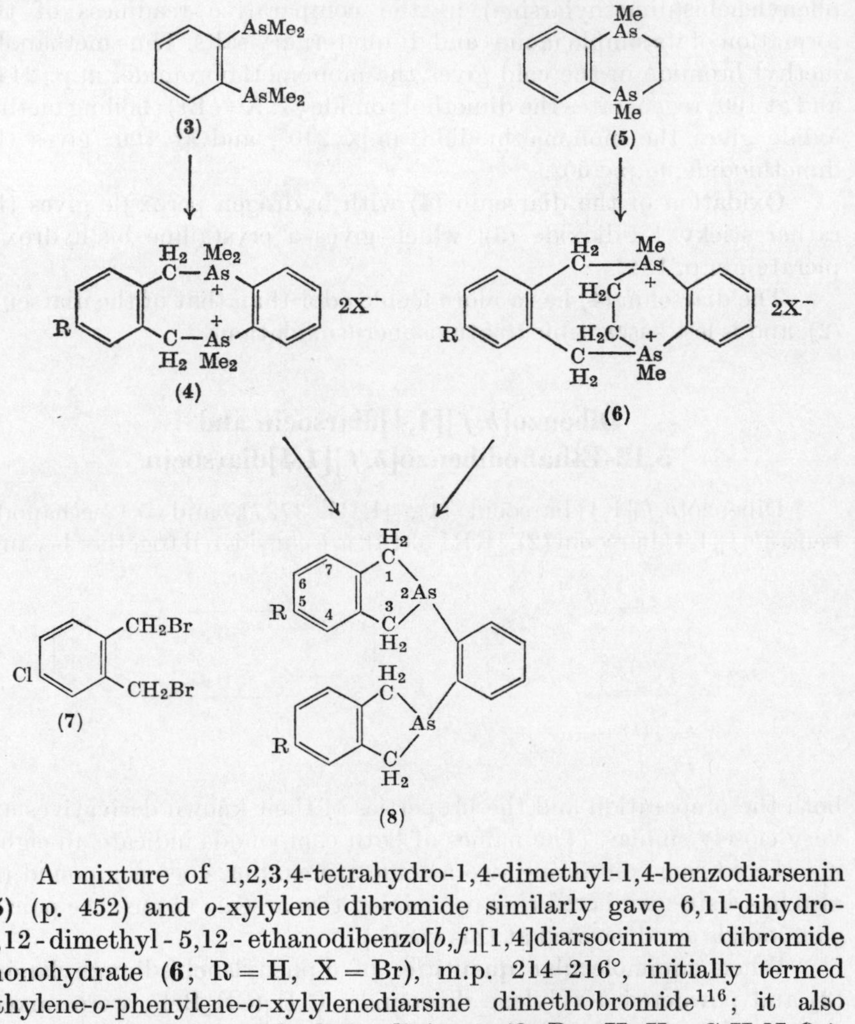

A mixture of 1,2,3,4-tetrahydro-1,4-dimethyl-1,4-benzodiarsenin (5) (p. 452) and o-xylylene dibromide similarly gave 6,11-dihydro-5,12 - dimethyl - 5,12 - ethanodibenzo[b,f][1,4]diarsocinium dibromide monohydrate (6; R = H, X = Br), m.p. 214–216°, initially termed ethylene-o-phenylene-o-xylylenediarsine dimethobromide[116]; it also gave a corresponding anhydrous dipicrate (6; R = H, X = $C_6H_2N_3O_7$), m.p. 221–222°.

A mixture of the two dibromides (4 and 6; R = H, X = Br), immersed at 160° and heated at the same rate as the separate dibromides, melted at 212–213° with softening at 207°.

The infrared spectrum (hexachlorobutadiene mull) of the dibromide (**4**; R = H, X = Br) showed strong sharp bands at 2976, 3413, 3344, and 2890 cm^{-1}, with intensities decreasing in this order, whereas that of the dibromide (**6**; R = H, X = Br) showed weak bands at 3333, 2967, and 3401 cm^{-1} (decreasing intensities) with a very faint indication of a band at 2882 cm^{-1}.

The structure and stereochemistry of the dibromides (**4** and **6**; R = H, X = Br) have interesting features. Certain reactions of these salts gave some indication that they might be dimeric, *i.e.*, in the case of the former dibromide, that it might be formed by the condensation of two molecules of the diarsine (**3**) with two of *o*-xylylene dibromide, giving a tetravalent cation having a 16-membered ring. An X-ray crystal structure investigation by Dr. W. Cochran showed conclusively, however, that this dibromide has the 'monomer' structure (**4**; R = H, X = Br); moreover crystals of this dibromide are isomorphous with those of the dibromide (**6**; R = H, X = Br), which therefore also has the monomer structure.[116]

A scale model of the cation of (**4**; R = H) shows that the distances apart of the arsenic atoms 5 and 12, and of the carbon atoms 6 and 11 of the methylene groups are almost identical, and that the two As—CH$_2$ bonds are thus almost parallel. Furthermore, the tetrahedral angle at these methylene groups would enable the attached benzene ring to be tilted above or below the plane of the As—CH$_2$ bonds, so that geometric isomerism might occur.

The *trans*- and the *cis*-cations are shown schematically in Figures 12 and 13, in which hydrogen atoms have been omitted. In each Figure, the benzene ring A, the arsenic atoms 5 and 12, and the methyl groups 5A and 12A are in the plane of the paper, but the methyl groups 5B and 12B are projecting towards, and the methylene groups 6 and 11 are projecting away from, the observer. In the *trans*-form (Figure 12), the benzene ring B projects away from A, but the tetrahedral angles at the methylene groups cause it to be almost parallel to A, but in a lower plane.

In the *cis*-form (Figure 13), the tetrahedral angles at the methylene groups cause the benzene ring B to adopt the alternative position, in which it is bent backward towards A, and is inclined at an angle to A, again below the plane of the paper.

Similar models of the cation (**6**; R = H) differ from those shown in Figures 12 and 13 only in that the methyl groups 5B and 12B now become methylene groups which, being linked, are rather closer together, and this constriction brings the methyl groups 5A and 12A slightly above the plane of the paper.

16

4-Chloro-*o*-xylylene dibromide (**7**) similarly combined with the diarsines (**3**) and (**5**) to give 8-chloro-5,6,11,12-tetrahydro-5,5,12,12-tetramethyldibenzo[*b,f*][1,4]diarsocinium dibromide monohydrate (**4**; R = Cl, X = Br), m.p. 241–242°, and 8-chloro-6,11-dihydro-5,12-dimethyl-5,12-ethanodibenzo[*b,f*][1,4]diarsocinium dibromide monohydrate (**6**; R = Cl, X = Br), m.p. 236°, respectively.

Although the above *cis–trans*-isomerism could theoretically be shown by each of the two dibromides, the crystalline salts in each case

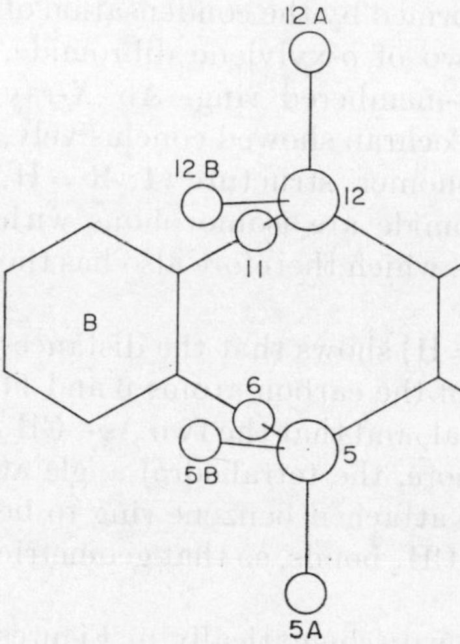

Figure 12 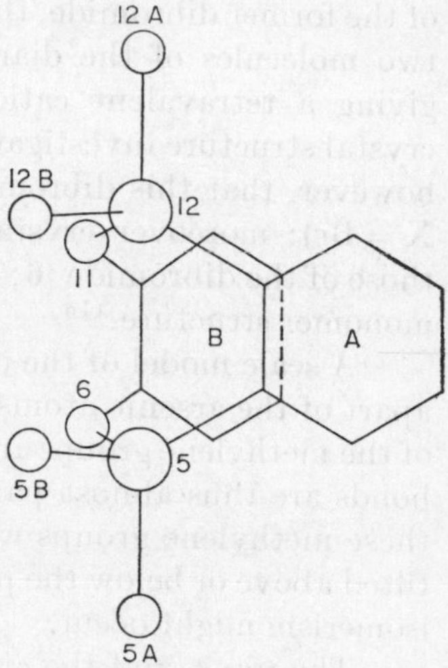   Figure 13

Figures 12 and 13. *trans*- (Figure 12) and *cis*-Isomers (Figure 13) of the cation of compound (**4**), as elucidated by Dr. W. Cochran. (Reproduced, by permission, from Emrys R. H. Jones and F. G. Mann, *J. Chem. Soc.*, **1955**, 405)

were apparently homogeneous; the X-ray analysis of the dibromide (**4**; R = H, X = Br) gave strong, but not conclusive, evidence that it had the *trans*-form (Figure 12).

The cations of the two dibromides (**4** and **6**; R = Cl, X = Br) are dissymmetric in both the *cis*- and the *trans*-isomers and should be resolvable into optically active forms. The possibility of any conversion *cis* ⇌ *trans* is negligible in the cation (**6**; R = Cl) especially, but fractional recrystallization of the di-(+)-camphorsulfonate and the (+)-3-bromo-8-camphorsulfonate gave no evidence of resolution.

The dibromides (**4** and **6**; R = H, X = Br), when heated just above their melting points at 20 mm until effervescence was complete, furnished a residue of *o*-phenylenebis-(2-isoarsindoline) (**8**; R = H), m.p. 129–131°.

The dibromides (**4** and **6**; R = Cl, X = Br) when similarly treated gave o-phenylenebis-[2-(5-chloroisoarsindoline)] (**8**; R = Cl), m.p. 191–195°.

The compound (**8**; R = H) formed only a monomethiodide, m.p. 180–181°, even under forcing conditions. The structure of (**8**; R = H) is

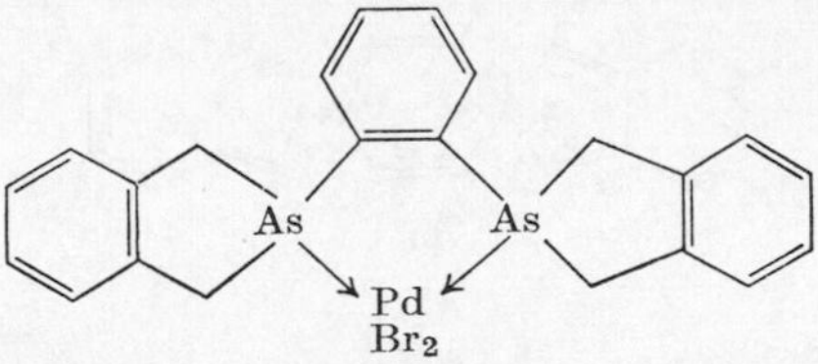

(**9**)

shown, however, by its ready formation of the crystalline o-phenylene-bis-(2-isoarsindoline)dibromopalladium (**9**), m.p. 336–337°, and by its cyclic diquaternization with 1,2-dibromoethane (cf. next Section).

The formation of the compounds (**8**; R = H or Cl) represents an entirely novel type of thermal decomposition of quaternary arsonium halides. It probably occurs by an intermediate dissociation of the dibromide to the original diarsine (**3** or **5**) and o-xylylene dibromide, followed by condensation of two molecules of the latter with one of the diarsine, with elimination of methyl (or ethylene) groups. This general mechanism is supported by the fact that an equimolecular mixture of the dibromide (**4**; R = H, X = Br) and o-xylylene dibromide, when subject to thermal decomposition under the same conditions, gave the compound (**8**; R = H) in increased yield.

The mechanism of this decomposition, and the evidence for the structure of the diarsines of type (**8**), have been discussed (Jones and Mann[116]).

Dispiro[isoarsindolium - 2,1' - [1,4]benzodiarsenin - 4',2'' - iso - arsindolium](2+) Salts

The bivalent cation named above, [RRI 6627], has the structure and numbering (**1**), being regarded systematically as a 1,4-benzodiarsenin unit having a 2-isoarsindole unit spirocyclically joined to each arsenic atom.

Salts of the cation (**1**) would obviously be unstable, and only hexahydro derivatives are known.

o-Phenylenebis-(2-isoarsindoline) (**2**) (cf. previous Section) when heated with 1,2-dibromoethane (one equivalent) at 125–130° undergoes cyclic diquaternization to form 1,1'',2',3,3',3''-hexahydrodispiro[isoarsindolium-2,1'-[1,4]benzodiarsenin-4',2''-isoarsindolium] dibromide monohydrate (**3**; X = Br), m.p. 223°, originally termed bis(isoars-indoline-*As*-spiro)ethylene-*o*-phenylenediarsonium dibromide (Jones and

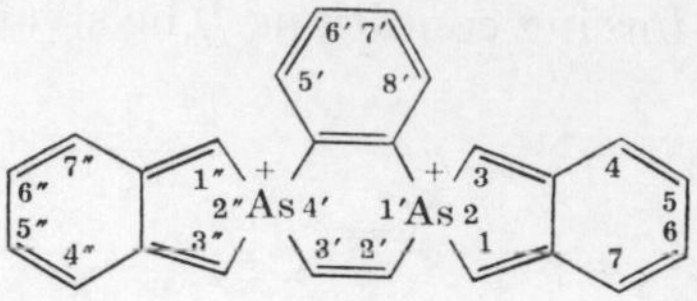

(**1**)

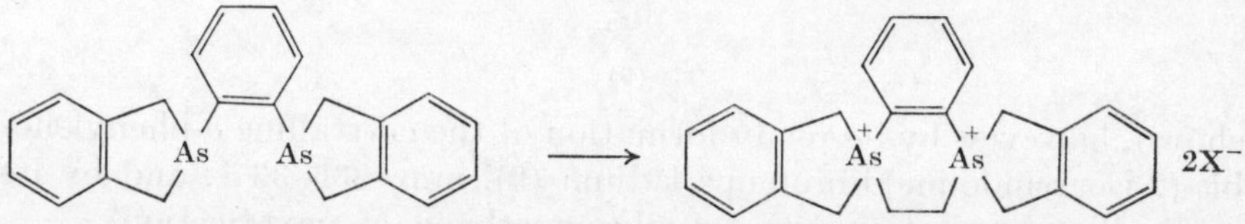

(**2**)   (**3**)   2X⁻

Mann[116]); it gives an anhydrous dipicrate (**3**; R = C$_6$H$_2$N$_3$O$_7$), m.p. 216–217°.

Arsanthrene

The parent compound arsanthrene* (**1**), [RRI 3262], is numbered as shown in accordance with the Ring Index and IUPAC rules; it has been termed diphenylenediarsine, diarsa-anthracene, and dibenzo-1,4-diarsenin, but these names should now be discarded. The existence of

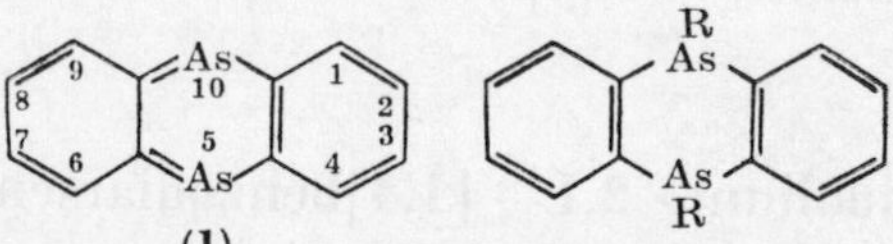

(**1**)   (**2**)

the true arsanthrene (**1**) is uncertain, and almost all its derivatives are of type (**2**); these are systematically regarded as substitution derivatives of 5,10-dihydroarsanthrene (**2**; R = H), and are known therefore as 5,10-disubstituted 5,10-dihydroarsanthrenes (**2**, where R is usually an alkyl or aryl group).

The arsanthrene ring system was first synthesized by Kalb,[117] utilizing the following series of reactions. *o*-Nitraniline was diazotized and converted by the Bart reaction into *o*-nitrophenylarsonic acid (**3**),

* In British and some other European literature, this word often becomes *arsanthren*.

which on reduction furnished *o*-aminophenylarsonic acid (**4**).* The latter was again diazotized and treated with oxophenylarsine, $C_6H_5As{=}O$, under the usual Bart reaction conditions to give *o*-arsono-diphenylarsinic acid (**5**). This acid, when mixed with concentrated hydrochloride acid containing a trace of hydriodic acid and then reduced with sulfur dioxide, gave a mixture of *o*-(dichloroarsino)diphenyl-arsinous chloride (**6**), m.p. 153–155°, and 5,10-dichloro-5,10-dihydro-arsanthrene (**7**), m.p. 182–183°. The conditions of this reduction and

NO2 / AsO(OH)2

(3)

NH2 / AsO(OH)2

(4)

O, OH / As / AsO(OH)2

(5)

Cl / As / AsCl2

(6)

Cl / As / As / Cl

(7)

cyclization were subsequently modified by Chatt and Mann[120] so that the yield of the trichloro compound (**6**) was reduced and that of the 5,10-dichloro compound (**7**) considerably increased. A very interesting by-product which arises at this stage, namely tri-*o*-phenylenediarsine (or 5,10-*o*-benzenoarsanthrene) is described later (p. 478).

Kalb investigated other possible routes to avoid this rather tedious synthesis. For example, he attempted to cyclize the arsinic acid (**5**) by dehydration with sulfuric acid to give arsanthrenic acid (**8**), alternatively named 5,10-dihydro-5,10-dihydroxyarsanthrene 5,10-dioxide, but the acid (**5**) was unaffected by sulfuric acid even at 150°; had this cyclization succeeded, the arsanthrenic acid could have been readily reduced in hydrochloric acid with sulfur dioxide to the 5,10-dichloro compound (**7**).

When the 5,10-dichloro-5,10-dihydroarsanthrene (**7**) was treated with aqueous sodium carbonate, it gave, not the corresponding di-hydroxy derivative, but the so-called 'oxide' (**9**), the anhydride of the 5,10-dihydroxy derivative, now termed 5,10-epoxy-5,10-dihydroars-

* Kalb's reduction of *o*-nitrophenylarsonic acid, using iron filings and hydrochloric acid, may give very erratic results, apparently depending on the condition of the iron. Reduction by iron filings in boiling aqueous sodium chloride solution gives uniformly good results (Rozina[118]; cf. also Mann and Stewart[119]).

anthrene (cf. p. 472). Reduction of the 5,10-dichloro compound (7) with zinc and hydrochloric acid, or of the oxide (9) with ethanolic phenyl-hydrazine, apparently gave the parent arsanthrene (1). This compound is of great interest because it represents one of the very few in which the =As— group does apparently replace the =CH— in an aromatic ring. Kalb describes this compound as forming orange-yellow plates, which on heating become orange-red at about 170° and greenish-yellow at about 290° and finally melt at 340°. Unfortunately the molecular weight was not recorded. It will be clear that a compound of constitution (1) would exist as a resonance hybrid, similar in type to anthracene. Although the existence of resonance in fused ring systems is often associated with visible color, this is not always the case, and in view of the above

O   OH
   As
   As
O   OH

(8)   (9)

colors it is not certain that Kalb's compound has the simple structure (1) or if it is a polymer based on this structure.

The dichloro compound (7) when treated with silver cyanide in boiling toluene gives the corresponding dicyano derivative, m.p. 165–167° (Plant et al.[121]).

Wieland and Rheinheimer[122] have attempted to nitrate 5,10-dichloro-5,10-dihydroarsanthrene in order subsequently to obtain amino derivatives of arsanthrene. The 5,10-dichloro compound resisted nitration by even energetic nitrating agents, but was converted into the above arsanthrenic acid (8), colorless crystals, m.p. above 360°. To overcome this difficulty, the arsinic acid (5) was nitrated to give 2-arsono-3′-nitrodiphenylarsinic acid (10), which on reduction with ferrous hydroxide furnished the 3′-aminoarsinic acid (11). This compound when reduced under the usual conditions with hydrochloric acid and sulfur dioxide gave 2-amino-5,10-dichloro-5,10-dihydroarsanthrene (12), the hydrochloride of which formed colorless crystals. Oxidation of this compound gave 2-aminoarsanthrenic acid (13) as colorless needles. It should be noted, however, that Wieland and Rheinheimer had no decisive evidence that the nitro group had entered the 3′-position in the arsinic acid (10) and that the amino group had therefore the 3′-position in the acid (11). Furthermore, if it is assumed that the latter acid has the constitution (11), it will be seen that on cyclization it could give either the 2-amino- or the 4-amino-5,10-dichloro-5,10-dihydro-arsanthrene, although if one can argue by analogy from Roberts and Turner's later work[123] on the cyclization of 3'-chloro-2-(dichloroarsino)-diphenyl ether to the corresponding phenoxarsine (p. 558), the formation of the 2-aminoarsanthrene is the more probable. It can be said, therefore,

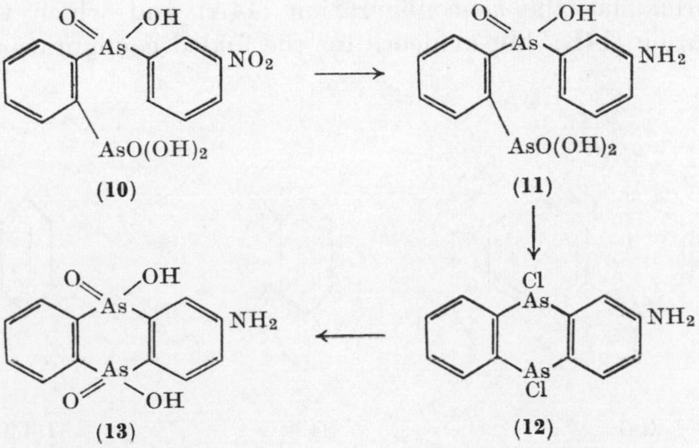

(10)   (11)

(13)   (12)

that the structures allocated to compounds (10)–(13) by Wieland and Rheinheimer are those which are most probable, although not certain.

The stereochemistry of 5,10-disubstituted 5,10-dihydroarsanthrenes is closely similar to that of 5,10-diethyl-5,10-dihydrophosphanthrene, which has been discussed in detail (pp. 186 ff.), and that of the arsanthrenes can therefore be more briefly described. Twenty-four years before isolation of the above phosphanthrene in two geometrically isomeric forms, Chatt and Mann[120] had reasoned that, if the arsenic atoms in the 5,10-dihydroarsanthrene system retained their 'natural' intervalency angle of *ca.* 98°, the molecule must be folded about the As—As axis and that this in turn should give rise to a *cis-* and a *trans-*form. The 5,10-dichloro derivative (7) was therefore converted into the stable, highly crystalline 5,10-dihydro-5,10-di-*p*-tolylarsanthrene (14), the *cis-* and the *trans-*form of which are shown schematically in (14A) and (14B), respectively, where T indicates the *p*-tolyl groups; in these diagrams, the arsenic and the tolyl groups are represented as being in the plane of the paper, but one *o*-phenylene group (shown in thick lines) projects towards, and the other *o*-phenylene group away from, the observer.

Chatt and Mann found that their 5,10-ditolyl compound, when initially prepared, formed white crystals, m.p. 133–158°; a slow and delicate process of fractional crystallization ultimately separated this product into the α-form, small leaflets, m.p. 178–179°, and the β-form,

large bipyramidal leaflets, m.p. 179–181°; a mixture of equal quantities
of the two forms melted at 144–158°. Each isomer showed the correct
molecular weight in boiling acetone. The two forms were, moreover,
surprisingly stable, for each isomer could be kept in the molten condition
at 190° for 10 minutes without any indication either of conversion into
the other form or of chemical decomposition. It is not known which of
these forms has the *cis*-configuration (14A) and which the *trans*-
configuration (14B), but evidence for the folded configurations is now
decisive.

(14)   (14 A)   (14 B)

The difference between these isomers extended also to their monomethiodides. Each monomethiodide had m.p. 178–179°, but a mixture
of equal quantities of the two compounds melted at 167–175°. The
α-monomethiodide crystallized from ethanol with one molecule of
ethanol of crystallization and from water in the anhydrous condition;
the β-monomethiodide crystallized from ethanol in the pure solvent-free
condition but from water as the monohydrate. Neither salt would form
a dimethiodide, although 'forcing' conditions were not applied. This
stability of the two monomethiodides may be the result of partial
deactivation of the tertiary arsenic atom by the positive pole of the
quaternized arsonium atom, combined with some degree of steric
hindrance. [It is noteworthy that 5,10-dihydro-5,10-diphenylphosph-
anthrene (p. 191) gives a dimethiodide, but tertiary phosphines in
general undergo quaternization more readily than their arsenic analogues.] Inductive deactivation by quaternized and other polar groups
has been discussed in some detail by Mann and J. Watson.[124]

The α- and the β-form of the 5,10-dihydro-5,10-di-*p*-tolylarsan-
threne, when treated with bromine, gave the same tetrabromide (15),
the identity of which as a single compound was further confirmed by
hydrolysis to the highly crystalline tetrahydroxide. This loss of isomeric
difference in the tetrabromide is to be expected. Wells[125] has found that
the trimethylstibine dihalides, $[(CH_3)_2SbX_2]$, have a trigonal bipyramid

structure, the antimony atom being at the center of an equilateral triangle, the methyl groups equatorially at its apices, and the halogen atoms at equal distances above and below the antimony atom on an axis perpendicular to the plane of the triangle; it follows that in these compounds, the C–Sb–C angle is 120°. There is evidence (p. 470) that tertiary arsine dihalides also have the trigonal bipyramid structure, and it follows that the C–As–C angle in the ring in the tetrabromide (15) should also be 120°; consequently the three rings would become uniplanar, and no basis for geometric isomerism would exist.

(15)   (16)

It should be added that when the tetrabromide was reduced back to the original 5,10-ditolyl derivative (14), an X-ray examination revealed the presence of only two isomers identical with those already obtained.

When the tetrabromide was treated with an excess of sodium sulfide, it gave elementary sulfur and 5,10-dihydro-5,10-di-p-tolylarsanthrene monosulfide. The latter could not be converted into a disulfide or a monosulfide monomethiodide, presumably because even the weak positive pole on the As$^+$–S$^-$ group was sufficient to inactivate the second arsenic atom.

When, however, the tetrabromide was treated with hydrogen sulfide, it gave the above monosulfide and also 5,10-dihydro-5,10-di-p-tolylarsanthrene dibromide. The properties of this compound indicated that the two bromine atoms were probably not linked covalently to one arsenic atom, but that the compound was a salt of structure (16). The evidence for the ionic structure is as follows. (a) The dibromide is soluble in hot water, ethanol, and other polar solvents, but not in ether or benzene. These are the properties of an electrovalent, and not a covalent compound. (b) The dibromide can be recrystallized unchanged from hot aqueous ethanol, whereas many tertiary arsine dibromides, R_3AsBr_2, give the corresponding hydroxybromide, $R_3As(OH)Br$, in these circumstances. (c) It reacts very slowly with hydrogen sulfide to give the above monosulfide, whereas most dibromides of type R_3AsBr_2 readily give the sulfide, R_3AsS, under these conditions. (d) Its high melting point, 298–300°, is in the region expected for a salt rather than a true covalent compound. In view of the considerable stability of the dibromide, it should be noted that if it had the ionic structure (**16**), it would of course exist as a resonance hybrid, which would enhance its stability; further-more, the tricyclic system would be planar.

Further light on the nature of such salts and on other aspects of arsanthrene chemistry was obtained by a detailed study of 5,10-dihydro-5,10-dimethylarsanthrene (**17**), a compound which was selected because the methyl groups should give maximum reactivity to the arsenic atoms (Jones and Mann[126]). This arsanthrene (**17**), prepared by the action of methylmagnesium iodide on the dichloro derivative (**7**), forms excellent crystals (90%), m.p. 191–192.5° after recrystallization from ethanol–chloroform.

Before discussing the reactions of this compound, it should be noted that a recent complete X-ray crystal analysis has shown (*a*) that the crystals consisted solely of one form and were not a *cis–trans*-mixture like the 5,10-di-*p*-tolyl compound (**14**), (*b*) that this form was folded about the As—As axis, the angle between the two wings of the 'butterfly' conformation being 117°, in good agreement with the earlier calculated

(17)   (17 A)   (17 B)

value of 121° (Mann[127]), and (*c*) that the two methyl groups were in the *cis*-position relative to one another, and were within the angle subtended by the two *o*-phenylene groups, as in (**17A**) where the arsenic and the methyl carbon atoms are depicted in the plane of the paper, with the *o*-phenylene group in thick outline projecting above this plane and the other projecting below (Kennard, Mann, Watson, Fawcett, and Kerr[128]). These results (1968) provided the first decisive evidence for the accuracy of the conformational deductions[120] of the 5,10-disubstituted 5,10-dihydroarsanthrene system made on a basis of intervalency angles and bond lengths in 1940, although a considerable amount of circumstantial evidence for this conformation had meanwhile accumulated.

The values of the intervalency angles and bond lengths in the arsanthrene (**17**), as determined by the *X*-ray crystal analysis, are shown in Fig. 14.

It is noteworthy that although the 5,10-dihydro-5,10-dimethyl-arsanthrene has the *cis*-conformation (**17A**) in the crystalline state, there is a possibility that in solution the molecule could 'flex' towards (**17B**), as described in detail for 5,10-dihydro-5,10-dimethylphosphan-threne (p. 187); this flexing could never be complete owing to the mutual obstruction of the methyl groups, and it would not involve *cis*–*trans*-interconversion.

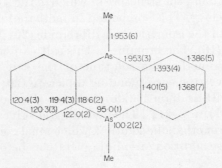

Figure 14. The intervalency angles (in degrees) and the interatomic distances (in Å) in 5,10-dihydro-5,10-dimethyl-arsanthrene. (Reproduced, by permission, from O. Kennard, F. G. Mann, D. G. Watson, J. K. Fawcett, and K. A. Kerr, *Chem. Commun.*, **1968**, 269)

On the chemical side,[126] the 5,10-dihydro-5,10-dimethylarsanthrene (**17**), when heated with methanolic methyl bromide in a sealed tube at 50° for 2 hours, gave the monomethobromide, m.p. 219°, but at 100° for 6 hours gave the dimethobromide, m.p. 302° (dimethopicrate, m.p. 282°); when heated with boiling methyl iodide it gave only the mono-methiodide, m.p. 222°, but with methyl *p*-toluenesulfonate at 180° for 4 hours it gave the dimetho-*p*-toluenesulfonate, m.p. 316°, which was converted into the dimethiodide, m.p. 263°. All these metho salts are colorless.

The stability of the ring system was again shown, for the dimethyl compound (**17**) when treated with a mixture of fuming nitric and concentrated sulfuric acids at 100° gave solely the corresponding diarsine di(hydroxynitrate) (**18**) (systematically termed 5,10-dihydro-5,10-dihydroxy-5,10-dimethylarsanthrenium dinitrate) which readily gave the corresponding di(hydroxypicrate).

The dihydrodimethylarsanthrene (**17**), when warmed with one equivalent of bromine in ethanol, gives the covalent 5,5-dibromide

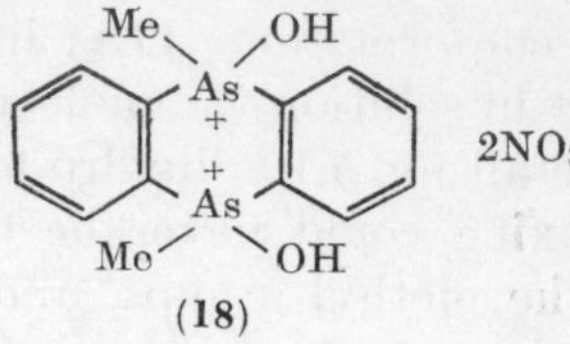

(**18**)

(**19**; X = Br), m.p. 256°, the structure of which in the solid state has been established (see below). This compound in methanol or ethanol gives a colorless solution of the isomeric ionic dibromide (**20**; X = Br), identical in structure with (**16**). The bromine in this salt can be estimated volumetrically, and a methanolic solution of the salt with methanolic sodium picrate gives the corresponding dipicrate (**20**; X = $C_6H_2N_3O_7$), m.p. 244°, and with sodium iodide gives the deep orange crystalline diiodide (**20**; X = I), m.p. 214°. The pure diiodide gives, however, a pale yellow methanolic or ethanolic solution, which must contain the colorless cation (**20**), because addition of sodium picrate as before gives the same

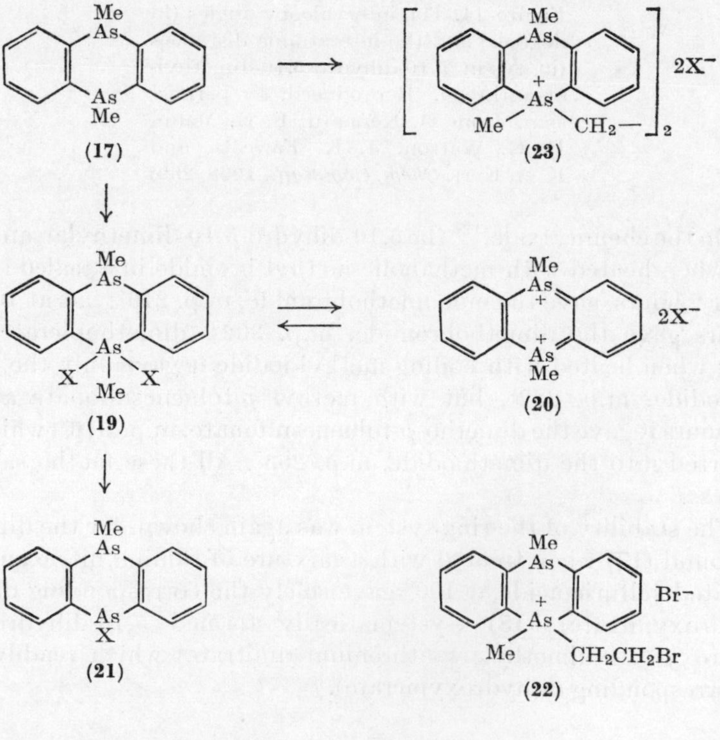

(**17**)   (**23**)

(**19**)   (**20**)

(**21**)   (**22**)

dipicrate (**20**; $X = C_6H_2N_3O_7$). Ionic iodides, however, would be colorless (cf. the above mono- and di-methiodides), whereas the orange-yellow color is characteristic of many tertiary arsine di-iodides of type R_3AsI_2. There is therefore very strong evidence that the diiodide in solution exists as a tautomeric equilibrium of the forms (**19**; $X = I$) and (**20**; $X = I$). This may also apply to the corresponding dibromide, but since tertiary arsine dibromides, R_3AsBr_2, are usually colorless, the

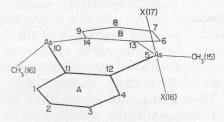

Figure 15. Diagrammatic structure of the compounds (**19**; X = Br or I), the normal numbering of the arsanthrene ring being extended as shown. The As(5) and As(10) atoms are depicted in a horizontal line in the plane of the paper, with the two o-phenylene groups directed slightly downwards, the group (A) towards and the other (B) away from the observer. The $CH_3(15)$ and the $CH_3(16)$ groups, and the two halogen atoms X attached to the atoms As(5), are in the plane of the paper. [From F. G. Mann, *J. Chem. Soc.*, **1963**, 4266, and adapted, by permission, from D. J. Sutor and F R. Harper, *Acta Cryst.*, **12**, 585 (1959)]

existence of the covalent form (**19**; $X = Br$) in equilibrium with the salt (**20**; $X = Br$) in solution would not be apparent. The closely comparable covalent compounds $(C_6H_5)_2CH_3AsI_2$ and $(C_6H_5)_2CH_3AsBr_2$ are deep orange and cream-colored.[126]

An X-ray crystal analysis of the dibromide (**19**; $X = Br$) and diiodide (**19**; $X = I$) (Sutor and Harper[129]) shows that these compounds have virtually identical structures corresponding to the covalent form (**19**).

An interesting feature of this structure of the dibromide (see Figure 15) is the strong evidence that the As(5) atom is in the same plane as the three carbon atoms to which it is linked, and that the Br(17)-As(5)-Br(18) atoms are almost linear. This trigonal-bipyramidal configuration of the As(5) atom, if undistorted, would tend to make the tricyclic system planar, but the As(10) atom retains its pyramidal configuration; the molecule is still folded about the As—As axis, but presumably because of these conflicting influences, the angle between the two o-phenylene groups is increased from the value of 117° in the arsanthrene (17) to a value of *ca*. 157°. The Br(18) and Br(17) atoms, *i.e.*, the atoms within and outside this angle, respectively, are tilted as shown at 14° from the perpendicular to the As—As axis, apparently to accommodate the folded structure. The closer proximity of the o-phenylene groups also increases slightly, but significantly, the arsenic–halogen distance within this angle compared with that 'outside' this angle; the inter-atomic distances are As–Br(18) 2.66, As–Br(17) 2.59; As–I(18) 2.98 and As–I(17) 2.80 Å. These lengths are greater than those in normal 'un-hindered' covalent compounds, for example, As–Br, 2.35 in $AsBr_3$[130] and $(CH_3)_2AsBr$[131]; As–I, 2.53 in AsI_3[132] and $(CH_3)_2AsI$.[131] This is to be expected,[125] for in the trimethylstibine dihalides, $(CH_3)_3SbX_2$, the two Sb–X lengths in each compound are of course equal but are of greater than normal covalent length (Mann[127]).

In view of the trigonal bipyramid structure of the five-covalent As(5) group, there is little doubt that both the arsenic groups in the tetrabromide (15) and the corresponding tetrahydroxide also have this structure, and that arsanthrene compounds of this type, and also those of the salt type (16) and (20), have a planar tricyclic system, whereas the simple arsanthrenes of type (10) and (17), and those of type (19), have a folded system.

Another arsanthrene compound similar in structure to (19; X = I) is described on p. 483.

The dibromide (19; X = Br), when heated at 0.1 mm, loses methyl bromide with the formation of the pale yellow 5-bromo-5,10-dihydro-10-methylarsanthrene (21; X = Br) (81%), which in boiling acetone with sodium iodide gives the corresponding deep yellow 5-iodo deriva-tive (21; X = I). These reactions can be reversed, for the 5-bromo compound (21; X = Br) when heated with an excess of methyl bromide at 100° gives the dibromide (19; X = Br), and when similarly heated with methyl iodide gives the diiodide (19; X = I); the mechanism of these reactions is probably simple eviction in the hot solution of the halogen atom in (21) to give the corresponding stable salt (20), which on cooling, however, crystallizes as the dihalide (19).[126]

The reaction of 1,2-dibromoethane and dimethylarsanthrene (17) depends on the conditions employed and is curiously varied. Equimolecular quantities of these compounds, when heated in a sealed tube at 100° for 24 hours or at 125° for 6 hours and then cooled, gave a moderate pressure of a gas (probably ethylene) and a residue of the dibromide (19; X = Br) (60% when pure). The mechanism of this process is uncertain. The probable intermediate is the monoquaternary salt (22), *i.e.*, 5-(2'-bromoethyl)-5,10-dihydro-5,10-dimethylarsanthrenium bromide, which under these conditions does not undergo cyclic quaternization with the (partially deactivated) As(10) atom; the positive charge on the As(5) atom apparently also weakens the C—Br bond of the bromoethyl group, with consequent liberation of ethylene and the formation of (19; X = Br), possibly through the intermediate formation of the tautomer (20; X = Br).

When the above mixture of reagents in methanol was heated at 100° for 9 hours, the 1,2-dibromoethane quaternized the As(5) atom of two molecules of (17) with the formation of 5,5'-ethylenebis-(5,10-dihydro-5,10-dimethylarsanthrenium) dibromide (23; X = Br), m.p. 195°; this in turn gave a dipicrate (23; X = $C_6H_2N_3O_7$), m.p. 237°, and a colorless diiodide, m.p. 185°.

When this dibromide was heated at 210°/0.1 mm for a few minutes, effervescence occurred, and the recrystallized residue furnished pale yellow mixed crystals, m.p. 142–147° (76%), of equimolecular amounts of (17) and of the bromo compound (21; X = Br). This composition was proved by analysis and as follows: (*a*) when equimolecular amounts of these components were recrystallized from ethanol–chloroform, the same crystals, m.p. 142–147°, were deposited; the m.p. was unaffected by admixture with the previous sample or by recrystallization from acetone; (*b*) the original crystals, in boiling methyl iodide, gave the dimethylarsanthrene monomethiodide and the less soluble diiodide (19; X = I), these products arising from (17) and from (21; X = Br), respectively.[126]

The diquaternization of 5,10-dihydro-5,10-dimethylarsanthrene (17) with 1,2-dibromoethane, 1,3-dibromopropane, and *o*-xylylene dibromide gives rise to new ring systems, which are described in the following Sections.

Arsanthrenes having alkyl groups in the benzo rings have been only slightly investigated.[126] Chloro-(*o*-dichloroarsinophenyl)-*p*-tolylarsine (24) on attempted distillation underwent cyclization to 5,10-dichloro-5,10-dihydro-2-methylarsanthrene (25; R = Cl), b.p. 188–193°/0.05 mm, a viscous liquid which with methylmagnesium iodide gave 5,10-dihydro-2,5,10-trimethylarsanthrene (25; R = CH_3), a gum which was not obtained pure. A solution of this product in cold methyl iodide, however, deposited the crystalline monomethiodide, m.p. 215–216°, probably the isomer (**26**); and a solution in cold methanolic methyl bromide slowly deposited the dimethobromide, m.p. 265°, which in turn gave a dimethopicrate, m.p. 258°. The trimethylarsanthrene (**25**; R = CH_3) gave

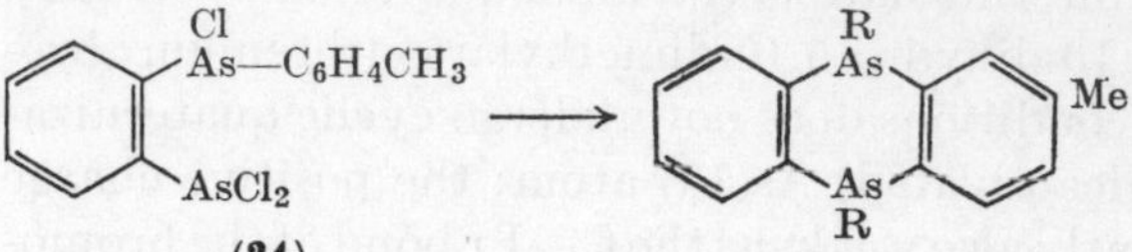

Cl
As—$C_6H_4CH_3$

$AsCl_2$

(**24**)

R
As

Me

As
R

(**25**)

Me
As

Me

$\overset{+}{As}$
Me_2

I⁻

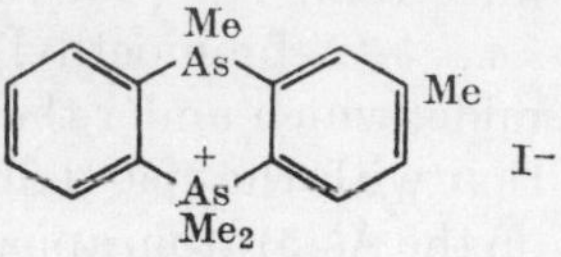

(**26**)

no well-defined products with 1,2-dibromoethane or with 1,3-dibromopropane; its derivative with o-xylylene dibromide is discussed on p. 484.

5,10-Epoxy-, 5,10-Epithio-, 5,10-Episeleno-, and 5,10-Epitelluro-5,10-dihydroarsanthrene

These compounds are discussed at this stage partly because they are the only known arsanthrene derivatives having a five-membered ring, and because they form a unique eutropic series; the four members have many properties in common and are more effectively discussed in one section than in separate sections according to their Group VI component.

5,10-Epoxy-5,10-dihydroarsanthrene (**1**), [RIS —], was recorded by Kalb[117] in 1921 but was not further investigated (p. 461); the

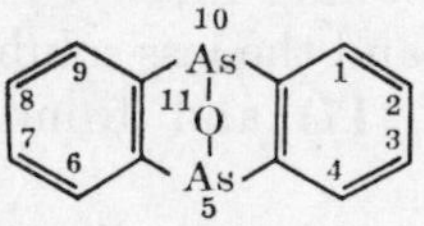

As
O
As

(**1**)

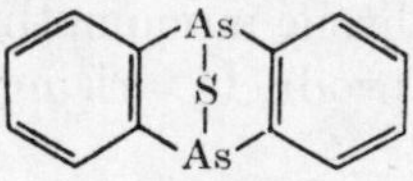

As
S
As

(**2**)

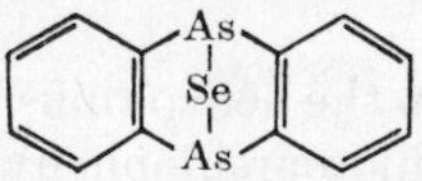

As
Se
As

(**3**)

As
Te
As

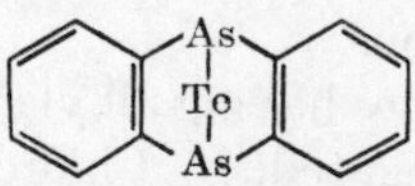

(**4**)

sulfur, selenium, and tellurium analogues have only recently been discovered (1968). A chemical and crystallographic investigation of these four compounds has been carried out by Allen, Coppola, Kennard, Mann, Motherwell, and D. G. Watson.[133]

Kalb showed that 5,10-dichloro-5,10-dihydroarsanthrene (5) could be very readily hydrolyzed, even when its ethereal solution was shaken with cold aqueous sodium carbonate, to give 'arsanthrene oxide', m.p. 196°, which he formulated as (1) (p. 462). This formulation, if correct, would indicate that the compound is systematically the anhydride of 5,10-dihydroarsanthrene-5,10-diol: in the absence of molecular-weight values, however, the possibility of a dimer, formed to relieve possible strain in the structure (1), could not be ignored.

Recent work[133] showed, however, that the compound (1) is monomeric in boiling benzene and in cyclohexane; this is confirmed by mass spectrography, which showed a molecular ion of m/e 318 (M for $C_{12}H_8As_2O$, 318). Moreover, a Dreiding type model of (1), constructed with very little strain using the approximate interatomic distances (Å) C–As 1.95, As–O 1.90, C–C 1.40, had a conformation shown diagrammatically as (6), in which the three intervalency angles at the arsenic atoms came reasonably close to the 'natural' angle of ca. 98°.

(5)   **(6)**   **(7)**

The comparative quantities used by Kalb in the preparation of the epoxyarsanthrene (1) require modification. A solution of the dichloroarsanthrene (5) (1 g) in ether (50 ml) is treated under nitrogen with aqueous N-Na_2CO_3 (50 ml) and vigorously shaken for 2 hours. The precipitated product (1), which is only slightly soluble in ether, is collected, washed with water, dried, and crystallized from benzene, ethanol, or cyclohexane; the yield being 0.7 g and the m.p. 197–199°.

The mechanism of this very ready hydrolysis of the 5,10-dichloroarsanthrene (5) to the epoxy compound (1), although unknown, must have certain points of interest. The 5,10-dichloroarsanthrene (5) has most probably the same configuration as the 5,10-dihydro-5,10-dimethylarsanthrene, (p. 466), namely, the cis-configuration with the two chlorine atoms within the angle subtended by the two o-phenylene groups [cf. formula (17A), p. 466]. This molecule could 'flex' over to the opposite position [cf. (17B), p. 466], and if meanwhile one chlorine atom had been replaced by an OH group, the electronic and stereochemical factors would promote loss of HCl and the closing (7) of the five-membered epoxy ring. This complete attainment of the '$anti$-cis'

configuration is of course impossible in the 5,10-dihydro-5,10-dimethyl-arsanthrene owing to the mutual obstruction of the stable methyl groups [cf. (**17B**), p. 466].

It should be recognized that epoxyarsanthrene is entirely different in type from the bridged compounds formed by the cyclic diquaterniza-tion of 5,10-dihydro-5,10-dimethylarsanthrene with ethylene and trimethylene dibromide (pp. 476, 477); these compounds, in which the 'bridge' across the arsenic atoms consists of CH_2 units, could be formed only when the dimethylarsanthrene has the normal *cis*-configuration [cf. (**17A**), p. 466], and the arsenic atoms in the final product must have the tetrahedral arsonium configuration.

When hydrogen sulfide is passed through an ethanolic solution of the epoxyarsanthrene (**1**) at room temperature, the colorless crystalline 5,10-epithio-5,10-dihydroarsanthrene (**2**), [RIS —], m.p. 184–185°, is rapidly deposited. It can also be prepared by shaking an ethereal solution of the dichloroarsanthrene (**5**) with aqueous sodium sulfide for 10–15 minutes and then evaporating the separated ethereal layer. It shows a normal molecular weight in dichloromethane at 30°.

Similarly, when hydrogen selenide in nitrogen is passed through an ethanolic solution of the epoxyarsanthrene (**1**), episeleno-5,10-dihydro-arsanthrene (**3**), [RIS —], is deposited as yellow crystals, m.p. 220° (sealed tube) after crystallization from ethanol–chloroform. The use of hydrogen telluride in nitrogen in a similar experiment gives a deposit of the orange-red crystalline 5,10-epitelluro-5,10-dihydroarsanthrene (**4**), [RIS —], m.p. >210° (dec.) (sealed tube) after sublimation at 170–180°/0.2 mm.

These compounds (**1**), (**2**), (**3**), and (**4**) have several properties in common, although insufficient material has precluded a full investiga-tion of the telluro compound (**4**).

All four compounds can be readily sublimed when heated under reduced pressure, and the epoxy compound (**1**) can be sublimed at room temperature at *ca.* 0.01 mm pressure.

The compounds (**1**), (**2**), and (**3**) are unaffected when their solutions in methyl iodide are boiled for 3–4 hours, and the compound (**1**) is unaffected when mixed with methyl *p*-toluenesulfonate and heated at 150° for 3 hours.

They are readily attacked by nucleophilic reagents, and compounds (**1**) and (**2**) when treated in the usual way with methylmagnesium iodide give 5,10-dihydro-5,10-dimethylarsanthrene in high yield.

All four compounds are oxidized in acetone solution with hydrogen peroxide (100-vol.) giving arsanthrenic acid (p. 462); this uniform eviction of the Group VI element with retention of the arsanthrene ring system is noteworthy. Solutions of the epoxy compound (1) in benzene, ethanol, or cyclohexane, when exposed to the air at room temperature, slowly deposit the insoluble arsanthrenic acid.

The mass spectra of the four compounds are similar, and show prominent molecular ions, m/e 318, 334, 381, and 430, respectively, thus confirming the composition and molecular weight of each compound: the molecular ions in the spectra of the compounds (3) and (4) are more

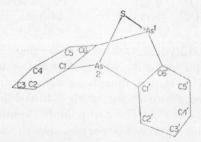

(8)

complex than those in the spectra of (1) and (2), owing to the natural isotopic distribution of selenium and tellurium. All members of the series show a similar fragmentation, the base peak in each spectrum being at m/e 227, attributed to the dibenzarsolyl ion (9-arsafluorenyl ion) (8). This arises from the loss of an arsenic atom and the bridging atom, followed by o,o'-coupling; the direct conversion of the molecular ion into the ion (8) is supported in each case by the appearance in the spectrum of the appropriate metastable peak.

Figure 16. The structure and conformation of 5,10-epithio-5, 10-dihydroarsanthrene (2). (Reproduced, by permission, from D. W. Allen, J. C. Coppola, O. Kennard, F. G. Mann, W. D. S. Motherwell, and D. G. Watson, *J. Chem. Soc., C,* in the press)[133]

Under suitable conditions, the epoxyarsanthrene (1) undergoes thermal decomposition to arsenious oxide and 5,10-o-benzenoarsanthrene (p. 478): the thermal decomposition of the other three members of the series has not been studied.

Crystals of the epoxyarsanthrene (**1**), obtained by recrystallization ('rhombs') or sublimation (fine needles), proved unsatisfactory for X-ray analysis, and the compound moreover decomposed on prolonged exposure to the X-rays. Excellent stable crystals of the compounds (**2**), (**3**), and (**4**) were, however, obtained.

The crystal structure of the 5,10-epithio-5,10-dihydroarsanthrene (**2**) was determined by a direct method of analysis that required no assumption as to the molecular structure. The crystal is orthorhombic, space group *Pnma* with 4 molecules in the unit cell; $a = 11.67$ Å, $b = 16.53$ Å, $c = 5.838$ Å. The molecule has a mirror plane through the arsenic and sulfur atoms which coincides with a crystallographic mirror plane. The conformation of the molecule (see Figure 16) is closely similar to the earlier suggested conformation (**6**).

The numbering of the atoms in Figure 16 is for crystallographic convenience, and differs from the chemical numbering (**1**). At a refinement $R = 8.5\%$, the annexed values were obtained.

Interatomic distances (Å)		*Intervalency angles* (°)	
As1—C6	1.936	C6–As–C6′	92.1
As2—C1	1.994	C1–As–C1′	92.7
As1—S	2.253	C6–As–S	92.6
As2—S	2.231	C1–As–S	91.7
C1—C6′	1.425	As–S–As	91.5

5,10-Episeleno-5,10-dihydroarsanthrene (**3**) also gave orthorhombic crystals, space group *Pnma* with 4 molecules in the unit cell; $a = 11.4$ Å, $b = 16.7$ Å, $c = 5.85$ Å. It is isomorphous with the tellurium compound (**4**), which has space group *Pnma*, with 4 molecules in the unit cell; $a = 11.3$, $b = 16.8$, $c = 6.1$ Å. The compounds (**3**) and (**4**) thus have the same structure, which approaches very closely to that of (**2**).

5,10-Ethanoarsanthrene

This compound, [RRI 4906], has the structure and numbering shown in (**1**); it is at present known only as the dimethyl quaternary salts (**2**).

The products obtained when an equimolecular mixture of 5,10-dihydro-5,10-dimethylarsanthrene and 1,2-dibromoethane is heated (*a*) without a solvent or (*b*) in methanol have been described (p. 471). However, when a mixture of the arsanthrene and two molar equivalents of 1,2-dibromoethane in methanol is heated at 100° for 24 hours, it gives

5,10-dimethyl-5,10-ethanoarsanthrenium dibromide (**2**; X = Br), m.p. 272° (63%).[126]

A methanolic solution of this dibromide gives (*a*) with methanolic sodium iodide the colorless diiodide (**2**; X = I), m.p. 248°, and (*b*) with

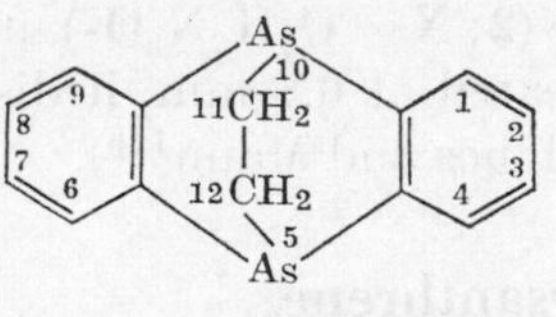

(1) 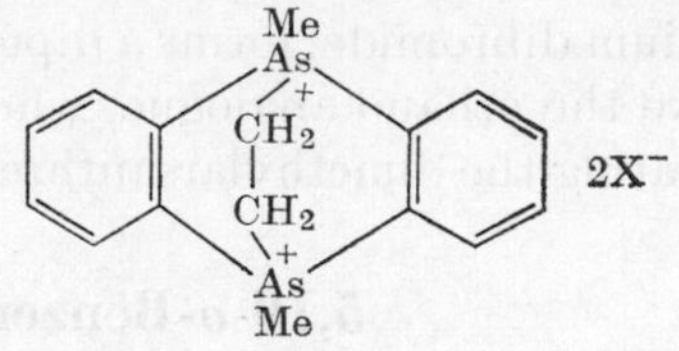   (2)

sodium picrate the dipicrate (**2**; X = C$_6$H$_2$N$_3$O$_7$), which crystallizes from ethanol–dimethylformamide as yellow crystals, m.p. 273°, of composition C$_{28}$H$_{22}$As$_2$N$_6$O$_{14}$,C$_3$H$_7$NO, having one molecule of dimethylformamide (Jones and Mann[126]).

The dibromide (**2**; X = Br) is a stable salt that can readily be recrystallized; the comparative difficulty of its preparation may indicate, however, some degree of strain in the cation (**2**), for the analogous propano derivative (next Section) is readily formed.

When dibromide is heated at 0.1 mm, there is smooth effervescence with the production of the 5,10-dihydro-5,10-dimethylarsanthrene, and the 'normal' decomposition to give methyl bromide and the compound (**1**) does not occur.

The compound (**1**) was originally termed ethylene-di-*o*-phenylene-diarsine, and the dibromide (**2**; X = Br) was termed 1,4-dimethylethyl-enedi-*o*-phenylenediarsonium dibromide.[126]

5,10-Propanoarsanthrene

The chemistry of this compound (**1**), [RRI 5274], is closely similar to that of the previous compound.

An equimolecular mixture of 5,10-dihydro-5,10-dimethylarsanthrene and 1,3-dibromopropane, when heated at 155–160° for 1.5 hours

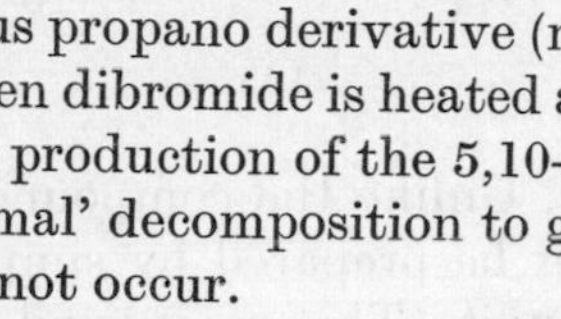

(1) 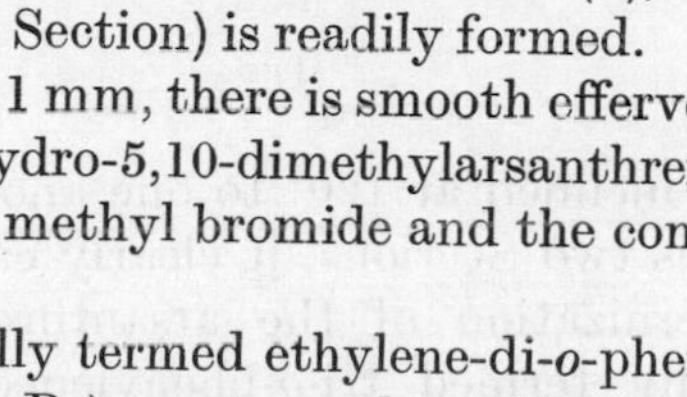   (2)

and cooled, gives a stone-hard mass of 5,10-dimethyl-5,10-propano-arsanthrenium dibromide (**2**; X = Br), m.p. 267–268° after crystallization from methanol. No other products, analogous to those formed in the condensation with 1,2-dibromoethane, were detected, The dibromide, initially termed 1,4-dimethyltrimethylene-*o*-phenylene-diarsonium dibromide, forms a dipicrate (**2**; X = $C_6H_2N_3O_7$), m.p. 252°, and, like the ethano analogue, when heated at 0.1 mm, it dissociates, regenerating the dimethylarsanthrene (Jones and Mann[126]).

5,10-*o*-Benzenoarsanthrene

This compound (**1**), [RRI 6138], is unique in type, for the arsenic atoms of the arsanthrene unit are bridged by an *o*-phenylene group, thus giving a highly symmetric compound having the three *o*-phenylene

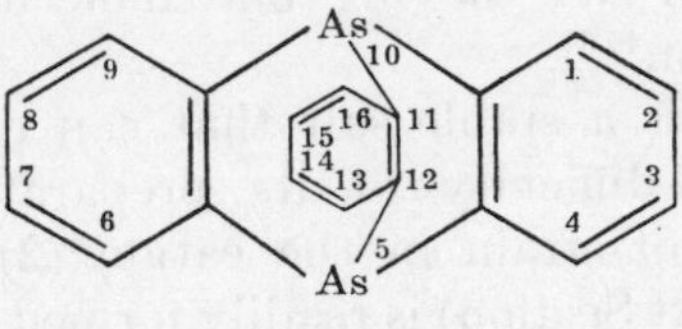

(1) 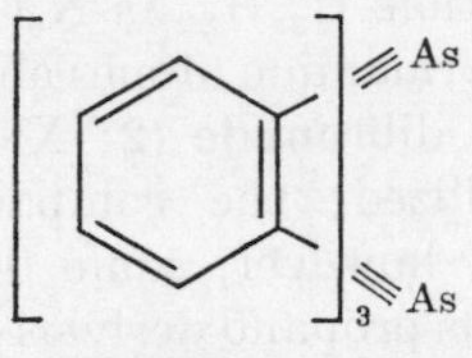   (1A)

groups inclined at 120° to one another. Unlike the compounds in the previous two Sections, it clearly cannot be prepared by simple cyclic diquaternization of the arsanthrene unit. The compound (**1**) was originally termed tri-*o*-phenylenediarsine and its formula is often conveniently drawn as (**1A**).

The compound was first synthesized and recognized by McCleland and Whitworth,[134] who diazotized *o*-aminodiphenylarsinic acid (**2**) and then coupled the product with oxophenylarsine, $C_6H_5As{=}O$, to give

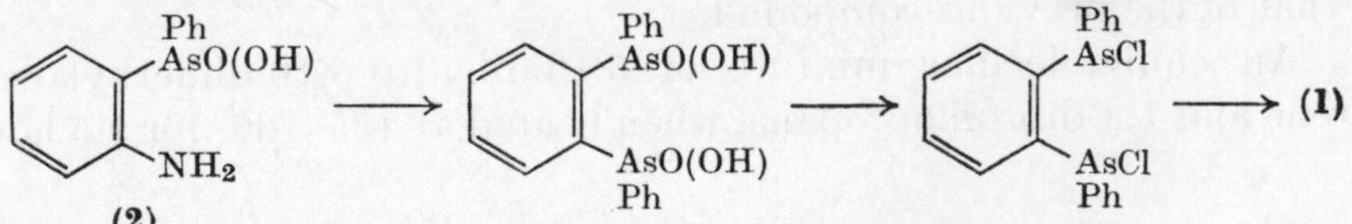

(2)   (3)   (4)

o-phenylenebis(phenylarsinic acid) (**3**). This was treated with phosphorus trichloride, being converted thereby into *o*-phenylenebis-(chlorophenylarsine) (**4**), which without isolation was heated under reduced pressure, with the formation of hydrogen chloride and the crystalline diarsine (**1**).

McCleland and Whitworth also showed that, although 'arsanthrene oxide' (pp. 461, 472) would distil at 15 mm pressure without appreciable decomposition, yet when heated in a current of carbon dioxide at pressures only slightly less than atmospheric, considerable decomposition occurred with formation of arsenious oxide and the diarsine (1); but note ready sublimation of the 'oxide' (p. 474).

The diarsine (1) is also formed as a by-product in Kalb's conversion of o-arsonodiphenylarsinic acid (p. 461) into 5,10-dichloro-5,10-dihydroarsanthrene.[117] A modification of this process (Chatt and Mann[120]) considerably decreases the yield of the intermediate o-(dichloroarsino)diphenylarsinous chloride but nearly doubles that of the tri-o-phenylenediarsine (1).

The diarsine (1) forms magnificent colorless crystals, m.p. 295–296°, which can be sublimed under reduced pressure. When boiled with nitric acid it gives the crystalline dioxide (5), and the corresponding bis-(hydroxynitrate) has not been recorded; the absence of nitration in these circumstances recalls the resistance of 5,10-dichloro-5,10-dihydroarsanthrene to nitration. The diarsine also readily adds bromine to give the tetrabromide (6), orange-red crystals, m.p. 255–256°, which is readily hydrolyzed, even by water, to give the dioxide (5). When heated under reduced pressure, the tetrabromide undergoes considerable decomposition, with the formation of bromine, the diarsine (1), o-dibromobenzene, and apparently 5,10-dibromo-5,10-dihydroarsanthrene.

In view of the ready formation of the dioxide (5), it is noteworthy that the diarsine (1) is unaffected when heated with methanolic methyl bromide or methyl iodide under nitrogen at 100° for 10 hours (Baker

(5)   (6)   (7)

and Mann[114]). When, however, the diarsine is heated with an excess of methyl p-toluenesulfonate at 180° for 4 hours, or even at 210° for 6 hours, only the monometho-p-toluenesulfonate (7; X = $C_7H_7SO_3$), m.p. 238°, is formed; it gives a monomethopicrate (7; X = $C_6H_2N_3O_7$), m.p. 226–227°. The disposition of the three valencies of each arsenic atom in the diarsine allows ready ingress of an oxygen atom (5), the arsenic atom becoming tetrahedral, but apparently the methyl group meets with considerable steric resistance. The fact that quaternization does not proceed beyond the monoquaternary salts (7) is understandable: the As=O links in the dioxide (5) must be very weakly polar (hence the apparent non-existence of the bis(hydroxynitrate), (a characteristic derivative of most tertiary arsine oxides), but the strong positive pole on the quaternary salt (7) would exert a powerful electronic pull by the three routes from the second arsenic atom, causing a degree of deactivation reinforced in effect by the steric hindrance. The structure of the tetrabromide (6) would prove of great interest.

5,12-o-Benzenodibenzo[b,f][1,4]diarsocin and 6,11:18,23-Di-o-benzenotetrabenzo[b,f,j,n][1,4,9,12]tetraarsacyclohexadecin

These two systems are conveniently discussed together, as they are both derivatives of arsanthrene and are formed together (as 2- or 4-hydro derivatives) in one operation by interaction of 5,10-dihydro-5,10-dimethylarsanthrene and o-xylylene dibromide.

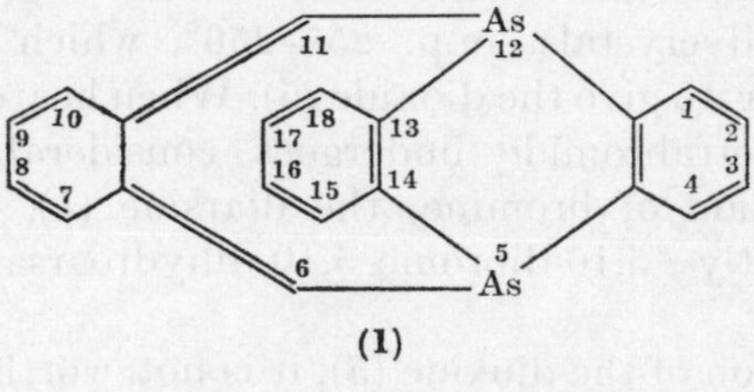

(1) 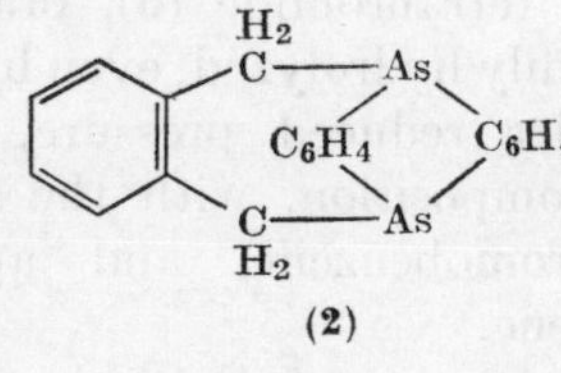   (2)

5,12-o-Benzenodibenzo[b,f][1,4]diarsocin, [RRI 6437], has the structure, numbering, and presentation shown in (1); although the suffix of this name indicates that the arsenic is part of an eight-membered ring, it is also part of the central six-membered ring of an arsanthrene

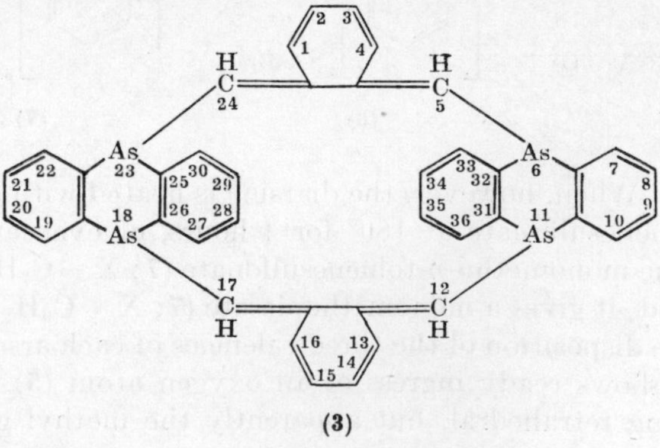

(3)

unit; hence its inclusion at this stage. Its 6,11-dihydro derivative has been termed di-o-phenylene-o-xylylenediarsine and conveniently represented as (2) (Jones and Mann[126]) but this formula is clearly useless for nuclear-substituted derivatives—at present unknown.

6,11:18,23 - Di - o - benzenotetrabenzo[b,f,j,n][1,4,9,12] - tetraarsa - cyclohexadecin, [RRI 7628], has the structure and numbering shown (3); its 5,12,17,24-tetrahydro derivative would consist of two arsanthrene units linked in the above positions by two o-xylylene units.

The compounds (1) and (3) have not been isolated: the known derivatives are hydrogenated and (usually) quaternized products.

When a powdered equimolecular mixture of 5,10-dihydro-5,10-dimethylarsanthrene (p. 466) and o-xylylene dibromide is heated at 160°

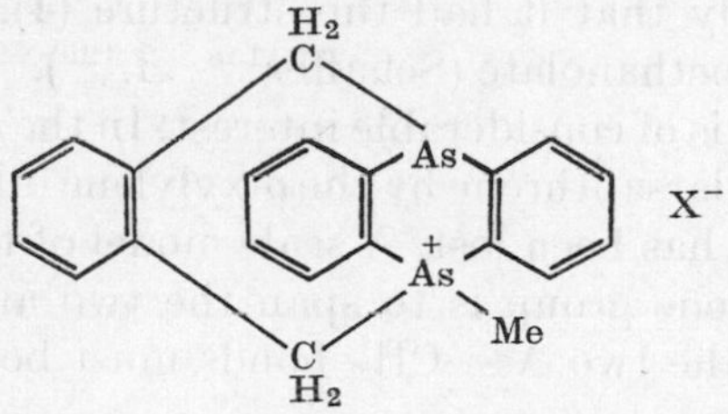

(4)

for 2 hours, two products are formed. The more soluble is 6,11-dihydro-5-methyl-5,12-o-benzenodibenzo[b,f][1,4]diarsocinium bromide (4; X = Br), m.p. 200°, earlier termed di-o-phenylene-o-xylylenediarsine monomothobromide,[126] in 20% yield. The less soluble is 5,12,17,24-tetrahydro-6,11:18,23-di-o-benzenotetrabenzo[b,f,j,n][1,4,9,12]tetraarsacyclohexadecinium tetrabromide (5; X = Br), in 39% yield.

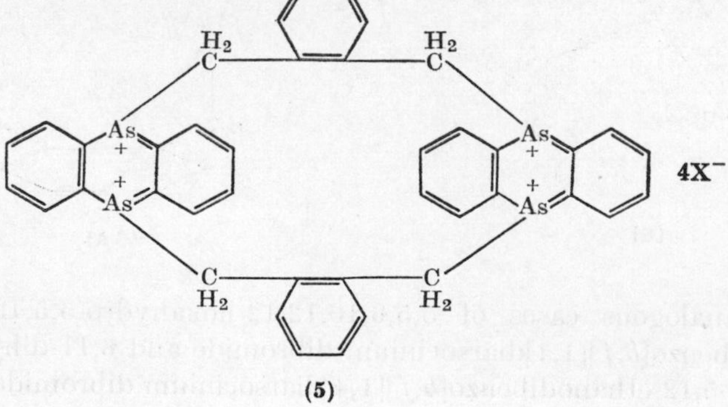

(5)

When the original mixture is heated at 130° for 1 hour, the yield of the monobromide (4) and the tetrabromide (5) are 65 and 8%, respectively, and when the mixture contains two equivalents of o-xylylene dibromide and is heated at 160° for 1 hour, only the tetrabromide (5) is formed, in 64% yield.[126]

The two compounds will be discussed in turn.

The monobromide (4; X = Br)

There are three not improbable structures for a compound of composition $(C_{21}H_{19}As_2)Br$, arising in the above circumstances. A sample of the bromide which had been recrystallized from methanol–ethanol and had the composition of a dimethanolate or a hydrate-ethanolate was subjected to an X-crystal structure investigation, which showed conclusively that it had the structure (4), and that it was a monohydrate-monoethanolate (Schaffer[135]; cf.[126]).

This structure is of considerable interest. In the cyclic diquaternization of the dimethylarsanthrene by the o-xylylene dibromide, a molecule of methyl bromide has been lost. A scale model of the cation (4) shows that if the o-xylylene group is to span the two arsenic atoms of the arsanthrene unit, the two As—CH$_2$ bonds must be almost parallel (as

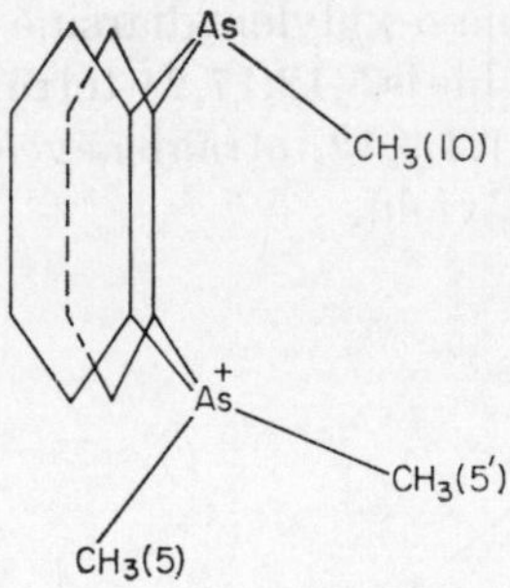

(6) 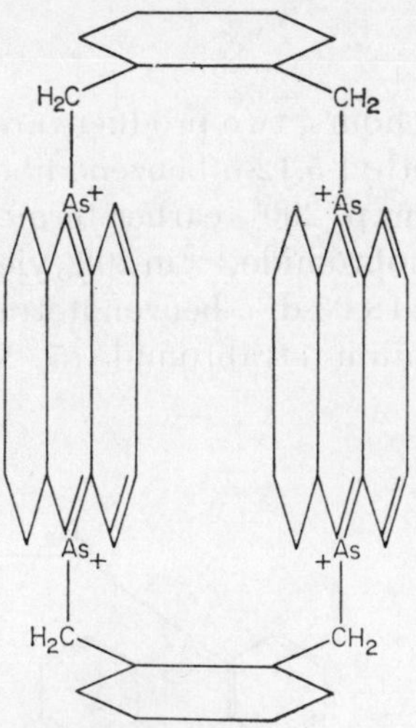   (5 A)

in the analogous cases of 5,5,6,10,12,12-hexahydro-5,5,12,12-tetra-methyldibenzo[b,f][1,4]diarsocinium dibromide and 6,11-dihydro-5,12-dimethyl-5,12-ethanodibenzo[b,f][1,4]diarsocinium dibromide (pp. 456,

458). But in the 5,10-dihydro-5,10-dimethylarsanthrene and, for example, its dimethobromide, the C–As–C intervalency angle, of *ca.* 98° and 107° in the two compounds respectively, does not allow any conformation in which the As—CH_3 bonds of the two arsenic atoms are parallel. However, in the monomethobromide, shown diagrammatically in (**6**), where $CH_3(5)$ and $CH_3(10)$ are the original methyl groups and $CH_3(5')$ is that inserted on methobromide formation, the intervalency angles at the As(10) and the As(5) atoms in the molecule are *ca.* 98° and *ca.* 107°, respectively, so that in this conformation the As—$CH_3(5')$ and the As—$CH_3(10)$ bonds are almost parallel. Consequently it would appear that the cation (**4**) can exist without undue strain only as the monoquaternary salt, in which the two As—CH_2 bonds have the spatial arrangement of the As—$CH_3(5')$ and the As—$CH_3(10)$ bonds in the diagram (**6**).

This reasoning must, however, be regarded with reserve, for it applies apparently only to the *trans*-form of dimethylarsanthrene and ignores the partial 'flattening' with bond deflection during monoquaternization.

The monobromide gave a cream-colored iodide (**4**; X = I), m.p. 197°, and a picrate, m.p. 211°.[126]

The tetrabromide (**5**; X = Br)

This salt, originally termed 5,5':10,10'-bis-*o*-xylylenebisarsanthrenium tetrabromide,[126] forms cream-colored crystals, m.p. 273°, and when treated in methanolic solution with sodium iodide gives deep orange crystals of the tetraiodide (**5**; X = I), m.p. 244°, strongly resembling the earlier 5,10-dimethylarsanthrenium diiodide (**20**) (p. 468). It also gives a tetrapicrate decahydrate, m.p. 232°, which has not been obtained in the anhydrous condition.

Reliable molecular-weight determinations on salts such as (**5**) cannot readily be obtained. The loss of all four methyl groups is clearly shown by analysis of the tetrabromide and tetraiodide, and the four halogen atoms are all ionic. If each arsanthrenium unit in (**5**) is planar, the As—CH_2 bonds must also be in this plane and must form an extension of each As—As axis; consequently the cation must have the 'dimer' structure (**5**), for one *o*-xylylene group could not in these circumstances span the two arsenic atoms of one arsanthrene unit. The cation (**5**) can therefore be shown diagrammatically as (**5A**), in which the two planar arsanthrene ring systems are depicted at right angles to the plane of the paper, only the quinonoid double bonds being shown: the four As—CH_2 groups are also in this plane. Models indicate that the parallel orientation of these As—CH_2 groups causes the benzene ring of each *o*-xylylene bridge to be tilted towards or away from the observer. This marked tilt by both rings could give rise to *cis–trans*-isomers, but the salts isolated were homogeneous, and most probably have the more symmetric *trans*-form. Calculations show that the two parallel arsanthrene units in (5A) should be *ca.* 2.9 Å apart. The shape and size of this cation may be associated with the stability of the tetrapicrate decahydrate.

5,10-Dihydro-2,5,10-trimethylarsanthrene,[126] when similarly heated with one equivalent of *o*-xylylene dibromide at 100° for 1.5 hours gives a highly hygroscopic tetrabromide (7; X = Br), which however when treated with methanolic sodium iodide gives a crystalline cream colored tetraiodide (7; X = I), m.p. 180–181° (70%), and a solvent-free tetrapicrate, m.p. 171–173°. Complete quaternization has occurred in

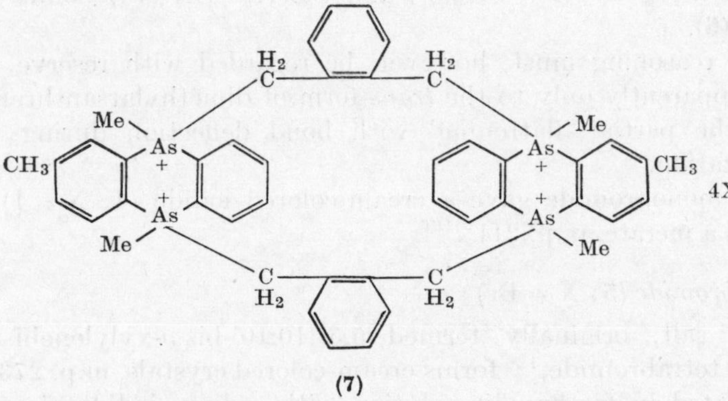

(7)

the formation of the initial tetrabromide without any loss of the methyl groups (possibly owing to the lower temperature of the reaction) and for reasons already stated it is very unlikely in these circumstances that a simple 1:1 combination of the reagents could have occurred; hence the assignment of the 'dimer' structure (7). The tetraiodide (7; X = I) should thus be called 5,6,11,12,17,18,23,24-octahydro-6,8,11,18,21,23-hexamethyl-6,11:18,23-di-*o*-benzenotetrabenzo[*b,f,j,n*][1,4,9,12]-tetra-arsacyclohexadecinium tetraiodide: it was originally termed 5,5′,10,10′-tetrahydro-2,2′,5,5′,10,10′-hexamethyl-5,5′:10,10′-bis-*o*-xylylenebis-arsanthrenium tetraiodide.[126]

The cation (7) could exist in several geometrically isomeric forms, of which one would be dissymmetric, but fractional recrystallization of the corresponding tetra-(+)-3-bromo-8-camphorsulfonate (7; X = $C_{10}H_{14}BrO_4S$) and the tetraantimonyl-(+)-tartrate (7; X = $C_4H_4O_7Sb$) gave no indication of optical resolution.

Application of modern spectroscopic methods to the halides of (5) and (7) might further elucidate their structures.

Seven-membered Ring Systems containing Carbon and Two Arsenic Atoms

1H-1,5-Benzodiarsepin

This system (1), [RRI 1817], numbered as shown, is at present known only as one type of derivative, prepared similarly to the analogous

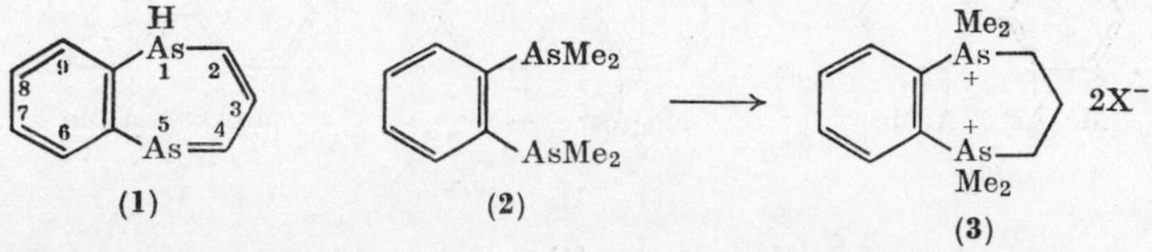

(1)   (2)   (3)

derivative of 1,4-benzodiarsenin (p. 451). When o-phenylenebis(dimethylarsine) (2) is heated with 1,3-dibromopropane (one equivalent) under nitrogen, with the temperature rising to 156–160°, cyclic quaternization occurs with the formation of 2,3,4,5-tetrahydro-1,1,5,5-tetramethyl-1H-1,5-benzodiarsepinium dibromide (3; X = Br), m.p. 258°; this salt is freely soluble in water, but very slightly soluble in boiling ethanol (Glauert and Mann[113]). It gives a dipicrate (3; X = $C_6H_2N_3O_7$), m.p. 260°, which is almost insoluble in boiling water, methanol, or ethanol. The dibromide (3; X = Br) was initially termed 1,5-dimethyl-o-phenylenetrimethylenediarsine dimethobromide.

The dibromide (3; X = Br) has not been further investigated. Its thermal decomposition would be of interest to see if the heterocyclic system would survive with only the loss of two molecules of methyl bromide, or if 1,3-dibromopropane would be lost with the re-formation of the diarsine (2).

5H-Dibenzo[d,f][1,3]diarsepin

This diarsine (1), [RIS 8626], is known only as derivatives of the 6,7-dihydro-5H-dibenzo[d,f][1,3]diarsepin (2).

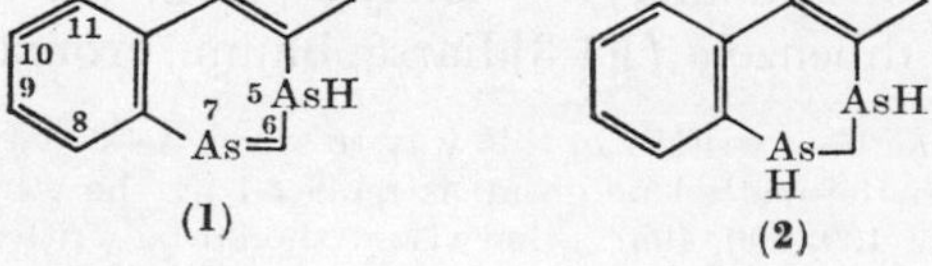

(1)   (2)

2,2′-Biphenylylenebisdimethylarsine (**3**), m.p. 46–46.5°, is readily prepared by the action of dimethylarsinous iodide, $(CH_3)_2AsI$, on 2,2′-dilithiobiphenyl. For steric reasons, the molecule of (**3**) is almost certainly non-planar; a scale model shows that the angle subtended by the planes of the two benzenoid rings cannot be less than *ca.* 25°, and the ultraviolet spectrum lacks the characteristic biphenyl band in the 252 mμ region. The available rotation of the two benzenoid rings about their common

Me₂As AsMe₂	Me₂As⁺ ⁺AsMe₂	MeAs AsMe
(**3**)	C 2X⁻ H₂	C H₂
	(**4**)	(**5**)

link, combined with that of the two As—aryl links, enables the diarsine (**3**) to adapt itself for cyclic diquaternization with a number of alkylene dihalides; for example, when heated with dibromomethane it gives 6,7-dihydro-5,5,7,7-tetramethyl-5*H*-dibenzo[*d*,*f*][1,3]diarsepinium dibromide* (**4**; X = Br), m.p. 224–226°, originally termed 5,5,7,7-tetramethyl - 5,7 - diarsonia - 1,2 : 3,4 - dibenzocycloheptadiene dibromide (Heaney, Heinekey, Mann, and Millar[33]); this salt gives the diiodide, m.p. 264°, and the dipicrate (**4**; X = $C_6H_2N_3O_7$), m.p. 283–285°.

This dibromide, when heated under nitrogen at 250° (air-bath) and 0.15 mm, loses methyl bromide and gives a distillate of 6,7-dihydro-5,7-dimethyl-5*H*-dibenzo[*d*,*f*][1,3]diarsepin (**5**), m.p. 95.5–96°. An aqueous solution of the dibromide, when shaken with silver oxide, filtered and evaporated, gives the syrupy dihydroxide (**4**; X = OH), which when similarly heated also gives the diarsine (**5**); this retention of the heterocyclic system is not shown by the homologous dihydroxides (pp. 489, 491), which undergo considerable disruption (Forbes, Heinekey, Mann, and Millar[136]).

The structure of the diarsine (**5**) is confirmed by its reaction with methyl iodide to re-form the diiodide (**4**; X = I).

An attempt to form a second methylene 'bridge' across the arsenic atoms in the diarsine (**5**), by heating the latter with methanolic dibromomethane at 100° for 72 hours, gave solely 5-(bromomethyl)-6,7-dihydro-5,7-dimethyl-5*H*-dibenzo[*d*,*f*][1,3]diarsepinium bromide (**6**; X = Br),

---

* Formula (**4**) has been written in this way to stress its structural similarity to its homologues in which the methylene group is replaced by the ethylene, propylene, or *o*-xylylene group (pp. 489, 490, 495); otherwise it should be written in the form of (**2**).

m.p. 210°, which gave the picrate (6; $X = C_6H_2N_3O_7$), m.p. 146–147°: the positive pole on one arsenic atom has apparently completely in-activated the tertiary arsenic atom, although the 'twisting' of the cation

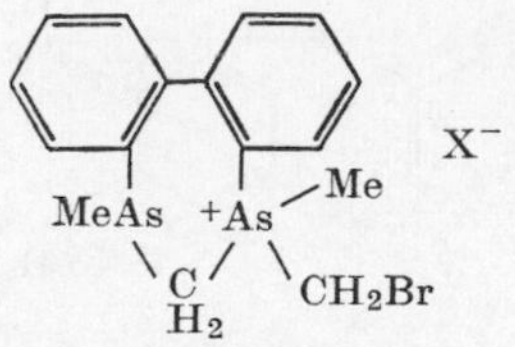

(6)

(6) may have placed this atom in a conformationally impossible position for quaternization by the bromomethyl group.

The 'twisted' conformation of the cation (4) must necessarily make the cation dissymmetric. Attempts to obtain optically active forms were ultimately successful when the dibromide (4; $X = Br$) was converted into the (+)-dibenzoyltartrate (4; $2X = C_{18}H_{12}O_8$). The diastereo-isomers of this salt differed markedly in their solubility in acetonitrile at room temperature. Repeated extraction of the mixture with cold acetonitrile left the insoluble (+)-diarsonium (+)-dibenzoyltartrate trihydrate, $[\alpha]_D$ +88.15° in aqueous solution; this salt in aqueous solution, when passed down a suitable ion-exchange column and evaporated under reduced pressure, gave the (+)-dibromide trihydrate, m.p. 228°, $[\alpha]_D$ +29.5°, $[M]$ +174°. Conversion of the racemic dibromide into the (−)-dibenzoyltartrate, followed by similar treatment, gave the (−)-dibromide trihydrate, $[\alpha]$ −30.3°, $[M]$ −179°. The cation (4) can apparently oscillate about the mean position, although this process is slow at room temperature. An aqueous solution of the (+)-dibromide kept at room temperature (ca. 20–22°) retained some of its activity after 4 weeks, but the activity of an aqueous solution at 70° fell from $[\alpha]$ +29.5° to +3.07° in 2 hours.[136] The racemization of this salt has been studied kinetically (Forbes, Mann, Millar, and Moelwyn-Hughes[137]).

Eight-membered Ring Systems containing Carbon and Two Arsenic Atoms

2,5-Benzodiarsocin

This ring system (1), [RRI 1886], is at present known only as one type of derivative. When ethylenebis(methylphenylarsine),

$$[—CH_2As(CH_3)C_6H_5]_2,$$

is added to one equivalent of o-xylylene dibromide under nitrogen, the mixture rapidly sets to a hard mass, which is then heated at 100° for

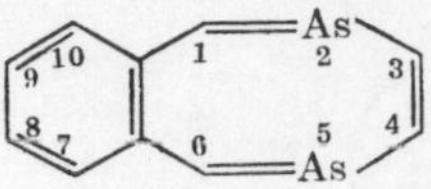

(1)

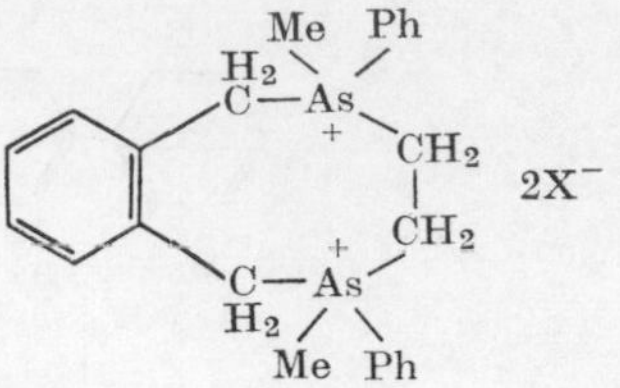

(2)

1 hour to ensure complete formation of 1,2,3,4,5,6-hexahydro-2,5-dimethyl-2,5-diphenyl-2,5-benzodiarsocinium dibromide (**2**; X = Br) This highly deliquescent salt is extracted with cold ethanol, leaving an insoluble yellow resinous material, probably formed by linear condensation of the two reactants. The extract, when treated with ethanolic sodium iodide, deposits the crystalline diiodide diethanolate (**2**; X = I), m.p. 160–161°, which was earlier termed 1,4-diphenyl-1,4-diarsa-6,7-benzocyclo-oct-6-ene dimethiodide (Jones and Mann[106]).

It is not surprising that the interaction of the above reactants gives rise to some resinous material, whereas the reaction of o-xylylene dibromide with o-phenylenebis(dimethylarsine) and with 1,2,3,4-tetrahydro-1,4-dimethyl-1,4-benzodiarsenin (p. 456) does not. In the last two compounds the tertiary arsine groups are held by their ring systems in the most favorable position for cyclic diquaternization, whereas the general rotation of the chain in ethylenebis(methylphenylarsine) could bring the arsine groups into positions at least as favorable to linear condenation as to cyclic condensation. The same factor applies to the condensation of this diarsine with 1,2-dibromoethane (p. 488), in which polymeric material is also formed.

Dibenzo[*e,g*][1,4]diarsocin

This fundamental compound (**1**), [RIS 8658], is known only as diquaternary salts of its 5,6,7,8-tetrahydro derivative.

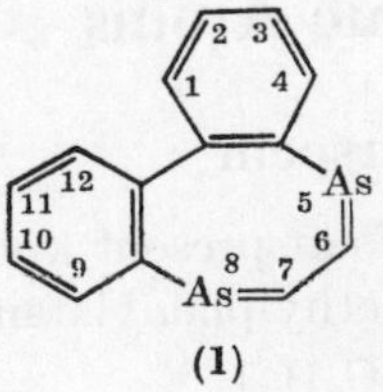

(1) 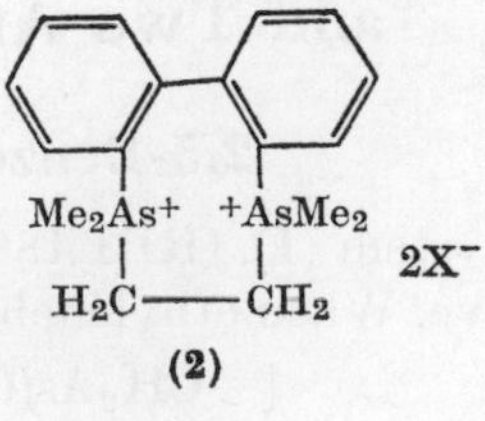   (2)

When 2,2'-biphenylylenebis(dimethylarsine) (p. 486) is heated with 1,2-dibromoethane, it readily undergoes diquaternization to give 5,6,7,8-tetrahydro-5,5,8,8-tetramethyldibenzo[e,g][1,4]diarsocinium dibromide, which can more conveniently be presented as (2; X = Br), m.p. 211–213°. It was initially termed As,As,As',As'-tetramethyl-As,As'-ethylene-As,As'-2,2'-diphenylylenediarsonium dibromide (Heaney, Mann, and Millar[138]). It gives a dipicrate (2; X = $C_6H_2N_3O_7$), m.p. 250–252°, and a diiodide, m.p. 222–223°.[136]

Salts of the cation (2) differ markedly in their chemical properties from those of the lower homologous 6,7-dihydro-5,5,7,7-tetramethyl-5H-dibenzo[d,f][1,3]diarsepinium cation (p. 486). The dibromide (2; X = Br), when heated at 0.1 mm and 210–215° (oil-bath), decomposes with the formation of a distillate of 9-methyl-9-arsafluorene (5-methyl-dibenzoarsole) (3; R = CH_3) containing some of the 9(5)-bromo derivative (3; R = Br); trimethylarsine is also formed and can be condensed

(3)

in a liquid-air trap and identified as its palladium dibromide derivative. The mixture of (3; R = CH_3) and (3; R = Br), when treated with methyl-magnesium bromide, gives the pure (3; R = CH_3) [methiodide, m.p. 206–207°, methopicrate, m.p. 214.5–215.5°, and diarsinedibromopalladium, yellow crystals (decomp. on heating)[33]].

This thermal decomposition in its general type is similar to that of 5,6,7,8-tetrahydro-5,5,8,8-tetramethyldibenzo[e,g][1,4]diphosphocinium dibromide (p. 198), but whilst formation of the monoarsine (3; R = CH_3) is apparently limited to the decomposition of the dibromide (2; X = Br) and the two higher homologues (pp. 491, 495), the 9-alkyl-9-phosphafluorene arises in the thermal decomposition of a wider range of diphosphonium dibromides (pp. 194, 198, 200, 201).

The production of the mixture of (3; X = CH_3) and (3; X = Br) recalls the production of a similar mixture of a methyl- and a bromo-arsine during the thermal decomposition of 1,2,3,4-tetrahydro-1,1,4,4-tetramethyl-1,4-benzodiarseninium dibromide (p. 452).

The dihydroxide (2; R = OH) decomposes when heated at 160° at 0.05 mm, regenerating 2,2'-biphenylylenebis(dimethylarsine).

The cation (2) is dissymmetric (like that of its lower and higher

17

homologue), and its di-(+)-3-bromo-8-camphorsulfonate (**2**; X = $C_{10}H_{14}BrO_4S$) after ten recrystallizations from ethanol yielded the optically pure (+)-diarsonium (+)-dibromocamphorsulfonate, $[M]_D$ +862° in aqueous solution, which when passed in solution down an ion-exchange column gave the (+)-dibromide (**2**; X = Br), $[M]$ +306°. The di-(−)-bromocamphorsulfonate, also prepared from the racemic dibromide, and similarly recrystallized from ethanol, gave the (−)-diarsonium di-(−)-bromocamphorsulfonate, $[M]_D$ −859°, which in turn gave the (−)-dibromide[137] (**2**; X = Br), $[M]$ −307°.

The (+)-dibromide (**2**; X = Br) has greater optical stability than its lower homologue (p. 487), for the activity of its aqueous solution remained unimpaired after the solution had been kept at room temperature for several months. The activity of an aqueous solution, kept in a sealed tube at 100° for 60 hours, fell to 9.84% of its original value, and when kept at 150° for 50 minutes similarly fell to 7.87%.

It is noteworthy that the 5,5,8,8-tetraethyl homologue of the dibromide (**2**) when heated gives 5-ethyldibenzarsole (Allen, Millar, and Mann[139]).

Note. Certain ring systems which contain two arsenic atoms as part of an eight-membered ring have been described earlier either because they also contain the two arsenic atoms as part of a six-membered ring, or because their formation and reactions are closely akin to such six-membered ring systems. They are:

Dibenzo[*b,f*][1,4]diarsocin and 5,12-Ethanodibenzo[*b,f*][1,4]-diarsocin (p. 455).

5,12-*o*-Benzenodibenzo[*b,f*][1,4]diarsocin (p. 480).

Nine-membered Ring Systems containing Carbon and Two Arsenic Atoms

5*H*-Dibenzo[*f,h*] [1,5]diarsonin

Our knowledge of the chemistry of this ring system (**1**), [RIS 8668], is very similar to that of benzo[*e,g*][1,4]diarsocin discussed in the previous Section.

Diquaternization of 2,2′-biphenylylenebis(dimethylarsine) (p. 486) with 1,3-dibromopropane readily gives the 6,7,8,9-tetrahydro-5,5,9,9-tetramethyl-5*H*-dibenzo[*f,h*][1,5]diarsoninium dibromide, conveniently depicted as (**2**; X = Br), and originally termed 5,5,9,9-tetramethyl-5,9-diarsonia-1,2:3,4-dibenzocyclononadiene dibromide (Heaney, Heinekey,

Mann, and Millar[33]). The hygroscopic dibromide has m.p. 261–262° and gives a dipicrate (2; X = $C_6H_2N_3O_2$), m.p. 201–203°.

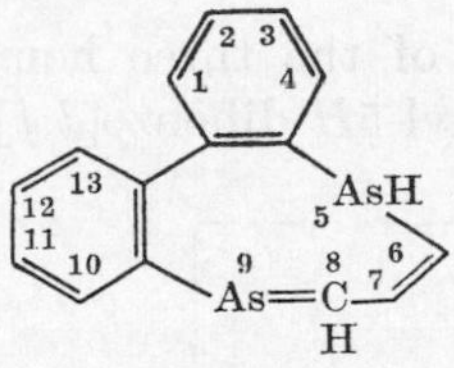

(1) 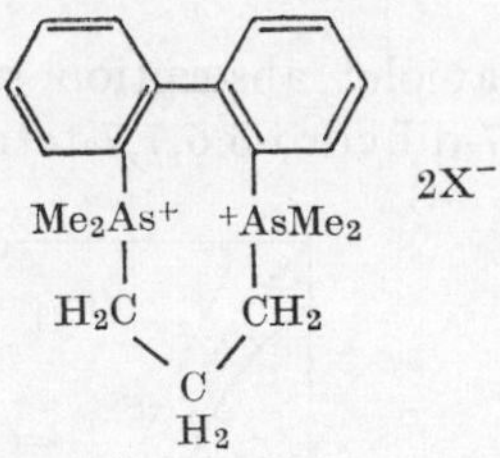   (2)

Thermal decomposition of the dibromide (2; X = Br), at 230° rising to 260° at 0.1 mm under nitrogen, gives 9-methyl-9-arsafluorene (3) contaminated with only a trace of the 9-bromo-9-arsafluorene; the second arsenic atom almost certainly forms the more volatile dimethyl-vinylarsine, $(CH_3)_2As$—CH=CH$_2$.[33] Some aspects of this type of thermal decomposition are discussed on pp. 496, 501, 194.

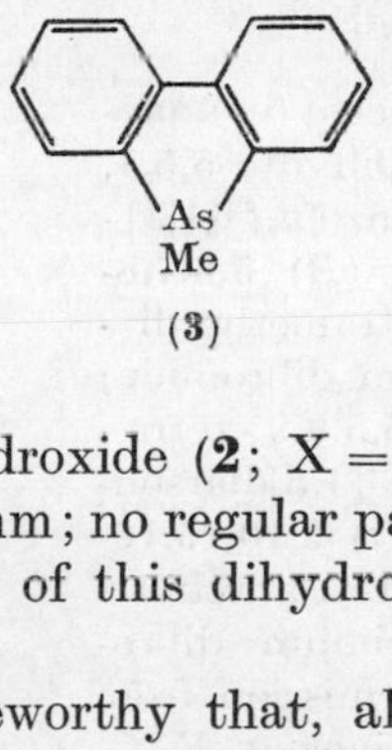

(3) 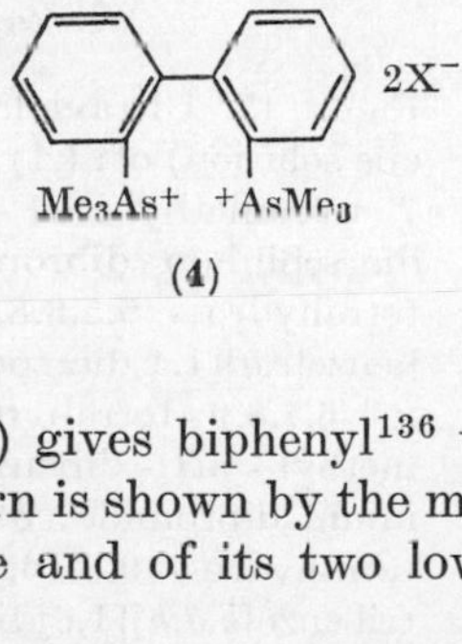   (4)

The dihydroxide (2; X = OH) gives biphenyl[136] when heated at 165° and 0.5 mm; no regular pattern is shown by the modes of thermal decomposition of this dihydroxide and of its two lower homologues (pp. 486, 489).

It is noteworthy that, although 2,2′-biphenylylenebis(dimethyl-arsine) dimethobromide (4; X = Br) regenerates the diarsine on thermal decomposition, the corresponding dihydroxide (4; X = OH)—like the dihydroxide (2; X = OH)—gives biphenyl.

The dibromide (2; X = Br) was converted into the di-(+)-3-bromo-8-camphorsulfonate (2; X = $C_{10}H_{14}BrO_4S$), which after five recrystallizations from ethanol gave the optically pure (+)-diarsonium di-(+)-sulfonate, m.p. 286–288°, $[M]_D$ +1047° in aqueous solution; it yielded the (+)-dibromide, m.p. 240°, $[M]$ +485°.

The racemic dibromide (2; X = Br) similarly furnished the (−)-diarsonium di-(−)-bromocamphorsulfonate, m.p. 290°, $[M]$ −1042°, which in turn gave the (−)-dibromide, m.p. 241°, $[M]$ −501°.[137]

An aqueous solution of the (+)-dibromide was heated at 150° for 3 hours without diminution of its optical activity; this very high optical stability indicates that the cation (**2**) does not 'twist' about the mean position.

The ultraviolet absorption spectra of the three homologous dibromides, 6,7-dihydro-5,5,7,7-tetramethyl-5H-dibenzo[d,f][1,3]diarse-

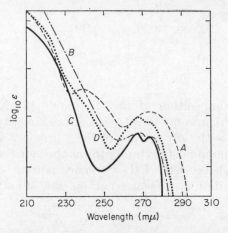

Figure 17. Ultraviolet spectra (in aqueous solution) of: (*A*) 6,7-Dihydro-5,5,7, 7 - tetramethyl - 5H - dibenzo[d,f][1,3] - diarsepinium dibromide; (*B*) 5,6,7,8-tetrahydro - 5,5,8,8 - tetramethyldi - benzo[e,g][1,4]diarsocinium dibromide; (*C*) 6,7,8,9 - tetrahydro - 5,5,9,9 - tetra - methyl - 5H - dibenzo[f,h][1,5]diarsoninium dibromide; and (*D*) 9,10,15,16-tetrahydro - 9,9,16,16 - tetramethyl - tribenzo[b,d,h][1,6]diarsecinium dibromide. (Reproduced, by permission, from M. H. Forbes, D. M. Heinekey, F. G. Mann, and I. T. Millar, *J. Chem. Soc.,* **1961**, 2762)

pinium dibromide (p. 486), 5,6,7,8-tetrahydro-5,5,8,8-tetramethyldibenzo[e,g][1,4]diarsocinium dibromide (p. 489), and the present dibromide (**2**; X = Br), and also 9,10,15,16-tetrahydro-9,9,16,16-tetramethyltribenzo[b,d,h][1,6]diarsecinium dibromide discussed on. p. 495 are shown together in the annexed Figure 17. It is noteworthy that all these spectra show no evidence of the characteristic biphenyl band in the 252 mμ region, although the optically active cations of the first two of the above compounds slowly racemize in aqueous solution.

9H-Tribenzo[d,f,h] [1,3]diarsonin

This compound (1), [RIS —], is unknown, but its hydro and its quaternized derivatives represent the first of a homologous series which

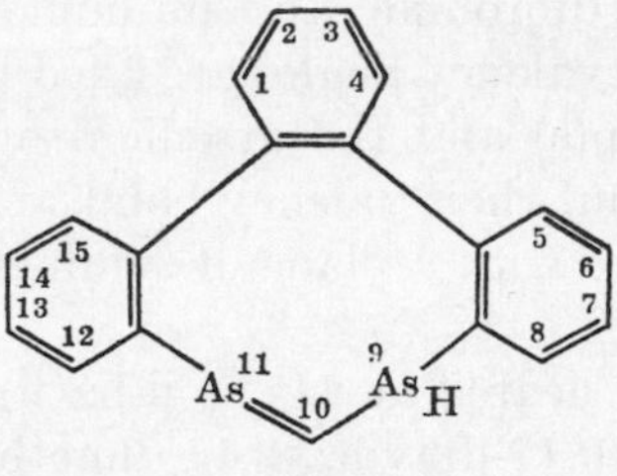

may be obtained by the cyclic diquaternization of o-terphenyl-2,2″-ylenebis(dimethylarsine) (2) with the customary alkylene dibromides. This process has a wide application because, as in 2,2′-biphenylylenebis-(dimethylarsine) (p. 489), the available rotation about the As—aryl and the aryl—aryl bonds allows the two tertiary arsenic atoms a wide and varied range of relative positions to accommodate the various dibromides. The fact that this rotation can bring the two arsine groups into close proximity is shown by the formation of the crystalline orange diarsine dibromopalladium (3), the molecular weight of which has been determined (Heaney, Mann, and Millar[138]).

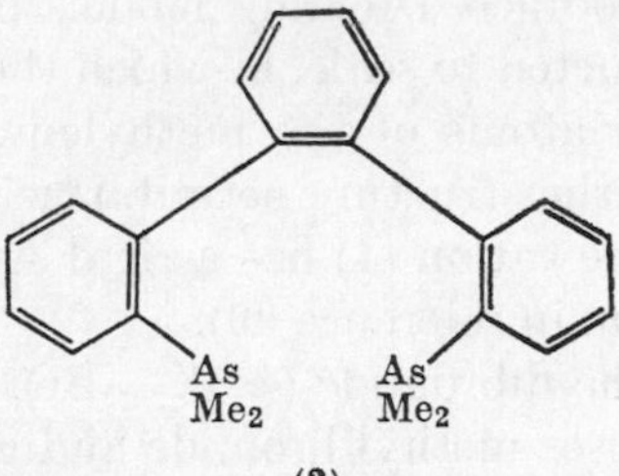

The diarsine (2) is one of the several arsines which arise when o-bromoiodobenzene in ether is treated with magnesium (3 equivalents) at 0°, and the product then treated with dimethylarsinous iodide, $(CH_3)_2AsI$.[138] It could probably be obtained by the action of this iodide on 2,2″-dilithio-o-terphenyl.

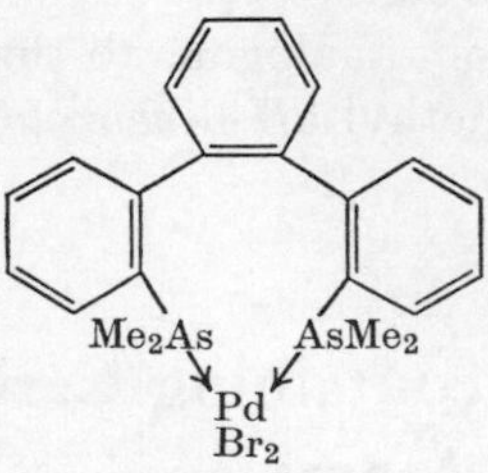

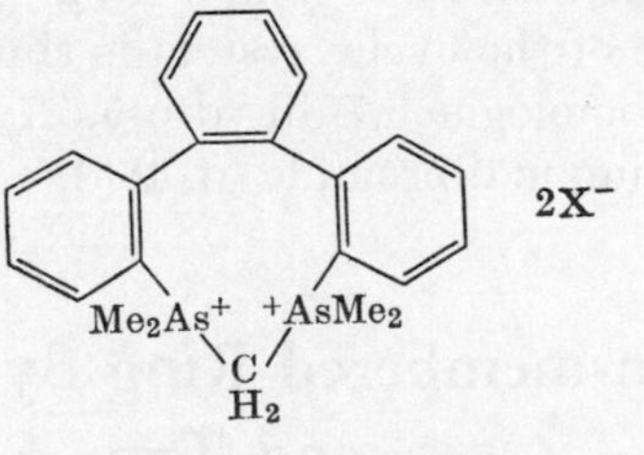

An equimolar mixture of the diarsine (2) and dibromomethane, when heated at 100° for 8 hours, gives 10,11-dihydro-9,9,11,11-tetra-methyl-9H-tribenzo[d,f,h][1,3]diarsoninium dibromide (4; X =Br), initially termed 7,7,9,9-tetramethyl-7,9-diarsonia-1,2:3,4:5,6-tribenzo-cyclononatriene dibromide (Mann, Millar, and Baker[99]). The dibromide, m.p. 258–264°, gives a corresponding dipicrate (4; $X = C_6H_2N_3O_7$) m.p. 266°.

Modified Dreiding models of the dibromide and its homologues, constructed to scale, in which the intervalency angles are fixed but the carbon atoms of the methylene group(s) and the arsenic atoms can (when the structure permits) twist about their valency bonds, indicate that the cation (4) has a rigid structure with a plane of symmetry (cf. diagram in reference 99).

The dibromide (4; $X = Br$), when heated at 245° (oil-bath) at 0.1 mm, loses methyl bromide and gives 10,11-dihydro-9,11-dimethyl-9H-tribenzo[d,f,h][1,3]diarsonin (5), m.p. 137–138°. This diarsine when treated in ethanol with potassium tetrabromopalladite gives a yellow

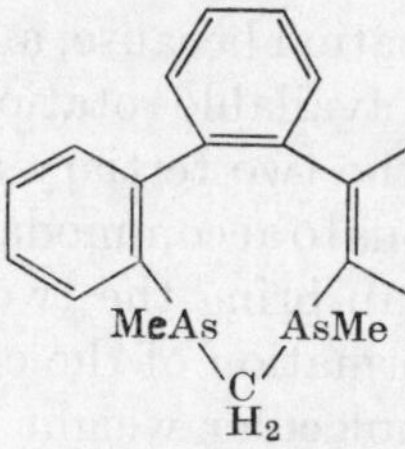

MeAs   AsMe
  C
  H$_2$

(5)

product of composition $(C_{21}H_{20}As_2,PdBr_2)_n$; this product is too insoluble for molecular-weight determinations, but if it is the monomer, $n = 1$, a palladium bridge has been formed across the arsenic atoms in (5).

This thermal decomposition of the dibromide (4; $X = Br$) without rupture of the cyclic system is thus precisely analogous to that of the lower homologue, 6,7-dihydro-5,5,7,7-tetramethyl-5H-dibenzo[d,f][1,3]-diarsepinium dibromide (p. 486).

Ten-membered Ring Systems containing Carbon and Two Arsenic Atoms

Dibenzo[b,d] [1,6]diarsecin

Derivatives of this compound (1), [RIS 10556], have been only very briefly studied.

2,2'-Biphenylylenebis(dimethylarsine) (p. 486) combines with 1,4-dibromobutane to give 5,6,7,8,9,10-hexahydro-5,5,10,10-tetra-methyl-dibenzo[b,d][1,6]diarsecinium dibromide (2), originally termed

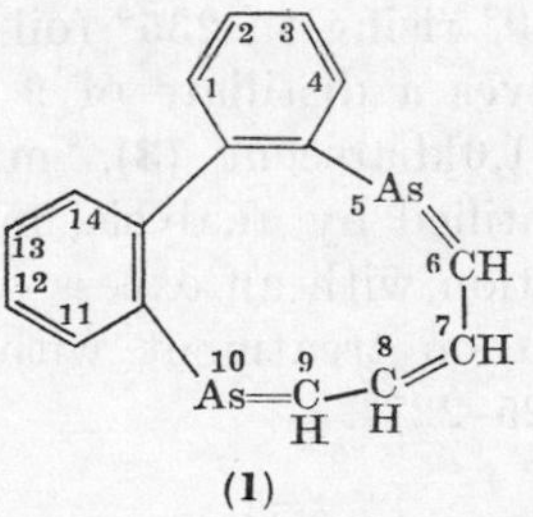

(1) 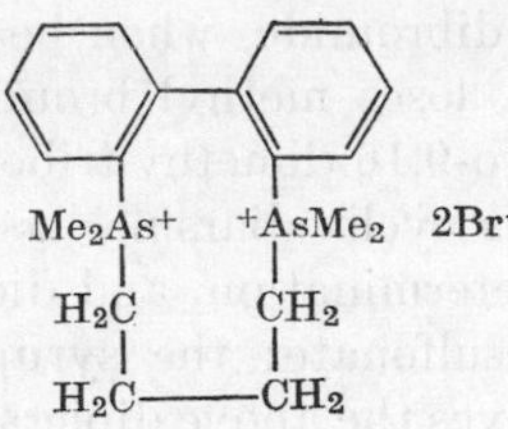   (2)

5,5,10,10-tetramethyl-5,10-diarsonia-1,2:3,4-dibenzocyclodecadiene dibromide (Forbes, Heinekey, Mann, and Millar[136]). This salt, m.p. 210°, does not crystallize as readily as its lower homologues already described.

The dibromide, when heated at 215° at 0.1 mm in nitrogen, decomposed with the formation of a mixture of 9-methyl-9-arsafluorene and 9-bromo-9-arsafluorene, and in this respect behaves as the corresponding eight-membered and nine-membered homologues (pp. 489, 491).

The thermal decomposition of the corresponding dihydroxide has not been investigated.

Tribenzo[b,d,h] [1,6]diarsecin

The parent compound (1), [RIS 9117], is numbered and presented as shown.

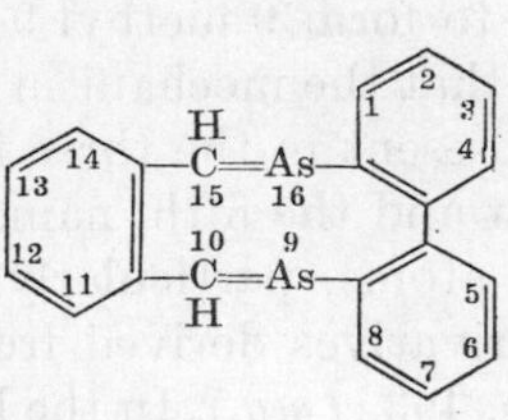

(1) 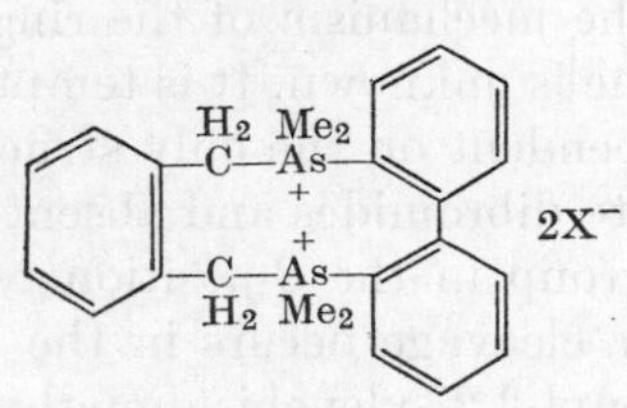   (2)

The only known type of derivative is the last of the series of compounds prepared by cyclic diquaternization of 2,2'-biphenylylenebis-(dimethylarsine) (p. 486). This diarsine, when heated with o-xylylene dibromide (one equivalent) in methanol at 100°, gives 9,10,15,16-tetra-hydro-9,9,16,16-tetramethyltribenzo[b,d,h][1,6]diarsecinium dibromide

(**2**; X = Br), initially termed 5,5,10,10-tetramethyl-5,10-diarsonia-1,2:3,4:7,8-tribenzocyclodecatriene dibromide (Heaney, Heinekey, Mann, and Millar[33]). The dibromide, m.p. 209–210°, gives a dipicrate, m.p. 225–227°.

The dibromide, when heated at 210° rising to 235° (oil-bath) at 0.05 mm, loses methyl bromide and gives a distillate of 9,10,15,16-tetrahydro-9,16-dimethyltribenzo[*b*,*d*,*h*][1,6]diarsecin (**3**), m.p. 110–111°. This cyclic diarsine has been identified by analysis, molecular-weight determination, and diquaternization with an excess of methyl *p*-toluenesulfonate: the syrupy product on treatment with sodium picrate gives the above dipicrate, m.p. 225–227°.

(**3**)

It is a striking fact, therefore, that of the five dibromides obtained by the cyclic diquaternization of 2,2′-biphenylylenebis(dimethylarsine) with dibromomethane, 1,2-dibromoethane, 1,3-dibromopropane, 1,4-dibromobutane, and *o*-xylylene dibromide, the first and the last, on thermal decomposition, lose methyl bromide but retain their cyclic diarsine system intact, whereas the three intermediates undergo cleavage and ring contraction to give 9-methyl-9-arsafluorene (5-methyldibenzarsole).

The mechanism of the ring contraction to form 9-methyl-9-arsafluorene is unknown. It is tempting to assert that the mechanism must be dependent on the only structural factor present in the three intermediate dibromides and absent from the first and the fifth, namely, a CH_2 group in the β-position to the arsenic atoms, particularly as a similar cleavage occurs in the analogous derivatives derived from *o*-terphenyl-2,2″-ylenebis(dimethylarsine) (cf. p. 493 *et seq.*). In the latter series, however, the distinction is not so sharp, but in the phosphorus analogues (pp. 194, 198, 200) the contraction is general.

The dibromide (**2**; X = Br) was converted into the di-(+)-3-bromo-8-camphorsulfonate (**2**; X = $C_{10}H_{14}BrO_4S$); repeated extractions with cold acetonitrile brought the [α] of the residue up to 94.7°, and recrystallization from water then gave the optically pure (+)-diarsonium di-(+)-bromocamphorsulfonate, having [α]$_D$ +103°, [M] +1110° in aqueous solution. This salt in turn gave the (+)-dibromide, m.p. 185–187°, having $[M]$ +557°. The di-(−)-3-bromo-8-camphorsulfonate on recrystallization gave the (−)-diarsonium di-(−)-bromocamphorsulfonate, having $[\alpha]$ −102.2°, $[M]$ −1109°, which furnished the optically pure (−)-dibromide, m.p. 189–190°, having $[M]$ −560°.[137]

An aqueous solution of the (+)-dibromide was heated in a sealed tube at 150° for 3 hours without change in its optical activity. The cation (2) therefore resembles that of the 6,7,8,9-tetrahydro-5,5,9,9-tetramethyl-5H-dibenzo[f,h][1,5]diarsonium cation (p. 492) in its very high optical stability.

The behavior of the dibromide (2; X = Br) on being heated differs strikingly from that of the analogous 9,9,16,16-tetraethyl dibromide, which undergoes apparently complete cleavage of the ring system, with the formation of 9-ethyl-9-arsafluorene and 2,2′-biphenylylenebis(diethylarsine) (Allen, Millar, and Mann[139]).

Tribenzo[e,g,i] [1,4]diarsecin

This parent compound, [RIS —], has the structure and numbering (1).

Its only known derivatives have been obtained by the diquaternization of o-terphenyl-2,2″-ylenebis(dimethylarsine) (p. 493) and 1,2-dibromoethane, giving 9,10,11,12-tetrahydro-9,9,12,12-tetramethyltri-

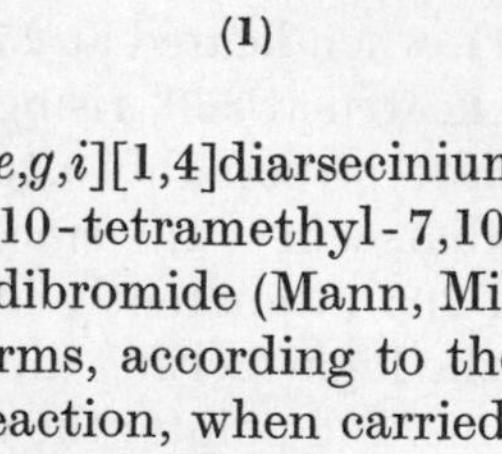

(1) 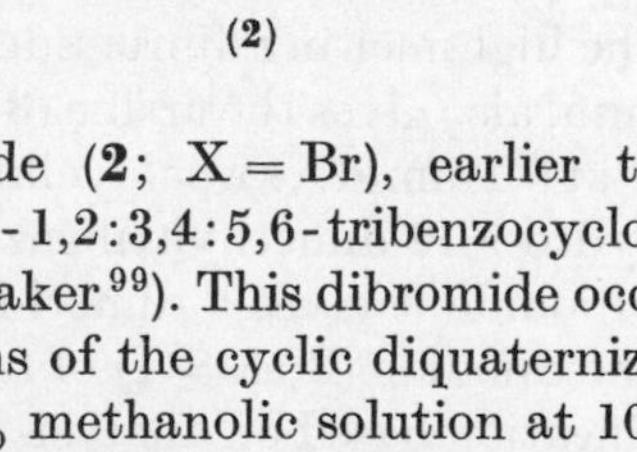   (2)

benzo[e,g,i][1,4]diarsecinium dibromide (2; X = Br), earlier termed 7,7,10,10-tetramethyl-7,10-diarsonia-1,2:3,4:5,6-tribenzocyclodeca-triene dibromide (Mann, Millar, and Baker[99]). This dibromide occurs in two forms, according to the conditions of the cyclic diquaternization. This reaction, when carried out in 5% methanolic solution at 100° for 10 hours, furnishes a dibromide, m.p. 201–202°, which can be recrystallized from methanol and gives a dipicrate (2; X = $C_6H_2N_3O_7$), m.p. 246°, and a diiodide (2; X = I) m.p. 256–257°, which can also be converted into the same dipicrate.

When, however, the reaction is carried out without a solvent, and with heating at 150–155°, it furnishes a 'high-melting' dibromide, m.p. 272–275°, of much lower solubility in organic solvents: this in turn gives a diiodide (2; X = I), m.p. 289–292°.

The special models (p. 494) show that the cation (2) has no element of symmetry, but that it can 'twist' or flex very considerably about the mean 'open' position (see diagram in reference 99). This model does not, however, provide any adequate explanation of the two forms of the dibromide (2; X = Br). Since the molecular weight of these salts cannot be readily determined, it is impossible to dismiss completely the (un-likely) possibility that one of them may be formed by condensation of two molecules of the diarsine with two of 1,2-dibromoethane, giving a cation having a 20-membered heterocyclic ring. It is noteworthy that the analogous dibromides formed by the condensation of the initial diarsine with dibromomethane, 1,3-dibromopropane, and o-xylylene dibromide give no indication of the existence of two forms.

The low-melting dibromide (2; X = Br), when heated at 200–205° at 14 mm under nitrogen, undergoes decomposition with the formation

(3)   (4)

of 9-methyl-9H-tribenz[b,d,f]arsepin (3), earlier termed 7-methyl-7-arsa-1,2:3,4:5,6-tribenzoheptatriene[99]; the properties of this arsine, m.p. 159–160°, have been discussed briefly on p. 440.

The high-melting dibromide (2; X = Br), when heated at 250–270° at 14 mm, also gives the arsine (3), but when heated at 200° rising slowly to 220° at 0.1 mm gives a pale yellow oily distillate, presumably 9,10,11,12-tetrahydro-9,12-dimethyltribenzo[e,g,i][1,4]diarsecin (4), since with methyl iodide it gives a dimethiodide, m.p. 296°, apparently identical with the diiodide (2; X = I) obtained from the high-melting dibromide (identification based on analysis and mixed melting points: the melting points of halides in this and analogous series are, however, affected by the temperature of immersion).

The decomposition of both dibromides to give the monoarsine (3) is markedly similar to the decomposition of 5,6,7,8-tetrahydro-5,5,8,8-

tetramethyldibenzo[e,g][1,4]diarsocinium dibromide to give 9-methyl-9-arsafluorene (p. 489).

Eleven-membered Ring Systems containing Carbon and Two Arsenic Atoms

9H-Tribenzo[f,h,j] [1,5]diarsundecin

This parent compound (1), [RIS —], has the structure and numbering shown. Our knowledge of its derivatives follows closely on that of the tribenzo[e,g,i][1,4]diarsecin of the previous Section.

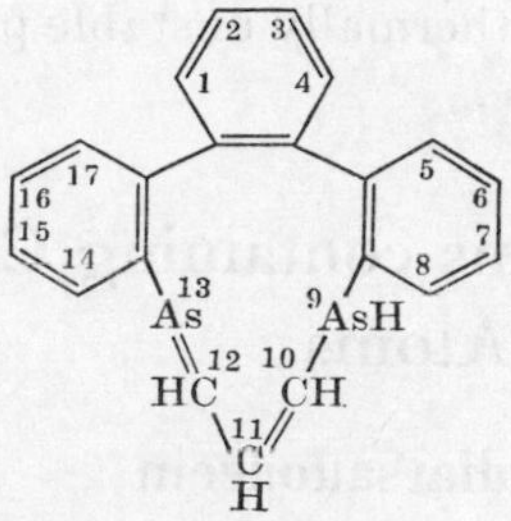

(1) 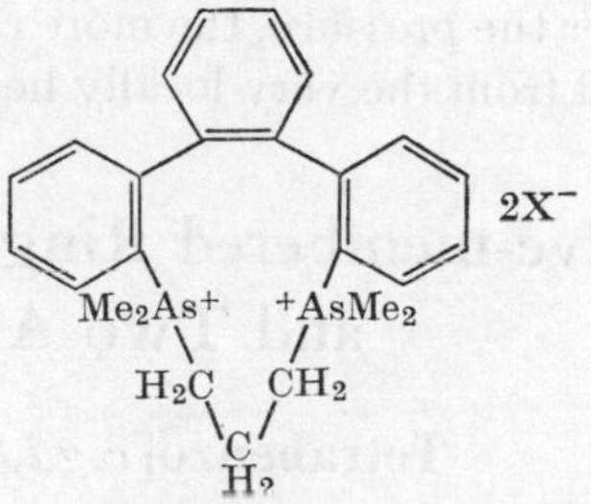   (2)

o-Terphenyl-2,2″-ylenebis(dimethylarsine) (p. 493) undergoes di-quaternization when heated with 1,3-dibromopropane at 120–125° for 6 hours to give 10,11,12,13-tetrahydro-9,9,13,13-tetramethyl-9H-tri-benzo[f,h,j][1,5]diarsundecinium dibromide (2; X = Br), initially termed 7,7,11,11-tetramethyl-7,11-diarsonia-1,2:3,4:5,6-tribenzocyclo undecatriene dibromide (Mann, Millar, and Baker[99]). The dibromide, m.p. 225–226°, gives a diiodide (2; X = I), m.p. 240–242°, which is almost insoluble in methanol and ethanol.

The special model (p. 494) shows that the cation (2) could exist in two isomeric forms, each having a plane of symmetry similar to that of the tribenzodiarsoninium dibromide (p. 494), and that each form shows a considerable degree of strainless 'twist' without losing its isomeric identity (cf. diagrams in reference 99). When, however, gentle pressure is applied to the central CH_2 group of either model, the ring will 'spring' over to the other isomeric form, indicating that the energy barrier between the two forms may be low. No decisive evidence for the existence of the two forms has been obtained, but it might well be masked by the very low solubility of the initial salts of the cation (2).

The thermal decomposition of the dibromide (2; X = Br) is affected by the conditions employed and may give a mixture of products of which in each case only one component has been identified. When heated at 220–225° at 14 mm, the dibromide gives a golden syrupy distillate, which with methyl iodide gives the dimethiodide of the original uncyclized o-terphenyl-2,2″-ylenebis(dimethylarsine). The dibromide, heated at 220° at 0.1 mm, gives a distillate which furnishes the crystalline 9-methyl-9H-tribenz[b,d,f]arsepin (p. 440). Finally, the dibromide, heated at 220° at 0.002 mm, gives a distillate which with methyl iodide forms apparently the diiodide (2; X = I), pale yellow crystals, m.p. 270°. This melting point is markedly different to that of the original diiodide and may possibly indicate a change of isomeric identity during the decomposition and/or distillation. The apparent effect of pressure on these decompositions may be due to the fact that the lower the pressure, the more rapidly a thermally unstable product is removed from the very locally heated zone.

Twelve-membered Ring Systems containing Carbon and Two Arsenic Atoms

Tetrabenzo[c,g,i,k] [1,6]diarsadodecin

This parent compound (1), [RIS —], is depicted and numbered as shown. Its derivatives form the fourth class of ring system obtained by

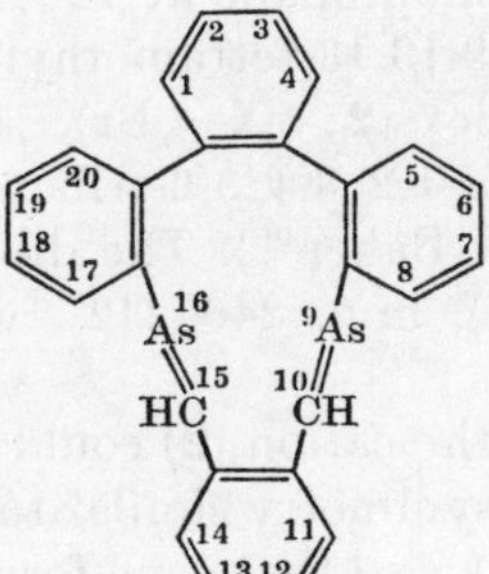

(1) 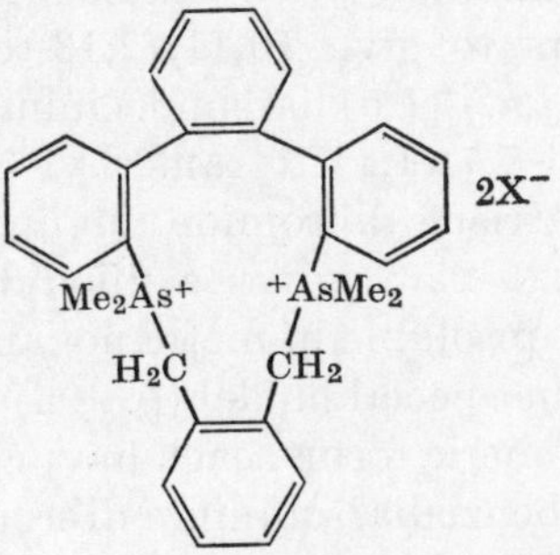   (2)

the cyclic diquaternization of o-terphenyl-2,2″-ylenebis(dimethylarsine) (p. 493). This diarsine rapidly reacts with one equivalent of o-xylylene dibromide in boiling xylene, depositing the crystalline 9,10,15,16-tetrahydro - 9,9,16,16 - tetramethyltetrabenzo[c,g,i,k][1,6]diarsadodeci-nium dibromide (2; X = Br), originally termed 7,7,12,12-tetramethyl-7,12-diarsonia-1,2:3,4:5,6:9,10-tetrabenzocyclododecatetraene dibromide (Mann, Millar, and Baker[99]).

This ready reaction of the diarsine with o-xylylene dibromide in boiling or even warm xylene contrasts markedly with its lack of reaction with dibromomethane under the same conditions. The dibromide (2; X = Br), moreover, is readily soluble in cold water, unlike the earlier members of this series.

This dibromide, m.p. 195–197°, gives a diiodide (2; X = I), m.p. 181–182°, soluble in hot methanol.

The model (p.494) shows that the cation (2) has a plane of symmetry, but is capable of 'twisting' about this mean position to an even greater extent than the diarsundecinium ion discussed in the previous Section.

The dibromide (2; X = Br), when heated at 185–190° at 14 mm for 4 hours gives only a trace of distillate, but the residue on recrystallization gives 9,10,15,16-tetrahydro-9,16-dimethyltetrabenzo[c,g,i,k][1,6]diarsa-dodecin (3), m.p. 67–71°, the decomposition of the dibromide having thus been limited to loss of methyl bromide.

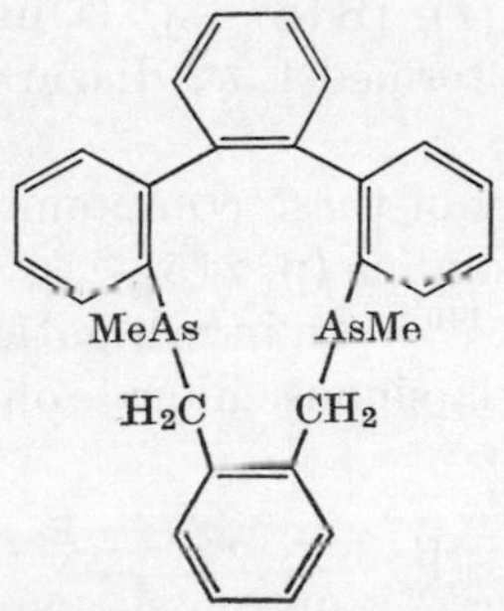

(3)

The diiodide (2; X = I), however, when heated at 245–250° at 14 mm, gives an oily distillate, which in ethanol deposits the crystalline 9-methyl-9H-tribenz[b,d,f]arsepin (p. 498).

The difference in the two types of decomposition may have been caused by the change in temperature or anion employed, or both. Very little work has been done on the possible influence of the anion on the mode of thermal decomposition of cyclic arsine methohalides.

The ring contraction to the above arsepin which the diiodide undergoes shows that a CH_2 group in the β-position to the arsenic atoms is not essential for this type of reaction, although the dibromides obtained by the diquaternization of 2,2'-biphenylylenebis(dimethyl-arsine) and of o-terphenyl-2,2''-ylenebis(dimethylarsine) with 1,2-dibromoethane and 1,3-dibromopropane all have a CH_2 group in this position and all undergo the ring contraction when heated (p. 496).

Note. In all monocyclic systems containing nitrogen and arsenic as the only heteroatoms, the numbering starts at the nitrogen atom. If the ring is hydrogenated, the normal terminations or suffixes to the names of rings containing arsenic do not apply: the names of fully hydrogenated rings containing 3, 4, or 5 atoms end in -arsiridine, -arsetidine, and -arsolidine, respectively, but those of hydrogenated larger rings retain the names of the corresponding unsaturated rings preceded by the position and number of hydrogen atoms, *e.g.* '2,3,4,5-tetrahydro'.

Five-membered Ring Systems containing Carbon and One Arsenic and Two Nitrogen Atoms

1*H*-1,3,2-Diazarsole and 1,3,2-Diazarsolidines

1*H*-1,3,2-Diazarsole (**1**), [RIS —], is unknown and its 2,3,4,5-tetrahydro derivative (**2**), termed 1,3,2-diazarsolidine, is known only as substituted products.

The known chemistry of these compounds is closely analogous to that of 1,3,2-diazaphospholidine (p. 215).

Scherer and Schmidt,[140] in an investigation of the action of arsenic trichloride and of dimethylarsinous chloride on various aliphatic amines,

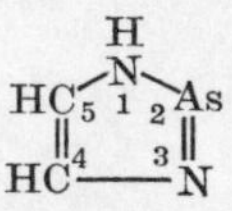

(1) 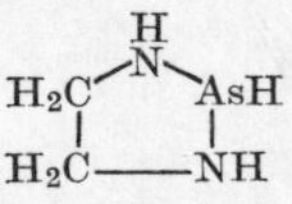   (2) 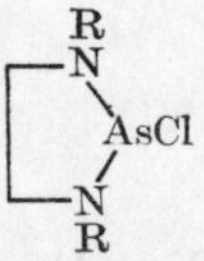   (3) 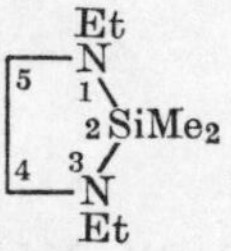   (4)

showed that arsenic trichloride reacted with N,N'-dimethylethylene-diamine, $[CH_3NHCH_2—]_2$, in ethereal solution to give 2-chloro-1,3-dimethyl-1,3,2-diazarsolidine (**3**; $R = CH_3$), b.p. 103–105°/14 mm, m.p. 19–21° (*ca.* 50% yield).

Using a different method, capable of wide utilization, Abel and Bush[141] have added arsenic trichloride in toluene to an equimolecular amount of 1,3-diethyl-2,2-dimethyl-1,3-diaza-2-silacyclopentane (**4**), [RIS 9742], also in toluene. The reaction mixture was boiled under reflux for 6 hours and then fractionally distilled, giving in turn (*a*) dichlorodimethylsilane, $Cl_2Si(CH_3)_2$, (*b*) toluene, and (*c*) 2-chloro-1,3-diethyl-1,3,2-diazarsolidine (**3**; $R = C_2H_5$), b.p. 81°/0.7 mm (42%). The infrared and nmr spectra of this compound and of several analogous compounds are discussed.

The compound (**3**; $R = CH_3$) was initially termed 2-chloro-1,3-dimethyl-1,3,2-diazoarsenolidine,[140] and (**3**; $R = C_2H_5$) was termed 1,3-diethyl-2-chloro-1,3-diaza-2-arsacyclopentane.[141]

Six-membered Ring Systems containing Carbon and One Arsenic and One Nitrogen Atom

Azarsenines

There is a possibility of three isomeric azarsenines, 1,2-azarsenine (**1**), 1,3-azarsenine (**2**) and 1,4-azarsenine (**3**). These three isomers,

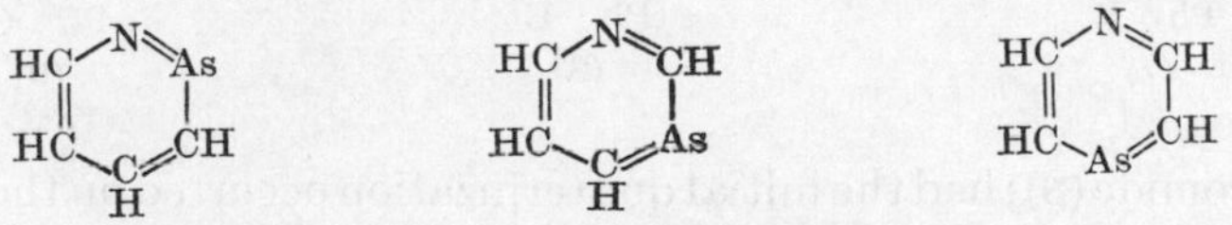

$$
\begin{array}{ccc}
\underset{\textstyle HC\underset{\parallel}{}}{}\overset{N}{\diagdown}As & \underset{HC}{}\overset{N}{\diagdown}CH & \underset{HC}{}\overset{N}{\diagdown}CH \\
HC\diagdown\underset{H}{C}{\diagup}CH & HC\diagdown\underset{H}{C}{\diagup}As & HC\diagdown As\diagup CH \\
(1) & (2) & (3)
\end{array}
$$

having a 'semi-aromatic' structure, have not been isolated, and apparently no derivative of (**1**) or (**2**) has been recorded. The isomer (**3**) has been isolated as its hexahydro derivatives.

1,4-Azarsenine

This system (**3**) [RRI 229] has been synthesized in the form of hexahydro-1,4-diphenyl-1,4-azarsenine (**4**).

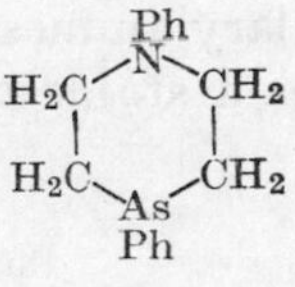

$$
\begin{array}{c}
Ph \\
| \\
N \\
H_2C \diagup \quad \diagdown CH_2 \\
| \qquad\qquad | \\
H_2C \diagdown \quad \diagup CH_2 \\
As \\
| \\
Ph
\end{array}
$$
(**4**)

Phenylarsinebis(magnesium bromide) (**5**) (p. 229) reacts in benzene with di-(2-bromoethyl)aniline (**6**) to give a mixture of the crystalline

$$C_6H_5As(MgBr)_2 \qquad C_6H_5N(CH_2CH_2Br)_2$$
(**5**) $\qquad\qquad$ (**6**)

hexahydro-1,4-diphenyl-1,4-azarsenine (**4**) and a resinous product, probably formed by extensive linear interaction of (**5**) and (**6**) (Beeby and Mann[142]).

The azarsenine (4), m.p. 96–97.5° (monopicrate, m.p. 172–173°), forms a monomethiodide, m.p. 181–182°, a monoethobromide, m.p. 179–179.5°, and a viscous mono-p-chlorophenacyl quaternary bromide, which gives a crystalline picrate, m.p. 93–94°. Diquaternary salts have not been obtained, even under forcing conditions.

There is no doubt that these monoquaternary salts have the structure (7), $i.e.$, that the tertiary arsine group, and not the amine group, has undergone quaternization. This is shown by the fact that the monoethobromide (7; R = C$_2$H$_5$, X = Br) forms a stable crystalline

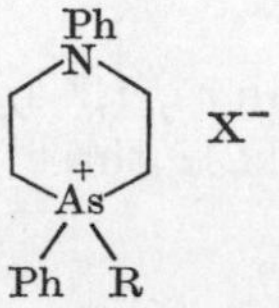

(7) 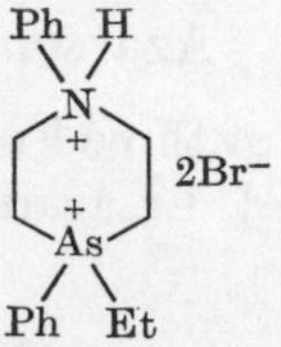   (8) 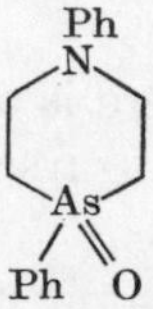   (9)

hydrobromide (8); had the initial quaternization occurred on the nitrogen atom, the tertiary arsine group could not have formed a stable hydrobromide. The general structure (7) is in harmony with the greater reactivity of a tertiary arsine with alkyl halides than of the analogous tertiary amine (Davies and Lewis,[143] Davies and Addis[144]).

The hexahydroazarsenine (4) when treated in solution either with very dilute hydrogen peroxide or with 'Chloramine-T' (N-chloro-N-sodio-p-toluenesulfonamide, CH$_3$C$_6$H$_4$SO$_2$NClNa) gives hexahydro-1,4-diphenylazarsenine 4-oxide dihydrate (9). The structure of this compound is confirmed by the findings that (a) aqueous 'Chloramine-T' readily converts tertiary arsines into the arsine oxides (Mann[145]) but does not affect tertiary alkylarylamines, and (b) the oxide (9) forms only a monopicrate but gives a stable dihydrochloride (10); had the

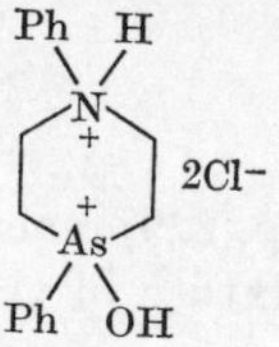

(10) 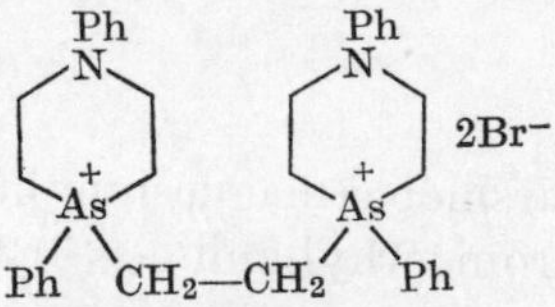   (11)

initial compound been the isomeric N-oxide, it would have formed only a monohydrochloride, in view of the neutral properties of the tertiary arsine group.

Both quaternization and oxidation thus occur preferentially at the arsenic atom. It is not surprising, therefore, that heating the azarsenine (4) with 1,2-dibromoethane in an attempt to achieve cyclic diquaterniza- tion gave only ethylenebis(hexahydro-1,4-diphenyl-1,4-azarseninium) dibromide[142] (11), m.p. 226–227° (dipicrate, m.p. 166–167°).

The inability of the quaternary arsonium salts of type (7) to undergo further quaternization at the nitrogen atom is almost certainly due to the strong positive charge on the arsenic atom causing an electron- attracting (inductive) effect through the —CH_2CH_2— groups and thus partially deactivating the already rather weak amine group. (For a discussion of many analogous examples, see Mann and J. Watson[124]). If this is so, a weaker positive charge on the arsenic atom should allow a correspondingly greater reactivity of the amine group. The arsenic atom in the As=O group of the 4-oxide (9) does carry a much weaker charge, and this compound, when heated with an excess of methyl p-toluene- sulfonate at 100°, gives the hexahydro-1-methyl-1,4-diphenyl-1,4- azoniaarsenine 4-oxide p-toluenesulfonate (12; $X = C_7H_7O_3S$), which with picric acid gives the 4-hydroxy dipicrate, m.p. 182–183°.

(12)   (13)   (14)

Treatment of the azarsenine (7) with more concentrated hydrogen peroxide gives the syrupy hexahydro-1,4-diphenyl-1,4-azarsenine 1,4- dioxide (13) which, however, with an excess of picric acid gives only a mono-hydroxypicrate, m.p. 179°; it is not known whether the oxygen atom attached to the nitrogen or to the arsenic has accepted the proton in the formation of this salt, but the significant fact is that both the oxide groups in (13) cannot simultaneously combine with a weak acid such as picric acid.

The diphenylazarsenine (4), heated in boiling hydriodic acid, gives the cream-colored hydriodide (m.p. 173–174°) of hexahydro-4-iodo-1- phenyl-1,4-azarsenine (14; R = I), which with aqueous sodium hydrogen carbonate gives the 4-hydroxy compound (14; R = OH), m.p. 116–116.5°. Previous examples of this cleavage of aryl groups from arsenic, with replacement by iodine, have been given on pages 371, 416, 428.

The above hydriodide of the 4-iodo derivative (**14**; R = I), when treated in hot benzene solution with iodine, gives the deep brown hydriodide (m.p. 136–138°) of hexahydro-4-triiodo-1-phenyl-1,4-azarsenine (**14**; R = I₃); this compound, when treated in aqueous suspension with sulfur dioxide, is reduced back to the hydriodide of (**14**; R = I).[112]

Benzo Derivatives of 1,4-Azarsenine

Theoretically there should be a monobenzo derivative (**1**) and a dibenzo derivative (**2**) of 1,4-azarsenine. The former system is still

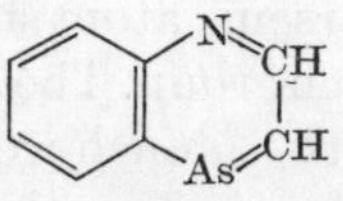

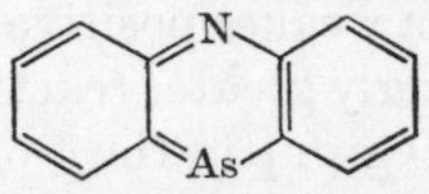

unknown, whereas a wide range of derivatives of the dibenzo derivative (**2**), known as phenarsazine, has been studied.

Phenarsazine

Phenarsazine [RRI 3257] has the structure and numbering shown in (**1**); it could therefore be termed dibenzo-1,4-azarsenine or 5-aza-10-arsaanthracene, and at one time it was commonly termed 'Adamsite' in the U.S.A. The earlier systems of numbering such as (**2**), formerly

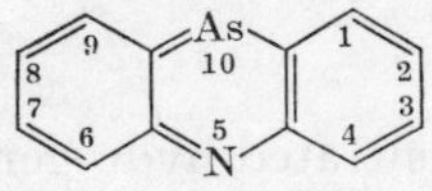

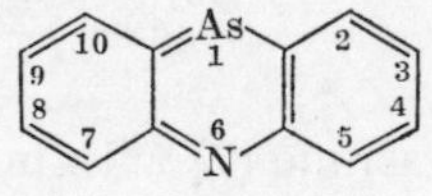

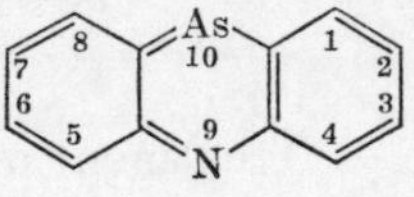

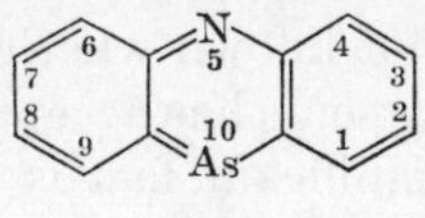

used in *Chemical Abstracts*, and (**3**), used in German literature, are obsolete.

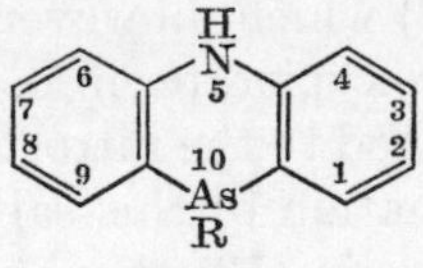

The existence of the very reactive phenarsazine (**1**) is reasonably established, but the vast majority of phenarsazine compounds are derivatives of 5,10-dihydrophenarsazine; since the reactions of most of these compounds involve substituents on the arsenic atom, it has become

customary in many cases to invert the formula (**1**) to (**1A**) and the
5,10-dihydrophenarsazine becomes (**4**; R = H).

A considerable number of such phenarsazine derivatives are known,
particularly because cyclization to form the phenarsazine ring system is
usually a simple and ready process. Consequently the present account is
not exhaustive: it aims at providing a general account of the preparation,
properties, and structure of the main types of phenarsazine derivative,
and it is followed by a brief record of those derivatives having one or
more alicyclic rings fused to the main phenarsazine ring system.

The first announcement of the condensation of diphenylamine and
arsenic trichloride to give 10-chloro-5,10-dihydrophenarsazine (**5**)
(known then as phenarsazine chloride) appeared in a German patent
specification,[146] but the first reasonably full description of the preparation and properties of compounds of this class was provided by Wieland
and Rheinheimer.[122] These authors showed that, if a mixture of diphenylamine and arsenic trichloride was boiled under reflux until
evolution of hydrogen chloride ceased (*ca.* 4 hours) and was then mixed
with boiling xylene, the solution on cooling deposited the crystalline
10-chloro compound (**5**), which after recrystallization or sublimation in
a vacuum had m.p. 193°. It is normally obtained as pale greenish-yellow
needles, which give a yellow powder.

Fischer[147] has stated that the stable rhombic form has m.p. 195°,
but that the compound also exists in two metastable forms, one monoclinic, m.p. 186°, and the other probably triclinic, m.p. 182°. Camerman
and Trotter,[148] when preparing crystals for an *X*-ray crystal structure
investigation (p. 516), found that the chloride separated from many
solvents, *e.g.*, xylene, as metastable yellow crystals containing solvent
of recrystallization; heating these crystals under reduced pressure gave
the solvent-free stable yellowish-green crystals, which when crushed
formed a yellow powder.

This compound has many of the chemical, and some of the physiological, properties of a diarylmonochloroarsine, such as diphenylarsinous

$$HN{<}(C_6H_4)_2{>}As{-}O{-}As{<}(C_6H_4)_2{>}HN$$

(**6**)

(**5**)

(**7**)

chloride, $(C_6H_5)_2AsCl$, and without doubt this similarity stimulated much of this earlier research into the phenarsazine series generally. The 10-chloro compound (5), for example, when in the form of dust or vapor, has an irritating effect on the mucous membrane similar to that of diphenylarsinous chloride (cf. p. 521).

On the chemical side, the 10-chloro compound when treated with alkalis gives 'phenarsazine oxide', to which the constitution (6) was allotted. This compound is thus systematically the anhydride of 5,10-dihydro-10-hydroxyphenarsazine (4; R = OH), named 10,10'-oxybis-(5,10-dihydrophenarsazine), recently renamed 10,10'-(5H,5'H-oxydiphenarsazine). It is very slightly soluble in most solvents and has a melting point 'above 350°', properties which would normally be associated with a more complex molecule.

When a boiling mixture of the 'oxide' (6) and methanol is treated with hydrogen sulfide, the phenarsazine sulfide [10,10'-thiobis(5,10-dihydrophenarsazine) or 10,10'-(5H,5'H-thiodiphenarsazine)], analogous to (6), is obtained as yellow crystals, m.p. 262°. Furthermore, acetic acid readily converts the 'oxide' into 10-acetoxy-5,10-dihydrophenarsazine (7), greenish leaflets, m.p. 223–224°.[122]

When the 10-chloro compound (5) is boiled with sodium methoxide in methanol, the 10-methoxy derivative (8), colorless needles, m.p. 194°,

$$HN{<}(C_6H_4)_2{>}As-N{<}(C_6H_4)_2{>}As-N{<}(C_6H_4)_2{>}AsCl$$

(9)

$$[HN{<}(C_6H_4)_2{>}As]_3N$$

(10)

(8)

is formed. A number of such compounds have been prepared. Furthermore, the 10-chloro compound, when heated with silver cyanide, gives the 10-cyano compound,[149] m.p. 227–228°.

The compound (8) in boiling methanol also reacts with hydrogen sulfide to give the above-mentioned sulfide.

It will be clear that if the elements of hydrogen chloride could be broken off the 10-chloro compound (5), the parent phenarsazine (1) would be obtained. Wieland and Rheinheimer showed, however, that when the chloro compound was boiled for 2 hours with pyridine (or with quinoline or dimethylaniline), a compound termed triphenarsazine chloride, whose composition corresponded with the structure (9), was obtained. When the 10-chloro compound in boiling xylene was treated with ammonia gas, triphenarsazin-10-ylamine (10) was obtained; this

reaction recalls the action of ammonia on molten 2-bromoethylphthal-
imide, which produces tris-(2-phthalimidoethyl)amine,

$$[C_6H_4(CO)_2N-CH_2CH_2]_3N.^{150}$$

However, the parent phenarsazine was obtained by heating the
10-methoxy compound (8) in N-methyldiphenylamine solution, through
which a current of carbon dioxide was passed. In these circumstances,
methanol is split off, and the phenarsazine (1) obtained as an orange-
yellow product, having a melting point 'about 310°'. As might be
expected, this compound is very reactive. Its solutions become de-
colorized very rapidly when exposed to air; this effect is apparently
caused, not by oxidation, but by absorption of water to give the 'oxide'
(6). It also readily combines with alcohols and phenols to give compounds
of type (8), with acetic and other acids to give products of type (7), and
with hydrogen chloride to give the 10-chloro compound (5). The last
addition is apparently the only one which has not been reversed.

The parent phenarsazine (1) is structurally of great interest, because
it is one of the few arsenic compounds having an 'aromatic' structure;
in this case, the structure analogous to that of anthracene.

When the 10-chloro compound (5) is treated in acetic acid with
hydrogen peroxide, it is converted into the colorless acid (11), m.p.
above 300°; since this is clearly an arsinic acid, it is termed phenazarsinic
acid (its systematic name is 5,10-dihydro-10-hydroxyphenarsazine 10-
oxide). It gives the N-acyl derivatives of m.p.: acetyl, 244–245°;

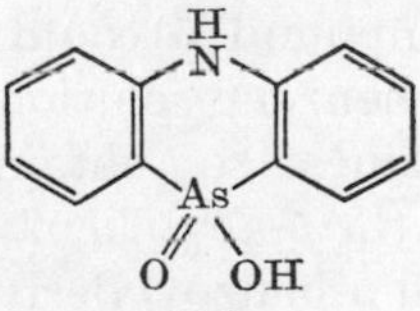

(11)

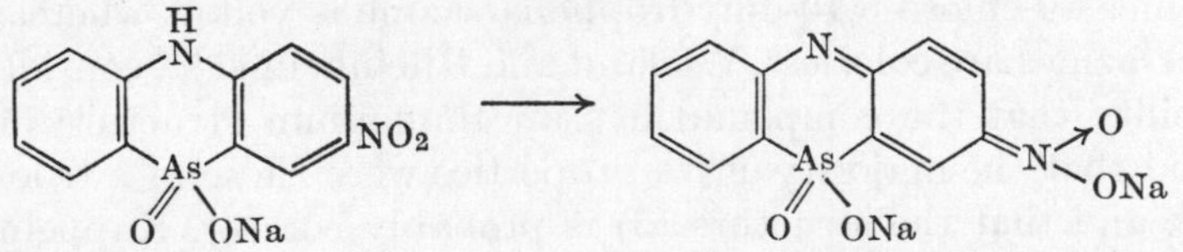

(12)   (13)

propionyl, 232°; benzoyl, 250°. The nitro derivatives of this acid are of
considerable interest. They can be prepared in two ways. If the 10-chloro
compound in acetic acid solution is treated cautiously with fuming nitric
acid, the following derivatives can be isolated: 4-nitro, scarlet needles,

m.p. 156°; 2-nitro, greenish-yellow leaflets; 2,8-dinitro, pale yellow needles, m.p. >300°. If these 10-chloro-nitro derivatives are now treated with hydrogen peroxide in acetic acid, the corresponding nitrophenazarsinic acids are obtained. Alternatively the latter can be obtained by the direct nitration of the phenazarsinic acid (**11**). The 2-nitrophenazarsinic acid forms two series of salts. For example, it gives a yellow monosodium salt, which is undoubtedly the normal sodium arsinate (**12**); with an excess of sodium hydroxide it will give a cherry-red disodium salt, which is almost certainly a derivative of the *aci*-form (**13**), in which the right-hand ring necessarily has the quinonoid structure.

These nitro compounds can be reduced by ferrous hydroxide to the corresponding aminophenazarsinic acids, since the arsinic acid group is not affected by this reducing agent. The aminophenazarsinic acids are colorless crystalline compounds having amphoteric properties; the arsinic acid group is not sufficiently strongly acidic to prevent them forming crystalline hydrochlorides, for instance.

Preparation of 10-chloro-5,10-dihydrophenarsazine, of the oxide (**6**) and of phenazarsinic acid was described independently by Schmidt[151] shortly after the appearance of Wieland and Rheinheimer's work.[122]

Burton and Gibson[152] have shown that the oxidation of 10-chloro-5,10-dihydrophenarsazine to the phenazarsinic acid can also be very conveniently carried out by using an aqueous solution of Chloramine-T (p. 504), a reaction which appears to be general also for dialkyl- and diaryl-monochloroarsines, R_2AsCl, which similarly give the corresponding arsinic acids, $R_2AsO(OH)$. These authors also showed that the NH group in the 10-chloro compound (**5**) could be readily acylated by the appropriate acyl chloride in benzene solution, and that the 5-acyl-10-chloro-5,10-dihydrophenarsazines so obtained could be similarly oxidized by Chloramine-T to the 5-acylphenazarsinic acids. In this way the 5-acetyl, 5-propionyl, and 5-benzoyl derivatives in each series were prepared.

Since 10-chloro-5,10-dihydrophenarsazine is yellow whereas many phenarsazines are colorless, Wieland and Rheinheimer[122] considered the possibility that the compound has an ammonium structure (**5a**), but decided that the majority of its properties were those of a true chloroarsine, and that the structure (**5**) is probably correct. Kappelmeier[153]

(5a)   (5b)

has suggested that the compound is an arsonium salt of structure (**5b**), a suggestion which was vigorously opposed by Gibson, Johnson, and Vining.[154] The unambiguous synthesis of the chloro compound (p. 513) and its X-ray crystal-structure analysis (p. 516) provide powerful support for the structure (**5**), and there is very little evidence that this compound undergoes ionization in solution.

When 10-chloro-5,10-dihydrophenarsazine is reduced in ethanolic acetone solution with hypophosphorous acid, it gives 10,10′-bis-5,10-dihydrophenarsazine;

$$HN{<}(C_6H_4)_2{>}As{-}As{<}(C_6H_4)_2{>}NH$$

recently renamed 10,10′-(5H,5′H-biphenarsazine),[155] m.p. 304–305°. The same compound is obtained when phenarsazinic acid in acetone solution containing a trace of iodine is similarly reduced with hypophosphorous acid. It is a reasonably stable compound, but in hot xylene solution undergoes ready oxidation back to phenarsazinic acid. It has been suggested that this compound, in view of its ready oxidation and its orange color, might exist partly (or wholly) as free radicals:

$$H{<}N(C_6H_4)_2{>}As{-}As{<}(C_6H_4)_2{>}NH \rightleftarrows 2HN{<}(C_6H_4)_2{>}As\cdot$$

The compound has been found to be diamagnetic, however, and hence the solid material must consist of the dimeric form with no detectable amount of the above free radical.[156]

With regard to the preparation of 5- and 10-substituted phenarsazines, it should be noted that Wieland and Rheinheimer[122] claimed that the phenarsazine cyclization was not peculiar to secondary diarylamines, because, they stated, N-methyldiphenylamine, $(C_6H_5)_2NCH_3$, would similarly condense with arsenic trichloride to give 10-chloro-5,10-dihydro-5-methylphenarsazine (**14**), yellowish-green needles, m.p. 203°,

(14)   (15)   (16)

although the yield was much lower than that of the unmethylated product (**5**). Lewis and Hamilton[157] showed that N-phenyl-1-naphthylamine condensed readily with arsenic trichloride when heated to 200° in the course of 2 hours, to give '3,4-benz-10-chloro-5,10-dihydrophenarsazine' (**15**; R = Cl) (p. 524). This compound formed yellow crystals, m.p. 219°, and had only a slight irritating action on the mucous membrane of the nose and throat. It is interesting that this compound, when treated with hydrobromic acid, gave the 10-bromo compound (**15**; R = Br), dark yellow needles, m.p. 227°, and with hydriodic acid gave the 10-iodo compound (**15**; R = I), red needles, m.p. 205°. A series of 10-alkoxy and other derivatives was also made. Lewis and Stiegler[158] claimed that dichloro-(2-chlorovinyl)arsine, $ClCH{=}CHAsCl_2$, in the above condensation, gave '3,4-benz-10-(2'-chlorovinyl)-5,10-dihydrophenarsazine' (**15**; R = CH=CHCl), bright yellow needles, m.p. 213°. For this condensation, the reactants were heated together for only 15 minutes; too long heating, too high a temperature, or the use of catalysts produced only a tar, while no reaction occurred in boiling xylene. These workers also claimed that the condensation of diphenylamine with dichloro-2-chlorovinylarsine gave 10-(2'-chlorovinyl)-5,10-dihydrophenarsazine (**16**), m.p. 186–187°, which had a very irritating action on the eyes and nose. The validity of these claims to have prepared either the 5-methyl derivative (**14**) or the 10-(2-chlorovinyl) derivatives of type (**15**) and (**16**) are discussed on pp. 514–516.

In 1926 there appeared the first[159] of a long series of papers[159, 155, 160–175] by Gibson with Burton and Johnson which have vastly increased our knowledge of the preparation, structure, and properties of phenarsazine derivatives. The interested reader is referred particularly to the first paper,[159] in which (in a footnote) a brief account is given of the various independent centers of research on phenarsazines which were operating during World War I; this account is valuable because it is clear that many of the early discoveries in phenarsazine chemistry were made independently by various workers, and it is now not possible to assert who first prepared (as distinct from recorded in open publication) many of the more simple phenarsazine derivatives. Much of the work recorded in Burton and Gibson's first paper in this series was in progress before Wieland and Rheinheimer's publication in 1921.[122]

For the preparation of 10-chloro-5,10-dihydrophenarsazine by the condensation of diphenylamine and arsenic trichloride, Burton and Gibson[159] recommend the use of *o*-dichlorobenzene as a solvent; on cooling, the 10-chloro compound crystallizes directly. They point out that this 10-chloro compound has a remarkable facility for separating from solution with solvent of crystallization; it will crystallize with a definite proportion of acetic acid, *sym*-tetrachloroethane, chlorobenzene, *o*-dichlorobenzene, acetone, carbon tetrachloride, or even arsenic trichloride. Furthermore, the 'oxide' or anhydride (**6**) can be prepared by heating diphenylamine with arsenious oxide in the presence of phosphoric anhydride above 130°.

It will be realized that although the diphenylamine–arsenic trichloride condensation places the constitution of 10-chloro-5,10-dihydro-
phenarsazine (5) beyond reasonable doubt, it does not provide an
absolute proof of this structure. The following alternative synthesis[159]
was developed by Burton and Gibson to provide this proof.

o-Nitraniline was diazotized and then coupled with o-bromophenyl-
arsenoxide under the usual conditions of the Bart reaction to give (2-
bromophenyl)(2′-nitrophenyl)arsinic acid (17). This compound was
reduced to the 2′-aminophenyl derivative (18), which when heated in
pentyl alcohol with potassium carbonate and copper powder underwent
the normal Ullmann condensation with the formation of phenazarsinic
acid (11), which on reduction gave the 10-chloro derivative (5). This
constitutes the first unambiguous demonstration of the structure of the
10-chloro compound (5).

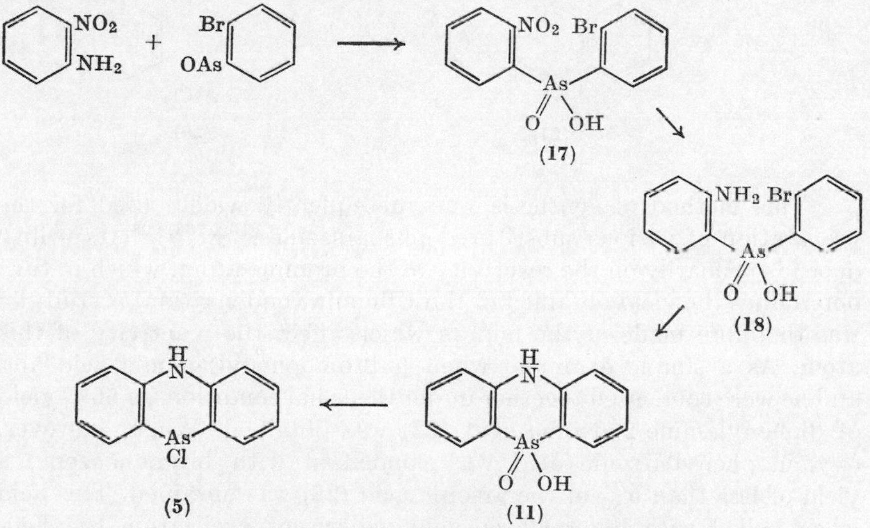

A third synthesis[161] was developed initially for the preparation of
10-chlorophenarsazines carrying carboxylic acid groups substituted in
the benzene rings. These derivatives cannot readily be prepared by the
arsenic trichloride condensation method because it is difficult to obtain
the necessary carboxydiphenylamines. For this purpose, Burton and
Gibson condensed (o-bromophenyl)arsonic acid with, for example, an-
thranilic acid by boiling the two reactants in nitrobenzene solution with
potassium carbonate and copper powder. This Ullmann condensation
thus furnished 2-carboxydiphenylamine-2′-arsonic acid (19), which when
reduced by sulfur dioxide (and a trace of iodine) in ethanolic solution

gave the dichloroarsine (**20**). However, this compound was unstable, but when either boiled in acetic acid solution, or alternatively boiled with sodium hydroxide solution and then treated with hydrochloric acid, it furnished 10-chloro-5,10-dihydrophenarsazine-4-carboxylic acid (**21**), yellow needles, m.p. 243°.

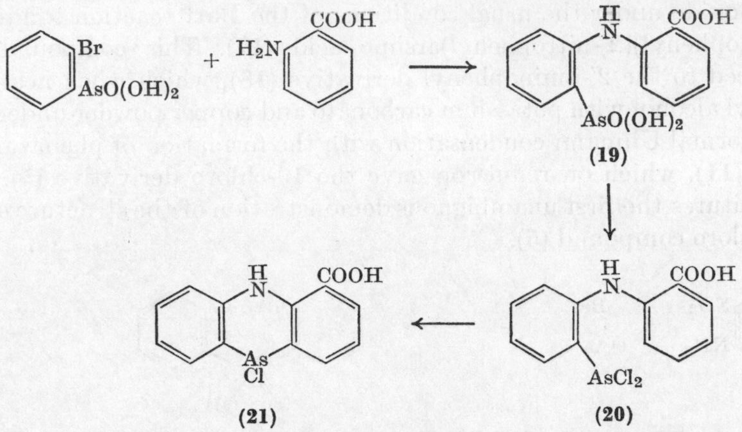

This method of synthesis was subsequently widely used for the preparation of various substituted phenarsazines.[162, 163, 168] Its utility depends primarily on the reactivity of the bromine atom, which in turn determines the yield obtained in the Ullmann condensation. A study[162] was therefore made of the factors which affect the reactivity of this atom. As a simple example, when (o-bromophenyl)arsonic acid and aniline were condensed together under the usual conditions, a 50% yield of diphenylamine-2-arsonic acid (**22**) was obtained. When, however, (o-aminophenyl)arsonic acid was condensed with bromobenzene, a yield of less than 5% of the arsonic acid (**22**) was obtained. This acid when boiled with hydrochloric acid underwent cyclization to phenazarsinic acid (**11**) and when reduced with sulfur dioxide in hydrochloric acid solution gave the 10-chloro compound (**5**). Since the phenazarsinic acid can be reduced to the 10-chloro compound, and the latter oxidized back to the acid, the relationship of these compounds was clearly demonstrated.

Burton and Gibson had earlier[159] cast doubts on the statement of Wieland and Rheinheimer[122] that arsenic chloride condensed with N-methyldiphenylamine to give the 5-methyl derivative (**14**). They repeated this experiment but obtained only the unmethylated 10-chloro-5,10-dihydro compound (**5**), and all their attempts to methylate this

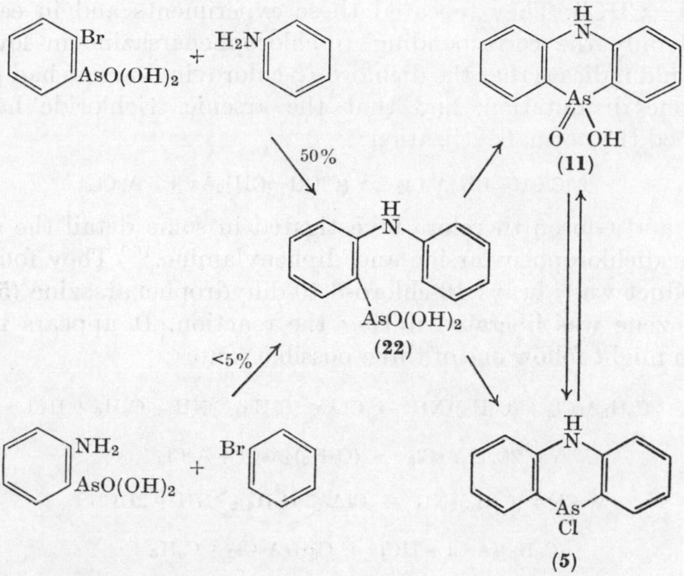

10-chloro compound or to synthesize the 5-methyl derivative by an indirect route had failed.[159] They concluded therefore that Wieland and Rheinheimer's product (14) was really a sample of 10-chloro-5,10-dihydrophenarsazine, the diphenylmethylamine having become demethylated in the course of the condensation. This conclusion received strong confirmation when it was found later that (o-bromophenyl)arsonic acid could be condensed with N-methylaniline to give methyldiphenylamine-2-arsonic acid (23), but that the latter could not be cyclized to

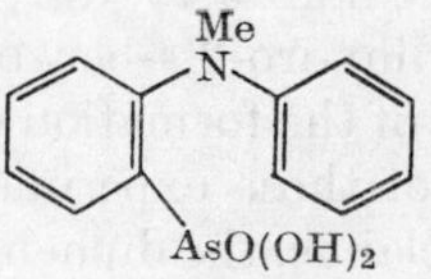

give either the 5-methylphenazarsinic acid or the 10-chloro-5-methylphenarsazine.[162] This result is in striking contrast to the ready cyclization of the unmethylated arsonic acid (22).

Burton and Gibson also questioned the accuracy of Lewis and Stiegler's claim[158] that dichloro-(2-chlorovinyl)arsine would condense with diphenylamine to give the 10-(2'-chlorovinyl)-5,10-dihydrophenarsazine (16) and with N-phenyl-1-naphthylamine to give the corresponding '3,4-benz-10-(2'-chlorovinyl)-5,10-dihydrophenarsazine' (15;

R = CH=CHCl). They repeated these experiments and in each case isolated only the corresponding 10-chlorophenarsazine in low yield. This would indicate that the dichloro-(2-chlorovinyl)arsine had possibly undergone dismutation, and that the arsenic trichloride had then performed the normal cyclization:

$$2\,(ClCH{=}CH)AsCl_2 \rightleftharpoons (ClCH{=}CH)_2AsCl + AsCl_3$$

Burton and Gibson therefore investigated in some detail the reaction between dichlorophenylarsine and diphenylamine.[160] They found that the product was always 10-chloro-5,10-dihydrophenarsazine (5), while free benzene was liberated during the reaction. It appears that the reaction might follow one of three possible routes:

$$C_6H_5AsCl_2 + (C_6H_5)_2NH \;\rightarrow\; ClAs{<}(C_6H_4)_2{>}NH + C_6H_6 + HCl \tag{1}$$

$$2C_6H_5AsCl_2 \;\rightarrow\; (C_6H_5)_2AsCl + AsCl_3 \tag{2a}$$

$$AsCl_3 + (C_6H_5)_2NH \;\rightarrow\; ClAs{<}(C_6H_4)_2{>}NH + 2HCl \tag{2b}$$

$$(C_6H_5)_2AsCl + HCl \;\rightarrow\; C_6H_5AsCl_2 + C_6H_6 \tag{2c}$$

$$C_6H_5AsCl_2 + (C_6H_5)_2NH \;\rightarrow\; C_6H_5As{<}(C_6H_4)_2{>}NH + HCl \tag{3}$$

$$C_6H_5As{<}(C_6H_4)_2{>}NH + HCl \;\rightarrow\; ClAs{<}(C_6H_4)_2{>}NH + C_6H_6$$

Of these routes, the reactions under (2) appear the most probable, with possibly both the diphenyl- and the monophenyl-chloroarsine reacting to some extent with the hydrogen chloride to liberate benzene.

Precisely similar results were obtained when dichlorophenylarsine was condensed with N-phenyl-p-toluidine, di-p-tolylamine, and N-phenyl-1-naphthylamine:[160] the products were 10-chloro-5,10-dihydro-1-methylphenarsazine, 10-chloro-5,10-dihydro-1,8-dimethylphenarsazine, and 10-chloro-5,10-dihydro-3,4-benzophenarsazine (15; R = Cl), respectively. No indication of the formation of a 10-phenylphenarsazine could be obtained in any of these experiments. It appears, therefore, that, in spite of the earlier claims, the diphenylamine–arsenic trichloride condensation cannot be modified to allow the direct preparation of 5,10-dihydrophenarsazines having alkyl or aryl groups substituted in either the 5- or the 10-position.[176]

The X-ray crystal structure determination of 10-chloro-5,10-dihydrophenarsazine (5) by Camerman and Trotter[148] fully confirm the results of Burton and Gibson.[159] For the X-ray work, the crystals which separated from xylene and contained 0.5 mole of solvent per mole of solute were heated at 200° and 0.001 mm, giving the stable solvent-free orthorhombic crystals of the compound (5). The main results of the investigation of these crystals are shown in Figures 18 (atomic distances) and 19 (intervalency angles). The C–As–C angle within the ring (97.0°)

closely approaches the 'natural' angle (*ca.* 98°) of a trivalent arsenic atom, as do the two C–As–Cl angles (96.6° and 95.7°). The heterocyclic ring is slightly distorted, with the C–N–C angle of 128°. The molecule is, however, folded about the N–As axis, the two *o*-phenylene groups subtending an angle of 169°; the C–Cl bond projects beyond this angle, which is sufficiently near 180° to make the occurrence of *cis-trans*-forms improbable. The As—Cl bond (2.30 Å) is significantly longer than

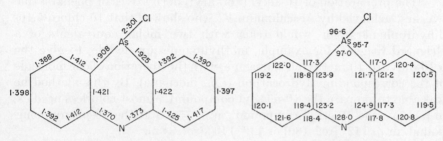

Figure 18   Figure 19

Figure 18, Atomic distances (in Å) and, Figure 19, Intervalency angles (in degrees), of 10-chloro-5,10-dihydrophenarsazine. (Adapted, by permission, from A. Camerman and J. Trotter, *J. Chem. Soc.*, 1965, 730)

that in arsenic trichloride (2.16 Å) but is near to that in diphenylarsonous chloride (2.26 Å). The As–C distance (mean, 1.917 Å) is shorter than the normal single-bond distance (*e.g.*, 1.990 Å in cacodyl disulfide), and the C–N distances (mean, 1.371 Å) are shorter than the normal single-bond distance (1.48 Å) but almost identical with that in aromatic amines (*e.g.*, *p*-nitroaniline, 1.371 Å). The authors conclude from these bond lengths that the compound (5) has an extended aromatic system involving interaction of the arsenic and nitrogen lone-pair electrons with the *o*-phenylene π-electrons, with possibly $d_\pi-p_\pi$ bonding between the π-electrons and vacant $4d$-orbitals of the arsenic atom. A later investigation by Fukuyo, Nakatsu, and Shimada has given closely similar results.[177]

The deep yellow 10-bromo-5,10-dihydrophenarsazine (4; R = Br) has been prepared by the action of arsenic tribromide on diphenylamine without a solvent,[146] and (better) in boiling *o*-dichlorobenzene;[160] also by treating the 10-chloro compound (5) or oxide (6), reduced in formic acid, with bromine also dissolved in this acid (Razuvaev[178]), and by the action of acetylenebis(magnesium bromide), $BrMgC\equiv CMgBr$, on the 10-chloro compound (5).[179]

Bromine (4 equivalents) reacts with (5) in hot acetic acid, disrupting the heterocyclic ring and forming di-(2,4-dibromophenyl)amine.[165]

The 10-iodo compound (**4**; R = I) has been obtained by heating the 10-acetoxy compound (**7**) in a dilute hydriodic acid–acetic acid mixture,[160] by treating the above reduced compounds (**5**) or (**6**) with ethanolic iodine,[178] and by the action of IMgC≡CMgI on the 10-chloro compound.[179]

10-Fluoro-5,10-dihydrophenarsazine (**4**; R = F) is prepared by the action of hydrofluoric acid on the 'oxide' (**6**).[180]

The preparation of 10-alkyl (and aryl) derivatives of phenarsazine was first achieved by Aeschlimann,[181] who showed that 10-chloro-5,10-dihydrophenarsazine would react with two molar equivalents of a Grignard reagent, for example, methylmagnesium iodide, to give the 5,10-dihydro-10-methylphenarsazine, with the liberation of one molecule of the corresponding hydrocarbon (*e.g.*, methane). By this method he was able to prepare the 10-methyl compound, almost colorless needles, m.p. 105°, the 10-ethyl compound, m.p. 75°, and the 10-phenyl compound, m.p. 142° (ref. 180) or 148–149° (ref. 182).

These compounds now contain true tertiary arsine groups and therefore unite with alkyl iodides to give the corresponding arsonium salts. The 10-methyl compound will combine with methyl iodide even in the cold to give the crystalline 5,10-dihydro-10,10-dimethylphenarsazinium iodide (**24**). When heated in a vacuum, this salt loses methyl iodide, regenerating 5,10-dihydro-10-methylphenarsazine.

(**24**)   (**25**)

The above 10-methyl, 10-ethyl, and 10-phenyl compounds were later prepared by the same method also by Seide and Gorski,[182] who in addition prepared the 5,10-dihydro-10-(1-naphthyl)phenarsazine, m.p. 154–155°. They showed that the 10-methyl and 10-phenyl compounds added chlorine to give arsine dichlorides such as (**25**), which on heating lost the 10-alkyl or 10-aryl group, giving 10-chloro-5,10-dihydrophenarsazine, presumably with the loss of methyl chloride and chlorobenzene, respectively.

The chemistry of quaternary salts of type (**24**) and their corresponding hydroxides, and also of the betaines obtained by addition of chloroacetic acid to the tertiary arsines followed by loss of hydrogen chloride, has been investigated by Razuvaev and his co-workers.[183, 184]

It is noteworthy that although 10-chloro-5,10-dihydrophenarsazine and phenazarsinic acid apparently do not react with nitrous acid to give the 5-nitroso derivatives, the 10-alkyl- and 10-aryl-5,10-dihydrophenarsazines do so if the conditions of nitrosation are carefully controlled. Thus Razuvaev, Godina, and Yemelyanova[185] have shown that the addition of hydrochloric acid to an ethanolic solution containing 5,10-dihydro-10-methylphenarsazine and sodium nitrite affords the 5-nitroso derivative, m.p. 108–110°; the 10-ethyl-5,10-dihydrophenarsazine similarly gives the 5-nitroso derivative as a viscous liquid, and the 5,10-dihydro-10-phenylphenarsazine gives the 5-nitroso derivative, m.p. 143–145°. These 5-nitroso derivatives are unstable, however, and in ethanolic or benzene solution are readily converted into resinous red products, which appear to contain both nitro derivatives and oxidation derivatives.

For an account of phenarsazines having methyl groups substituted in the benzene ring, see in particular references 164, 166, 167, 169, and 170; for the chloro derivatives see reference 172; for the nitro derivatives see references 166 and 175; for those having a carboxylic acid group see reference 161; and for those having an acetyl group in the benzene ring see reference 173.

For an account of the condensation of the 10-chloro compound (5) with 2-mercaptoethanol in the presence of ethanolic potassium hydroxide to give 5,10-dihydro-10-(2′-hydroxyethylthio)phenarsazine, m.p. 164–166°, see Rueggeberg, Ginsburg, and Cook.[186]

Many 10-substituted 5,10-dihydrophenarsazines on reduction (particularly in the presence of acids, for example, in hot formic acid solution) give intensely colored solutions of the 'meriquinoid' derivatives, which have been investigated in some detail by Razuvaev and his co-workers.[185, 187–191] The production of these colored products is sometimes dependent on the temperature, the color becoming more intense as the temperature rises and fading again as the solution cools, a property that suggests dissociation to free radicals. In other cases, the colored product slowly absorbs atmospheric oxygen with formation of the original 10-substituted 5,10-dihydrophenarsazine or of the corresponding phenazarsinic acid. The structure of the intermediate colored 'meriquinoid' compound remains uncertain, and it is probable that the processes involved in the color formation may vary with the particular phenarsazine derivative investigated.

Mass spectrograms of compounds (4; R = Cl) and of (4; R = CH$_3$) show that the former first loses hydrogen chloride with formation of the phenarsazine ion radical corresponding to (1), which then undergoes further fragmentation with loss of arsenic to form carbazole; the carbazole is shown by its own peak and by those of its known fragmentation products. The methyl compound (4; R = CH$_3$), on the other hand, readily loses the methyl group with the formation of the 5-hydro-phenarsazine ion radical, which then breaks down as above to carbazole; the loss of the methyl group and one hydrogen atom to form the phenarsazine ion is much less pronounced than the formation of this ion from the compound (4; R = Cl) (Buu-Hoï, Mangane, and Jacquignon[192]).

Table 4 gives the melting points of, and chief references to, the simpler phenarsazine compounds.

Table 4. Derivatives of 5,10-dihydrophenarsazine

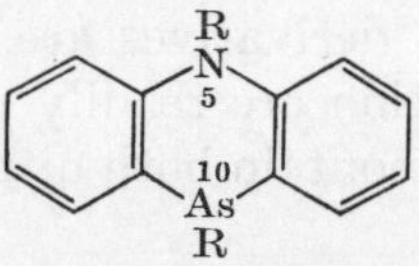

R-5	R-10	M.p. (°)	Ref.
H	Cl	193	122
H	Br	217–218	160
H	I	222–224	160, 178, 211
H	F	285	153, 180
H	CH₃*	105	181, 182
H	C₂H₅†	75	181
H	C₆H₅‡	148–149	181, 182
H	CN	246–247	206
H	CN	227–228	149
H	—SCN§	235–237¶	193, 154, 206
H	OCH₃	194	122
H	OC₆H₅	200–202	206
H	OCOCH₃‖	223–224	122, 154
H	OCOC₂H₅	135–136	152
CH₃CO	Cl	229–230	152
C₂H₅CO	Cl	135–136	152
C₆H₅CO	Cl	180–181	152
CH₃CO	CH₃	154	181

* Methiodide,[181] m.p. 259–268°.
† Methiodide,[181] m.p. 229–235°.
‡ Methiodide,[181] m.p. 158°.
§ Two forms, yellow needles and red plates.[193]
‖ Picrate,[154] deep red crystals, m.p. 170–174°.
¶ Dependent on rate of heating.[193]

It will be obvious that Burton and Gibson's third synthesis of the phenarsazine system by the use of (o-bromophenyl)arsonic acid opens up a wide field for the preparation of complex phenarsazines, particularly those having other rings fused on to the phenarsazine system itself (p. 524 *et seq.*). A derivative of another type is obtained by condensing benzidine with two equivalents of (*o*-bromophenyl)arsonic acid, reducing the diarsonic acid so obtained to the bis(dichloroarsine), and then cyclizing as usual to produce 2,2'-bi-(10-chloro-5,10-dihydro-phen-arsazine)[163] (26).

(26)

The irritant properties of 10-chloro-5,10-dihydrophenarsazine (5) had made it a potential chemical warfare compound in the First World War, and many papers (often of a very general nature) subsequently appeared on the detection and determination of this compound in the air, in water, and in food. These papers tended to increase in number in the years immediately before, and also for some time after the onset of, the Second World War. Only a selection can be cited here. For data regarding its toxic properties and those of other irritant compounds, and for an assessment of its comparative value as a chemical warfare agent, see Mielenz[194] and others.[195] For an account of the size of the particles of (5) in smoke, particularly compared with diphenylarsinous chloride, see Redlinger.[196] The detection, occurrence, and safety measures to be observed in manipulating the 10-chloro compound have been described by Leroux,[197] with a considerable number of references to this subject.

Mohler and Pólya[198] recorded the ultraviolet spectra of (5) and of many other irritant compounds, and later Mohler[199] discussed the detection and estimation of (5) and various other chemical warfare agents based on absorption spectroscopy.

Other noteworthy discussions of this subject were contributed by Hennig,[200] Leipert,[201] Delga,[202] Cox,[203] Hoogeveen,[204] and Fenton.[205]

The claims made for the physiological action of phenarsazine probably cover a wider field of effect and application than those for any other heterocyclic system and only a brief summary can be given here. A number of recent Japanese patent specifications bear on this subject. Nagasawa *et al.*[206] make rather sweeping claims that the 10-substituted derivatives (4; R = Cl, CN, SCN, CH$_3$, C$_6$H$_5$, OCH$_3$, and OC$_6$H$_5$) and others are effective fungicides. Two other specifications[207, 208] claim that various compounds of the 'oxide' type (6) and the 10,10'-biphen-arsazine type (p. 511) are effective as insecticides. Fukunaya *et al.*[209]

18

claim that the 10-acetoxy- and the 10-chloroacetoxy-, 10-(dichloro-acetoxy)-, and 10-(trichloroacetoxy)-5,10-dihydrophenarsazines and the four corresponding 5-substituted derivatives can be used as bactericides for agricultural purposes.

Nakanishi *et al.*[180, 210, 211] assert that the 10-fluoro- and 2,10-difluoro-5,10-dihydrophenarsazine and many similar derivatives are effective antifouling agents in marine paints, being stable, insoluble in water and most organic solvents, easily dispersed in organic vehicles, and, in contrast to their effect on marine organisms, having negligible oral or cutaneous toxicity to mammals. Nagasawa[212] also claims as essential ingedients in antifouling paints 5,10-dihydro-10-[(dimethyl-thiocarbamoyl)thio]phenarsazine [4; R = —SC(S)N(CH$_3$)$_2$], the corre-sponding 10-(ethylxanthyl) derivative [4; R = —SC(S)OC$_2$H$_5$], and the simpler derivatives (4; R = SCN, OCH$_3$, and CN).

'Phenarsazine monothiobenzoate' [≡10-(benzoylthio)-5,10-di-hydrophenarsazine] (4; R = —SCOC$_6$H$_5$) is reported as a seed disin-fectant,[213] and the 10-(2-benzothiazolylthio) derivative as both a fungicide and a vulcanization accelerator.[214]

For toxicity studies on rats, mice, and guinea-pigs, and also on humans, see Punte *et al.*,[215, 216] and for anthelmintic efficiency studies, see Gordon and Lipson.[217]

The effect of 10-chloro-, 10-bromo-, and other nuclear-substituted 5,10-dihydrophenarsazines on isocitric dehydrogenase activity has been studied in detail by Lotspeich and Peters.[218]

5,10-*o*-Benzenophenarsazine

This very interesting compound (1), [RIS —], is presented and numbered as shown, although it may be convenient to invert this

(1)   (1A)

structural presentation to bring the more reactive arsenic atom into the lower position. The structure can be indicated briefly as (1A). This 'tryptycene' type of structure is similar to that of 5,10-*o*-benzenoarsan-threne (p. 478), and the compound is fundamentally different from those phenarsazines having extra rings fused to the *o*-phenylene groups (pp. 524-532).

It has been synthesized (Earley and Gallagher [218a]) by the action of
o-chlorophenylmagnesium bromide on 10-chloro-5,10-dihydrophen-
arsazine (2) to give 10-(o-chlorophenyl)-5,10-dihydrophenarsazine (3),

(2)   (3)

colorless needles, m.p. 110° (76%). An ethereal solution of (3) containing
lithium diethylamide, when boiled under reflux for 5 days, gave the
5,10-o-benzenophenarsazine (1), colorless cubes, m.p. 233° (48%). It
can be readily sublimed. The identities of the compounds (1) and (3)
have been confirmed by their analysis, their molecular weights (mass
spectra), and their nmr spectra. The compound (1) has a characteristic
ultraviolet spectrum, which shows no fine structure and only intense
end absorption (λ 215 mμ) with an inflection at 250–260 mμ.

The compound (1) has considerable stability. Attempts to prepare
quaternary salts by the action of methyl iodide (alone or mixed with
silver tetrafluoroborate), methyl p-toluenesulfonate, or triethyloxonium tetrafluoroborate, $Et_3O^+BF_4^-$, were unsuccessful. Reaction of
compound (1) with bromine gives an unstable brown addition product,
which when heated reverts to (1) and when hydrolyzed gives the 10-
oxide; this oxide is also formed by the action of boiling nitric acid on the
parent (1).

These properties of the arsenic atom in the phenarsazine (1) are
closely similar to those of the arsenic atoms in 5,10-o-benzenoarsan-
threne: the 5,10-dioxide of the latter compound can be obtained either
by hydrolysis of the tetrabromide or by the direct action of nitric acid.
Moreover, even under 'forcing' conditions (methyl p-toluenesulfonate
at 210° for 6 hours) only one of the arsenic atoms undergoes quaternization (p. 479).

The compound (1), when treated with Raney nickel in boiling
ethanol, loses arsenic with formation of triphenylamine, a reaction that
supports the assigned structure.

The phenarsazine (1) is also unaffected by boiling ethanol containing
sodium ethoxide, but when it is heated with sodamide in hexamethyl-
phosphoramide, $OP[N(CH_3)_2]_3$, one of the As–C bonds is broken, with

the formation of 5,5'-diphenyl-10,10'(5H,5'H)-oxydiphenarsazine) (**4**), m.p. 284–285° (80%); this compound is systematically the anhydride of

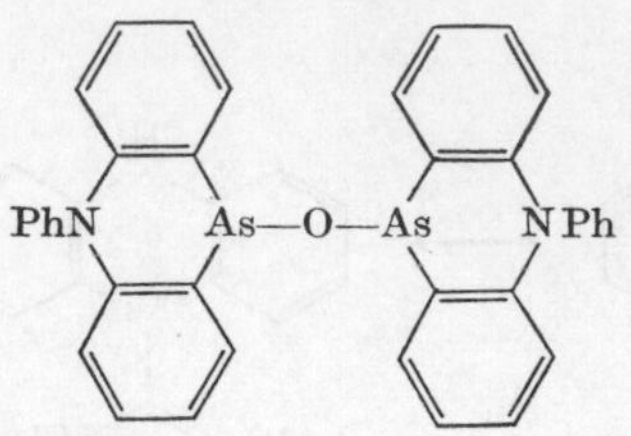

(**4**)

5,10-dihydro-10-hydroxy-5-phenylphenarsazine. The mechanism of this reaction is unknown, but it may involve the intermediate formation of 5,10-dihydro-5-phenyl-10-sodioaminophenarsazine, which during the working up is hydrolyzed to the 10-hydroxy compound, which in turn gives the anhydride (**4**). It is noteworthy that sodamide also converts the 10-(*o*-chlorophenyl) compound (**3**) into the anhydride (**4**), possibly by initial conversion into the phenarsazine (**1**) and thence into the anhydride.

Phenarsazines having extra fused carbocyclic systems

The following are briefly recorded, with the current numbering and presentation of the parent compounds. The presentation of these compounds in the Ring Index, following the IUPAC rules, entails in some compounds the nitrogen atom of the heterocyclic system being shown uppermost and in others the arsenic atom being thus shown.

Unless otherwise stated, the phenarsazines have been prepared by condensation of arsenic trichloride with the given diarylamine, by direct fusion without a solvent, or in boiling *o*-dichlorobenzene. References to their early production and names are given.

Benzo[c]*phenarsazine* (**1**), [RRI 4903]. From *N*-phenyl-1-naphthylamine, giving the canary-yellow 7-chloro-7,12-dihydro derivative (**2**), m.p. 219° (Lewis

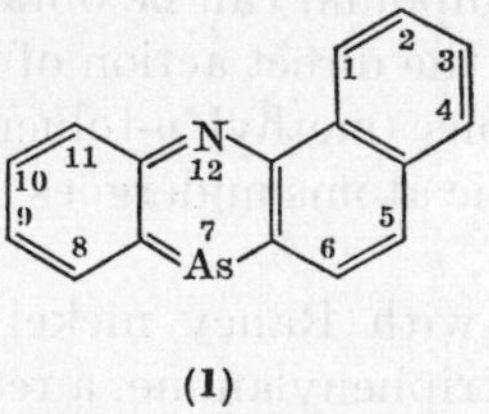

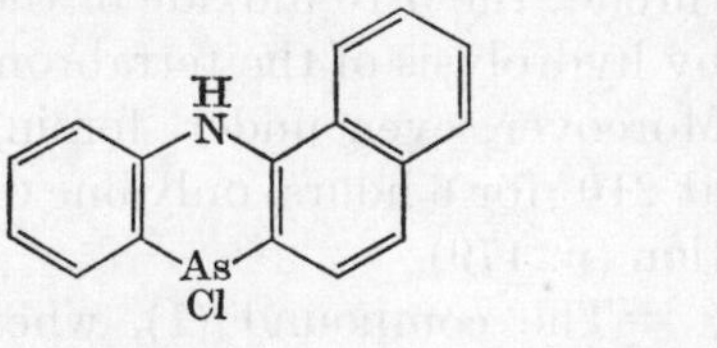

and Hamilton[157]); many derivatives, including the corresponding phenazarsinic acid, have also been prepared. Historically, this was the first monobenzophenarsazine. *N*-*p*-Tolyl-1-naphthylamine gives the 7-chloro-7,12-dihydro-9-methyl derivative, deep yellow, m.p. 252–255° (Burton and Gibson[155]).

Benzo[b]phenarsazine (**3**), [RRI 4901]. From *N*-phenyl-1-methyl-2-naphthyl-amine, giving the 12-chloro-5,12-dihydro-6-methyl derivative (**4**; R = CH₃),

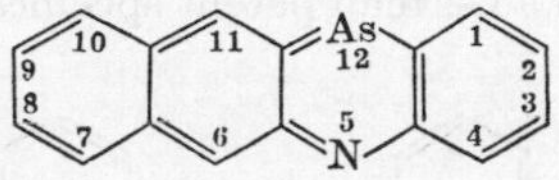

(3) 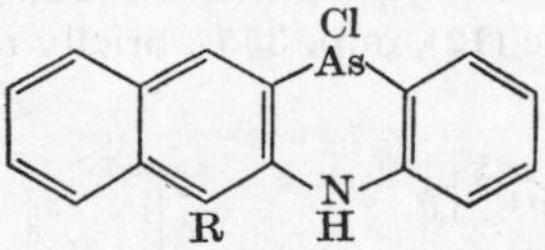   (4)

m.p. 227–228° (Buu-Hoï *et al.*[219]); the methyl group prevents condensation at the otherwise more reactive 1-position of the naphthalene ring. The parent compound (**4**; R = H) has apparently not been prepared.

Benzo[a]phenarsazine (**5**), [RRI 4902]. From *N*-phenyl-2-naphthylamine, giving the yellow 12-chloro-7,12-dihydro derivative (**6**; R = H), m.p. 249–250°;

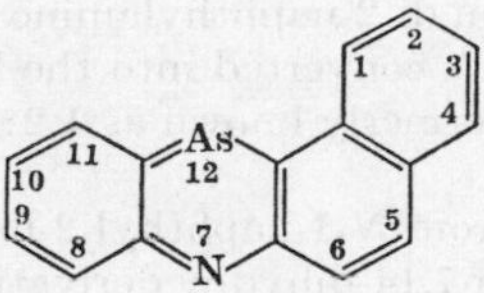

(5) 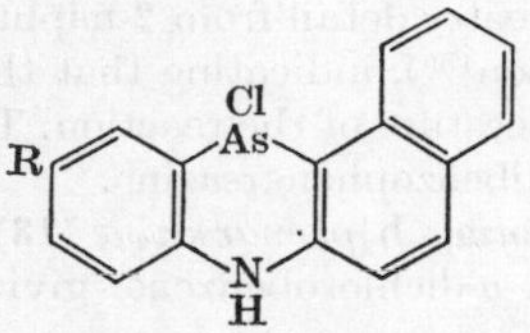   (6)

the *N-p*-tolylamine gives the 10-methyl derivative (**6**; R = CH₃), m.p. 266–267° (Burton and Gibson[155]). The compound (**5**) was earlier known as α-benzophen-arsazine.

Dibenzo[a,i]phenarsazine (**7**), [RRI 6133]. Apparently known only as the 14-chloro-7,9,10,11,12,14-hexahydro derivative (**8**), m.p. 260°, from 5,6,7,8-tetrahydro-*N*-(2′-naphthyl)-2-naphthylamine (Buu-Hoï *et al.*[220]); gives 14-methyl derivative, m.p. 200°, and 14-ethyl derivative, m.p. 146°. The system was initially known as 1,2:7,8-dibenzophenarsazine.

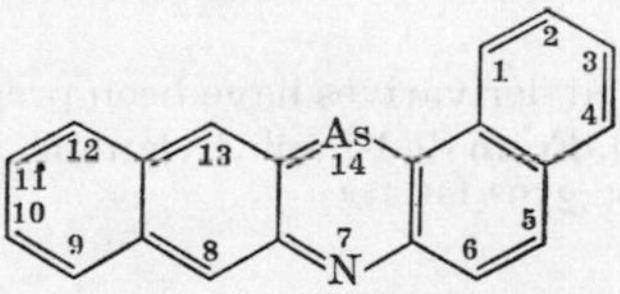

(7) 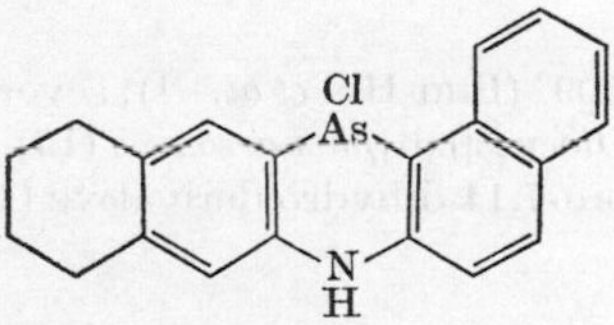   (8)

Dibenzo[b,h]phenarsazine (**9**), [RRI 6134]. Apparently known only as the 7-chloro-7,9,10,11,12,14-hexahydro derivative (**10**), m.p. 264°, from 5,6,7,8-tetrahydro-2-*N*-(1′-naphthyl)-2-naphthylamine; gives 7-methyl derivative, m.p.

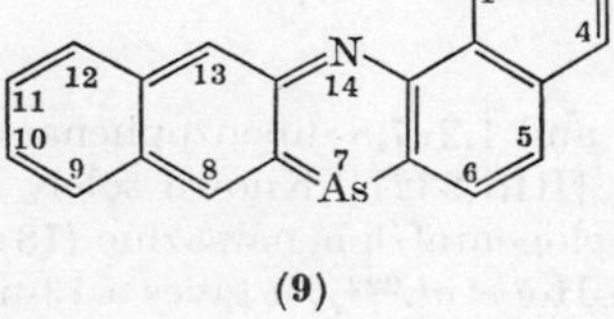

(9) 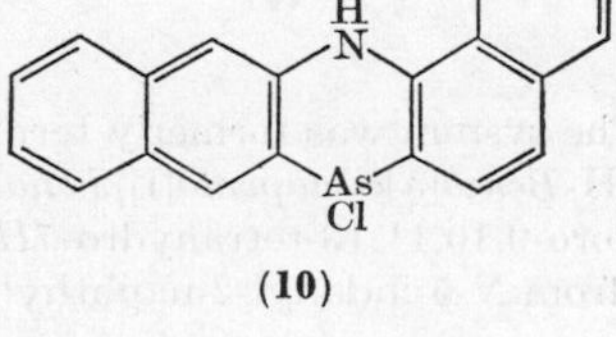   (10)

226°, and 7-ethyl derivative, m.p. 157° (Buu-Hoï *et al.*[220]). These compounds were formerly termed 2,3:6,7-dibenzophenarsazines.

Dibenzo[a,j]*phenarsazine* (**11**). [RRI 6136]. The yellow 14-chloro-7,14-dihydro derivative (**12**), m.p. 355°, briefly mentioned in a German patent specification,[146]

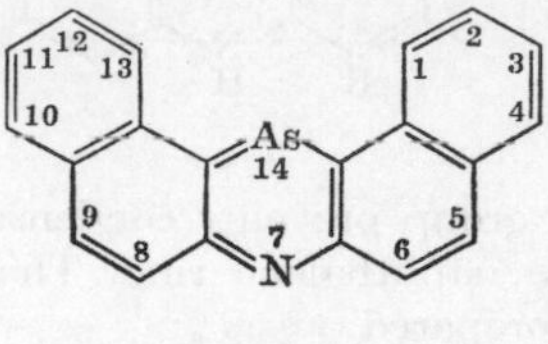

(11) 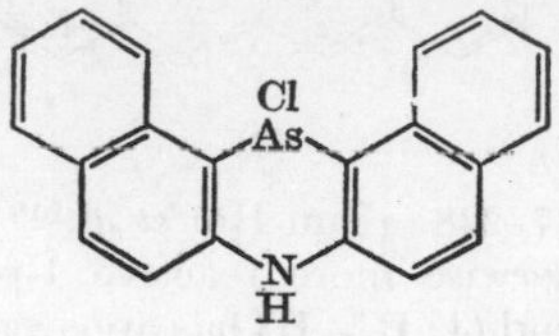   (12)

and in greater detail from 2-naphthylamine or from di-2-naphthylamine (Burton and Gibson[159]), indicating that the former amine is converted into the latter at the temperature of the reaction. The system was formerly known as 1,2:8,9- and 3,4:5,6-dibenzophenarsazine.

Dibenzo[a,h]*phenarsazine* (**13**), [RRI 6135]. From N-1-naphthyl-2-naphthyl-amine in o-dichlorobenzene, giving the 14-chloro-7,14-dihydro derivative (**14**),

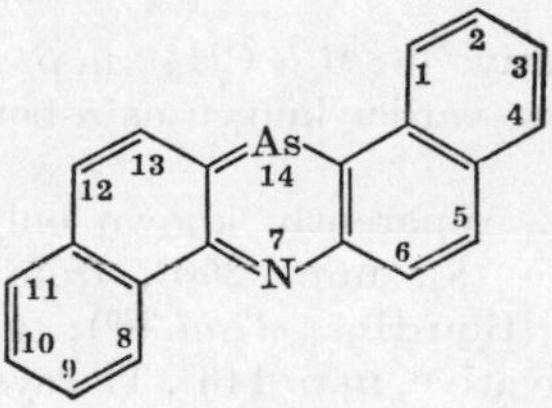

(13) 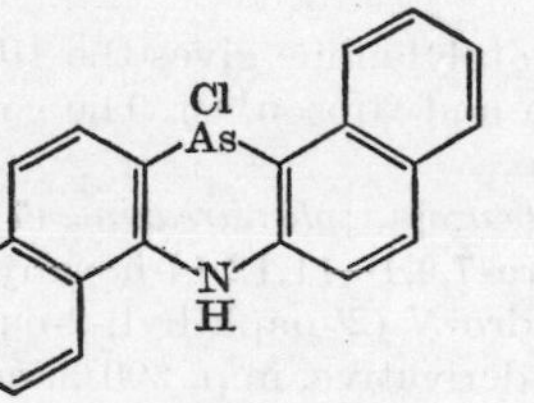   (14)

m.p. 309° (Buu-Hoï *et al.*[221]); several substituted derivatives have been prepared.

Dibenzo[c,h]*phenarsazine* (**15**), [RRI 6137]. From di-1-naphthylamine, giving 7-chloro-7,14-dihydro derivative (**16**), m.p. 278–279°.[146, 159]

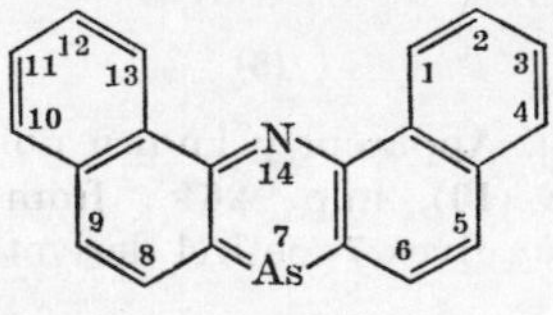

(15) 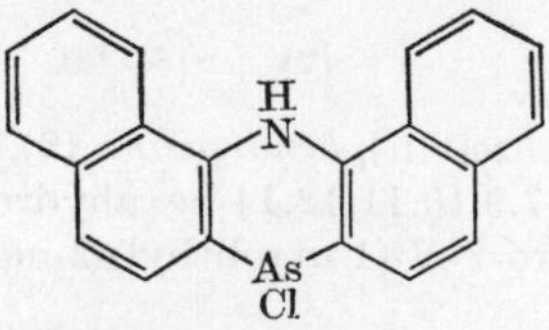   (16)

The system was formerly termed 3,4:6,7- and 1,2:7,8-dibenzophenarsazine.

7H-Benzo[a]*cyclopenta*[i]*phenarsazine* (**17**), [RIS 9321]. Known solely as the 13-chloro-9,10,11,13-tetrahydro-7H-benzo[a]cyclopenta[i]phenarsazine (**18**), m.p. 290°, from N-5-indanyl-2-naphthylamine (Buu-Hoï *et al.*[222]); it gives a 13-methyl derivative, m.p. 207°. Compound (18) was initially termed 10-chloro-5,10-dihydro-7,8-cyclopenteno-1,2-benzophenarsazine.

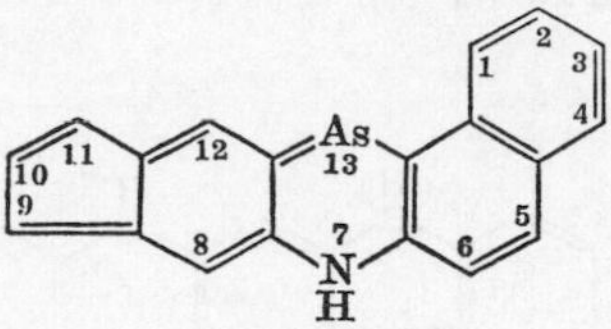

(17) 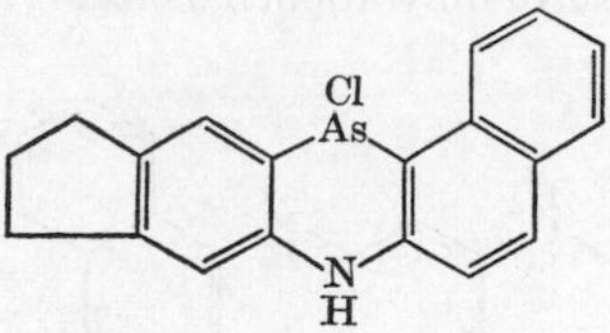   (18)

7H-*Benzo*[h]*cyclopenta*[b]*phenarsazine* (19), [RIS 9322]. Similarly known solely as the 7-chloro 9,10,11,13-tetrahydro-7H-benzo[h]cyclopenta[b]phenarsazine (20), m.p. 278°, from N-5-indanyl-1-naphthylamine;[222] this was initially termed 10-chloro-5,10-dihydro-2,3-cyclopenteno-6,7-benzophenarsazine.

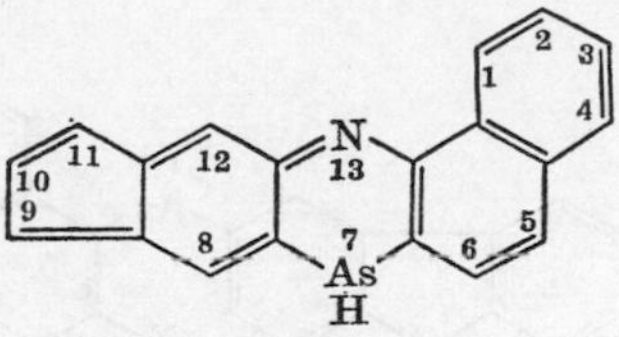

(19) 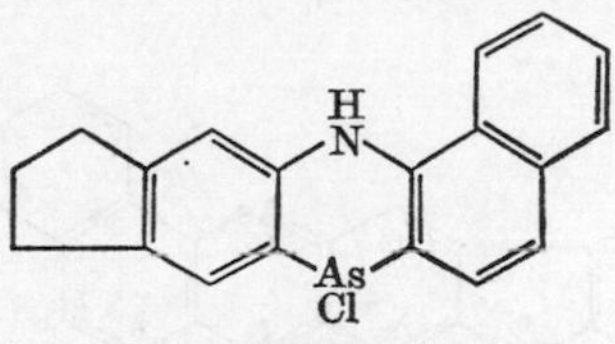   (20)

Indeno[1,7-b,c]*phenarsazine* (21), [RRI 5987]. This system has been synthesized by reducing o-(5-acenaphthylamino)phenylarsonic acid [22; R = AsO(OH)2] in hot ethanolic hydrochloric acid with sulfur dioxide to the dichloroarsine (22; R = AsCl2), which undergoes cyclization to the orange-red 7-chloro-4,5,7,12-tetrahydroindeno[1,7-b,c]phenarsazine (23), m.p. 241°, formerly termed 7-chloro-7,12-dihydroisoacenaphthabenzarsazine (Gibson and Johnson[167]).

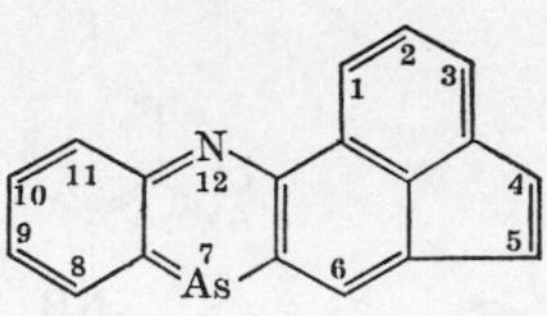

(21) 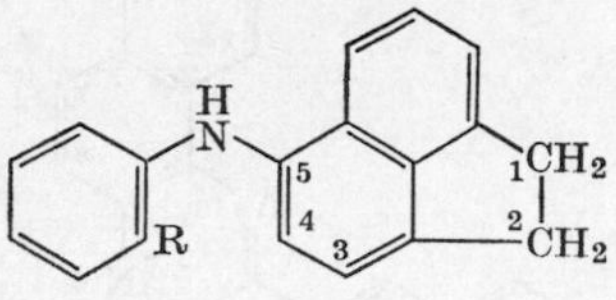   (22)

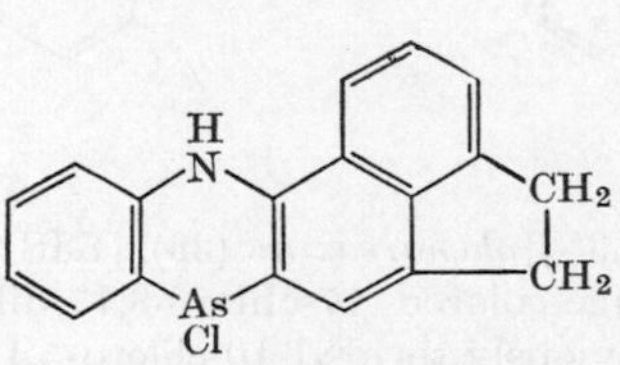

(23)

13H-*Benz*[h]*indeno*[1,2-b]*phenarsazine* (**24**), [RRI 6803]. From 2-(1-naphthyl-amino)fluorene, giving 7-chloro-7,15-dihydro-13*H*-benz[*h*]indeno[1,2-*b*]phenars-azine, chars above 300°; this wa searlier termed 3,4-benzo-10-chloro-2′,3′,7,8-indeno-5,10-dihydrophenarsazine (Buu-Hoï and Roger [223]).

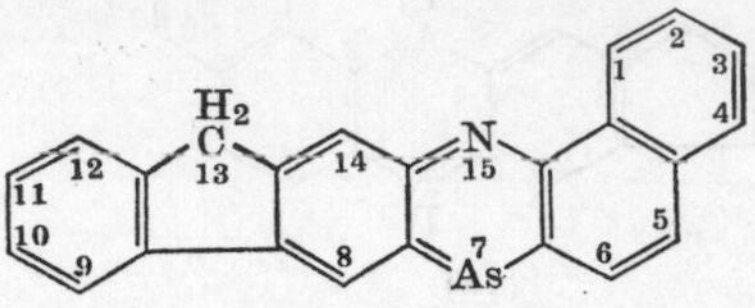

(**24**) 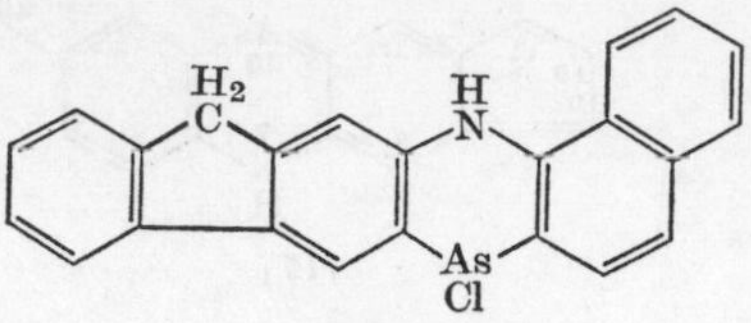   (**25**)

9H-*Benz*[a]*indeno*[2,1-i]*phenarsazine* (**26**), [RRI 6804]. Prepared from 2-(2-naphthylimino)fluorene, giving 15-chloro-7,15-dihydro-9*H*-benz[*a*]indeno[2,1-*i*]-phenarsazine (**27**), chars above 300°, formerly termed 1,2-benzo-10-chloro-2′,3′,7,8-indeno-5,10-dihydrophenarsazine.[223]

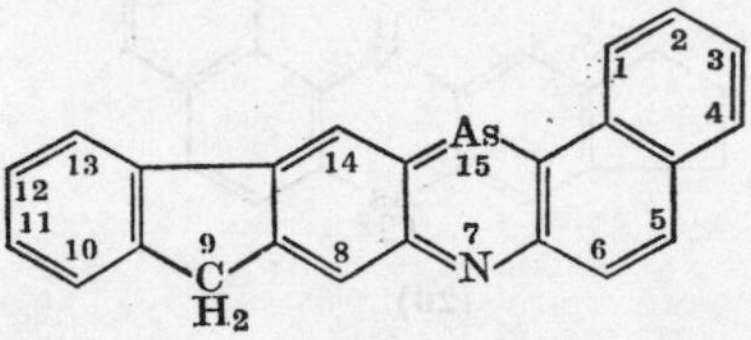

(**26**) 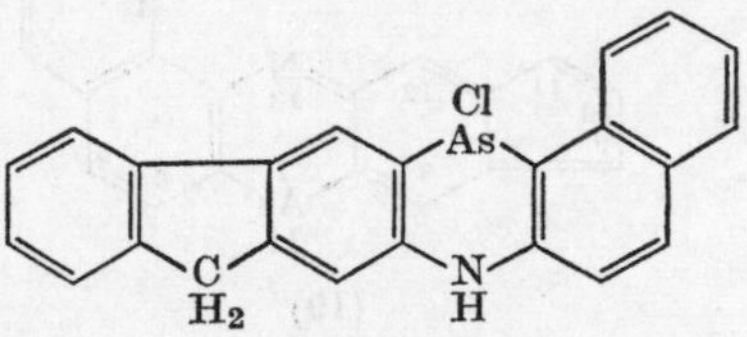   (**27**)

Benzo[a]*naphtho*[2,3-j]*phenarsazine* (**28**), [RIS 9517]. From 2-(2-naphthyl-amino)anthracene, giving the deep yellow 16-chloro-7,16-dihydrobenzo[*a*]-naphtho[2,3-*j*]phenarsazine (**29**), m.p. 304–305°, formerly termed 10-chloro-5,10-dihydro-1,2-benzonaphtho(2′,3′:8,9)phenarsazine (Buu-Hoï et al.[224]).

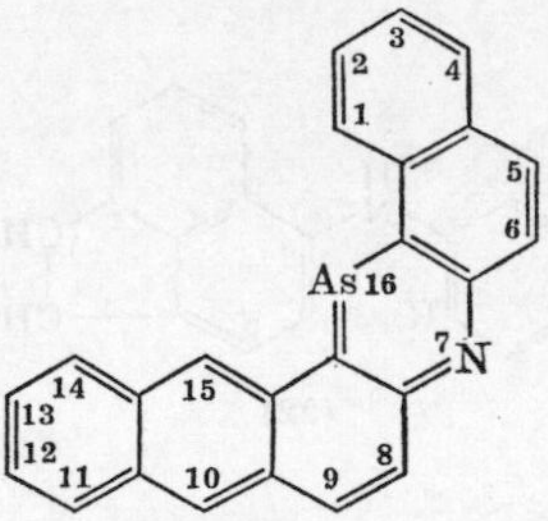

(**28**) 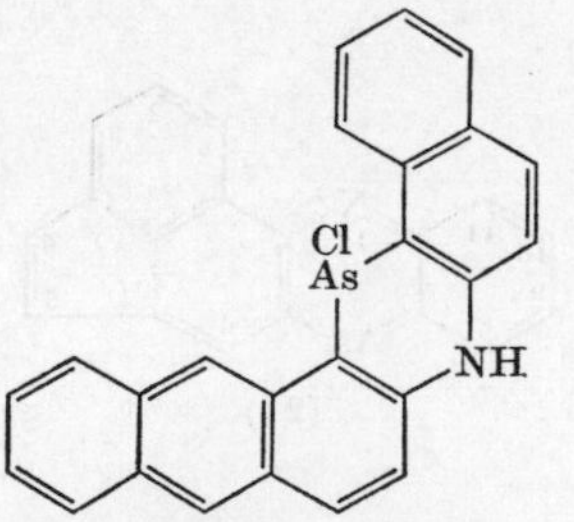   (**29**)

Dinaphtho[2,3-a:2′,3′-j]*phenarsazine* (**30**), [RIS 9583]. From di-2-anthranyl-amine, giving the orange-colored 17-chloro-8,17-dihydro-dinaphtho[2,3-*a*:2′,3′-j]phenarsazine (**31**), previously termed 10-chloro-5,10-dihydronaphtho[2′,3′:1,2]-2″,2″:8,9]phenarsazine (Barrett and Buu-Hoï[225]).

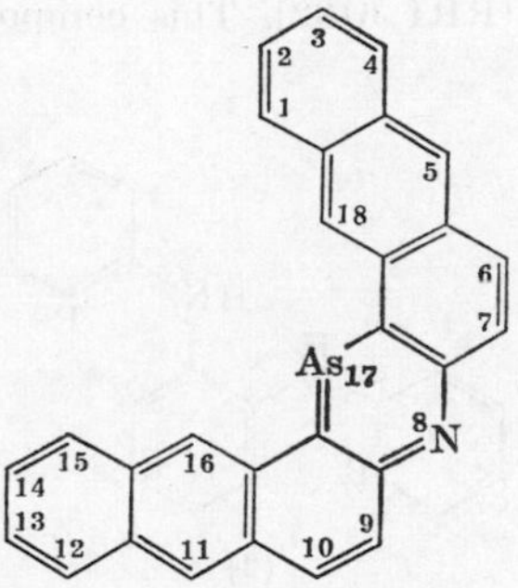

(30) 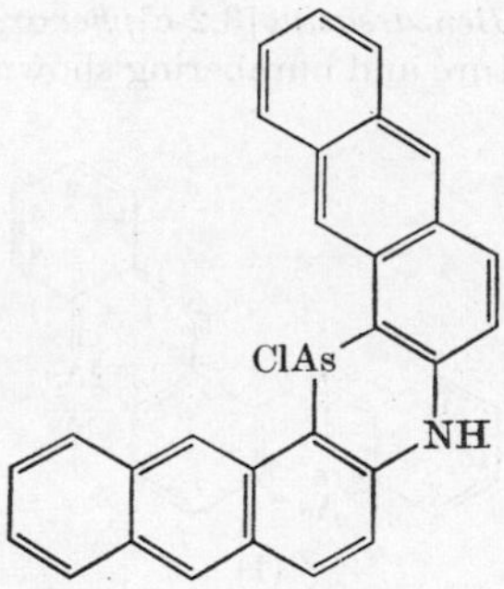   (31)

Benzo[h]*phenaleno*[1,9-bc]*phenarsazine* (**32**), [RIS 14189]. From 1-(1′-naphthylamino)pyrene in *o*-dichlorobenzene, giving the orange 7-chloro-7,14-dihydrobenzo[*h*]phenaleno[1,9-*bc*]phenarsazine (**33**), m.p. not stated (Buu-Hoï et al.[226]).

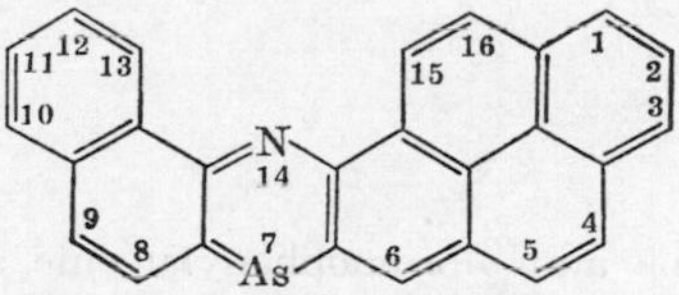

(32) 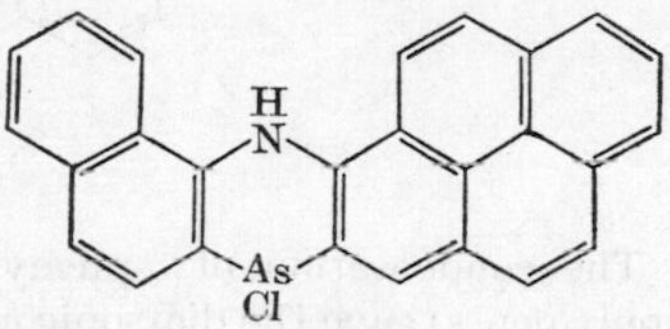   (33)

Benzo[a]*phenaleno*[1,9-hi]*phenarsazine* (**34**), [RIS 14190]. Similarly from 1-(2′-naphthylamino)pyrene, giving the orange 7-chloro-7,14-dihydrobenzo[*a*]phenaleno[1,9-*hi*]phenarsazine[226] (**35**), m.p. 321–322°.

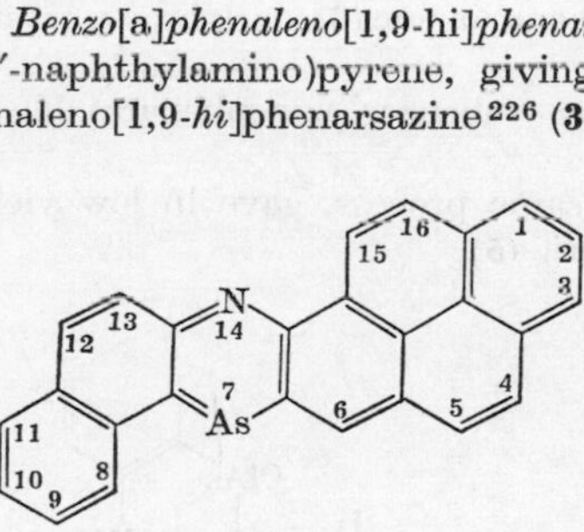

(34) 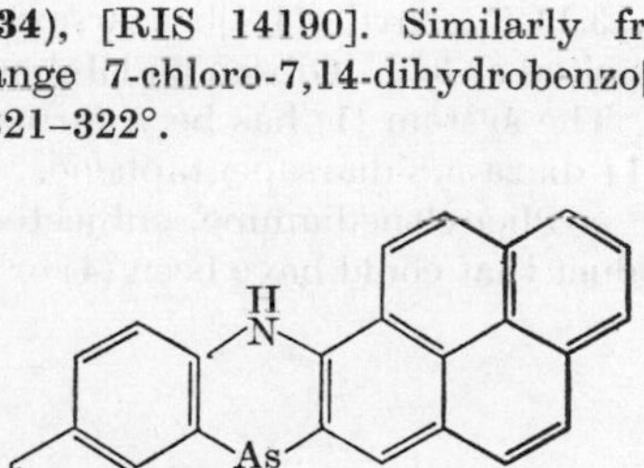   (35)

Most of the above compounds prepared by Buu-Hoï and his co-workers were required for investigation of their possible therapeutic properties.

Phenarsazines having extra fused heterocyclic rings

One particular type of ring system in which two phenarsazine units are fused together, can potentially arise by the application of the *o*-bromophenylarsonic acid synthetic method (p. 514) to the three phenylenediamines. Gibson and Johnson,[163] however, obtained decisive results only from the *o*-diamine.

[1,4]*Benzarsazino*[3,2-c]*phenarsazine* (**1**), [RRI 6132]. This compound has the structure and numbering shown.

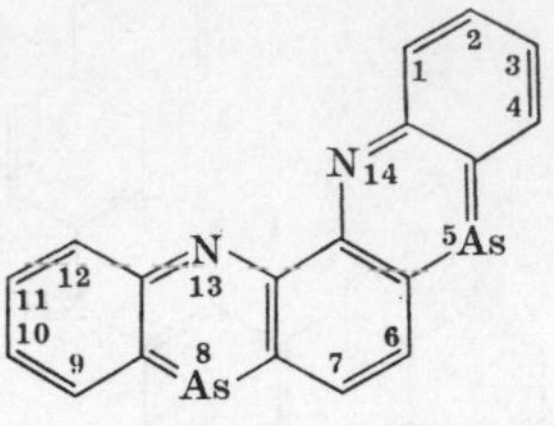

(**1**) 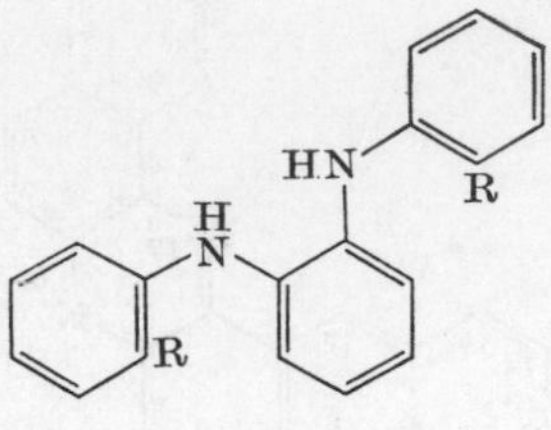   (**2**)

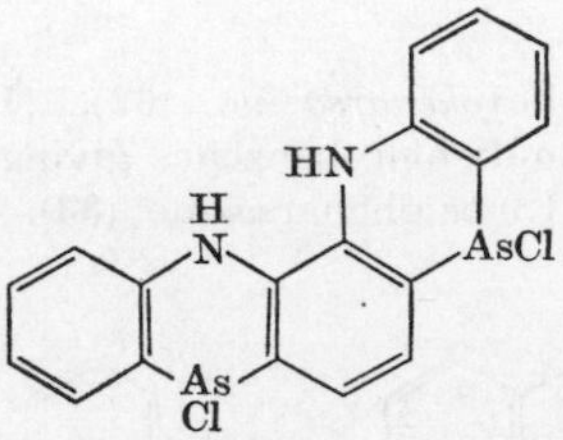

(**3**)

The condensation of o-phenylenediamine and o-bromophenylarsonic acid (2 equivalents) gave the diarsonic acid [**2**; R = AsO(OH)$_2$], which proved difficult to purify. Reduction in ethanolic hydrochloric acid with sulfur dioxide gave the bis(dichloroarsine) (**2**; R = AsCl$_2$), which on dicyclization gave 5,8-dichloro-5,8,13,14-tetrahydro[1,4]benzarsazino[3,2-c]phenarsazine (**3**), dark brown crystals, unmelted at 320° (Gibson and Johnson[163]).

The system (**1**) has been termed 13,14,5,8-isobenzarsazinephenarsazine and 13,14-diaza-5,8-diarsapentaphene.

m-Phenylenediamine, subjected to the same process, gave in low yield a product that could have been (**4**) or the isomeric (**5**).

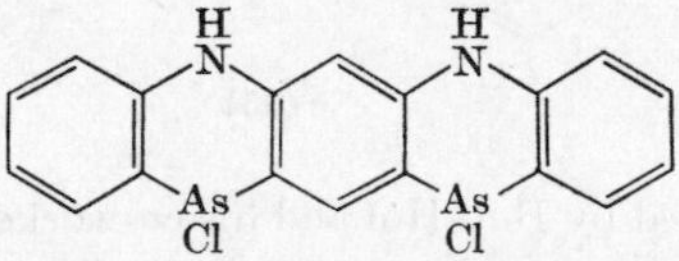

(**4**) 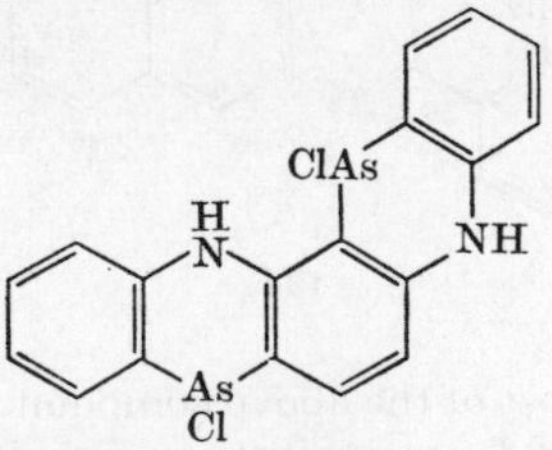   (**5**)

p-Phenylenediamine gave unsatisfactory products.

Pyrido[3,2-a]*phenarsazine* (**6**), [RIS 9034]. From 6-anilinoquinoline (**7**; R = C$_6$H$_5$) with arsenic trichloride in o-dichlorobenzene; on the basis of the greater reactivity of the 5-position in the quinoline system than of the 7-position, the

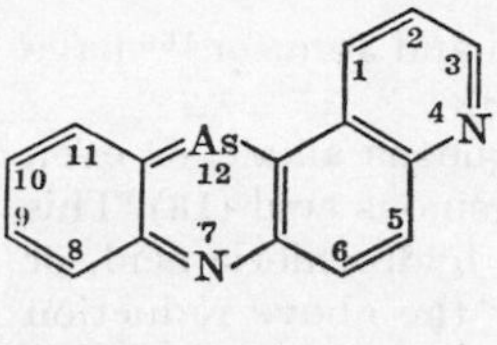

(6) 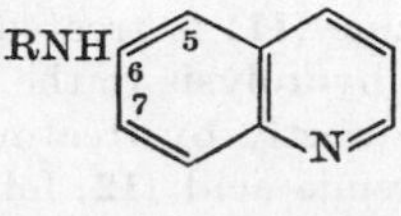   (7) 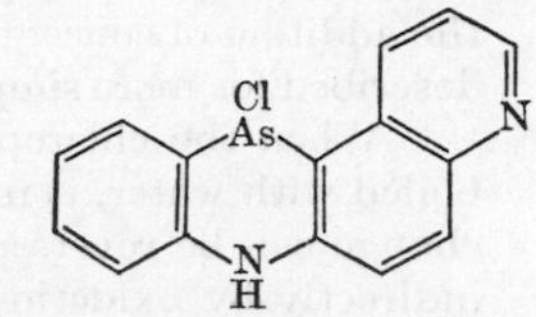   (8)

angular (rather than the linear) cyclization is accepted, giving the yellow hydro-chloride of 12-chloro-7,12-dihydropyrido[3,2-a]phenarsazine (8), m.p. 265–266°; the free base could not be isolated. The compound (8) was termed 9-chloro-9,10-dihydro-10,4′-diaza-9-arsa-1,2-benzanthracene (Buu-Hoï et al.[227]).

The use of 6-p-toluidinoquinoline similarly gave the hydrochloride of the 10-methyl derivative of (8), m.p. 290–291°; this base also was not isolated.

Quino[8,7-b][1,4]benzarsazine (9), [RRI 4899]. This compound (9) can be considered here, as it is isomeric with the above compound (6) and contains the phenarsazine system.

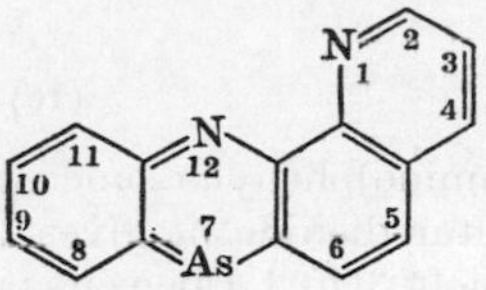

(9) 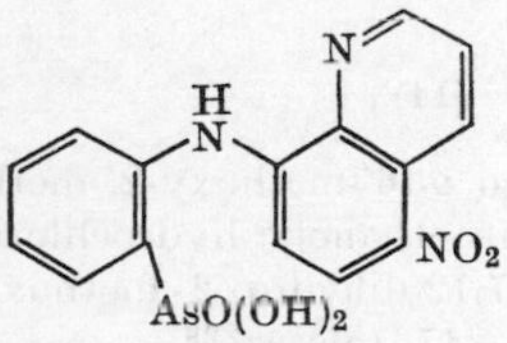   (10)

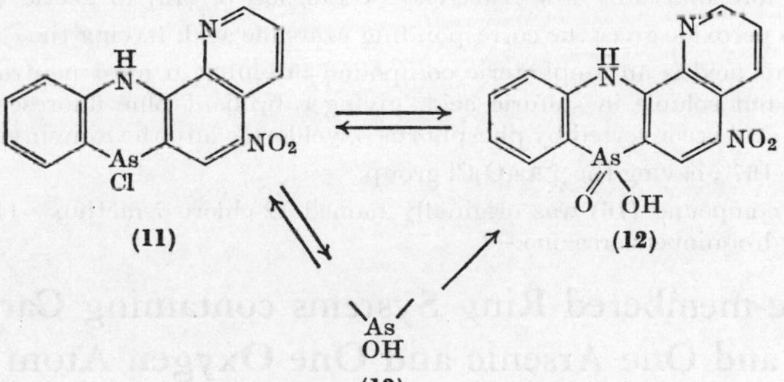

(11)   ⇌   (12)

(13)

8-Amino-5-nitroquinoline condenses with (o-bromophenyl)arsonic acid to give o-(5′-nitro-8′-quinolylamino)phenylarsonic acid (10). This acid, when reduced in ethanolic hydrochloric acid with sulfur dioxide, readily cyclizes to give the deep red crystalline hydrochloride (m.p. 258–260°) of 7-chloro-7,12-dihydro-5-nitro-quino[8,7-b][1,4]benzarsazine (11) (Slater[228]). The base was not isolated. The hydrochloride, when oxidized in acetic acid with hydrogen peroxide gives the phenazarsinic acid (12), a process that can be reversed by the further application of the HCl–SO₂ reduction. The acid (12) is an amphoteric compound, and aqueous solutions of its alkali salts show all the marked color changes on dilution and on the addition of concentrated sodium hydroxide that Gibson and Johnson[166] have described for more simple 2-nitrophenazarsinic acids (p. 509).

When the chlorophenarsazine (11) is treated with aqueous alkali, or even boiled with water, it undergoes hydrolysis to the phenazarsinous acid (13). This change can be reversed either directly by treatment with hydrochloric acid, or indirectly by oxidation to azarsinic acid (12) followed by the above reduction to the arsazine (11). The compound (11) was initially termed 12-chloro-10-nitro-5,12-dihydroquinbenzarsazine.[228]

Quino[4,3-b][1,4]*benzarsazine* (14), [RRI 4900]. This compound does not contain the phenarsazine ring system, but is conveniently considered here, as it is isomeric with the compounds (8) and (11).

(14)   (15)   (16)

When o-(6′-methoxy-2′-methyl-4′-quinolylamino)phenylarsonic acid (15) is reduced in ethanolic hydrochloric acid with sulfur dioxide, it gives the yellow 7-chloro-7,12-dihydro-2-methoxy-6-methylquino[4,3-b][1,4]benzarsazine (16), m.p. 245–247° (Slater[229]).

This compound can be recrystallized from hot water, and the chlorine atom has therefore unusually low reactivity. Oxidation of (16) in acetic acid with hydrogen peroxide gives the corresponding azarsinic acid, having the $>$As(O)OH group; this acid is an amphoteric compound insoluble in most neutral organic solvents, but soluble in sulfuric acid, giving a brilliant blue fluorescence. The azarsinic acid is converted by phosphorus oxychloride into the azarsinyl chloride, m.p. 165–167°, having the $>$As(O)Cl group.

The compound (16) was originally named 12-chloro-7-methoxy-11-methyl-5,12-dihydroquinbenzarsazine.[229]

Five-membered Ring Systems containing Carbon and One Arsenic and One Oxygen Atom

1,2-Oxarsole and 1,2-Oxarsolanes

The compound (1) is 1,2-oxarsole, [RIS —], its 2,3-dihydro derivative (2) is 1,2-oxarsol-4-ene, and the 2,3,4,5-tetrahydro derivative (3) is 1,2-oxarsolane.

(1)   (2)   (3)

Only derivatives of (3) are at present known, and their investigation (Hands and Mercer[230]) has been on the same lines as that of 1,2-oxaphospholane (p. 259).

A mixture of (3-hydroxypropyl)triphenylarsonium iodide (4) and sodium hydride in tetrahydrofuran, when boiled under reflux for 4

$$[\mathrm{HO(CH_2)_3\overset{+}{As}Ph_3}]\ \mathrm{I^-}$$

(4)   (5)   (6)

hours, gives 2,2,2-triphenyl-1,2-oxarsolane (5), m.p. 154°. That this product is in fact the cyclic compound (5) and not the betaine-like zwitterion (6) is shown by the nmr spectrum, which shows a triplet at τ 6.29 ($J_{4,5} = 6$ c/sec), a triplet at τ 7.30 ($J_{3,4} = 8$ c/sec), and a complex multiplet at ca. τ 8.0 (each multiplet 2H); moreover, the infrared spectrum shows ν_{max} 1055 cm^{-1}, assigned to the AsOCH$_2$ unit.

A methanolic solution of the arsolane (5), when treated with hydriodic acid at 22°, rapidly deposits the original iodide (4).

When the oxarsolane (5) is heated at 160–180°, it gives triphenylarsine and allyl alcohol. The mechanism of this degradation probably involves an intramolecular elimination reaction:

$$\mathrm{(5) \rightleftharpoons (6) \rightleftharpoons Ph_3\overset{+}{As}-CH_2-\overset{-}{CH}-CH_2OH \longrightarrow Ph_3As + CH_2{=}CH-CH_2OH}$$

This rearrangement of the intermediate zwitterion (6) is similar to that suggested for the intermediate (7) in the reaction between methylenetriphenylarsorane and benzophenone (Trippett[231]).

$$\mathrm{Ph_3\overset{+}{As}-CH_2 \atop \bar{O}-CPh_2} \longrightarrow \mathrm{Ph_3\overset{+}{As}-\overset{-}{CH} \atop HO-CPh_2} \longrightarrow \mathrm{Ph_3As + CHPh_2CHO}$$

(7)

1,6-Dioxa-5-arsaspiro[4.4]nonane

This spirocyclic compound (1), [RIS 7889], is unknown, but the existence of a very interesting type of derivative (2; where R is an aryl group) is now well established (Braunholtz and Mann[61]).

The three homologous acids, phenylarsinylidenediacetic acid (3),[61] 3,3'-phenylarsinylidenedipropionic acid (4), and 4,4'-phenylarsinylidenedibutyric acid (R. C. Cookson and Mann[58]) show significant differences

(1)   (2)

on oxidation. The acids (3) and (5) give normal arsine oxides of composition corresponding to (6) and (7), respectively. The acid (4), however,

$C_6H_5As(CH_2CO_2H)_2$ $C_6H_5As(CH_2CH_2CO_2H)_2$ $C_6H_5As(CH_2CH_2CH_2CO_2H)_2$

(3)   (4)   (5)

$C_6H_5\overset{\|}{\underset{O}{As}}(CH_2CO_2H)_2$ $C_6H_5\overset{\|}{\underset{O}{As}}(CH_2CH_2CH_2CO_2H)_2$

(6)   (7)

on oxidation gives a crystalline product, m.p. 235°, of composition $C_{12}H_{13}AsO_4$; the corresponding p-chlorophenyl acid (as 4) similarly gives a product, m.p. 223°, of composition $C_{12}H_{12}AsClO_4$. These products can be recrystallized from water, but a cold aqueous solution has pH 4.5, indicating some hydrolysis (see also p. 395 for their preparation).

The evidence that these products have the structure (2), *i.e.*, that they are 5-phenyl-(and 5-p-chlorophenyl)-1,6-dioxa-5-arsaspiro[4.4]-nonane-2,7-dione, respectively (2; $R = C_6H_5$ or C_6H_4Cl), can be summarized as follows. (*a*) The presence of only four atoms of oxygen in the molecule, unlike the five atoms of (6) and (7), is clearly shown by analysis. (*b*) The infrared spectra of the product shows no OH groups, but powerful CO absorption at 1681 (2; $R = C_6H_5$) and 1684 cm^{-1} (2; $R = C_6H_4Cl$); this is consistent with structure (2), particularly as the As—O bond may have some polar character. The spectra do not show the strong absorption band in the 815–845 cm^{-1} region, shown in the spectra of (6) and (7) and attributed to the As=O group. (*c*) The products in aqueous solution can be titrated smoothly with dilute sodium hydroxide, a sharp end-point being obtained potentiometrically without an intermediate 'step', and the equivalent weight thus obtained agrees closely with the theoretical value. (*d*) The compounds cannot have the isomeric acid anhydride structure (8), for such anhydrides show two CO absorption bands in the 1870–1820 and 1800–1750 cm^{-1} regions.

(8)   (9)

The structure (2) is thus comparable with that of a normal γ-lactone, and the phenyl compound was initially termed di-(2-carboxy-ethyl)phenylarsine dihydroxide dilactone, to indicate its relationship to the hydroxy compound (9) produced by hydration of the initial arsine oxide.

The formation of compounds of type (2) is an example of the very ready formation of five-membered rings in the arsenic series; other examples are given in each of the next three Sections.

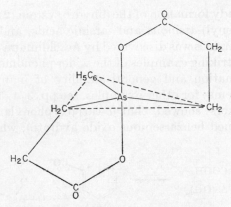

Figure 20. Probable structure of 5-phenyl - 1,6 - dioxa - 5 - arsaspiro[4.4] -nonane-2,7-dione (2; R = C₆H₅). (Reproduced, by permission, from J. H. Braunholtz and F. G. Mann, *J. Chem. Soc.*, 1957, 3285)

The configuration of compounds of type (2) is unknown: it is probable that coplanarity of the bonds from arsenic to both CH₂ groups and the aryl group will be derived from trigonal sp^2-hybridization, while both cyclic oxygen atoms will be collinear with the arsenic atom if the remaining pair of arsenic electrons undergo pd-hybridization (see Figure 20). This structure would be analogous to the trigonal-bipyramidal structure of the tertiary stibine dihalides[125] and of various phosphorus and arsenic compounds.

2,1-Benzoxarsole

The parent compound (1), [RRI 1108], is unknown, but the ring system has in the past been termed 2,1-benzoarsenole and 2-oxa-1-arsenaisoindene. The known members of this series are systematically

1,3-dihydro derivatives, and since most reactions occur at the arsenic atom, it may be convenient to invert this dihydro formula, giving (2).

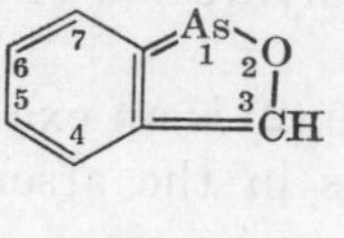

(1) 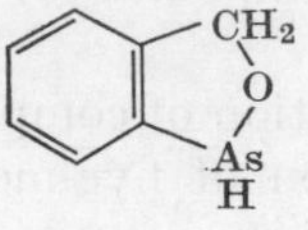   (2)

The very ready formation of the dihydro system (2) from derivatives of (o-carboxyphenyl)-arsinous and -arsinic acids, and from the corresponding chloroarsines was discovered by Aeschlimann and McCleland.[89] It affords some striking examples of the wider phenomenon, namely, the very ready formation and general stability of many five-membered arsenic ring systems; for other examples, see pp. 374, 541.

These workers showed that o-carboxyphenylarsonous acid (3) (which they termed benzarsenious oxide hydrate), when heated at 70°

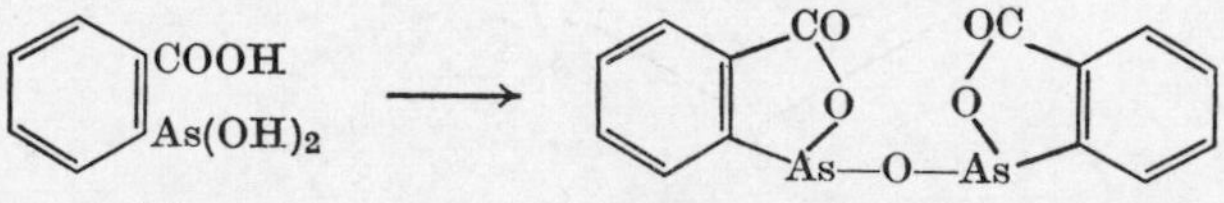

(3)   (4)

in a vacuum, lost water with the formation of the anhydride (4) of 1,3-dihydro-1-hydroxy-2,1-benzoxarsol-3-one. A simpler example of this cyclization was obtained by heating o-carboxyphenylarsonous dichloride (5) at low pressure, whereupon hydrogen chloride was lost with the formation of the 1-chloro-1,3-dihydro-2,1-benzoxarsol-3-one (6). This compound formed colorless crystals, m.p. 145°, b.p. ca. 220°/15 mm, and

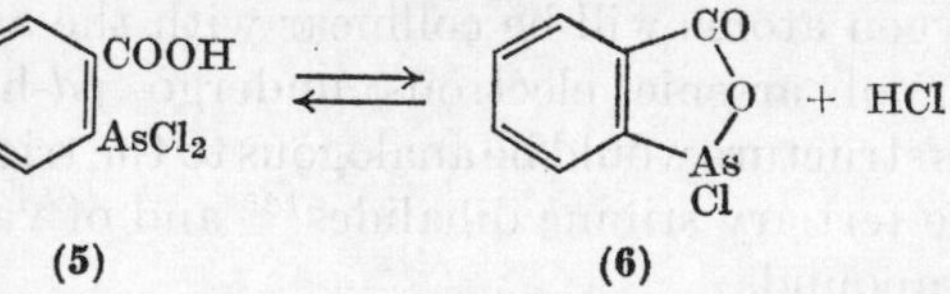

(5)   (6)

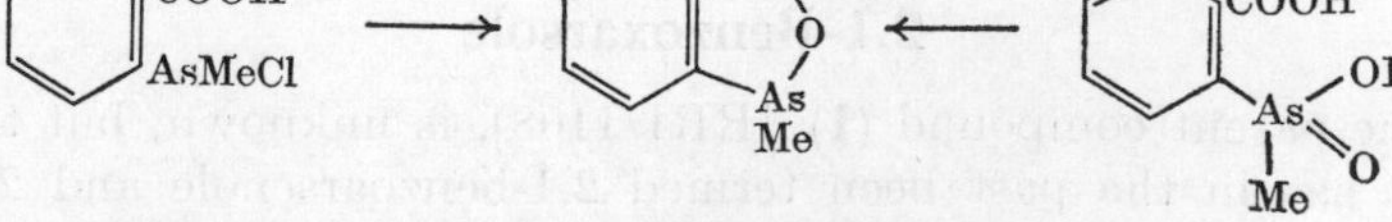

(7)   (8)   (9)

when treated in benzene with hydrogen chloride was converted back into the dichloroarsine (5).

The latter method of ring formation can also be applied readily to the corresponding arsinous acids. Thus, when (o-carboxyphenyl)methylarsinous chloride (7), m.p. 141°, is heated in a vacuum it loses hydrogen chloride with the formation of 1,3-dihydro-1-methyl-2,1-benzoxarsol-3-one (8), m.p. 106°; this change can also be readily reversed by the action of hydrogen chloride dissolved in benzene. The compound (8), however, can also be prepared from o-carboxyphenylmethylarsinic acid (9), which on reduction even in aqueous solution gives the arsine (8).

Precisely similar relationships, with one notable exception, hold in the corresponding phenyl series. Thus, (o-carboxyphenyl)phenylarsinous chloride (10), m.p. 163°, when heated in a vacuum, gives 1,3-dihydro-1-phenyl-2,1-benzoxarsol-3-one (11), m.p. 133°, a change that

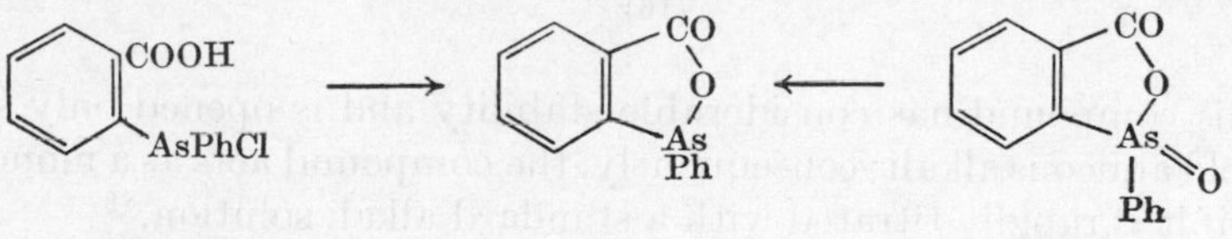

can be reversed as before. In this series, however, the o-carboxyphenylphenylarsinic acid is apparently unknown, since it passes spontaneously over to the 1,3-dihydro-1-phenyl-2,1-benzoxarsol-3-one 1-oxide (12); the latter in turn can be readily reduced to the arsine (11).

This spontaneous formation of the benzoxarsole ring shown in the above (o-carboxyphenyl)phenyl compound (12) occurs even more readily in the di-(o-carboxyphenyl)arsine series of compounds. For instance, di-(o-carboxyphenyl)arsinous chloride (13) is unknown, since

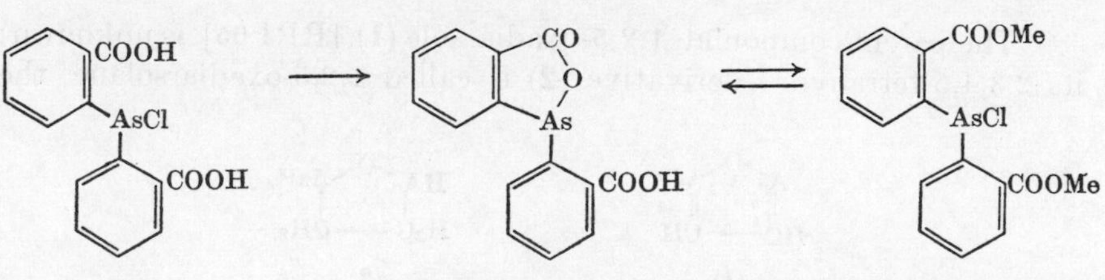

it passes spontaneously into 1-(o-carboxyphenyl)-1,3-dihydro-2,1-benzoxarsol-3-one (14), m.p. 251–255°. When the latter is boiled with methanolic hydrogen chloride, the benzoxarsole ring opens on esterification, with the formation of bis-[(o-methoxycarbonyl)phenyl]arsinous chloride (**15**), m.p. 184°. When this dimethyl ester is hydrolyzed either with ethanolic potassium hydroxide or even with boiling aqueous hydriodic acid (under the conditions of a Zeisel demethylation), the compound (**14**) is again obtained.

As would be expected from the above reactions, di-(o-carboxy-phenyl)arsinic acid is unknown, since it also undergoes spontaneous conversion into the benzoxarsolone 4-oxide (**16**). The benzoxarsole ring

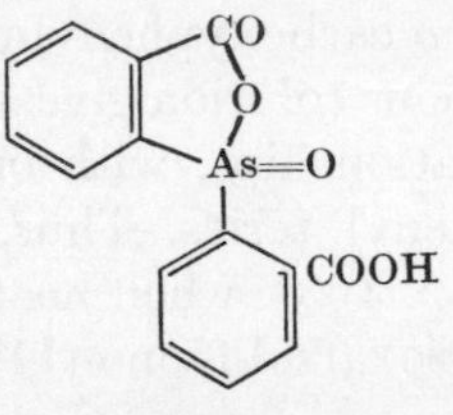

(**16**)

in this compound has considerable stability and is opened only slowly by cold aqueous alkali; consequently, the compound acts as a monobasic acid if it is rapidly titrated with a standard alkali solution.[89]

The formation of the 1,3-dihydro-2,1-benzoxarsol-3-ones is thus closely analogous to that of the γ-lactones, which can be similarly prepared by the loss of water from the γ-hydroxy carboxylic acids or of hydrogen chloride from the γ-chloro carboxylic acids.

Five-membered Ring Systems containing Carbon, Two Arsenic Atoms and One Oxygen Atom

1,2,5-Oxadiarsole and 1,2,5-Oxadiarsolanes

The parent compound, 1,2,5-oxadiarsole (**1**), [RRI 65], is unknown; its 2,3,4,5-tetrahydro derivative (**2**) is called 1,2,5-oxadiarsolane, the

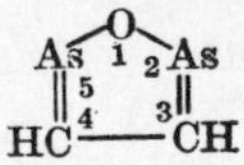

(**1**) 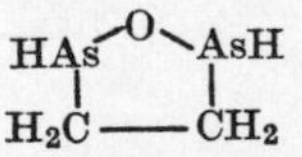   (**2**)

2,5-disubstituted derivatives of which are known.

When ethylenebis(phenylarsinous chloride) (**3**) is hydrolyzed with boiling aqueous sodium hydroxide, a deposit of 2,5-diphenyl-1,2,5-oxadiarsolane (**4**), m.p. 94°, is obtained. The compound is clearly the anhydride of ethylenebis(phenylarsinous acid), and its formation even in the presence of the excess of alkali indicates the strong tendency of many arsenic compounds to form five-membered ring systems (cf. pp. 375, 574).

$$
\begin{array}{cccc}
& \text{Ph} & & \text{Ph} & \text{O} & & \text{Ph} & \text{O} \\
\text{H}_2\text{C—AsCl} & & \text{H}_2\text{C—As} & & \text{H}_2\text{C—As} & & \text{H}_2\text{C—As—OH} \\
| & \longrightarrow & | \quad \text{O} & \longrightarrow & | \quad \text{O} & \longrightarrow & | \\
\text{H}_2\text{C—AsCl} & & \text{H}_2\text{C—As} & & \text{H}_2\text{C—As} & & \text{H}_2\text{C—As—OH} \\
\text{Ph} & & \text{Ph} & & \text{Ph} \;\; \text{O} & & \text{Ph} \;\; \text{O} \\
\\
(3) & & (4) & & (5) & & (6)
\end{array}
$$

The compound (4), when oxidized with hydrogen peroxide in aqueous acetone, yields 2,5-diphenyl-1,2,5-oxadiarsolane 2,5-dioxide (5), m.p. 191°. This compound is similarly the anhydride of ethylenebis(phenylarsinic acid) (6), m.p. 200°. When the dioxide (5) is dissolved in warm hydrochloric acid it deposits the acid (6) on cooling; this acid can be obtained more directly by similarly dissolving the compound (4) in hot dilute nitric acid (Jones and Mann[106]).

The compounds (4) and (5) were initially termed s.-ethylenebis-(phenylarsinous) anhydride and s.-ethylenebis(phenylarsinic) anhydride, respectively.[100]

2,1,3-Benzoxadiarsole

The parent compound is 2,1,3-benzoxadiarsole (1) [RRI 1038], but the known member is a 1,3-disubstituted 1,3-dihydro-2,1,3-benzoxadiarsole (2).

$$
\begin{array}{cc}
(1) & (2)
\end{array}
$$

Kalb's synthesis[117] of o-aminophenylarsonic acid (3) has been described (p. 461); by treating the diazotized acid with sodium arsenite he obtained o-phenylenediarsonic acid (4). By reducing this acid in hot concentrated hydrochloric acid solution with sulfur dioxide he obtained only the 'oxychloride', i.e., 1,3-dichloro-1,3-dihydro-2,1,3-benzoxadiarsole (5), a stable compound which he was unable to convert into the tetrachloro compound (6); the formation of the compound (5) illustrates again the stability of many five-membered arsenic ring systems. Chatt and Mann[107] ultimately converted the compound (5) by the action of thionyl chloride into o-phenylenebis(dichloroarsine) (6), which when

NH$_2$ AsO(OH)$_2$ → AsO(OH)$_2$ AsO(OH)$_2$ →

(3) (4)

Cl—As—O—As—Cl (5) → AsCl$_2$ AsCl$_2$ (6) → AsMe$_2$ AsMe$_2$ (7)

(5) (6) (7)

treated with methylmagnesium iodide gave the corresponding o-phenyl-enebis(dimethylarsine) (7), a strongly chelating compound which they were the first to use for the study of the structure of metallic coordination compounds. A more recent synthesis of the diarsine is based on the action of dimethylsodioarsine on o-dichlorobenzene (Feltham, Kasenally, and Nyholm[232]). The application of the diarsine (7) for the synthesis of non-metallic heterocyclic arsenic systems has been described (pp. 452, 485).

An X-ray crystal-structure investigation of 1,3-dichloro-1,3-dihydro-2,1,3-benzoxadiarsole (5) (Cullen and Trotter[233]) shows that the bicyclic ring system is planar, but that the two chlorine atoms project one above, and the other below, this plane so that they are in the trans-position about the pyramidal arsenic atoms. Figure 21 gives the bond lengths (in Å) and the intervalency angles.

Cullen and Trotter point out that the As—Cl bond length is only slightly greater than those reported for AsCl$_3$ (2.16) and (CH$_3$)$_2$AsCl (2.18), and that the As–O distances are almost identical with those of (CH$_3$)$_3$As (1.98) and (C$_6$H$_5$)$_2$AsBr (1.99), but that the As–O distances are shorter than those in arsenic trioxide (1.78) and in the arsenate ion (1.75) (all bond lengths in Å units).

The intervalency angles Cl–As–C (97°) and Cl–As–O (104°), i.e., those external to the five-membered ring, are within the normal range of those in unrestricted molecules, but the formation of the five-membered ring reduces the O–As–C angle to 77°, and increases the As–O–As angle to 151° (the 'normal' oxygen valency value is 105°, and the As–O–As angle in As$_4$O$_6$ is 128°). These values are particularly noteworthy in view of the high stability of the molecule and suggest that the system may have some degree of aromatic character. Cullen and Trotter[233] write: 'The large oxygen valency angle suggests that the oxygen is approaching a state of sp hybridization, and that the two lone pairs on the oxygen are in approximately p_π orbitals. If this is the case, then $d_\pi-p_\pi$ bonding could occur using unoccupied $4d$ orbitals on the arsenic atoms, accounting for the shorter As–O distance in the oxychloride in comparison with that in the oxide and the arsenate ion. This

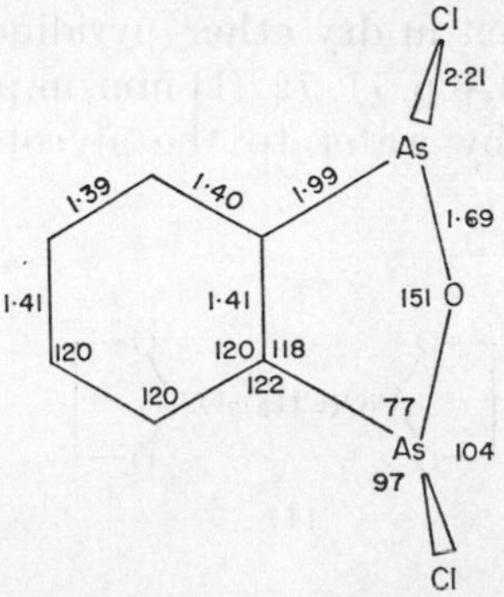

Figure 21. Bond lengths (in Å) and intervalency angles (in degrees) of 1,3-dichloro-1,3-dihydro-2,1,3-benzoxadiarsole. [Reproduced, by permission of the National Research Council of Canada, from W. R. Cullen and J. Trotter, *Can. J. Chem.*, **40**, 1113 (1962)]

$d_\pi-p_\pi$ bonding might also be expected to occur between the arsenic atoms and the phenylene π-electrons, resulting in an extended aromatic system.'

The compound (5) was originally termed o-phenylenediarsinoxy-chloride.[117]

Five-membered Ring Systems containing Carbon and One Arsenic Atom and Two Oxygen Atoms

1,3,2-Dioxarsole and 1,3,2-Dioxarsolanes

1,3,2-Dioxarsole (1), [RRI 63], is encountered only as derivatives of the 4,5-dihydro compound (2), named 1,3,2-dioxarsolane.

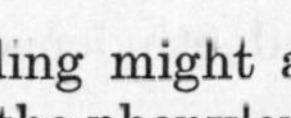

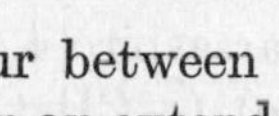

1,2-Glycols react readily with arsenic acid and with alkyl- and aryl-arsonic acids to give spirocyclic compounds (pp. 544 ff.). The conditions for obtaining the monocyclic system (2), and the chemistry of such compounds, have been elucidated mainly by Kamai and his collaborators.

Kamai and Khisamova[234, 235] have shown that arsenic trichloride and ethylene glycol react in dry ether–pyridine to give 2-chloro-1,3,2-dioxarsolane (3; R = Cl), b.p. 71–72°/11 mm, m.p. 44–45°; this compound is readily hydrolyzed by water to the glycol and arsenious acid. A

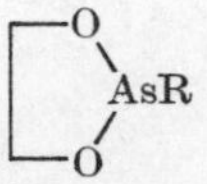

(3) 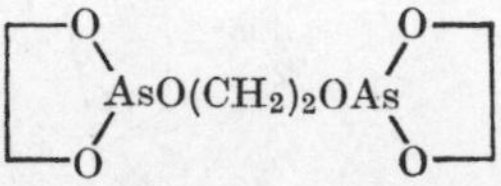   (4) 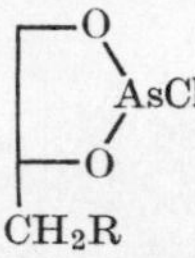   (5)

second product of the above reaction is 2,2′-ethylenedioxybis(1,3,2-dioxarsolane) (4), a viscous liquid, b.p. 166–167°/4 mm, which can be prepared directly by the action of (3; R = Cl) on glycol in ether–pyridine, and by heating arsenious oxide with glycol at 140–150°. When the compound (4) is heated with $AsCl_3$, it regenerates (3; R = Cl), and when it is treated with bromine in carbon tetrachloride it furnishes 1,4,6,9-tetraoxa-5-arsaspiro[4.4]nonan-5-ol (p. 544) and ethylenedioxybis(dibromoarsine), $(—CH_2OAsBr_2)_2$, b.p. 86°/11 mm.

More importantly, the compound (3; R = Cl) reacts with alcohols in pyridine to give the 2-alkoxy-1,3,2-dioxarsolanes, of which the methoxy (3; R = OCH_3), ethoxy, propoxy, butoxy and cyclohexyloxy members are recorded. These compounds are also readily hydrolyzed by water; they do not form methiodides or sulfides, but the ethoxy compound gives an As-dibromo derivative, b.p. 86°/11 mm.

Arsenic trichloride similarly reacts with 3-substituted propylene glycols to give compounds of type (5); thus 3-chloropropane-1,2-diol gives 2-chloro-4-chloromethyl-1,3,2-dioxarsolane (5; R = Cl), b.p. 103–104°/11 mm, and 3-methoxy-1,2-propanediol gives 2-chloro-4-(methoxymethyl)-1,3,2-dioxarsolane (5; R = OCH_3), b.p. 103–104°/11 mm; several such alkoxy derivatives have been described (Kamai and Chadeva[236]).

Arsenic trichloride reacts readily with pinacol in ether–pyridine to give 2-chloro-4,4,5,5-tetramethyl-1,3,2-dioxarsolane (6; R = Cl), m.p. 121°, which with butanol in boiling ether–pyridine gives the 2-butoxy compound[237] (6; R = OC_4H_9), b.p. 170–190°/12 mm, m.p. 94°.

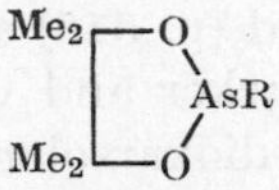

Me₂—O
        \
         AsR
        /
Me₂—O

(6)

O
 \
  AsNMe₂
 /
O

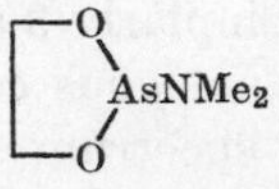

(7)

The chloro compound (**3**; R = Cl) reacts readily with dimethylamine in ice-cooled benzene solution to give N,N-dimethyl-1,3,2-dioxarsolane-2-amine (**7**), b.p. 101–102°; a byproduct of this reaction is 2,2'-oxydi-1,3,2-dioxarsolane[238] (**8**), b.p. 172–175°/18 mm.

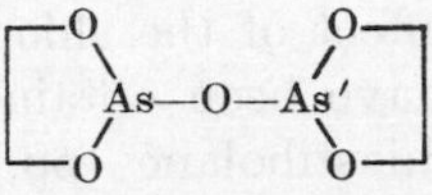

O   O
 \   /
  As—O—As'
 /   \
O   O

(8)

O
 \
  AsC₆H₄NO₂
 /
O

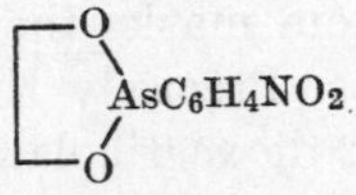

(9)

It is noteworthy that p-nitrophenylarsonous acid,

$$p\text{-}NO_2C_6H_4As(OH)_2,$$

when heated with glycol at 80° under reduced pressure, gives 2-p-nitrophenyl-1,3,2-dioxarsolane (**9**), m.p. 119–121°, and 1-naphthyl-arsonous acid similarly gives the 1-naphthyl analogue, b.p. 169–170°/4 mm. 3-Alkoxypropane-1,2-diols react with p-nitrophenylarsonous acid to give the 4-alkoxymethyl derivatives of (**9**) (Kamai and Chadeva[239]).

Phenylarsonous dichloride, $C_6H_5AsCl_2$, reacts with ethylene glycol in ether–pyridine at 0° and after 1 hour's boiling gives (**3**; R — C₆H₅), b.p. 122–123°/10 mm; it reacts similarly with various 3-alkoxypropane-1,2-diols to give the corresponding 4-(alkoxymethyl)-2-phenyl-1,3,2-arsolanes. p-Tolylarsonous dichloride gives analogous derivatives.[240]

Certain substituted 1,2-diols will condense with arsonic acids to give the ring system (**2**) if the product is stabilized by special factors. Thus when an acetic acid solution of carboxymethylarsonic acid, $HO_2C \cdot CH_2 \cdot As(O)(OH)_2$ and (+)-tartaric acid is boiled for a few minutes, the crystalline 2-carboxymethyl-1,3,2-dioxarsolane-4,5-dicarboxylic acid 2-oxide (**10**) is deposited, and further reaction to form a spirocyclic compound does not occur; moreover, meso-tartaric acid will not even form a compound of type (**10**), the configuration of the two asymmetric carbon atoms apparently inhibiting the formation of the ring system

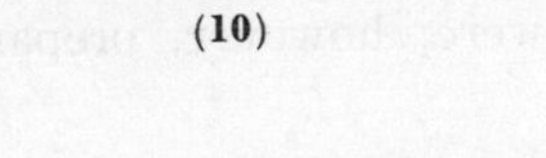

HOOC—O
        \ //O
         As
        / \
HOOC—O   CH₂COOH

(10)

HOOC—O
        \ //O
         As
        / \
HOOC—O   CMe=CHCOOH

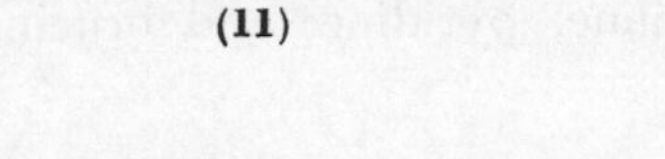

(11)

(Englund[241]). Similarly 3-arsonocrotonic acid (p. 545) and (+)-tartaric acid give the analogous compound (11) (Backer and van Oosten[242]).

The nmr spectrum of 2-chloro-1,3,2-dioxarsolane (3; R = Cl) resembles that of 2-chloro-1,3,2-dithiarsolane (p. 574), in that both show a single sharp line for the CH_2 absorption (Foster and Fyfe[243]). Since the inversion times of the pyramidal arsine (AsH_3) and trimethylarsine are long compared with the probable limit of resolution obtainable in this investigation, it is unlikely that the observed equivalence of the ring CH_2 groups in the above two compounds is the result of inversion. It is suggested that the large size of the arsenic atom compared with the chlorine atom masks the stereochemical effect of the chlorine on the CH_2 groups. Precisely the same results have been obtained with 2-chloro-1,3,2-dioxastibolane and -1,3,2-dithiastibolane (pp. 613, 629) and have been similarly interpreted.[243] The nmr spectra of N,N-diethyl-1,3,2-dioxaphospholane-2-amine and other substituted 1,3,2-dioxaphospholanes show on the other hand a multiplet splitting of the CH_2 protons (p. 274).

1,4,6,9-Tetraoxa-5-arsaspiro[4.4]nonane

This spirocyclic compound (1), [RRI 848], can be prepared in the form of its derivatives very readily. Englund[241] has recorded that when arsenic acid, H_3AsO_4, is dissolved in ethylene glycol with gentle warming,

(1)   (2)

the solution when cooled deposits 1,4,6,9-tetraoxa-5-arsaspiro[4.4]-nonan-5-ol (2; R = OH) originally termed di(ethylenedioxy)arsenic acid, m.p. 120° after crystallization from ethanol. The compound is stable in air but is immediately hydrolyzed in water, and consequently when titrated with alkali acts merely as a solution of arsenic acid. Its molecular weight, determined cryoscopically in bromoform, is twice that required for the above formula, but this is very probably due to association in solution. Its pyridine and brucine salts have been recorded.

If arsenic acid is dissolved in three equivalents of ethylene glycol by gentle warming, later cooling gives a solid mixture of the compound (2; R = OH) and of the tris(ethylenedioxy)arsenic acid (3), which Englund[241] was unable to separate by recrystallization; the pure aniline, pyridine, and brucine salts of (3) were, however, prepared.

The modern name for the acid (**3**) is tris-[1,2-ethanediolato(2-)]arsenic(v) acid, [RIS —]; it is analogous to the tris[pyrocatecholato(2-)]arsenic(v) acid described on p. 551.

(**3**)   (**4**)

Carboxymethylarsonic acid, $HO_2C—CH_2—AsO(OH)_2$, when dissolved in warm methanolic glycol deposits the compound (**2**; $R = CH_2CO_2H$), m.p. 142°, in 60% yield. It is immediately hydrolyzed in water to its components. It also exists as a dimer in freezing bromoform.

When a solution of the arsonic acid in glycol is heated above 130°, carbon dioxide is evolved, and distillation of the residue gives the compound (**2**; $R = CH_3$), b.p. 135–136°/15 mm, which solidifies in the receiver. It is also rapidly hydrolyzed by water.

Pinacol forms similar compounds even more readily than ethylene glycol, and the products have rather greater stability. Pinacol and arsenic acid in acetone readily give 2,2,3,3,7,7,8,8-octamethyl-1,4,6,9-tetraoxa-5-arsaspiro[4.4]nonan-5-ol (**4**; $R = OH$; $R^1 = R^2 = R^3 = R^4 = CH_3$), m.p. 131°. This acid dissolves in ice-cold water without hydrolysis and can be titrated as such with dilute alkali and phenolphthalein as an indicator; at room temperature the acid undergoes steady hydrolysis.

Pinacol and carboxymethylarsonic acid similarly give the compound (**4**; $R = CH_2CO_2H$; $R^1 = R^2 = R^3 = R^4 = CH_3$), m.p. 188°.

Pinacol in hot ethanol reacts with 2-arsonoacrylic acid (**5**) and with 3-arsonocrotonic acid (**6**) to give analogous compounds (**4**; $R = H_2C=\overset{|}{C}—CO_2H$; $R^1 = R^2 = R^3 = R^4 = CH_3$), m.p. 173–174°, and (**4**;

(**5**)   (**6**)

$R = CH_3—\overset{|}{C}=CH—CO_2H$; $R^1 = R^2 = R^3 = R^4 = CH_3$), m.p. 198–200° (Backer and van Oosten [242]).

Salmi, Merivuori, and Laaksonen [244] have described the condensation of alkyl- and aryl-arsonic acids with several 1,2-glycols (**7**) to give compounds of type (**4**), and also with α-hydroxy carboxylic acids to give similar compounds of type (**8**), i.e., derivatives of 1,4,6,9-tetraoxa-3,7-dioxo-5-arsaspiro[4.4]nonane.

Compounds of type (8) are prepared by the azeotropic distillation of the arsonic acid and the α-hydroxycarboxylic acids in benzene or

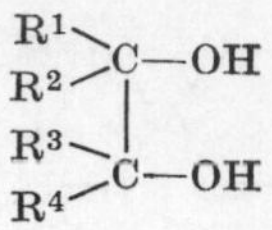

(7) 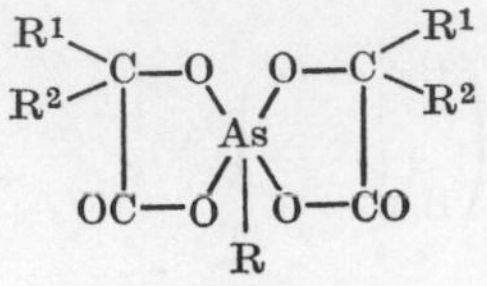   (8)

toluene or, if the arsonic acid is insoluble in benzene, by boiling the components with acetic anhydride.

Examples of the compounds obtained[244] are given in Table 5; most of the solid products form canary-yellow crystals.

Table 5. Some spirans of types (4) and (8)

Type	R	R^1	R^2	R^3	R^4	M.p. (°)	B.p. (°/mm)
(4)	CH_3	H	H	H	H	—	110–111/4
	CH_3	CH_3	H	H	H	—	115.5–116/3
	CH_3	CH_3	CH_3	CH_3	CH_3	—	131–132/3
	n-C_4H_9	H	H	H	H	—	140.5–141.5/3
	n-C_4H_9	CH_3	H	H	H	—	142.6–143.4/5
	n-C_4H_9	CH_3	CH_3	CH_3	CH_3	—	169.0–170/4
	C_6H_5	H	H	H	H	105.5	
	C_6H_5	CH_3	CH_3	CH_3	CH_3	176.0	
(8)	C_6H_5	CH_3	CH_3	—	—	138.2	
	C_6H_5	CH_3	C_2H_5	—	—	91.4	
	$C_6H_5CH_2$	CH_3	CH_3	—	—	169	
	$C_6H_5CH_2$	CH_3	C_2H_5	—	—	161	

The above ready condensations of arsonic acids with 1,2-glycols and with α-hydroxy carboxylic acids can give rise to more complex systems, some of which are very briefly recorded below.

11,13,24,25-*Tetraoxa*-12-*arsapentaspiro*[4.0.4.1.1.4.0.4.1.1]*pentacosane*. This compound (1; R = H), [RRI 6553], is formed by the condensation of arsonic acids with bi(cyclopentane)-1,1′-diol. The derivatives[242] (1; R = CH$_2$=$\overset{|}{C}$—CO$_2$H) and (1; R = CH$_3\overset{|}{C}$=C—COOH) have m.p.s 208–210° and 162–162.5°, respectively.[242]

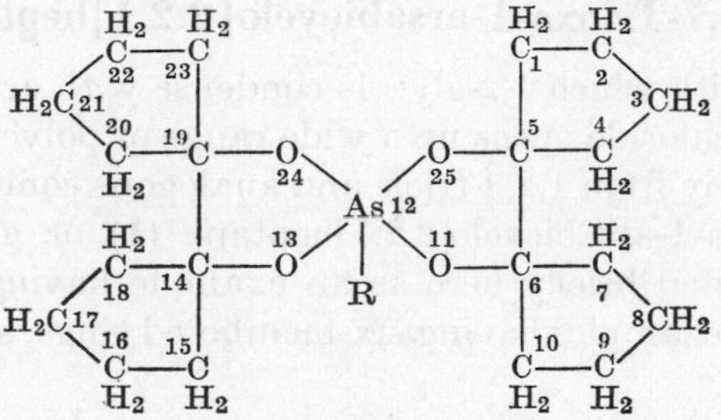

H₂ H₂ ... (structure 1)

(1)

13,15,28,29-*Tetraoxa*-14-*arsapentaspiro*[5.0.5.1.1.5.0.5.1.1]*nonacosane*. This compound (**2**; R = H), [RRI 6599], is similarly prepared[242] from bi(cyclohexane)-1,1'-diol. The derivative (**2**; R = CH₃C̶=CHCO₂H) has m.p. 233–234°.

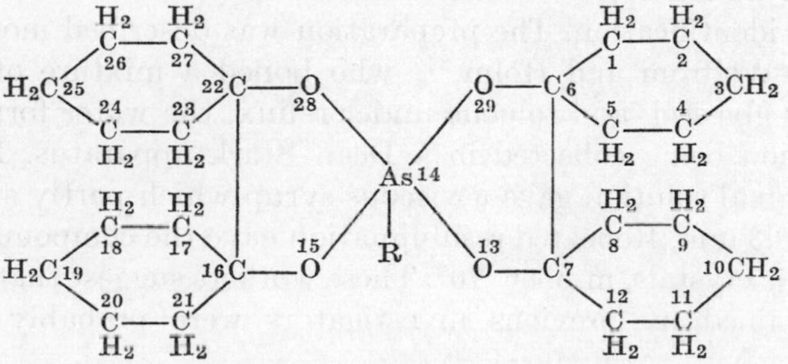

(2)

7,9,17,18-*Tetraoxa*-8-*arsatrispiro*[5.1.1.5.2.2]*nonadecane*. This compound (**2**; R = H), [RRI 4114], is represented by derivatives obtained by the condensation of arsonic acids and α-hydroxy carboxylic acids; in these derivatives the 16- and 19-CH₂ groups of formula (**3**) become CO groups. The compounds having R = CH₃,

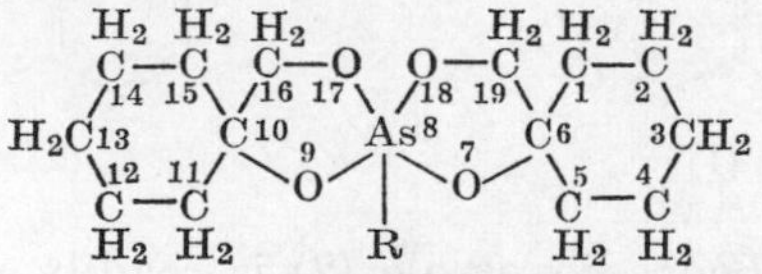

(3)

n-C₄H₈, and C₆H₅CH₂ have m.p.s 213°, 194°, and 175°, respectively[244]; those having R = CH₂=C̶CO₂H and CH₃C̶=CHCO₂H have m.p.s 173–174° and 198–200°, respectively.[242]

2,6,7-Trioxa-1-arsabicyclo[2.2.1]heptane

The ease with which 1,2-glycols condense with arsenious acid and with arsenic trichloride opens up a wide range of polycyclic compounds obtained similarly from 1,2,3-triols and analogous compounds.

2,6,7-Trioxa-1-arsabicyclo[2.2.1]heptane (1) or glyceryl arsenite, [RRI 847], is cited briefly here as an example having five-membered rings; for other examples having six-membered rings, see p. 570.

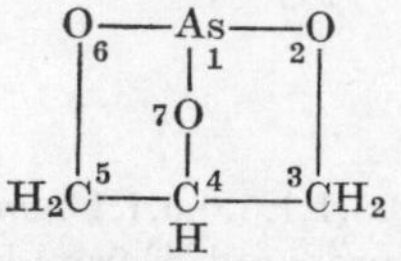

(1) 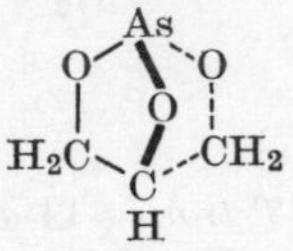   (1A)

This compound (1) was mentioned by Pictet and Bon,[245] and later by Pascal and Dupire,[246] who recorded physical constants but no analytical identification. The preparation was described more recently (1961) by Wolfrom and Holm,[247] who boiled a mixture of arsenious oxide, dry glycerol, and toluene under reflux, the water formed in the condensation being collected in a Dean–Stark apparatus. Distillation of the residual solution gave a viscous syrup which partly sublimed at 120–140°/0.3 mm. Repeated resublimation gave the compound (1), very hygroscopic crystals, m.p. 66–70°. These workers suggest that the glassy solids obtained by previous investigators were probably polymeric material.

The conformation of (1) is more clearly depicted as (1A).

1,3,2-Benzodioxarsole

This system (1) [RRI 1037], is encountered almost solely as its 2-substituted derivatives.

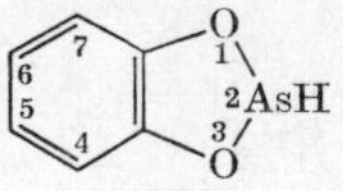

(1) 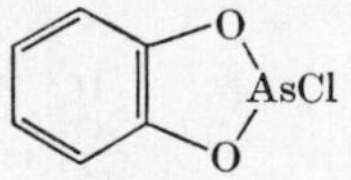   (2)

2-Chloro-1,3,2-benzodioxarsole (2) is readily prepared by heating arsenic trichloride and pyrocatechol under reflux; when the vigorous reaction is complete, the cooled product deposits the compound (2), m.p. 125–127° (93% yield) (Funk and Köhler[248]). In an older method the components were heated first in an oil-bath and then in a sealed tube at 200°; the product had m.p. 133° (61% yield) (Kamai et al.[237]).

When an excess of pyrocatechol is used in the first method the main product is 2,2'-(o-phenylenedioxy)di-(1,3,2-benzodioxarsole) (3), b.p. 269°/3 mm, sometimes termed tri-o-phenylene diarsonite.

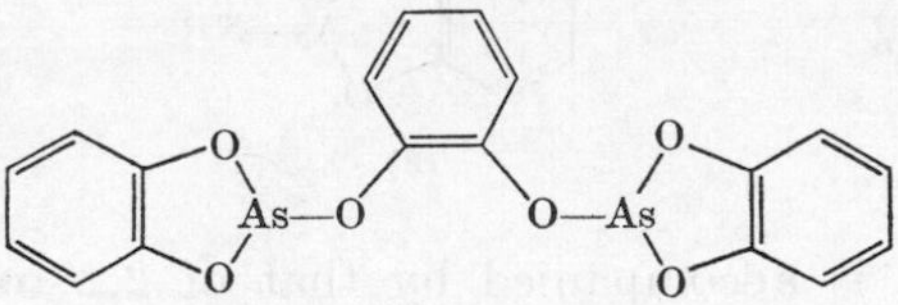

(3) 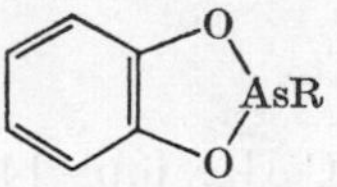   (4)

When a mixture of arsenic trichloride, pyrocatechol, and phenol is heated, the intermediate product is undoubtedly (2), which then condenses with the reactive phenol to give 2-phenoxy-1,3,2-benzdioxarsole (4; R = OC_6H_5), m.p. 98–100°. This compound can be regarded as a double ester of arsenious acid with pyrocatechol and phenol.

2-Phenyl-1,3,2-benzodioxarsole (4; R = C_6H_5), m.p. 83°, was obtained by Michaelis[249] by interaction of the lead salt of pyrocatechol, $C_6H_4O_2Pb$, and phenylarsonous dichloride, $C_6H_5AsCl_2$. It can be prepared more readily by the action of phenylarsonous dichloride on pyrocatechol in ether–pyridine at 0°, followed by 1 hour's boiling; the compound (4; R = C_6H_5) has b.p. 179–181°/10 mm, m.p. 85–86°, and the compound (4; R = $C_6H_4CH_3-p$) has b.p. 185–186°/9 mm, m.p. 106–107.[240]

Most compounds in this series are derivatives of the chloro compound (2), which incidentally shows an unusual ability to form 1:1 crystalline complexes with pyridine, quinoline, dioxane, and other compounds (Funk and Köhler[250]).

The compound (2) is much less susceptible to hydrolysis than its 2-chloro-1,3,2-dioxarsolane analogue (p. 542) and requires six hours' boiling with water for complete hydrolysis to pyrocatechol and arsenious acid. However, when it is treated with ammonia in boiling benzene or xylene, it forms, according to the solvent, 2-amino-1,3,2-benzodioxarsole (5), m.p. 128°, di-(1,3,2-benzodioxarsol-2-yl)amine (6), m.p. 138°, and tri-(1,3,2-benzodioxarsol-2-yl)amine (7), m.p. 149°.

With piperidine in chilled benzene it forms 2-piperidino-1,3,2-benzodioxarsole (8), b.p. 142°/3 mm, and with dimethylamine and

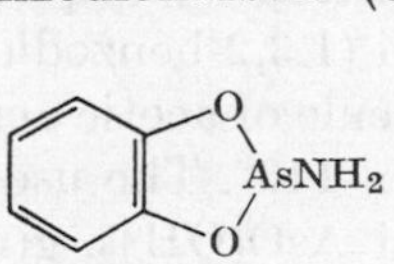

(5) 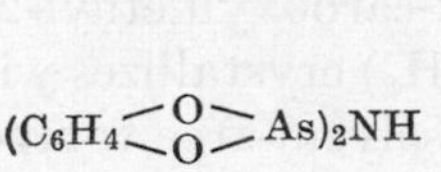   (6) 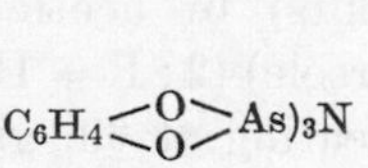   (7)

diethylamine the corresponding 2-dialkylamino derivatives (9). The formation of the two compounds (9; R = CH$_3$, b.p. 146°/26 mm;

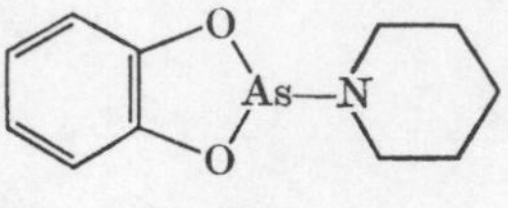

(8) 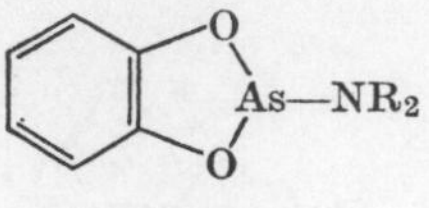   (9)

R = C$_2$H$_5$, b.p. 149°/17 mm) is accompanied by that of 2,2'-oxydi-(1,3,2-benzodioxarsole) (10), m.p. 147°.

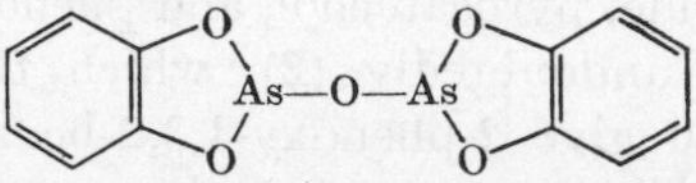

(10) 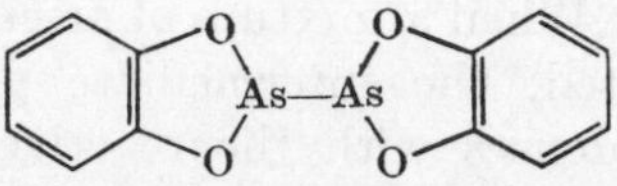   (11)

The compound (2) in a boiling ethanol–pyridine–benzene mixture gives the 2-ethoxy derivative (4; R = OC$_2$H$_5$), m.p. 150°, which is recorded as reacting vigorously with water; the 2-butoxy derivative, similarly prepared,[250] has b.p. 140–141°/14 mm.

A xylene solution of the compound (2), when boiled with copper-bronze, gives 2,2'-bi-(1,3,2-benzodioxarsole) (11), m.p. 99° (Legler[251]).

2,2'-Spirobi[1,3,2-benzodioxarsole]

Derivatives of this system (1), [RRI 4115], have the hydrogen of the AsH group in (1) replaced by alkyl or aryl groups (2). Alternatively, the compound (1) may act as an acid by ionization of the central hydrogen atom.

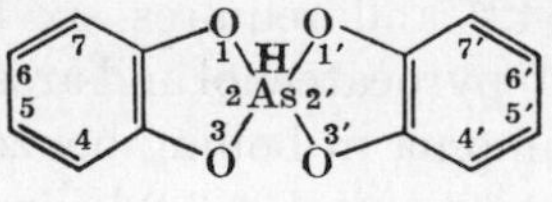

(1) 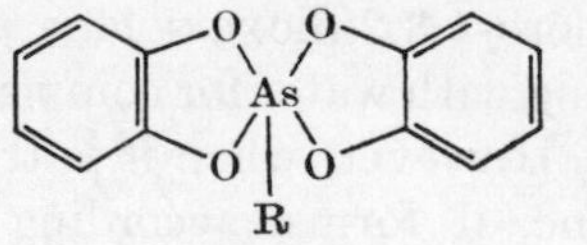   (2)

Englund[252] has shown that when carboxymethylarsonic acid, HO$_2$C—CH$_2$AsO(OH)$_2$, is briefly heated with pyrocatechol (2 equivalents) in acetic acid, 2-carboxymethyl-2,2'-spirobi-(1,3,2-benzodioxarsole) (2; R = HO$_2$C—CH$_2$) crystallizes with a molecule of acetic acid; heating at 60° gives the solvent-free compound, m.p. 146°. The use of (methoxycarbonylmethyl)arsonic acid, CH$_3$O$_2$C—CH$_2$AsO(OH)$_2$, gives the analogous compound (2; R = CH$_3$O$_2$C—CH$_2$), m.p. 117°.

Salmi *et al.*[244] have carried out the same condensation with methyl-, *n*-butyl-, and benzyl-arsonic acids to give the yellow crystalline compounds (**2**), where R is CH_3 (m.p. 152°), $n\text{-}C_4H_9$ (m.p. 83°), or $C_6H_5CH_2$ (m.p. 147°).

A preliminary account of an *X*-ray crystal structure investigation of the ionized potassium salt $K[As(O_2C_6H_4)_2]$ has been given by Skapski.[253]

Tris[pyrocatecholato(2−)]arsenic(v) Acid

Two different types of arsenic derivative, in which one and two pyrocatechol units, respectively, are linked to the arsenic atom to form a five-membered ring, have been described (pp. 548, 550). The present compound, most probably of structure (**3**), has three such units of pyrocatechol; it is acidic and for many years was called tripyrocatechyl-arsenic acid [RRI —]: its modern name is tris[pyrocatecholato(2−)]-arsenic(v) acid. It will be seen, however, that decisive evidence for the structure of the acid or its anion is still lacking.

Weinland and Heinzler[254] showed that if pyrocatechol was added to a boiling aqueous solution of arsenic acid, H_3AsO_4, the solution on cooling deposited colorless crystals to which they gave the structure (**1**). It is noteworthy that the same compound was obtained if the molecular ratio of pyrocatechol to arsenic acid was 1:1, 2:1, 3:1, or 1:2. With regard to this structure, the following points should be noted. (*a*) The structure ignores the strong tendency of pyrocatechol to form cyclic or

$$H_3[O\text{---}As\text{---}(O\cdot C_6H_4O^-)_3]4H_2O$$

(**1**)

(**2**)   (**3**)

chelated rings. (*b*) Although Weinland and Heinzler prepared a number of simple salts of the acid, the latter was acting as a monobasic acid in the formation of all these salts and never as a tribasic acid.

Reihlen, Sapper, and Kall[255] drew attention to the monobasicity of this acid, but also pointed out that only four molecules of water could be detached from the acid, the elements of the fifth molecule remaining tenaciously in the acid and it salts. Consequently, they allotted the structure (2) to the acid. In this structure the acid is mono-basic, two molecules of pyrocatechol are chelated to the arsenic, but the third is linked by only one bond, and the one molecule of water is coordinated to the arsenic atom, which is thus showing a coordination number of six. They also demonstrated that a similar stable acid (and salts) could be obtained in which the coordinated molecule of water was replaced by a molecule of pyridine—also firmly attached.

Rosenheim and Plato[256] considered that all three pyrocatechol residues were chelated to the six-coordinated arsenic atom, and that the

Table 6. Some salts of tris[pyrocatecholato(2−)]arsenic(v) acid

Salt	Rotation, $[M]_D$*	Solvent
	(+)-Series	
$H[As(C_6H_4O_2)_3]5H_2O$	+2002.6°	Aqueous acetone
$K[As(C_6H_4O_2)_3]1H_2O$	+2218.0	Aqueous acetone
	(−)-Series	
$H[As(C_6H_4O_2)_3]5H_2O$	−2187.5°	Aqueous acetone
$K[As(C_6H_4O_2)_3]1H_2O$	−2097.1	Aqueous acetone
$NH_4[As(C_6H_4O_2)_3]1H_2O$	−2483.4	Aqueous acetone
$\frac{1}{2}Ba[As(C_6H_4O_2)_3]5H_2O$	−2325.5	Water

* The rotations are cited as given by the authors, although five significant figures are clearly not warranted by their measured rotations.

compound was consequently tripyrocatechylarsenic acid pentahydrate of structure (3); they admitted, however, that only four of the molecules of water of crystallization could be readily removed, the fifth being firmly linked. If the acid has the structure (3), it should be resolvable into optically active forms because it is of the same general stereochemical type as, for example, potassium cobalt(III) trioxalate. They were able, therefore, to resolve the acid through its cinchonine salt, which furnished the (−)-rotatory acid, and through the cinchonidine and quinine salts, which furnished the (+)-rotatory acid.

Table 6 gives the composition and rotation of some of the compounds isolated by Rosenheim and Plato.[256] The optically active anion undergoes only slow racemization in neutral aqueous solution and has a surprising optical stability in the presence of caustic alkalis, a short boiling with aqueous sodium hydroxide leaving the rotation almost unchanged. The anion is very sensitive to acids, however, which causes immediate racemization; this would be expected because acids would certainly readily break the chelated pyrocatechol ring.

The structure of the acid still remains uncertain in one respect, however. It is to be noted that all the salts isolated by Rosenheim and Plato—salts of metals, ammonia, and alkaloids—contained at least one molecule of water. Furthermore the structure (2) put forward by Reihlen, Sapper, and Kall would also allow optical activity if the unchelated pyrocatechyl group and the molecule of water occupied the *cis*- or 1,2-positions in the six-coordination octahedron.

This work has been extended by Larkins and Jones,[257] who have prepared similar sodium salts from 3-methylpyrocatechol (4; R = CH$_3$, R′ = H), 4-methylpyrocatechol (4; R = H, R′ = CH$_3$), 4-chloropyrocatechol (4; R = H, R′ = Cl) and also from naphthalene-2,3-diol (5),

(4)   (5)   (6)

which had been earlier used by Weinland and Seuffert.[258] The free acids of type (3) could not be isolated from 4-methyl- and 4-chloro-pyrocatechol, although salts with cinchonine and quinine were obtained: moreover, the free acid (6) was isolated as a tan-colored monohydrate and a white sesquihydrate, although a pentahydrate had been reported by Weinland and Seuffert.

Optical resolution was obtained by recrystallization of the cinchonine and quinine salts, which were then converted into the corresponding sodium salts, for which the following rotations in aqueous solutions at 30° are recorded.

Na[(3-CH$_3$C$_6$H$_3$O$_2$)$_3$As],2H$_2$O	$[\alpha]_D$	−385°	+395°
Na[(4-CH$_3$C$_6$H$_3$O$_2$)$_3$As],1H$_2$O	$[\alpha]_D$	−26	+32
Na[(4-ClC$_6$H$_3$O$_2$)$_3$As],5H$_2$O	$[\alpha]_D$	−21	+22
Na[(C$_{10}$H$_6$O$_2$)$_3$As],7H$_2$O	$[\alpha]_D$	−20	+19

It was suggested that the low values for rotations may have been caused by alkaloid contamination of the sodium salts.[257]

The striking feature however is the persistence of at least one molecule of water in the above salts. Following a suggestion put forward (Porte, Gutowsky, and Harris[259]) to account for a similar persistence in potassium trioxalatorhodium(III), K$_3$[Rh(C$_2$O$_4$)$_3$],4$\frac{1}{2}$H$_2$O, Larkins and Jones[257] envisage the possibility that in the tris(pyrocatecholato)-arsenate ion, one molecule of water is co-ordinated through its oxygen

19

(7)

to the arsenic atom, and is also hydrogen-bonded to the (otherwise) free
phenolic group which is formed in this process (7). This is, however,
closely akin to the earlier structure (2), with hydrogen-bonding added
for additional stability.

Larkins and Jones[260] have also studied the degree of retention of
the normal phenolic properties of the pyrocatechol and the 3-methyl-
pyrocatechol in the above optically active sodium salts. They find that
both bromination and coupling with diazotized o-anisidine occur without
appreciable change of optical activity or, therefore, of configuration.
They conclude that these reactions occur with the complex anion as a
whole, rather than with 'momentarily disengaged ligand molecules'.
Clearly, however, the precise structure of the complex anion must be
determined before this conclusion can be accepted; other mechanisms
may perhaps be suggested to explain these results.

The optically active tripyrocatechylarsenic acid has been un-
successfully employed to resolve a dissymmetric tetraarylarsonium salt
(Mann and J. Watson[261]). This salt, p-bromophenyl-p-chlorophenyl-
phenyl-4-o-xylylarsonium   (−)-tripyrocatechylarsenate, is the only
anhydrous salt which has been recorded. However, it has so great a
molecular weight that the addition of one molecule of water would
hardly affect the carbon and hydrogen content by which the salt was
identified. In an attempt (also unsuccessful) to resolve a complex gold
compound (Davis and Mann[262]), the salt di-[4-methyl-o-phenylenebis-
(diethylphosphine)]gold(I) (−)-tripyrocatechylarsenate was found to
crystallize with two molecules of ethanol. The role of the one molecule
of water (or ethanol) in the acid and its salts therefore still awaits
elucidation.

Six-membered Ring Systems containing Carbon
and One Arsenic and One Oxygen Atom

4H-1,4-Oxarsenin and 1,4-Oxarsenanes

This compound (1), [RRI 230], is unknown but its 2,3,5,6-tetrahydro
derivative, termed 1,4-oxarsenane (2), has been isolated as 4-substituted
products (Beeby and Mann[142]).

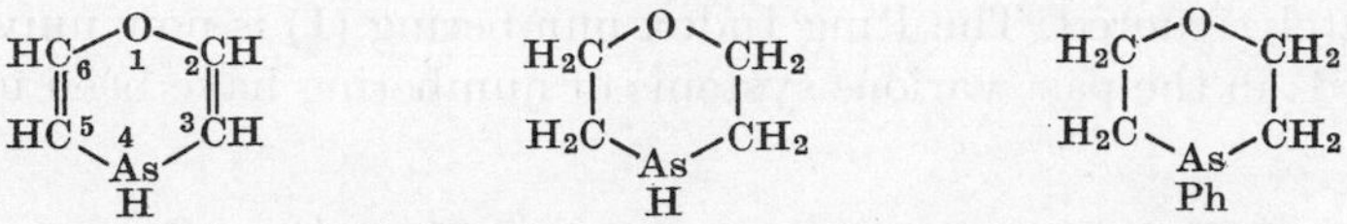

(1)   (2)   (3)

Di-(2-bromoethyl) ether in benzene was added to an ether–benzene solution of phenylbis(magnesium bromide), $C_6H_5As(MgBr)_2$ (one equivalent) (p. 229), and the reaction mixture was then boiled, cooled, and hydrolyzed. Evaporation of the organic layer left a viscous residue, almost certainly formed by extensive linear condensation of the above reactants. This residue, heated in nitrogen at 0.1 mm, underwent progressive decomposition, giving a middle fraction, b.p. 150–165°, of crude 4-phenyl-1,4-oxarsenane (3).

For purification, this product was oxidized by hydrogen peroxide to the 4-oxide (4), which with picric acid gave the 4-hydroxy-4-phenyl-1,4-oxarsenanium picrate (5; $X = C_6H_2N_3O_7$), m.p. 123° after recrystallization. This pure salt was decomposed by cold hydrochloric

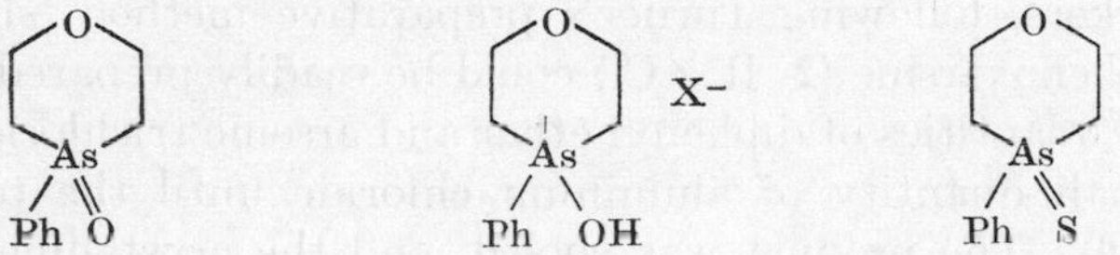

(4)   (5)   (6)

acid, the liberated picric acid being extracted with ether, and the aqueous solution of the chloride (5; X = Cl), when reduced with sulfur dioxide gave the pure oxarsenane (3), b.p. 149–151°/18 mm. It gave a methiodide, m.p. 162–162.5°, and an orange dichlorobis(oxarsenane)palladium, $(C_{10}H_{13}AsO)_2PdCl_2$, m.p. 182°, crystallizing from dioxane with one molecule of solvent.

A portion of the aqueous solution of the chloride (5; X = Cl), when basified with ammonia and treated with hydrogen sulfide, gave 4-phenyl-1,4-oxarsenane 4-sulfide (6), m.p. 101.5–102°, the molecular weight of which confirms the above structure.

The compound (3) was originally termed tetrahydro-4-phenyl-oxarsine.[142]

Phenoxarsines

The name phenoxarsine [RRI 3258] is applied to the system (1), and the systematic but longer name, dibenz-4H-1,4-oxarsenin, is fortunately ignored. The Ring Index numbering (1) is now universally accepted: in the past various systems of numbering have been used (as

(1)   (2)

in the case of phenarsazine), and the reader when consulting earlier papers should first note carefully the notation employed.

The majority of phenoxarsine derivatives are 10-substituted phenoxarsines (2), and it may be convenient in this series also to invert the formula, as shown, and so have the reactive arsenic atom at the bottom.

The phenoxarsines, like the phenarsazines, owe their development very largely to the conditions obtaining in World War I. Their first record in open publication was by Lewis, Lowry, and Bergeim,[263] who made a number of derivatives by a method initiated by Turner[264] and sent to them by Sir William Pope in a Chemical Warfare Communication. These workers, following Turner's preparative method, showed that 10-chlorophenoxarsine (2, R = Cl) could be readily prepared by boiling equivalent quantities of diphenyl ether and arsenic trichloride together with a small quantity of aluminum chloride until the temperature reached 200°. The product was cooled, and the crystalline 10-chloro-phenoxarsine was then filtered off. If the mother-liquor were then again boiled, a second crop could be obtained, the total yield being 46%.

The 10-chloro compound thus obtained had an irritating action on the mucous membrane, particularly if the compound was dissolved in a volatile solvent, although Lewis, Lowry, and Bergeim[263] state that it is less irritating to the nose, throat, and skin than is 10-chloro-5,10-di-hydrophenarsazine.

10-Chlorophenoxarsine is a stable compound not readily affected by boiling water or by 10% aqueous sodium hydroxide solution. The chlorine atom, however, can be speedily replaced, for treatment in methanol solution with the appropriate sodium salt gives 10-bromo-phenoxarsine, m.p. 128°, 10-iodophenoxarsine, m.p. 144°, and 10-thio-cyanatophenoxarsine (2; R = —CNS), m.p. 129°; these three compounds all form yellow crystals.

When the 10-chloro compound is heated with alkalis or with sodium methoxide or ethoxide, or treated with ethanolic ammonia, it gives the 'oxide', $O{<}(C_6H_4)_2{>}As{-}O{-}As{<}(C_6H_4)_2{>}O$, m.p. 182°; this compound is really the anhydride of 10-hydroxyphenoxarsine (2; R = OH) and would now be termed 10,10'-oxydiphenoxarsine; it is soluble in all the common organic solvents, but insoluble in water or alkalis. When treated with the hydrogen halides, it is converted into the corresponding 10-halogenophenoxarsine. This is the best method of preparing the pure 10-bromo compound. When the 10-chloro compound in ethanol is treated with hydrogen sulfide, it forms the corresponding sulfide, $O{<}(C_6H_4)_2{>}As{-}S{-}As{<}(C_6H_4)_2{>}O$, termed 10,10'-thiodiphenoxarsine, pale yellow needles, m.p. 161°, insoluble in all common solvents except acetic acid. It is interesting that these solubility relations are the reverse of those in the corresponding phenarsazine compounds, where the 'oxide', $HN{<}(C_6H_4)_2{>}As{-}O{-}As{<}(C_6H_4)_2{>}NH$, is insoluble in most liquids but the corresponding sulfide is moderately soluble.

When the 10-chlorophenoxarsine in aqueous suspension is treated with bromine, addition occurs at the arsenic atom, and if the mixture is then boiled to effect hydrolysis phenoxarsinic acid (3), m.p. 219°, is obtained. This acid is systematically named 10-hydroxyphenoxarsine 10-oxide.

(3)

When the above 'oxide' in ethanolic solution is reduced with phosphorous acid, it gives 10,10'-biphenoxarsine,

$$O{<}(C_6H_4)_2{>}As{-}As{<}(C_6H_4)_2{>}O.$$

Lewis, Lowry, and Bergeim [263] record this compound as yellow needles, m.p. 159°, which when exposed to air slowly gives a mixture of the 'oxide' and phenoxarsinic acid, but which when dissolved in organic solvents undergoes rapid oxidation to the acid. Blicke, Weinkauff, and Hargreaves [265] prepared this compound under an atmosphere of carbon dioxide and record m.p. 176–177°.

The statement by Lewis, Lowry, and Bergeim [263]—that when the 10-chlorophenoxarsine is treated with ethylmagnesium iodide, it furnishes the 10-ethylphenoxarsine, as white needles, m.p. 218°—is in error (see p. 561): the compound they isolated but analyzed only for arsenic was almost certainly phenoxarsinic acid.

These properties, and the method of preparation, leave the structure of 10-chlorophenoxarsine reasonably certain. The following unambiguous synthesis, worked out by Turner and Sheppard, [266] provides, however, a complete proof of this structure. 2-Aminodiphenyl ether was diazotized and converted by the Bart reaction into 2-phenoxyphenylarsonic acid (**5**), which when reduced in the usual way with sulfur dioxide in the presence of hydrochloric acid and a trace of iodide gave 2-phenoxyphenylarsonous dichloride (**6**). This compound formed an orange-red oil, which on attempted distillation at 10 mm pressure lost hydrogen chloride very readily with cyclization to 10-chlorophenoxarsine (**7**) (57%).[267]

Roberts and Turner[267] have investigated the relative ease of cyclization of [2-(2'-chlorophenoxy)phenyl]arsonous dichloride and the 3'- and 4'-chloro isomers. The 2'-chloro compound thus gave 4,10-dichlorophenoxarsine, m.p. 99°, and the 4'-chloro compound (**8**) gave 2,10-dichlorophenoxarsine (**9**), m.p. 144–145°. As an alternative route to the latter, the [2-(4'-chlorophenoxy)phenyl]arsonic acid (**10**), from which the dichloroarsine (**8**) had been prepared, was warmed with sulfuric acid when cyclization readily occurred to give 2-chlorophenoxarsinic acid (**11**), which when reduced in the usual way also furnished

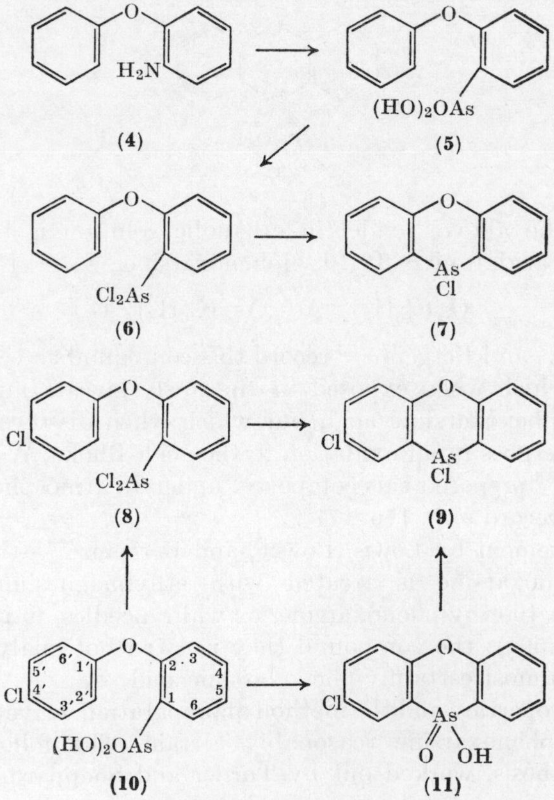

2,10-dichlorophenoxarsine (9). In the case of [2-(3′-chlorophenoxy)-phenyl]arsonous dichloride (12), cyclization might theoretically occur to give either 3,10-dichlorophenoxarsine (13), or 1,10-dichlorophenoxarsine, or a mixture of both. In practice, this cyclization occurred very readily and was found to give only one product, m.p. 125°, which was identified as the 3,10-dichloro isomer by the following synthesis. 2,4-Dichloro-nitrobenzene (10) was condensed with potassium phenoxide in phenol solution to give 5-chloro-2-nitrodiphenyl ether (15). This compound was then reduced to the 2-amino compound, which when diazotized and subjected to the Bart reaction gave 4-chloro-2-phenoxyphenylarsonic acid, which in turn on reduction gave the arsonous dichloride (16). The

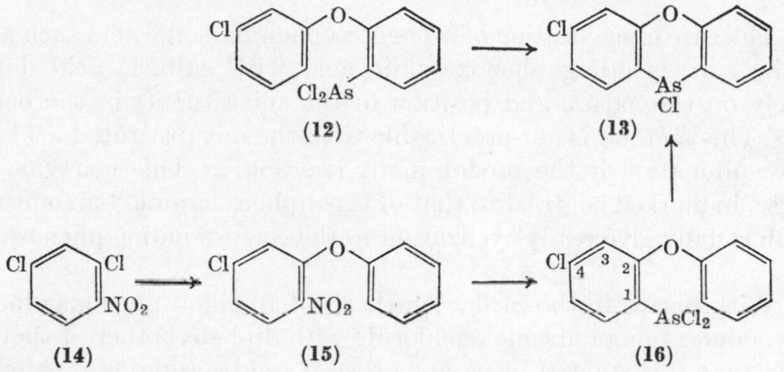

(12) → (13)

(14) → (15) → (16)

latter then underwent the usual smooth cyclization, and the product, which was necessarily 3,10-dichlorophenoxarsine, was found to be identical with that obtained by the cyclization of the isomeric dichloro-arsine (12), and also with that obtained by reduction of the phenoxarsinic acid prepared by the cyclization of [2-(3′-chlorophenoxy)phenyl]-arsonic acid.

These workers[267] found that under comparable conditions the unsubstituted dichloroarsine (6) underwent cyclization more rapidly than the 3′-chloro compound (12), and the latter in turn reacted more readily than the 2′-chloro compound and the 4′-chloro compound (8) which underwent cyclization at approximately equal rates.

The effect of the nature and position of substituents upon the ease of ring closure in compounds of the type (8) and (12) was subsequently examined in detail by Mole and Turner.[268] Their results, which were based on a quantitative study of the relative speeds of the ring closure in the various compounds examined, should be consulted by the interested reader. In the course of this work they isolated new derivatives of 10-chlorophenoxarsine recorded in Table 7.

Table 7. 10-Chlorophenoxarsine derivatives prepared
by Mole and Turner [268]

Compound	M.p. (°)
10-Chloro-3-methylphenoxarsine	140–141
10-Chloro-4-methylphenoxarsine	90–91
10-Chloro-1,3-dimethylphenoxarsine	138–139
10-Chloro-1,4-dimethylphenoxarsine	146–147
10-Chloro-2,4-dimethylphenoxarsine	130–131
10-Chloro-2-methoxyphenoxarsine	108–109
2-Bromo-10-chlorophenoxarsine	172–173

The ease of cyclization of a phenoxyphenylarsonic acid such as (**10**) to the corresponding phenoxarsinic acid with sulfuric acid depends largely on the nature and position of the substituents in the benzene rings. This method is not practicable with the unsubstituted acid itself, since sulfonation is the predominant reaction.[267] This behavior is, of course, in marked contrast to that of the diphenylamine-2-arsonic acids, which usually give ready cyclization to the corresponding phenazarsinic acid.

With regard to the earlier synthesis of 10-chlorophenoxarsine, *i.e.*, the condensation of arsenic trichloride with diphenyl ether, it should be noted that this reaction does not proceed unless aluminum chloride is present.[266] In this respect it is in marked contrast to the condensation of arsenic trichloride with diphenylamine (pp. 507 ff.), which usually proceeds vigorously to form the corresponding phenarsazine without the presence of catalysts. In one other respect, however, these two condensations have a feature in common. Aeschlimann[269] has shown that phenylarsonous dichloride, $C_6H_5AsCl_2$, reacts with diphenyl ether in the presence of aluminum chloride to give, not the expected 10-phenylphenoxarsine, but 10-chlorophenoxarsine. This is strictly analogous to Burton and Gibson's discovery[160] that phenylarsonous dichloride reacts with diphenylamine to give 10-chloro-5,10-dihydrophenarsazine. In the phenoxarsine series, however, one might expect the formation of the 10-chloro compound to proceed more readily, because the aluminum chloride would promote the dismutation:

$$2C_6H_5AsCl_2 \rightleftarrows (C_6H_5)_2AsCl + AsCl_3$$

If the arsenic trichloride thus formed were condensing readily with the diphenyl ether, the production of the 10-chlorophenoxarsine would proceed smoothly. In the phenarsazine series, there is no aluminum chloride to hasten the attainment of the above equilibrium and hence the 10-chloro derivative is formed more slowly and (over a reasonable period of time) in low yield.

Aeschlimann[269] attempted to prepare the 10-phenylphenoxarsine by heating 10-chlorophenoxarsine in benzene solution with aluminum chloride in the hope that a condensation of the Friedel–Crafts type would ensue, but the experiment proved fruitless.

The action of Grignard reagents on 10-chlorophenoxarsine has been investigated by Roberts and Turner,[270] and also by Aeschlimann.[269, 181] It is noteworthy that the former were unable to prepare 10-methylphen-

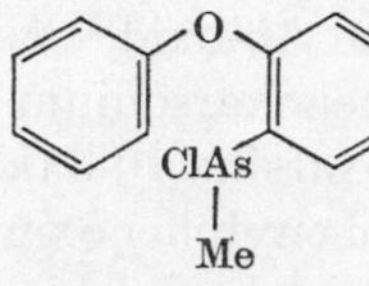

(17) 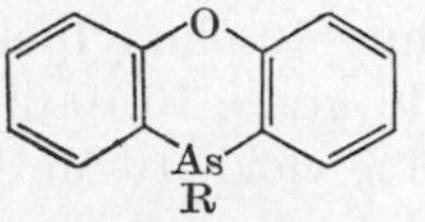   (18)

oxarsine (18; R = CH₃) by the cyclization of the monochloro arsine (17): apparently the presence of the methyl group completely inhibits the cyclization which occurs so readily in the dichloroarsine. 10-Methyl- and 10-ethyl-phenoxarsine can be readily prepared by the action of the appropriate Grignard reagent in ethereal solution on 10-chlorophenoxarsine; this reaction does not proceed so readily with phenylmagnesium bromide, and this Grignard reagent has to be boiled with the 10-chlorophenoxarsine in benzene solution for 4 hours to ensure conversion into the 10-phenylphenoxarsine.

Aeschlimann[269, 181] records the following properties for these three derivatives. 10-Methylphenoxarsine is an oil, b.p. 185°/20 mm, 198–200°/40 mm. When a solution in aqueous hydrogen peroxide is evaporated and the residue recrystallized, the dihydroxide (19), m.p. 94° is obtained. The arsine readily gives a methiodide, systematically 10,10-dimethylphenoxarsonium iodide, m.p. 220° or 225° dependent on the rate of heating, and it also readily combines similarly with bromoacetic acid to give 10-(carboxymethyl)-10-methylphenoxarsonium bromide (20). 10-Ethylphenoxarsine is also an oil, b.p. 194°/20 mm, which could not be obtained in crystalline form. It also gives a dihydroxide, a

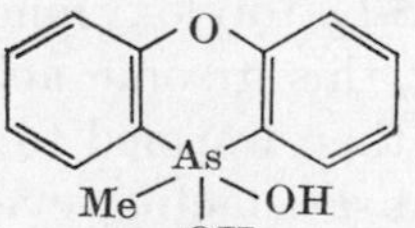

(19) 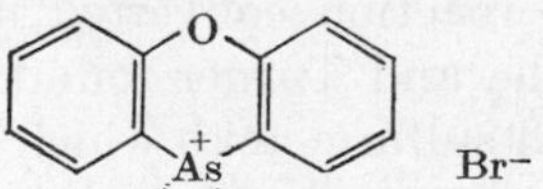   (20) 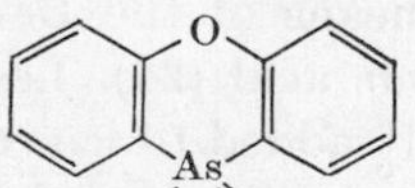   (21)

methiodide, m.p. 186° or 193° dependent on the rate of heating, an ethiodide, m.p. 193°, and a bromoacetic acid addition product analogous to (20). When 10-ethylphenoxarsine is boiled in carbon disulfide solution with one equivalent of sulfur, or when hydrogen sulfide is passed through an aqueous solution of the dihydroxide, 10-ethylphenoxarsine 10-sulfide (21) is readily formed as colorless crystals, m.p. 109°.

10-Phenylphenoxarsine forms colorless needles, m.p. 107°. It gives a methiodide, m.p. 175°, and a very hygroscopic oxide, m.p. 184°.

Quaternary salts can also be prepared by the action of Grignard reagents on the 10-alkyl(or aryl)phenoxarsine 10-oxides (Blicke and Cataline[35]). Thus 10-phenylphenoxarsine 10-oxide with phenylmagnesium bromide gives 10,10-diphenylphenoxarsonium bromide, m.p. 229–230°, in 73% yield. 10-Methylphenoxarsine 10-oxide with phenylmagnesium iodide gives 10-methyl-10-phenylphenoxarsonium iodide, m.p. 170–171°, in 87% yield, but this salt is obtained in only 27% yield by the interaction of 10-phenylphenoxarsine 10-oxide and methylmagnesium iodide.

Derivatives of 10-chlorophenoxarsine having substituents in the benzene rings can be prepared by the use of suitably substituted diphenyl ethers. This has already been illustrated in the preparation of the dichloro derivatives (9) and (13), and the preparation of phenoxarsines having an extra fused benzo group is discussed on p. 568.

One of the most interesting features of the phenoxarsines is their stereochemistry. If the arsenic atom of the phenoxarsine system has a C–As–C angle of *ca.* 98°, and the oxygen atom has a C–O–C angle of *ca.* 112° (cf. dioxane and trioxymethylene which have this angle[271]), the tricyclic phenoxarsine system might be folded about the O–As axis. This type of folding to a 'butterfly' conformation has already been discussed in connection with the 5,10-dihydrophosphanthrenes (p. 186) and the 5,10-dihydroarsanthrenes (p. 464).

If this folding occurred in the phenoxarsine system, the three rings could, of course, not be planar and a suitably substituted phenoxarsine should be susceptible to optical resolution.

This point has been investigated by Lesslie and Turner.[272] For the preparation of a suitable compound, *o*-nitrophenyl *p*-tolyl ether (22) was reduced to the *o*-amino compound (23), which was then diazotized and by means of the Bart reaction converted into 2-(*p*-tolyloxy)phenylarsonic acid (24). Lesslie and Turner found that this arsonic acid, if slowly added to ice-cold sulfuric acid, which was then warmed to 100° for 5 minutes, cooled, and poured on ice, underwent satisfactory cyclization to 2-methylphenoxarsinic acid (25). On oxidation, the latter gave the 2-carboxy derivative (26), which when reduced as usual by sulfur

dioxide in the presence of hydrochloric acid, gave 10-chlorophenox-
arsine-2-carboxylic acid (27). This compound, when treated with methylmagnesium iodide, gave 10-methylphenoxarsine-2-carboxylic acid (28).

This acid was converted into its strychnine salt, which on recrystallization ultimately gave complete optical resolution. The 10-
methyl acid (28) was thus obtained in the dextrorotatory form having

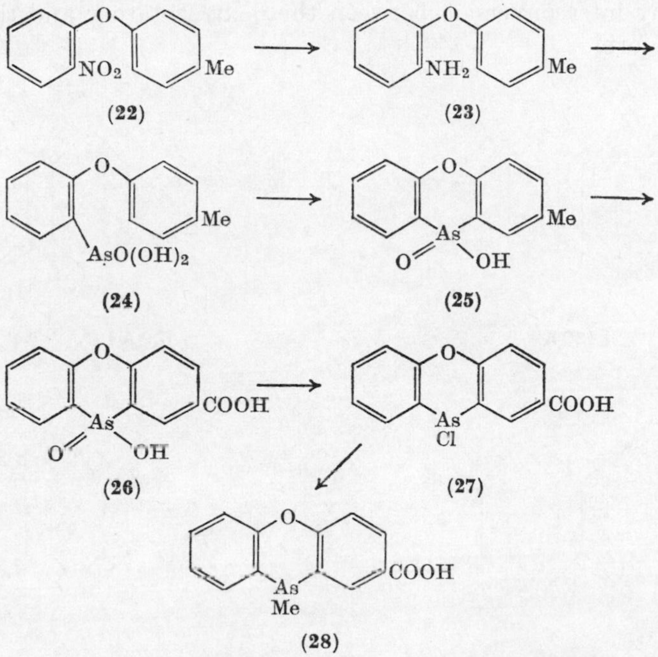

(22)   (23)

(24)   (25)

(26)   (27)

(28)

$[\alpha]_{5791}^{20}$ +95.8°, $[\alpha]_{5461}^{20}$ +111.5°, and in the laevorotatory form having
$[\alpha]_{5791}^{20}$ −96.0°, $[\alpha]_{5461}^{20}$ −111.7°, all in ethanolic solution. This acid had
considerable optical stability; thus an ethanolic solution could be
boiled for several hours without racemization, and a solution in N-NaOH
solution could be heated at 100° (sealed tube) for 6 hours without
diminution of the optical activity.

Lesslie and Turner point out that there are two possible explanations
of the optical activity of this acid. (a) If the molecule is folded about the
O–As axis [as in (29A), in which the oxygen and the arsenic and the
methyl group are depicted as lying in the plane of the paper, while the
carboxyphenylene group projects above the plane of the paper, and the
unsubstituted phenylene group projects below this plane], then the acid
has no element of symmetry irrespective of the disposition of the methyl

group. There may be an additional factor here, suggested later by
Mislow, Zimmerman, and Melillo [273] and discussed in detail with reference to the 5,10-dihydrophosphanthrenes (p. 187), that the molecule
(**29A**) might 'flex' over to the form (**29A′**) and that the two forms might
exist in equilibrium in solution. If (**29A**) represents the dextrorotatory
form, (**29B**) would represent the laevorotatory form and this might also
be in equilibrium with (**29B′**). This 'flexing' motion, however, does not
entail any interconversion between the dextrorotatory and the laevorotatory form.

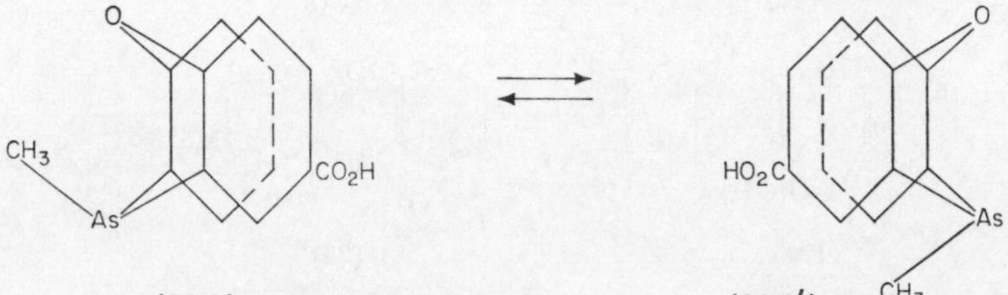

(b) If the tricyclic system were rigidly planar, the methyl group
would lie above or below this plane and would thus deny any element of
symmetry to the molecule. The dextrorotatory and laevorotatory forms
would thus have the methyl group on opposite sides of this plane, and
any oscillation between these positions would be a racemization process.
In the acid (**28**), the position of the methyl group might be stabilized by
the disposition of the ring system, but it is unlikely that this stabilization
is so great that oscillation of the methyl group above and below the
plane of the rings under the above vigorous conditions is entirely
prevented; yet this condition would be essential in view of the great
optical stability of the acid.

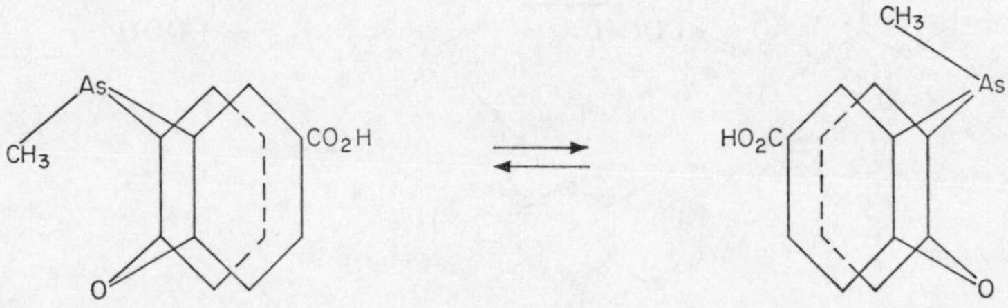

The second explanation (b) is therefore highly improbable, and the
first explanation (a), based on the folded molecule (**29**), is almost certainly

correct. There is, however, at present no decisive evidence for either explanation, and an X-ray crystal structure examination of 10-chloro-phenoxarsine (**2**; R = Cl) would be of great value.

It is noteworthy, nevertheless, that when the acid was dissolved in methyl iodide, racemization occurred, slowly at first and then more rapidly as conversion into the methiodide, *i.e.*, 2-carboxy-10,10-di-methylphenoxarsonium iodide, proceeded. Similarly, racemization occurred with ethyl iodide, but more slowly owing to the lower reactivity of ethyl iodide and the more sluggish formation of the ethiodide. Now, if the acid had a *rigidly* folded structure as in (**29A**), the quaternary arsonium salt should also be devoid of any element of symmetry, and it would have been expected that the optical activity would have been retained. Lesslie and Turner point out, however, that since the formation of the methiodide or ethiodide is accompanied by the acquisition of a positive charge by the arsenic atom, the effective size of the latter must undergo simultaneous diminution, since the arsenic atom in an arsonium salt is probably little larger than an oxygen atom. They suggest, there-fore, that it is almost inconceivable that a folded molecule such as (**29A**) could withstand the agitation accompanying this process without complete loss of optical activity.

In view of these results, Lesslie and Turner[274] have prepared the quaternary arsonium salts, 2,10,10-trimethylphenoxarsonium iodide (**30**), and the corresponding 2,10-dimethyl-10-phenyl derivative, but

(**30**)

neither of these compounds gave any indication of optical resolution. It should be noted, however, that in all these four quaternary salts, the arsenic atom has now almost certainly become tetrahedral, and the intervalency angle at the arsenic atom within the ring has now increased from about 97° to about 109°. Although this value is still far short of 120°, it may in the phenoxarsine compounds weaken the rigidity of the folded structure. (No such weakening could be detected, however, in the arsanthrene compounds described on p. 464, but in view of the absence of the oxygen atom they are not necessarily strictly comparable).

Lesslie and Turner[275] also have resolved the 10-ethylphenoxarsine-2-carboxylic acid, which had $[\alpha]_{5791} \pm 119°$ and $[\alpha]_{5461} \pm 139°$, and which also had great optical stability in organic solvents and in hot sodium hydroxide solution. With methyl iodide, however, it formed the corresponding methiodide, the process again being accompanied by complete racemization.

The 10-phenylphenoxarsine-2-carboxylic acid has also been investigated. Lesslie and Turner [276] were unable to prepare this compound by the action of phenylmagnesium bromide on the 10-chloro acid (**27**). Consequently they reduced the 2-methylphenoxarsinic acid (**25**) to the 10-chloro compound, which reacted readily with the above Grignard reagent to give 2-methyl-10-phenylphenoxarsine (**31**). On oxidation,

this compound gave 10-phenylphenoxarsine-2-carboxylic acid 10-oxide (**32**) and the latter then underwent smooth reduction to the required 10-phenylphenoxarsine-2-carboxylic acid (**33**). This acid was resolved through its 1-phenylethylamine salts, and obtained in the two forms having $[\alpha]_{5791} \pm 223°$ and $[\alpha]_{5461} \pm 261°$; these values are notably higher than those of the 10-methyl and 10-ethyl homologues. The 10-phenyl acid also possessed great optical stability and was unaffected when heated in ethanolic solution in a closed tube at 100° for 4 hours. Oxidation to the oxide (**32**) caused complete racemization, however.

The oxide (**32**) has been resolved by Lesslie. [277] It is so weakly acidic that salts with many amines dissociate in solution; fractional recrystallization of the morphine salt, however, ultimately gave the (+)-acid, $[\alpha]_{5791}^{20} +36.4°$, and the (−)-acid, $[\alpha] -29.7°$, each measured in dilute aqueous ammonia solution. The acid in this solution slowly racemized at room temperature, possibly because of the hydration of the As=O group to the As(OH)$_2$ group: its half-life period at 50° and 80° was 35 and 6.4 minutes, respectively. The oxide (**32**) has two potential sources of dissymmetry, namely, the asymmetric arsenic atom, and the 'folding' of the molecule, although the latter factor would probably be negligible in this compound. Reduction of the active oxide gave, as expected, the inactive arsine (**33**).

The optical properties of another compound in this series, namely, p-(2-chloro-10-phenoxarsinyl)benzoic acid (**37**), have been investigated by Lesslie.[278] It had been suggested by Campbell[279] that one of the factors contributing to the optical instability of p-(2-methyl-10-phenoxstibinyl)benzoic acid (p. 623) may have been the heavy polar p-carboxyphenyl group in the 10-position, which might encourage inversion of the pyramidal configuration of the antimony atom. The analogous phenoxarsine (**37**) was therefore prepared and resolved into optically active forms in order to determine whether it also showed a similar optical instability.

The acid was prepared by treating 2,10-dichlorophenoxarsine (**34**) with p-tolylmagnesium bromide, whereby 2-chloro-10-p-tolylphenoxarsine (**35**) was obtained. This compound, on oxidation, gave the acid

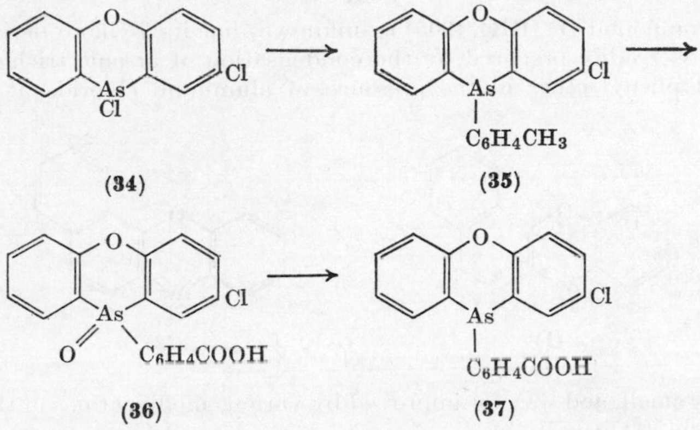

(**34**)   (**35**)

(**36**)   (**37**)

oxide (**36**), which, on reduction, gave the required compound (**37**). This acid was resolved by the fractional crystallization of its strychnine and cinchonine salts, and obtained optically pure, having $[\alpha]_{5461}^{20}$ +187.5° and −187.1° in chloroform. The acids were optically unaffected by boiling in ethanol or chloroform solution for 1 hour, or by boiling in aqueous 0.5 N-sodium hydroxide for 3 hours, and it is clear, therefore, that in this series the optical stability is not weakened by the 10-p-carboxyphenyl group.

Phenoxarsine derivatives, like those of phenarsazine, have attracted wide claims for their action on various pests. Nagasawa et al.[280] claim as agricultural fungicides phenoxarsines having in the 10-position the following substituent groups: Cl, I, CH_3O, C_6H_5, —SCN, CN, CH_3CO_2, Cl_3CCO_2, ClCH=CH, and many others; Shvetsova-Shilovskaya et al.[281] claim as insecticides 10-substituted phenoxarsines having the groups:

$(CH_3)_2NCS_2$, $(C_2H_5)_2$, NCS_2, $ROCS_2$ (where $R = CH_3$, C_2H_5, or C_3H_7), and also, of lower potency, $(CH_3O)_2PS_2$ and similar groups.

Strycker and Dunbar[282] have treated 10-chloro(and bromo)phenoxarsine with the sodium salt of various substituted xanthic acids and thioalkanoic acids, giving the corresponding 10-acyloxy derivatives, stated to be useful as parasiticides, toxicants, or herbicides. Similar treatment of 10-chlorophenoxarsine with sodium trifluoroacetate, has given the 10-(trifluoroacetoxy)phenoxarsine, having the $10\text{-}F_3CCO_2$ substituent; this compound, and the analogues having the $CF_3CF_2CO_2$ and $CF_3CF_2CF_2CO_2$ substituents, are claimed to give control of 'Southern army worm', nematodes, crab grass, and other plants and to be useful as wood-preservatives and in paints and adhesives (Dunbar[283]).

7*H*-Benzo[*c*]phenoxarsine

This compound (1) [RRI 4905] is unknown, but its 7-chloro derivative (2), m.p. 168°, is readily prepared by the condensation of arsenic trichloride with 1-naphthyl phenyl ether in the presence of aluminum chloride at 180–250°.

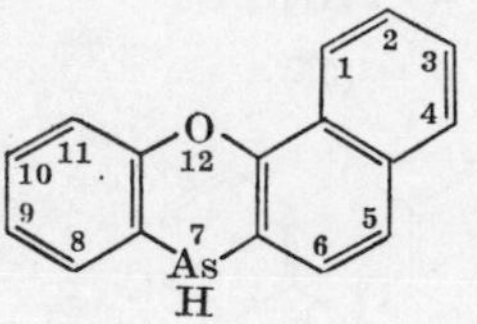

(1) 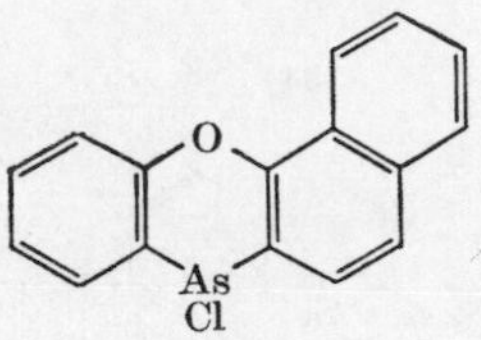   (2)

The yield is small, and was not improved by various modifications of the method (Aeschlimann[269]).

The ring system has also been prepared by condensing phenol with 1-chloronaphthyl-2-arsonic acid in nitrobenzene containing potassium carbonate and copper foil, to give 1-phenoxynaphthyl-2-arsonic acid (3), m.p. 322°. This acid

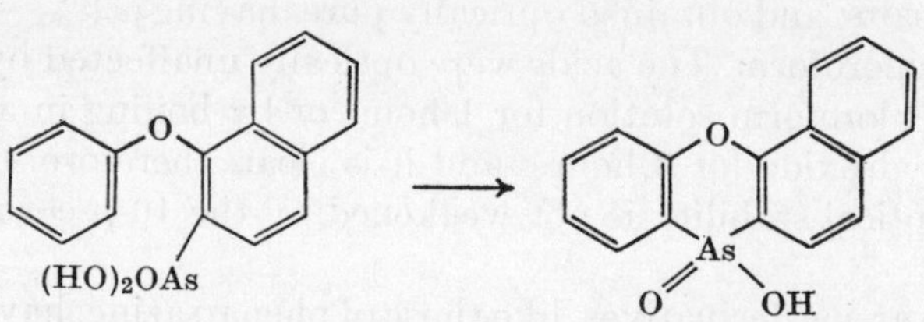

(3)   (4)

in boiling acetic acid readily cyclizes to 7*H*-benzo[*c*]phenoxarsinic acid (4), m.p. 319°. The use of *p*-cresol gives the corresponding 9-methyl-arsinic acid (Bowers and Hamilton[284]).

The system (1) was initially termed α-benzophenoxarsine.[269]

12H-Benzo[a]phenoxarsine

The system (1) [RRI 4904] has been prepared as the previous system, phenol being condensed with 2-chloronaphthyl-2-arsonic acid to give 1-phenoxynaphthyl-1-arsonic acid (2), m.p. 211–295°, which in boiling acetic acid gives 12H-benzo[a]-phenoxarsinic acid[284] (3), m.p. 278–280°.

The system (1) was earlier termed γ-benzophenoxarsine.

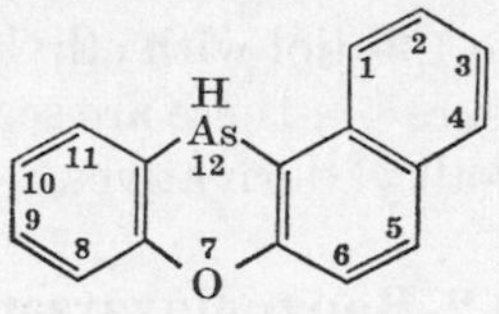

(1)

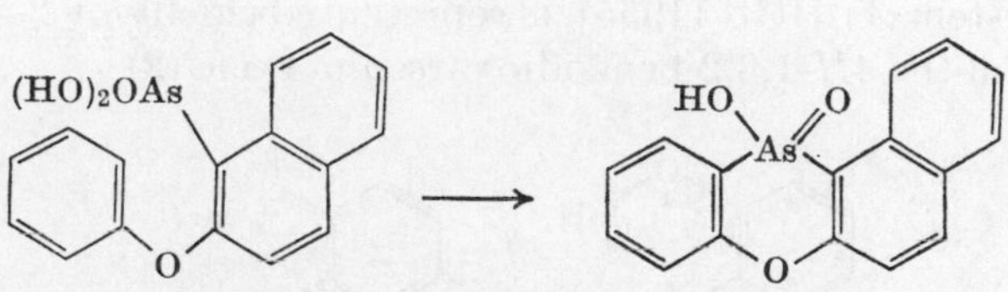

(2)   (3)

Six-membered Ring Systems containing Carbon and One Arsenic and Two Oxygen Atoms

4H-1,3,2-Dioxarsenin and 1,3,2-Dioxarsenanes

The compound 4H-1,3,2-dioxarsenin (1), [RRI 183], is represented solely by its 5,6-dihydro derivative, named 1,3,2-dioxarsenane (2).

The 2-chloro-1,3,2-dioxarsenane (3; R = Cl), a liquid, b.p. 66–67°/10 mm, was prepared by Kamai and Chadaeva[236, 285] by treating propane-1,3-diol in ether–pyridine at 0° with arsenic trichloride, followed by

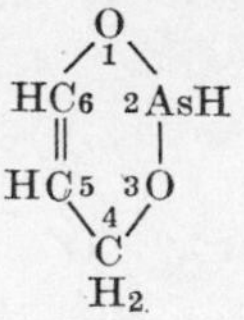

(1) 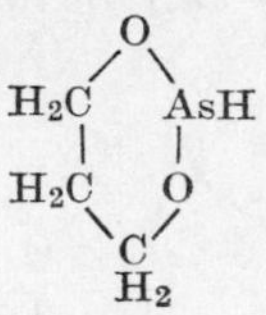   (2) 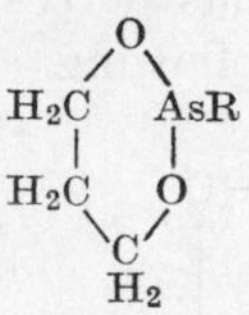   (3)

1 hour's boiling: a by-product is 2,2'-(trimethylenedioxy)bis-(1,3,2-dioxarsenane) (4), b.p. 164–167°/3 mm.

2-Phenyl-1,3,2-dioxarsenane (3; R = C_6H_5), b.p. 135–136°/12 mm, is similarly prepared by using phenylarsonous dichloride, $C_6H_5AsCl_2$, whereas the 2-ethyl compound (3; R = C_2H_5), b.p. 74–75°/10 mm, is

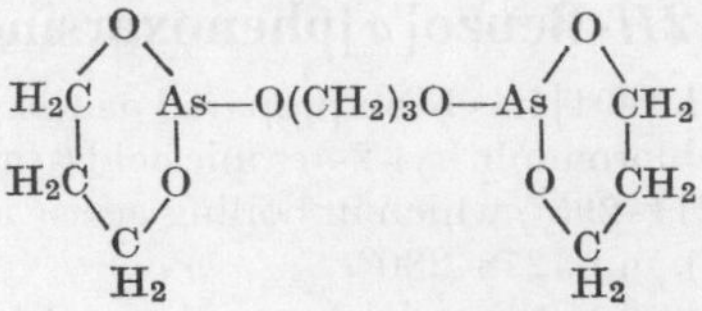

$$\text{(4)}$$

best prepared by heating the 1,3-diol with ethyloxoarsine, C_2H_5AsO, at 90–100° under reduced pressure[288]: there are scattered references in the literature to various other 2-alkyl derivatives.

4H-1,3,2-Benzodioxarsenin

This system (1), [RIS 11955], is represented chiefly by 2-substituted derivatives of the 4H-1,3,2-benzodioxarsenin-4-one (2).

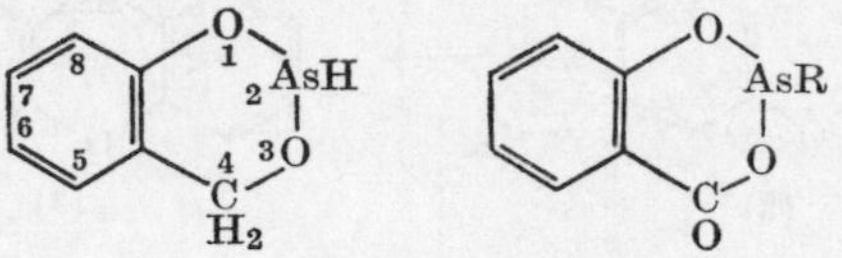

$$\text{(1)}\qquad\qquad\text{(2)}$$

When a solution of salicylic acid in arsenic trichloride is boiled for 2 hours, cooled, and then added slowly to stirred hexane, 2-chloro-4H-1,3,2-benzodioxarsenin-4-one (2; R = Cl) (63%), m.p. 120–122°, is precipitated (Funk and Köhler[248]).

2,6,7-Trioxa-1-arsabicyclo[2.2.2]octane

This compound, [RIS 10019], has the structure (1), depicted alternatively as (1A), and thus contains the 1,3,2-dioxarsenane ring system; analogous derivatives can be prepared by the same general method as the following.

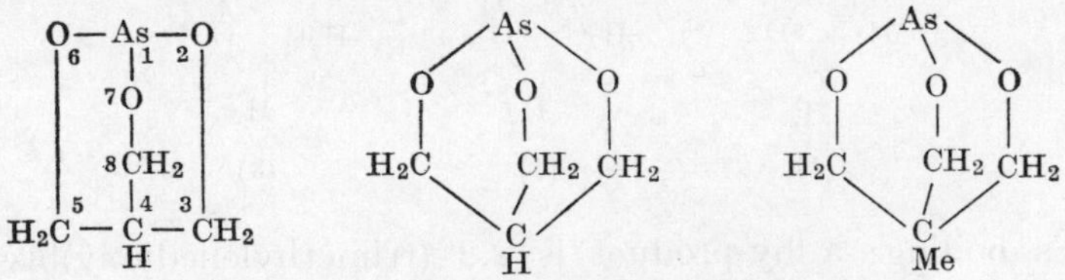

$$\text{(1)}\qquad\qquad\text{(1A)}\qquad\qquad\text{(2)}$$

Verkade and Reynolds[289] have shown that when solutions of (a) arsenic trichloride in tetrahydrofuran and (b) 2-(hydroxymethyl)-2-

methylpropane-1,3-diol, $CH_3C(CH_2OH)_3$, are mixed under nitrogen with vigorous stirring, the reaction mixture furnishes 4-methyl-2,6,7-trioxa-1-arsabicyclo[2.2.2]octane (2), m.p. 41–42° (38%) after sublimation at room temperature at 1 mm to discard polymeric material.

The compound (2) was initially numbered from the bridgehead C atom and therefore termed 1-methyl-4-arsa-3,5,8-trioxabicyclo[2.2.2]-octane.[289]

The compound is readily affected by moisture, and attempts to prepare the 1-oxide and 1-sulfide were unsuccessful; in these respects it is markedly different from its stable phosphorus analogue (p. 312).

Six-membered Ring Systems containing Carbon and Two Arsenic and Two Oxygen Atoms

1,4,2,5-Dioxadiarsenin

The above system (1), [RRI 161], is known only as derivatives of the tetrahydro compound (2), known as 1,4,2,5-dioxadiarsenane. In these

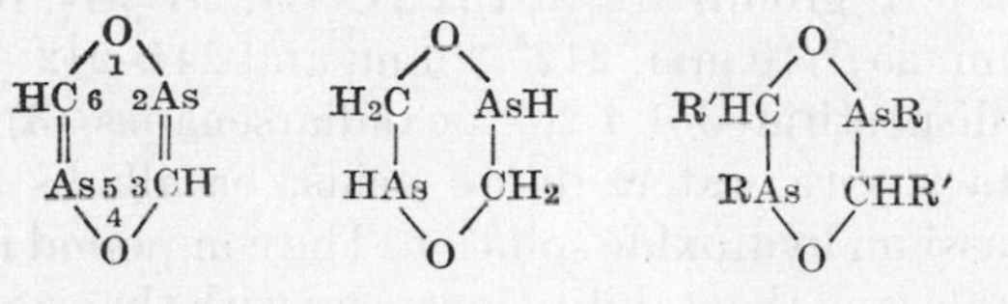

(1)   (2)   (3)

derivatives (3), discovered and studied by Palmer and Adams,[290] the substituents R may be alkyl or aryl, but R′ may be only hydrogen or an alkyl group.

These workers showed that when, for example, phenylarsine, $C_6H_5AsH_2$, maintained under carbon dioxide with ice-cooling, was treated with dry hydrogen chloride and acetaldehyde (2 equivalents), and then after 1–2 days the by-products (mainly ethanol) were distilled off, the residue was the almost pure 3,6-dimethyl-2,5-diphenyl-1,4,2,5-dioxadiarsenane (3; $R = C_6H_5$, $R′ = CH_3$). This residue could be distilled, but at the lowest pressure employed (2 mm) certain other members of the series underwent some decomposition.

When, however, phenylarsine was similarly treated with an excess of an alkyl or aryl aldehyde in the presence of a small quantity of concentrated hydrochloric acid, a ready reaction occurred with the formation of the 'bis-α-hydroxy-tertiary arsine' (4) in 60–95% yield. Pure compounds of type (4) having an alkyl group R′, when treated with dehydrating agents such as acetyl chloride or anhydride, or anhydrous hydrogen chloride, undergo condensation with the formation of the 1,4,2,5-dioxadiarsenane (**3**) as indicated above; Palmer and Adams therefore consider that compounds of type (**4**) are intermediates in the formation of those of type (**3**).

$$
\begin{array}{c}
\text{CHR'OH} \quad \text{HOCHR'} \qquad\qquad \text{O---CHR'} \\
\diagup\qquad\qquad\diagdown \qquad\qquad\qquad\qquad \diagup\qquad\qquad\diagdown \\
\text{RAs} \qquad\qquad + \qquad \text{AsR} \longrightarrow \text{RAs} \qquad\qquad \text{AsR} \; + \; 2\text{R'CH}_2\text{OH} \\
\diagdown\qquad\qquad\diagup \qquad\qquad\qquad\qquad \diagdown\qquad\qquad\diagup \\
\text{CHR'OH} \quad \text{HOCHR'} \qquad\qquad \text{CHR'---O}
\end{array}
$$

$$\qquad\qquad\qquad (4) \qquad\qquad\qquad\qquad\qquad\qquad\qquad\qquad (3)$$

It is noteworthy that compounds of type (**4**), where R' is an aryl group, are unaffected by dry hydrogen chloride and hence 3,6-diaryl derivatives of (**3**) cannot be obtained. Conversely, when phenylarsine is treated with paraformaldehyde and concentrated hydrochloric acid, the product (**4**; $R = C_6H_5$, $R' = H$) cannot be isolated; distillation of the reaction mixture causes some decomposition and the production of 2,5-diphenyl-1,4,2,5-dioxadiarsenane (**3**; $R = C_6H_5$, $R' = H$).

The boiling points of the compounds (**3**), where $R = C_6H_5$ throughout and the two R' groups are H, CH_3, C_2H_5, n-C_3H_7, respectively, are 215–216°/9 mm, 257°/10 mm, 212°/2 mm, and 241–242°/2 mm.

The 2,5-disubstituted 1,4,2,5-dioxadiarsenanes (**3**) are unaffected by long contact with water, dilute acids, or alkalis and even 10% ethanolic potassium hydroxide solution. The compound in which $R' = H$ is the least stable, and the stability increases with the size of the R' group. This is shown by the fact that the compound (**3**; $R = C_6H_5$, $R = H$) oxidizes very rapidly in air with the production of C_6H_5AsO, but the vigor of the oxidation decreases as the group R' increases in size.

Iodine in ethereal solution breaks these compounds down to phenylarsonous diiodide, $C_6H_5AsI_2$, and the aldehyde; phosphorus pentachloride similarly gives $C_6H_5AsCl_2$.

It is recorded that the compound (**3**; $R = C_6H_5$, $R' = CH_3$), *i.e.*, $C_{16}H_{18}As_2O_2$, reacts with hexachloroplatinic acid to give a colorless compound, m.p. 130–131°, of composition $C_{16}H_{18}As_2O_2,H_2PtCl_6$, and with cupric chloride to give a colorless compound, m.p. 150–152°, of composition $C_{16}H_{18}As_2O_2,CuCl_2$. The fact that both these compounds are colorless indicates that the metals have been reduced to the Pt(II) and the Cu(I) condition, respectively, probably with coordination to the diarsine portion of the molecule. It is difficult, however, to reconcile these processes with the above compositions. Since the first compound was identified solely by a platinum analysis, and the second by an arsenic analysis, the behavior of compounds of type (**3**) with various metal halides might repay a much fuller study.

Five-membered Ring Systems containing Carbon and Two Arsenic and One Sulfur Atom

[1,2,3]Benzothiadiarsolo[3,2-*b*][1,2,3]benzothiadiarsole

Our knowledge of the above compound (**1**), [RRI 4116], is based on the work of Barber,[291, 292] who converted *o*-iodophenylarsonic acid

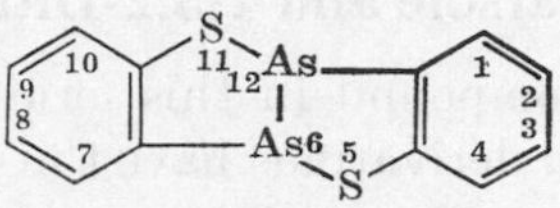

(1)

(**2**; R = I) by the action of sodium sulfite and a trace of cupric acetate into *o*-sulfonophenylarsonic acid (**2**; R = SO$_3$H); this acid, treated with a mixture of phosphorus trichloride and pentachloride with subsequent

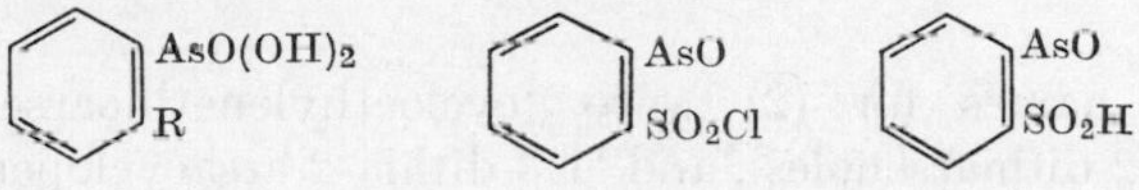

partial hydrolysis, gave oxo-*o*-(chlorosulphonyl)phenylarsine (**3**), which on reduction with sodium sulfite yielded oxo-*o*-sulphinophenylarsine (**4**).

The acid (**4**), when reduced in hot acetic acid by hypophosphorous acid and potassium iodide, gave pale yellow crystals, m.p. 177–178°, the C, H, As, and S content and the molecular weight of which agreed with those required for the compound (**1**).

o-Thiocyanatophenylarsonic acid (**2**; R = SCN), when similarly reduced with hypophosphorous acid or with sodium hyposulfite, gave the same compound.

Barber points out that the composition and molecular weight of the crystals would apply also to an isomeric compound of structure (**5**),

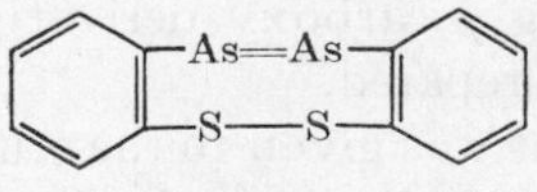

(5)

but that the general properties of his product differ markedly from those reasonably to be expected of a compound of structure (**5**).

It is highly probable that the structure (**1**) is correct. An *X*-ray crystal analysis, to establish the presence or absence of a center of symmetry, would be decisive.

Five-membered Ring Systems containing Carbon and One Arsenic and Two Sulfur Atoms

1,3,2-Dithiarsole and 1,3,2-Dithiarsolanes

The fundamental compound in this series is 1,3,2-dithiarsole (**1**), [RRI 64], but the known derivatives have the 4,5-dihydro ring system, named 1,3,2-dithiarsolane (**2**), and almost invariably carrying a 2-substituent as in (**3**).

$$
\underset{(1)}{\begin{array}{c} HC_5 \overset{S}{\underset{1}{\diagdown}} \overset{2}{} AsH \\ \| | \\ HC_4 \underline{}_{3} S \end{array}}
\qquad
\underset{(2)}{\begin{array}{c} H_2C \overset{S}{\diagdown} AsH \\ | | \\ H_2C \underline{} S \end{array}}
\qquad
\underset{(3)}{\begin{array}{c} H_2C \overset{S}{\diagdown} AsR \\ | | \\ H_2C \underline{} S \end{array}}
$$

Earlier names for (**2**) were 'cycloethylenethioarsenites', '4,5-dihydro-1,3,2-dithiarsenoles', and '1,3-dithia-2-arsacyclopentanes'.

The ring system (**2**) was first synthesized by Cohen, King, and Strangeways,[293] who showed that a mixture of 1,2-ethanedithiol (**4**) and *p*-(acetylamino)phenyloxoarsine (**5**) in ethanol, when boiled for 10 minutes and cooled, deposited the colorless 2-(*p*-acetylaminophenyl)-1,3,2-dithiarsolane (**6**; $R = CH_3CONH$), m.p. 155°. It was soluble in

$$
\underset{(4)}{\begin{array}{c} CH_2SH \\ | \\ CH_2SH \end{array}}
+ \underset{(5)}{OAs\!\!\bigcirc\!\!NHCOCH_3}
\longrightarrow
\underset{(6)}{\begin{array}{c} H_2C - S \\ | \diagdown \\ As - \bigcirc - R \\ | \diagup \\ H_2C - S \end{array}}
$$

most organic solvents and dissolved unchanged in glacial acetic acid; it was, however, readily decomposed by aqueous alkalis and the mineral acids. The corresponding *p*-carboxy derivative (**6**; $R = CO_2H$), m.p. 223–224°, was similarly prepared.

Considerable impetus was given to the study of compounds having the ring structure (**2**) by the discovery by Peters and his co-workers in the early stages of World War II that certain 1,2-dithiols, but especially 2,3-dimercapto-1-propanol (**7**) would condense readily with dichloroarsines and thus provide an effective treatment for contamination in particular with primary Lewisite (2-chlorovinylarsonous dichloride) (8). Much work on the chemical and physiological aspects of this subject was done by British and American workers both during and after the war. Peters, Stocken, and Thompson,[294] giving a brief account (1945) of the basis of this subject, pointed out that 2,3-dimercapto-1-propanol (7)—termed BAL (British Anti-Lewisite) by the American workers— reacted with primary Lewisite (8) to give 2-(2'-chlorovinyl)-4-(hydroxy-methyl)-1,3,2-dithiarsolane (9). They stated that various 1,2- and

$$CH_2SH$$
$$|$$
$$CHSH$$
$$|$$
$$CH_2OH$$

$$\begin{matrix} Cl \\ Cl \end{matrix} AsCH=CHCl$$

$$\begin{matrix} H_2C—S \\ | \\ HC—S \\ | \\ CH_2OH \end{matrix} AsCH=CHCl$$

(7)   (8)   (9)

1,3-dithiols were condensed with dichlorarsines to give compounds of type (9) and the higher six-membered homologue (p. 583), respectively, but that BAL in particular 'proved highly effective in stopping the toxic action of Lewisite upon the pyruvate oxidase system in brain'.

Peters and Stocken[295] later showed that the drug named Matharside or Mapharsen—(3-amino-4-hydroxyphenyl)oxoarsine hydrochloride (10) —condensed with an equivalent of BAL to give 2-(3'-amino-4'-hydroxy-phenyl)-4-(hydroxymethyl)-1,3,2-dithiarsolane (11), m.p. 172–173°.

(7) + OAs⟨benzene ring⟩OH / NH₂,HCl ⟶ H₂C—S / H₂C—S As⟨benzene ring⟩OH / NH₂,HCl with CH₂OH

(10)   (11)

The two methods of synthesizing the ring system (2), namely by condensing a 1,2-dithiol with an arsonous dichloride ($RAsCl_2$) or an oxoarsine (RAsO), are thus illustrated; later workers showed that an arsonic acid, $RAs(O)(OH)_2$, could also be used if the dithiol (in excess) or ammonium mercaptoacetate was present to reduce the arsonic acid or its intermediate condensation product (p. 577).

Experimental details of the preparation of various 1,2-dithiols and of their condensations with dichloroarsines in chloroform–pyridine at 0–5° have been given by Stocken,[296] who records the derivatives of (12) shown in Table 8.

Table 8. 2-Substituted and 2,4-disubstituted 1,3,2-dithi-
arsolanes (12)

R	R'	B.p. (°/mm)	M.p. (°)
ClCH=CH	H	120/0.4	35
ClCH=CH	CH$_3$	127/0.6	—
CH$_3$	CH$_2$OH	158/1.5	—
C.CH=CH	CH$_2$OH	165/2	—
C$_6$H$_5$	CH$_2$OH	—	97–98
ClCH=CH	CH$_3$OCH$_2$	145/0.6	—
p-ClC$_6$H$_4$	CH$_3$OCH$_2$	—	178–180

A brief account of the chemistry, biochemical action, toxicology, pharmacology, and medical applications of 1,2-dithiols such as BAL

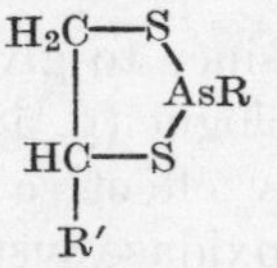

(12) 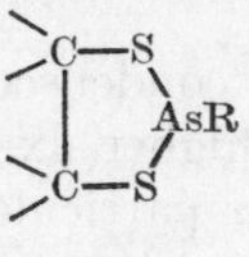   (13)

has been given by Waters and Stock.[297] The authors state that the available evidence (1945) indicates 'that dithiols added to protein previously treated with arsenic probably compete successfully for the arsenic by formation of compounds of type (13)'. Furthermore this evidence 'focused attention on the theory that the toxicity of trivalent arsenicals is largely due to their binding of essential thiol groups in enzyme proteins'. Essentially the same conclusions were given by Stocken.[296]

A much fuller account of the physiological action and general therapy of 1,2-dithiols was given later (1949) in 'Reactions of British Antilewisite with Arsenic and Other Metals in Living Systems' (Stocken and Thompson[298]), in which the use of BAL in the treatment of other metal intoxications, such as those due to Hg, Au, Pb, Cd, and Sb, is also discussed. A later account (1964) of the inhibition of the pyruvate oxidase system by certain arsenical compounds, and its reversal by dithiol derivatives, has been given by Dixon and Webb.[299]

Friedheim and Vogel[300] record that the hydrochloride (11) and its base are stable compounds, and that solutions of the hydrochloride in propylene glycol can be sterilized by one hour's heating at 100°. The base was found to possess spirochetocidal action when tested against rabbit syphilis, and the hydrochloride to have trypanocidal action when tested against *T. equiperdum* in mice.

Friedheim and Berman[301] have shown that the trichomonacidal action of Mapharsen (**10**) is significantly increased when it is condensed with BAL to give the compound (**11**). On the other hand, condensation of 4-(p-arsenophenyl)butyric acid, $OAsC_6H_4(CH_2)_3CO_2H$, with BAL to give 2-(p-3-carboxypropylphenyl)-4-hydroxymethyl-1,3,2-dithiarsolane was not accompanied by the development of any significant trichomonacidal action. These workers suggest that this is probably because 'the strongly polar anionic character of the butyric acid residue precludes the degree of lipid solubility necessary for a trichomonocidal effect'.

Other substituted 1,2-ethanedithiols have been employed in the standard 1,3,2-dithiarsolane ring closure. Friedheim[302, 303] in patent specifications also cites the use of 1,2-propanedithiol (**14**; R = CH_3) and 3-ethoxy-1,2-propanedithiol (**14**; R = $CH_2OC_2H_5$) to give 4-methyl- and 4-(ethoxymethyl)-1,3,2-dithiarsolane, respectively, and Petrun'kin[304] has used sodium 2,3-dimercaptopropanesulfonate ('Unithiol') (**14**;

$$CH_2SH$$
$$|$$
$$CHSH$$
$$|$$
$$R$$

$$H_2C—S$$
$$\quad\quad \backslash$$
$$\quad\quad\quad AsR$$
$$\quad\quad /$$
$$H_2C—S$$
$$|$$
$$CH_2SO_3Na$$

$$HOOC—CH—SH$$
$$|$$
$$HOOC—CH—SH$$

$$HOOC—CH—S$$
$$\quad\quad\quad\quad \backslash$$
$$\quad\quad\quad\quad\quad As—⟨\ ⟩—R$$
$$\quad\quad\quad\quad /$$
$$HOOC—CH—S$$
$$\quad\quad\quad\quad\quad\quad R'$$

(**14**)   (**15**)   (**16**)   (**17**)

R = CH_2SO_3Na) to give the sodium salts of various 2-substituted (1,3,2-dithiarsolan-4-yl)methanesulfonates (**15**). Rhone-Poulenc S.A.[305] have used 2,3-dimercaptosuccinic acid (**16**) to prepare various substituted 2-phenyl-1,3,2-dithiarsolane-4,5-dicarboxylates (**17**), in which R is a carbamoyl (NH_2CO), sulfamoyl (NH_2SO_2), or ureido (NH_2CONH) group, and R' is H, Cl, OH, or NH_2, or an alkyl, hydroxyalkyl, or alkoxy group; it is claimed that these products can be used as anthelmintics, filaricides, and trypanocides. Friedheim[306] has also used the acid (**16**) to make various 2-substituted derivatives of type (**17**).

In another patent specification, Friedheim[307] recommends the use of ammonium mercaptoacetate, $HSCH_2CO_2NH_4$ for reducing arsonic acids before their condensation with dithiols, instead of using an excess of the dithiol itself. It should be noted that in Friedheim's work on the synthesis of compounds of type (**3**) and their larger-ring analogues, and the corresponding antimony compounds (p. 628), the members having the group p-(4',6'-diamino-1',3',5'-triazin-2'-ylamino)phenyl (**18**) joined to the arsenic (or antimony) are usually recorded.

Nagasawa *et al.*[308, 309] in patent specifications describe various simple compounds of type (**12**), 'all effective bactericides', and others such as (**19**), 'useful as agricultural germicides'.

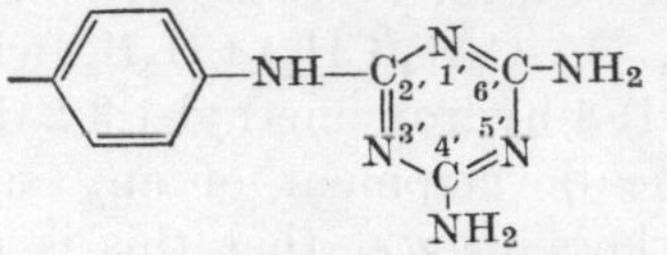

$$\text{(18)}$$

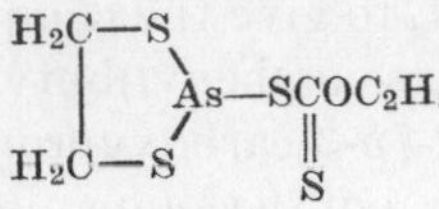

$$\text{(19)}$$

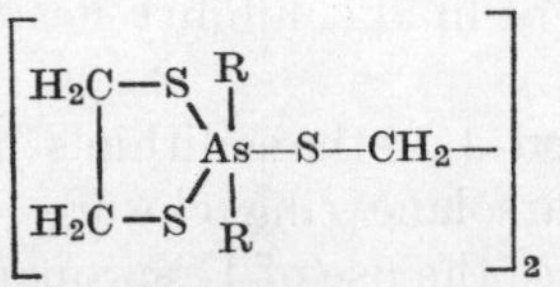

$$\text{(20)}$$

Epstein and Willsey[310] have used methyl- and ethyl-arsinic acid, $R_2AsO(OH)$, in the ring closure and have shown that a benzene solution of 1,2-ethanedithiol and the arsinic acid (3 and 2 equivalents, respectively) on azeotropic distillation give the corresponding 2,2'-ethylene-dithiobis-(2,2-dialkyl-1,3,2-dithiarsolane) (**20**; $R = CH_3$ or C_2H_5), the products being 'effective as post-emergent herbicides'.

Some simple and very interesting derivatives of the parent compound (**2**) have been obtained by Rueggeberg, Ginsburg, and Cook.[311] When 1,2-ethanedithiol was added to a solution of arsenic trichloride in carbon tetrachloride, hydrogen chloride was instantly liberated, and the endothermic reaction caused a marked fall in the temperature of the solution, the crystalline 2-chloro-1,3,2-dithiarsolane (**21**), m.p. 37.5–38°, separating. In an attempt to oxidize this compound to the corresponding arsinic acid, air was drawn through an aqueous-ethanolic solution of (**21**) at 80° for 2 hours, but the product was 2,2'-ethylene-dithiobis-(1,3,2-dithiarsolane) (**22**). This compound is probably formed

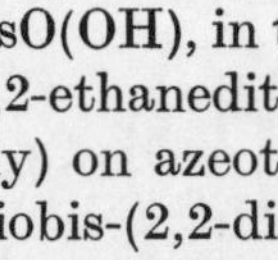

$$\text{(21)} \qquad \text{(22)} \qquad \text{(23)}$$

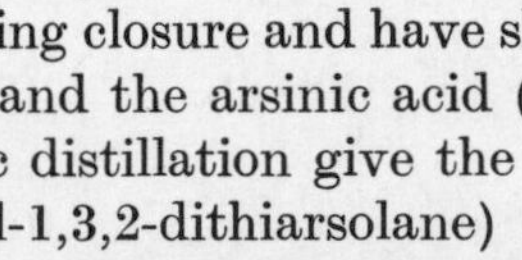

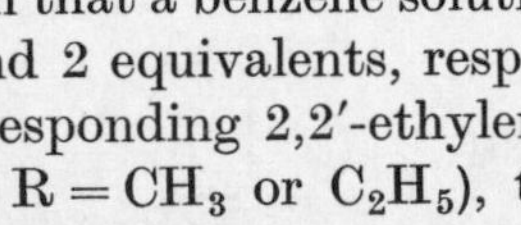

because some of the chloro compound (**21**) undergoes hydrolysis with the liberation of ethanedithiol, which then readily condenses with two equivalents of the unhydrolyzed compound (**21**) to form the dicyclic compound (**22**).

The compound (**21**) is obviously of great synthetic value by virtue of its reactive chlorine atom, which for instance could readily be replaced by alkyl or aryl groups by application of the appropriate Grignard or lithium derivative. When a benzene solution of (**21**) and of mercapto-acetic acid, $HS.CH_2CO_2H$ (one equivalent) is boiled under reflux, the crystalline 1,3,2-dithiarsolan-2-ylacetic acid (**23**), m.p. 77–80°, is readily formed.[311] For the nmr spectrum of (**21**) see page 544.

1,3,2-Benzodithiarsole

This compound, [RRI 1039], of structure and numbering (**1**), is encountered—as one would expect—as the 2-substituted derivatives (**2**). These derivatives were earlier named 1,3-dithia-2-arsaindane.

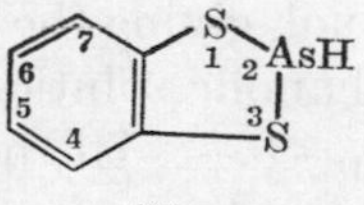

(**1**) 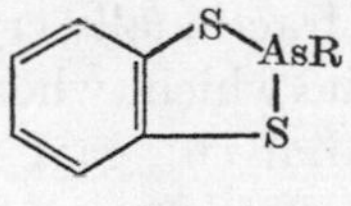   (**2**)

Compounds of type (**2**) were first mentioned by Friedheim,[302, 303, 307] who recorded their preparation by the condensation of o-benzenedithiol with arylarsonous dichlorides ($RAsCl_2$), oxoarsines ($RAsO$), or arsonic acids [$RAsO(OH)_2$]. For example, the use of 3-amino-4-hydroxyphenyl-arsonic acid, preferably reduced with ammonium mercaptoacetate, and o-benzenedithiol gave 2-amino-4-(1',3',2'-benzodithiarsol-2'-yl)phenol (**3**).

The stereochemistry of the 1,3,2-benzodithiarsole ring system has been studied by Campbell[312] in order to investigate the stability of the

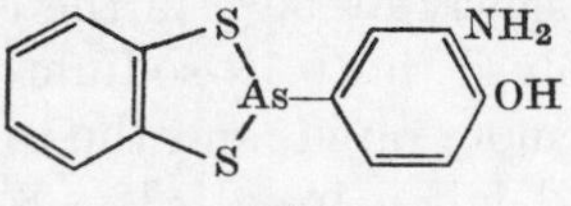

(**3**)

pyramidal disposition of the bonds of the tervalent arsenic atom: a compound of type (**2**) is particularly suitable, as the bicyclic system is undoubtedly planar.

For this purpose 3,4-toluenedithiol (**4**) in warm methanol containing sodium acetate was condensed with p-carboxyphenylarsonous dichlor-ide (**5**), giving p-(5'-methyl-1',3',2'-benzodithiarsol-2'-yl)benzoic acid

(6), yellow crystals, m.p. 200–202°. This compound is dissymmetric, for the pyramidal configuration of the arsenic atom will direct the *p*-carboxyphenyl group above or below the plane of the remainder of the molecule. Optical resolution of the compound (6) would therefore give a

(4)   (5)   (6)

product, the optical stability of which would depend entirely on the pyramidal stability of the bonds of the arsenic atom.

The acid (6) was soluble in most organic solvents and dissolved in aqueous sodium hydrogen carbonate, from which on acidification it was reprecipitated unchanged.

The salt formed with (+)-α-methylbenzylamine (= 1-phenylethyl-amine) was fractionally crystallized from ethanol, giving the (+)-amine–(−)-acid salt, which, when decomposed in ethanolic solution by dilute sulfuric acid at <0°, gave the (−)-acid (6), $[\alpha]_D$ −8.7°. The quinine salt, recrystallized from ethanol–chloroform, gave the quinine–(+)-acid, which when similarly treated yielded the (+)-acid, $[\alpha]_D$ +8.9° (both rotations in chloroform solution at 20°).

The rotation was unaffected when these acids were recrystallized from boiling ethanol or heated in ethanolic solution at 110° for 2 hours. The optical stability of the acid (6) is therefore comparable with that of 2-amino-9-phenyl-9-arsafluorene (p. 380) under similar conditions and shows the considerable stability of the pyramidal configuration of the tervalent arsenic atom, *i.e.*, there is no oscillation of the *p*-carboxyphenyl group between the two corresponding positions above and below the plane of the bicyclic system under these conditions.

The rotation of the (−)-acid, when dissolved in pyridine at room temperature, decreased slowly by 50% in the course of 6 weeks. When the (−)-acid was dissolved in 0.1N-sodium hydroxide the partial racemization was much more rapid, and the rotation of the acid, when recovered as before, had fallen from $[\alpha]_D$ −8.7° to −3.2°. This rapid racemization was probably caused by the fission of the As—S bonds by

$$RAs(SR')_2 + 2H_2O \underset{H^+}{\overset{OH^-}{\rightleftharpoons}} RAs(OH)_2 + 2R'SH$$
(7)

aqueous alkali and their reunion by acids, a process which in the dithio-arsenites (7) has been noted by Klement and May[313] and studied by Cohen, King, and Strangeways.[293]

Six-membered Ring Systems containing Carbon and One Arsenic and One Sulfur Atom

4H-1,4-Thiarsenin and 1,4-Thiarsenane

The parent compound in this series, 4H-1,4-thiarsenin (1), [RRI 231], is unknown; the 2,3,5,6-tetrahydro derivative (2) is named 1,4-thiarsenane.

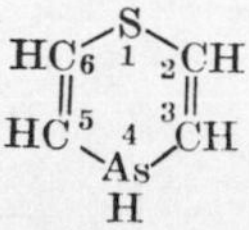

(1) 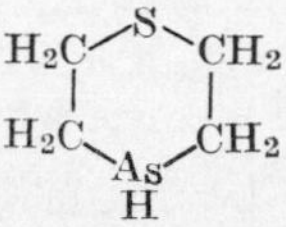   (2) 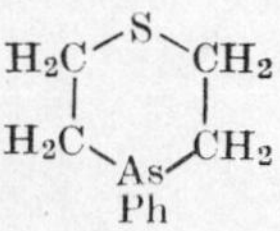   (3)

Job, Reich, and Vergnaud[314] have shown that a benzene solution of phenylarsinebis(magnesium bromide), $C_6H_5As(MgBr)_2$ (cf. pp. 229, 415), when boiled with 2,2' dichlorodiethyl sulfide, gives 4-phenyl-1,4-thiarsenane (3), m.p. 38°, b.p. 134°/14 mm.

They record a monomethiodide, m.p. 226° (almost certainly an arsonium and not a sulfonium salt), and also a white mercuric chloride adduct, m.p. 181°, and a yellow mercuric iodide adduct, m.p. 153°.

The compound (3) has apparently not been further studied, possibly because the above synthesis entails the use of 'mustard gas'.

Phenothiarsine

Phenothiarsine [RRI 3260] has the structure, numbering and presentation shown in (1). Comparatively little is known about this

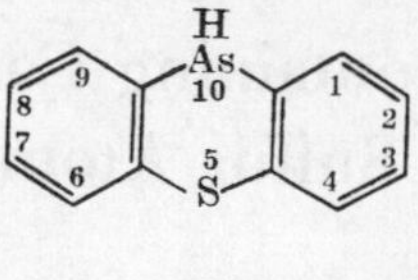

(1) 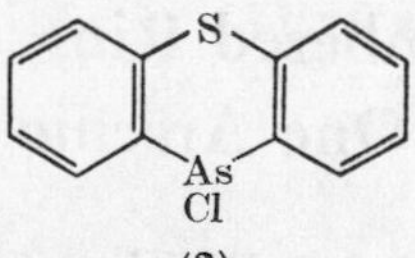   (2)

system, but since most reactions will clearly concern chiefly the tertiary arsine group, the 10-substituted derivatives may most conveniently be indicated by the inverted formula (2).

10-Chlorophenothiarsine (**2**) has been synthesized by Roberts and Turner[270] by the following method, which is strictly parallel to that used by Turner and Sheppard[266] for synthesizing the phenoxarsine analogue. Potassium phenyl sulfide was condensed with *o*-chloronitrobenzene by heating with copper bronze at 170–185° for 2 hours, giving 2-nitrodiphenyl sulfide (**3**; R′ = NO$_2$). This was reduced to the 2-amino

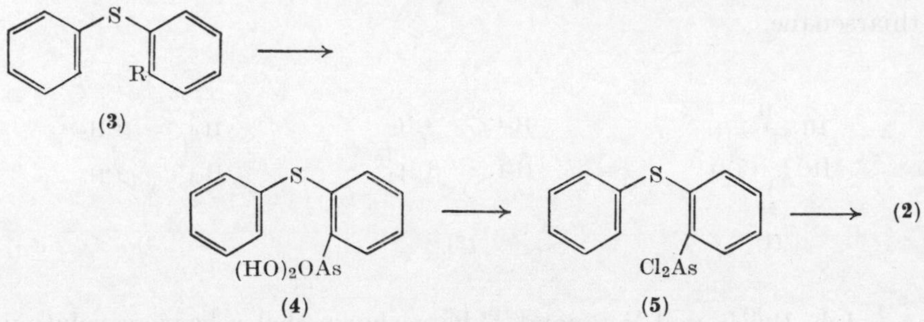

(**3**)

(HO)$_2$OAs

(**4**)

$\longrightarrow$

Cl$_2$As

(**5**)

$\longrightarrow$ (**2**)

derivative (**3**; R′ = NH$_2$), which was then diazotized and converted by the Bart reaction into 2-(phenylthio)phenylarsonic acid (**4**), m.p. 192°–194°. All attempts to cyclize this with sulfuric acid to the arsinic acid failed. It was therefore reduced to the 2-(phenylthio)phenylarsonous dichloride (**5**); this dark oily compound could not be purified, so it was hydrolyzed to the arsine oxide, which on recrystallization gave white crystals, m.p. 187–189°, and this pure oxide was converted back into the dichlorarsine (**5**). This cyclized at 200° in a stream of carbon dioxide, and the 10-chlorophenothiarsine (**2**) was obtained as sulfuryellow crystals, m.p. 129–130°. It is noteworthy that the maximum yield in this cyclization was 20%, whereas the yield of the 10-chlorophenoxarsine from the corresponding 2-phenoxyphenylarsonous dichloride under the same conditions was 57% (p. 558).

Six-membered Ring Systems containing Carbon and One Arsenic and Two Sulfur Atoms

4*H*-1,3,2-Dithiarsenin and the 1,3,2-Dithiarsenanes

4*H*-1,3,2-Dithiarsenin (**1**), [RRI 184], is the parent compound of this series, but the ring system is encountered, as one would expect, only as 2-substituted 1,3,2-dithiarsenanes (**2**), which is the name given to the 5,6-dihydro derivatives.

Stocken[296] has shown the 1,3-propanedithiol condenses with 2-chlorovinylarsonous dichloride, $ClCH{=}CHAsCl_2$, in chloroform–pyridine at 0–5° to give 2-(2'-chlorovinyl)-1,3,2-dithiarsenane (**2**; R = $ClCH{=}CH$), b.p. 150°/0.4 mm, and that 2-hydroxy-1,3-propanedithiol, $HSCH_2CH(OH)CH_2SH$, condenses similarly with this arsonous

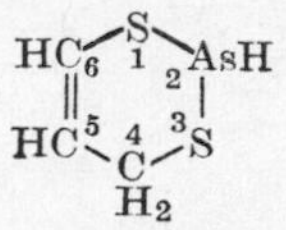

(1) 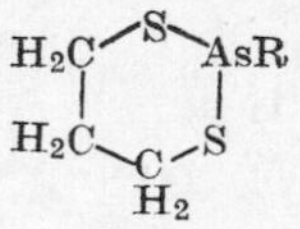   (2) 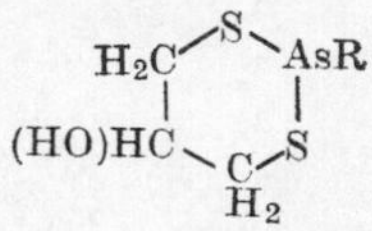   (3)

dichloride to give 2-(2'-chlorovinyl)-1,3,2-dithiarsenan-5-ol (**3**; R = $ClCH{=}CH\cdot$), b.p. 166°/0.8 mm, and with phenylarsonous dichloride, $C_6H_5AsCl_2$, to give 2-phenyl-1,3,2-dithiarsenan-5-ol (**3**; R = C_6H_5), b.p. 190°/1 mm.

Similar derivatives have been recorded by Friedheim.[303, 307]

Seven-membered Ring System containing Carbon and One Arsenic and Two Sulfur Atoms

2,4,3-Benzodithiarsepin

This compound (**1**), [RRI 1796], is unknown, and the ring system is found only as 3-substituted 1,5-dihydro-2,4,3-benzodithiarsepins (**2**).

The preparation of compounds of type (**2**) has been noted in patent specifications by Friedheim[303, 307] and follows closely on his work on 1,3,2-benzodithiarsoles (p. 579).

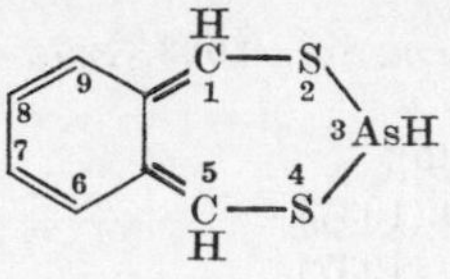

(1) 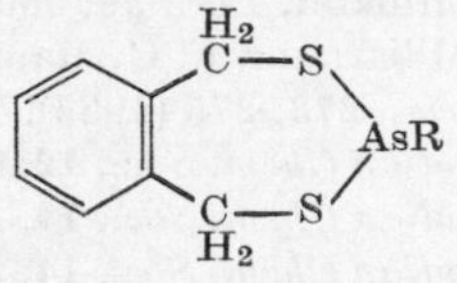   (2) 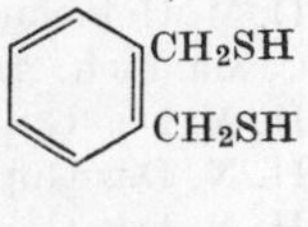   (3)

o-Xylene-αα'-dithiol (**3**) will condense with arylarsonous dichlorides, aryloxoarsines, or arylarsonic acids (the acids in the presence of ammonium mercaptoacetate or an excess of the dithiol for reductive purposes) to give the 3-aryl-1,5-dihydro-2,4,3-benzodithiarsepin. A

particular example quoted by Friedheim is the condensation with 3
amino-4-hydroxyphenylarsonic acid to give 2-amino-4-(1′,5′-dihydro-
2′,4′,3′-benzodithiarsepin-3′-yl)phenol (4).

(4)

References (Part II, Arsenic)

[1] F. C. Leavitt, T. A. Manuel, and F. Johnson, *J. Amer. Chem. Soc.*, **81**, 3163 (1959).
[2] F. C. Leavitt, T. A. Manuel, F. Johnson, L. U. Matternas, and D. S. Lehman, *J. Amer. Chem. Soc.*, **82**, 5099 (1960).
[3] F. C. Leavitt and F. Johnson (Dow Chemical Co.), U.S. Pat. 3,116,307 (Dec. 31, 1963); *Chem. Abstr.*, **60**, 6872 (1964).
[4] E. H. Braye, W. Hübel, and I. Caplier, *J. Amer. Chem. Soc.*, **83**, 4406 (1961).
[5] K. W. Hübel, E. H. Braye, and I. H. Caplier (to Union Carbide Corporation), U.S. Pat. 3,151,140 (Sept. 29, 1964); *Chem. Abstr.*, **61**, 16097 (1964).
[6] J. B. Hendrickson, R. E. Spenger, and J. J. Sims, *Tetrahedron Lett.*, **1961**, 477.
[7] J. B. Hendrickson, R. E. Spenger, and J. J. Sims, *Tetrahedron*, **19**, 707 (1963).
[8] D. A. Brown, *J. Chem. Soc.*, **1962**, 929.
[9] W. B. McCormack, personal communication.
[10] G. Grüttner and E. Krause, *Ber.*, **49**, 437 (1916).
[11] J. J. Monagle, *J. Org. Chem.*, **27**, 3851 (1962).
[12] R. C. Evans, F. G. Mann, H. S. Peiser, and D. Purdie, *J. Chem. Soc.*, **1940**, 1209.
[13] W. Steinkopf, I. Schubart, and J. Roch, *Ber.*, **65**, 409 (1932).
[14] H. N. Das-Gupta, *J. Indian Chem. Soc.*, **15**, 498 (1938).
[15] E. Wiberg and K. Mödritzer, *Z. Naturforsch.*, **12b**, 135 (1957).
[15a] H. Schmidt and F. Hoffmann, *Ber.*, **59**, 560 (1926).
[16] D. M. Heinekey, I. T. Millar, and F. G. Mann, *J. Chem. Soc.*, **1963**, 725.
[17] C. Mannich, *Arch. Pharm.*, **273**, 275 (1935).
[18] H. N. Das-Gupta, *J. Indian Chem. Soc.*, **12**, 627 (1935).
[19] H. N. Das-Gupta, *J. Indian Chem. Soc.*, **14**, 231, 349 (1937).
[20] H. N. Das-Gupta, *J. Indian Chem. Soc.*, **14**, 397, 400 (1937).
[21] E. E. Turner and F. W. Bury, *J. Chem. Soc.*, **123**, 2489 (1923).
[22] H. N. Das-Gupta, *J. Indian Chem. Soc.*, **15**, 495 (1938).
[23] Emrys R. H. Jones and F. G. Mann, *J. Chem. Soc.*, **1958**, 1719.
[24] D. R. Lyon and F. G. Mann, *J. Chem. Soc.*, **1945**, 30.
[25] D. R. Lyon, F. G. Mann, and G. H. Cookson, *J. Chem. Soc.*, **1947**, 662.
[26] M. H. Beeby, G. H. Cookson, and F. G. Mann, *J. Chem. Soc.*, **1950**, 1917.
[27] F. G. Mann and H. R. Watson, *J. Chem. Soc.*, **1957**, 3945.
[27a] G. Ferguson and E. W. Macaulay, *Chem. Commun.*, **1968**, 1288.

27b K. A. Jensen, *Z. anorg. Chem.*, **250**, 268 (1943).

27c G. S. Harris and F. Inglis, *J. Chem. Soc.*, A, **1967**, 497.

27d F. G. Mann and E. J. Chaplin, *J. Chem. Soc.*, **1937**, 527.

28 F. G. Holliman and F. G. Mann, *J. Chem. Soc.*, **1945**, 45.

29 J. A. Aeschlimann, N. D. Lees, N. P. McCleland, and G. N. Nicklin, *J. Chem. Soc.*, **127**, 66 (1925).

30 G. H. Cookson and F. G. Mann, *J. Chem. Soc.*, **1949**, 2895.

31 F. F. Blicke, O. J. Weinkauff, and G. W. Hargreaves, *J. Amer. Chem. Soc.*, **52**, 780 (1930).

32 R. J. Garascia and I. V. Mattei, *J. Amer. Chem. Soc.*, **75**, 4589 (1953).

33 H. Heaney, D. M. Heinekey, F. G. Mann, and I. T. Millar, *J. Chem. Soc.*, **1958**, 3838.

34 D. M. Heinekey and I. T. Millar, in an Appendix to D. Sartain and M. R. Truter, *J. Chem. Soc.*, **1963**, 4414.

35 F. F. Blicke and E. L. Cataline, *J. Amer. Chem. Soc.*, **60**, 423 (1938).

36 B. N. Feitelson and V. Petrow, *J. Chem. Soc.*, **1951**, 2279.

37 R. J. Garascia, A. A. Carr, and T. R. Hauser, *J. Org. Chem.*, **21**, 252 (1956).

38 E. Urbschat (to Farbenfabriken Baeyer), U.S. Pat. 2,644,005 (June 30, 1953); *Chem. Abstr.*, **48**, 5879 (1954).

39 G. Wittig and E. Benz, *Chem. Ber.*, **91**, 873 (1958).

40 I. G. M. Campbell and R. C. Poller, *J. Chem. Soc.*, **1956**, 1195.

41 D. Sartain and M. R. Truter, *J. Chem. Soc.*, **1963**, 4414.

42 D. M. Burns and J. Iball, *Proc. Roy. Soc.*, A, **227**, 200 (1955).

43 L. Horner and H. Fuchs, *Tetrahedron Lett.*, **1962**, 203; **1963**, 1573.

44 G. Wittig and A. Maercker, *Chem. Ber.*, **97**, 747 (1964).

45 G. Wittig and D. Hellwinkel, *Angew. Chem.* **74**, 76, 782 (1962); *Int. Ed. Engl.*, **1**, 53, 598 (1962).

46 G. Wittig and E. Kochendörfer, *Chem. Ber.*, **97**, 741 (1964).

47 F. G. Mann and E. J. Chaplin, *J. Chem. Soc.*, **1937**, 527.

48 G. Wittig and D. Hellwinkel, *Chem. Ber.*, **97**, 769 (1964).

48a D. Hellwinkel, *Chem. Ber.*, **98**, 576 (1965); *Angew. Chem.*, **77**, 378 (1965); *Int. Ed. Engl.*, **4**, 356 (1965).

48b D. Hellwinkel and G. Kilthau, *Annalen*, **705**, 66 (1967).

49 E. V. Zappi, *Bull. Soc. chim. France*, **19**, 151, 290 (1916).

50 G. Grüttner and M. Wiernik, *Ber.*, **48**, 1473 (1915).

51 W. Steinkopf, H. Donat, and P. Jaeger, *Ber.*, **55**, 2579 (1922).

52 I. Gorski, W. Schpanski, and L. Muljav, *Ber.*, **67**, 730 (1934).

53 E. V. Zappi and H. Degiorgi, *Bull. Soc. chim. France*, **49**, 366 (1931).

54 E. V. Zappi and L. M. Simonin, *Ciencia Mex.*, **3**, 160 (1942); *Chem. Abstr.*, **37**, 2009 (1943).

55 A. Tzschach and W. Lange, *Chem. Ber.*, **95**, 1360 (1962).

56 W. Steinkopf and A. Wolfram, *Ber.*, **54**, 848 (1921).

57 Ger. Pat. 313,876 (1919).

58 R. C. Cookson and F. G. Mann, *J. Chem. Soc.*, **1947**, 618.

59 R. P. Welcher, G. A. Johnson, and V. P. Wystrach, *J. Amer. Chem. Soc.*, **82**, 4437 (1960); Ger. Pat., 1,149,004 (May 22nd, 1963).

60 M. J. Gallagher and F. G. Mann, *J. Chem. Soc.*, **1962**, 5110.

61 J. T. Braunholtz and F. G. Mann, *J. Chem. Soc.*, **1957**, 3285.

62 H. Tschamler and R. Leutner, *Monatsh.*, **83**, 1502 (1952).

63 H. Tschamler, *Spectrochim. Acta*, **6**, 95 (1953).

20

[64] G. J. Burrows and E. E. Turner, *J. Chem. Soc.*, **119**, 426 (1921).

[65] F. G. Holliman and D. A. Thornton, unpublished work: D. A. Thornton, 'Heterocyclic Derivatives of Arsenic', Dissertation submitted for the Ph.D. Degree, The University of Cape Town, 1958.

[66] E. Roberts, E. E. Turner, and F. W. Bury, *J. Chem. Soc.*, **1926**, 1443.

[67] R. C. Cookson and F. G. Mann, *J. Chem. Soc.*, **1949**, 67.

[68] F. G. Mann and A. J. Wilkinson, *J. Chem. Soc.*, **1957**, 3336.

[69] F. G. Mann and B. B. Smith, *J. Chem. Soc.*, **1952**, 4544.

[70] J. A. C. Allison, J. T. Braunholtz, and F. G. Mann, *J. Chem. Soc.*, **1954**, 403.

[71] H. L. Hergert and E. F. Kurth, *J. Amer. Chem. Soc.*, **75**, 1622 (1953).

[72] P. I. Ittyerah and F. G. Mann, *J. Chem. Soc.*, **1956**, 3179.

[73] F. G. Holliman, F. G. Mann, and D. A. Thornton, *J. Chem. Soc.*, **1960**, 9.

[74] F. G. Mann and J. Watson, *J. Chem. Soc.*, **1947**, 505.

[75] F. G. Holliman and F. G. Mann, *J. Chem. Soc.*, **1947**, 1634.

[76] F. G. Mann and A. J. Wilkinson, *J. Chem. Soc.*, **1957**, 3346.

[77] J. T. Braunholtz and F. G. Mann, *J. Chem. Soc.*, **1955**, 381.

[78] F. G. Holliman and F. G. Mann, *J. Chem. Soc.*, **1943**, 547.

[79] F. G. Holliman and F. G. Mann, *J. Chem. Soc.*, **1942**, 737.

[80] E. L. Anderson and F. G. Holliman, *J. Chem. Soc.*, **1950**, 1037.

[81] J. Colonge and P. Boisde, *Bull. Soc. chim. France*, **1956**, 1337.

[82] G. Kamai, *Ber.*, **66**, 1779 (1933); *J. Gen. Chem. U.S.S.R.*, 4, 184 (1934).

[83] W. J. Pope and A. W. Harvey, *J. Chem. Soc.*, **79**, 828 (1901).

[84] F. G. Holliman and F. G. Mann, *J. Chem. Soc.*, **1943**, 550.

[85] W. Gump and H. Stoltzenberg, *J. Amer. Chem. Soc.*, **53**, 1428 (1931).

[86] C. L. Hewett, L. J. Lermit, H. T. Openshaw, A. R. Todd, A. H. Williams, and F. N. Woodward, *J. Chem. Soc.*, **1948**, 292.

[87] R. E. Davies, H. T. Openshaw, F. S. Spring, R. H. Stanley, and A. R. Todd, *J. Chem. Soc.*, **1948**, 295.

[88] K. Mislow, A. Zimmerman, and J. T. Melillo, *J. Amer. Chem. Soc.*, **85**, 594 (1963).

[89] J. A. Aeschlimann and N. P. McCleland, *J. Chem. Soc.*, **125**, 2025 (1924).

[90] E. Sakellarios, *Ber.*, **59**, 2552 (1926).

[91] Emrys R. H. Jones and F. G. Mann, *J. Chem. Soc.*, **1958**, 294.

[92] R. N. Jones, *J. Amer. Chem. Soc.*, **67**, 2127 (1945).

[93] G. H. Cookson and F. G. Mann, *J. Chem. Soc.*, **1949**, 2888.

[94] P. J. Wheatley, *J. Chem. Soc.*, **1962**, 3733.

[95] J. T. Braunholtz and F. G. Mann, *J. Chem. Soc.*, **1953**, 1817; **1954**, 651.

[96] J. T. Braunholtz and F. G. Mann, *J. Chem. Soc.*, **1954**, 651.

[97] F. G. Mann, I. T. Millar, and B. B. Smith, *J. Chem. Soc.*, **1953**, 1130.

[98] M. H. Beeby, F. G. Mann, and E. E. Turner, *J. Chem. Soc.*, **1950**, 1923.

[99] F. G. Mann, I. T. Millar, and F. C. Baker, *J. Chem. Soc.*, **1965**, 6342.

[100] A. Tzschach and G. Pacholke, *Chem. Ber.*, **97**, 419 (1964).

[100a] K. Hedberg, E. W. Hughes and J. Waser, *Acta Cryst.*, **14**, 369 (1961).

[101] A. Tzschach and G. Pacholke, *Z. anorg. allg. Chem.*, **336**, 270 (1965).

[102] K. Issleib and F. Krech, *Z. anorg. allg. Chem.*, **328**, 21 (1964).

[103] W. Steinkopf and G. Schwen, *Ber.*, **54**, 1437 (1921).

[104] K. Issleib, A. Tzschach, and R. Schwarzer, *Z. anorg. allg. Chem.*, **338**, 141 (1965).

[105] R. N. Collinge, R. S. Nyholm, and M. L. Tobe, *Nature (London)*, **201**, 1322 (1964).

106 Emrys R. H. Jones and F. G. Mann, *J. Chem. Soc.*, **1955**, 401.

107 J. Chatt and F. G. Mann, *J. Chem. Soc.*, **1939**, 610, 1622.

108 F. G. Mann and M. J. Pragnell, unpublished work; M. J. Pragnell, 'Some Heterocyclic Derivatives of Phosphorus and Arsenic: De-alkylation by the Diarylphosphide Ion', Dissertation submitted for the Ph.D. Degree, The University of Cambridge, 1965.

109 F. G. Mann and F. C. Baker, *J. Chem. Soc.*, **1957**, 1881.

110 S. C. Nyburg and J. Hilton, *Acta Cryst.*, **8**, 358 (1955).

111 F. G. Mann and A. Senior, *J. Chem. Soc.*, **1954**, 4476.

112 C. G. Krespan, *J. Amer. Chem. Soc.*, **83**, 3432 (1961).

113 R. H. Glauert and F. G. Mann, *J. Chem. Soc.*, **1950**, 682.

114 F. G. Mann and F. C. Baker, *J. Chem. Soc.*, **1952**, 4142.

115 J. Chatt and F. G. Mann, *J. Chem. Soc.*, **1939**, 1622.

116 Emrys R. H. Jones and F. G. Mann, *J. Chem. Soc.*, **1955**, 405.

117 L. Kalb, *Annalen*, **423**, 39 (1921).

118 D. Sh. Rozina, *Zh. prikl. Khim.*, **23**, 1110 (1950); *Chem. Abstr.*, **46**, 10121 (1952).

119 F. G. Mann and F. H. C. Stewart, *J. Chem. Soc.*, **1954**, 4127.

120 J. Chatt and F. G. Mann, *J. Chem. Soc.*, **1940**, 1184.

121 L. J. Goldsworthy, W. H. Hook, J. A. John, S. G. P. Plant, J. Rushton, and L. M. Smith, *J. Chem. Soc.*, **1948**, 2208.

122 H. Wieland and W. Rheinheimer, *Annalen*, **423**, 1 (1921).

123 E. Roberts and E. E. Turner, *J. Chem. Soc.*, **127**, 2004 (1925).

124 F. G. Mann and J. Watson, *J. Org. Chem.*, **13**, 502 (1948).

125 A. F. Wells, *Z. Krist.*, **99**, 367 (1938).

126 Emrys R. H. Jones and F. G. Mann, *J. Chem. Soc.*, **1955**, 411.

127 F. G. Mann, *J. Chem. Soc.*, **1963**, 4266.

128 O. Kennard, F. G. Mann, D. G. Watson, J. K. Fawcett, and K. A. Kerr, *Chem. Commun.*, **1968**, 269.

129 D. J. Sutor and F. R. Harper, *Acta Cryst.*, **12**, 585 (1959).

130 M. W. Lister and L. E. Sutton, *Trans. Faraday Soc.*, **37**, 393 (1941).

131 H. A. Skinner and L. E. Sutton, *Trans. Faraday Soc.*, **40**, 164 (1944).

132 O. Hassell and H. Viervoll, *Acta. Chem. Scand.*, **1**, 149 (1947).

133 D. W. Allen, J. C. Coppola, O. Kennard, F. G. Mann, W. D. S. Motherwell, and D. G. Watson, *J. Chem. Soc.*, *C*, in the press.

134 N. P. McCleland and J. B. Whitworth, *J. Chem. Soc.*, **1927**, 2753.

135 W. Schaffer, *Acta Cryst.*, **9**, 401 (1956).

136 M. H. Forbes, D. M. Heinekey, F. G. Mann, and I. T. Millar, *J. Chem. Soc.*, **1961**, 2762.

137 M. H. Forbes, F. G. Mann, I. T. Millar and E. A. Moelwyn-Hughes, *J. Chem. Soc.*, **1963**, 2833.

138 H. Heaney, F. G. Mann, and I. T. Millar, *J. Chem. Soc.*, **1957**, 3930.

139 D. W. Allen, I. T. Millar, and F. G. Mann, *J. Chem. Soc.*, *C*, **1967**, 1869.

140 O. J. Scherer and M. Schmidt, *Angew. Chem.*, **76**, 787 (1964); *Int. Ed. Engl.*, **3**, 702 (1964).

141 E. W. Abel and R. P. Bush, *J. Organometal. Chem.*, **3**, 245 (1965).

142 M. H. Beeby and F. G. Mann, *J. Chem. Soc.*, **1951**, 886.

143 W. Cule Davies and W. P. G. Lewis, *J. Chem. Soc.*, **1934**, 1599.

144 W. Cule Davies and H. W. Addis, *J. Chem. Soc.*, **1937**, 1622.

145 F. G. Mann, *J. Chem. Soc.*, **1932**, 958.

146 F. Bayer and Co., Ger. Pat. 281,049 (1914).

[147] R. Fischer, *Mikrochemie*, **12**, 257 (1932).

[148] A. Camerman and J. Trotter, *J. Chem. Soc.*, **1965**, 730.

[149] E. Gryszkiewicz-Trochimowski (Grischkievitch-Trochimovski), L. Mateyak, and Zablotscki, *Bull. Soc. Chim. France*, [4], **41**, 1323 (1927); *Roczniki Chem.*, **7**, 230 (1927); *Chem. Abstr.*, **22**, 760 (1928).

[150] E. Ristenpart, *Ber.*, **29**, 2526 (1896).

[151] J. H. Schmidt, *J. Amer. Chem. Soc.*, **43**, 2449 (1921).

[152] H. Burton and C. S. Gibson, *J. Chem. Soc.*, **125**, 2275 (1924).

[153] C. P. A. Kappelmeier, *Rec. Trav. chim.*, **49**, 57 (1930); **50**, 44 (1931).

[154] C. S. Gibson, J. D. A. Johnson, and D. C. Vining, *Rec. Trav. chim.*, **49**, 1006 (1930).

[155] H. Burton and C. S. Gibson, *J. Chem. Soc.*, **1926**, 2241.

[156] F. L. Allen and S. Sugden, *J. Chem. Soc.*, **1936**, 440.

[157] W. L. Lewis and C. S. Hamilton, *J. Amer. Chem. Soc.*, **43**, 2218 (1921).

[158] W. L. Lewis and H. W. Stiegler, *J. Amer. Chem. Soc.*, **47**, 2546 (1925).

[159] H. Burton and C. S. Gibson, *J. Chem. Soc.*, **1926**, 450.

[160] H. Burton and C. S. Gibson, *J. Chem. Soc.*, **1926**, 464.

[161] H. Burton and C. S. Gibson, *J. Chem. Soc.*, **1927**, 247.

[162] C. S. Gibson and J. D. A. Johnson, *J. Chem. Soc.*, **1927**, 2499.

[163] C. S. Gibson and J. D. A. Johnson, *J. Chem. Soc.*, **1928**, 2204.

[164] C. S. Gibson and J. D. A. Johnson, *J. Chem. Soc.*, **1929**, 767.

[165] L. A. Elson, C. S. Gibson, and J. D. A. Johnson, *J. Chem. Soc.*, **1929**, 1080.

[166] C. S. Gibson and J. D. A. Johnson, *J. Chem. Soc.*, **1929**, 1229.

[167] C. S. Gibson and J. D. A. Johnson, *J. Chem. Soc.*, **1929**, 1473.

[168] C. S. Gibson and J. D. A. Johnson, *J. Chem. Soc.*, **1929**, 1621.

[169] C. S. Gibson and J. D. A. Johnson, *J. Chem. Soc.*, **1929**, 2743.

[170] C. S. Gibson and J. D. A. Johnson, *J. Chem. Soc.*, **1930**, 1124.

[171] C. S. Gibson, E. S. Hiscocks, J. D. A. Johnson, and J. L. Jones, *J. Chem. Soc.*, **1930**, 1622.

[172] L. A. Elson and C. S. Gibson, *J. Chem. Soc.*, **1931**, 294.

[173] L. A. Elson and C. S. Gibson, *J. Chem. Soc.*, **1931**, 2381.

[174] C. S. Gibson and J. D. A. Johnson, *J. Chem. Soc.*, **1931**, 2518.

[175] C. S. Gibson and J. D. A. Johnson, *J. Chem. Soc.*, **1931**, 3270.

[176] C. S. Gibson, *J. Amer. Chem. Soc.*, **53**, 376 (1931).

[177] M. Fukuyo, K. Nakatsu, and A. Shimada, *Bull. Chem. Soc. Japan*, **39**, 1614 (1966); *Chem. Abstr.*, **65**, 14552 (1961).

[178] I. G. Rasuvaiev (also transliterated Razuvaev and Rasuwajew), *Ber.*, **62**, 605 (1929); *J. Russ. Phys.-Chem. Soc.*, **61**, 13 (1929).

[179] V. V. Shtishevskii and A. I. Voronina, *J. Gen. Chem. (U.S.S.R.)*, **7**, 2406 (1937); *Chem. Abstr.*, **32**, 2134 (1938).

[180] I. Kageyama and S. Nakanishi, Brit. Pat., 861,500 (Feb. 22, 1961); *Chem. Abstr.*, **55**, 21152 (1961).

[181] J. A. Aeschlimann, *J. Chem. Soc.*, **1927**, 413.

[182] O. Seide and J. Gorski, *Ber.*, **62**, 2186 (1929).

[183] G. A. Razuvaev and V. G. Malinovski, *J. Gen. Chem. U.S.S.R.*, **5**, 570 (1935); *Chem. Abstr.*, **29**, 6895 (1935).

[184] G. A. Razuvaev, V. S. Malinowski, and S. E. Arkina, *J. Gen. Chem. U.S.S.R.*, **5**, 575 (1935); *Chem. Abstr.*, **29**, 6895 (1935).

[185] G. A. Razuvaev (Rasuwajew), D. A. Godina, and T. I. Yemelyanova (Jemeljanowa), *Ber.*, **65**, 666 (1932).

186 W. H. C. Rueggeberg, A. Ginsburg, and W. A. Cook, *J. Amer. Chem. Soc.*, **68**, 1860 (1946).

187 G. A. Razuvaev, *Ber.*, **62**, 1208 (1929).

188 G. A. Razuvaev and W. Malinowski, *Ber.*, **62**, 2675 (1929).

189 G. A. Razuvaev and A. Benediktow, *Ber.*, **63**, 343 (1930).

190 G. A. Razuvaev, *Rec. Trav. Chem.*, **50**, 900 (1931).

191 G. A. Razuvaev and W. Malinowski, *Ber.*, **66**, 463 (1933).

192 Ng. Ph. Buu-Hoï, M. Mangane, and P. Jacquignon, *J. Heterocyc. Chem.*, **3**, 149 (1966).

193 P. G. Sergeev and I. M. Gorskiï, *J. Gen. Chem. U.S.S.R.*, **1**, 263 (1931); *Chem. Abstr.*, **26**, 2195 (1932).

194 W. Mielenz, *Gasschutz u. Luftschutz*, **2**, 10 (1932).

195 Anon., *Gasschutz u. Luftschutz*, **2**, 264 (1932).

196 L. Redlinger, *Gasschutz u. Luftschutz*, **2**, 263 (1932).

197 L. Leroux, *Rev. Hyg. Med. prév.*, **57**, 81 (1935).

198 H. Mohler and J. Pólya, *Helv. Chim. Acta*, **19**, 1222, 1239 (1936).

199 H. Mohler, *Protar*, **7**, 78 (1941); *Chem. Abstr.*, **35**, 4868 (1941).

200 H. Hennig, *Gasschutz u. Luftschutz*, **7**, 18 (1937).

201 T. Leipert, *Wien. Klin. Wochenschr.*, **51**, 549 (1938).

202 J. Delga, *J. Pharm. Chim.* [9], **1**, 73 (1940).

203 H. E. Cox, *Analyst*, **64**, 807 (1939).

204 A. P. J. Hoogeveen, *Chem. & Ind. (London)*, **1940**, 550.

205 P. F. Fenton, *J. Chem. Educ.*, **21**, 488 (1944).

206 M. Nagasawa, M. Yoshido, and T. Totsuka, Jap. Pat. 6399 (1957); *Chem. Abstr.*, **52**, 10484 (1958).

207 M. Nagasawa, R. Kubota, and F. Mamamota, Jap. Pat. 2299 (1958); *Chem. Abstr.*, **53**, 3588 (1959).

208 M. Nagasawa, R. Kubota, and F. Yamamota, Jap. Pat. 9500 (1958); *Chem. Abstr.*, **54**, 4635 (1960).

209 K. Fukunaga, K. Takita, M. Matsuo, and W. Yamatani, Jap. Pat. 8549 (1959); *Chem. Abstr.*, **54**, 7056 (1960).

210 I. Fujiyama and S. Nakanishi, Jap. Pat. 10,771 (1959); *Chem. Abstr.*, **54**, 17911 (1960).

211 I. Kageyama and S. Nakanishi, U.S. Pat. 3,041,188 (1962).

212 M. Nagasawa, U.S. Pat. 3,214,281 (1965).

213 E. Urbschat and P. E. Frohberger, U.S. Pat. 2,767,114 (1956); *Chem. Abstr.*, **51**, 5354 (1957).

214 I. Biro and M. Parkany, *Magy. Kem. Foly.*, **68**, 437 (1962); *Chem. Abstr.*, **58**, 7009 (1963).

215 C. L. Punte, T. A. Ballard, and J. T. Weimer, *Amer. Ind. Hyg. Ass. J.*, **23**, 194 (1962).

216 C. L. Punte, P. J. Gutentag, E. J. Owens, and L. E. Gongiver, *Amer. Ind. Hyg. Ass. J.*, **23**, 199 (1962).

217 H. McL. Gordon and M. Lipson, *J. Counc. Sci. Ind. Res.*, **13**, 173 (1940).

218 W. D. Lotspeich and R. A. Peters, *Biochem. J.*, **49**, 704 (1951).

218a R. A. Earley and M. J. Gallagher, *J. Chem. Soc., C*, in the press.

219 Ng. Ph. Buu-Hoï, Hiong-Ki-Wei, and R. Royer, *Compt. rend.*, **220**, 361 (1945); *Chem. Abstr.*, **41**, 5534 (1947).

220 Ng. Ph. Buu-Hoï and P. Jacquignon, *J. Chem. Soc.*, **1951**, 2964.

221 Ng. Ph. Buu-Hoï, Hiong-Ki-Wei, and R. Royer, *Compt. rend.*, **220**, 50 (1945); *Chem. Abstr.*, **40**, 2838 (1946).

222 Ng. Ph. Buu-Hoï, P. Jacquignon, and D. Lavit, *J. Chem. Soc.*, **1956**, 2593.

223 Ng. Ph. Buu-Hoï and R. Royer, *Bull. Soc. Chim.* (*France*), **1946**, 379.

224 Ng. Ph. Buu-Hoï, G. Saint-Ruf, P. Jacquignon, and G. C. Barrett, *J. Chem. Soc.*, **1958**, 4308.

225 G. C. Barrett and Ng. Ph. Buu-Hoï, *J. Chem. Soc.*, **1958**, 2946.

226 Ng. Ph. Buu-Hoï, O. Roussel, and L. Petit, *J. Chem. Soc.*, **1963**, 956.

227 Ng. Ph. Buu-Hoï, R. Royer, and M. Hubert-Habart, *J. Chem. Soc.*, **1956**, 2048.

228 R. H. Slater, *J. Chem. Soc.*, **1931**, 1938.

229 R. H. Slater, *J. Chem. Soc.*, **1931**, 107.

230 A. R. Hands and A. J. H. Mercer, *J. Chem. Soc.*, (*C*), **1967**, 1099.

231 S. Trippett, *Quart. Rev.*, **17**, 438 (1963).

232 R. D. Feltham, A. Kasenally, and R. S. Nyholm, *J. Organometal. Chem.*, **7**, 285 (1967).

233 W. R. Cullen and J. Trotter, *Can. J. Chem.*, **40**, 1113 (1962).

234 G. Kamai and Z. L. Khisamova, *Dokl. Akad. Nauk SSSR*, **76**, 535 (1951); *Chem. Abstr.*, **45**, 10190 (1951).

235 G. Kamai and Z. L. Khisamova, *Zh. Obschchei Khim.*, **23**, 1323 (1953); *Chem. Abstr.*, **48**, 3889 (1954).

236 G. Kamai and N. A. Chadaeva, *Dokl. Akad. Nauk SSSR*, **81**, 837 (1951); *Chem. Abstr.*, **47**, 3792 (1953).

237 G. Kamai and Z. L. Khisamova, *Zh. Obschchei Khim.*, **24**, 816 (1954); *Chem. Abstr.*, **49**, 8093 (1955).

238 G. Kamai and Z. L. Khisamova, *Zh. Obsch. Khim.*, **24**, 821 (1954); *Chem. Abstr.*, **49**, 8094 (1955).

239 G. Kamai and N. A. Chadaeva, *Zh. Obsch. Khim.*, **23**, 1431 (1953); *Chem. Abstr.*, **48**, 3890 (1954).

240 G. Kamai and N. A. Chadaeva, *Izv. Kazan. Filiala Akad. Nauk SSSR, Ser. Khim. Nauk*, **1955**, 19; *Chem. Abstr.*, **52**, 292 (1958).

241 B. Englund, *J. Prakt. Chem.*, **120**, 179 (1929).

242 H. J. Backer and R. P. van Oosten, *Rec. Trav. Chim.*, **59**, 41 (1940).

243 R. Foster and C. A. Fyfe, *Spectrochim. Acta*, **21**, 1785 (1965).

244 E. J. Salmi, K. Merivuori, and E. Laaksonen, *Suomen Kemistilehti*, **19B**, 102 (1946); *Chem. Abstr.*, **41**, 5440 (1947).

245 A. Pictet and A. Bon, *Bull. Soc. chim. France*, **33**, 1139 (1905).

246 P. Pascal and A. Dupire, *Compt. rend.*, **195**, 14 (1932).

247 M. L. Wolfrom and M. J. Holm, *J. Org. Chem.*, **26**, 273 (1961).

248 H. Funk and H. Köhler, *J. Prakt. Chem.*, [iv], **13**, 322 (1961).

249 A. Michaelis, *Annalen*, **320**, 271 (1902).

250 H. Funk and H. Köhler, *J. Prakt. Chem.* [iv], **14**, 226 (1961).

251 H. Legler, Ger. Pat. 536,081 (1931).

252 B. Englund, *Ber.*, **59**, 2669 (1926).

253 A. G. Skapski, *Chem. Commun.*, **1966**, 10.

254 R. Weinland and J. Heinzler, *Ber.*, **52**, 1316 (1919).

255 H. Reihlen, A. Sapper, and G. A. Kall, *Z. Anorg. Allgem. Chem.*, **144**, 218 (1925).

256 A. Rosenheim and W. Plato, *Ber.*, **58**, 2000 (1925).

257 T. H. Larkins, Jr., and M. M. Jones, *Inorg. Chem.*, **2**, 142 (1963).

258 R. Weinland and H. Seuffert, *Arch. Pharm.*, **266**, 455 (1928).

259 A. L. Porte, H. S. Gutowsky, and G. M. Harris, *J. Chem. Phys.*, **34**, 66 (1961).

260 T. H. Larkins, Jr., and M. M. Jones, *J. Inorg. Nucl. Chem.*, **25**, 1487 (1963).

261 F. G. Mann and J. Watson, *J. Chem. Soc.*, **1947**, 505.

262 M. Davis and F. G. Mann, *J. Chem. Soc.*, **1964**, 3791.

263 W. L. Lewis, C. D. Lowry, and F. H. Bergeim, *J. Amer. Chem. Soc.*, **43**, 891 (1921).

264 E. E. Turner, Dissertation submitted for the D.Sc. Degree, London University, 1920.

265 F. F. Blicke, O. J. Weinkauff, and G. W. Hargreaves, *J. Amer. Chem. Soc.*, **52**, 780 (1930).

266 E. E. Turner and A. B. Sheppard, *J. Chem. Soc.*, **127**, 544 (1925).

267 E. Roberts and E. E. Turner, *J. Chem. Soc.*, **127**, 2004 (1925).

268 J. D. C. Mole and E. E. Turner, *J. Chem. Soc.*, **1939**, 1720.

269 J. A. Aeschlimann, *J. Chem. Soc.*, **127**, 811 (1925).

270 E. Roberts and E. E. Turner, *J. Chem. Soc.*, **1926**, 1207.

271 P. W. Allen and L. E. Sutton, *Acta Cryst.*, **3**, 46 (1950), personal communication from W. Shand.

272 M. S. Lesslie and E. E. Turner, *J. Chem. Soc.*, **1934**, 1170.

273 K. Mislow, A. Zimmerman, and J. T. Melillo, *J. Amer. Chem. Soc.*, **85**, 594 (1963).

274 M. S. Lesslie and E. E. Turner, *J. Chem. Soc.*, **1935**, 1051.

275 M. S. Lesslie and E. E. Turner, *J. Chem. Soc.*, **1935**, 1268.

276 M. S. Lesslie and E. E. Turner, *J. Chem. Soc.*, **1936**, 730.

277 M. S. Lesslie, *J. Chem. Soc.*, **1939**, 1050.

278 M. S. Lesslie, *J. Chem. Soc.*, **1949**, 1183.

279 I. G. M. Campbell, *J. Chem. Soc.*, **1947**, 4.

280 M. Nagasawa, S. Imamiya, and M. Fukuda (to Ihara Agricultural Chemical Co.), Jap. Pat. 6400 (Aug. 17, 1957); *Chem. Abstr.*, **52**, 10484 (1958).

281 K. D. Shvetsova-Shilovskaya, N. N. Mel'nikov, E. I. Andreeva, L. P. Bocharova, and Y. N. Sapozhkov, *Zh. Obshch. Khim.*, **31**, 845 (1961); *Chem. Abstr.*, **55**, 23551 (1061).

282 S. J. Strycker and J. E. Dunbar (to Dow Chemical Co.), U.S. Pat. 3,038,921 (June 12, 1962); *Chem. Abstr.*, **57**, 12540 (1962).

283 J. E. Dunbar, U.S. Pat. 3,036,107 (May 22, 1962); *Chem. Abstr.*, **57**, 13806 (1962).

284 G. W. Bowers and C. S. Hamilton, *J. Amer. Chem. Soc.*, **58**, 1573 (1936).

285 G. Kamai and N. A. Chadaeva, *Izv. Akad. Nauk SSSR, Otd. Khim. Nauk*, **1952**, 908; *Chem. Abstr.*, **47**, 10470 (1953).

286 G. Kamai and N. A. Chadaeva, *Dokl. Akad. Nauk SSSR*, **86**, 71 (1952); *Chem. Abstr.*, **47**, 6365 (1953).

287 G. Kamai, Z. L. Khisamova and N. A. Chadaeva, *Dokl. Akad. Nauk SSSR*, **89**, 1015 (1953); *Chem. Abstr.*, **48**, 6391 (1954).

288 G. Kamai and N. A. Chadaeva, *Zh. Obshch. Khim.*, **26**, 2468 (1956); *Chem. Abstr.*, **51**, 4932 (1957).

289 J. G. Verkade and L. T. Reynolds, *J. Org. Chem.*, **25**, 663 (1960).

290 C. S. Palmer and R. Adams, *J. Amer. Chem. Soc.*, **44**, 1356 (1922).

291 H. J. Barber, *J. Chem. Soc.*, **1930**, 2047.

292 H. J. Barber, *J. Chem. Soc.*, **1930**, 2725.

293 A. Cohen, H. King, and W. I. Strangeways, *J. Chem. Soc.*, **1931**, 3043.

294 R. A. Peters, L. A. Stocken, and R. H. S. Thompson, *Nature (London)*, **156**, 616 (1945).

[295] R. A. Peters and L. A. Stocken, *Biochem. J.*, **41**, 53 (1947).

[296] L. A. Stocken, *J. Chem. Soc.*, **1947**, 592.

[297] L. L. Waters and C. Stock, *Science*, **102**, 601 (1945).

[298] L. A. Stocken and R. H. S. Thompson, *Physiol. Rev.*, **29**, 168 (1949).

[299] M. Dixon and E. C. Webb, 'Enzymes,' 2nd edn., Longmans Green and Co., London, 1964, p. 343.

[300] E. A. H. Friedheim and H. J. Vogel, *Proc. Soc. Exp. Biol. Med.*, **64**, 418 (1947).

[301] E. A. H. Friedheim and R. L. Berman, *Proc. Soc. Exp. Biol. Med.*, **65**, 180 (1947).

[302] E. A. H. Friedheim, Brit. Pat. 655,435 (July 1951); *Chem. Abstr.*, **47**, 144 (1953).

[303] E. A. H. Friedheim, U.S. Pat. 2,659,723 (Nov. 1953); 2,664,432 (Dec. 1953); *Chem. Abstr.*, **47**, 144 (1953); **49**, 1816 (1955). The chemical examples given in the two patents are the same, but the claims differ.

[304] V. E. Petrun'kin, *Ukr. Khim. Zh.*, **22**, 608 (1956); *Chem. Abstr.*, **51**, 5693 (1957).

[305] Rhone-Poulenc S.A., French Pat. M854 (Nov. 1961): *Chem. Abstr.*, **58**, 3460 (1963).

[306] E. A. H. Friedheim, U.S. Pat. 3,035,052 (May 1962); *Chem. Abstr.*, **57**, 11240 (1962).

[307] E. A. H. Friedheim, U.S. Pat. 2,772,303 (Nov. 1956); *Chem. Abstr.*, **51**, 5836 (1957).

[308] M. Nagasawa, K. Nagamizu, and F. Yamamoto (for Ihara Agric. Chem. Co.), Jap. Pat. 8700 (Sept. 1958); *Chem. Abstr.*, **54**, 5000 (1960).

[309] M. Nagasawa, K. Nagamizu, S. Imamiya, and T. Maeda (for Ihara Agric. Chem. Co.), Jap. Pat. 17,132 (1960); *Chem. Abstr.*, **55**, 17661 (1961).

[310] P. F. Epstein and G. P. Willsey (to Stauffer Chem. Co.), U.S. Pat. 3,152,159 (Oct. 1964); *Chem. Abstr.*, **61**, 14713 (1964).

[311] W. H. C. Rueggeberg, A. Ginsburg, and W. A. Cook, *J. Amer. Chem. Soc.*, **68**, 1860 (1946).

[312] I. G. M. Campbell, *J. Chem. Soc.*, **1956**, 1976.

[313] R. Klement and A. May, *Ber.*, **71**, 890 (1938).

[314] A. Job, R. Reich, and P. Vergnaud, *Bull. Soc. Chim. France*, [4], **35**, 1404 (1924).

PART III

Heterocyclic Derivatives of Antimony

Five-membered Ring Systems containing Carbon and One Antimony Atom

Stibole and Stibolanes

The chemistry of stibole (**1**) [RRI 151, in which the Sb atom is shown uppermost] closely follows that of the corresponding arsole

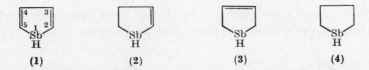

(**1**)   (**2**)   (**3**)   (**4**)

(p. 357). Stibole (**1**) should theoretically have two isomeric dihydro derivatives, the 2-stibolene (**2**) and the 3-stibolene (**3**), and the tetrahydro derivative, stibolane (**4**). In practice, stibole (**1**) is known only as its heavily substituted derivatives, the two stibolenes are unknown, and stibolane is known (as expected) only as its 1-substituted derivatives.

Stibole (**1**) has been isolated only as the pentaphenylstibole (**5**); its preparation was carried out in the same way and by the same workers as for the phosphorus (p. 13) and the arsenic analogue (p. 357) and thus requires only brief mention. Leavitt and his co-workers [1, 2, 3] have shown that an ethereal solution of 1,4-dilithio-1,2,3,4-tetraphenylbutadiene (prepared by the action of lithium on diphenylacetylene) when added at 0° to ethereal phenylstibonous dichloride gives pentaphenylstibole (52%), yellow needles, m.p. 160°.

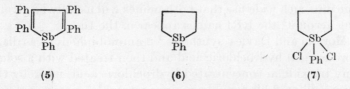

(**5**)   (**6**)   (**7**)

Braye, Hübel, and Caplier [4, 5] carried out this reaction at −40° and obtained the stibole (**5**) (21%), m.p. 162–170°, decomposing at 220°; atmospheric oxidation converted it into the faintly yellow 1-oxide, m.p. 250–255°.

1-Phenylstibolane (**6**) was obtained by the interaction of the di-Grignard agent of 1,4-dibromobutane, $(CH_2CH_2MgBr)_2$, and phenyl-stibonous dichloride under hydrogen (Grüttner and Krause,[6] 1916). It is a colorless oily liquid, b.p. 156–158°/20–22 mm, having an unpleasant odor and readily becoming cloudy by atmospheric oxidation. When treated in carbon tetrachloride solution with chlorine, it gives the 1,1-dichloride (**7**), colorless crystals, m.p. 150°; the 1,1-dibromide, similarly prepared, also forms colorless crystals, m.p. 149°.

The stibolane system (**4**) was earlier known as tetramethylene-stibine and as cyclotetramethylenestibine.

Dibenzostibole

The naming of the ring system (**1**) has been even more confused than those of the analogous phosphorus and arsenic systems (pp. 72,

(1)   (1a)

373). Morgan and Davies,[7] who first (1930) prepared compounds having the ring system (**1**), revived a term originally used by von Hofmann[8] for the three isomeric monoaminobiphenyls and termed them 'xenyl-amines'. On this basis, the univalent biphenylyl radicals, $C_6H_5C_6H_4$—, were termed 'xenyl' radicals, and the bivalent biphenylylene radicals, —$C_6H_4C_6H_4$—, were termed 'xenylene' radicals. Consequently, Morgan and Davies termed the chloro derivative of (**1**) 'xenylenestibine chloride'.

The Ring Index name for the compound (**1**) is now dibenzostibole [RRI 3065] with the numbering as shown. In the meantime, however, the alternative name 9-stibafluorene, with the numbering given in (**1A**), has come into such wide use that both names will usually be given in the following account; the RRI names appear in the Index.

In Morgan and Davies' synthesis,[7] 2-aminobiphenyl was diazotized in the presence of hydrochloric acid and then treated with a solution of antimony trioxide in concentrated hydrochloric acid, whereby the pale yellow crystalline 2-biphenyldiazonium tetrachloroantimonate (**2**) was precipitated. [This type of antimony 'double salt' was discovered by May[9] (1912) and later studied by Brucker and Nikiforova[10, 11] (1949).] A suspension of these crystals in ice-water was first stirred with aqueous sodium hydroxide solution until decomposition of the diazonium salt was complete and evolution of nitrogen had ceased, and was then warmed with dilute hydrochloric acid to precipitate the 2-biphenylylstibonic acid (3); the latter, after purification from inorganic antimony compounds, was obtained as colorless crystals from ethanol.

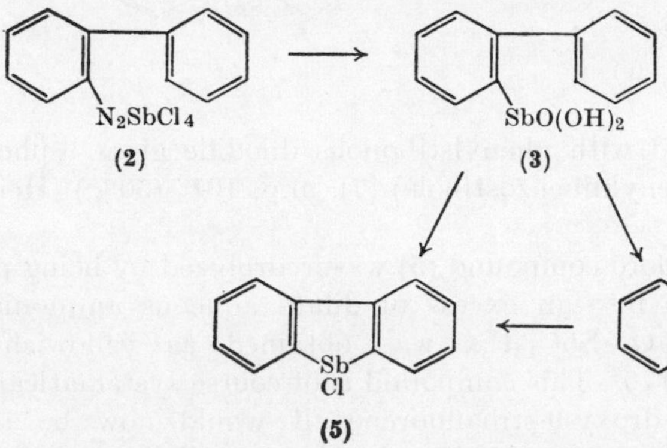

When a solution of the stibonic acid (3) in methanolic hydrochloric acid containing a trace of potassium iodide was treated at room temperature with sulfur dioxide, ready reduction occurred to form 2-biphenylylstibonous dichloride (4), colorless needles, m.p. 76°. This compound was further characterized by treatment with an acetone solution of sodium iodide, whereby the corresponding diiodide was obtained as dark yellow needles, m.p. 95–96°.

The 9-chloro-9-stibafluorene (5-chlorodibenzostibole) (5) was obtained by cyclization of either the dichloride (4) or the stibonic acid (3). When the dichloride (4) was heated at 100°/25 mm pressure, hydrogen chloride was lost and the 9-chloro derivative (5) obtained as greenish white crystals, m.p. 209°. More satisfactory results were obtained, however, when the stibonic acid (3) was dissolved in concentrated sulfuric acid and the solution warmed at 100° for 10 minutes and then poured into much cold water. The precipitated material was dissolved in methanolic hydrochloric acid and reduced as before with sulfur dioxide, yielding the 9-chloro-9-stibafluorene. This compound was also characterized by treatment with sodium iodide in acetone solution, giving lemon-yellow needles of 9-iodo-9-stibafluorene, m.p. 222°.

It should be noted that when the sulfur dioxide reductions cited above are carried out, care has to be taken to ensure that an excess of the dioxide is not used, otherwise the reduction proceeds too far, with the separation of elementary antimony. In arsenic compounds this extreme reduction to elemental arsenic is apparently never encountered.

The 9-stibafluorene ring system (1) can also be synthesized by converting 2,2'-diiodobiphenyl into the 2,2'-dilithio derivative (6),

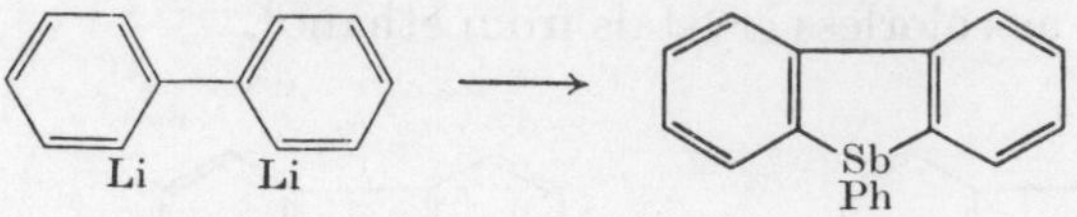

(6)   (7)

which when treated with phenylstibonous diiodide gives 9-phenyl-9-stibafluorene (5-phenyldibenzostibole) (7), m.p. 101° (50%) (Heinekey and Millar[12]).

When the 9-chloro compound (5) was hydrolyzed by being poured in acetone solution into an excess of dilute aqueous ammonia, the 'oxide', $C_{13}H_8Sb$—O—$SbC_{13}H_8$, was obtained as yellowish-white crystals, m.p. 177–179°. This compound is of course systematically the anhydride of 9-hydroxy-9-stibafluorene; it would now be termed 9,9'-oxydi-(9-stibafluorene) or 5,5'-oxybis(dibenzostibole).

An ethereal suspension of 9-iodo-9-stibafluorene, when treated with methylmagnesium iodide, gave the tertiary stibine, 9-methyl-9-stibafluorene (5-methyldibenzostibole) (8), colorless crystals, m.p. 57°. It is

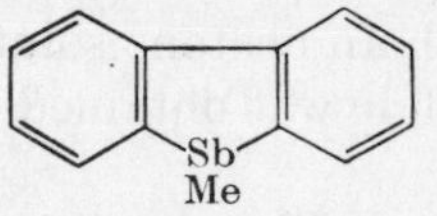

(8) 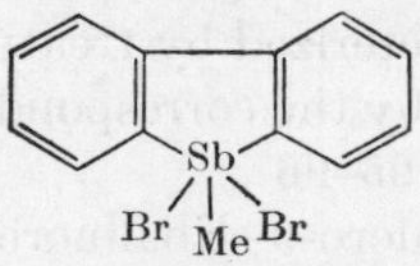   (9)

noteworthy that this stibine did not combine with methyl iodide, although it united readily with one molecular equivalent of bromine to give the stibine dibromide (9), a pale yellow microcrystalline powder, m.p. 207°.

Certain properties of 2-biphenylylstibonous dichloride (4) and its derivatives deserve discussion, because they illustrate the extraordinary ease with which organic derivatives of antimony may in certain cases undergo dismutation, and they also provide another route for cyclization to the 9-stibafluorene ring system. Morgan and Davies[7] found that when the stibonous dichloride (4) was digested with warm dilute aqueous sodium hydroxide, it underwent dismutation with the formation of antimony trioxide and the 'oxide' (10). The latter is clearly the anhydride of bis-(2-biphenylyl)hydroxystibine, $(C_6H_5C_6H_4)_2Sb(OH)$, and this is confirmed by the fact that on treatment with hot ethanolic hydrochloric acid it gives bis-(2-biphenylyl)stibinous chloride (11), colorless crystals, m.p. 125.5°, which can be converted in the usual way into the corresponding monoiodide, yellow crystals, m.p. 156–157°. The anhydride

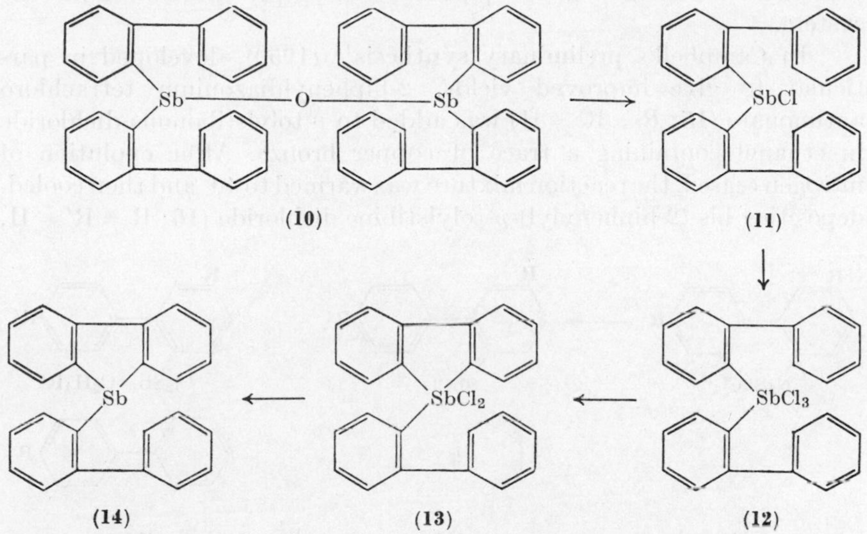

(10) is in fact best purified by this conversion into the chloride (11), for when an acetone solution of the latter is added to a large volume of dilute aqueous ammonia, the pure anhydride (10) can be readily isolated as large colorless needles, m.p. 157°.

The bis-(2-biphenylyl)stibinous chloride (11) in chloroform solution readily combined with one molecular equivalent of chlorine to give the trichloro derivative (12), colorless needles, m.p. 177°, and this compound, when heated at 50°/20 mm underwent cyclization with loss of hydrogen chloride to form 9-(2-biphenylyl)-9-stibafluorene dichloride (13), colorless crystals, m.p. 212°. This product could be reduced to the tertiary stibine by boiling in ethanolic suspension with zinc dust for 2 hours, and the 9-(2-biphenylyl)-9-stibafluorene [5-(2-biphenylyl)-dibenzostibole] (14) obtained as colorless needles, m.p. 106–107°.

The stereochemistry of the 9-stibafluorene ring system has been examined in considerable detail by Campbell, with the prime object of investigating the optical resolution of suitably substituted 9-stibafluorenes and, if this were successful, the evidence that the degree of optical stability of these products would provide regarding the configurational stability of the pyramidal 3-covalent antimony atom. At the time of these investigations (1950–1959) there was considerable circumstantial, but no decisive, evidence that the fused tricyclic ring of the

9-stibafluorene (dibenzostibole) system was planar. The later decisive evidence by X-ray crystal-structure analysis that the corresponding arsenic system is flat (Sartain and Truter,[13] 1963) (p. 381) is also strong evidence by analogy for the planar condition of the 9-stibafluorene system.

In Campbell's preliminary synthesis[14] (1950), developed in particular to give improved yields, 2-biphenyldiazonium tetrachloro antimonate (15; $R = R' = H$) was added to p-tolylstibonous dichloride in ethanol containing a trace of copper bronze. After evolution of nitrogen ceased, the reaction mixture was warmed to 40° and then cooled, depositing bis-(2-biphenylyl)-p-tolylstibine dichloride (16; $R = R' = H$,

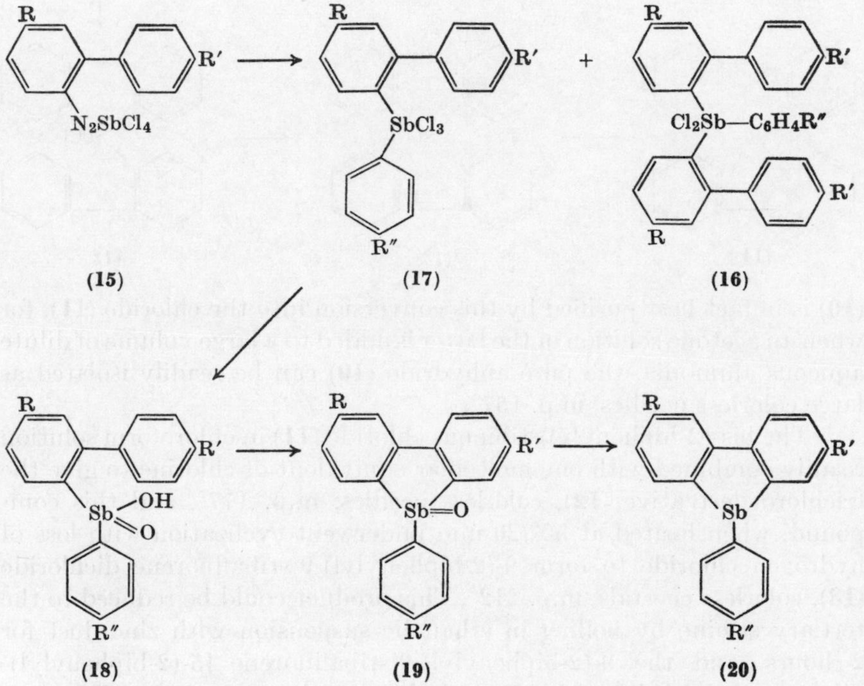

(15)   (17)   (16)

(18)   (19)   (20)

$R'' = CH_3$) in small amount; on concentration, the filtered solution deposited 2-biphenylyl-p-tolylstibine trichloride (17; $R = R' = H$, $R'' = CH_3$) in greater amount. Direct cyclization of the latter could not be achieved; it was therefore added in ethanol–acetone to cold aqueous sodium acetate for hydrolysis to the stibinic acid (18; $R = R' = H$, $R'' = CH_3$). This acid, when dried and heated with acetic anhydride containing a trace of sulfuric acid (an earlier method of Morgan and Davies,[15] p. 608), underwent cyclization to the 9-p-tolyl-9-stibafluorene oxide (**19**; R = R′ = H, R″ = CH₃). Reduction with stannous chloride in hydrochloric acid then gave the tertiary stibine (**20**; R = R′ = H, R″ = CH₃).

This synthetic method, starting with 5-bromo-2-biphenylylamine and ethyl *p*-(dichlorostibino)benzoate gave ethyl *p*-(3-bromo-9-stibafluoren-9-yl)benzoic acid (**20**; R = Br, R′ = H, R″ = CO₂C₂H₅); hydrolysis gave the acid, which however with alkaloids formed salts that were too slightly soluble for fractional crystallization.

The same route, starting with ethyl 2′-amino-4-biphenylcarboxylate and *p*-tolylstibonous dichloride, gave 9-*p*-tolyl-9-stibafluorene-2-carboxylic acid (5-*p*-tolyldibenzostibole-3-carboxylic acid) (**20**; R = H, R′ = COOH, R″ = CH₃). Fractional recrystallization of the salts which this acid gave with (+)- and (−)-α-methylbenzylamine (= 1-phenylethylamine), followed by cautious removal of the amine, gave the free acids, $[\alpha]_D^{20}$ ±245° in pyridine; they were optically stable at 20° but underwent slow racemization at 40°. The acids were rapidly racemized by hydrochloric acid unless it was very dilute.

The optical activity of this acid could not be ascribed with certainty to the pyramidal configuration of the antimony atom, without definite evidence that the fused tricyclic system was flat. This system might, however, have had a 'skew' configuration in which the two *o*-phenylene groups, in order to accommodate the antimony atom, are twisted so that they are not coplanar but remain coaxial. In this case, a compound of type (**20**), but in which R = R′ = H and which therefore carries only the 9-*p*-substituted phenyl group, would be dissymmetric and capable of showing optical activity.

To investigate this factor, Campbell[16] prepared the 9-*p*-carboxyphenyl compound (**21**) by the previous synthetic route. This acid formed

(**21**)   (**22**)

crystalline salts with quinine, ephedrine and (+)-α-methylbenzylamine (= 1-phenylethylamine), but fractional crystallization of these salts gave no evidence of resolution. In contrast, 9-*p*-tolyl-9-stibafluorene-2-amine

(5-p-tolyldibenzostibole-3-amine) (**22**), which would be dissymmetric if the tricyclic system were flat and the antimony atom pyramidal, was readily resolved by crystallization of its (+)- and (−)-hydrogen tartrates, and gave the (+)-amine, $[\alpha]_D^{22}$ +250.5° ± 1°, and the (−)-amine, $[\alpha]_D^{23}$ −248.0° ± 2°, in benzene.

The active amine (**22**) underwent racemization even in warm solutions and could not be recrystallized without some loss in activity. The racemization was also very susceptible to traces of an adventitious catalyst, and a detailed kinetic study was therefore curtailed. The racemization of one batch of the (+)-base in benzene was, however, observed at 22°, 25.4°, 30°, and 40°, giving rate constants of 1.15, 1.61, 2.02, and 5.16 × 10^{-2} hour^{-1}, respectively, leading to an energy of activation of 15 kcal/mole.

The failure to resolve the compound (**21**) is of course only negative evidence; in the light of present knowledge, however, the attempted resolution was almost certainly doomed to failure.

Since both the resolved compounds (**20**; R = H, R′ = COOH, R″ = CH₃) and (**22**) had the salt-forming group in the 2-position and moreover had closely similar specific rotations, Campbell and Morrill[17] investigated the effect of change in position of the polar groups on the rotatory power and the optical stability, and the isolation of an active 9-stibafluorene suitable for the study of its racemization rate, in particular a compound whose racemization was not affected by adventitious catalysts.

The resolution of three compounds was investigated. For the first, p-(2-methyl-9-stibafluorenyl)benzoic acid (**23**; R = CH₃), ephedrine

(23)   (24)

proved the most effective reagent, but experimental difficulties prevented isolation of the optically pure acid, the highest rotation obtained being $[\alpha]_D$ +78.0°. This value is in marked contrast to the specific rotation $[\alpha]_D$ +245° of the positional isomer 9-p-tolyl-9-stibafluorene-2-carboxylic acid (**20**; R = H, R′ = COOH, R″ = CH₃).

The second compound was the methoxy analogue (**23**; R = OCH$_3$), the 2-methyl group of the previous acid being replaced by the more polar methoxyl group. This acid was also best resolved as the ephedrine salt, and the liberated acid had [α]$_D$ +153.0° ± 1°; the replacement had virtually doubled the specific rotation. The solubility of this acid in suitable solvents, although greater than that of the first acid, was still too low for the required racemization studies. A preliminary examination of the acid in pyridine indicated a half-life of 41.3 hours at 22° and 26.3 hours at 38°.

A third acid, selected for its probable greater solubility, was (9-*p*-tolyl-2-stibafluorenyloxy)acetic acid (3-carboxymethoxy-5-*p*-tolyl-dibenzostibole) (**24**) prepared by the interaction of 4-[(ethoxycarbonyl)-methoxy]-2-biphenylyldiazonium tetrachlorostibonate and *p*-tolylsti-bonous dichloride, followed by the usual cyclization. Resolution of this acid (**24**) with ephedrine gave the optically pure acids, [α]$_D$ ±112.0° in 0.3% (g/ml) pyridine solution, the rotation being considerably influenced by the concentration. The racemization of portions of the same batch of acid, observed in pure pyridine at the same concentration, and at 30°, 38°, 44°, and 46° were 3.8, 6.5, 10.4, and 12.7 × 10^{-4} min^{-1}, respectively, giving an energy of activation of 14.4 ± 0.5 kcal/mole, slightly lower than that of the compound (**22**).

The theoretical significance of these results, in conjunction with those obtained with the compound (**22**), are discussed by the authors.[17]

Campbell and White[18] have investigated the 9-stibafluorene 9-oxides for possible optical resolution. These compounds (**19**) were apparently obtained in the penultimate stage of the many syntheses of the 9-stibafluorenes (p. 600), but (except for one compound[16]) were not purified and identified. The exception was 9-[*p*-(ethoxycarbonyl)-phenyl]-9-stibafluorene 9-oxide (**25**), which after crystallization from

Sb=O

COOC$_2$H$_5$

(25)

Sb—OH

R

R'

(26)

ethyl acetate had m.p. 172–173° and gave an excellent analysis for the oxide.

The cyclization of the two stibinic acids (**26**; R = OCH$_3$, R′ = COOC$_2$H$_5$) and (**26**; R = OCH$_2$COOC$_2$H$_5$, R′ = CH$_3$) was now carefully examined, and in each case the product was the 9,9-dihydroxide, (**27**; R = OCH$_3$, R′ = COOC$_2$H$_5$) and (**27**; R = OCH$_2$COOC$_2$H$_5$, R′ = CH$_3$), respectively. Every compound was identified by analysis and by reduction to the corresponding known 9-stibafluorene. Neither hydroxide was affected by prolonged heating at 100° under reduced pressure.

By an alternative preparative method, bromine (one molar equivalent) was added to the appropriate 9-stibafluorenes in carbon tetrachloride to form the 9,9-dibromides (**28**; R = Br, R′ = H, R″ = CO$_2$C$_2$H$_5$),

(**27**)   (**28**)

(**28**; R = Br, R′ = H, R″ = CH$_3$) and (**28**; R = H, R′ = CO$_2$H, R″ = CH$_3$), each crystallizing from carbon tetrachloride with 1 mole of solvent. The first two of these dibromides, when treated with 10% ethanolic sodium hydroxide at 60° for 10 minutes, gave the corresponding very stable monohydrated 9,9-dihydroxides, whilst the third dibromide gave an insoluble and apparently polymeric material, which was also formed when the parent 9-*p*-tolyl-9-stibafluorene-2-carboxylic acid was oxidized in acetic acid by hydrogen peroxide.

This property of the 9-substituted 9-stibafluorenes of forming stable 9,9-dihydroxides instead of 9-oxides is common to many simpler tertiary stibines. Triphenylstibine forms a dihydroxide, C$_6$H$_5$Sb(OH)$_2$, which is stable up to its m.p., 212°, and the oxide is unknown. The recorded analyses of many tertiary stibine 'oxides' show that they are hydrated and are probably dihydroxides. The resolution of a dissymmetric 9-substituted 9-stibafluorene 9-oxide would have to be performed under anhydrous conditions throughout to avoid the intermediate formation of the 9,9-dihydroxide and the accompanying racemization.

Stiboranes

Wittig and Hellwinkel[19] have prepared some stiboranes, or antimony(v) compounds, containing the 9-stibafluorene (5-dibenzostibole)

nucleus, analogous in type to the phosphoranes (p. 88) and arsoranes (p. 382) recorded earlier by Wittig and his co-workers, although the stiboranes are more restricted in range.

9-Chloro-9-stibafluorene (5) was converted by the action of phenyl-lithium into 9-phenyl-9-stibafluorene (5-phenyldibenzostibole) (7). This compound in tetrahydrofuran when treated with Chloramine-T gave the corresponding N-(p-tolylsulfonimide) (p. 504), which without isolation was treated in turn with 2,2'-dilithiobiphenyl to form bis-(2,2'-biphenylylene)phenylstiborane (5-phenyl-5,5'-spirobi[dibenzosti-bole]) (29), m.p. 211–212° (39%). When an ethanolic solution of (29)

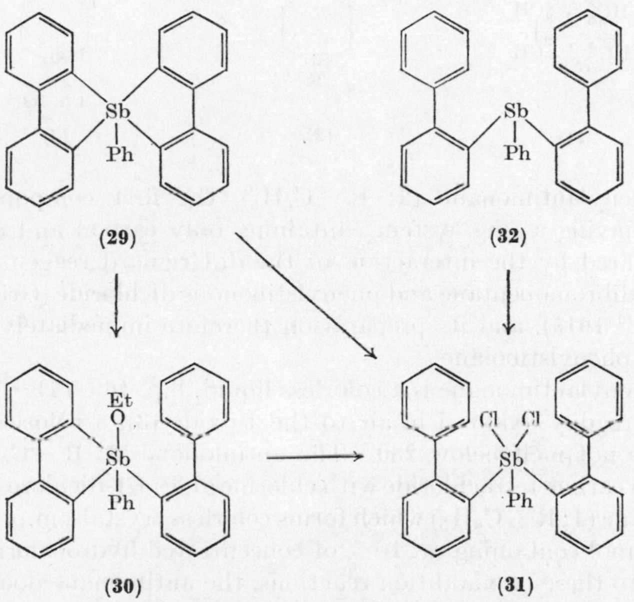

(29)   (32)

(30)   (31)

was boiled for several hours, one of the heterocyclic rings underwent fission to form 9-(2-biphenylyl)-9-ethoxy-9-phenyl-9-stibafluorene [5-(2-biphenylyl)-5-ethoxy-5-phenyldibenzostib(v)ole] (30), m.p. 149–151° (22%). This fission of the heterocyclic systems goes to completion when a suspension of the compound (29) or (30) is boiled with concentrated hydrochloric acid, giving in each case dichlorobis-(2-biphenylyl)phenylstiborane (31), m.p. 244.5–245.5° (87%, 90%). To confirm the identity of (31), phenylstibonous diiodide, $C_6H_5SbI_2$, was condensed with 2-lithiobiphenyl to give bis-(2-biphenylyl)phenylstibine (32), m.p. 116–118°, which when treated in chloroform with chlorine gave the compound (31).

Six-membered Ring Systems containing Carbon and One Antimony Atom

Antimonin and Antimonane

Antimonin (**1**) [RRI 287, in which it is shown with the Sb atom uppermost] is known only as 1-substituted derivatives of hexahydro-antimonin, known as antimonane (**2**; R = H).

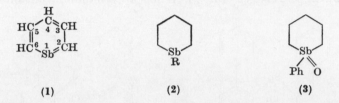

(1)   (2)   (3)

1-Phenylantimonane (**2**; R = C_6H_5), the first compound to be isolated having a ring system containing only carbon and antimony, was prepared by the interaction of the di-Grignard reagent obtained from 1,5-dibromopentane and phenylstibonous dichloride (Grüttner and Wiernik,[20] 1915), and its preparation therefore immediately preceded that of 1-phenylstibolane.

1-Phenylantimonane is a colorless liquid, b.p. 169–171°/18–20 mm, which is readily oxidized in air to the 1-oxide (**3**), a colorless powder that does not melt below 280°. The antimonane (**2**; R = C_6H_5) when treated in carbon tetrachloride with chlorine gives 1,1-dichloro-1-phenyl-antimonane (**4**; R = C_6H_5) which forms colorless crystals, m.p. 141–142°, from ethanol containing *ca.* 10% of concentrated hydrochloric acid. In contrast to these two addition reactions, the antimonane does not give

(4)   (5)

a mercuric chloride addition product and is unaffected when its solution in ethyl iodide is boiled for 3 hours. A methiodide has apparently not been recorded.

1-Methylantimonane (**2**; R = CH_3) was prepared much later by Steinkopf, Schubart, and Roch,[21] by interaction of the di-Grignard reagent from 1,5-dichloropentane and methylstibonous dichloride. It is a colorless liquid, b.p. 73–73.5°/17 mm, which when treated in petroleum solution with chlorine gives the 1,1-dichloro-1-methylantimonane (4; R = CH₃). This compound when heated to 160–185° loses methyl chloride with the formation of 1-chloroantimonane (5), b.p. 110–111°/13 mm. This decomposition is precisely similar to that shown by 1,1-dichloro-1-methylarsenane (p. 392) and is indeed a reaction of frequent occurrence with tertiary methylarsine dichlorides.

The antimonanes (2; R = C₆H₅ and CH₃) have unpleasant odors.

Earlier names for the antimonane system have been cyclopentamethylenestibine and stibacyclohexane.

Dibenz[b,e]antimonin

Dibenz[b,e]antimonin (1) [RRI 3614] has the structure and numbering shown; it is unknown, and the system is found in its 5-substituted

(1)   (2)

5,10-dihydro derivatives (2). The system (1) has in the past been named acridostibine and stibacridine, but these names, and earlier different notations of the ring system, should now become obsolete.

5-Substituted 5,10-dihydrodibenz[b,e]antimonins have been synthesized by Morgan and Davies[15] by a series of reactions closely parallel to those used earlier by Gump and Stoltzenberg[22] in their original synthesis of acridarsine derivatives (p. 420). For this purpose, o-benzylaniline (3) in hydrochloric acid solution was first diazotized and then treated with a solution of antimony trioxide in hydrochloric acid and with glycerol to check subsequent foaming. The agitated ice-cold mixture was then treated with aqueous 5N sodium hydroxide until evolution of nitrogen was complete. Colored impurities and free antimony oxide were first precipitated by saturating the solution with carbon dioxide, and the (o-benzylphenyl)stibonic acid (4) was then precipitated by hydrochloric acid. The stibonic acid was thus obtained as a colorless amorphous powder, sufficiently pure for subsequent synthetic purposes; the pure acid was obtained by recrystallization from an ethanol–dichloromethane mixture as pale needles, which were only slowly soluble in hot dilute aqueous sodium hydroxide.

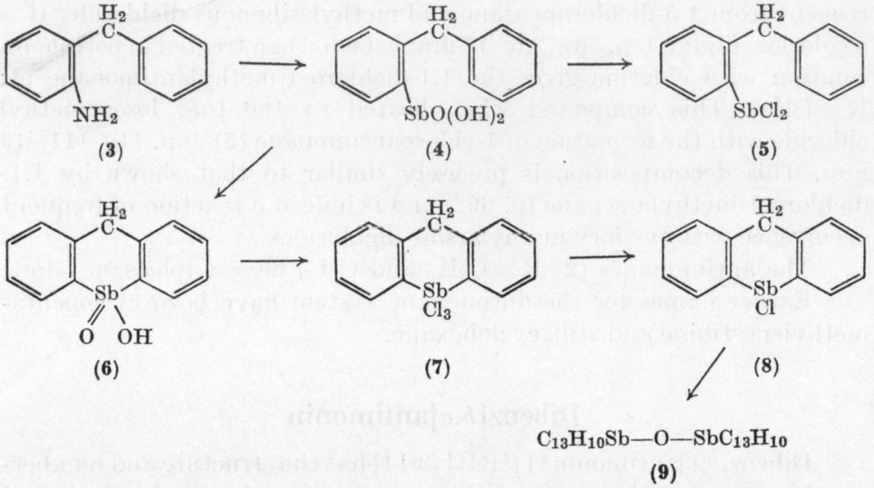

(3) → (4) → (5)

(6) → (7) → (8)

$$C_{13}H_{10}Sb-O-SbC_{13}H_{10}$$
(9)

A number of derivatives of this acid were prepared. For example, when a mixture of the acid with methanolic hydrochloric acid was reduced in the usual way with sulfur dioxide, the (o-benzylphenyl)-stibonous dichloride (5) was obtained as colorless needles, m.p. 129–130°. This compound in acetone solution when treated with sodium iodide readily gave the diiodide as golden-yellow crystals, m.p. 95°. When a similar solution of the dichloride was poured into ice-cold dilute aqueous ammonia, (o-benzylphenyl)oxostibine, $C_6H_5CH_2C_6H_4$—SbO, was obtained as a microcrystalline powder, m.p. 82–83°, which when treated with hydrobromic acid gave the dibromide as colorless plates, m.p. 221°.

Attempts to cyclize the dichloride (5) by heating it in a vacuum and so to obtain 5-chloro-5,10-dihydrodibenz[b,e]antimonin (8) failed. Attempts to cyclize the stibonic acid (4) by means of hot concentrated sulfuric acid also failed. Ultimately, cyclization was achieved by mixing the stibonic acid (4) with a considerable quantity of acetic anhydride containing 3–4% (by volume) of sulfuric acid, and then heating it at 100° for 3 hours. When the mixture was poured into warm water, the crude acid (6) was precipitated as a pale brown powder. This acid was washed, dried, and then treated with ice-cold hydrochloric acid, whereby the trichloride (7) was obtained. The latter was dissolved in methanol, reduced with sulfur dioxide, and finally treated with hydrochloric acid, the 5-chloro compound (8) then being isolated as colorless needles, m.p. 105°.

The 5-chloro derivative was characterized by direct conversion into the 5-iodo compound, yellow needles, m.p. 160–162°, and by hydrolysis with dilute aqueous ammonia to the anhydride (**9**) of the 5-hydroxy compound; the anhydride, in turn, when treated with hydrobromic acid, gave the 5-bromo derivative, colorless crystals, m.p. 112°. The compound (**9**) would now be termed 5,5'-oxybis(10H,10'H-dibenz[b,e]-antimonin).

The 5-iodo derivative in ethereal solution, treated with methyl-magnesium iodide, gave the 5,10-dihydro-5-methyldibenz[b,e]anti-monin (**10**), colorless crystals, m.p. 101°. This tertiary stibine, when

(**10**)   (**11**)

treated in dichloromethane solution with chlorine, gave the tertiary stibine dichloride, $C_{13}H_{10}{>}Sb(CH_3)Cl_2$, as colorless crystals which melted at 177–178° with evolution of gas. A similar addition of iodine gave a crude diiodide but all attempts to recrystallize this compound caused decomposition with the formation of methyl iodide and of 5,10-dihydro-5-iododibenz[b,e]antimonin. The crude dibromide decomposed similarly.

Attempts were made to prepare the 5-phenyl derivative by treating the 5-iodo compound with phenylmagnesium bromide, but the crude product underwent such ready oxidation that it was stabilized by addition of chlorine, and the 5,5-dichloro-5,10-dihydro-5-phenyldibenz-[b,e]antimonin (**11**) was isolated as colorless crystals, m.p. 221–223°.

The stibonous dichloride (**5**) showed the same ready dismutation under the influence of alkalis as that shown by 2-biphenylylstibonous dichloride (p. 598), and a similar series of derivatives was obtained. Thus, when the dichloride (**5**) in acetone solution was added to an excess of dilute aqueous ammonia, antimony oxide was split off and the anhydride (**12**) of bis-(o-benzylphenyl)hydroxystibine was formed as colorless needles, m.p. 117°. This anhydride, when treated in acetone solution with ice-cold concentrated hydrochloric acid, gave bis-(o-benzylphenyl)stibinous chloride (**13**), m.p. 87.5°. Addition of chlorine to the latter compound then furnished the trichlorostibine (**14**), colorless plates, m.p. 129°, which when heated at 150–180° at 18–20 mm slowly cyclized with loss of hydrogen chloride and the formation of 5-(o-benzylphenyl)-5,5-dichloro-5,10-dihydrodibenz[b,e]antimonin (**15**). It was difficult to purify the latter directly and it was therefore hydrolyzed to the corresponding 5-oxide (**16**), which after purification and treatment with hydrochloric acid gave the pure dichloride (**15**) as a microcrystalline

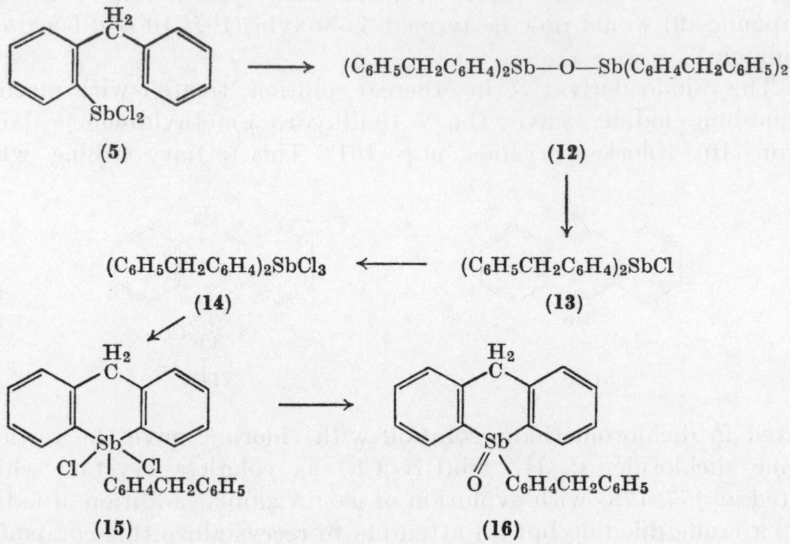

H2
C
SbCl2
(5)
(C6H5CH2C6H4)2Sb—O—Sb(C6H4CH2C6H5)2
(12)
(C6H5CH2C6H4)2SbCl3
(14)
(C6H5CH2C6H4)2SbCl
(13)
H2
C
Sb
Cl
Cl
C6H4CH2C6H5
(15)
H2
C
Sb
O
C6H4CH2C6H5
(16)

powder, m.p. 220–224° (dec.). No attempt to reduce this dichloride to the tertiary stibine is recorded.[15]

Five-membered Ring Systems containing Carbon and One Antimony and Two Oxygen Atoms

1,3,2-Dioxastibole and 1,3,2-Dioxastibolanes

1,3,2-Dioxastibole (**1**), [RRI 106], is known only as its 4,5-dihydro derivatives, the 1,3,2-dioxastibolanes, and these are encountered mainly as their 2-substituted products (**2**).

HC
O
SbH
HC
O
5
1
2
4
3
H2C
O
SbR
H2C
O

(1) **(2)**

When chlorodiethoxystibine (**3**) is heated with two equivalents of 1-butanol at 110–115°, transesterification occurs with the formation of dibutoxychlorostibine (**4**; R = H), b.p. 147–148°/10 mm; the use of 3-methyl-1-butanol similarly gives chlorodi-(3-methylbutoxy)stibine

(4; R = CH₃), b.p. 140–141°/3 mm (Arbusov and Samoĭlova[23]). This reaction contrasts with that of $(C_2H_5O)_2AsCl$, in which under these conditions the more reactive chlorine would attack the butanol.

(EtO)₂SbCl   ⟶   (CH₃CHRCH₂CH₂O)₂SbCl   ⟶

$$\begin{array}{c} \text{O} \\ | \quad \diagdown \\ \quad \quad \text{SbCl} \\ | \quad \diagup \\ \text{O} \end{array}$$

(3)   (4)   (5)

$$\text{Me} \begin{array}{c} \quad \text{O} \\ \diagdown \quad \diagdown \\ \quad \quad \text{SbCl} \\ \diagup \quad \diagup \\ \quad \text{O} \end{array} \quad \longrightarrow \quad \text{Me} \begin{array}{c} \quad \text{O} \\ \diagdown \quad \diagdown \\ \quad \quad \text{SbOEt} \\ \diagup \quad \diagup \\ \quad \text{O} \end{array}$$

(6)   (7)

This process can be repeated, for when the compound (4; R = CH₃) is heated with ethylene glycol for 1 hour at 100° and the volatile portion of the reaction mixture is then removed, the residue consists of 2-chloro-1,3,2-dioxastibolane (5), m.p. 190–194° (from ethanol). The use of 1,2-propanediol similarly gives 2-chloro-4-methyl-1,3,2-dioxastibolane (6), m.p. 112–116°. Both these compounds undergo hydrolysis on exposure to air.

The compound (6) when heated in a $C_2H_5OH-C_2H_5ONa$ solution gives 2-ethoxy-4-methyl-1,3,2-dioxastibolane, m.p. 130–133° (from ethanol) (Dubrovina-Samoĭlova[24]).

Some very interesting derivatives of 1,3,2-dioxastibolane have been described by Nerdel, Buddrus, and Höher,[25] (1964). Antimony penta-chloride reacts with sodium acetate to give the hygroscopic crystalline sodium antimony(v) hexaacetate (8). This compound oxidizes 2-hydroxy

Na[Sb(OOCCH₃)₆]   Na[Sb(O—CHR—CHROH)₆]
     (8)   (9)

ketones to 1,2-diketones, e.g., acetoin to biacetyl, and benzoin to benzil. It reacts with certain glycols to give the corresponding hexaalkoxides (9); thus ethylene glycol gives the salt (9; R = H) and 2,3-butanediol gives (9; R = CH₃).

The hexaacetate reacts, however, with pinacol in absolute dioxane at 90° with eviction of the acetate groups and the formation of sodium tris-[2,3-dimethyl-2,3-butanediolato(2−)] antimonate(v) (10) (needles, 97%). With meso-hydrobenzoin and the salt (8) in dioxane, 1–5 minutes' boiling gives incomplete eviction of the acetate groups, with formation

of sodium bis(acetato)bis-[1,2-diphenyl-1,2-ethanediolato(2−)]antimon-
ate(v) (**11**) (needles, 69%).

$$\text{Na}\left[\text{Sb}\left(\begin{array}{c}\text{O—CMe}_2\\|\\\text{O—CMe}_2\end{array}\right)_3\right]$$

$$\text{Na}\left[(\text{CH}_3\text{COO})_2\text{Sb}\left(\begin{array}{c}\text{O—CHPh}\\|\\\text{O—CHPh}\end{array}\right)_2\right]$$

(10)   (11)

Stiboranes of comparable structure can also be obtained by using
triphenylstibine oxide, $(C_6H_5)_3SbO$, prepared by the hydrogen peroxide
oxidation of triphenylstibine. A mixture of the oxide and pinacol, when
heated quickly to 110° and then at 120–130° for not more than $1\frac{1}{2}$
minutes, gives 4,4,5,5-tetramethyl-2,2,2-triphenyl-1,3,2-dioxastibolane
(**12**; R = R′ = CH₃), m.p. 92° (32%), which when treated in chloroform

$$\begin{array}{c}\text{R′RC—O}\\|\qquad\quad\text{SbPh}_3\\\text{R′RC—O}\end{array}$$

$$\begin{array}{c}\text{Me}_2\text{C—O}\\|\qquad\quad\text{SbPh}_2\\\text{Me}_2\text{C—O}\quad|\\\qquad\qquad\text{Br}\end{array}$$

(12)   (13)

solution with bromine gives 2-bromo-4,4,5,5-tetramethyl-2,2-diphenyl-
1,3,2-dioxastibolane (**13**), m.p. 133° (81%).

When, however, the mixture of the stibine oxide and pinacol is
heated at 170° for 15 minutes, methanolic extraction of the cold product
gives 2,2′ - oxybis - [4,4,5,5 - tetramethyl - 2,2 - diphenyl - 1,3,2 - dioxa -
stibolane] (**14**; R = R′ = CH₃), m.p. 175°, which when treated with

$$\begin{array}{c}\qquad\quad\text{Ph}\qquad\text{Ph}\\\text{R′RC—O}\ |\qquad\ |\ \text{O—CRR′}\\|\qquad\ \text{Sb—O—Sb}\qquad|\\\text{R′RC—O}\ |\qquad\ |\ \text{O—CRR′}\\\qquad\quad\text{Ph}\qquad\text{Ph}\end{array}$$

(14)

'damp' methanolic hydrogen chloride breaks down with the formation
of trichlorodiphenylstibine(v), $(C_6H_5)_2SbCl_3,H_2O$.

2,3-Butanediol, when heated with the oxide for 30 seconds at 60°,
yields the 4,4′,5,5′-tetramethyl derivative (**14**; R = H, R′ = CH₃),
m.p. 161.5° (70%), whereas *meso*-hydrobenzoin in dioxane yields
2,2,2,4,5-pentaphenyl-1,3,2-dioxastibolane (**12**; R = H, R′ = C₆H₅),
m.p. 168°.

It is noteworthy that the production of (13) and of the compounds (14) entails eviction of a phenyl group from the antimony atom without the heterocyclic rings being affected.

The identified products of thermal decomposition of the above compounds can be listed: (10), acetone; (11), benzaldehyde; (12; R = R' = CH$_3$), acetone and triphenylstibine; (13), acetone and bromobenzene; (14; R = H, R' = CH$_3$), acetone; (12; R = H, R' = C$_6$H$_5$), benzaldehyde, benzil, and triphenylstibine.

The nmr spectrum of the chloro compound (5) is closely similar to that of the corresponding dithio compound (p. 629).

2,6,7-Trioxa-1-stibabicyclo[2.2.1]heptane

This system, [RRI —], has the structure and numbering shown in (1); it consists therefore of two fused 1,3,2-dioxastibolane rings having the Sb—O—C portion in common. There is now no doubt that it forms

(1)   (2)

the fundamental heterocyclic system of the class of compound exemplified by potassium antimonyl tartrate or tartar emetic. This salt is obtained by treating a boiling aqueous solution of potassium hydrogen (+)-tartrate with antimony oxide, and it can be readily crystallized from water, in which its solubility is 8.3 g/100 ml at 25° and 33 g/100 ml at 100°. Its composition corresponds to the formula K[SbC$_4$H$_4$O$_7$],$\frac{1}{2}$H$_2$O, and its aqueous solution gives the normal reactions of the K$^+$ ion but none of the reactions of the Sb^{3+} ion.

The salt has been known for very many years and its structure has been the object of several investigations, but most of the suggested structures, being certainly incorrect, can now be discarded. Brief mention should be made, however, of the work of Reihlen and Hezel[26] (1931), who put forward the structure (2), in which the dotted line represents a 'residual valence'. They established the following points: (a) The half-molecule of water in the potassium salt has no structural significance, for the barium salt when treated with phenethylamine sulfate gives a crystalline salt of composition C$_6$H$_5$C$_2$H$_4$NH$_2$,HSbC$_4$H$_4$O$_7$, and p-nitrophenethylamine gives a salt of corresponding composition. (b) A molecular model indicates that the structure (2), when composed of (+)-tartaric units, would be strainless; it is significant, therefore, that attempts to prepare an analogous salt from potassium hydrogen *meso*-tartrate failed. (*c*) Barium antimonyl (+)-tartrate when treated with pyridine sulfate gave a pyridine antimonyl tartrate; when this compound

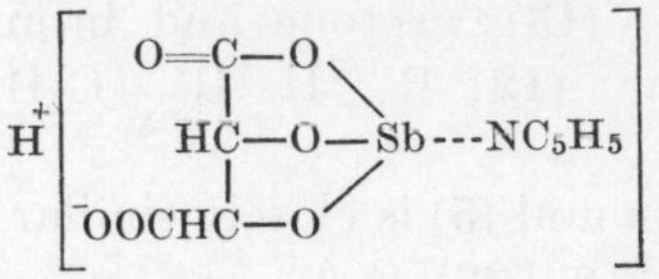

(3) 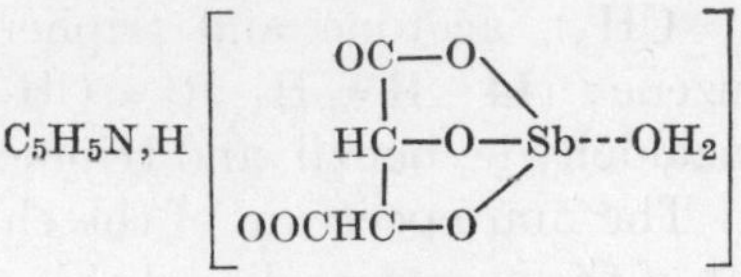   (4)

was dissolved in water, its optical rotation increased rapidly to that of an equimolecular solution of tartar emetic. Consequently the solid salt was considered to be an acid of constitution (3), which in aqueous solution changed to the pyridinium salt (4) that had regained the same cation as (2).

Reihlen and Hezel's structure (2) came very near to the truth, for a recent *X*-ray crystal analysis by Grdenić and Kamenar[27] (1965) shows that the potassium salt prepared from the (±)-tartrate (chosen to facilitate the analysis) has the structure (5): its systematic name, based

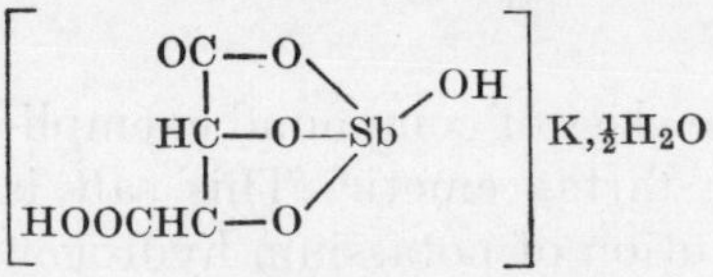

(5) 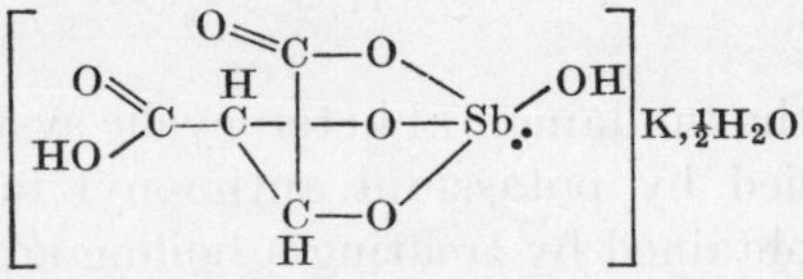   (6)

on (1), is therefore potassium 3-carboxy-1-hydroxy-5-oxo-2,6,7-trioxa-1-stibabicyclo[2.2.1]heptanoate, [RIS —], a name which is unlikely to displace the incorrect potassium antimonyl tartrate or the more homely tartar emetic except perhaps in indexes of specialized systematic nomenclature.

Grdenić and Kamenar consider that the structure (5), whose spatial arrangement is better depicted as (6), shows that the anion does not belong to the antimonyl series, but is a derivative of the antimonate anion $[Sb(OH)_4]^-$. They consider that the configuration of the antimony complex is that of a deformed trigonal bipyramid, four apices of which are occupied by oxygen atoms while one equatorial apex is occupied by the unshared electron pair: the bipyramid should be deformed 'just in the direction which approximates to a square pyramid'. Further, 'the final values of the O–Sb–O angles will show whether a deformed trigonal bipyramid or a deformed square pyramid is the more convenient for the description of the actual co-ordination round the antimony atom'.

For a study of the action of (+)-, (±)-, and *meso*-tartaric acids on antimony oxide, and for cryoscopic and infrared spectral investigations of tartar emetic, see Girard[28, 29] and Girard and Lecomte.[30]

For an account of the therapeutic action of tartar emetic, see Schubert[31] and also Bang and Hairston.[32]

1,3,2-Benzodioxastibole

This ring system (1), [RRI 1102], must be one of the earliest recognized carbon–antimony cyclic systems, for Rosing[33] (1858) refers

(1) — benzene ring fused to a five-membered ring containing O(1)–Sb(2)H–O(3); ring positions numbered 4, 5, 6, 7.

(2) — benzene ring fused to O–SbR–O five-membered ring.

(3) — benzene ring fused to O–Sb—O—C₆H₄COOH—O five-membered ring.

Sb—O—C$_6$H$_4$COOH

(1)   **(2)**   **(3)**

briefly to the 2-hydroxy derivative (2; R = OH), which Causse[34] (1898) examined in some detail. Brown and Austin[35] (1942) modified Causse's preparation of this compound; a solution of antimony trichloride in an excess of saturated aqueous sodium chloride (containing undissolved crystalline chloride) was neutralized with 2N-Na$_2$CO$_3$ solution until the precipitated antimony hydroxide just failed to dissolve, and was then added to a solution of pyrocatechol also in saturated brine; the precipitated product (2; R = OH) was collected, washed repeatedly by vigorous stirring with water, and obtained as microcrystals (70%) which were unaffected by heating to 300°.

If antimony trifluoride is added to pyrocatechol, each in aqueous solution, the 2-fluoro derivative (2; R = F) (55%) is precipitated; this compound when boiled with 2N-Na$_2$CO$_3$ solution is quantitatively converted into (2; R = OH).

The 2-hydroxy compound is insoluble in water, aqueous ammonia, and most organic solvents.

Causse[36] subsequently showed that an aqueous 'solution' of antimony trichloride, *i.e.*, one containing hydrochloric acid instead of sodium chloride, reacted with aqueous pyrocatechol to give the crystalline 2-chloro derivative (2; R = Cl), which was insoluble in the usual solvents. This compound was also prepared by the interaction of the trichloride and pyrocatechol in boiling benzene (Funk and Köhler[37]). It reacts vigorously with acetic anhydride to give pyrocatechol diacetate, C$_6$H$_4$(OOCCH$_3$)$_2$. Causse[36] also, by modifying the above conditions, prepared the 2-bromo (**2**; R = Br) and 2-iodo (**2**; R = I) derivatives from antimony tribromide and triiodide, respectively.

The compound (**2**; R = Cl) when boiled with water for 6 hours undergoes hydrolysis only to the hydroxy compound (**2**; R = OH), whereas the arsenic analogue, 2-chloro-1,3,2-benzodioxarsole, similarly treated, undergoes complete hydrolysis to pyrocatechol and arsenious and hydrochloric acids (p. 549).

When the moist 2-hydroxy compound is added portionwise to a hot aqueous solution of sodium salicylate, N-NaHCO$_3$ solution being added to keep the reaction mixture alkaline, the solution deposits the crystalline sodium salt of 2-(o-carboxyphenoxy)-1,3,2-benzodioxastibole [p-(1,3,2-benzodioxastibol-2-yloxy)benzoic acid] (**3**) (30%); use of m- and p-hydroxybenzoic acids gives the respective isomers, whereas use of thiosalicylic acid gives the 2-(o-carboxyphenylthio) derivative (50%).[35]

The simpler 2-phenoxy derivative[37] (**2**; R = OC$_6$H$_5$) is obtained by adding (**2**; R = Cl) to a benzene solution of phenol and ammonia; it is a stable compound, not decomposing below 300°.

The 2-chloro derivative (**2**; R = Cl) combines with pyridine to give a compound of composition C$_6$H$_4$ClO$_2$Sb,2C$_5$H$_5$N, m.p. 55–57°, and with quinoline one of composition C$_6$H$_4$ClO$_2$Sb,1.5C$_9$H$_7$N, m.p. 90°; it also reacts readily with piperidine to give the 2-piperidino derivative (**4**), which crystallizes with one molecule of piperidine (Funk and Köhler[38]).

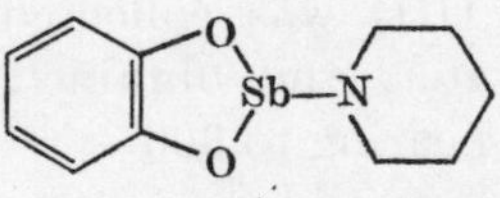

(4)

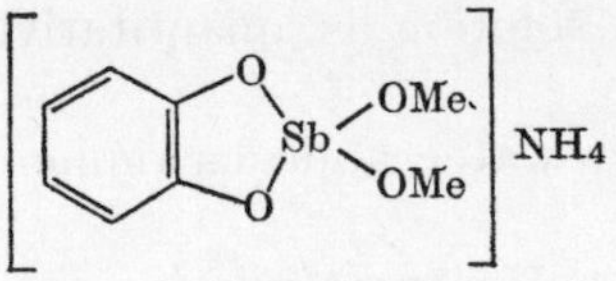

(5)

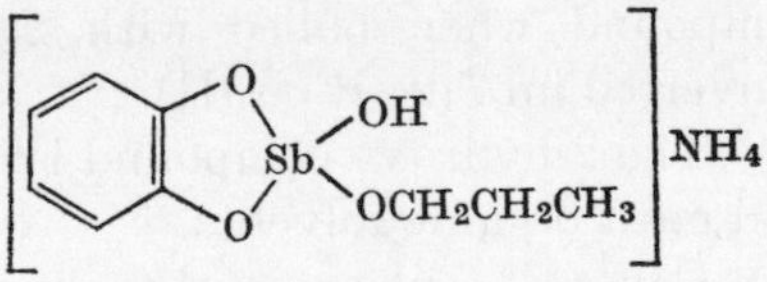

(6)

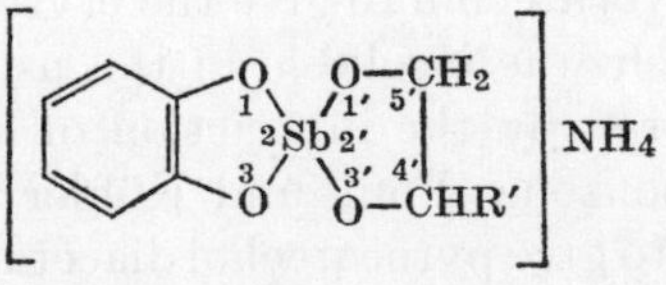

(7)

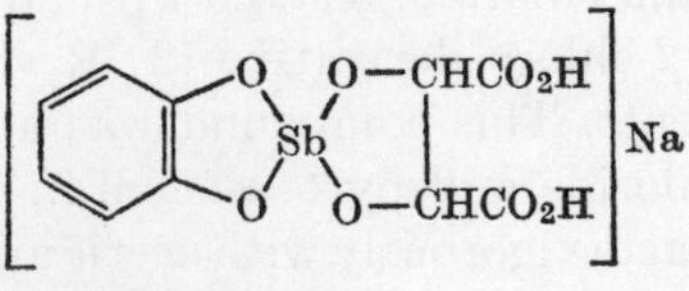

(8)

Wheeler and Banks[39] have claimed that when the compound (**2**; R = OH) is added to aqueous methanol containing ammonia, which is then heated to 80–90° for 15 minutes, the filtered cooled solution deposits the dimethoxy salt (**5**), whereas the use of aqueous 1-propanol gives the hydroxy-propoxy salt (**6**); similarly ethylene glycol gives the spirocyclic salt (**7**; R′ = H), and glycerol gives the similar salt (**7**; R′ = CH₂OH), isolated as the ammonium and the diethylammonium salts. The compound (**2**; R = OH) when added to an aqueous solution of tartaric acid and sodium hydrogen carbonate at 80°, affords the crystalline sodium salt (**8**). These salts are all soluble in water (the compounds **5**, **6**, and **7** giving nearly neutral solutions) and they may be recrystallized from water, methanol, or aqueous acetone.

Wheeler and Banks[40] also claim that the hydroxy compound (**2**; R = OH) combines with a variety of amino-alcohols, giving zwitterionic products of which (**9**) is an example.

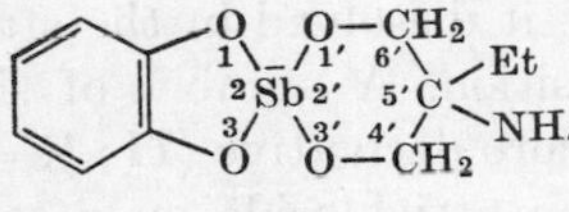

(**9**)

The compounds (**7**), (**8**), and (**9**) all contain novel heterocyclic systems which could prove of considerable interest. Unfortunately the compounds (**5**), (**6**), (**7**), and (**9**), which are all hydrated, have been identified solely by analyses for Sb and N, and no other evidence for their structure has been provided. The compound (**8**) is not hydrated, but its composition is based solely on its Sb content. Each paper contains a minimum of experimental detail.

If the above structures of these anions are established, the compounds would presumably be termed:

(**5**) Ammonium 2,2-dimethoxy-1,3,2-benzodioxastib(v)olate.

(**6**) Ammonium 2-hydroxy-2-propoxy-1,3,2-benzodioxastib(v)olate.

(**7**; R = H) Ammonium spiro[1,3,2-benzodioxastib(v)ole-2,2′-1′,3′,2′-dioxastib(v)ol]ate.

(**7**; R = CH₂OH) Ammonium 4′-(hydroxymethyl)spiro[1,3,2-benzodioxastib(v)ole-2,2′-1′,3′,2′-dioxastib(v)ol]ate.

(**8**) Sodium 4′,5′-dicarboxyspiro[1,3,2-benzodioxastib(v)ole-2,2′-1′,3′,2′-dioxastib(v)ol]ate.

(**9**) 5′-Ammonio-5′-ethylspiro[1,3,2-benzodioxastib(v)ole-2,2′-1′,3′,2′-dioxaantimon(v)in]ate.

21

Derivatives of 1,3,2-benzodioxastibole having substituents in the aromatic system could clearly be obtained by using the appropriate substituted pyrocatechol. Thus Causse and Bayard [41] showed that, when solutions of pyrogallol and of antimony trichloride, each in saturated brine, were mixed, a white flocculent precipitate of 2,4-dihydroxy-1,3,2-benzodioxastibole (10) was formed, and that this slowly changed to small dense crystals. Later, Causse [42] prepared this compound by the action of

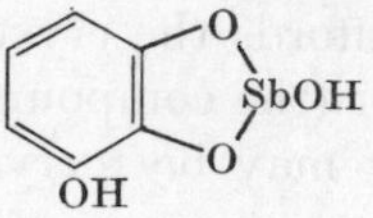

(10) 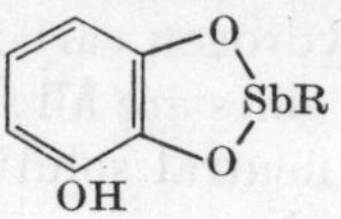   (11)

pyrogallol on a solution of antimony trioxide in aqueous tartaric acid; this method became widely used subsequently. Causse and Bayard [41] found that the compound (10) was insoluble in water and the usual neutral organic solvents; it dissolved in the strong mineral acids, but with complete decomposition. A number of simple derivatives were prepared. [42] Thus, the chloro derivative (11; R = Cl) was best prepared by the action of antimony trichloride on a solution of pyrogallol in methanol. The bromo derivative was similarly prepared and could be crystallized from dilute hydrobromic acid. The iodo, fluoro, and hydrogen oxalato derivatives were also prepared. All these compounds formed colorless crystals very similar in their general physical properties. Gallic acid gave 2-chloro-4-hydroxy-1,3,2-benzodioxastibole-6-carboxylic acid

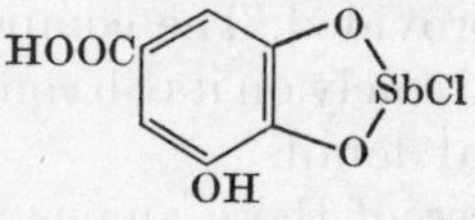

(12)

(12), and the potassium salt and the methyl ester were also obtained; the methyl ester was prepared by the action of methyl gallate on antimony trioxide in tartaric acid solution.

Christiansen [43] has widely extended the preparation of compounds of this class by using the antimony in the form of sodium antimonyl tartrate, which is more convenient than the less soluble potassium salt. A number of derivatives similar to (3) were obtained by boiling the appropriate 1,2-dihydric phenol with the sodium antimonyl tartrate in aqueous solution, although pyrocatechol did not react under these conditions. For this purpose he employed pyrogallol, gallic acid, its amide and various esters, (3,4,5-trihydroxybenzoyl)glycollic acid and its amide, (5'-carboxy-2',3'-dihydroxyphenoxy)acetic acid, and 2,3,4-trihydroxybenzaldehyde and its phenylhydrazone. The products were insoluble in water and the usual solvents and could not be purified by recrystallization; their composition depended to some extent on the conditions of their preparation, but analysis indicated that compounds approximated (usually closely) in composition to derivatives of type (10). It is stated that all these products were trypanocidally active.

Earlier names for the compound (2; R = OH) were antimonial catechol, pyrocatechyl antimonite, and o-phenylene hydrogen stibinate; the last name emphasizes its dual-ester nature.

2,2'-Spirobi[1,3,2-benzodioxastibole]

This parent compound (1), [RIS 10684], with the quinquevalent antimony atom as depicted, is unknown, but the hydrogen atom can be

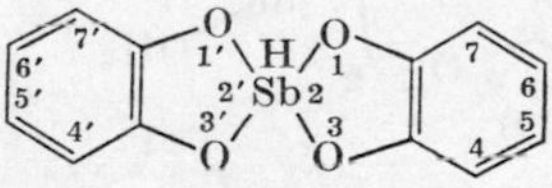

(1) 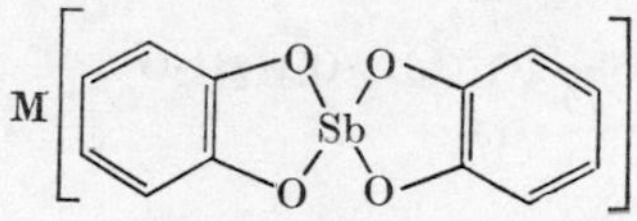   (2)

considered to separate as a proton, and the resulting anion to give salts of type (2), where M is a univalent cation, and the negative charge of the anion is represented formally as being on the antimony atom. The anion (2) is therefore very similar in type to those shown on p. 616.

A solution of pyrocatechol in aqueous ammonia reacts with antimony trioxide to give the ammonium salt (2; M = NH$_4$), very stable white crystals, which are soluble in water; the identity of these crystals is placed beyond reasonable doubt by their analysis (Rosenheim and Baruttschisky[44]). A much wider range of salts of the corresponding bismuth anion have been prepared (p. 640).

A compound which is probably of the type (2) has been prepared from the sodium salt of pyrocatechol-3,5-disulfonic acid and has been employed as a drug under the name of Fouadin (or Fuadin). It is, however, often depicted as having the structure (3). [The stability of

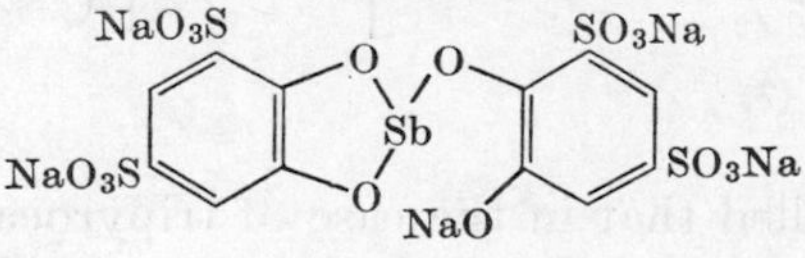

(3)

the ammonium salt (2; $M = NH_4$) precludes it having an 'open' structure of this type.]

Fouadin has been employed for the treatment of schistosomiasis and filariasis; for a general account of its therapeutic application, see Schubert[31] and Bang and Hairston.[32]

The possible existence of compounds having three pyrocatechol units linked to one antimony atom can conveniently be briefly noted here. The history of such compounds runs closely parallel to that of the apparently similar arsenic compounds (pp. 551 ff.).

Weinland and Scholder[45] stated that antimony pentoxide and pyrocatechol reacted to form 'tripyrocatechylantimonic acid', which formed yellowish-green crystals easily soluble in water. To this acid they attributed the structure (4), precisely similar to that which

$$H_3[O{=}Sb{-}(O \cdot C_6H_4 \cdot O){-}]_3 \, 6H_2O$$

(4)

$$KH_2[O{=}Sb{-}(O \cdot C_6H_4 \cdot O)]_3 \, \tfrac{1}{2}H_2O$$

(5)

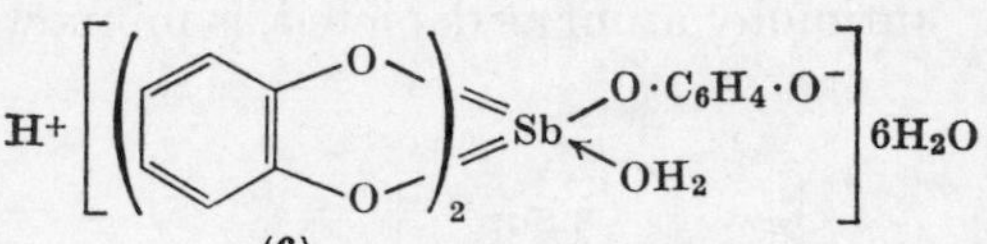

(6)

Weinland and Heinzler[46] had earlier given to the corresponding arsenic acid. They prepared a number of salts, such as the potassium salt (5), and showed that the acid and its alkali salts even in hot aqueous solution were decomposed only very slowly by hydrogen sulfide.

Reihlen, Sapper, and Kall[47] pointed out that all the salts prepared by Weinland and Scholder, with the exception of some complex mercury derivatives, were salts such as (5) in which the acid was acting as if it were monobasic and not tribasic. Consequently, they suggested the structure (6) for the acid, and in support of this formula they showed that—as in the case of the arsenic analogue—the coordinated water molecule could be replaced by a coordinated pyridine molecule, stable sodium and potassium pyridinotripyrocatechyl antimonates (7) being thus obtained.

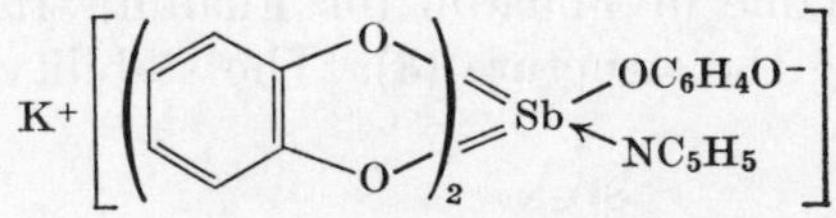

(7) 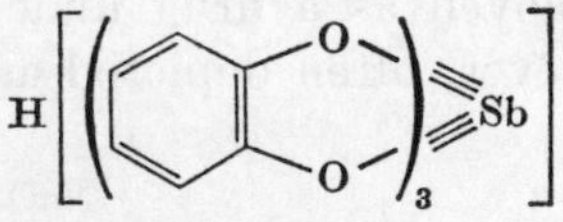   (8)

It will be recalled that in the case of tripyrocatecholarsenic acid, now termed tris[pyrocatecholato(2−)]arsenic(v) acid (p. 551), to which structures analogous to (4) and (6) had been attributed, Rosenheim and Plato[48] went one step further and asserted that all three pyrocatechol units were chelated (*i.e.*, joined by two bonds) to the arsenic atom, and their optical resolution of salts of this acid provided considerable, although not decisive, evidence for this structure; the position and linkage of the one tenaciously held molecule of water remained uncertain. By analogy, one might suspect that the antimonic acid might have the analogous structure (8); Rosenheim and Plato were aware of this possibility but state that so far the resolution of the antimonic acid has not been achieved. Consequently, there is at present no concrete evidence to support structure (8). The analogous salt having three 2,3-dimethyl-2,3-butanediolato groups linked to the antimony is, however, a well-defined stable compound (p. 611).

Six-membered Ring Systems containing Carbon and One Antimony and One Oxygen Atom

Phenoxantimonin

This compound has the structure and numbering shown in (1), [RRI 3411], but it is known only as 10-substituted derivatives which are

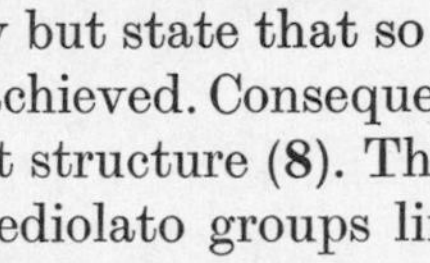

(1) 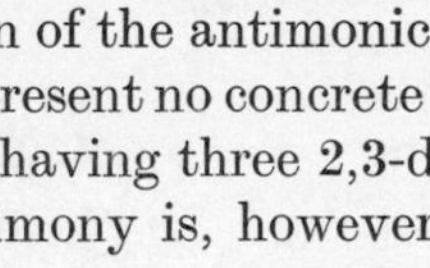   (2)

conveniently shown as (2). The system was initially named phenoxstibine.

The synthesis of the phenoxantimonin ring system has been investigated by Campbell,[49] who found that methods based on those successfully used for the analogous arsenic derivatives failed in the antimony series. For example, diazotization of 2-amino-4'-methyldiphenyl ether followed by the usual Bart reaction gave only a 5–7% yield of the 2-(*p*-tolyloxyphenyl)stibonic acid, although by a similar reaction Lesslie and Turner[50] had obtained a high yield of the corresponding arsonic acid (p. 562). Furthermore, attempts made to increase this yield by the modifications employed by Morgan and Davies[7, 15] (pp. 596, 607) gave no improvement.

The following synthesis was ultimately successful. 2,2'-Dibromo-4-methyldiphenyl ether (3) was converted into the double Grignard reagent (4) in about 80% yield, and this reagent was then treated with

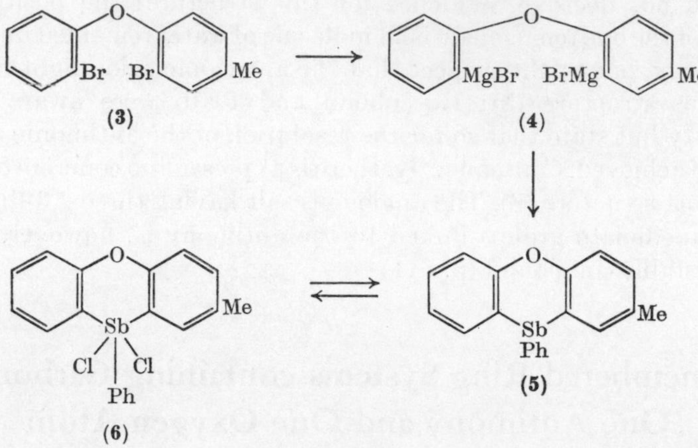

phenylstibonous diiodide, $C_6H_5SbI_2$, with the formation of 2-methyl-10-phenylphenoxantimonin (5). This compound was initially obtained only as a crude viscous oil, b.p. 200–210°/0.5 mm. It was therefore converted into the dichloride (6), which when recrystallized from petroleum formed colorless crystals, m.p. 140–142°. This was converted back into the phenoxantimonin (5) by treating it in ethanolic solution with aqueous ammonia until a precipitate of the oxide began to appear; hydrogen sulfide was then passed through the mixture to reduce the oxide, and the 2-methyl-10-phenylphenoxantimonin (5) ultimately obtained as colorless crystals, m.p. 62–63°. The overall yield of the pure stibine (5), based on the phenylstibonous diiodide used, was 30–35%.

The prime object of the above synthesis was an investigation of the stereochemistry of the phenoxantimonin ring system. Campbell[49] points out that physical measurements show that the molecules of the antimony trihalides are pyramidal, with an intervalency angle of about 100°. This value is very close to that of arsenic, which is about 97° (p. 463). Consequently, if this intervalency angle persisted in the phenoxantimonin series, the molecule (5), for example, would be folded about the O–Sb axis in the same manner as Lesslie and Turner's phenoxarsine compounds[50,51,52] were folded about the O–As axis (p. 562), and Chatt and Mann's arsanthrene compounds[53] were folded about the As–As axis (p. 464), and Davis and Mann's phosphanthrene compounds[54] were folded about the P–P axis (p. 186). Consequently, a phenoxantimonin of type (5) should be resolvable into optically active forms if it possessed an acidic or basic group for salt formation with a suitable optically active base or acid.

Attempts made for this purpose to oxidize the 2-methyl group in the phenoxantimonin (5) to a carboxylic group proved impracticable. The di-Grignard reagent (4) was therefore treated with (p-cyanophenyl)-stibonous diiodide to give 10-(p-cyanophenyl)-2-methylphenoxantimonin (7), m.p. 82°, but attempted hydrolysis of the cyano group caused disruption of the heterocyclic ring system. The use of (p-bromophenyl)-stibonous diiodide gave 10-(p-bromophenyl)-2-methylphenoxantimonin (8), but the bromophenyl group in this compound would not form a Grignard reagent whereby the bromine might ultimately have been replaced by a carboxylic group. Finally, the di-Grignard reagent (4) was treated with p-(ethoxycarbonyl)phenylstibonous dichloride, C_2H_5OOC—C_6H_4—$SbCl_2$, and 10-p-(ethoxycarbonyl)phenyl-2-methyl-phenoxantimonin obtained, although never in yields greater than 10%. This ester, when hydrolyzed with ethanolic sodium hydroxide, ultimately furnished the 10-p-carboxyphenyl-2-methylphenoxantimonin [p-(2'-methyl-10'-phenoxantimoninyl)benzoic acid] (9). The heterocyclic ring in all these compounds proved very stable toward alkalis but was very sensitive to acids; sulfur dioxide in the presence of hydrochloric acid could cause fission of the ring and even elimination of elementary antimony unless the conditions were carefully controlled.

The acid (9) was resolved into its optically active forms by means of strychnine, which gave diastereoisomeric salts of widely different solubilities.[49] The (+)-acid was thus isolated having $[\alpha]_D$ +77.5° in chloroform, +89.6° in benzene, and +87.6° in ethanol; the (−)-acid had $[\alpha]_D$ −77.2° in chloroform and −89.9° in benzene. This acid did not possess the considerable optical stability of the phenoxarsine derivatives

CN	Br	COOH
(7)	(8)	(9)

(p. 563). Thus,[49] although the acid was optically unaffected in boiling chloroform and benzene, it underwent slow racemization in boiling ethanol. Furthermore, 1 hour's boiling of the (+)-acid in aqueous sodium hydroxide solution reduced the activity to $[\alpha]_D$ +55.3°, while after the ammonium salt of the (+)-acid had been kept in aqueous solution at room temperature for 14 days it furnished the free acid having only $[\alpha]_D$ +12.5°.

The physical properties of the 10-substituted 2-methylphenoxantimonins synthesized by Campbell[49] are collected in Table 1.

Table 1. Derivatives of 10-substituted 2-methylphenoxantimonins

10-Substituent	Stibine, m.p. (°)	Stibine dichloride, m.p. (°)
Phenyl	62–63	140–142
p-Cyanophenyl	82	180–181
p-Bromophenyl	116	202
p-(Ethoxycarbonyl)phenyl	136–137	—
p-Carboxyphenyl (±)	201	—
p-Carboxyphenyl (+) and (−)	192	—

It should be added that the (±)-acid (9) has been tested for trypanocidal activity by administration by intravenous injection in mice infected with *Trypanosoma equiperdum*. These tests showed that the acid possessed low toxicity but also very little trypanocidal activity.

Six-membered Ring Systems containing Carbon and One Antimony and Two Oxygen Atoms

4H-1,3,2-Dioxantimonin

This system (1), [RRI —], is known only as substitution products of its 5,6-dihydro derivative, named 1,3,2-dioxantimonane (2), which itself has been very little studied.

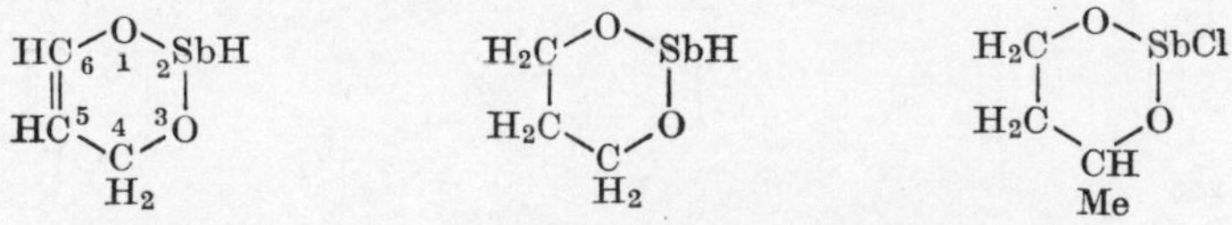

(1)   (2)   (3)

A disubstituted derivative of (2) can be obtained by a transesterification process, precisely similar to that used for the preparation of 1,3,2-dioxastibolanes (p. 610). Bis-(3-methylbutoxy)stibinous chloride,

$[(CH_3)_2CHCH_2CH_2O]_2SbCl$, when heated with 1,3-butandiol gives 2-chloro-4-methyl-1,3,2-dioxantimonane (**3**), m.p. 160–165°, with the regeneration of 3-methyl-1-butanol (Dubrovina-Samoĭlova[24]). The use of other 1,3-diols would presumably afford a number of similar disubstituted derivatives of the dioxantimonane (**2**).

Five-membered Ring Systems containing Carbon and One Antimony, One Oxygen and One Sulfur Atom

1,3,2-Oxathiastibole and 1,3,2-Oxathiastibolanes

1,3,2-Oxathiastibole (**1**), [RRI 99], occurs only as 2-substituted products of the 4,5-dihydro derivative, 1,3,2-oxathiastibolane (**2**).

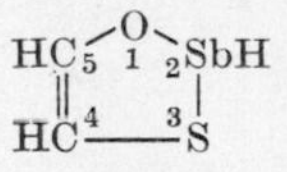

(1) 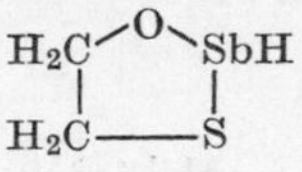   (2) 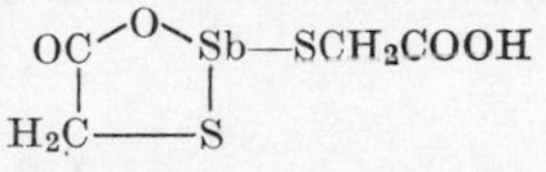   (3)

Klason and Carlson[55] have shown that mercaptoacetic acid, $HSCH_2COOH$, (three equivalents), when added to a solution of antimony trichloride in dilute hydrochloric acid, gives 2-(carboxymethylthio)-5-oxo-1,3,2-oxathiastibolane (**3**), colorless crystals, only moderately soluble in water. Ramberg[56] prepared this compound by boiling a $\frac{2}{3}$N-aqueous solution of mercaptoacetic acid with antimony trioxide, and also prepared a number of salts in all of which the compound (**3**) is acting as a monobasic acid. Holmberg[57] confirmed the production of Klason and Carlson's compound, and refuted the earlier claim of Rosenheim and Davidson[58] that antimony formed a mercaptoacetic acid derivative of composition $Sb(SCH_2COOH)_3,12H_2O$ (identified solely by an Sb analysis), an error subsequently admitted by Rosenheim.[59] It is noteworthy, however, that arsenic does form a compound $As(SCH_2COOH)_3$, which has been described by Rosenheim and Davidsohn[58] and by Klason and Carlson.[55]

There is very little doubt that the acid (**3**), having a sharp m.p. 201–202°, does contain the heterocyclic ring as depicted (**3**).

Formation of cyclic compounds of this type is shown by other α-mercapto carboxylic acids; for example, 2-mercaptopropionic acid gives a similar compound (4), m.p. 192° (Volmar and Weil[60]). As would be expected, these cyclic compounds are formed only by the free mercapto acids. If the latter are esterified, normal non-cyclic derivatives of

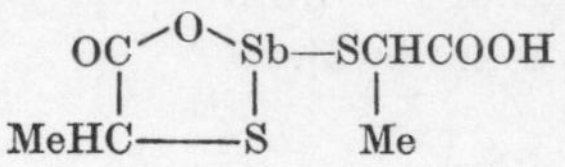

$$OC \overset{O}{\diagdown} Sb-SCHCOOH$$

MeHC——S   Me

(4)

$$Sb(SCH_2COOC_2H_5)_3$$

(5)

antimony are formed: for example, the ethyl ester of mercaptoacetic acid reacts with anhydrous antimony trioxide to give the compound (5).

The sodium salt of (3), which is readily soluble in water, has been used as a drug to combat kala azar, schistosomiasis, and filariasis.

The use of the compound (3) and many similar 2-substituted 1,3,2-oxathiastibolanes having a CH_2 or a CO group in the 5-position, has been claimed for stabilizing acrylonitrile–vinyl chloride copolymers, particularly against the development of color with age or on heating. The patent specification gives no chemical information regarding these stabilizers (Longemann, Malz, Rachwalsky, and Süling[61]).

1,3,2-Benzoxathiastibole

This system (1), [RRI —], is apparently known only as a 2-substituted derivative. Brown and Austin[35] claim that when an ethanolic

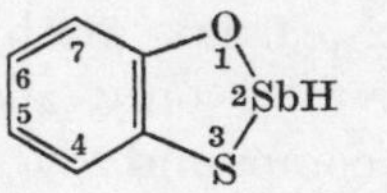

(1) 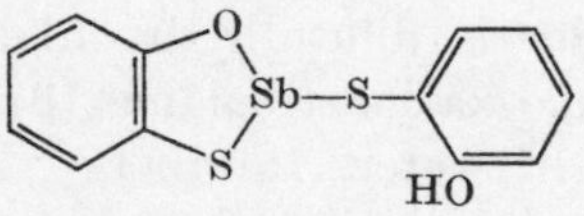   (2)

solution of monothiocatechol (o-mercaptophenol) is boiled with fresh precipitated arsenious oxide for 6 hours, working up the reaction mixture yields 2-(o-hydroxyphenylthio)-1,3,2-benzoxathiastibole (2), faintly yellow crystals from ethanol, not melting below 250° (25% yield on the thiol used). Furthermore, all attempts to prepare the simpler 2-hydroxy or 2-halogeno derivatives of (1) by using monothiocatechol for the ring formation also gave the compound (2), termed 'antimonial monothio-catechol o-hydroxythiophenol'. The compound (2) does not react with tartar emetic (cf. p. 630).

The structure (2) was allocated to this compound, however, solely on the basis of an Sb determination, and it clearly requires more rigorous confirmation.

Five-membered Ring Systems containing Carbon and One Antimony and Two Sulfur Atoms

1,3,2-Dithiastibole and 1,3,2-Dithiastibolanes

1,3,2-Dithiastibole (**1**), [RRI 113], is the fundamental member in this series, but it is encountered almost solely as 2-substituted products of its 4,5-dihydro derivative, 1,3,2-dithiastibolane (**2**; R = H). This

(1)   (2)   (3)

reduced system has earlier received several other names, such as 1,3-dithia-2-stibacyclopentane, and cyclo-2,5-dithia-3,4-dimethylenestibine, the latter indicating also an earlier notation of the ring constituents.

1,2-Ethanedithiol (one equivalent), when added to antimony trichloride dissolved in cold concentrated hydrochloric acid, which is then heated on a water-bath for 15–20 minutes, gives 2-chloro-1,3,2-dithiastibolane (**2**; R = Cl), colorless crystals, m.p. 124° (from benzene). Oxidation with hydrogen peroxide gives an amorphous powder, whose composition indicates that it is 2-chloro-2,2-dihydroxy-1,3,2-dithiastibolane (**3**) (Clark[62]).

p-Tolylstibonous dichloride also reacts with the dithiol to give 2-*p*-tolyl-1,3,2-dithiastibolane (**2**; R = $C_6H_4CH_3$-*p*), dimorphous, each form having m.p. 90°.

The chloro compound (**2**; R = Cl) reacts with mercaptoacetic acid in pyridine solution to form the crystalline pyridine salt, m.p. 101°, of the 2-(carboxymethylthio) derivative, (**2**; R = SCH_2COOH); the pure free acid could not be isolated.[62]

Friedheim in a series of patent specifications[63-66] has recorded a number of 2-substituted 1,3,2-dithiastibolanes, formed by the condensation of various 1,2-dithiols with arylstibonous dihalides, $RSbX_2$, or with the corresponding sodium arylstibonates, $RSbO_3HNa$; for good yields, however, these stibonates require the presence of ammonium or potassium mercaptoacetate in order to reduce them to the corresponding intermediate oxostibines (p. 575).

The general conditions for the preparation of the 1,3,2-dithiastibolanes and the similar but larger ring systems which follow are almost identical with those given by Friedheim for the analogous arsenic compounds (pp. 577, 579) and preparative directions common to the arsenic and antimony members in each series are usually given in his specifications.

He records the condensation of sodium p-aminophenylstibonate with 2,3-dimercapto-1-propanol (BAL, p. 575) to form 2-(p-amino-phenyl)-4-(hydroxymethyl)-1,3,2-dithiastibolane (**4**); and the p-acet-amido-, the 4′-acetamido-2′-chloro-, the p-hydroxy-, the 3′-amino-4′-hydroxy- and the 3′-acetamido-4′-hydroxy-phenyl analogues, prepared

(4)   (5)

from the corresponding phenylstibonates, are also recorded. Other members, similar in type to (**4**), but having the CH_2OH group replaced by the CH_3 or the COOH group, are obtained by using 1,2-propane-dithiol and 2,3-dimercaptopropionic acid, respectively.

In the 1,3,2-dithiastibolane series and the following analogous ring systems, Friedheim also frequently uses the sodium salt of p-(4′,6′-diamino-s-triazin-2′-ylamino)phenylstibonic acid (**5**) (p. 578) for linking this substituted phenyl group to the antimony atom, with the aid of the usual reducing agent.

An X-crystal study of the structure of 2-chloro-1,3,2-dithiastibolane (**2**; R = Cl) has proved of considerable interest (Bush, Lindley, and Woodward[67]). The unit cell contains four molecules, two of which are mirror images of the other two. The ring system is non-planar (Figure 1), showing two enantiomorphs, the atoms being numbered as in (**1**). The ring system can be regarded as folded about the S(1)–S(3) axis, the plane on one side of the axis being defined by the Sb–S(1)–S(3), and that on the other side defined by S(1), S(3), and the midpoint between C(4) and C(5); the angle subtended by these planes is 168°. The angle between the Cl—Sb bond and the line joining the Sb atom to the midpoint of S(1)–S(3) is 99°.

The Sb, S(1), S(3), and C(5) atoms are in fact almost coplanar, the C(5) atom deviating by only 0.1 Å from the Sb, S(1), S(3) plane; the C(4) atom is well above this plane. The mean intervalency angles are Cl–Sb–S(1), 94.9°; Cl–Sb–S(3), 98°; Sb–S(1)–C(5), 100.3°; Sb–S(3)–C(4), 96.2°; (S1)–C(5)–C(4), 111.2°; S(3)–C(4)–C(5), 111.6°. The interatomic distances (Å) are Sb–Cl, 2.46; Sb–S(1), 2.40; Sb–S(3), 2.41; S(1)–C(5), 1.84; S(3)–C(4), 1.84; C(4)–C(5), 1.49.

Foster and Fyfe[69] have recorded that the nmr spectra of 2-chloro-1,3,2-dioxastibolane (p. 613) and of 2-chloro-1,3,2-dithiastibolane (**2**; R = Cl); both show a single sharp line for the CH_2 absorption. The antimony atom has undoubtedly the pyramidal configuration, but the inversion time of, for example, trimethylstibine is long compared with

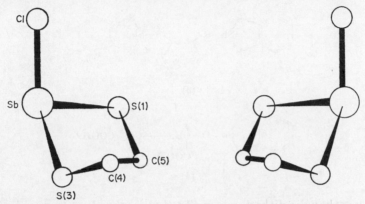

Figure 1. The two enantiomorphs of crystalline 2-chloro-1,3,2-dithiastibolane (**3**; R – Cl). (Adapted, by permission, from M. A. Bush, P. F. Lindley and P. Woodward, *Chem. Commun.*, **1966**, 149)

the probable limit of resolution in the nmr spectrum. Foster and Fyfe consequently consider that the observed equivalence of the CH_2 groups in the above compounds is unlikely to result from inversion at the antimony atom: they consider that the larger size of the antimony atom than of the chlorine atom masks the stereochemical effect of the chlorine on the CH_2 groups. The same considerations apply to the corresponding chloroarsenic compounds (pp. 544, 574).

1,3,2-Benzodithiastibole

This system (**1**), [RRI 1106], is encountered, as one would expect, almost exclusively as its 2-substituted derivatives (**2**).

Brown and Austin[35] have shown that a benzene solution of antimony trichloride and *o*-benzenedithiol, $C_6H_4(SH)_2$, when boiled for 3 hours

(1)

(2)

and cooled, deposits 2-chloro-1,3,2-benzodithiastibole (**2**; R = Cl),
m.p. 174–175° (40%); antimony tribromide similarly gives the 2-bromo
compound, m.p. 162–163° (33%). An ethanolic solution of the dithiol,
boiled with freshly precipitated antimony oxide, gives the 2-hydroxy
compound (**2**; R = OH). Antimony trifluoride in water, treated with
the dithiol, gives a yellow product, the analysis of which indicated that
it was 2,2'-o-phenylenedithiobis-(1,3,2-benzodithiastibole) (**3**).

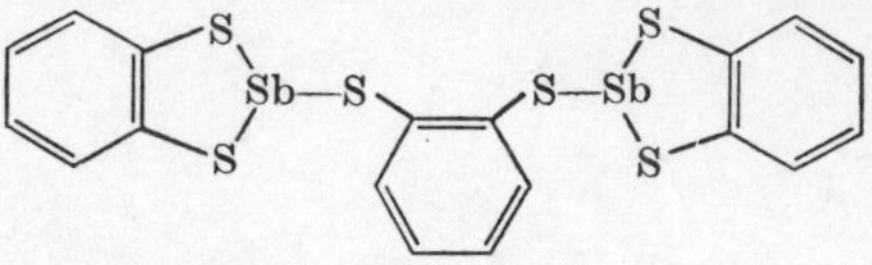

(**3**)

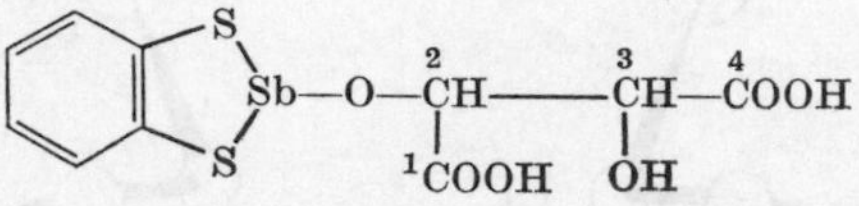

(**4**)

The ethanolic dithiol, when added to a stirred aqueous solution of
potassium antimonyl tartrate, gives the yellow crystalline potassium
salt (60%) of 2-(1',3',2'-benzodithiastibol-2'-yloxy)-3-hydroxysuccinic
acid (**4**); a solution in N-Na$_2$CO$_3$ solution, made just acid to litmus with
N-HCl gives the free acid (**4**).

The compounds (**2**; R = OH), (**3**), and the potassium salt of (**4**)
do not melt below 300°.

Friedheim [63-66] has recorded in patent specifications that sodium
(p-aminophenyl)stibonate reacts with the dithiol in acetic acid solution
containing potassium mercaptoacetate to give 2-p-aminophenyl-1,3,2-
benzodithiastibole (**2**; R = C$_6$H$_4$NH$_2$-p). He also records the 2-p-
(4'6'-diamino-s-triazin-2'-ylamino)phenyl derivative.

Six-membered Ring Systems containing Carbon and One Antimony and Two Sulfur Atoms

4H-1,3,2-Dithiantimonin and 1,3,2-Dithiantimonanes

4H-1,3,2-Dithiantimonin, [RRI 223], has the numbering and
presentation shown in (**1**); its 5,6-dihydro derivative, 1,3,2-dithi-
antimonane (**2**), is always encountered as its 2-substituted products.

Friedheim[63-66] has recorded the preparation of various substituted 1,3,2-dithiantimonanes under precisely the conditions employed for

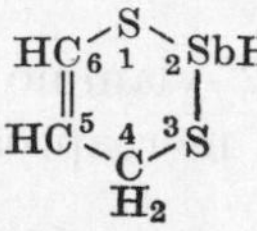

HC$_6$ 1 $_2$SbH / HC^5 $_4$ $_3$S / C / H$_2$ (S top)

(1)

H$_2$C S SbH / H$_2$C C S / H$_2$

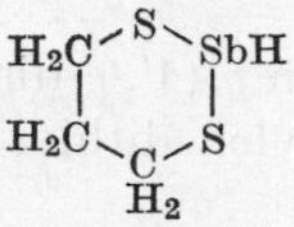

(2)

1,3,2-dithiastibolanes (p. 627), *i.e.*, by direct condensation of a 1,3-dithiol with an arylstibonous dihalide, or by indirect action of a 1,3-dithiol on an arylstibonic acid in the presence of a reducing agent such as ammonium mercaptoacetate. As examples of his products, 1,3-dimercapto-2-propanol and sodium (3-amino-4-hydroxyphenyl)stibonate gave 3-(3'-amino-4'-hydroxyphenyl)-5-hydroxy-1,3,2-dithiantimonane (3); 2,2-bis(mercaptomethyl)-1,3-propanediol, $(HOCH_2)_2C(CH_2SH)_2$, and

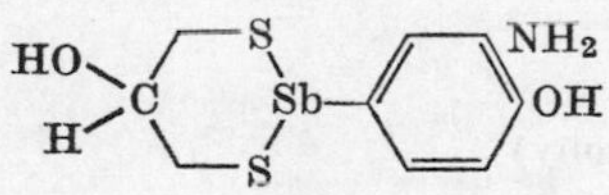

HO C Sb S NH$_2$ OH / H S

(3)

HOCH$_2$ C Sb S NHCOCH$_3$ / HOCH$_2$ S

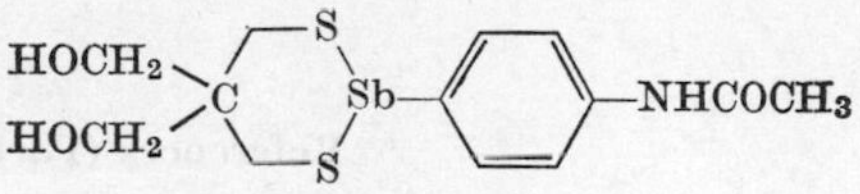

(4)

sodium *p*-(acetylamino)phenylstibonate gave 2-*p*-(acetylamino)phenyl-5,5-bis(hydroxymethyl)-1,3,2-dithiantimonane (4).

The ring system (2) was earlier termed 1,3-dithia-2-stibacyclohexane and 1,3,2-dithiastibinane.

Seven-membered Ring Systems containing Carbon and One Antimony and Two Sulfur Atoms

2,4,3-Benzodithiastibepin

This fundamental system (1) [RRI 1814] will clearly be encountered mainly as 3-substituted derivatives of the 1,5-dihydro-2,4,3-benzo-dithiastibepin (2).

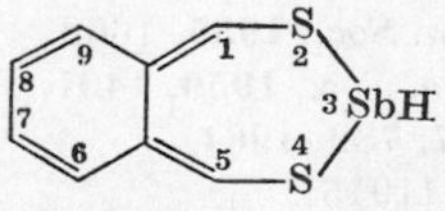

9 8 1 S 2 / 7 6 5 3 SbH / S 4

(1)

S SbH S

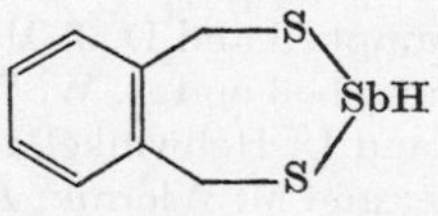

(2)

The preparation of these derivatives has been recorded by Fried-heim[63-66] using the same method as that employed for 1,3,2-dithi-antimonane, but using o-xylene-αα'-dithiol, $C_6H_4(CH_2SH)_2$, instead of a 1,3-dithiol.

Thus sodium p-(4',6'-diamino-s-triazin-2'-ylamino)phenylstibonate reacts with o-xylenedithiol and potassium mercaptoacetate in acetic

(3)

acid to give 3-p-(4',6'-diamino-s-triazinyl-2'-amino)phenyl-1,5-dihydro-2,4,3- benzodithiastibepin (3).

References (Part III, Antimony)

1 F. C. Leavitt, T. A. Manuel, and F. Johnson, *J. Amer. Chem. Soc.*, **81**, 3163 (1959).
2 F. C. Leavitt, T. A. Manuel, F. Johnson, L. U. Matternas, and D. S. Lehman, *J. Amer. Chem. Soc.*, **82**, 5099 (1960).
3 F. C. Leavitt and F. Johnson (to Dow Chemical Co.), U.S. Pat. 3,116,307 (Dec. 31, 1963); *Chem. Abstr.*, **60**, 6872 (1964).
4 E. H. Braye, W. Hübel, and I. Caplier, *J. Amer. Chem. Soc.*, **83**, 4406 (1961).
5 K. W. Hübel, E. H. Braye, and I. H. Caplier (to Union Carbide Corporation), U.S. Pat. 3,115,140 (Sept. 29, 1964); *Chem. Abstr.*, **61**, 16097 (1964).
6 G. Grüttner and E. Krause, *Ber.*, **49**, 437 (1916).
7 G. T. Morgan and G. R. Davies, *Proc. Roy. Soc. London, A*, **127**, 1 (1930).
8 A. W. von Hofmann, *Proc. Roy. Soc. London, A*, **12**, 389, 576 (1862–1863).
9 P. May, *J. Chem. Soc.*, **101**, 1037 (1912).
10 A. B. Bruker and N. M. Nikiforova, *Zh. Obshch. Khim.*, **18**, 1133 (1948); *Chem. Abstr.*, **43**, 1737 (1949).
11 A. B. Bruker, *Zh. Obshch. Khim.*, **18**, 1297 (1948); *Chem. Abstr.*, **43**, 4647 (1949).
12 D. M. Heinekey and I. T. Millar, *J. Chem. Soc.*, **1959**, 3101.
13 D. Sartain and M. R. Truter, *J. Chem. Soc.*, **1963**, 4414.
14 I. G. M. Campbell, *J. Chem. Soc.*, **1950**, 3109.
15 G. T. Morgan and G. R. Davies, *Proc. Roy. Soc. London, A*, **143**, 38 (1933).
16 I. G. M. Campbell, *J. Chem. Soc.*, **1952**, 4448.
17 I. G. M. Campbell and D. J. Morrill, *J. Chem. Soc.*, **1955**, 1662.
18 I. G. M. Campbell and A. W. White, *J. Chem. Soc.*, **1959**, 1491.
19 G. Wittig and D. Hellwinkel, *Chem. Ber.*, **97**, 789 (1964).
20 G. Grüttner and M. Wiernik, *Ber.*, **48**, 1473 (1915).
21 W. Steinkopf, I. Schubart, and J. Roch, *Ber.*, **65**, 409 (1932).

[22] W. Gump and H. Stoltzenberg, *J. Amer. Chem. Soc.*, **53**, 1428 (1931).

[23] B. A. Arbusov and D. Samoïlova, *Izv. Akad. Nauk SSSR, Khim. Nauk*, **1955**, 435; *Bull. Acad. Sci. USSR, Div. Chem. Sci.*, **1955**, 385; *Chem. Abstr.*, **50**, 6298 (1956).

[24] O. D. Dubrovina-Samoïlova, *Zh. Obshch. Khim.*, **28**, 2933 (1958); *Chem. Abstr.*, **53**, 9034 (1959).

[25] F. Nerdel, J. Buddrus, and K. Höher, *Chem. Ber.*, **97**, 124 (1964).

[26] H. Reihlen and E. Hezel, *Annalen*, **487**, 213 (1931).

[27] D. Grdenić and B. Kamenar, *Acta Cryst.*, **19**, 197 (1965).

[28] M. Girard, *Compt. Rend.*, **239**, 1638 (1954); *Chem. Abstr.*, **49**, 6758 (1955).

[29] M. Girard, *Bull. Soc. Chim. France*, **1955**, 571; *Chem. Abstr.*, **49**, 10708 (1955).

[30] M. Girard and J. Lecomte, *J. Phys. Radium*, **17**, 9 (1956); *Chem. Abstr.*, **50**, 10532 (1956).

[31] J. Schubert, *Amer. J. Trop. Med.*, **28**, 121 (1948).

[32] F. B. Bang and N. G. Hairston, *Amer. J. Hyg.*, **44**, 348 (1946).

[33] A. Rosing, *Compt. Rend.*, **46**, 1139 (1858).

[34] H. Causse, *Compt. Rend.*, **114**, 1072 (1892).

[35] H. P. Brown and J. A. Austin, *J. Amer. Chem. Soc.*, **63**, 2054 (1941).

[36] H. Causse, *Compt. Rend.*, **125**, 954 (1897).

[37] H. Funk and H. Köhler, *J. Prakt. Chem.*, **13**, 322 (1961).

[38] H. Funk and H. Köhler, *J. Prakt. Chem.*, **14**, 226 (1961).

[39] L. M. Wheeler and C. K. Banks, *J. Amer. Chem. Soc.*, **70**, 1264 (1948).

[40] L. M. Wheeler and C. K. Banks, *J. Amer. Chem. Soc.*, **70**, 1266 (1948).

[41] H. Causse and C. Bayard, *Compt. Rend.*, **115**, 507 (1892).

[42] H. Causse, *Ann. Chim. Phys.* [vii], **14**, 526 (1898).

[43] W. G. Christiansen, *J. Amer. Chem. Soc.*, **48**, 1365 (1926).

[44] A. Rosenheim and I. Baruttschisky, *Ber.*, **58**, 891 (1925).

[45] R. F. Weinland and R. Scholder, *Z. Anorg. Chem.*, **127**, 343 (1923).

[46] R. F. Weinland and J. Heinzler, *Ber.*, **52**, 1316 (1919).

[47] H. Reihlen, A. Sapper, and G. A. Kall, *Z. Anorg. Chem.*, **144**, 218 (1925).

[48] A. Rosenheim and W. Plato, *Ber.*, **58**, 2000 (1925).

[49] I. G. M. Campbell, *J. Chem. Soc.*, **1947**, 4.

[50] M. S. Lesslie and E. E. Turner, *J. Chem. Soc.*, **1934**, 1170.

[51] M. S. Lesslie and E. E. Turner, *J. Chem. Soc.*, **1935**, 1268.

[52] M. S. Lesslie and E. E. Turner, *J. Chem. Soc.*, **1936**, 730.

[53] J. Chatt and F. G. Mann, *J. Chem. Soc.*, **1940**, 1184.

[54] M. Davis and F. G. Mann, *J. Chem. Soc.*, **1964**, 3770.

[55] P. Klason and T. Carlson, *Ber.*, **39**, 732 (1906).

[56] L. Ramberg, *Ber.*, **39**, 1356 (1906).

[57] B. Holmberg, *Z. Anorg. Chem.*, **56**, 385 (1907).

[58] A. Rosenheim and I. Davidsohn, *Z. Anorg. Chem.*, **41**, 231 (1904).

[59] A. Rosenheim, *Z. Anorg. Chem.*, **57**, 359 (1908).

[60] Y. Volmar and E. Weil, *Comp. Rend.*, **207**, 534 (1938).

[61] H. Longemann, H. Malz, H. Rachwalsky, and C. Süling, French Pat. 1,395,398 (to Farbenfabriken Bayer A.G.) (March 1st, 1965); *Chem. Abstr.*, **63**, 18337 (1965).

[62] R. E. D. Clark, *J. Chem. Soc.*, **1932**, 1826.

[63] E. A. H. Friedheim, Brit. Pat. 655,435 (July 1951); *Chem. Abstr.*, **47**, 144 (1953).

[64] E. A. H. Friedheim, U.S. Pat., 2,659,723 (Nov. 1953); 2,664,432 (Dec. 1953); *Chem. Abstr.*, **47**, 144 (1953); **49**, 1816 (1955).

[65] E. A. H. Friedheim, U.S. Pat., 2,772,303 (Nov. 1956); *Chem. Abstr.*, **51**, 5836 (1957).

[66] E. A. H. Friedheim, U.S. Pat., 3,035,052 (May 1962); *Chem. Abstr.*, **57**, 11240 (1962).

[67] M. A. Bush, P. F. Lindley, and P. Woodward, *Chem. Commun.*, **1966**, 149.

[68] M. A. Bush, P. F. Lindley, and P. Woodward, *J. Chem. Soc., A*, **1967**, 221.

[69] R. Foster and C. A. Fyfe, *Spectrochim. Acta*, **21**, 1785 (1965).

PART IV

Heterocyclic Derivatives of Bismuth

Five-membered Ring Systems containing Carbon and One Bismuth Atom

Dibenzobismole

The ring system (1) is termed bismole [RIS —], but simple substituted derivatives are apparently unknown. The system (2), [RIS —],

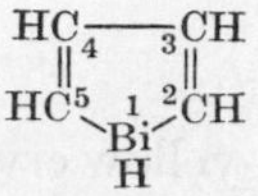

HC$_4$———$_3$CH   HC^5 $_1$ ^{2}CH   Bi H

(1) 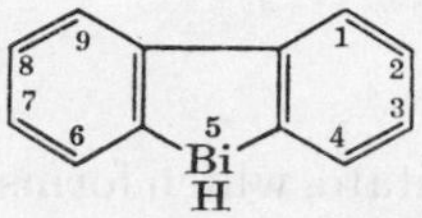   (2) 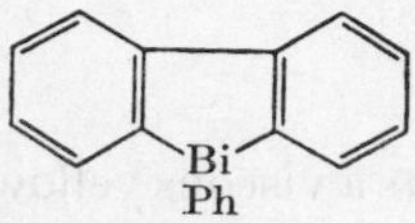   (3)

is thus dibenzobismole, of which one derivative has been recorded (Wittig and Hellwinkel[1]).

When 2,2'-dilithiobiphenyl in ether is added to diiodophenylbismuthine, $C_6H_5BiI_2$, in dry ether–tetrahydrofuran, and the reaction mixture is boiled, cooled, and hydrolyzed, it yields 5-phenyldibenzobismole (3), m.p. 167–168° (72%). This compound (not unexpectedly) does not react with chloramine-T (p. 504) in boiling dioxane solution.

The compound (3) has been termed phenyl-2,2'-biphenylylenebismuthine.[1]

Hellwinkel and Bach[2] (1969) have extended this line of research by showing that when ether–petroleum solutions of dichlorotriphenylbismuthine, $(C_6H_5)_3BiCl_2$, and 2,2'-dilithiobiphenyl are mixed at room temperature and, after 1 hour, cooled to −70°, orange crystals of 5,5,5-triphenyldibenzobismole (4) (64%) are deposited: it can be recrystallized from ether under nitrogen and then has m.p. 130–136° (dec.).

When an ethereal suspension of this 'bismuthorane' is chilled to −70° and treated with a saturated ethereal solution of hydrogen chloride, the heterocyclic ring is split with the formation of 2-biphenylyltriphenylbismuthonium chloride (5; X = Cl), which decomposes at 131–133°. The addition of saturated aqueous solutions of sodium nitrate or sodium tetraphenylborate to an aqueous solution of the chloride causes deposition of the nitrate (5; X = NO$_3$), m.p. 99–101°, and the tetraphenylborate [5; X = B(C$_6$H$_5$)$_4$], m.p. 202–214° (dec.), respectively. These two salts have noteworthy stability.

An ethereal suspension of the triphenyl compound (4), when
treated with phenyllithium, also in ether, and then cooled to −70°,

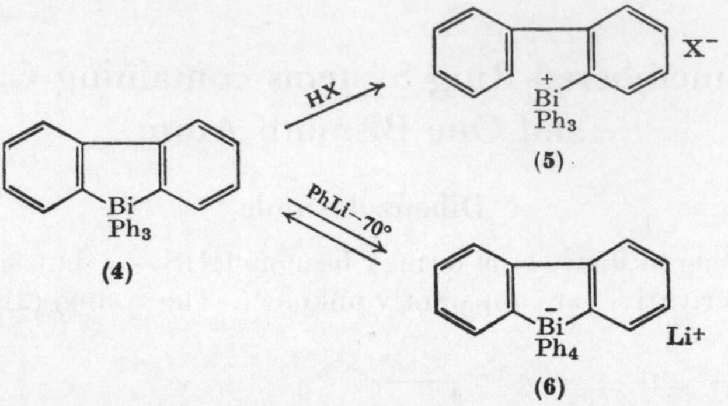

deposits a viscous yellow precipitate, which forms lemon-yellow crystals:
these dissociate into their components if the temperature is allowed to
rise towards 20°. These crystals have not been analyzed, but Hellwinkel
and Bach consider that they are the ionic lithium 5,5,5,5-tetraphenyl-
dibenzobismolate (6), being apparently analogous in type to lithium
hexaphenylbismuthate, $Li^+[Bi(C_6H_5)_6]^-$, which Hellwinkel and Kilthau[3]
had earlier prepared.

These authors term the compounds (4) and (5) 2,2′-biphenylylene-
triphenylbismuth and 2-biphenylyltriphenylbismuthonium chloride,
respectively.

Six-membered Ring System containing Carbon and One Bismuth Atom

Bismin and Bisminane

Bismin (1), [RRI 274, which depicts the bismuth uppermost], is the
unknown systematic precursor of the hexahydro derivative, bisminane
(2; R = H). This in turn is known apparently only as the 1-ethylbis-
minane (2; R = C_2H_5), which Grüttner and Wiernik[4] prepared by inter-

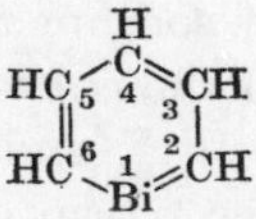

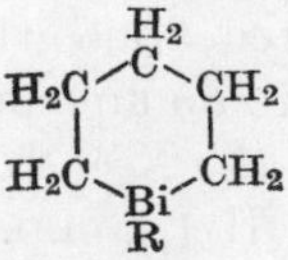

action of ethylbismuth dibromide, $C_2H_5BiBr_2$, and the di-Grignard reagent obtained from 1,5-dibromopentane, $CH_2(CH_2CH_2MgBr)_2$, in ethereal solution under hydrogen. It was obtained as a malodorous, faintly yellow, oily liquid, b.p. 108–112°/18–20 mm, very susceptible to oxidation and hence rapidly becoming cloudy on exposure to air. Grüttner and Wiernik record that a piece of paper which has absorbed the bisminane will on exposure to the air burst into flame within a few seconds with the production of thick yellowish smoke.

Grüttner's synthesis of 1-ethylbisminane is of course precisely similar to that which he and his co-workers employed for the preparation of the phosphorus, arsenic and antimony analogues (pp. 100, 391, 606).

The bisminane system (2) has previously been termed pentamethylenebismuthene, cyclopentamethylenebismuth, and bismacyclohexane.[4]

Five-membered Ring Systems containing Carbon, One Bismuth Atom and Two Oxygen Atoms

1,3,2-Benzodioxabismole

This compound (1), [RIS 9909], has been obtained as the 2-hydroxy derivative (2), a compound very similar to its antimony analogue (p. 615)

in its stability and insolubility in most liquids. The study of (2) and of its aryl-substituted derivatives was pursued in some detail by French chemists in the early years of this century.

Richard[5] showed that a bismuth salt in aqueous solution reacted with pyrocatechol to give the lemon-yellow insoluble 2-hydroxy-1,3,2-benzodioxabismole (2). He obtained similar compounds from homocatechol (4-methylpyrocatechol), protocatechuic acid (3,4-dihydroxybenzoic acid), pyrogallolcarboxylic acid (2,3,4-trihydroxybenzoic acid), gallic acid (3,4,5-trihydroxybenzoic acid), and methyl gallate. The formation of these compounds, and the fact that 1,3- and 1,4-dihydric phenols did not give these characteristic yellow derivatives, supported the structure (2) which he had given to the pyrocatechol product.

Thibault in a series of papers also investigated the products from gallic acid,[6, 7, 8] pyrogallolcarboxylic acid,[9] and protocatechuic acid.[10]

These were sometimes prepared by the action of the aqueous 1,2-dihydric phenol on pure hydrated bismuth oxide, since the anhydrous oxide gave a very slow action; the reaction mixture at room temperature was shaken occasionally over a period of several days. From those derivatives having a nuclear COOH group, he isolated crystalline alkali and ammonium salts, and in some cases the amide and the p-tolylamide. His product from protocatechuic acid was identical with Richard's in appearance and composition.

It is significant that p-hydroxybenzoic acid gave a compound of composition $(C_7H_5O_3)_3Bi$, colorless needles which were easily decomposed by water; 2,4-dihydroxybenzoic acid gave a compound corresponding in composition to $C_7H_4O_4{>}BiOH$, also colorless needles.[11]

The accurate estimation of bismuth by precipitation with gallic acid as the insoluble dihydrated carboxy-dihydroxy derivative of (2) has been described (Dick and Mihai[12]); the precipitate, of composition $C_7H_9BiO_8$, (43.08% Bi), is collected, thoroughly washed, dried, and weighed.

2,2'-Spirobi[1,3,2-benzodioxabismole]

This compound (1), [RIS 12966], is not known with a 5-covalent bismuth atom as shown, but (as in the antimony analogue, p. 619) the

(1)   (2)

hydrogen can be regarded as forming a proton, and consequently salts of type (2), in which M^+ represents a univalent cation, are readily formed.

Rosenheim and Baruttschisky[13] have shown that an aqueous solution of pyrocatechol containing sodium carbonate, when boiled under nitrogen with bismuth carbonate, gives the crystalline monohydrated sodium salt (2; M = Na), which with barium chloride solution deposits the corresponding barium salt; they have also isolated a crystalline solvent-free pyridine salt (2; M = C_5H_5NH). A saturated solution of pyrocatechol in aqueous ammonia reacts with bismuth

$$M[Bi(C_6H_4O_2)_2]C_6H_4(OH)_2$$

(3)

carbonate to give a salt of composition (3; M = NH_4) with a molecule of water; this compound, when boiled in aqueous solution with an excess of the carbonate, gives the 'normal' ammonium salt (2; $M = NH_4$). Pyrocatechol in aqueous potassium carbonate similarly gives with bismuth carbonate an anhydrous salt of composition (3; $M = K$). All these compounds form yellow crystals, and their composition has been determined by analyses for bismuth, carbon, and (where appropriate) nitrogen, sodium, or potassium.

Two structural points arise here. Compounds of the type (3) differ by only two hydrogen atoms from the corresponding tris[pyrocatecholato(2−)]bismuthate salt (4). The evidence for this type of compound

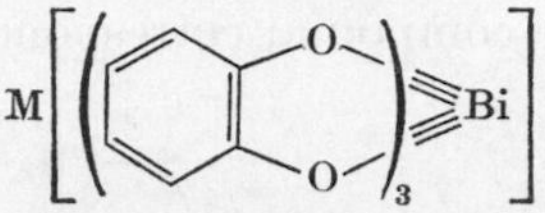

(4) 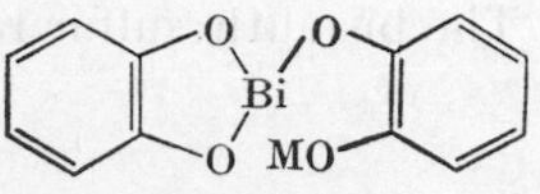   (5)

is strong in the arsenic series (p. 551) and indecisive in the antimony series (p. 620). If it does exist in the bismuth series, the ready shedding of the third pyrocatechol unit would be expected. There is, however, no decisive evidence for the existence of salts of the type (4).

Secondly, the existence of the stable pyridine and ammonium salts of composition $M[Bi(C_6H_4O_2)_2]$, where M is the pyridinium or ammonium cation, is strong evidence that these salts have the spirocyclic anion (2) and not the 'open' structure (5), since the phenolic group would not form stable salts with such bases.

The application of modern spectroscopic methods would undoubtedly clarify the structures of many of the bismuth compounds noted in this and neighboring Sections.

Five-membered Ring Systems containing Carbon, One Bismuth Atom and Two Sulfur Atoms

1,3,2-Dithiabismole and 1,3,2-Benzodithiabismole

Our knowledge of these two related systems is so scanty that they can be considered together.

1,3,2-Dithiabismole (1), [RRI 69], is recorded only as derivatives of its 4,5-dihydro product, 1,3,2-dithiabismolane (2).

1,3,2-Benzodithiabismole, [RRI —], has the structure (3), giving 2-substituted derivatives (4).

$$
\begin{array}{cccc}
\underset{\text{HC}}{\overset{\text{HC}}{}}\overset{S}{\underset{S}{\rangle}}\text{BiH} & \underset{\text{H}_2\text{C}}{\overset{\text{H}_2\text{C}}{}}\overset{S}{\underset{S}{\rangle}}\text{BiH} & & \\
(1) & (2) & (3) & (4)
\end{array}
$$

Friedheim records in two patent specifications[14, 15] the preparation of various compounds having the above ring systems, but no decisive evidence is given even for the composition of his products, and their chemical identity remains uncertain. Thus when a methanolic solution of 2,3-dimercapto-1-propanol (BAL, p. 575) is added to a solution of bismuth oxide in hydrochloric acid at 40°, a yellow precipitate is formed. 'The bismuth:sulfur ratio in the compound corresponds to the

$$
\begin{array}{ccc}
\text{(5)} & & \text{(6)}
\end{array}
$$

formula' (5), *i.e.*, to 2-chloro-4-(hydroxymethyl)-1,3,2-dithiabismolane.[14] When, however, an aqueous solution of sodium bismuth citrate is treated with ammonia until its pH is 8–9 and is then added to 2,3-dimercapto-1-propanol (two molar equivalents) in aqueous methanol, a yellow precipitate is formed, in which the ratio of bismuth to sulfur 'is 1:3, corresponding to the probable formula' (6).[14] The use of 2,3-dimercaptosuccinic acid is claimed to give a compound of analogous type.[15] Other similar examples are given.

When ethanolic *o*-benzenedithiol is added to an equimolar quantity of bismuth hydroxide in hydrochloric acid, the yellow precipitate 'thus obtained has the probable formula' (4; R = Cl), *i.e.*, 2-chloro-1,3,2-benzodithiabismole.[14]

References (Part IV, Bismuth)

1 G. Wittig and D. Hellwinkel, *Chem. Ber.*, **97**, 789 (1964).
2 D. Hellwinkel and M. Bach, *Annalen*, **720**, 198 (1969).
3 D. Hellwinkel and G. Kilthau, *Annalen*, **705**, 66 (1967).
4 G. Grüttner and M. Wiernik, *Ber.*, **48**, 1473 (1915).
5 E. Richard, *J. Pharm. Chim.* [vi], **12**, 145 (1900).
6 P. Thibault, *J. Pharm. Chim.* [vi], **14**, 487 (1901).
7 P. Thibault, *Ann. Chim. Phys.* [vii], **25**, 268 (1902).
8 P. Thibault, *Bull. Soc. Chim. France* [iii], **29**, 531 (1903).

[9] P. Thibault, *Bull. Soc. Chim. France* [iii], **29**, 680 (1903).

[10] P. Thibault, *Bull. Soc. Chim. France* [iii], **31**, 176 (1904).

[11] P. Thibault, *Bull. Soc. Chim. France* [iii], **31**, 36 (1904).

[12] J. Dick and F. Mihai, *Stud. Cercet. Sti., Chim. (Baza Cercet. Sti., Timisoara), Ser. I*, **2**, 97 (1955); *Chem. Abstr.*, **50**, 15329 (1956).

[13] A. Rosenheim and I. Baruttschisky, *Ber.*, **58**, 891 (1925).

[14] E. A. H. Friedheim, Brit. Pat. 712,828 (Aug. 4, 1954); *Chem. Abstr.*, **49**, 15946 (1955).

[15] E. A. H. Friedheim, Brit. Pat. 716,647 (Oct. 13, 1954); *Chem. Abstr.*, **49**, 12530 (1955).

Author Index

Numerals in square brackets are reference numbers for the Parts denoted by P, As, Sb, or Bi. Other numerals are page numbers where these references are cited either by the author's name and/or by the reference number. Thus, the first item denotes that E. W. Abel is an author of references numbered 305 and 493 in the phosphorus Part cited on pages 219 and 327 and also of reference numbered 141 in the arsenic Part cited on pages 502 and 503.

Bibliographic details will be found on pages immediately preceding the appropriate colored card pages.

Abel, E. W., 219, 327 [P. 305, 493]; 502, 503 [As. 141]

Acton, E. M., 321, 325 [P. 481]

Adams, R., 571, 572 [As. 290]

Addis, H. W., 504 [As. 144]

Aeschlimann, A. E., 373, 374, 421, 518, 520, 536, 538, 560, 561, 568 [As. 29, 89, 181, 269]

Aguiar, A. M., 176–8, 180, 196, 198, 316–19 [P. 249–54, 276, 476]

Aguiar, H., 176, 177 [P. 249–51]

Aksnes, G., 277, 278 [P. 402, 405]

Algor, T., 28 [P. 44]

Allen, D. W., 79, 80, 148, 149, 194, 200, 201, 294, 295 [P. 122, 123, 127, 128, 217, 218, 220, 221]; 472, 473, 475, 490, 497 [As. 133, 139]

Allen, P. W., 562 [As. 271]

Allison, J. A. C., 404 [As. 70]

Anderson, E. L., 415 [As. 80]

Anderson, G. W., 282 [P. 418]

Anderson, W. A., 28 [P. 42]

Andreeva, E. I., 567 [As. 281]

Anon., 521 [As. 195]

Anschütz, L., 252, 254, 280–4, 303 [P. 341, 410–12, 417, 448]

Anschütz, R., 303, 306 [P. 446, 446a, 447, 453]

Appel, R., 330, 331 [P. 495]

Arbuzov, A. E., 281, 282, 326, 327 [P. 415, 420, 491]

Arbuzov, B. A., 31, 46, 48, 272, 299 [P. 64, 67, 390]; 611 [Sb. 23]

Arkina, S. E., 518 [As. 184]

Armitage, D. A., 327 [P. 493]

Arnold, H., 245, 321, 325, 327 [P. 330, 485, 489]

Asta-Werke A. G. Chemische Fabrik, 321–5 [P. 482–4]

Atherton, F. R., 303–6 [P. 445, 454, 455]

Aufderhaar, E., 54 [P. 81]

Austin, J. A., 615, 616, 626, 629 [Sb. 35]

Autenrieth, W., 217, 221 [P. 300, 301, 309]

Awl, R. A., 245 [P. 332]

Ayres, D. C., 271, 283 [P. 386]

Bach, M., 637, 638 [Bi. 2]

Backer, H. J., 544, 546, 547 [As. 242]

Bär, F., 118 [P. 184]

Bailey, W. J., 171 [P. 243]

Baker, B. R., 321, 325 [P. 481]

Baker, F. C., 181, 186, 230 [P. 255, 265]; 441, 449, 453, 455, 479, 497–500 [As. 99, 109, 144]

Baksova, R. A., 274 [P. 396]

Ballard, T. A., 522 [As. 215]

Balon, W. J., 35, 61 [P. 54]

Balszuweit, A., 214 [P. 295]

Banas, E. M., 265 [P. 373]

Bang, F. B., 615, 620 [Sb. 32]

Banks, C. K., 617 [Sb. 39, 40]

Barber, H. J., 573 [As. 291, 292]

Barket, T. P., 28, 42–4, 60 [P. 43, 61]

Barow, R. L., 307 [P. 458]

Subject Index

Heterocyclic compounds of As, Bi, P, and Sb are listed alphabetically under the name of the ring system, derivatives being collected thereafter as inverted entries, *e.g.* Acridarsine, 5-chloro-, or Dibenzarsole, 5-methyl-. Hydrogenated forms of the ring system are often given as separate main entries, derivatives again being listed in inverted form, *e.g.* Acridarsine, 5,10-dihydro- as a main entry with 5-chloro- as a sub-entry.

Other compounds are listed under the first letter of the full name (i.e. not in inverted form), e.g. 2-Aminoethanol, Carboxymethylarsonic acid.

The following abbreviations are used: condensn. (condensation), cpd. (compound), cyclizn. (cyclization), decomp. (decomposition), deriv. (derivative), prep. (preparation), redn. (reduction), subst. (substituted), synth. (synthesis).

23*